Ensuring Global Food Safety

Ensuring Global Food Safety

Exploring Global Harmonization

SECOND EDITION

Edited by

Aleksandra Martinović

University of Donja Gorica, Centre of Excellence-FoodHub, Podgorica, Montenegro

Sangsuk Oh

Department of Food Science and Technology, Ewha Womans University, Seoul, Korea

Huub Lelieveld

Global Harmonization Initiative (GHI), Vienna, Austria

Global Harmonization Initiative

ELSEVIER

ACADEMIC PRESS

An imprint of Elsevier

Academic Press is an imprint of Elsevier
125 London Wall, London EC2Y 5AS, United Kingdom
525 B Street, Suite 1650, San Diego, CA 92101, United States
50 Hampshire Street, 5th Floor, Cambridge, MA 02139, United States
The Boulevard, Langford Lane, Kidlington, Oxford OX5 1GB, United Kingdom

Library of Congress Cataloging-in-Publication Data
A catalog record for this book is available from the Library of Congress

British Library Cataloguing-in-Publication Data
A catalogue record for this book is available from the British Library

ISBN: 978-0-12-816011-4

For information on all Academic Press publications visit our website at https://www.elsevier.com/books-and-journals

Publisher: Charlotte Cockle
Acquisitions Editor: Nancy J. Maragioglio
Editorial Project Manager: Charlotte Rowley
Production Project Manager: Vijayaraj Purushothaman
Cover Designer: Miles Hitchen

Typeset by TNQ Technologies

Contents

26. Testing for food safety using human competent liver cells (HepG2): a review

Firouz Darroudi

27. Capacity building

Larry Keener and Tatiana Koutchma

28. Capacity building: building analytical capacity for microbial food safety

Debdeep Dasgupta, Mandyam C. Varadaraj and Paula Bourke

List of contributors

Fadwa Al-Taher, VDF FutureCeuticals, Inc., Momence, IL, United States

Veslemøy Andersen, Global Harmonization Initiative (GHI), Vienna, Austria

Alejandro Ariosti, National Institute of Industrial Technology (INTI) — Plastics Center, Buenos Aires, Argentina; Department of Food Science, Faculty of Pharmacy and Biochemistry, University of Buenos Aires (UBA), Buenos Aires, Argentina

Elisabeth J. Arundell, The New South Wales Department of Primary Industries, Orange, NSW, Australia

Adina Alexandra Baicu, University of Agronomic Sciences and Veterinary Medicine of Bucharest, Romania

Gustavo V. Barbosa-Cánovas, Center for Nonthermal Processing of Food, Washington State University, Pullman, WA, United States

Daniela Bermúdez-Aguirre, Center for Nonthermal Processing of Food, Washington State University, Pullman, WA, United States

Fehmi Kerem Bilgin, İzmir Bakirçay University, Faculty of Law, Menemen, İzmir, Turkey

Paula Bourke, School of Biosystems and Food Engineering, University College Dublin, Dublin, Ireland

Hans Bouwmeester, Division of Toxicology, Wageningen University and Research, Wageningen, the Netherlands

Adelia C. Bovell-Benjamin, Food and Nutritional Sciences, Tuskegee University, Tuskegee, AL, United States

Julie Larson Bricher, Quiddity Communications, Inc., McMinnville, OR, United States

Frank F. Busta, University of Minnesota, Minneapolis, St. Paul, MN, United States

Kezban Candoğan, Faculty of Engineering, Department of Food Engineering, Ankara University, Ankara, Turkey

Melissa M. Card, Institute for Food Laws & Regulations, MSU, Michigan State University's College of Law, United States

Yifan Cheng, Department of Food Science, Cornell University, Ithaca, NY, United States

M.B. Cole, Head, School of Agriculture Food and Wine. University of Adelaide, Urrbrae, SA, Australia

Pamela L. Coleman, Mérieux NutriSciences, Chicago, IL, United States

Firouz Darroudi, Global Harminization Initiaitve (GHI), Section of Genetic Toxicology and Genomics, Oegstgeest, The Netherlands

Debdeep Dasgupta, Department of Microbiology, Surendranath College-Kolkata, Kolkata, West Bengal, India

H.K.S. De Zoysa, Department of Bioprocess Technology, Faculty of Technology, Rajarata University of Sri Lanka, Anuradhapura, North Central Province, Sri Lanka; Department of Biology, University of Naples Federico II, Naples, Italy

Ahmad Din, National Institute of Food Science & Technology, University of Agriculture, Faisalabad, Pakistan

Hazel Farrell, The New South Wales Department of Primary Industries, Taree, NSW, Australia

Anthony J. Fontana, Mérieux NutriSciences, Chicago, IL, United States

Neal D. Fortin, Institute for Food Laws and Regulations, Michigan State University, East Lansing, MI, United States

Beatriz Gonçalves Franco, Center for Nonthermal Processing of Food, Washington State University, Pullman, WA, United States

L.G.M. Gorris, Food Safety Expert, Food Safety Futures, Nijmegen, The Netherlands

Jaap C. Hanekamp, University College Roosevelt, Middelburg, the Netherlands; Environmental Health Sciences, University of Massachusetts Amherst, Amherst, MA, United States; HAN-Research, Zoetermeer, the Netherlands

Helen Nonye Henry-Unaeze, Department of Food, Nutrition and Home Science, Faculty of Agriculture, University of Port Harcourt, East-West Road Choba, Rivers, Nigeria

Alison Imlay, The New South Wales Department of Primary Industries, Silverwater, NSW, Australia

Heinz-Dieter Isengard, University of Hohenheim, Institute of Food Science and Biotechnology, Stuttgart, Germany

Lauren S. Jackson, U.S. Food and Drug Administration, Division of Food Processing Science & Technology, Bedford Park, IL, United States

Sewon Jeong, BiofoodCRO, Seoul, Korea

Katy A. Jones, FoodLogiQ, Durham, NC, United States

Frans W.H. Kampers, Wageningen UR, Wageningen, the Netherlands

Larry Keener, International Product Safety Consultants, Seattle, WA, United States

Ji Yeon Kim, Department of Food Science and Technology, Seoul National University of Science and Technology, Seoul, Korea

Thea King, The New South Wales Department of Primary Industries, Silverwater, NSW, Australia

Tatiana Koutchma, Agriculture and Agri Foods, Canada

Oran Kwon, Department of Nutritional Science and Food Management, Graduate Program in System Health Science and Engineering, Ewha Womans University, Seoul, Korea

Joe Lederman, FoodLegal, Australia

Huub Lelieveld, Global Harmonization Initiative (GHI), Vienna, Austria

Rebeca López-García, Logre International Food Science Consulting, Mexico

Alida Mahmudova, Bona Mente Consulting LLC Law Company, Azerbaijan

Bernard Maister, Intellectual Property Unit, University of Cape Town, Cape Town, South Africa

Carmen I. Moraru, Department of Food Science, Cornell University, Ithaca, NY, United States

Sangsuk Oh, Department of Food Science and Technology, Ewha Womans University, Seoul, Korea

Margherita Paola Poto, K. G. Jebsen Centre for the Law of the Sea, UiT, Tromsø, Norway

Jamuna Prakash, Global Harmonization Initiative, Austria

Syed S.H. Rizvi, Department of Food Science, Cornell University, Ithaca, NY, United States

V.D. Sattigeri, Food Safety and Analytical Quality Control Laboratory, Central Food Technological Research Institute, Mysuru, Karnataka, India

Bert Schwitters, Independent Researcher

Craig Shadbolt, The New South Wales Department of Primary Industries, Silverwater, NSW, Australia

Xian-Ming Shi, MOST-USDA Joint Research Center for Food Safety, School of Agriculture and Biology, State Key Lab of Microbial Metabolism, Shanghai Jiao Tong University, Shanghai, China

Ga Young Shin, Center for Nonthermal Processing of Food, Washington State University, Pullman, WA, United States

Mungi Sohn, Food Science and Biotechnology, College of Life Sciences, Kyung Hee University, Republic of Korea

Cynthia M. Stewart, Silliker Food Science Center, South Holland, IL, United States

Juanjuan Sun, Food Law, Nantes University of France, Center for Coordination and Innovation of Food Safety Governance, Renmin University, Beijing, China

Elizabeth A. Szabo, The New South Wales Department of Primary Industries, Silverwater, NSW, Australia

John Szpylka, Mérieux NutriSciences, Chicago, IL, United States

John Y.H. Tang, Universiti Sultan Zainal Abidin, Terengganu, Malaysia

Matthew D. Taylor, The New South Wales Department of Primary Industries, Taylors Beach, NSW, Australia

The International Commission on Microbiological Specifications for Foods, www.icmsf.org

Halide Gökçe Türkoğlu, İzmir Bakirçay University, Faculty of Law, Menemen, İzmir, Turkey

Altinay Urazbaeva, Studying Advanced Master Program in European, International Business Law, Leiden University

M.A.J.S. van Boekel, Food Quality & Design Group, Wageningen University & Research, Wageningen, the Netherlands

Bernd van der Meulen, GHI, Prof. Comparative Food Law, Renmin University of China School of Law, University of Copenhagen, European Institute for Food Law, Amsterdam, The Netherlands

Mandyam C. Varadaraj, Department of Human Resource Development, Central Food Technological Research Institute, Mysore, Karnataka, India

Yuriy Vasiliev, Stavropol Branch, North Caucasus Civil Service Academy, Russia

Viduranga Y. Waisundara, Australian College of Business & Technology - Kandy Campus, Peradeniya Road, Kandy, Central Province, Sri Lanka

Odel Yun LI, Shanghai Jiao Tong University, Shanghai Legislative Research Institute, Shanghai, China

Chapter 1

Introduction: Ensuring global food safety: A public health priority and a global responsibility

Julie Larson Bricher
Quiddity Communications, Inc., McMinnville, OR, United States

Only if we act together can we respond effectively to international food safety problems and ensure safer food for everyone.
Dr. Margaret Chan, former Director-General, World Health Organization.

Dr. Margaret Chan's words served as the opening quote of this book during its first publication nearly a decade ago. We have seen improvements in global health outcomes in the past 10 years, in part due to political and community pressure to implement evidence-based and scientifically informed health and food safety policies and legislation on a global scale. Chan's words ring especially true today: It was only through community effort that these improved outcomes were achieved—and it is only through continued community effort that we can ensure that safe food in adequate supply is the reality for all the world's people.

The march toward globalization appears inexorable, even as the trend remains politically controversial on the world stage. The International Monetary Fund defines globalization as "the process through which an increasingly free flow of ideas, people, goods, services, and capital leads to the integration of economies and societies" (IMF, 2006). At its core, globalization is a process driven by free trade economics and an ideal driven by the promise of greater societal benefits for all peoples of the world.

Proponents put forward that an economy without borders spurs greater market competition and therefore economic freedom, driving down prices and increasing availability and variety of affordable goods and services for a greater number of people. In turn, globalization promises further benefits, such as increases in productivity, access to new technologies and information streams, and higher living, environmental, and labor standards for those in both developed and developing countries. Critics charge that inherent economic and infrastructure inequalities that exist between developed and developing nations preclude less developed and poorer nations from fully realizing these benefits.

Whatever the measurable positive benefits experienced by some countries in recent years, there remain tangible challenges not only brought on by the rapid acceleration of globalization in the world economy but the impact of global climate change on the planet's food supply. Perhaps there are no statistics more compelling than those of the 2018 World Resources Institute's report, "Creating a Sustainable Food Future," which projects that the human population is expected to grow from 7 billion in 2010 to 9.8 billion in 2050. The demand for food is estimated to increase by more than 50% and demand for animal-based foods by nearly 70% (WRI, 2018). According to the report, major changes to the global food system—by farmers, food companies, consumers, and governments—will be necessary to mitigate looming food shortages worldwide.

In addition, nearly 2 decades into the 21st century, the challenges of ensuring food security, food safety, and nutrition on a global scale continue to grow in complexity. Recent statistics show that the levels of world hunger, malnutrition, and food- and waterborne diseases are among the most critical global public health issues facing the international community. For example:

- According to the Food and Agriculture Organization (FAO) of the United Nations, 10.9% of the world's population are undernourished, down from 14.5% in 2005. This percentage still represents roughly 770 million people (FAO, 2018).

Ensuring Global Food Safety. https://doi.org/10.1016/B978-0-12-816011-4.00012-4

- Globally, 22.7% of children under five who experience undernourishment suffer from stunted growth (FAO, 2018).
- The World Health Organization (WHO) reports that more than 1000 children under five die daily from diarrheal disease caused by inadequate access to water sanitation (WHO, 2014).
- In 2015, foodborne diarrheal disease agents alone were the cause of death for more than 230,000 people (WHO, 2018; WHO 2015).
- Worldwide, nearly 1 in 10 people fall ill from all foodborne diseases, which equates to 33 million healthy life years lost and results in the deaths of approximately 420,000 people (WHO, 2015). Children account for one-third of deaths from foodborne diseases.
- In developed countries, one in three consumers get a foodborne disease associated with microbes or their toxins every year. This does not include other foodborne diseases associated with naturally occurring or man-made chemical contaminants, such as aflatoxin, acrylamide, furan, or dioxin (Schlundt, 2008).

The WHO Initiative to Estimate the Global Burden of Foodborne Diseases identifies the rapid globalization of food trade as a worldwide trend that has introduced an increased potential for contaminated food to adversely affect greater numbers of people (WHO, 2015). As the food supply chain becomes more integrated, the potential for massive foodborne illness outbreaks caused by pathogens, chemicals, viruses, and parasites increases—as do the difficulties in controlling foodborne infections, morbidity, disability, and mortality.

Rapid globalization also has exposed critical gaps in national and international capabilities to assure adequate levels of food safety and quality. Disparities related to national infrastructural and technological capacities and international food production, distribution and handling standards, and law have become more visible as global commerce becomes more interconnected. As a result, WHO and other food-related international public health, development, and standard-setting bodies have targeted these gaps as priority items and are working together to reinforce the need to use an integrated international food safety regulatory system in the era of "one global market."

To be effective, such a system must include advancing the use of risk analysis and management to better direct resources toward areas of high risk, providing a scientific basis for international food safety action, moving from conventional "vertical" legislation within nations to more "horizontal" rulings among nations to attain harmonization of standards and reduce barriers to trade, and building capacity to promote the availability and use of new food safety technologies, testing and preventing strategies that will reduce the public health risks of foodborne disease around the globe.

In the second edition of this volume, *Ensuring Global Food Safety—A Public Health Priority and a Global Responsibility*, members of the Global Harmonization Initiative (GHI) once again contribute to the world dialogue, discussing tools for promoting harmonization of scientific methods, standards, and regulations. Established in 2004, GHI is a network of international scientific organizations and individual scientists that aims to achieve objective consensus on the science of food regulations and legislation to ensure the global availability of safe and wholesome food products for all consumers.

With support and participation of its individual members and member organizations, the GHI's Working Groups have conducted a series of meetings at which members have formulated approaches to critically (re-)evaluate the scientific evidence used to underpin existing global regulations in the areas of product composition, processing operations, and technologies or measures designed to prevent foodborne illness. Each chapter is reflective of outcomes of these discussions and progress in developing strategies to find the shortest route to achieving global harmonization in concert with international public health and food safety authorities, including the WHO, FAO, the Codex Alimentarius Commission (CAC), and the International Organization for Standardization (ISO).

GHI's overarching objective is to provide regulators, policymakers, and public health authorities with a foundation for sound, sensible, science-based international regulations in order to eliminate hurdles to scientific advancement in food safety technology. For example, there is no question that the more that the avenues of global trade narrow, the higher the probability of traffic jams in worldwide commerce. Barriers to trade in the form of differing—and sometimes conflicting—country-by-country import/export rules and requirements can and do make it difficult for food businesses to get traction in overseas markets.

Food safety concerns are frequently cited by individual nations as underpinning the justification for their legislative acts and rulemaking—and for erecting trade barriers and other measures that have the impact of curtailing free trade. Unfortunately, in some cases, the science used to inform and bolster food safety policymaking is insufficient, inconsistent, or contradictory, creating a roadblock to the promulgation of laws that have a clear and evident benefit to protecting public health.

National differences in food safety regulations and laws also trigger a red light to the advances offered by science and technology. Though many food companies throughout the world have invested significant monies to food safety and nutrition technology research and development efforts, industry is understandably hesitant to apply newly developed capabilities on an international scale in an uncertain, maze-like regulatory environment.

GHI anticipates that elimination of regulatory differences will make it more attractive for the private sector to invest in food safety and nutrition research and development, consequently strengthening the competitiveness of each nation's food industry and of the industries supplying the food sector. Harmonizing global regulations will aid in the uptake and application of new technologies and encourage the food industry to invest in technologies to ensure the safety, quality, and security of the global food supply.

Ultimately, "globalizing" food safety regulations and laws based on sound science can only serve to help bridge public health gaps and create opportunities for all stakeholders to realize the big picture benefits promised by economic globalization, including measurable global reductions in morbidity and mortality associated with foodborne disease; increases in food availability to combat malnutrition and enhance food security for consumers worldwide; and decreases in poverty rates among less-developed or impoverished nations through capacity building that enables full participation in the global economy.

For public health agencies responsible for overseeing the safety of the international food supply, harmonization of food safety and quality standards and regulations will bring a higher level of confidence that risk reduction strategies and food safety measures are effective and that decisions taken are based on science and not on underlying political agendas that may be in conflict with public health goals. Harmonization will also ensure that available resources are allocated where they have the highest impact on the most pressing food disease—related problems.

To paraphrase WHO Director General Chan, it is only through collective action that we can fully embrace our global responsibility to respond effectively to the challenges of ensuring food security, food safety, and nutrition for everyone. As the authors in this volume attest, meeting that global responsibility requires cooperation, collaboration, and consensus building if we are to achieve harmonization of food regulations and standards, and thereby accomplish even greater gains in global public health.

References

Food and Agriculture Organization, 2018. World Food and Agriculture - Statistical Pocketbook 2018. http://www.fao.org/publications/card/en/c/CA1796EN.

International Monetary Fund, 2006. Glossary of Selected Financial Terms. http://www.imf.org/external/np/exr/glossary/showTerm.asp#91.

Schlundt, J., 2008. Food safety: a joint responsibility. In: 14th World Congress of Food Science and Technology. Shanghai, China. October 20, 2008.

World Health Organization, 2014. Preventing Diarrhoea through Better Water, Sanitation and Hygiene. https://www.who.int/water_sanitation_health/publications/gbd_poor_water/en.

World Health Organization, 2015. WHO Estimates of the Global Burden of Foodborne Diseases: Foodborne Disease Burden Epidemiology Reference Group, 2007—2015. http://www.who.int/iris/handle/10665/199350.

World Resources Institute, 2018. Synthesis Report: Creating a Sustainable Future: A Menu of Solutions to Feed Nearly 10 Billion People by 2050. Full report to be published in 2019. www.wri.org/publication/creating-sustainable-food-future.

Chapter 2

Safety and security: the costs and benefits of traceability and transparency in the food chain

Katy A. Jones
FoodLogiQ, Durham, NC, United States

The food supply chain is one of the most important business aspects of food companies and restaurants. It is critical to every operation and provides organizations with the opportunity to build and deliver their brand promise. An efficient, well-managed food supply chain can help to improve operational efficiency, mitigate risk, improve brand reputation, and increase (or maintain) consumer confidence in the products being served to customers.

But these benefits are only achievable if supply chains are kept safe and secure. This requires supply chains that are monitored and tracked using strong processes supported with advanced technologies. As a food company or restaurant, it is critical to track and monitor the supply chain to reduce the overall risk to the brand—and to the customers. In this chapter, we will look into the costs of foodborne outbreaks not only to society but to the impact of the business as well.

2.1 The burden of foodborne outbreaks

2.1.1 Each year 48 million people get sick from foodborne illnesses, 128,000 are hospitalized, and 3000 die (Centers for Disease Control and Prevention, 2015)

Many people are under the assumption that foodborne illness is more of a problem in developing countries where regulations are not as strict, but it hits closer to home more often than we think. The Centers for Disease Control and Prevention (CDC) in the United States of America closely monitor all cases of foodborne illness that come through the United States, and the statistics are not pretty.

It's nearly impossible to eradicate foodborne illnesses. There are too many factors that go into it, some out of our control, but more can be done for prevention. When organizations employ enhanced traceability programs, combined with effective safety plans, auditing and corrective actions, the risks of contamination can be much lower. Depending on where you are, companies should be able to monitor exactly where their food was, is, and will be, maintaining total visibility across the supply chain—and react quickly when a food safety issue hits.

Additionally, when foodborne illnesses occur, traceability can reduce exposure. As soon as the adulterated food is identified and the root cause is identified, it can be recalled to prevent further illness and used to analyze the medical action needed to rectify the situation.

2.1.2 About 23% of U.S. food recalls cost the food industry over $30 million and 14% cost organizations over $50 million (Grocery Manufacturers Association, 2011)

These statistics are staggering. Executing a recall on food products can be a manufacturer's worst nightmare simply because of the time and money lost. Recalling products is an essential part of maintaining public health, but it can be stressful.

Ensuring Global Food Safety. https://doi.org/10.1016/B978-0-12-816011-4.00028-8

Consider one *E. coli* outbreak in Germany (Centers for Disease Control and Prevention, 2014). It's difficult to contain, and in this instance, there was not strong traceability throughout the supply chain. 3800 people were affected worldwide, 47 died, and European Union farmers lost €417 million ($611 million) (Grieshaber, 2011). The holistic cost of this incident shows that traceability is essential to healthy food industries.

And according to a study from researchers at John Hopkins Bloomberg School of Public Health, a single foodborne disease outbreak at a fast-casual establishment could cost between $6330 and $2.1 million in lost revenue, fines, and lawsuits. And this is only on the financial side.

2.1.3 Reducing foodborne illnesses by just 1% would prevent nearly 500,000 Americans from getting sick each year

If we were able to reduce foodborne illness by 1%, nearly 500,000 Americans could avoid sickness. This will require a more strategic approach from individuals and food organizations alike.

Reducing foodborne illnesses so drastically will require a new perspective on how we view foodborne illnesses and mitigate risks. In 2013, the United States spent about $40 million on treating the problem but not preventing it. Through the adoption of traceability supported by technology, organizations can shift the focus toward prevention, saving money, and lives.

Reducing foodborne illnesses will not be an easy task, and it relies on individual organizations making decisions that impact society as a whole. As they employ traceability, they will start to make an impact and take us closer to reducing these issues.

2.2 The food supply chain: increasing risk

Food supply chains are growing increasingly complex, global, and fresh. Consumers are driving for more fresh format concepts and want healthier options overall. Yet that profile of supply chain in theory has increased risk.

Fresh produce items like cilantro, cucumbers, cantaloupes, and peppers that are often eaten raw cause more foodborne illness than any other single category of food, according to a study by the Center for Science in the Public Interest (CSPI) (Food Safety News, 2015). The nonprofit food safety group reviewed 10 years of outbreak data to determine which foods are most often linked to outbreaks of foodborne disease and identify trends in illnesses. Over the period studied, fresh produce caused 629 outbreaks and almost 20,000 illnesses.

But that does not mean Americans should avoid fruits and vegetables, CSPI says. While the number of outbreaks and illnesses is large, on a pound-for-pound basis fresh produce is safer than many other foods.

Over the period studied, there was a total of 193,754 illnesses reported from 9626 outbreaks. Of the total number of reported outbreaks, the CDC was able to identify both the food source and the contaminant in fewer than 40%. CSPI only reviewed the 3485 solved outbreaks.

The report also found that seafood caused more illnesses per pound consumed than any other food category, while fruits, vegetables, and dairy caused the fewest illnesses per pound consumed.

2.3 Working toward traceability and transparency

Now that we have established the burden of the issue and the associated risk in the food supply chain, we ask ourselves what are the real costs associated with a failure in transparency and traceability?

Food regulatory requirements have shifted in nature; now, rather than primarily focusing on responding to food safety incidents, there is an increased emphasis on food safety prevention. Fortunately, there are a variety of tools to help food and beverage manufacturers ensure that they are compliant with food safety regulations; the key is preparing for food safety audits in advance. Audits can occur in-house by a dedicated team or by an external auditor, with the purpose being to identify areas for improvement to processes and systems.

Under U.S. Food and Drug Administration (FDA) guidelines and regulations under the Food Safety Modernization Act, food and beverage manufacturers must have a food safety plan in place that includes oversight and management of preventive controls established in each manufacturing facility. Regulatory audits and audit reports must be submitted to the FDA. In 2019, the FDA announced the New Era of Smarter Food Safety initiative that encompasses four pillars that include tech-enabled traceability, smarter tools for prevention, adapting to new business models in retail, and food safety culture. The blueprint released also includes proposed Section FSMA 204 rulemaking to harmonize the key data elements and critical tracking events needed for enhanced traceability (FDA New Era of Smarter Food Safety).

In Europe, the European Food Safety Authority (EFSA), established in 2002, oversees the regulation of the food supply chain. The organization's mission is "to deliver independent, high-quality and timely scientific advice on risks in the food chain from farm to fork in an integrated manner and to communicate on those risks in an open manner to all interested parties and the public at large."

The exact same law that established EFSA, Regulation EC/178/2002, also established the basis for food traceability in Europe. Under the law, any food produced in Europe or imported into Europe is subject to an incredibly high standard for traceability. The regulation requires that both food manufacturers and distributors demonstrate "the ability to trace and follow food, feed, and ingredients through all stages of production, processing, and distribution" (European Commission, 2019).

Although going through an audit can be a stressful event, a passing result will assure you and your team that your company has achieved a satisfactory level of food safety. A successful audit also lets consumers know that your company prioritizes their wellbeing.

2.4 The costs associated to a lack of traceability

With a managed, improved, and efficient food supply chain in place, businesses can begin to reap other benefits. This can include financial gains, such as an increase in credit request approvals and a reduction in insurance premium costs.

2.4.1 Issues with labeling and brand claims

Providing proof of your product claims is by itself enough to boost your business' reputation in the industry and worldwide. With access to robust supply chain data, companies can proactively remove any issues that may negatively impact your brand in the future. By identifying quality issues in advance, you can protect your brand commitment and your business' reputation.

For example, you can communicate across your supply chain to prevent poor or tainted products from being delivered to your consumer. To examine the real potential return on investment that supply chain transparency can achieve, let us look at an example:

A consumer packaged goods company that produces nut-free granola bars has been notified that a lot from its oat supplier actually contained traces of walnut, when the allergen claim on the packaging does not list tree nuts. The issues have only been identified after the batch has been sent out to retailers and put on shelves. The batch of 1500 cases had an average sale price of $40, totaling $60,000 worth of sales. $20,000 of that sales figure was sunk manufacturing costs (the cost of the raw ingredients, manufacturing, and labor).

Without Supply Chain Traceability: The manufacturer is slow to respond to the crisis.
It takes time for staff to identify if the contaminated batch has been sent to figure out exactly which stores were sent the tainted product. This equates to a labor cost of $1,000. Once the shipment locations have been identified, the manufacturer has the opportunity to get the sales back quickly if they can replace the product. There is a 6% logistics cost of $3,000 to recall the products and a $9,000 cost to expedite delivery of the new product. All of this is on top of the $20,000 cost of manufacturing the new product. This is a total cost of $33,000. However, because the product defect was not caught before it went on sale, the manufacturer's reputation is severely damaged. Consumers no longer trust their product, and they see a 20% drop in sales. The company's share price also falls as a result.
With Supply Chain Traceability: With enhanced traceability in place, the situation is a different story. Corporate food safety can use data to instantly see that the contaminated batch has only been sent to three store locations. The $1,000 labor cost is significantly reduced.
Companies that utilize traceability can also benefit from an average 30% reduction in the direct costs associated with a recall. So instead of the logistical costs totaling $12,000, the costs only total $8000. That is a saving of $4000.
Because the company is able to quickly respond to the issue and pull the tainted product from the exact store locations, the damage to their brand is greatly reduced. They may see a small reduction in sales but not near the damage caused from a long, drawn-out recall played out in the news media and social media.

2.4.2 Restrictions to market access

With globalization comes the opportunity for businesses to enter new markets across the world with relative ease. A traceable supply chain can help to ease compliance regulation on a global scale. With a traceable supply chain, you can ensure that all trading partners meet or exceed the minimum acceptable standards for markets across the globe.

Enhanced traceability connects every stage of the supply chain; the manufacturer was able to identify where the nuts entered the manufacturing process. The issue was the fault of a well-known supplier, not the manufacturer and it is the supplier that suffers the reputational damage. The manufacturer still sees a slight decrease in sales, but it is able to save its reputation and its share price.

2.5 Benefits beyond food safety

We all know the important role food supply chain traceability plays in food safety. Being transparent about where your food comes from helps put customers' minds at ease—and also can help resolve a food safety issue more quickly, should one arise. But what about the other benefits food supply chain traceability? Can being transparent about your supply chain provide an additional return of investments for your business, beyond safety?

Today's consumers are more concerned than ever about what they are eating and where their food is coming from. With this trend toward awareness and transparency, businesses can capitalize on additional benefits, simply by telling your customers more about your processes. Here is how:

2.5.1 Enhanced credibility

Being open about your supply chain shows you have nothing to hide. This transparency will resonate with customers and help build the credibility of your business as a trusted place to bring their business. Improving your credibility can also help to further establish your branding and help you stand out from the competition. Your customers will be more likely to choose you knowing they can trust where your food comes from.

2.5.2 Transparent marketing

Supply chain traceability allows you to create a "farm-to-fork" story that can be a very effective marketing tool. Consumers like to know they are putting good, clean ingredients in their body and they like to know they are supporting local farmers at the same time. By using your supply chain as a marketing tool, you can attract a health-conscious audience. While many companies like to claim their food is "farm fresh," traceability allows you to authenticate your sources.

2.5.3 Increased reliability for consumers

Meticulously tracing your supply chain helps to ensure consistent quality of suppliers across each chain or franchise of your business. This means the food your customers eat is always exactly the same, regardless of which location they visit. This level of reliability and consistency helps solidify your brand and lets your customers know they can trust your product. Rather than worrying your food may be "hit or miss" depending on location, they will know what level of quality they can expect.

2.6 More operational efficiency

You depend on your supply chain to maintain efficiency in the daily operations of your business. Traceability helps to improve communication between you and your suppliers, which keeps everyone on the same page and helps improve efficiency. When your processes are running smoothly, your relationships with your suppliers, employers, and customers all benefit.

The benefits of food supply traceability go well beyond food safety. With the opportunity to improve your business' branding, marketing, reputation, and processes, traceability is an investment that can transform your entire business.

References

Centers for Disease Control and Prevention, 2014. Questions about the 2011 *E. coli* Outbreak in Germany (accessed 19.10.10.). www.cdc.gov/ecoli/germany.html.

Centers for Disease Control and Prevention, 2015. Surveillance for Foodborne Disease Outbreaks United States, 2013: Annual Report.

European Commission, 2019. Food Law General Requirements (accessed 19.10.16.). https://ec.europa.eu/food/safety/general_food_law/general_requirements_en>.

Food Safety News, 2015. Fresh Produce Responsible for Most Foodborne Illnesses in the U.S (accessed 19.10.10.). www.foodsafetynews.com/2015/12/report-fresh-produce-responsible-for-most-foodborne-illness-outbreaks/.

Grieshaber, K., 2011. 2 New *E. coli* Deaths as EU Holds Emergency Meeting. The Post and Courier.

Grocery Manufacturers Association, 2011. Capturing Recall Costs: Measuring and Recovering the Losses.

FDA New Era of Smarter Food Safety, 2021. (Accessed 26 October 2021).

Chapter 3

Food regulation around the world

Bernd van der Meulen[1], Melissa M. Card[2], Ahmad Din[3], Neal D. Fortin[4], Alida Mahmudova[5], Bernard Maister[6], Halide Gökçe Türkoğlu[7], Fehmi Kerem Bilgin[7], Joe Lederman[8], Margherita Paola Poto[9], V.D. Sattigeri[10], Mungi Sohn[11], Juanjuan Sun[12], Altinay Urazbaeva[13] and Yuriy Vasiliev[14]

[1]GHI, Prof. Comparative Food Law, Renmin University of China School of Law, University of Copenhagen, European Institute for Food Law, Amsterdam, The Netherlands; [2]Institute for Food Laws & Regulations, MSU, Michigan State University's College of Law, United States; [3]National Institute of Food Science & Technology, University of Agriculture, Faisalabad, Pakistan; [4]Institute for Food Laws and Regulations, Michigan State University, East Lansing, MI, United States; [5]Bona Mente Consulting LLC Law Company, Azerbaijan; [6]Intellectual Property Unit, University of Cape Town, Cape Town, South Africa; [7]İzmir Bakirçay University, Faculty of Law, Menemen, İzmir, Turkey; [8]FoodLegal, Australia; [9]K. G. Jebsen Centre for the Law of the Sea, UiT, Tromsø, Norway; [10]Food Safety and Analytical Quality Control Laboratory, Central Food Technological Research Institute, Mysuru, Karnataka, India; [11]Food Science and Biotechnology, College of Life Sciences, Kyung Hee University, Republic of Korea; [12]Food Law, Nantes University of France, Center for Coordination and Innovation of Food Safety Governance, Renmin University, Beijing, China; [13]Studying Advanced Master Program in European, International Business Law, Leiden University; [14]Stavropol Branch, North Caucasus Civil Service Academy, Russia

Chapter 3.1

Introduction

Bernd van der Meulen

GHI, Prof. Comparative Food Law, Renmin University of China School of Law, University of Copenhagen, European Institute for Food Law, Amsterdam, The Netherlands

3.1.1 Purpose of this chapter

Against the background of global harmonization through scientific consensus, this chapter provides an inventory of approaches to the regulation of food and related issues in a variety of jurisdictions around the world. To each jurisdiction, a separate section is dedicated. Each section has been written by an author well versed in the jurisdiction at issue. The sections can be read as independent texts.

3.1.2 Food law

We have labeled the rules and regulations that apply to the food sector "food law." This label can cover two closely related but distinguishable phenomena. It may relate to a branch of law recognized within a legal system that is labeled "food law" or the outcome of an analysis of the legal system from the perspective of the food sector. If the latter, one can speak of a functional feed of law.[1] In this chapter, we approach the topic from this functional perspective.

1. On this concept, see: Bernd van der Meulen, The Functional Field of Food Law. The Emergence of a Functional Discipline in the Legal Sciences, European Institute for Food Law working paper 2018/02. Available at < http://www.food-law.nl/Working-papers/>.

Ensuring Global Food Safety. https://doi.org/10.1016/B978-0-12-816011-4.00001-X

3.1.3 Framework of analysis

The sections differ considerably due to differences in the subject matter, in the availability of data and in background and style of the authors. For the purpose of comparative analysis, some points of attention have been chosen. These include the presence or absence of a branch of food law; the most important sources of food law; game changing events that triggered development or reform; the role of science through risk analysis; the role of the Codex Alimentarius in national systems; the institutional framework; underlying principles and concepts; the role of product specific requirements ("standards"); authorization requirements; food safety limits; process requirements on hygiene and incident management; labeling; human right to food; and jurisdiction specific elements. Where available, literature in English has been indicated.[2]

In this chapter, the sections are grouped by region. The chapter opens with a brief section on International food law—the Codex Alimentarius in particular—because this has been chosen as benchmark in many jurisdictions. Then follows the section on the United States of America. The United States of America were relatively early in developing modern food law. Many other jurisdictions have taken the United States' experiences and examples into account in their systems. The chapter concludes with a reflection from the perspective of global harmonization.

Further reading

Van der Meulen, B.M.J. (2018). The Functional Field of Food Law. The Emergence of a Functional Discipline in the Legal Sciences, European Institute for Food Law working paper 2018/02. Available at: http://www.food-law.nl/Working-papers/.

Chapter 3.2

International food law

Bernd van der Meulen

GHI, Prof. Comparative Food Law, Renmin University of China School of Law, University of Copenhagen, European Institute for Food Law, Amsterdam, The Netherlands

3.2.1 Codex Alimentarius

Between 1961 and 1963 the Food and Agriculture Organization (FAO) and the World Health Organization (WHO) established the Codex Alimentarius Commission (CAC). Over the years, the CAC established specialized committees. These committees are hosted by member states all over the world. Some 188 members (187 countries and the EU), representing about 99% of the world's population, participate in the work of Codex Alimentarius (Van der Meulen, 2018).

Food standards are established through an elaborate procedure of international negotiations (FAO/WHO, 2016). All standards taken together are called "*Codex Alimentarius.*" In Latin, this means "food code" It can be seen as a virtual book filled with food standards. The food standards represent models for national legislation on food.

By July 2015[3] the total output of the Codex Alimentarius stood at: 191 commodity standards, 73 guidelines, 51 codes of practice, and 17 maximum levels for contaminants in foods; over 3770 maximum limits for food additive in foods covering 301 different additives, 4347 maximum residue limits for pesticide residues covering 196 pesticides, and 610 maximum residue limits of veterinary drugs in foods covering 75 veterinary drugs. Finally, the Codex Alimentarius includes requirements of a horizontal nature on labeling and presentation and on methods of analysis and sampling (FAO/WHO, 2016; Masson-Matthee, 2007; Van der Meulen, 2018).

---

2. For many jurisdictions, important data are available at: < https://www.fas.usda.gov/regions >.
3. FAO/WHO, *Understanding Codex*, Rome 2016, available at: < www.fao.org/3/a-i5667e.pdf >, p. 10.

3.2.2 Procedural manual

The "constitution" of the Codex Alimentarius is the Procedural Manual. The Procedural Manual not only provides the procedures and format for setting Codex Standards and Guidelines, but also some general principles and definitions (Table 3.2.1). The principles relate, among other things, to the scientific substantiation of the work of Codex Alimentarius and the use of risk analysis for food safety (Table 3.2.2).

3.2.3 Standards

The work of the CAC has resulted in a vast collection of internationally agreed food standards that are presented in a uniform format. Most of these standards are of a vertical (product specific) nature. They address all principal foods, whether processed, semiprocessed, or raw. Standards of a horizontal nature are often called "general standards," like the General Standard for the Labeling of prepackaged Foods.[4]

According to this general standard, the following information shall appear on the labeling of prepackaged foods:

- the name of the food; this name shall indicate the true nature of the food;
- list of ingredients (in particular if one of a list of eight allergens is present);
- net contents;
- name and address of the business;
- country of origin where omission could mislead the consumer;
- lot identification;
- date marking and storage instructions;
- instructions for use.

3.2.4 Codes

In addition to the formally accepted standards, the Codex includes recommended provisions called codes of practice or guidelines. There is, for example, a "Code of Ethics for International Trade in Food,"[5] and a set of hygiene codes like the "Recommended International Code of Practice General Principles of Food Hygiene" and the "Hazard Analysis and Critical Control Point (HACCP) System and Guidelines for its Application" (Table 3.2.3).

3.2.5 Legal force

The Codex standards do not represent legally binding norms. They present models for national legislation. Member states undertake to transform the Codex standards into national legislation. However, no sanctions apply if they do not honor this undertaking.

By agreeing on nonbinding standards, the participating states develop a common language. All states and other subjects of international law will mean the same thing when they meet to negotiate about food (i.e., food as defined in the Codex). The same holds true for "milk" and "honey" and all the standards that have been agreed upon. The notion of HACCP has been developed—and is understood—within the framework of Codex Alimentarius.[6] In this way, the Codex Alimentarius provides a common frame of reference, but there is more.

TABLE 3.2.1 Some definitions in the Codex Alimentarius Procedural Manual.

Food means any substance, whether processed, semiprocessed or raw, which is intended for human consumption, and includes drink, chewing gum, and any substance which has been used in the manufacture, preparation, or treatment of "food" but does not include cosmetics or tobacco or substances used only as drugs.
Food hygiene comprises conditions and measures necessary for the production, processing, storage, and distribution of food designed to ensure a safe, sound, wholesome product fit for human consumption.

4. CODEX STAN 1–1985. The Codex texts on food labeling are available at: < www.fao.org/ag/humannutrition/foodlabel/78292/en/>.
5. CAC/RCP 20–1979 (Rev. 1–1985).
6. Recommended International Code of Practice General Principles of Food Hygiene CAC/RCP 1–1969. A collection of the Codex texts regarding food hygiene is available at: < http://www.fao.org/3/a-a1552e.pdf >.

TABLE 3.2.2 Some principles in the Codex Alimentarius Procedural Manual.

Statements of principle concerning the role of science in the codex decision-making process and the extent to which other factors are taken into account
1. The food standards, guidelines, and other recommendations of codex Alimentarius shall be based on the principle of sound scientific analysis and evidence, involving a thorough review of all relevant information, in order that the standards assure the quality and safety of the food supply.
2. When elaborating and deciding upon food standards, codex Alimentarius will have regard, where appropriate, to other legitimate factors relevant for the health protection of consumers and for the promotion of fair practices in food trade.
3. In this regard it is noted that food labeling plays an important role in furthering both of these objectives.
4. When the situation arises that members of codex agree on the necessary level of protection of public health but hold differing views about other considerations, members may abstain from acceptance of the relevant standard without necessarily preventing the decision by codex.

TABLE 3.2.3 The principles of HACCP according to Codex Alimentarius.

Principle 1	Conduct a hazard analysis.
Principle 2	Determine the critical control points (CCPs).
Principle 3	Establish critical limit(s).
Principle 4	Establish a system to monitor control of the CCP.
Principle 5	Establish the corrective action to be taken when monitoring indicates that a particular CCP is not under control.
Principle 6	Establish procedures for verification to confirm that the HACCP system is working effectively.
Principle 7	Establish documentation concerning all procedures and records appropriate to these principles and their application.

The mere fact that national specialists on food law enter into discussion on these standards will influence them in their work at home. Civil servants drafting a piece of legislation will look for examples. As regards food, they will find examples in abundance in the Codex. In these subtle ways, the Codex Alimentarius is likely to have a major impact on the development of food law in many countries even without a strict legal obligation to implement.

It turns out more than once that soft law has a tendency to solidify. Once agreements are reached, parties tend to put more weight on them than was initially intended. This is true for Codex standards as well which, following several developments, eventually acquire at least a quasibinding force.

3.2.6 WTO/SPS

The World Trade Organization[7] (WTO) tries to remove barriers to trade. To achieve this, several measures have been taken. Tariff barriers were reduced and to the extent that this was successful nontariff barriers became more of a concern. The basic treaty addressing trade in goods is the General Agreement on Tariffs and Trade (GATT). The GATT recognizes that certain exceptions to free trade can be necessary to protect higher values like health and (food) safety.

7. Established 1 January 1995 by the Agreement Establishing the World Trade Organization as the result of the so-called Uruguay round of trade negotiations and signed in Marrakesh on 15 April 1994 (WTO Agreement). The WTO is the institutional continuation of the General Agreement on Tariffs and Trade 1947 (GATT).

In food trade, differences in technical standards like packaging requirements may cause problems, but it is mostly concerns about food safety, human health, animal, and plant health that induce national authorities to take measures which may frustrate the free flow of trade. To address these concerns, two WTO treaties were concluded: the Agreement on Technical Barriers to Trade (the TBT Agreement) and the Agreement on the Application of Sanitary and Phytosanitary Measures (the SPS Agreement).

The SPS Agreement was drawn up to ensure that countries only apply measures to protect human and animal health (sanitary measures) and plant health (phytosanitary measures) based on the assessment of risk, or in other words, based on science. The SPS Agreement incorporates, therefore, safety aspects of foods in trade. The TBT Agreement covers all technical requirements and standards (applied to all commodities), such as labeling, that are not covered by the SPS Agreement. Therefore, the SPS and TBT Agreements can be seen as complementing each other.

To a certain extent the WTO is a supranational organization. The treaties concluded between its members are binding. There is the Dispute Settlement Understanding, providing an arbitration procedure to resolve conflicts. If a party wants to present a conflict, a Dispute Settlement Body (DSB) is formed to arbitrate on the basis of WTO law. If a party does not agree with the decision of the DSB, it can take the case to an Appellate Body (AB). While the WTO does not have powers to enforce decisions taken in this arbitration procedure, it can condone economic sanctions by the winning party if the decision reached is not implemented by the party found at fault. These sanctions usually take the form of additional import levies on goods from the state found at fault. If the levies are condoned by the DSB (or the AB), setting them does not in itself constitute an infringement of WTO obligations.

As follows from the above, the SPS Agreement is very important from a food safety point of view. The SPS Agreement recognizes and further elaborates on the right of the parties to this agreement to take sanitary and phytosanitary measures necessary for the protection of human, animal, or plant life or health. The measures must be scientifically justified, and they may not be discriminating, nor constitute disguised barriers to international trade.

If the measures are in conformity with international standards, no scientific proof of their necessity is required. These measures are by definition considered to be necessary. The most important international standards regarding SPS are set by the so-called "three sisters" of the SPS Agreement: The Codex Alimentarius Commission, the International Office of Epizootics (OIE[8]), and the Secretariat of the International Plant Protection Convention (IPPC). The standards on food and on food safety are mainly to be found in the Codex Alimentarius.[9]

3.2.7 Conclusion

The inclusion of the Codex Alimentarius in the SPS Agreement greatly enhances its significance. WTO members who follow Codex standards are liberated from the burden of having to prove the necessity of their sanitary and phytosanitary measures. National measures not based on Codex have to be proven to be science based.

References

FAO/WHO, 2016. Understanding Codex. Rome. www.fao.org/3/a-i5667e.pdf.

Masson-Matthee, M.D., 2007. The Codex Alimentarius Commission and its standards. In: An Examination of the Legal Aspects of the Codex Alimentarius Commission. Asser Press, The Netherlands.

van der Meulen, B.M.J., 2018. Codex Alimentarius: The Impact of the Joint FAO/WHO Food Standards Program on EU Food Law. European Institute for Food Law Working Paper 2018/04. http://www.food-law.nl/Working-papers/.

Further reading

Codex Alimentarius Commission, 2018. Procedural Manual. FAO/WHO, Rome. www.fao.org/documents/card/en/c/I8608EN.

8. The abbreviation follows the French spelling.
9. The Agreement on Technical Barriers to Trade (TBT Treaty) has similar articles.

Chapter 3.3

United States of America

Neal D. Fortin
Institute for Food Laws and Regulations, Michigan State University, East Lansing, MI, United States

3.3.1 Introduction

And chalk and alum and plaster are sold to the poor for bread.

—Alfred Lord Tennyson, *Maud* (1886).

3.3.1.1 What is food law?

The history of adulteration and misrepresentation of food is as old as trade in food (Hart, 1952). Consequently, the history of food law designed to thwart adulteration and misrepresentation runs as far back as the history of commerce itself (Hart, 1952).

The earliest food law in the United States extends back into the colonial years and the adoption of British common law. The essence of the common law was plain and direct: (1) Do not poison food and (2) Do not cheat (Hutt, 1960).

Under the common law, adulterated food consisted of food that was unfit for human consumption or contained some deleterious substance, whereby rendering it dangerous to health (*Ibid.*) Food labels were infrequent, so correspondingly there was no common-law offense of mislabeling. However, the common-law offense of false representation of merchandise for sale is similar to the present day offense of false labeling (*Ibid.*)

For most of recorded history (throughout the world), food law revolved around these two pillars: regulation against adulteration and misrepresentation. Food production and consequently food law has grown more complex, but these fundamental pillars remain. Food law remains as the body of public law designed to prevent adulteration and mislabeling.

Among U.S. academics, it is vogue today to use "food law" expansively to include any area of law or even policy that is related to food. For example, crop insurance law affects the price of farm commodities used for food, ergo food law; animal welfare law applies to livestock used for food, ergo food law; water quality issues arising from manure in animal husbandry are part of food production, ergo food law. This broad context reminds us that our food supply is connected to agriculture, the environment, and social issues. Nonetheless, this expansive definition diverges from tradition, and this scope is so capacious that it becomes amorphous (Fortin, 2017).

Moreover, the morality, ethics, and the philosophy of what our food production systems *should* look like—those fields of discussion fundamentally are not food law. For example, a law banning *foie gras* production is a law related to food. The discussion of the controversy over force feeding ducks and geese is a moral, ethical, and philosophical concern. However, this is hardly a reason to classify it as a food law issue.

Therefore, for clarity, consistency, and practicality, "food law" in this section will use the traditional meaning of food law, which is regulatory law regarding food.[10] That is, the laws and regulations enforced by the government concerning the safety, labeling, and honest presentation of food.

3.3.1.2 The history of U.S. food law

The earliest food law recognized the importance of protecting commerce as well as consumers. Merchants pushed for the establishment of food inspection laws because they recognized the marketing problems created by inferior goods, and they wished to create a level playing field (Janssen, 1975). Honest dealings are important in creating and preserving the markets.

10. "Regulatory law" is the term loosely used to describe the laws and regulations enforced by administrative agencies of the executive branch of government.

The early 19th century was the height of *laissez faire* capitalism in the United States. This economic philosophy called for deregulation of business but not ensuring safe, pure food.[11] At the same time, adulteration of food increased.[12] This degradation of the food supply was increasingly documented as scientists found new ways to detect various forms of adulteration (Beck, 1846; Byrn, 1852; Hoskins, 1861; Felker, 1880; Richards, 1886; Batrershall, 1887). In response, the period from 1865 to 1900 was one of increased state food legislation (Hart, 1952). In these early days through the late 1800s, nearly all of the early food laws in the United States were state and local.

The United States is a federation. Some powers are assigned to the federal or national government, but other powers are reserved to the individual states. Of these, regulation for health and safety is a power primarily held by the states. However, with the increase of national and international commerce, the need for federal regulation of food also increased (authorized under the power of the federal government to regulate commerce). The enactment of federal food law began at the end of the 19th century largely in reaction to scandals about food fraud and adulteration.

> *We face a new situation in history. Ingenuity, striking hands with cunning trickery, compounds a substance to counterfeit an article of food. It is made to look like something it is not; to taste and smell like something it is not; to sell like something it is not, and so deceive the purchaser.*
>
> US Congressional Record, 49 Congress I Session 1886

For example, in 1883, the United States Congress enacted a law to prevent the importation of adulterated tea. The oleomargarine statute followed in 1896, which was passed because of the dairy industry's objections to the sale of fats colored to look like butter.[13] In 1890, Congress passed a meat inspection act to facilitate the export sale of meat (Hutt and Hutt, 1984). A live cattle inspection law followed in 1891 (Hutt and Hutt, 1984). In 1899, Congress authorized the Secretary of Agriculture to inspect and analyze any imported food, drug, or liquor when there was reason to believe there was a danger (*Ibid.*)

3.3.1.3 The evolution of food law through scandal and tragedy

Public support for passage of a federal food and drug law grew as journalists exposed in shocking detail the frauds and dangers in the food industry, such as the use of poisonous preservatives and dyes in food. A final catalyst for change was the 1905 publication of Upton Sinclair's *The Jungle*. Sinclair's portrayal of nauseating practices and unsanitary conditions in the meatpacking industry captured the public's attention. On June 30, 1906, President Theodore Roosevelt signed both the Pure Food and Drug Act[14] and the Meat Inspection Act[15] into law.

Not long after passage of the Pure Food and Drug Act, legislative efforts began to expand and strengthen the law. For example, leaders in the food industry called for more stringent product quality standards to stop dubious competitors by creating a level playing field. Likewise, consumers wanted stronger safety standards and fair dealing. However, major revision of the 1906 Act stalled until a precipitous tragedy occurred. The agonizing deaths of more than 100 children from elixir of sulfanilamide spurred the passage of the Federal Food, Drug, and Cosmetic Act of 1938. Sulfanilamide is an effective antibiotic but bitter tasting. A manufacturer added diethylene glycol to sweeten the elixir. The company was unapologetic that it had not tested the glycol for toxicity because the law required none. That a company would put profits first and forgo even a simple study to safeguard the lives of children remains a shocking reminder of the need for preventive regulation of food and drug manufacturers.

The food laws continued to evolve based upon the concerns and issues of the times. The following are just some prominent examples: the FDCA has been amended more than 100 times since 1938. In the 1950s, concerns over new and increasing use of synthetic food additives and pesticides were high, and the public dreaded cancer. Consequently, the Food Additives Amendment of 1958 was enacted, requiring the evaluation of food additives to establish safety. The amendment includes the Delaney Clause, which forbids the use of any substance as a food additive if it is found to cause cancer in laboratory animals or humans (FDCA sec. 409(c) (3) (A)).

---

11. This was not called the era of the "robber barons" for nothing.
12. For example, in the period around 1880, over 73% of the milk in Buffalo, New York, was watered; 41% of the samples of ground coffee in New York were adulterated; and 71% of the olive oil in New York and Massachusetts were adulterated (Hart, 1952).
13. Margarine was patented in 1869 (Hutt and Hutt, 1984).
14. 21 U.S.C. § one *et seq.*
15. 21 U.S.C. § 601 *et seq.*

In the 1980s, growing public interest in nutrition and the food industry's desire to make health claims led to passage of the Nutrition Labeling and Education Act (NLEA) in 1990. A series of high profile foodborne illness outbreaks—2006 (*E. coli*—fresh spinach), 2007 (melamine—pet food and infant formula), 2008 (salmonellosis—tomatoes/peppers), 2009 (salmonellosis—peanut paste), and 2010 (salmonellosis—eggs)—pushed forward the Food Safety Modernization Act in 2011, the largest overhaul of U.S. food law since 1938.

3.3.2 The food regulatory system

Federal responsibility for the direct regulation of food in the United States has primarily been delegated to the U.S. Food and Drug Administration (FDA) and the U.S. Department of Agriculture (USDA). However, a number of other agencies are involved on certain aspects of food and with certain types of food. Just as the statutes were written to address specific problems at particular points in history, the delegation of food regulation developed piecemeal to address specific concerns as they arose. The delegation, thus, represents a conglomeration of many roughly grouped pieces rather than an organization by design.

3.3.2.1 Food and Drug Administration

The FDA regulates all domestic and imported food sold in interstate commerce except for meat, poultry, catfish, egg products, and most alcoholic beverages. The agency inspects food production establishments and food warehouses, and collects and analyzes samples for physical, chemical, and microbial contamination. The FDA is also responsible for reviewing safety of food and color additives before marketing, animal drugs for safety to animals that receive them (as well as humans who eat food produced from the animals), and the safety of animal feeds used in food-producing animals. The agency also establishes rules for good manufacturing practices, such as sanitation and hazard analysis and critical control point (HACCP) programs.

3.3.2.2 U.S. Department of Agriculture Food Safety and Inspection Service

The USDA FSIS oversees domestic and imported meat and poultry and related products, such as meat- or poultry-containing stews, soups, and frozen foods. The agency also oversees processed egg products, such as liquid, frozen, and dried pasteurized egg.

FSIS inspects food animals for diseases before and after slaughter and inspects meat and poultry slaughter and processing plants and "egg product" processing plants (egg breaking and pasteurizing operations). The agency is also responsible for ensuring foreign meat and poultry processing plants that export these products to the United States meet U.S. standards.

3.3.2.3 Centers for Disease Control and Prevention

The Centers for Disease Control and Prevention investigate sources of foodborne disease outbreaks together with local, state, and other federal officials. The CDC also maintains a nationwide system of foodborne disease surveillance, develops and advocates public health policies to prevent foodborne diseases, and conducts research to help prevent foodborne illness.

3.3.2.4 U.S. Environmental Protection Agency

The Environmental Protection Agency oversees the regulation of drinking water standards and pesticide safety. The EPA also regulates toxic substances and wastes to prevent their entry into the environment and food chain. The EPA establishes requirements for the safe use of pesticides and also sets tolerance levels for pesticide residues in foods.

3.3.2.5 Alcohol and Tobacco Tax and Trade Bureau

The Alcohol and Tobacco Tax and Trade Bureau of the U.S. Department of Treasury has jurisdiction over the labeling of alcoholic beverages under the Federal Alcohol Administration Act, 27 USC § 201 et seq. The TTB oversees all alcoholic beverages except nonmalt beverages containing less than 7% alcohol. The TTB enforces the food safety laws governing alcoholic beverages and investigates adulteration of alcoholic products, sometimes with help from FDA.

3.3.2.6 U.S. Customs and Border Protection

The U.S. Customs and Border Protection oversees all importation into the United States and works with other federal regulatory agencies to ensure that all goods entering and exiting the United States do so according to U.S. laws and regulations.

3.3.2.7 Federal Trade Commission

The Federal Trade Commission oversees advertising. The FTC enforces a variety of laws that protect consumers from unfair, deceptive, or fraudulent practices, including deceptive and unsubstantiated advertising.

3.3.2.8 Other federal agencies

Five additional federal agencies and units become involved with food in some way as well. For example, the USDA has a number of programs that, while nonregulatory by nature, can affect food regulation. The USDA Agricultural Marketing Service (AMS) provides voluntary standardization, grading, and market news services for specific agricultural commodities. The Agricultural Research Service (ARS) is the main scientific research arm of USDA. The USDA Economic Research Service (ERS) provides economic analysis relating to agriculture, food, the environment, and rural development. The USDA Grain Inspection, Packers, and Stockyards Administration (GIPSA) provides grading and standardization programs for grains and related products and regulates and maintains fair trade practices in the marketing of livestock.

The U.S. Codex Office is the point of contact in the United States for the Codex Alimentarius Commission and its activities. The Department of Commerce, National Marine Fisheries Service (NMFS), provides voluntary inspection and certification of fish operations, and administers grades and standards for fish and fish products (similar to the AMS grading and standards programs). The NMFS also inspects and certifies fishing vessels, seafood processing plants, and retail facilities for federal sanitation standards.

3.3.2.9 State and local governments

State and local governments also play a prominent role in food safety regulation in the United States. The combined food-related budget of the above-mentioned federal agencies amounts to only a tiny fraction of the total federal government budget, and the number of federal food authorities is relatively small. Correspondingly, most food safety regulatory work is performed by state and local officials, who far outnumber the federal food regulatory staff.

State and local governments employ food inspectors, sanitarians, microbiologists, epidemiologists, food scientists, and more. Their duties and areas of responsibility are defined by state and local laws. For example, some of these officials monitor only one kind of food, such as milk or seafood. Many work within a specified geographical area, such as a county or a city. Others regulate only one type of food establishment, such as restaurants or meat-packing plants.

State food laws generally model the federal laws but, with some exceptions, are not required to be the same as federal law. Only in meat inspection does federal law broadly preempt state authority. State meat and poultry inspection programs must be assessed by the USDA FSIS to determine whether the state inspection programs are at least equal to the federal program. If a state chooses to end its inspection program or cannot maintain the FSIS equivalent standard, the FSIS assumes responsibility for inspection in that state.

3.3.3 Major federal laws

3.3.3.1 Food, Drug, and Cosmetic Act

The Federal Food, Drug, and Cosmetic Act of 1938 was adopted to correct the imperfections of the 1906 Act and with amendments remains the main food law today. The 1938 Act established a comprehensive set of standards by which food safety could be regulated. More than 100 amendments and revisions after 1938 extended the act's coverage and authority. Most states have enacted primary food laws that are largely the same as the federal law.

3.3.3.2 Federal Meat Inspection Act

Federal Meat Inspection Act of 1906 was substantially amended by the Wholesome Meat Act (1967). The FMIA requires USDA to inspect all cattle, sheep, swine, goats, and horses when slaughtered and processed into products for human consumption. The primary goals of the law are to prevent adulterated or misbranded livestock and products from being sold as food, and to ensure that meat and meat products are slaughtered and processed under sanitary conditions. The Food and Drug Administration is responsible for all other or exotic meats, such as venison and bison.

3.3.3.3 Poultry Products Inspection Act

Poultry Products Inspection Act provides for the inspection of poultry and poultry products and regulates the processing and distribution of poultry to prevent the movement or sale of poultry products that are adulterated or misbranded.

3.3.3.4 Egg Products Inspection Act

Egg Products Inspection Act (EPIA) provides for the inspection of certain egg products, restrictions on the certain qualities of eggs, and uniform standards for eggs. EPIA otherwise regulates the processing and distribution of eggs and egg products.

3.3.3.5 The regulations

While the above statutes or acts are written by Congress, regulations, also called administrative rules, are promulgated by federal agencies to implement and interpret the statutes. Regulations are codified (organized) in the Code of Federal Regulations (C.F.R.). Titles 7, 9, and 21 contain most of the laws regulating foods. However, Titles 5, 15, 16, 19, 27, 42, and 49 contain other matters that relate to food in an indirect manner.

3.3.4 Principles and concepts

3.3.4.1 Codex Alimentarius

The Codex Alimentarius Commission (CAC) was created in 1963 as a joint body organized by the FAO and the WHO. The CAC develops and publishes international food standards, codes of practice, guidelines, and recommendations. These published standards are collectively referred to as the Codex Alimentarius, or simply Codex.

The primary U.S. food safety agencies, FDA, USDA, and EPA, have always looked on Codex Alimentarius as an important organization. These agencies have invested in and long been active participants in Codex. In addition, the agencies routinely consider the applicable Codex standard when drafting new food regulations. Moreover, the FDA is required by law to review all food standards adopted by the Codex Alimentarius Commission and consider them for adoption (21 C.F.R. 130.6). The EPA is required by law to adopt Codex maximum residue levels for pesticides or to publish for public comment a notice explaining the EPA's reasons for departing from the Codex level (Food, Drug, and Cosmetic Act sec. 408(b) (4); 21 U.S.C. 346a(b) (4)).

Many U.S. food safety standards are the same as Codex Alimentarius standards. Nevertheless, the U.S. food safety agencies typically do not adopt Codex standards verbatim or completely. Often the statutes passed by Congress create conflicts with Codex standards. For example, the Delaney clause results in a U.S. ban on certain food colorants that are permitted under Codex standards.[16] The statute is binding on the FDA; thus, the agency cannot adopt the Codex standards for those colorants banned under the Delaney clause.

3.3.4.2 Standards

Detailed standards are typically put into regulations by the agencies based on a mandate or authorization by statute from Congress. Although there has been some movement away from use of prescriptive standards in recent years, U.S. food law nonetheless contains thousands of standards relating to food. Standards remain essential for informing regulated parties of expected performance.

The standards include everything from maximum pesticide residue limits to requirements to make a "low fat" claim. For example, standards of identity prescribe the composition and labeling of specific food products, such as ketchup and pork sausage. Pesticide regulations control usage and prescribe maximum residue levels. Food additive regulations provide

16. The Delaney clause sets a "zero risk" standard for food additives and food colors that induce cancer responses in test animals, even if the risk to humans is inconsequential because the carcinogenic potential is weak and/or human exposure is very low. *See*, Federal Food, Drug, and Cosmetic Act, sec. 409(c) (3) (A).

listing of permitted food additives and their allowed uses and maximum quantities. Defect action levels list the maximum permitted nonsafety-related defects tolerated in various foods. Good manufacturing practice rules set required sanitation (hygiene) requirements in food manufacturing facilities. In addition to these broad categories of standards, specific food categories have specific additional standards, such as infant formula and low acid canned foods.

3.3.4.3 Authorization requirements

Before beginning production or holding of food for sale in the United States, the facility must be licensed by the applicable state agency and registered with the Food and Drug Administration, or in the case of meat processing, registered with the USDA Food Safety Inspection Service. Typically, a prelicensing inspection is required to assure that minimum requirements are met.

For imported food, prior notice of import is required for FDA-regulated food products. In addition, the foreign manufacturing facility must be registered with the FDA. For USDA-regulated products, the establishment must be authorized by the foreign government and certified by the USDA as meeting U.S. requirements.

Labels for most meats and alcoholic beverages must be approved by FSIS and TTB, respectively, before being offered for sale. In contrast, the FDA does not approve labels.

Food additives must be approved by the FDA before use in food. New food additives require a petition to FDA with a corresponding dossier of evidence of safety for the intended use. Once approved, the food additive and its limits are published in a regulation and available for any food manufacturer to use. Uses of substances that are generally recognized by nearly all scientific experts in the field as being safe (GRAS) for that use are exempt from the food additive requirements. There is no GRAS provision, however, for food colorings, and all must be approved by FDA before use.

The regulatory distinctions between the three legally separate categories—food additives, color additives, and GRAS status—are among the most confused in U.S. food law. GRAS status, for example, exempts a food ingredient from the food additive review requirements, but the safety standard for GRAS is *more* stringent than the standard for food additives. Nonetheless, the food additive petition process is time consuming and slow; therefore, GRAS determination has become a preferred method of food substance acceptance. Another common confusion is that although the GRAS status exempts a substance's use from the food additive petition requirements, GRAS status does not exempt the use from the food safety standards for a food additive. To give an example, vinegar is a substance often added to food for its functional purpose as an acidifier, and vinegar falls under the food additive definition in the law, but then it is exempt from being considered a food additive by GRAS status. Nonetheless, the level of safety required for a GRAS use is the same as for a food additive.

3.3.4.4 Food from genetically engineered organisms

Food from recombinant DNA organisms, also called genetically modified organism (GMOs), falls into three different regulatory categories: animals, plants with GRAS status, and plants without GRAS status. All GMO animals require a new drug application and approval by the FDA before use in food. GMO plants containing a novel substance, or a non-GRAS use of a nonnovel substance, require a food additive petition and approval by FDA before use in food. On the other hand, GMO plants without novel substances generally do not require FDA approval before use in food so long as the use of those plants in food is GRAS. Nonetheless, the FDA strongly request a voluntary consultation with the agency before any use in food. This consultation is voluntary because nothing in the law requires a review, but practical reasons make the consultation mandatory. (In addition, *all* new GMO plants require a permit from the USDA Animal and Plant Health Inspection Service before being grown.)

3.3.4.5 Process requirements

Nearly all food produced in the U.S. must comply with a science-based, risk control system. Juices, seafood, and meat must comply with HACCP rules. Most other FDA-regulated food must comply with the Hazard Analysis and Risk-Based Preventive Controls Rule (21 C.F.R. § 117) or the Produce Safety Standard (21 C.F.R. § 112).

In addition, some foods must comply with their own specialized rules, specifically low acid canned foods, acidified food, bottled drinking water, infant formula, dietary supplements, and eggs. The FDA has the authority to establish additional specific food safety performance standards when needed.

3.3.4.6 Risk analysis

In the 1970s, the U.S. FDA developed rudimentary tools of quantitative risk assessment for food safety. Risk assessments tools became more sophisticated and quantitative risk assessment eventually became a common feature of food safety decision-making at the FDA, EPA, and USDA.

In the 1970s and early 1980s, the conceptual framework of risk analysis evolved and was described by the National Research Council (NRC) in its 1983 report, the "Red Book" (NRC, 1983). Under this framework, risk analysis consists of three processes: risk assessment, risk management, and risk communication. In 2007, this NRC framework formed the basis for the global standard for risk analysis (Codex Alimentarius, 2007).

As framed by both NRC and Codex, the risk assessment process should be distinguished—but not completely separated—from risk management. Continual communication between risk assessors (scientists) and risk managers (regulators and legislators) is vital if politicization of the science is to be avoided. For instance, policy judgments often frame the risk assessment and should be explicit to both risk assessment and risk management. Therefore, the often-misunderstood recommendation to distinguish risk assessment from risk management should not be interpreted as a recommendation for separation organizationally.

Typically, in the United States, the Congress begins risk management and enacts a statute that directs and empowers the FDA, USDA, or EPA to act. The statute defines the parameters that the agency must follow, but usually the agency is left to conduct both risk assessment and risk management (within the given limits), decide on the technical details, and put the details down in a regulation. Congress is empowered to review the final regulations and can override them if desired, but this is rare.

This structure puts the ultimate risk management decisions with the legislature, and the food safety agencies are often constrained in risk management choices by the mandates of their enabling statutes (written by Congress). For example, the Delaney clause sets a rigid risk management scheme for potential carcinogens in food additives and color additives that is more stringent than necessary for a quantitative risk assessment for negligible risk. The political decisions of the U.S. Congress are typically where other factors beyond science are weighed, such as society's dread of cancer and other societal concerns.

3.3.4.7 Powers of enforcement

The enforcement powers of the food safety agencies have evolved over time in response to scandals and crises just as the rest of the laws have evolved. The enforcement tools rest on the foundational prohibitions of adulteration and mislabeling. Other prohibited acts have been added to the list over time. The prohibited acts are criminal offenses punishable by fines and imprisonment.

In addition to the criminal penalties, there are a variety of civil actions that can be initiated. The FDA can order a food product's recall. And, in certain limited circumstances, the FDA has the authority to debar (prohibit) a person from conducting a food business.

The agencies also have the ability to detain food products (seizures or detentions). The FDA can suspend a facility's registration, which effectively stops all food sales from that facility. In addition, the agencies can also seek court injunctions to compel compliance with the law. Moreover, regular criminal charges (such as fraud) sometimes are brought in addition to the food law violations.

The agencies are also authorized to use adverse publicity as an enforcement tool. The lost sales from adverse publicity can be more punishing than any fines that can be imposed. More than one company has gone out of business due the bad publicity from a foodborne illness outbreak.

The USDA has additional powers related to its inspector authority. The agency can withdraw a specific product's inspection mark (which prohibits the product's sale), suspend inspection (effectively shutting down that facility during the suspension), and withdraw inspection (which puts the facility out of business).

3.3.5 Labeling

3.3.5.1 The affirmative requirements

U.S. food labeling law is primarily intended to provide consumers with the information needed to make informed choices. The main affirmative labeling requirements for packaged food are the following:

- Statement of identity of the food
- Net contents

- Ingredient statement
- Responsible party name and address
- Nutrition Facts
- Major allergen identification

The details of the rules concerning the above six affirmative requirements are extensive. The regulations specify everything from the size of the typeface to the location and wording.

There are many additional labeling rules, but most are requirements on voluntary statements on the label. For instance, placing the term "organic" on the label is voluntary, but the use of the term triggers many mandatory rules covering the use.

3.3.5.2 Labeling of ingredients and food additives

All ingredients must be listed in the ingredient statement by common or usual name; for example, sodium benzoate and acetic acid.

When the function of an ingredient is as a chemical preservative, the function must be included in the ingredient statement; for example, "BHT (preservative)."

Color additives are defined by law as a category separate from food additives. The function of color additives must be declared in the ingredient statement. This rule applies whether or not the coloring is from a natural source or synthetic, so, for example, "beet juice (coloring)" when the juice is added for color.

3.3.5.3 "Major" allergen labeling

The Food Allergen Labeling and Consumer Protection Act requires that certain major allergens must have the source of the allergen identified on the label. The law's design is twofold: to focus attention on the allergens that cause the vast majority of allergic reactions in the United States and to identify the source for ingredients that might otherwise not be recognized by consumers (for example, casein might not be recognized as a milk protein). The identification of the allergen source may be either in the ingredient statement, such as "casein (milk)," or listing in a "Contains" statement immediately after the ingredient statement.

The major allergens specified in the law are the following:

- eggs,
- fish,
- crustacea,
- milk,
- tree nuts,
- wheat,
- peanuts, and
- soybeans.

It is worth noting that in addition to the calling out of the source of major allergens, other allergens are still required to be listed in the ingredient statement. Thus, while not listed as major allergens under U.S. law, sesame seed and molluscan shellfish must be declared in the ingredient statement.

3.3.5.4 Nutrition facts labeling

Since 1990, the United States has required Nutrition Facts labeling on most foods. The requirement came about largely due to consumer demand for more information to make healthy choices. The requirements are quite detailed and numerous. The FDA lists mandatory nutrient information (e.g., calories, sodium) and voluntary nutrient information (e.g., vitamins A and C). Only the FDA-listed mandatory and voluntary nutrients may be included.

3.3.5.5 Nutrient level claims

Nutrient content claims are labeling statements about the level of a nutrient in food (e.g., low fat and sodium free). Nutrient content claims are strictly controlled. No nutrient content claim may be made unless expressly defined by the FDA in regulation with one exception for claims based on certain statements by scientific bodies of the U.S. government. Surrounding each permitted claim are details on conditions for authorized use.

3.3.5.6 Health claims

A health claim is any claim made on the label or labeling that characterizes the relationship of any substance to a disease or health-related condition. 21 C.F.R. § 101.14(a) (1). Before 1990, all health claims were considered illegal drug claims. There are now three categories of permitted health claims on food:

- NLEA or significant scientific agreement claims
- FDMA or authoritative statement claims
- Qualified claims

3.3.5.6.1 NLEA or significant scientific agreement claims

The NLEA of 1990 amended the Food, Drug, and Cosmetic Act to allow health claims for foods. These claims must be approved by the FDA through publication of a regulation. The standard of proof for these claims is "significant scientific agreement," which is a high hurdle. An example of an NLEA approved claim is "Diets low in sodium may reduce the risk of high blood pressure, a disease associated with many factors."

3.3.5.6.2 FDAMA or authoritative statement claims

The FDA Modernization Act of 1997 (FDAMA) further amended the FD&C Act to permit health claims based on an "authoritative statement" linking a nutrient to a disease made by a scientific body of the U.S. government, such as the National Academy of Sciences. An example of an accepted FDAMA claim is "Diets containing foods that are good sources of potassium and low in sodium may reduce the risk of high blood pressure and stroke."

3.3.5.6.3 Qualified health claim

In December 2002, FDA announced the availability for companies to petition the FDA to authorize qualified health claims. These claims require a qualifying statement explaining how the quality or strength of evidence falls below the requirement for an FDA authorizing regulation for a health claim. An example of an accepted qualified claim is "Scientific evidence suggests but does not prove that eating 1.5 ounces per day of most nuts [such as name of specific nut] as part of a diet low in saturated fat and cholesterol may reduce the risk of heart disease. [See nutrition information for fat content.]"

3.3.5.6.4 Medicinal claims

The term "medicinal claim" is not defined in U.S. law. However, the area of medicinal claims overlaps with the U.S. regulated areas of health claims, drug claims, and medical food claims.

A health claim is defined as any claim made on the label or labeling that expressly or by implication characterizes the relationship of any substance to a disease or health-related condition. 21 C.F.R. § 101.14(a) (1). In comparison, a drug claim is a claim to diagnose, cure, mitigate, treat, or prevent disease. A structure–function claim on a product other than food is also a drug claim (FDCA § 201(g) (1) (B), and (C)).

U.S. law does not expressly prohibit drug claims on food, but all drug claims require FDA approval. However, as a practical matter, essentially, no claims to cure, mitigate, treat, or prevent disease are approved for foods. Food health claims must stay restricted to reducing the risk of disease. That said, the line between these two categories can sometimes be murky.

A "medical food" is a food formulated to be consumed or administered under the supervision of a physician and which is intended for the specific dietary management of a disease or condition for which distinctive nutritional requirements (Orphan Drug Act § 5(b)). This is a narrow category, and medical food claims must be based on therapeutic or chronic medical needs, such as foods for those with genetic metabolic disorders.

3.3.6 Conclusion

The history of U.S. food law reveals how it evolved piecemeal in response to the needs of the day. In the beginning, the protection of commerce was a primary factor. In the last 100 plus years, food safety scandals have been a significant force toward change. Sadly, all too often this has been outrageous tragedy.

The result today is a complex conglomeration of requirements and provisions. It is important to recognize the history that culminated in change in the law. History instructs us that the legislatures will follow rather than lead food safety reforms. Scientists play a critical role in increasing awareness of food safety risks. Public outrage also plays a role.

In addition, the food industry plays a leadership role. Enlightened self-interest points to stringent but fair food safety regulation as necessary to preserve and grow food trade and profits. Most important, because legislators are reactive, the food industry must be proactive. Major food law reform usually occurs only when all—scientists, the public, and food industry leadership—are galvanized by current events.

3.3.6.1 Sources for more information

A free guide to United States food laws and regulations is available at < http://libguides.lib.msu.edu/c.php?g=212832 >.

The federal agencies also maintain repositories of food law and guidance documents. Some key ones are the following:

- Food and Drug Administration: www.fda.gov
- USDA Food Safety Inspection Service: www.fsis.usda.gov
- Environmental Protection Agency: www.epa.gov
- Alcohol and Tobacco Tax and Trade Bureau: www.ttb.gov
- Federal Trade Commission: www.ftc.gov
- Centers for Disease Control: www.cdc.gov

Free online locations for reference to the federal statutes include the following:

- Cornell's LII: www.law.cornell.edu/uscode
- Office of the Law Revision Counsel: http://uscode.house.gov
- U.S. Government Printing Office (GPO): www.gpo.gov/fdsys/browse/collectionUScode.action?collectionCode=USCODE

References

Batrershall, J.P., 1887. Food Adulteration and its Detection.

Beck, L.C., 1846. Adulteration of Various Substances Use in Medicine and the Arts.

Byrn, M.L., 1852. Detection of Fraud and Protection of Health: A Treatise on the Adulteration of Food and Drink.

Codex Alimentarius, 2007. Working Principles for Risk Analysis for Food Safety for Application by Governments. CAC/GL 62−2007.

Delaney Clause, Federal Food, Drug, and Cosmetic Act, Sec. 409(c)(3)(A)).

Egg Products Inspection Act.

Federal Alcohol Administration Act, 27 USC § 201 et seq.

Federal Meat Inspection Act of, 1906 Federal Meat Inspection Act of 1906.

Felker, P.H., 1880. What the Grocers Sell Us: A Manual for Buyers.

Food Additives Amendment of 1958.

Fortin, N.D., 2017. What is food law?. In: Foreword, Food Regulation: Law, Science, Policy, and Practice, second ed. Wiley.

Hart, F.L., 1952. A history of the adulteration of food before 1906. Food Drug Cosmet. Law J. 7, 5−22.

Hoskins, T.H., 1861. What We Eat: An Account of the Most Common Adulterations of Food and Drink.

Hutt, P.B., 1960. Criminal prosecution for adulteration and misbranding of food at common law. Food Drug Cosmet. Law J. 15, 382−398.

Hutt, P.B., Hutt II, P.B., 1984. A history of government regulation of adulteration and misbranding of food. Food Drug Cosmet. Law J. 39, 2−72.

Janssen, W.F., 1975. America's first food and drug laws. Food Drug Cosmet. Law J. 30, 665−672.

National Research Council, 1983. Risk Assessment in the Federal Government: Managing the Process. National Academies Press, Washington, DC.

Poultry Products Inspection Act.

Richards, E.H., 1886. Foods Materials and Their Adulterations.

Sinclair, U., 1905. The Jungle.

Wholesome Meat Act of 1967 (Pub. L. 90-201), 1967.

Further reading

Food and Drug Administration, 2013a. Current Good Manufacturing Practices and Hazard Analysis and Risk-Based Preventive Controls Rule. 21 C.F.R. § 117.

Food and Drug Administration, 2013b. Standards for the Growing, Harvesting, Packing, and Holding of Produce for Human Consumption. 21 C.F.R. § 112.

Chapter 3.4

Canada

Melissa M. Card
Institute for Food Laws & Regulations, MSU, Michigan State University's College of Law, United States

3.4.1 Introduction

Canadian food law has evolved, and continues to evolve, based on societal concerns. For example, the Safe Food for Canadians Act was introduced due to the societal concern that Canada's Federal Legislation was not modernized or effectively protecting Canadians from unsafe foods. This chapter will cover various topics of Canadian food law exemplifying the evolution of food law due to societal concerns.

3.4.1.1 Sources of legislation for food

Food law touches all Canadians' lives. Canadian food laws and regulations ensure that the food products that Canadians consume, import, and export are safe for consumers. This section will discuss the sources or legislation for food in Canada, game changing events in Canada leading to evolved food law, the addressees of food law, and Canada's role in Codex Alimentarius.

The Food and Drugs Act is the primary source of food law in Canada. The Food and Drugs Act enacts safety, compositional, nutritional, and labeling requirements for food, as well as sets out inspection provisions and creates offences for violations of the Food and Drugs Act and its related regulations.[17] The Food and Drug Regulations, which prescribe standards for food content and labeling, were promulgated under the Food and Drugs Act.[18]

To protect Canadian families from potentially unsafe food, the Safe Food for Canadians Act, S-11, was adopted by the Senate on October 17, 2012, and passed by the House of Commons on November 20, 2012. On November 22, 2012, it received Royal Assent. This Act consolidated the authorities of the Fish Inspection Act, the Canada Agricultural Products Act, the Meat Inspection Act, and the food provisions of the Consumer Packaging and Labeling Act.

In addition to the federal statutes mentioned above, each Canadian Province and Territory has enacted additional legislation that impacts food that is manufactured or sold in that Province or Territory. The additional legislation is designed to mirror Federal legislation, but Provincial statutes and regulations sometimes take a different approach.

3.4.1.2 Game changing event(s)

The Food and Drugs Act was enacted in late 1920, introducing provisions to prevent the false or misleading labeling of food. During the 1960s, the Government of Canada had approved the drug thalidomide, resulting in the deaths of thousands of infants and severe birth defects in others when the drug was taken by women in early stages of pregnancy. As a result of the health adversities that stemmed from the drug thalidomide, the federal government introduced a new bill to strengthen the legislation. The new legislation required manufacturers to provide evidence of efficacy prior to the drug being sold to the public.

The enactment of Safe Food for Canadians Acts was another meaningful change in Canadian Food Law. The Safe Food for Canadians Act consolidates food provisions administered and enforced by the Canadian Food Inspection Agency. The Food and Drugs Act will continue to exist separately, providing overarching protection for consumers from any foods that

17. *Food Regulation in Canada: Spring Semester 2016, Module One Reading Materials*, The Institute for Food Laws and Regulations (on file with author).

18. There are other federal statutes that apply to food products. Important federal statutes include the following: Agriculture and Agri-Food Administrative Monetary Penalties Act, Canada Agricultural Products Act, Canada Grain Act, Canadian Food Inspection Agency Act, Consumer Packaging and Labeling Act, Fish Inspection Act, Health of Animals Act, and Plant Protection Act.

are unsuitable for consumption, including those marketed exclusively within provinces. The Safe Food for Canadians Act focuses on (1) improved food safety oversight to better protect consumers; (2) streamlined and strengthened legislative authorities; and (3) enhanced international market opportunities for Canadian industry.

3.4.1.3 Addressees of food law and the role of risk analysis

The Safe Food for Canadians Act provides additional regulatory powers to the Federal government with respect to "food commodities," including the power to require licensing and/or registration of participants in the food commodity supply chain including food exports, new prohibitions such as prohibitions against tampering and selling recalled products, and a broadened arbitration and complaints adjudication regime to address disputes arising under the statute.[19] The food industry has the burden of coming into compliance. An outcome of these approaches taken under the Safe Food for Canadians Act will be a food safety system capable of continuous improvement based on science, global trends, and best practices and be focused on science-based and risk-based prevention and control of potential hazards as well as robust and responsive.

3.4.1.4 Role of the Codex Alimentarius

Canada is one of the 188 member governments involved in the Codex Alimentarius Commission, and has been a member of Codex since its creation in 1963.[20] An interdepartmental committee consisting of senior officials from Health Canada, the Canadian Food Inspection Agency, the Pest Management Regulatory Agency, Foreign Affairs and International Trade, and Agriculture and Agri-Food Canada manages the Codex Program in Canada.

The most recent strategic plan for Canada's involvement in the Joint FAO/WHO Food Standards Program focused on enhancing the work of the Codex Alimentarius Commission. For example, there is increased recognition of the importance of international regulatory cooperation under the Cabinet Directive on Streamlining Regulation. Generally, the strategic objectives relate to four themes: (1) enhance Canada's influence on Codex deliberations and outcomes; (2) promote the use of Codex standards as the basis for national policies and regulations; (3) enhance strategic and functional management of Canada's domestic Codex program; and (4) promote processes to enhance the efficiency and responsiveness of the Codex Alimentarius Commission.

3.4.2 Institutional

This section will examine sanctions prescribed by federal legislation for noncompliance with Canada's food laws and the process by which enforcement decisions are made by the federal agencies involved. Bear in mind that noncompliance with the relevant provincial legislation can result in sanctions as well. This section will also cover the main authorities involved in food law and the model of Canadian food law.

3.4.2.1 Authorities in food law

The Canadian Food Inspection Agency (the "CFIA") was created in 1997 to respond to food safety. The CFIA reports to Parliament through the Minister of Agriculture and Agri-Food. While CFIA has the responsibilities to provide inspection services relating to food safety, economic fraud, trade-related requirements, and animal and plant health programs, Health Canada maintains the responsibilities for establishing food safety policy, standard-setting, risk assessment, analytical testing, researching, and auditing. Even though there has been consolidation of food-related inspection activities in the CFIA, other Federal departments and agencies play a role in Canada's food industry. Additionally, each Province or Territory has a complement of government departments and agencies that are tasked with regulating agriculture, health, and food industries in that Province or Territory.

3.4.2.2 Is an FDA or EFSA model applied?

The Food and Drugs Act concerns the production, transport across provinces and territories, import, export, and sale of food, drugs, medical devices, and cosmetics. This Act's purpose is to ensure that these products are safe for consumers that

---

19. *Institute for Food Laws & Regulations*, Food Regulations in Canada, Spring Semester 2106 Reading Materials, Module 13 Food Safety (on file with author).

20. The World Health Organization and the Food and Agriculture Organization of the United Nations established the Codex Alimentarius Commission in 1963 to develop international food standards to protect consumer health and to facilitate fair trading practices in foods.

manufacturers disclose the ingredients of these products to consumers, and that drugs are effective and are not sold as food or cosmetics. The Act also sets out inspection and enforcement provisions and creates offences for violations of the Food and Drugs Act and its related regulations.

3.4.2.3 Powers of enforcement

The CFIA's compliance and enforcement activities can occur within the countries of Canada's trading partners; at or near the Canadian border, domestically in food; at animal and plant product processing facilities; at points of distribution and retail sale; or at food service locations.[21] Regulated parties are responsible for complying with all acts and regulations that apply to them. A party who fails to comply with federal legislation and regulations relating to the packaging, labeling, and/ or safety of food will face penalties under Canadian law. Government enforcement includes searches, seizures, and recalls.

3.4.3 Principles and concepts

Definitions play an important role in the principles and concepts of Canadian food laws. Definitions determine how products are classified, but also determine whether products are violative of the Food and Drugs Act. This section delves into the principles and concepts relating to food law.

3.4.3.1 Food law principles

A product's classification has implications for how the product will be regulated. The classification of a product is based on statutory definitions. "Food" as defined by the Food and Drugs Act means "Any article manufactured, sold, or represented for use as a food or drink for human beings, including chewing gum, and any ingredients that may be mixed with food for any purpose whatever." Two separate definitions for the term "food" exist, the common sense definition and the statutory definition. The common sense definition is a product that one would consume for nutrition, taste, or aroma. The definition of food in Food and Drugs Act is more expansive. The statutory definition includes items one may not normally classify as food. For instance, most would not classify chewing gum as food. In addition, the use of the term "ingredients" in the statutory definition of food is vague. Does the term "ingredients" refer to plastic particles that migrate into an end product? Probably. The expansive statutory definition of food is imperative for public policy. Generally, no premarket authorization is needed for food; therefore, an expansive statutory definition helps to ensure food safety.

3.4.3.2 Important concepts

The Food and Drugs Act prohibits the sale of foods beyond products that are adulterated. Under the Food and Drugs Act section 412, "[n]o person shall sell an article of food that (a) has in or on it any poisonous or harmful substance; (b) is unfit for human consumption; (c) consists in whole or in part of any filthy, putrid, disgusting, rotten, decomposed, or diseased animal or vegetable substance; (d) is adulterated; or (e) was manufactured, prepared, preserved, packaged, or stored under unsanitary conditions." (CAN) Food and Drugs Act R.S., 1985, c. F-27, s. 4; 2005, c. 42, s. 1; 2012, c. 19, s. 412. There are exceptions to this provision: "[a] food does not have a poisonous or harmful substance in or on it for the purposes of paragraph (1) (a)—or is not adulterated for the purposes of paragraph (1) (d)—by reason only that it has in or on it a pest control product as defined in subsection 2(1) of the Pest Control Products Act, or any of its components or derivatives, if the amount of the pest control product or the components or derivatives in or on the food being sold does not exceed the maximum residue limit specified under section 9 or 10 of that Act."

3.4.4 Authorization requirements

Change is constant, especially when it comes to food. New technologies bring unknown risks requiring different, or more stringent, regulatory oversight. This section covers the regulatory processes for products that do not have a history of safe use as a food. Specifically, this section delves into the requirements for food additives, food irradiation, novel foods, and genetically modified foods.

---

21. *Compliance and Enforcement Activities*, Canadian Food Inspection Agency, *available at*: < http://www.inspection.gc.ca/about-the-cfia/accountability/compliance-and-enforcement/eng/1299846323019/1299846384123 >.

3.4.4.1 Food additives

The Food and Drug Regulations, and associated Marketing Authorizations, regulated food additives[22] in Canada. Canada employs a "positive listing" approach to food additives. Section B.16.007 of the Food and Drug Regulations provides that "[n]o person shall sell a food containing a food additive other than a food additive provided for in sections B.01.042, B.01.043, and B.25.062." (CAN) Food and Drug Regulations Section B.16.100 provides the list of approved food additives that can be found in the Tables. These tables set out the specific conditions explaining how food additives may be used in food. Frequently, if an additive is approved to be used in one type of food for one process, that does not mean that the additive can be used in another food unless the Food and Drug Regulations state that.

If the Lists provided in Section B.16.007 of the Food and Drug Regulations do not permit a particular use for a food additive, then the manufacturer is required to file a food additive submission to the Minister in a form, manner, and content that is satisfactory to him. The submission must contain a detailed description of the food additive, its proposed use, the results of safety tests, information on the effectiveness of the food additive for its intended use, data to indicate the residues that may remain in or upon the finished product, a proposed maximum limit for residues of the food additive in or upon the finished food, proposed labeling, and a sample of the food additive in the form in which it is proposed to be used in foods. The submission is then evaluated by scientists from Health Canada's Food Directorate, Health Products, and Food Branch. The scientists focus their testing on safety issues such as toxicological aspects, and relevant microbiological and/or nutritional factors of the food additive. The scientists must determine through their evaluation that the food additive is of suitable quality, effective for its intended purpose, and, when used according to the Lists, must not pose a hazard to the health of the consumer.

3.4.4.2 Food irradiation

The CFIA is responsible for the enforcement of the regulations relating to irradiated food products under the Food and Drugs Act. The CFIA establishes inspection and testing programs to verify compliance by both domestic producers and importers. The CFIA takes action if irradiated products are found on the Canadian marketplace that have not been approved for sale in Canada or are improperly labeled.

The safety and labeling of food irradiation are subject to federal controls. Division 26 (Part B) of the Food and Drug Regulations includes a table that lists the foods that may be irradiated and sold in the Canadian marketplace. Only the following irradiated foods are permitted for sale in Canada: (1) potatoes; (2) onions; (3) wheat, flour, whole wheat flour; and (4) whole or ground spices and dehydrated seasoning preparations. If the Table in Division 26 (Part B) of the Food and Drug Regulations does not include a particular food that may be irradiated, then the manufacturer is required to file submission. Division 26 sets out the submission requirements for those stakeholders seeking approval to irradiate a new food.

3.4.4.3 Novel foods/genetically modified foods

Foods that have been produced through new processes are called novel foods. Often novel foods do not have a history of "safe use as food," or have been modified by genetic modification. Under the Food and Drugs Act and Food and Drug Regulations, before a novel food can be sold in the Canadian marketplace it must be assessed by Health Canada. Companies that want to sell an unapproved novel food must file premarket notification, containing detailed scientific data, for Health Canada's Food Directorate to review. Health Canada's Food Directorate scientists will assess whether the novel food meets Canadian and international standards, and is totally safe to use.

Foods that have been genetically altered are often called genetically modified foods, GM foods, or genetically engineered foods. Under Division 28 of Part B of the Food and Drug Regulations, manufacturers and importers who wish to sell or advertise a genetically modified food in Canada, must file premarket notification to Health Canada for a premarket safety assessment. The regulatory process for assessing a new genetically modified food includes (1) presubmission consultation; (2) premarket notification; (3) scientific assessment; (4) requests for additional information; (5) summary report of findings; (6) preparation of food rulings proposal; (7) letter of no objection; and (8) decision document on Health Canada Website.

22. A food additive is any chemical substance that is added to food during preparation or storage and either becomes a part of the food or affects its characteristics for the purpose of achieving a particular technical effect. Division 16 of the Food and Drug Regulations specifically defines food additives.

3.4.5 Food safety limits

Food law evolves in response to societal concerns. For example, as technology improved, consumer and government agencies became concerned with manufacturers making products with pesticides, chemicals, and other materials that increased safety concerns for consumers. This section discusses the laws and regulations regarding residue limits for veterinary drugs, chemical contaminants, and pesticides.

3.4.5.1 Residues limits of veterinary drugs

A Maximum Residue Limit[23] is a level of residue that can remain in the tissue or product derived from a food-producing animal that has been treated with a veterinary drug and the residue is considered to pose no adverse health effects if ingested daily by humans over a lifetime. The Maximum Residue Limits are listed in the List of Maximum Residue Limits for Veterinary Drugs in Foods. This List includes Column I (Veterinary Drug); Column II (Name of the Substance for Drug Analysis Purposes); Column III (Foods); and Column IV (Maximum Residue Limits p.p.m.).

3.4.5.2 Chemical contaminants and/or microorganisms limits

Health Canada's Bureau of Chemical Safety, Food Directorate, is responsible for the assessment of risks associated with exposure to foodborne chemical contaminants.[24] The development of maximum levels for chemical contaminants in retail foods is a risk management technique that is used to reduce exposure to a particular contaminant. When establishing the maximum level of a particular contaminant in food, the primary concern is safety to human health; however, availability, nutritional value, and importance of the food in the Canadian diet are also considered. Maximum levels for chemical contaminants are listed in either the *List of Contaminants and Other Adulterating Substances in Foods* or *List of Maximum Levels for Various Chemical Contaminants in Foods*.

3.4.5.3 Pesticide residue limits

The Pest Control Products Act sets the requirements for maximum residue limits for pesticide products or for the products' components or derivatives. Under the Pest Control Products Act,[25] after any required evaluations and consultations have been completed, if the Minister considers that the health and environmental risks and the value of a pest control product to be acceptable, the Minister shall register the product or amend its registration in accordance with the regulations. When the Minister is making a decision regarding the registration of a pest control product, the Minister shall specify any maximum residue limits for the product or for its components or derivatives when appropriate.

3.4.6 Process requirements

In Canada, one in eight Canadians, a total of about 4 million annually, is affected by a foodborne illness. Of these 4 million Canadians that are affected by foodborne illnesses, there are about 11,600 hospitalizations and 238 deaths.[26] These statistics demonstrate that foodborne illnesses are a significant public health burden. Luckily, the burden associated with foodborne illnesses is largely preventable through food safety measures. This section covers certain Canadian food safety measures. This section will discuss how business processes are regulated; HACCP requirements; traceability requirements; and recall obligations.

23. *Maximum Residue Limits*, Government of Canada, *available at*: < https://www.canada.ca/en/health-canada/services/drugs-health-products/veterinary-drugs/maximum-residue-limits-mrls.html. >.

24. *Health Canada's Maximum Levels for Chemical Contaminants in Foods*, Government of Canada, *available at*: < https://www.canada.ca/en/health-canada/services/food-nutrition/food-safety/chemical-contaminants/maximum-levels-chemical-contaminants-foods.html >.

25. *Pest Control Products Act*, Justice Laws Website, *available at*: http://laws-lois.justice.gc.ca/eng/acts/P-9.01/page-3.html#h-9.

26. *Yearly Food-Borne Illness Estimates for Canada*, Government of Canada, *available at*: < https://www.canada.ca/en/public-health/services/food-borne-illness-canada/yearly-food-borne-illness-estimates-canada.html >.

3.4.6.1 Business hygiene processes

Effective hygiene controls are imperative to preventing foodborne illness,[27] foodborne injury, and food spoilage. The CFIA's General Principles of Food Hygiene, Composition and Labeling (GPFHCL) is designed to serve as a guideline for Canadian food manufacturers. The GPFHCL is meant to assist food manufacturers in establishing manufacturing practices that maintain food safety and meet regulatory requirements.

3.4.6.2 HACCP requirements

Hazard Analysis Critical Control Point (HACCP) is a widely recommended approach to food safety that is systematic and preventive. In Canada, HACCP is mandatory for federally registered meat and poultry establishments, but it is not mandatory for federally registered dairy, processed product, egg, honey, maple, and hatchery establishments. However, the CFIA strongly recommends that these establishments adopt HACCP. Developing, implementing, and maintaining an HACCP system is industry's responsibility.

3.4.6.3 Traceability requirements

Traceability is the ability to track any food through all its stages of production, processing, and distribution. Food traceability is an increasingly important element of food safety especially with the improvements in technology. The Safe Food for Canadians Act gives the CFIA the ability to develop innovative regulations to trace and to take action on potentially unsafe food commodities.

3.4.6.4 Recall obligations

The CFIA manages about 350 food recalls each year through the Office of Food Safety and Recall within the CFIA. In cases where the product poses a serious health risk, the CFIA issues a public warning, advising consumers through the media. Food recalls are centrally coordinated with CFIA staff across Canada and external partners. The CFIA categorizes food recalls into three different classes: Class I, Class II, and Class III.[28]

3.4.7 Labeling

Food labeling regulations cause explosive debates between government departments, consumers, and industry. Incorrect food labeling can have significant economic effects on industry, like leading to a recall, but also can send a consumer to the hospital if a product does not include proper allergen labeling. This section will focus on food labeling regulations that are designed to manage the expectations of consumers and industry.

3.4.7.1 Introduction

Health Canada and the CFIA share responsibility for the development of Canadian food labeling requirements. Health Canada is responsible for establishing policies, regulations, and standards that relate to the health, safety, and nutritional quality of food.[29] CFIA creates regulations related to misrepresentation, labeling, advertising, composition, grade, and packaging.[30]

---

27. *General Principles of Food Hygiene, Composition, and Labeling*, Canadian Food Inspection Agency, *available at*: < http://www.inspection.gc.ca/food/non-federally-registered/safe-food-production/general-principles/eng/1352919343654/1352920880237?chap=0 - s8c2 >.

28. "Class I" is a situation in which there is a reasonable probability that the use of, or exposure to, a violative product will cause serious adverse health consequences or death. "Class II" is a situation in which the use of, or exposure to, a violative product may cause temporary adverse health consequences or where the probability of serious adverse health consequences is remote. "Class III" is a situation in which the use of, or exposure to, a violative product is not likely to cause any adverse health consequences.

29. *See* (CAN) Canadian Food Inspection Agency Act, S.C. 1997, c. 6, s. 11(4).

30. (CAN) Canadian Food Inspection Agency Act, S.C. 1997, c. 6, ss. 11(2) and 11(3) (b).

3.4.7.2 Definition

The term "label," as it relates to food, is defined by four federal acts, which include (1) the Food and Drugs Act, (2) the Consumer Packaging and Labeling Act, (3) the Meat Inspection Act, and (4) the Canada Agricultural Products Act.[31] There are two common requirements for the legal definition of "label" among the four federal acts.[32] The first common requirement is that "a label must contain several elements or claims that will convey information to the consumer."[33] The second common requirement is that "those elements or claims must be attached or be in very close proximity or belong to the product."[34] While there are four federal acts that contain definitions for the term label, as it pertains to food, there is no legislative definition that exists for the term labeling.[35]

3.4.7.3 Mandatory labeling requirements

The Food and Drugs Act and the Consumer Packaging and Labeling Act set out the mandatory requirements for food labels. Under the Food and Drugs Act, every food label must be clear and its information readily discernible to the purchaser, be in English and French, have specified elements on a principal display panel, and not have these required elements on the bottom of the food product.[36]

For all prepackaged foods in Canada, the following elements must be on the label: (1) the ingredients or components by their common name; (2) a declaration of net quantity in metric units of measurement; (3) a declaration of products that have certain compositional characteristics; (4) the identity and principal place of business of the person by or for whom the food was manufactured or produced; (5) durable life or packaging date for products with less than 91 days expected shelf life; (6) a list of ingredients, shown in descending order of their proportion, and (7) the Nutrition Facts Table.[37]

3.4.7.4 Prohibited elements

Under the following statutes, there are four types of prohibited elements or claims that may not appear on labels: Food and Drugs Act, Consumer Packaging and Labeling Act, Canada Agricultural Products Act, Meat Inspection Act, Fish Inspection Act, Trademarks Act, and Competition Act.[38] Prohibited claims include (1) claims that a food is a treatment; preventive; or cures diseases, disorders, or abnormal physical states; (2) claims that refer to a product's compliance with the Food and Drugs Act or the Food and Drug Regulations; (3) claims that are scandalous, obscene, or immoral; and (4) claims that are associated with certain entities or organizations for which no permission for use has been obtained.[39]

Additionally, there are two broad prohibitions that apply to all information provided in food labeling or advertising[40]:

- Subsection 5(1) of the Food and Drugs Act states: "No person shall label, package, treat, process, sell, or advertise any food in a manner that is false, misleading, or deceptive or is likely to create an erroneous impression regarding its character, value, quantity, composition, merit, or safety."
- Subsection 7(1) of the Consumer Packaging and Labeling Act states: "No dealer shall apply to any prepackaged product or sell import into Canada or advertise any prepackaged product that has applied to it a label containing any false or misleading representation that relates to or may reasonably be regarded as relating to that product."

Subsection 7(2) of the CPLA defines "false and misleading representation" to include "any representation in which expressions, words, figures, depictions, or symbols are used that may likely deceive a consumer with respect to the net quantity of a prepackaged product, any representation that implies or may reasonably be regarded as implying that an ingredient is present when it is not, and any representation that would likely deceive a consumer with respect to the type, quality, performance, function, origin or method of manufacture, or production of a prepackaged product."

---

31. *See* Halsbury's Laws of Canada (2014), at 122 n. 1.
32. *Id.* at 121. *See* Halsbury's Laws of Canada (2014), at 123 n. 3 (stating that the common requirements among the four federal acts have been retained in the Safe Food for Canadians Act).
33. *Id.*
34. *Id.*
35. *Id.*
36. *See* (CAN) Food and Drug Regulations, C.R.C., c. 870. s. A.01.016, s. B.01.012(2), s. B.01.005.
37. *See* Halsbury's Laws of Canada (2014), at 130 and accompanying footnotes.
38. *See* Halsbury's Laws of Canada (2014), at 124.
39. *Id.*
40. *General Principles for Labeling and Advertising*, Canadian Food Inspection Agency, < http://www.inspection.gc.ca/food/labelling/food-labelling-for-industry/general-principles/eng/1392324632253/1392324755688?chap=0 >.

3.4.7.5 Labeling of food additives

The Food and Drug Regulations require food additives to be declared by an acceptable common name in the list of ingredients of prepackaged foods, or as per section B.01.008.2(4) (d). Food additives may be listed at the end of the list of ingredients in any order. Health Canada's lists of permitted food additives are always acceptable common names for food additives.[41]

3.4.7.6 Allergen labeling

Under Section B.01.010.1(2) of the Food and Drug Regulations, all prepackaged foods must declare food allergens and gluten in the list of ingredients for products in which those allergens and/or gluten are present. The source of food allergens and gluten must be declared in one of two ways: (1) in the list of ingredients [B.01.010.1(2) (a), FDR] or (2) in a "food allergen source, gluten source, and added sulfites statement" [B.01.010.1(2) (b), FDR]. A cross-contamination statement may also be declared by food manufacturers and importers when, despite all reasonable measures, there is the unintended presence of food allergens in the food. Cross-contamination statements are not a substitute for Good Manufacturing Practices.[42] Cross-contamination statements must not be used when an allergen, or allergen-containing ingredient, is deliberately added to a food—in such situations, food manufacturers are required to include the mandatory food allergen and gluten declaration.

3.4.7.7 Nutrition labeling

Due to the interconnectedness of diet and chronic diseases, nutrition labeling in Canada is designed to provide consumers with adequate information allowing consumers to make informed choices about the foods they purchase.[43] For example, most prepackaged food's must carry a nutrition facts table.[44] The Nutrition Facts Table gives consumers information about calories, 13 core nutrients (serving of stated size, energy value, amount of fat, amount of saturated fatty acids, amount of transfatty acid, the sum of the saturated fatty acids and transfatty acids, amount of cholesterol, amount of sodium, amount of carbohydrate, amount of fiber, amount of sugars, amount of protein, amount of vitamin A, vitamin C, calcium, and iron), and percent daily value of nutrients.[45] All of the information in the Nutrition Facts Table is based on an amount of food. This amount is always found at the top of the Nutrition Facts Table. To ensure that nutrition labeling is based on current scientific findings, new Canadian nutrition labeling regulations took effect as of December 14, 2016, and companies have until December 14, 2021, to be in compliance with the new regulations.

Under the Food and Drug Regulations Section B.01.401, if the products meet the requirements stated in B.01.402 to B.01.406 and B.01.467 to B.01.469, then the products are exempted from the nutrition label requirement. For example, the requirements for nutritional labeling are not required for a fresh vegetable or fruit or any combination of fresh vegetables or fruits without any added ingredients, an orange with added food color, or a fresh vegetable or fruit coated with mineral oil, paraffin wax, petrolatum, or any other protective coating (*see* B1.402(2)).

3.4.7.8 Nutrition content claims, health claims, and medicinal claims

Since 2003, two additional types of food label claims—nutrient content claims and health claims—are permitted for products meeting specific conditions. The Food and Drug Regulations Section B.01.513 positively lists permitted uses of nutrient content claims and diet-related health claims. No other claims are permitted on food labels under the Food and Drug Regulations Section B.01.502(1).[46]

---

41. *Lists of Permitted Food Additives*, Government of Canada, *available at*: < https://www.canada.ca/en/health-canada/services/food-nutrition/food-safety/food-additives/lists-permitted.html >.

42. *Food Allergen Cross Contamination (or Precautionary) Statements*, Canadian Food Inspection Agency, *available at*: < http://www.inspection.gc.ca/food/labelling/food-labelling-for-industry/allergens-and-gluten/eng/1388152325341/1388152326591?chap=4 > (stating that a cross contamination statement is a declaration on the label of a prepackaged product that alerts consumers of the possible presence of an allergen in the food).

43. Nutrition labeling is information found on the labels of prepackaged foods. The legislated information includes: the Nutrition Facts table, and ingredient list, optional nutrition claims.

44. *See* (CAN) Food and Drug Regulations, C.R.C., c. 870. s. A B.01.401 (1).

45. *See id.*

46. Part D of the Food and Drug Regulations regulates permitted nutrient content claims with respect to vitamins and minerals; these claims not covered in the table in B.01.513.

3.4.8 Human right to food

The right to food is a fundamental human right.[47] Article 11.1 of the International Covenant of Economic, Social, and Cultural Rights states that parties recognize "the right of everyone to an adequate standard of living for himself and his family, including adequate food, clothing, and housing, and to the continuous improvement of living conditions"; additionally, Article 11.2 states that parties recognize that more immediate and urgent steps may be needed to ensure "the fundamental right to freedom from hunger and malnutrition."[48] Canada has a legal obligation to respect, protect, and fulfill the right to food because Canada signed the International Covenant on Economic, Social and Cultural Rights in 1976. In 2015, the Trudeau government charged the Department of Agriculture and Agri-Food with leading the development of A Food Policy for Canada (NFP). In a letter to Mr. MacAulay, the Minister of Agriculture and Agri-Food, the Prime Minster of Canada Justin Trudeau instructed MacAulay to deliver on top priorities concerning food and agriculture. One of those top priorities was to "develop a food policy that promotes healthy living and safe food by putting more healthy, high-quality food, produced by Canadian ranchers and farmers, on the tables of families across the country."

References

Codex Alimentarius, Government of Canada. Available at: https://www.canada.ca/en/health-canada/services/food-nutrition/international-activities/codex-alimentarius.html.

Food Additives, Government of Canada. Available at: https://www.canada.ca/en/health-canada/services/food-nutrition/food-safety/food-additives.html - a2.

Food Irradiation, Government of Canada. Available at: https://www.canada.ca/en/health-canada/services/food-nutrition/food-safety/food-irradiation.html.

Lists of Permitted Food Additives, Government of Canada. Available at: https://www.canada.ca/en/health-canada/services/food-nutrition/food-safety/food-additives/lists-permitted.html.

Pest Control Products Act, Justice Laws Website. Available at: http://laws-lois.justice.gc.ca/eng/acts/P-9.01/page-3.html - h-9.

Further reading

A Guide to Canadian Trade-Marks, Alexander Holburn Beaudin + Lang LLP. http://www.ahbl.ca/wp-content/uploads/2013/02/Guide-to-Canadian-Trade-Marks-2.pdf.

A Proposal from Food Secure Canada for the National Food Policy, Five Big Ideas for a Better Food System. Available at: https://foodsecurecanada.org/policy-advocacy/five-big-ideas-better-food-system.

Acceptable Disease Risk Reduction Claims and Therapeutic Claims, Canadian Food Inspection Agency. Available at: http://www.inspection.gc.ca/food/labelling/food-labelling-for-industry/health-claims/eng/1392834838383/1392834887794?chap=7.

Approved Products, Government of Canada. Available at: https://www.canada.ca/en/health-canada/services/food-nutrition/genetically-modified-foods-other-novel-foods/approved-products.html.

Canada's Strategic Framework for Participation in the Joint FAO/WHO Food Standards Program, Government of Canada. Available at: https://www.canada.ca/en/health-canada/services/food-nutrition/international-activities/codex-alimentarius/activities/canada-strategic-framework-participation-joint-who-food-standards-program.html.

CFIA at A Glance, Canadian Food Inspection Agency. Available at: http://www.inspection.gc.ca/about-the-cfia/organizational-information/at-a-glance/eng/1358708199729/1358708306386.

Compliance and Enforcement Operational Policy, Canadian Food Inspection Agency. Available at: http://www.inspection.gc.ca/about-the-cfia/accountability/compliance-and-enforcement/operational-policy/eng/1326788174756/1326788306568.

Diana Bronson, Coming Up with a Canadian Food Policy, TED Talk at TEDxConcordia. Available at: https://foodsecurecanada.org/resources-news/news-media/coming-canadian-food-policy.

Food Allergen Cross Contamination (or Precautionary) Statements, Canadian Food Inspection Agency. Available at: http://www.inspection.gc.ca/food/labelling/food-labelling-for-industry/allergens-and-gluten/eng/1388152325341/1388152326591?chap=4.

Food Labelling Changes, Government of Canada. Available at: https://www.canada.ca/en/health-canada/services/food-labelling-changes.html.

Food Regulation in Canada: Spring Semester, 2016a. Module One Reading Materials, The Institute for Food Laws and Regulations (On File with Author).

Food Regulation in Canada: Spring Semester, 2016b. Module Six Lecture Video, The Institute for Food Laws and Regulations (On File with Author).

Genetically Modified (GM) Foods and Other Novel Foods, Government of Canada. Available at: https://www.canada.ca/en/health-canada/services/food-nutrition/genetically-modified-foods-other-novel-foods.html.

Halsbury's Laws of Canada 2014.

---

47. Universal Declaration of Human Rights, *available at*: < http://www.ohchr.org/EN/UDHR/Documents/UDHR_Translations/eng.pdf >.

48. Substantive Issues Arising in the Implementation of the International Covenant on Economic, Social, and Cultural Rights: General Comment 12, *available at*: < http://www.fao.org/fileadmin/templates/righttofood/documents/RTF_publications/EN/General_Comment_12_EN.pdf >.

Health Canada's Maximum Levels for Chemical Contaminants in Foods, Government of Canada. Available at: https://www.canada.ca/en/health-canada/services/food-nutrition/food-safety/chemical-contaminants/maximum-levels-chemical-contaminants-foods.html.
Health Produces and Food Branch Inspectorate, Government of Canada. Available at: https://www.canada.ca/en/health-canada/corporate/about-health-canada/branches-agencies/health-products-food-branch/health-products-food-branch-inspectorate.html.
Institute for Food Laws & Regulations, Food Regulations in Canada, Spring Semester, 2016. Reading Materials, Module 13 Food Safety (On File with Author).
Labelling Legislative' Framework, June 3, 2018. Canadian Food Inspection Agency. http://www.inspection.gc.ca/food/labelling/labelling-legislative-framework/eng/1387771371233/1387771427304.
Learn About, Food Secure Canada. Available at: https://foodsecurecanada.org/food-policy/learn-about.
Maximum Residue Limits (MRLs), Government of Canada. Available at: https://www.canada.ca/en/health-canada/services/drugs-health-products/veterinary-drugs/maximum-residue-limits-mrls.html.
McKeown J., Scandalous, Obscene or Immoral Trademarks. GSNH. Available at: http://www.gsnh.com/scandalous-obscene-immoral-trademarks/.
Melissa, M.C., John, F.A., 2015. Just A Spoonful of sugar will land you six feet underground: should the food and drug administration revoke added sugar's gras status? Food & Drug L.J 70, 395–398.
Minister of Agriculture and Agri-Food Mandate Letter, November 12, 2015. Available at: https://pm.gc.ca/eng/minister-agriculture-and-agri-food-mandate-letter.
Pest Control Products (pesticides) Acts and Regulations, Government of Canada. Available at: https://www.canada.ca/en/health-canada/services/consumer-product-safety/pesticides-pest-management/public/protecting-your-health-environment/pest-control-products-acts-and-regulations-en.html.
Quality Claims, Canadian Food Inspection Agency. Available at: http://www.inspection.gc.ca/food/labelling/former-food-labelling-for-industry/former-composition-and-quality-claims/eng/1518720087614/1518720088627?chap=3.
Safe Food for Canadians Act, Canadian Food Inspection Agency. Available at: http://www.inspection.gc.ca/about-the-cfia/acts-and-regulations/regulatory-initiatives/sfca/eng/1338796071420/1338796152395.
Safe Food for Canadians Act: An Overview, Canadian Food Inspection Agency. Available at: http://www.inspection.gc.ca/about-the-cfia/acts-and-regulations/regulatory-initiatives/sfca/overview/eng/1339046165809/1339046230549
Safe Food for Canadians Act: What It Means for Consumers, Canadian Food Inspection Agency. Available at: http://www.inspection.gc.ca/about-the-cfia/acts-and-regulations/regulatory-initiatives/sfca/consumers/eng/1339044365103/1339044455501.
Safe Food for Canadians Action Plan, Canadian Food Inspection Agency. Available at: http://www.europarl.europa.eu/meetdocs/2009_2014/documents/envi/dv/envi20131014_canada_presentation_/envi20131014_canada_presentation_en.pdf.
Section 1 – Food Safety Enhancement Program Description: Food Safety Enhancement Program Manual, Canadian Food Inspection Agency. Available at: http://www.inspection.gc.ca/food/safe-food-production-systems/food-safety-enhancement-program/program-manual/eng/1345821469459/1345821716482?chap=2-s1c2.
The Establishment-based Risk Assessment Model, Canadian Food Inspection Agency. Available at: http://www.inspection.gc.ca/about-the-cfia/accountability/inspection-modernization/era-model/eng/1487771637766/1487771638453.
The Regulation of Genetically Modified Food, Government of Canada. Available at: https://www.canada.ca/en/health-canada/services/science-research/reports-publications/biotechnology/regulation-genetically-modified-foods.html.
The War Amputations of Canada; Report: Thalidomide Task Force. Available at: https://thalidomide.ca/wp-content/uploads/2018/01/synopsis-war-amps-report.pdf.

Chapter 3.5

The road to harmonization in Latin America

Rebeca López-García
Logre International Food Science Consulting, Mexico

3.5.1 Introduction

Latin America is a very complex region that faces diverse challenges due to a disparity of social, economic, and cultural conditions. Each country and even each subregion has its own strengths, weaknesses, and challenges, so it is difficult to portray the whole region in just a few pages without making unavoidable generalizations. Latin American countries are most definitely not strangers to the "globalization processes" and have been quickly gaining a position in global markets with unique products. In addition, Latin America represents a huge market that is very attractive for companies around the world and commercial activities within the region have increased through the participation in several free trade agreements. These new market opportunities have helped shape the region's industry and food regulations and have sparked the interest in actively participating in International Organizations such as the *Codex Alimentarius*.

According to the FAO Regional Office for Latin America and the Caribbean, this was the first region to commit to the complete eradication of hunger through the 2025 Hunger-Free Latin America and the Caribbean Initiative. This renewed political commitment is based on the full conviction that eradicating in the region is an achievable target. Thus, the region is at the forefront of the global fight against hunger and has made the most progress in reducing the percentage and total number of people suffering from hunger in the past 20 years. Public policies implemented by Governments coupled with economic growth, reduced hunger considerably from 14.7% in 1990–93 to 7.9% in 2010–2013. Child malnutrition was also halved in the same period, while the total number of undernourished people decreased from 59 million in 1990 to 47 million in 2013. Food security policies and programs, as well as the region's role as one of the World's main food producers, place Latin America and the Caribbean at the forefront of the global fight against hunger (FAO, 2019). However, the region's diversity and the vastness of food produced present unique challenges to ensure food safety and for harmonization in itself. Additionally, the level of maturity and sophistication of the regulatory infrastructures and technical capabilities is extremely different. Thus, harmonization poses unique challenges.

Latin American countries have undergone a process that sought to promote a policy shift from protecting national industries through openly protectionist policies toward an open market, free commerce system that seeks to foster competition within the "global market" framework. The shift to free markets has not come without some opposition. Of course, there is the responsibility of the State to defend the health and safety of its national consumers. However, most Latin American countries are subject to private sector pressures and are struggling to find the balance between encouraging more open commerce while making sure the products being imported are safe. In primary production, the shift from subsistence agriculture to competitive productive systems has faced a lot of cultural resistance, and in many countries, land reform has fractionated land to a point where it is almost impossible to compete without proper cooperation.

A new complication has emerged in the region as it has undergone a transition from trying to address basic nutrition needs and control food safety in general to address the severe obesity crisis that has spread worldwide with the subsequent public health issues. Thus, a lot of energy and effort has been concentrated in developing public policies and proper intervention strategies to address these issues and help mitigate the future impact. According to a survey of policy initiatives on diet, health, and nutrition, performed by the Organization for Economic Cooperation and Development, overall, the major strategies taken by different member countries focus mainly on two activities: first, increasing information on diet and health to consumers to enable them to make informed food choices and second, promoting increased consumption of fruit and vegetables, particularly among children (Fulponi, 2009). These strategies, particularly the adoption of different labeling systems, have created new trials for harmonization in the last few years.

3.5.2 Steps toward harmonization

Harmonization is not new to the region. In fact, long before other regions of the world began to imagine a framework of food standards beyond their national borders, in 1924, in Buenos Aires, at the first Latin American Congress of Chemistry, a Commission composed by two delegates from each country represented in the Congress proposed the development of a *Codex Alimentarius Sudamericanus*. This Commission accomplished their objective, and in 1930 at the following Congress in Montevideo, they presented a Code that had 154 articles and was considered for adoption by all countries in Latin America. The Code contained definitions of food products and general dispositions. Unfortunately, then, as now, turning a proposal into a reality was not easy. Although many countries in Latin America have adopted "modern" and well thought out food legislation, as a region, Latin America did not achieve an early effective regional regulation of food. In the following Latin American Congresses of Chemistry, there was much discussion on the same topic always with the vision of developing a Latin American Code. During the sixth Congress in Caracas in 1955, after much discussion, there was vote for a new Commission that was composed by the official representative of each country. In addition, a group of specialists in Bromatology was formed. This group was charged with the project, and after working for 3 years, the Commission presented a new document that was unanimously approved. The Revised Latin American Food Code was published in Spanish in 1960. This document is highly relevant since it represents the regional efforts toward harmonization. In addition, its value is that in combination with the European Legislation, it served as a source for the *Codex Alimentarius* (Acosta and Marrero, 1985; Nader and Vitale, 1998). Despite these very valuable efforts, with so much diversity in the region and with the problems Latin America has had to face, it is not surprising that food regulation has also been challenging.

The Codex Alimentarius Coordinating Committee for Latin America (CACCLA) was created in 1976 with the mandate to define the region's challenges and needs for food regulations as well as inspection systems, strengthen the inspection infrastructure, and recommend the establishment of international standards for products of interest to the region; particularly, products that in the Committee's judgment could have commercial potential in international markets; establish regional regulations for products that are traded almost exclusively in regional markets; identify important challenges unique to the region; and promote the coordination of all food regulatory activities promoted by international organizations, local government, and nongovernmental organizations (Acosta and Marrero, 1985).

3.5.3 The challenges of regional food regulation

According to the Report of the 20th Session of the FAO/WHO Coordinating Committee for Latin America and the Caribbean (CAC, 2017), delegations generally agreed with the top six critical and emerging food safety and quality issues that have been identified in the past, namely: regulatory landscape; antimicrobial resistance (AMR); contaminants/residues/additives; new technologies; climate change; and capacity development. However, delegations noted that the analysis was not representative of the LAC region as it only reflected the responses of 12 countries (approximately 30% of the membership). Thus, participation in this type of international efforts to promote harmonization remains an issue with international organizations such as FAO making a continuous effort to promote involvement of all members.

The following additional critical and emerging issues/needs were mentioned:

- addressing the increase of obesity among children and adults;
- setting Maximum Residue Limits (MRLs) for minor crops;
- evaluating impact of EU regulations on endocrine disruptors;
- developing capacity development and promoting participation of all relevant stakeholders;
- strengthening collaboration among food laboratories, e.g., laboratory networking;
- promoting participation of the food sector to the achievement of the Sustainable Development Goals (SDG), e.g., reduction of poverty and hunger, good health and wellbeing, decent work and economic growth, responsible consumption and production; changes in food habits; and urbanization;
- establishing risk communication strategies, i.e., how risk managers should appropriately communicate food safety risk to the public;
- evaluating viruses in food;
- addressing water contamination as it affects the safety of food;
- considering food fraud;
- addressing distribution of food through new channels such as internet sale of food.

In addition, the need for harmonization was addressed throughout the meeting with delegations mentioning that it was important to

- set priorities at national and regional level;
- catalog national food safety legislation in order to allow for its analysis and comparison with that of other countries and with Codex standards;
- have information on the status of food safety systems in the countries of the region, including related capacity development needs, in order to prioritize interventions;
- know the status of implementation of food legislation at country level;
- develop national food safety policies;
- review and update national legislations; and analyze the impact of food safety legislation;
- strengthen regional coordination and collaboration;
- avoid fragmentation at the national level in the implementation of food legislation;
- increase involvement of private sector;
- have strong intersectorial collaboration, i.e., One Health approach; to develop MRLs for more pesticides;
- have more FAO and WHO work to assist countries to produce data for exposure assessment of chemicals;
- establish Codex guidelines for the prevention of noncommunicable diseases.

3.5.4 Regional intentions for improvement: the Pan American Commission of Food Safety (COPAIA 7)

Harmonizing regulatory requirements to assure safe and good quality foods and promote trade is of increasing interest to all countries in the Americas. In general, in the region, there are no structures and processes to achieve these objectives in a harmonized manner. Work toward regional structures and processes in the Americas has been strengthened and promoted via the activities of the Pan American Health Organization (PAHO), which is one of six regional organizations of the WHO, and the FAO of the United Nations. Under this framework, PAHO has supported establishment of a hemispheric Commission for Food Safety (COPAIA). Training and research links between the COPAIA and academic institutions in the Americas may be especially useful to promote collaborations and leverage resources for food safety (FAO, 2002). The most recent meeting (COPAIA 7) was held in Asunción, Paraguay, in 2016 where the members made recommendations in the different priority areas that included risk analysis, AMR, and food quality (COPAIA 7, 2016). In the area of risk analysis, the members identified the need to strengthen risk analysis capabilities taking into account the social and economic determinants for food safety. Specific emphasis is needed regarding the implementation of risk-based inspection and surveillance programs in addition to risk communication capabilities. Regarding AMR, the Commission acknowledged the need to improve awareness and knowledge on the subject; reinforce surveillance and research; strengthen prevention and infection control programs; optimize the use of antimicrobial drugs in human and veterinary medicine; and ensure sustainable investments in the fight against AMR in public and private human and agricultural health sectors. In relation to food quality, the Commission recognized that the prevalence of noncommunicable chronic illness is rising in the region and it is the main cause of death causing a high social and economic impact. Thus, the Commission urged countries to generate regulatory frameworks at a national level that are harmonized internationally; develop and implement food and nutrition education programs with a focus on safe and healthy foods; consider integrating programs of nutritional quality and food safety; and analyze, together with International Organizations, the availability of human and financial resources to implement these strategies.

3.5.5 General regulatory structure

There are many general strategies to group different Latin American countries in blocks. These divisions are made based on geographical location, cultural background, language spoken, level of development, etc. The following statements are based on an informal division based in part in a geographical division as well as common typical regulatory structure. In general, Caribbean countries have individual laws and regulations for consumer products. Some of the islands are associated with the European Union or the United States and follow their regulatory structure, while others have decided to adopt Codex standards as their own. Central American countries have either individual laws and regulations or harmonized through participation in a trade block (Central American Common Market) and, again, in many cases have opted for the adoption of Codex standards. Most countries in South America have individual laws and regulations for products. However, five have entered a common market arrangement known as MERCOSUR. In addition, even when countries such Bolivia, Chile, Colombia, Ecuador, the Falkland Islands, French Guiana, Guyana, Paraguay, Peru, and Suriname have existing trade agreements, they maintain individual laws and standards for products.

3.5.6 Trade agreements

There are several associations and trade agreements within the region. Each of these agreements varies widely in terms of their scope and the degree of harmonization.

3.5.6.1 Pacific Alliance

The Pacific Alliance is a trade bloc that aims to become the largest in Latin America. It is composed of Chile, Colombia, Mexico, and Peru. It was originally formed following the Lima declaration in April 2011. The objective of this regional integration is to foster economic growth, development, and competitiveness of the member economies. Through a deep integration moving progressively toward free movement of goods, services, investment and people (SAGARPA, 2019). While meeting in Santiago (December 2018), high-level officials responsible for agriculture in Chile, Mexico, and Peru agreed to create the Agricultural Council of the Pacific Alliance, a forum that will drive integration agreements and increase regional coordination. Thus, as integration continues, there will be potential for further harmonization in this area (IICA, 2018).

3.5.6.2 NAFTA/USMCA

The United States—Mexico—Canada Agreement (USMCA) is the result of the renegotiation of the North American Free Trade Agreement (NAFTA). On September 30, 2018, the governments of the United States, Canada, and Mexico announced they had reached a trilateral free trade agreement (in principle) concluding 13 months of negotiations. The revisions that new agreement does not seem to have an impact on specific food law harmonization issues as the reform is more relevant to market access of different products and commodities (USMCA, 2018).

The Trade Agreement between Mexico, the United States, and Canada has had a profound effect in Mexican regulations. So, even when technically speaking this is not a Latin American Agreement, it is still of extreme importance to the region. The original agreement includes text on sanitary and phytosanitary measures, modeled after the Uruguay Round Agreement on sanitary and phytosanitary measures with recommendations that the three countries pursue equivalence of their respective sanitary and phytosanitary standards. This recommendation was established to assist in avoiding trade disputes among the three regarding the preparation and processing of food products that are traded. The idea is that the countries pledge to harmonize food production processes to "the extent feasible" and that measures do not become disguised trade restrictions. To avoid barriers to trade, the agreement encourages countries to use relevant international standards, if existent, when developing their SPS measures. However, each country is permitted to adopt a standard more stringent than international standards to achieve an appropriate level of desired protection of human, animal, or plant health if the standard is based upon scientific principles. The signatories have agreed in the past to work toward "equivalent" SPS measures without reducing national levels of desired, appropriate protection. It would be expected that these efforts persist with the renewed agreement. Equivalency recognizes that different methods may be used to reach the same level of protection. Each country agreed to accept the others' SPS measures as equivalent, provided the exporter shows that its SPS measures meet the importer's desired level of protection as long as it is based on risk assessment techniques (Looney, 1995).

3.5.6.3 Andean Community

The Andean Community (Peru, Ecuador, Colombia, and Bolivia) is a South American organization that was founded to encourage industrial, agricultural, social, and trade cooperation. In 2005, this organization signed an agreement with MERCOSUR. Through this, the Andean Community gained four new associate members, Argentina, Brazil, Paraguay, and Uruguay. Among the objectives of this association is to facilitate the participation in the regional integration process, with a view to the gradual formation of a Latin American common market. To present, the Technical Committees have been able to finalize a few common (harmonized) Andean standards, with several more in the project stage. The Andean Group Member Countries are also working on the harmonization of health and consumer safety requirements for processed foods, pharmaceutical products, and cosmetics. Members are considering the creation of *Ad Hoc* Committees to consider standards-related aspects of security, health, consumer protection, the environment, and national defense of Andean Group members. The Andean countries have adopted the ISO/IEC guidelines related to standardization and conformity assessment procedures. With respect to the adoption and/or development of Andean standards, Decision 376 sets out an order of preference from which these should be drawn, proceeding from international standards, to regional standards, to national standards of Member Countries, followed by those of nonmember countries and lastly, to those of private standards organizations (OAS, 1998).

3.5.6.4 CARICOM

The Caribbean Common Market (CARICOM) is an organization that aims at the eventual integration of its members and economies and the creation of a common market. Its members include Antigua and Barbuda, Belize, Grenada, Montserrat, St. Vincent and the Grenadines, Turks and Caicos Islands, The Bahamas, British Virgin Islands, Guyana, St. Kitts and Nevis, Suriname, Barbados, Dominica, Jamaica, Saint Lucia, and Trinidad and Tobago. As signatories to the WTO, CARICOM countries are expected to harmonize national and regional food safety standards with Codex standards in the import and export of food products, and to adopt the WTO approach to food safety.

The Caribbean Food Safety Initiative (CFSI) was designed by the CARICOM Secretariat, the U.S. Department of Agriculture (USDA), the Food and Drug Administration (FDA), and the Inter-American Development Bank (IDB). The purpose of this initiative was to develop a model approach to assist countries in meeting their WTO Sanitary and Phytosanitary obligations (FAO, 2002).

3.5.6.5 Central American Customs Union (Unión Aduanera Centroamericana)

This group is composed of five Central American economies (Guatemala, Honduras, El Salvador, Nicaragua, and Costa Rica) and is well integrated. This integration was the consequence of 40 years under the Central American Common Market (CACM); this is a far-reaching trade agreement that, among other things, ensures tariff-free exchange for 99.9% of the native products within the region. This Union provides common regulation in many areas from services to product registry to dispute resolution. Standards are now harmonized and grouped in Technical Regulations (*Reglamentos Técnicos Centroamericanos*) that are fully based on Codex Standards.

3.5.6.6 MERCOSUR

Argentina, Brazil, Paraguay, and Uruguay created MERCOSUR in March 1991 with the signing of the Treaty of Asuncion. MERCOSUR was originally created with the ambitious goal of creating a common market similar to the European Union. Venezuela became a full member in 2006. Similar to the EU, MERCOSUR has different legislative and technical organizations that create legislation and standards that are voted on and if passed are supposed to be adopted into national law. Although many hundreds of standards have been created and adopted by MERCOSUR, individual country adoption may be challenging. If a MERCOSUR standard is not available for a specific issue, each country can adopt its own dispositions until a regional standard is established. Food law harmonization has been conducted by the SGT-3 (Technical Regulations Work Subgroup) under the responsibilities of the Food Committee. The process followed for adoption includes the following (de Figuereido, 2000):

1. Elaboration of the proposals to be discussed jointly by governmental and private institutions.
2. Submission of the proposals to all member countries during an ordinary meeting of the SGT3 - Food Commission.
3. Discussion of the proposal by the specific ad hoc group.
4. Approval of the proposal by consensus.
5. Elaboration of a MERCOSUR project of technical regulation.
6. Internal discussion of the project within each member state by all interested parties
7. Approval of the project by the SGT3.
8. Submission of the harmonized project to the GMC for approval as a resolution.

In order to supplement the scientific knowledge required to set food standards, Codex Alimentarius standards, guidelines, and recommendations as well as EU directives and the US FDA regulations are consulted. The process is well established. However, with the participation of four (now five) countries with different laws, habits, idiosyncrasies, and interests, harmonization has not progressed as originally planned.

In 2014, renewed efforts to harmonize and simplify processes were started originally with dietary supplements and their ingredients followed by work on processed foods which is still ongoing. Part of this process is the rationalization of the procedures for getting new food products and ingredients registered in the various markets. However, as stated before, the regulations concerning public health issues are still controversial and lack harmonization throughout the region.

3.5.7 Conclusions

Although food law harmonization is highly desirable to set a level playing field for global food trade. Its implementation, in reality, may be hampered by many challenges. In the Latin American region, these challenges are further complicated by the diverse level of development of national food control systems. Each country must first face the challenges of their own internal system before participating in a more regional approach. This development could be facilitated by the adoption of already internationally recognized standards such as the Codex Alimentarius. However, it is important to consider that each country has its own idiosyncrasies and needs, and thus, the level of adoption and the activities for implementation may still be very different in addition to the wide variety of free trade agreements and blocks present in the region. In addition, the nutrition transition and the wide variety of public policy strategies that have been adopted by different countries pose new harmonization challenges, particularly for food labeling. The process of regional harmonization will still take some time and will depend on technical assistance and the use of sound risk assessment activities that are already available through international organizations. In some cases, harmonization activities may already be well-defined in paper. However, the challenge still resides in making them a reality.

References

Acosta, A., Marrero, T., 1985. Bol of SanPPanmn. Normalización de alimentos y salud para América Latina y el Caribe. 4. Labor del Comité Coordinador Regional de la Comisión del Codex Alimentarius, vol. 99, pp. 642–652 (6).

Codex Alimentarius Commission, 2017. Report of the Twentieth Session of the FAO/WHO Coordinating Committee for Latin America and the Caribbean. http://www.fao.org/fao-who-codexalimentarius/sh-proxy/en/?lnk=1&url=https%253A%252F%252Fworkspace.fao.org%252Fsites%252Fcodex%252FMeetings%252FCX-725-20%252FREPORT%252FREP17_LACe.pdf. (Accessed 27 February 2019).

COPAIA 7, 2016. Report of the 7th Meeting of the Pan-American Commission on Food Safety. Asunción, Paraguay. http://www.panaftosa.org/copaia7/dmdocuments/COPAIA7_Recomendaciones_[020816]_english.pdf. (Accessed 27 February 2019).

De Figuereido Toledo, M.C., 2000. Southern common market standards. In: Rees, N., Watson, D. (Eds.), International Standards for Food Safety. Springer, U.S, pp. 79–94.

FAO, January 28–30, 2002. New approaches to consider in capacity building and technical assistance - building alliances. FAO/WHO Glob. Forum Food Saf. Regul. Marrakech, Morocco.

FAO, 2019. FAO in Latin America and the Caribbean. http://www.fao.org/americas/acerca-de/en/. (Accessed 27 February 2019).

Fulponi, L., 2009. Policy initiatives concerning diet, health and nutrition. In: OECD Food, Agriculture and Fisheries Working Papers, No. 14. OECD Publishing. https://doi.org/10.1787/221286427320.

IICA, 2018. Chile, Mexico and Peru Agree to Create the Agricultural Council of the Pacific Alliance, with Support from IICA. IICA Institutional Communication.

Nader, A., Vitale, G., September 1998. Legislación alimentaria - al alcance de la mano. Revista Alimentos Argentinos No. 8. S.A.G.P. y A. Dirección de Promoción de la Calidad Alimentaria, Buenos Aires, Argentina. http://www.alimentosargentinos.gov.ar/0-3/revistas/r_08/08_06_codex.htm. (Accessed 20 March 2009).

Organization of American States, February 1998. Standards and the Regional Integration Process in the Western Hemisphere. Trade Section. http://www.sedi.oas.org/DTTC/TRADE/PUB/STUDIES/STAND/stand2.asp. (Accessed 20 March 2009).

SAGARPA, 2019. Pacific Alliance. International Trade Negotiations. http://www.sagarpa.mx/English/InternationalTradeNegotiations/TradeAgreements/PA/Pages/default.aspx. (Accessed 25 February 2019).

United States-Mexico-Canada Agreement (USMCA), 2018. https://usmca.com/. (Accessed 20 February 2019).

Further reading

Panamerican Health Organization, 2008. Agriculture and Health: Alliance for Equity and Rural Development in the Americas. In: 15th Inter-American Meeting at Ministerial Level on Health and Agriculture. Rio de Janeiro, Brazil, 11–12 June 2008.

World Health Organization, 2005. International Health Regulations, second ed. (Switzerland).

Chapter 3.6

European Union

Bernd van der Meulen

GHI, Prof. Comparative Food Law, Renmin University of China School of Law, University of Copenhagen, European Institute for Food Law, Amsterdam, The Netherlands

3.6.1 Introduction

The European Union (EU) is a supranational international organization. The Institutions of the European Union only enjoy those competences that the Member States have entrusted them through the founding treaties. Initially, food was addressed in the context of the creation of an internal market with free movement of goods. Increasingly, protection of health and consumer interests play an important role.

In the 1990s, the EU was struck by a series of animal health and food safety outbreaks including the BSE crisis. From 2000 onward, the food legal system in the EU was subjected to a fundamental reform. The framework regulation that forms the corner stone of this reform is Regulation (EC) 178/2002 popularly known as the General Food Law (GFL) (Van der Meulen and Wernaart, 2020). By consequence, within EU law, "food law" is recognized as a branch of law in its own right.

3.6.2 Institutional

The core institutions in the EU are the European Parliament (representing the people), the Council of Ministers (representing the Member States), and the European Commission (the day-to-day administration of the EU). The right of initiative in legislation is with the Commission. The EU has legislation addressing the legislature in the Member States with a view to harmonizing national law (so-called directives) and legislation directly addressing people and businesses in the entire area of the EU (so-called regulations). Food law mainly consists of regulations.

The Court of Justice of the EU (CJEU) has the final say on the interpretation of EU law. National courts can ask the CJEU to provide interpretations by way of preliminary rulings. The European Commission can take Member States to court if they fail in their duties (Borchardt, 2018). Decisions of EU Institutions and failures to act can be contested at the CJEU by interested parties.

The staff of the European Commission is organized into Directorates General (DGs) each headed by one of the commissioners. Food law is the domain of DG Sante (Health and Food Safety). A special office within DG Sante (Health and Food Audits and Analysis—previously known as Food and Veterinary Office) supervises the quality of food safety inspection in the Member States and in third countries. Independent risk assessment is provided by the European Food Safety Authority (EFSA). EFSA is an agency independent from the European Commission.

3.6.3 Enforcement and incident management

According to Article 17(2) GFL "*Member States shall enforce food law, and monitor and verify that the relevant requirements of food law are fulfilled by food and feed business operators at all stages of production, processing, and distribution. For that purpose, they shall maintain a system of official controls and other activities as appropriate to the circumstances, including public communication on food and feed safety and risk, food and feed safety surveillance, and other monitoring activities covering all stages of production, processing, and distribution. Member States shall also lay down the rules on measures and penalties applicable to infringements of food and feed law. The measures and penalties provided for shall be effective, proportionate, and dissuasive.*"

The Official Controls Regulation (Regulation (EU) 2017/625) (OCR) further details Member States' obligations and provides powers of inspection and measures to deal with food safety incidents (Van der Meulen, 2018c). The OCR attributes control powers to the competent authorities in the Member States. The role of the national legislature is limited to designating the competent authority. The rest follows automatically from the direct applicability of the regulation. According to Article 14 OCR "methods and techniques for official controls" shall include (among others) an inspection of equipment, animals, goods, ingredients, processing aids, traceability, labeling, presentation, hygiene and assessment of procedures, examination of documents, verification of measurements and test results, sampling, analyses, and tests. Official controls may include interrogations. The OCR speaks in this context of "interviews" with operators and their staff. Additionally, the OCR provides the power to perform "any other activity required to identifying cases of noncompliance."

In case of noncompliances, the competent authority can order treatments on goods, order the alteration of labels, or corrective information to be provided to consumers; restrict or prohibit the placing on the market; order the operator to increase the frequency of own controls; order the recall, withdrawal, removal, and destruction of goods, authorize, where appropriate, the use of the goods for purposes other than those for which they were originally intended; order the isolation or closure, for an appropriate period of time, of all or part of the business of the operator concerned, or its establishments, holdings or other premises; order the cessation for an appropriate period of time of all or part of the activities of the operator concerned and, where relevant, of the internet sites it operates or employs; and order the slaughter or killing of animals.

3.6.4 Principles and concepts

The General Food Law provides some general concepts, obligations, requirements, and principles of food law. Food law should aim at the protection of human life and health and (other) consumers' interests (Article 5). In protecting life and health, it should be science based, that is to say, based on risk analysis (Article 6). When scientific risk assessment is inconclusive, the precautionary principle justifies temporary measures to be taken to protect from possible risks (Article 7). The authority responsible for risk assessment is the EFSA (Article 22). Risk management measures are taken by the European Commission, the legislature, or the Member States.Where international standards—like the Codex Alimentarius—exist or their completion is imminent, they shall in general be taken into consideration in the development or adaptation of food law (Article 5(3) GFL). In design and structure, there is a strong resemblance between the Codex

Alimentarius and EU food law (Van der Meulen, 2018b). The definition of food for example is tailored on the Codex and also the principle of HACCP as elaborated in the Codex is incorporated in EU food law (see hereafter). The legion of product standards that is available in the Codex has less influence on EU legislation as product specific legislation has been largely abandoned in the EU (see below).

Food businesses are responsible to ensure compliance. Member states are responsible for enforcement (Article 17). Food shall not be placed on the market if it is unsafe (Article 14) (Van der Meulen, 2018a).

The EU definition of "food" is clearly tailored after the Codex definition. Article 2 of GFL reads as follows:

For the purposes of this Regulation, "food" (or "foodstuff") means any substance or product, whether processed, partially processed, or unprocessed, intended to be, or reasonably expected to be ingested by humans.

"Food" includes drink, chewing gum, and any substance, including water, intentionally incorporated into the food during its manufacture, preparation, or treatment. It includes water after the point of compliance as defined in Article 6 of Directive 98/83/EC and without prejudice to the requirements of Directives 80/778/EEC and 98/83/EC.

"Food" shall not include

(a) feed;
(b) live animals unless they are prepared for placing on the market for human consumption;
(c) plants prior to harvesting;
(d) medicinal products within the meaning of Council Directives 65/65/EEC (1) and 92/73/EEC (2);
(e) cosmetics within the meaning of Council Directive 76/768/EEC (3);
(f) tobacco and tobacco products within the meaning of Council Directive 89/622/EEC (4);
(g) narcotic or psychotropic substances within the meaning of the United Nations Single Convention on Narcotic Drugs, 1961, and the United Nations Convention on Psychotropic Substances, 1971;
(h) residues and contaminants.

3.6.5 Standards

Prior to the foundation of the international organization now known as the European Union, legal definitions of food products (often referred to as "standards" in the EU also as "vertical legislation") played an important role in food law in the Member States. The purpose of these product standards was to protect consumers from fraudulent practices. These standards constituted important barriers to the creation of a single market in the EU. At first, attempts were made to harmonize food standards through EU directives. Success of this attempt was limited. The CJEU found a way out of the deadlock. In its case law[49] it established a principle of mutual recognition. This principle entails that only in exceptional situations—for example, when public health is at stake—national standards can be used to ban foreign products from the market. In normal situations, products that conform to the legal requirements in the Member State of origin must be accepted on the market. As a consequence, product standards lost much of their relevance. Modern EU food law is mainly horizontal in nature. This means that legal requirements address foods in general or at least broad categories of foods, not individual foods.

3.6.6 Authorization requirements

Foods that have a history of safe use in the EU can be freely used. For an increasing number of categories of foods market access has been made subject to prior authorization. These include food additives (substances not normally consumed as a food that are added for a technological purpose), food enzymes, flavorings, genetically modified organisms, and (other) novel foods. Foods are considered novel if they do not have a history of safe use in the EU prior to 1997.

Authorization requires an application by an interested party providing scientific evidence of safety. Additional requirements may apply such as that consumers are not misled and that there is a technological need. The application is assessed by EFSA. The European Commission decides taking account of EFSA's opinion and opinions of the Member States (represented through the Standing Committee on Plants, Animals, Food, and Feed: SCoPAFF).

Authorized substances are placed on a positive list. In the case of food additives, they are given a number with the suffix "E" to indicate EU authorization. The number in principle conforms to the INS number of the Codex Alimentarius.

Authorization usually is generic. This means that any business can place the authorized product on the market. In case of GMOs, the authorization is specific. It only applies to the applicant.

49. CJEU 20 February 1979, Rewe-Zentral AG v Bundesmonopolverwaltung für Branntwein, Case 120/78 ECLI:EU:C:1979:42 (Cassis de Dijon).

3.6.7 Food safety limits

In the EU, legal limits are set for an ever-increasing number of substance/product combinations on food contaminants (Regulation 315/93), chemical substances (Regulation 1881/2006), residues of plant protection products (Regulation (EU) 1107/2009), residues of veterinary drugs (Regulation (EC) 470/2009), pathogenic microorganisms (Regulation 2073/2005), and radio activity (Regulation (EURATOM) 2016/52).

3.6.8 Process requirements

It has been realized that in order to ensure food safety processes must be under control in production as well as in trade. Practices aimed at the prevention of food safety risks are known as "hygiene." At the heart of EU legislation on food hygiene is the so-called HACCP system: Hazard Analysis and Critical Control Points (Regulation 853/2004). This system requires food businesses to make such an analysis of their processes that they know where hazards may occur, how to recognize them, and how to deal with them in order to maintain food safety. Application of the system must be well documented. The EU concept of HACCP conforms to the Codex.

In trade a requirement of traceability applies (Article 18 GFL). Food businesses must record where their inputs come from and where their products go. If a food safety incident occurs, this information must enable the authorities to swiftly identify the origin of the problem and its dispersal in order to eliminate the cause and take care of the consequences.

Finally, businesses that have reason to believe that a food they have brought to the market may not be in conformity with food safety requirements are under obligation to withdraw it from the food chain and recall is from consumers (Article 19 GFL).

3.6.9 Labeling

A large part of food legislation addresses the information food businesses provide to consumers regarding their product through advertising and—mainly—labeling. The most important codification of these rules is to be found in Regulation (EU) 1169/2011 on food information to consumers. Labeling means "any words, particulars, trademarks, brand name, pictorial matter, or symbol relating to a foodstuff and placed on any packaging, document, notice, label, ring, or collar accompanying or referring to such foodstuff." Labeling may not be misleading.

All prepackaged food products must be labeled in a language that is easily understood. Usually, this means in the national language of the Member State. Other information is mandatory, restricted, or forbidden.

There are 12 required (mandatory) pieces of information, the most important of which are the name of the food; the list of ingredients; the quantity of certain ingredients or categories of ingredients; the presence of allergens; in the case of prepackaged foodstuffs, the net quantity; the date of minimum durability or, in the case of foodstuffs which, from the microbiological point of view, are highly perishable, the "use by" date; the name or business name and address of the manufacturer or packager, or of a seller established within the Union; and a nutrition declaration (Tables 3.6.1 and 3.6.2.

Specific labeling requirements demand that the presence of additives, novel ingredients, and GMOs be mentioned on the label. For food additives, the function must be mentioned (such as preservative). They can be indicated either by their name or by E-number.

Nutrition and health claims are dealt with in a separate regulation (Regulation (EC) 1924/2006). Nutrition claims must conform to the annex to this regulation. The annex states among other things that the expression "light" may be only used in case of a reduction of at least 30% certain nutrients or energy. Health claims, e.g., claims about the effects of a certain food on health must be approved and science based. Foods bearing health claims are sometimes called "functional foods."

Claims that suggest that a food has medicinal properties are forbidden (Article 7 Regulation (EU) 1169/2011) (Van der Meulen and Bremmers, 2015).

3.6.10 Human right to food/food security

Respect for human rights is a condition for a country to become a Member State of the EU. All EU Member States are state party to the International Covenant on Economic, Social, and Cultural Rights ICESCR). However, the right to food has not been included in the EU Charter of Fundamental Rights. In EU food law, no reference is made to the right to food nor to Article 11 ICESCR (Van der Meulen and Wernaart, 2019). In the EU and its Member State, the Right to Food is generally not considered to be an enforceable right (CEDR, 2005; Wernaart, 2013).

TABLE 3.6.1 Allergens.

Substances or products causing allergies or intolerances according to regulation 1169/2011	
(1) Gluten	(8) Nuts
(2) Crustaceans	(9) Celery
(3) Eggs	(10) Mustard
(4) Fish	(11) Sesame
(5) Peanuts	(12) Sulfur dioxide
(6) Soy	(13) Lupin
(7) Milk	(14) Molluscs

TABLE 3.6.2 nutrition declaration according to regulation (EU) 1169/2011.

Energy	kJ/kcal
Fat	g
Of which	
- Saturates,	g
- Mono-un saturates,	g
- Polyunsaturates,	g
Carbohydrate	g
Of which	
- Sugars,	g
- Polyols,	g
- Starch,	g
Fiber	g
Protein	g
Salt	g
Vitamins and minerals	The units specified in point 1 of part A of Annex XIII

References

Borchardt, K.-D., 2018. The ABC of EU Law. Publication office of the European Union (Open access). https://publications.europa.eu/s/kIT8.

CEDR, the European Council for Rural Law, on the Right to Food http://www.cedr.org/congresses/roros/roros.php. (Open access).

Meulen, B.van der, Bremmers, H., 2015. The prohibition of medicinal claims: food in fact but medicinal product in law?. In: Wageningen Working Paper in Law and Governance 2015/03 https://doi.org/10.2139/ssrn.2605881 (open access).

Meulen, B.van der, Wernaart, B., 2019. Food and Agricultural Organisation (FAO) and Codex Alimentarius Commission: the impact of the right to food and food standards on EU food law, chapter 5. In: Wessel, R.A., Odermatt, J. (Eds.), Research Handbook on the European Union and International Organizations. Edwar Elgar Publishing. https://www.e-elgar.com/shop/research-handbook-on-the-european-union-and-international-organizations.

Wernaart, B., 2013. The enforceability of the human right to adequate food. In: A Comparative Study, European Institute for Food Law Series, vol. 8. Wageningen Academic Publishers. http://www.food-law.nl/Books/.

van der Meulen, B.M.J., Wernaart, B. (Eds.), 2020. EU Food Law Handbook. Wageningen Academic Publishers. http://www.food-law.nl/Books/.

van der Meulen, B., 2018a. The safe food principle. A critical reflection on the key concept of EU food safety law. In: European Institute for Food Law Working Paper 2018/08 (Open access). http://www.food-law.nl/Working-papers/.

van der Meulen, B., 2018b. Codex Alimentarius. The impact of the joint FAO/WHO food standards programme on EU food law. In: European Institute for Food Law working paper 2018/04. http://www.food-law.nl/Working-papers/ (Open access).

van der Meulen, B.M.J., 2018c. Enforcement of EU Agri-food law. Regulation (EU) 2017/625 on official controls and other official activities performed to ensure the application of food and feed law, rules on animal health and welfare, plant health and plant protection products. ERA Forum 1−19. https://doi.org/10.1007/s12027-018-0532-5 (Open access).

Further reading

All EU legislation is available at: https://eur-lex.europa.eu and can easily be found through their number comprising a year and an identifier.

Chapter 3.7

Turkey

Halide Gökçe Türkoğlu and Fehmi Kerem Bilgin

İzmir Bakirçay University, Faculty of Law, Menemen, İzmir, Turkey

3.7.1 Introduction

3.7.1.1 Historical evolution of food law in Turkey

Contemporary Turkish food law took its shape and characteristics through the adoption of the European Union (EU) norms. Thus, Turkish food law displays substantial similarities with European food law.

Yet, the origins of regulating the field of food in Anatolia[50] go back to antiquity. Many archaeological findings pertaining to food regulation have been found in Anatolian lands. For example, it was discovered that a Hittite column dating 3500 years back bears the phrases "thou shall not poison your neighbor's meat" and "thou shall not fool your neighbor" (Yalçın, 2008).

Cultural values and practices show that food safety and food hygiene were important in ancient Greece which was the site of important studies on hygiene, plant-based medicines, and healthy dietary practices (Notermans and Powell, 2005).

The influence of teachings of Greece continued through ancient Rome. In the Roman Republic, the maintenance of roads, the preservation of public peace, and the organization of public games fell under the jurisdiction of the *aedilis curulis* that was established in 494 B.C. Among the functions of this authority was also the inspection of marketplaces. The *aediles* checked the authenticity of weights used in marketplaces, the quality of the products sold, and also secured the storage and conservation of wheat in preparation for periods of scarcity. They could impose fines on sellers who sold defective products or committed fraud (Türkoğlu, 2011).

During the reign of the Ottoman sultan Bayezid II (1481−1512), laws setting out quality standards and fixing prices for various foodstuffs were enacted (Tayar, 2010). Under Ottoman rule, the general control over shopkeepers was carried out by constabularies called *muhtesipler* or *ihtisap ağaları*. These officials supervised the prices and quality of goods sold in bazaars and marketplaces, inspected the cleanliness of shops, and checked the authenticity of weights. Their activities included the prevention of the sale of adulterated food (Yıldırım, 2011−12).

The development of food control based on considerations of hygiene started in the Ottoman Empire alongside efforts to establish a modern health organization and a modern municipal organization. The control of foodstuffs and sales restrictions for public health reasons improved particularly on the institutional and administrative levels in periods marked by cholera epidemics that broke out in Istanbul in the years 1831, 1847, 1865, and 1894−1895.

During the second constitutional period (*ikinci meşrutiyet*) (1908−1918), the public health control of bovines was carried out by the Veterinary Inspectorship of the Province of Istanbul. The *Regulation on Health Administration of Provinces* (1913) enumerated the duties of health directors to be appointed to each province regarding the inspection of

50. Ancient term referring to the western part of Asia comprising most of what is Turkey today.

places of food production and sale. A regulation (1914) regarding the duties of the Health Inspection Commission under the General Directorship of Health required inspectors to perform sanitary controls of foodstuffs and beverages, as well as of slaughterhouses and tanneries (Yıldırım, 2011–12).

Moreover, a *By-Law on Cow Barns, Dairies, and Milks* (1914) issued by the Health Directorate of the Municipality conditioned the establishment of new cow barns within the municipal boundaries of Istanbul on obtaining a permit from district municipalities. Besides these developments in the field of food control, a relative awareness, as regards the necessity to control food manufacturers and sellers in order to prevent the spread of contagious diseases, emerged around the beginning of the 20th century. For example, dairymen were required to present a health report to the municipality every 6 months. However, attempts to establish legal regulations in this field mostly remained at the stage of planning (Yıldırım, 2011–12).

Despite the adoption of detailed food regulations after the foundation of the Republic of Turkey (1923), this field could not be subjected to a global and articulated legislation for a long time. Major food laws adopted during the first half of the 20th century were the following: the *Law on Spirit and Alcoholic Beverages Monopoly*,[51] the *Law on Animal Health Constabulary*,[52] *Municipality Law*,[53] the *Public Health Preservation Law*,[54] the *Animal Health Constabulary Regulation*,[55] the *Spirit and Spirituous Beverages Monopoly Law*,[56] the *Regulation Indicating the Particular Qualities of Foodstuffs and of Commodities and Materials Concerning Public Health* and,[57] the *Regulation Indicating the Particular Qualities of Foodstuffs and of Commodities and Materials Concerning Public Health*.[58] Due to difficulties encountered in updating this latter regulation, official authorities strived to supplement the existing food legislation with food standards adopted by the Turkish Standards Institute (*Türk Standardları Enstitüsü*) during 1960s (Artık et al., 2017).

Various laws entrusting the field of food services to different administrative authorities such as the Ministry of Agriculture and Rural Affairs, the Ministry of Health and Social Assistance, the Ministry of Industry and Trade, the Turkish Standards Institute, and municipalities failed to distinguish the spheres of responsibility and powers of these with sufficient clarity.

Despite official reports and projects pointing at the necessity to remedy the confusion of powers, the lack of coordination between different authorities, and the disorder in food legislation, a concrete step in this direction could only be made with the enactment of the *Statutory Decree Regarding the Production, Consumption and Control of Foodstuffs*[59] with the adoption by the EC - Turkey Association Council of Decision No. 1/95 on implementing the final phase of the Customs Union.[60] However, while constituting an important advance in the harmonization of Turkish food legislation with European standards, this *Statutory Decree* was not sufficient to bring a satisfactory solution to institutional and normative problems.

51. Law on Spirit and Alcoholic Beverages Monopoly [*İspirto ve Meşrubat-ı Küuliye İnhisarı Hakkında Kanun*] (Law N°: 790, Date of Adoption: 22/03/1926, Date of Official Gazette: 03/04/1926, Number of Official Gazette: 338).

52. Law on Animal Health Constabulary [*Hayvan Sağlık Zabıtası Hakkında Kanun*] (Law N° 1234, Date of Adoption: 03/05/1928, Date of Official Gazette: 14/05/1928, Number of Official Gazette: 888).

53. Municipality Law [*Belediye Kanunu*] (Law N°: 1580, Date of Adoption: 03/04/1930, Date of Official Gazette: 14/04/1930, Number of Official Gazette: 1471).

54. Public Health Preservation Law [*Umumî Hıfzıssıhha Kanunu*] (Law N° 1593, Date of Adoption: 24/04/1930, Date of Official Gazette: 06/05/1930, Number of Official Gazette: 1489).

55. Animal Health Constabulary Regulation [*Hayvan Sağlık Zabıtası Nizamnamesi*] (Date of Council of Ministers Decree: 09/08/1931, N° 11656, Date of Official Gazette: 17/09/1931, Number of Official Gazette: 1901).

56. Spirit and Spirituous Beverages Monopoly Law [*İspirto ve İspirtolu İçkiler İnhisarı Kanunu*] (Law N°: 4250, Date of Adoption: 08/06/1942, Date of Official Gazette: 12/06/1942, Number of Official Gazette: 5130).

57. Regulation Indicating the Particular Qualities of Foodstuffs and of Commodities and Materials Concerning Public Health [*Gıda Maddelerinin ve Umumî Sıhhati İlgilendiren Eşya ve Levazımın Hususî Vasıflarını Gösteren Nizamname*] (Date of Council of Ministers Decree: 11/08/1942, N° 2/18542, Date of Official Gazette: 07/09/1942, Number of Official Gazette: 5204).

58. Regulation Indicating the Particular Qualities of Foodstuffs and of Commodities and Materials Concerning Public Health [*Gıda Maddelerinin ve Umumî Sağlığı İlgilendiren Eşya ve Levazımın Hususî Vasıflarını Gösteren Tüzük*] (Date of Council of Ministers Decree: 04/08/1952, N° 3/15481, Date of Adoption: 18/10/1952, Number of Official Gazette: 8236).

59. Statutory Decree Regarding the Production, Consumption and Control of Foodstuffs [*Gıdaların Üretimi, Tüketimi ve Denetlenmesine Dair Kanun Hükmünde Kararname*] (Statutory Decree N° 560, Date of Adoption: 24/06/1995, Date of the Enabling Law: 08/06/1995, N° 4113, Date of Official Gazette: 28/06/1995, Number of Official Gazette: 22327).

60. Decision No 1/95 of The EC - Turkey Association Council of 22 December 1995 on implementing the final phase of the Customs Union (96/142/EC), *Official Journal of the European Communities*, L 35, 13/02/1996, pp. 1–46.

In the efforts to align with EU norms, Regulation (EC) No 178/2002[61] was decisive with the enactment of the *Law on the Adoption through Modification of the Statutory Decree Regarding the Production, Consumption, and Control of Foodstuffs*.[62]

However, noting serious deficiencies concerning food safety, veterinary, and phytosanitary policy, the *Turkey 2006 Progress Report* of the Commission of the European Communities revealed that substantial reforms were still necessary. The report noted that the "food, feed, and veterinary package" had still not been adopted and that progress had remained limited in transposition and implementation in the field of general foodstuffs policy. Moreover, no progress was observed on subjects such as veterinary policy, registration of bovine and caprine animals, financing of veterinary inspection and controls, rules for placing on the market of food and feed, genetically modified organisms, and novel foods. As regards the institutional plane, the report emphasized that a clear definition of competences between the central authorities and municipalities had to be established and that the Ministry of Agriculture and Rural Affairs had to be strengthened at central and local levels.[63]

Successive progress reports concluded that institutional and normative progress in the field of food safety, veterinary, and phytosanitary policy remained limited. It was obvious that the harmonization of Turkish food law with EU standards required a comprehensive and systematic general regulation. This situation was redressed with the adoption of the *Veterinary Services, Plant Health, Food, and Feed Law*.[64]

3.7.1.2 Fundamental legislation

The *Veterinary Services, Plant Health, Food, and Feed Law*, hereafter "Food Statute," constitutes the fundamental statute of Turkish food law. The statute indicates its aim as protecting and ensuring food and feed safety, public health, plant and animal health, animal improvement, and welfare. This aim is attained by "taking into consideration" consumer interests and the protection of the environment (Article 1). The scope of the statute is quite wide.[65] Primary production activities for personal consumption and foodstuffs prepared for personal consumption are excluded from the scope of the statute (Article 2 § 2).

In addition, the *Turkish Food Codex By-Law*[66] issued on the basis of the *Food Statute* by the Ministry of Food, Agriculture, and Livestock constitutes a conceptual matrix for specific food regulations.[67]

Two by-laws issued simultaneously by the Ministry of Food, Agriculture, and Livestock which are considered as falling within the scope of the "horizontal food codex" are the *Turkish Food Codex Food Labeling and Informing Consumers By-Law*[68] and the *Turkish Food Codex Nutrition and Health Claims By-Law*.[69] The aim of the former is "to determine the rules pertaining to the protection at the highest level of consumers as regards food information including differences of

61. Regulation (EC) No 178/2002 of the European Parliament and of the Council of 28 January 2002 laying down the general principles and requirements of food law, establishing the European Food Safety Authority and laying down procedures in matters of food safety, *Official Journal of the European Communities*, L 031, 01/02/2002, pp. 1–24.

62. Law on the Adoption through Modification of the Statutory Decree Regarding the Production, Consumption and Control of Foodstuffs [*Gıdaların Üretimi, Tüketimi ve Denetlenmesine Dair Kanun Hükmünde Kararnamenin Değiştirilerek Kabulü Hakkında Kanun*] (Law N$\underline{o}$ 5179, Date of Adoption: 27/05/2004, Date of Official Gazette: 05/06/2004, Number of Official Gazette: 25483).

63. Commission of the European Communities, SEC (2006) 1390, Commission Staff Working Document, *Turkey 2006 Progress Report*, COM (2006) 649 final, Brussels, 8.11.2006, pp. 44–46.

64. Veterinary Services, Plant Health, Food and Feed Law [*Veteriner Hizmetleri, Bitki Sağlığı, Gıda ve Yem Kanunu*] (Law N$\underline{o}$: 5996, Date of Adoption: 11/06/2010, Date of Official Gazette: 13/06/2010, Number of Official Gazette: 27610).

65. The following subjects fall within the scope of the statute: (1) all phases of production, processing and distribution of food, substances and materials in contact with food and of feed; (2) controls of residues of plant protection products, veterinary medicinal products and other residues as well as controls of contaminants; (3) struggle against epidemics or contagious animal diseases, harmful organisms in plants and plant products; (4) welfare of farm and laboratory animals as well as of domestic and ornamental animals; (5) issues of zootechnics; (6) veterinary medicinal products and plant protection products; (7) veterinary and phytosanitary services; (8) entrance and exit procedures of alive animals and products; (9) official controls and sanctions pertaining to these issues (Article 2 § 1).

66. Turkish Food Codex By-Law [*Türk Gıda Kodeksi Yönetmeliği*] (Date of Official Gazette: 29/12/2011, Number of Official Gazette: 28157 bis 3).

67. This by-law comprises three main groups of rules: (1) rules pertaining to the determination of minimum technical and hygienic standards for food, substances and materials in contact with food; (2) rules pertaining to the determination of principles regarding the horizontal and vertical food codex for pesticide residues and veterinary medicinal residues, food additives, flavorings and aromatic food components, contaminants, packaging, labeling, sampling, analysis methods, transportation and storage; (3) rules pertaining to the determination of special provisions regarding geographical indications (Article 2).

68. Turkish Food Codex Food Labeling and Informing Consumers By-Law [*Türk Gıda Kodeksi Gıda Etiketleme ve Tüketicileri Bilgilendirme Yönetmeliği*] (Date of Official Gazette: 26/01/2017, Number of Official Gazette: 29960 bis).

69. Turkish Food Codex Nutrition and Health Claims By-Law [*Türk Gıda Kodeksi Beslenme ve Sağlık Beyanları Yönetmeliği*] (Date of Official Gazette: 26/01/2017, Number of Official Gazette: 29960 bis).

perception and necessities of information" (Article 1). The by-law covers general rules, requirements, and responsibilities relating to food information as well as information procedures and measures guaranteeing the right to information of consumers (Article 2 § 1). The aim of the latter by-law is to regulate the nutrition and health claims used in labeling, presenting, and advertising of foodstuffs provided to final consumers and mass consumption places (Article 2).

3.7.1.3 Risk analysis

By imposing an obligation to base "acts pertaining to food, feed, and plant health on risk analysis, in order to ensure the maximum protection of human health and life," the *Food Statute* gives a central position to risk analysis (Article 26 § 1).[70] While defining the concept of risk as "the functional relationship between the danger that may produce a negative effect on health and its severity" (Article 3 § 1, *54*), the statute configures risk analysis as a process consisting of the components of risk assessment, risk management, and risk communication (Article 3 § 1, *55*).

The *Food Statute* foresees the establishment of risk analysis "commissions" having different working areas in order to make independent scientific risk assessment. These commissions, which formulate risk analysis conclusions of an advisory character, consist of representatives of research institutions, research institutes, university faculty concerned with the subject and of other specialists if necessary. A "risk assessment unit" established by the Ministry of Agriculture and Rural Affairs oversees the commissions and may cooperate with similar national and international organizations (Article 26 § 2).

In order to implement these provisions, the *Statutory Decree on the Organization and Duties of the Ministry of Food, Agriculture, and Livestock*[71] established the General Directorate of Food and Control as one of the service units within the central organization of the Ministry. Currently, the duties of the General Directorate of Food and Control are governed by the *Presidential Decree on the Organization of the Presidency of the Republic.*[72] Duties of the General Directorate of Food and Control include "determining the principles of risk management for ensuring plant and animal health as well as food and feed safety, carrying out risk assessment and ensuring risk communication" (Article 413 § 1, *o*). The Presidency of the Department of Risk Assessment was established as the subunit of the General Directorate of Food and Control entrusted with the carrying out of risk assessment activities.[73]

70. For a more technical study, *see* Çopuroğlu, G., Kasımoğlu Doğru, A., & Ayaz, N. D. (2015). Türk Gıda Mevzuatında Risk Analizi. *Etlik Veteriner Mikrobiyoloji Dergisi*, *26* (1), 23–28.

71. Statutory Decree on the Organization and Duties of the Ministry of Food, Agriculture and Livestock [*Gıda, Tarım ve Hayvancılık Bakanlığının Teşkilat ve Görevleri Hakkında Kanun Hükmünde Kararname*] (Statutory Decree N° 639, Date of Adoption: 03/06/2011, Date of the Enabling Law: 06/04/2011, N° 6223, Date of Official Gazette: 08/06/2011, Number of Official Gazette: 27958 bis). However, this statutory decree was abrogated by the Statutory Decree on the Modification of Certain Laws and Statutory Decrees for Accommodating Constitutional Amendments [*Anayasada Yapılan Değişikliklere Uyum Sağlanması Amacıyla Bazı Kanun ve Kanun Hükmünde Kararnamelerde Değişiklik Yapılması Hakkında Kanun Hükmünde Kararname*] (Statutory Decree N° 703, Date of Adoption: 02/07/2018, Date of the Enabling Law: 10/05/2018, N° 7142, Date of Official Gazette: 09/07/2018, Number of Official Gazette: 30473 bis 3).

72. Presidential Decree on the Organization of the Presidency of the Republic [*Cumhurbaşkanlığı Teşkilatı Hakkında Cumhurbaşkanlığı Kararnamesi*] (Presidential Decree N° 1, Date of Official Gazette: 10/07/2018, Number of Official Gazette: 30474).

73. Specific duties of the Presidency of the Department of Risk Assessment as regards animal health and welfare, plant health as well as feed and food safety are enumerated in Article 13 of the Directive on the Duties of the Central Organization of the Ministry of Food, Agriculture and Livestock [*Gıda, Tarım ve Hayvancılık Bakanlığı Merkez Teşkilatı Görev Yönergesi*] (Date: 18/12/2014, N° 13805938/MEV-2011-127/54) as follows: (*a*) to establish committees and commissions as to different subjects for effectuating risk assessments according to scientific principles; (*b*) to carry out the secretariat of committees and commissions that will effectuate risk assessments; (*c*) to collect the data that will constitute the basis of risk assessment and to analyze such data; (*d*) to formulate advisory scientific opinions; (*e*) to carry out risk assessments and to formulate opinions as to the determination and definition of characteristics of urgent risks; (*f*) to provide scientific and technical assistance in crisis situations, if requested; (*g*) to cooperate with similar national and international organizations carrying out risk assessments when necessary; (*h*) to ensure coordination with all intra- and extra-ministerial parties concerned with risks; (*i*) to ensure the coordination of intraministerial risk communication in order to ensure the trustworthy, impartial and correct informing of the public and concerned parties; (*j*) to make all kinds of risk assessment researches and projects or to have them made; (*k*) to fulfill similar duties given by the General Directorate.

In order to regulate the formation and operation of risk assessment commissions foreseen by the *Food Statute* (Article 26 § 2), the Ministry of Food, Agriculture, and Livestock prepared the *By-law on the Working Procedures and Principles of Risk Assessment Committees and Commissions*.[74] To date, seven commissions have been established to date by the Ministry in regard to risk assessments relating to food and feed safety.[75]

Turkish food law attributes an essential role to scientific commissions in the field of risk assessment. The above cited by-law provides that the members of the scientific commissions are independent and that they shall not receive any instruction from any office, authority, or person (Article 5 § 5). Commissions are able to make risk assessments upon request of the Ministry or on their own initiative (Article 8 § 1, *a*). Yet, the members of the commissions are appointed by the minister among candidates determined upon open call (Article 7 § 3). Scientific commissions have to submit their advisory opinions to the Ministry and are not permitted to share these directly with the public (Article 8 § 1, *c*). Members of the commissions are not permitted to disclosure any information, documents, or secrets they have obtained in the course of their duty, even after they have quit their function (Article 15 § 2). Even though the scientific commissions have the main responsibility regarding risk assessments, the Presidency of the Department of Risk Assessment is empowered to independently formulate advisory scientific opinions regarding animal health and welfare, plant health, as well as feed and food safety.

The *Food Statutes* require that risk management and risk communication shall be carried out by the Ministry taking into consideration the results of risk assessment, scientific data, other related factors, and the precautionary principle (Article 26 § 3). In the EU, the system is different with there being a clear distinction between the political (European Commission etc.) and scientific (EFSA) authorities.[76] The overlap of risk assessment and risk management in the Turkish system is problematic.

Turkey has acquired a considerable technical knowledge in the field of food safety. However, substantial improvements are necessary as regards the transparency of risk analysis processes and the publication of risk assessment data.

3.7.1.4 Codex Alimentarius and Turkish food law

A member of the FAO since 1948, Turkey acceded the Codex Alimentarius Commission (CAC) in 1963. The country is represented in CAC meetings by officials of the Ministry of Agriculture and Forestry.[77] The General Directorate of Food and Control serves as codex contact point and coordinates with the National Food Codex Commission (*Ulusal Gıda Kodeksi Komisyonu*). The *Codex Alimentarius* standards constitute a fundamental reference in the preparation of Turkish food legislation, especially in the fields of contaminants, additives, and residues.[78]

Turkey is also bound to act in conformity with the WTO agreements including the WTO Agreement on the Application of Sanitary and Phytosanitary Measures.[79]

It must be noted that difficulties are encountered in the harmonization of Turkish food legislation with the standards of the *Codex Alimentarius*, as is the case for many countries. Nevertheless, the harmonization process is gaining momentum owing to requests by commercial operators involved in international trade as well as by consumers for access to safe and healthy food (Artık et al., 2017).

---

74. By-Law on the Working Procedures and Principles of Risk Assessment Committees and Commissions [*Risk Değerlendirme Komite ve Komisyonlarının Çalışma Usul ve Esasları Hakkında Yönetmelik*] (Date of Official Gazette: 24/12/2011, Number of Official Gazette: 28152).

75. The seven commissions in question are the following: (1) Commission on Plants Useable as Food (2012); (2) Contaminants Commission (2013); (3) Feed Commission (2015); (4) Biohazards Commission (2016); (5) Phytosanitary Codex Alimentarius Commission (2017); (6) Specific Food Ingredients, Claims and Novel Foods Codex Alimentarius Commission (2017) and (7) Food Additives Codex Alimentarius Commission (2018). *See*. Tarım ve Orman Bakanlığı Gıda ve Kontrol Genel Müdürlüğü (2018) *Gıda ve Kontrol Verileri* (General Directorate of Food and Control of the Ministry of Agriculture and Forestry, *Food and Control Data*). Available at < https://www.tarimorman.gov.tr/sgb/Belgeler/SagMenuVeriler/GKGM.pdf >, ss. 31–33.

76. EFSA, "Risk assessment versus risk management: What's the difference?" < http://www.efsa.europa.eu/en/press/news/140416 >.

77. As a consequence of the Statutory Decree numbered 703 (02/07/2018) and the Presidential Decree numbered 1 (10/07/2018) aiming at the realization of a presidential system in Turkey, the Ministry of Food, Agriculture and Livestock and the Ministry of Forestry and Water Affairs were merged to form the Ministry of Agriculture and Forestry.

78. *See* the information document prepared by the Ministry of Agriculture and Forestry on its relations with the Codex Alimentarius Commission: < https://www.tarimorman.gov.tr/Konular/Gida-Ve-Yem-Hizmetleri/Gida-Hizmetleri/Kodeks >. For example, the EU standards for aflatoxin limits constituted a major obstacle to the exportation of almonds, hazelnuts, pistachios and dried figs from Turkey to Europe. In order to handle this problem concerning various countries, studies on maximum levels of aflatoxins for these products were carried out under the auspices of the Codex Committee on Contaminants in Foods by working groups in which Turkey played an important role. *See* CAC, Codex Committee on Contaminants in Foods. (2007). *Discussion Paper on Aflatoxin in Dried Figs*. CX/CF 07/1/20. Available at < http://www.fao.org/tempref/codex/Meetings/CCCF/CCCF1/cf01_20e.pdf > and *Discussion Paper on Maximum Levels for Total Aflatoxins in "ready-to-eat" Almonds, Hazelnuts and Pistachios*. CX/CF 07/1/9. Available at < http://www.fao.org/tempref/codex/Meetings/CCCF/CCCF1/cf01_09e.pdf >.

79. *See*: < https://www.wto.org/english/thewto_e/coher_e/wto_codex_e.htm >.

3.7.2 Fundamental institutional framework

3.7.2.1 Major authorities

The principal authority in the field of food in Turkey is the Ministry of Agriculture and Forestry. The structure, duties, and powers of the Ministry are found in the *Presidential Decree on the Organization of the Presidency of the Republic*. The duties of the Ministry are varied.[80] The Ministry is composed of central, provincial, and foreign organizations. The central organization of the Ministry comprises 21 "service units."[81] There are also five "related establishments"[82] and five "dependent establishments."[83]

While placing food safety services principally within the competence of the Ministry of Agriculture and Forestry, the *Food Statute* provides certain powers to the Ministry of Health in the field of food. These two Ministries are expected to share some powers in conjunction with other establishments as, for example, in the case of contagious animal diseases and animal by-products not intended for human consumption. In certain areas such as waters, supplementary foods and dietary foods for special medical purposes the fields of competence of the Ministry of Agriculture and Forestry and the Ministry of Health are separated.

3.7.2.2 Enforcement powers

The examination, inspection, control, and sanction powers in the field of food law are stipulated in the *Food Statute*. In this regard, important powers are conferred upon the Ministry of Agriculture and Forestry.

The Ministry is empowered to take all necessary measures in the event of an outbreak of an animal disease subject to compulsory notification (Article 4 § 1, *b*).[84] Moreover, as regards operations pertaining to animal by-products not intended for human consumption, measures relating to the prevention of threats to human and animal health and to the prevention of environmental damage shall be taken by the Ministry of Agriculture and Forestry, the Ministry of Health, and the Ministry of Internal Affairs (Article 6 § 1).

The Ministry is responsible for taking appropriate measures against the introduction into the country of organisms harmful to plants or plant products and against their spread within the country. If there is a suspicion regarding harmful organisms or the appearance of harmful organisms in epidemical form, the Ministry may take any measures against their spread. Such measures may comprise the prohibition or restriction of seeding or planting, the prohibition of transport and sale of plants, plant products or other materials, as well as the destruction of these (Article 15 § 1, *b*).

If it receives notification of a direct or indirect risk relating to animal and plant health or to food and feed, the Ministry is required to take all necessary measures. The owner of the animal or the keeper of the animal on behalf of the owner, the owner of the plant and plant products, the food or feed business operator is required to apply the measures and decisions taken by the Ministry. In case a serious risk to human, animal, and plant health appears and the existing measures prove to be insufficient, administrative measures restricting or prohibiting the placing on the market, the use, and introduction into the country of live animals or products may be applied (Article 25 §§ 1–3).

---

80. The duties in question are as follows: (1) the conduct of research for the improvement of plant and animal production as well as aquacultural production, the improvement of the agricultural sector and the formation of agricultural policies; (2) the protection of food production, safety and reliability, rural development, soil, water resources and biodiversity; (3) the organization and awareness raising of farmers, the effective management of incentives, the regulation of agricultural markets, the determination of general policies relating to agriculture and livestock; (4) the protection, development, management, redress and maintenance of forests; struggle against desertification and soil erosion; afforestation and pasture redress; (5) the protection of nature, national parks, natural parks, natural monuments, wetlands, wildlife; (6) coordination of national water management; (7) follow-up of international studies on subjects falling within its field of activity and making preparations on the national plane for contributing to such studies.

81. Nine of these service units are concerned with food safety: (1) General Directorate of Food and Control; (2) General Directorate of Plant Production; (3) General Directorate of Livestock; (4) General Directorate of Fisheries and Aquaculture; (5) General Directorate of European Union and Foreign Relations; (6) Presidency of Tobacco and Alcohol Department; (7) Presidency of Sugar Department; (8) General Directorate of Agricultural Research and Policies; (9) General Directorate of Water Management.

82. The related establishments are the following: (1) General Directorate of Agricultural Products Office; (2) General Directorate of Agricultural Enterprises; (3) General Directorate of Tea Enterprises; (4) General Directorate of Meat and Dairy Institution; (5) Institution for Supporting Agriculture and Rural Development.

83. The dependent establishments are as follows: (1) General Directorate of Atatürk Forest Farm; (2) State General Directorate of Water Works; (3) General Directorate of Forestry; (4) Water Institute of Turkey; (5) General Directorate of Meteorology.

84. These measures may include the establishment of protection and surveillance zones; controlling, sampling, diagnosing and other examinations for investigating the disease and preventing the spread of the disease; vaccination, isolation, cull, and destruction of animals; cordoning for restricting or prohibiting the movements of animals or humans; cessation of artificial insemination and animal improvement activities; the destruction of animal products, feed, tools, equipment, and other contaminated materials that may cause the spread of the disease.

Even in the case of scientific uncertainties, the Ministry may, if there is a threat of possible harmful effects, order provisional cessation of production, marketing or consumption, placing on the market, recall and similar precautionary measures until additional scientific data enabling a comprehensive risk assessment are gathered (Article 26 § 5).

The Ministry is empowered to take measures such as quarantining, blocking the introduction, or placing on the market of an item if there exists any risk to human, plant, and animal health. Exported live animals and products returned for various reasons are submitted to official control by the Ministry. Live animals and products determined to be in conformity with the legislation are allowed to enter the country. Those which are not in conformity may be reexported, placed under quarantine, subjected to special treatment, allowed entry to be used for purposes other than their principal purpose of use, or subjected to culling and destruction (Article 34 § 3, and § 8).

The *Food Statute* contains detailed provisions regarding sanctions for various unlawful acts. These sanctions are grouped in the *Statute* as sanctions relating to animal health, animal welfare, and zootechnics; veterinary health products; plant health, plant protection products, food and feed; and hygiene and official controls (Article 36–41). While part of these sanctions are for contravening preventive and protective measures adopted by the Ministry, others are for actions violating provisions regarding certain activities. Furthermore, sanctions are provided for not keeping certain records, not having certain documents, and for forging documents. Most of the sanctions are administrative fines of various amounts. Apart from these, there are other sanctions such as the prohibition of exercising activity for those conducting unauthorized activities, the annulment of permits and approvals of those acting in contravention to the provisions of the *Statute*, and the confiscation of products which do not fulfill the legal standards. The sanctions under the *Food Statute* are applied by provincial and communal agriculture directors. In case of urgency, administrative sanctions other than administrative fines may be applied by control officers (Article 42).

3.7.3 Standards

Turkish food law is linked to various national and international food quality and safety standards. Preparing "Turkish standards" for all kinds of materials, products, services, and procedures at the national level is assigned to the Turkish Standards Institute.[85] Within the Institute, these standards are prepared by specialized boards and adopted by a Technical Board.

On the international level, the HACCP and traceability systems are prominent. Owing to a growing awareness of food safety in recent years, compliance with international standards gained importance in Turkey. In order to raise their competitiveness in national and international markets, Turkish food companies were impelled to restructure their production activities so as to conform to these international standards. In addition, despite "compliance costs," national incentive policies encourage companies to integrate food standards (Koç et al., 2018).

With the issuance of the *Turkish Food Codex By-Law*[86] in 1997, the establishment of HACCP system became compulsory for food businesses. Moreover, the Turkish Standards Institute adopted on March 3, 2003, the TS 13001 standard titled "Management of Food Safety based on HACCP - Rules Relating to the Management System for Food Producing Establishments and Suppliers."

Coming after a multitude of by-laws, the *Turkish Food Codex By-Law on Materials and Articles in Contact with Food*[87] adopts the principle of traceability for all phases of production, processing, and distribution in conformity with the ISO 22000 standard. However, serious problems are still faced due to structural deficiencies. Foodstuffs threatening human health are placed on the market in the absence of rigorous control mechanisms (Atlı et al., 2010). The situation is aggravated by underground food businesses operating without giving due consideration to hygiene norms.

85. Turkish Standards Institute Establishment Law [*Türk Standardları Enstitüsü Kuruluş Kanunu*] (Law N°: 132, Date of Adoption: 18/11/1960, Date of Official Gazette: 22/11/1960, Number of Official Gazette: 10661). However, most provisions of this law were abrogated, and its name was changed as "Law on Certain Regulations Relating to the Turkish Standards Institute" by the Statutory Decree numbered 703 (02/07/2018). At present time, the establishment, organization, duties and powers of the Turkish Standards Institute are governed by the Presidential Decree on the Organization of Institutions and Establishments Dependent, Related, Connected to Ministries and of other Institutions and Establishments [*Bakanlıklara Bağlı, İlgili, İlişkili Kurum ve Kuruluşlar ile Diğer Kurum ve Kuruluşların Teşkilatı Hakkında Cumhurbaşkanlığı Kararnamesi*] (Presidential Decree N° 4, Date of Official Gazette: 15/07/2018, Number of Official Gazette: 30479).

86. Turkish Food Codex By-Law [*Türk Gıda Kodeksi Yönetmeliği*] (Date of Official Gazette: 16/11/1997, Number of Official Gazette: 23172 bis).

87. Turkish Food Codex By-Law on Materials and Articles in Contact with Food [*Türk Gıda Kodeksi Gıda ile Temas Eden Madde ve Malzemelere Dair Yönetmelik*] (Date of Official Gazette: 05/04/2018, Number of Official Gazette: 30382).

3.7.4 Authorization requirements

3.7.4.1 Food subject to authorization

Under the *Biosafety Law*[88] and the *By-Law Regarding Genetically Modified Organisms and their Products*,[89] operations relating to genetically modified organisms and their products are subject to regulation. The importation, exportation, release for experimental purposes, placement on the market of GMOs, and their products as well as the use of genetically modified microorganisms in confined areas are dependent upon a risk assessment effectuated according to scientific principles. Both for the first importation of each GMO and its product and for the GMO and its product developed within the country, an application must be made to the Ministry by the gene owner, the importer or the interested party. As regards GMO and GMO products, the *Biosafety Law* prohibits their placement on the market without authorization, their use in contravention to the decisions of the Ministry, their use in infant food and infant formulae, as well as in follow-on food and follow-on formulae. The production of genetically modified plants and animals is also prohibited.

The *Turkish Food Codex Food Additives By-Law*[90] indicates both food additives, which may be placed on the market directly and food containing food additives, in two separate annexed lists. Food additives or food containing food additives, which are not on the lists, cannot be placed on the market. Although the sale of porcine products is not prohibited in Turkey, the by-law prohibits explicitly the use of food additives of porcine origin in food, food additives, food enzymes, and flavorings.

The *Food Statute* provides that "procedures and principles pertaining to novel food and feed shall be determined by the Ministry." (Article 21 § 4) As yet, no legislation of any kind has been adopted in this field. As stated above, a Specific Food Components, Declarations, and Novel Foods Commission has been established by the Ministry for, among other purposes, evaluating the safety of novel food. However, no activity of this Commission in this field is made public.

3.7.4.2 Procedural aspects

Applications pertaining to GMOs should indicate their purpose. The result of a particular application cannot constitute a precedent for other applications. All applications falling within the scope of the *Biosafety Law* are subject to risk assessment based on scientific principles and socio-economical assessment separately. Risk management principles are established on the basis of the results of these assessments. The applicant is responsible for the preparation and implementation of a detailed risk management plan (Article 4).

A special by-law establishes a common application procedure for the evaluation and authorization of food additives, food enzymes, flavorings, source materials of flavorings, and flavoring food components.[91] The General Directorate of Food and Control shall decide on applications within the framework of the relevant legislation. In the event of acceptance, the specific list in the by-law shall be updated through the insertion of the name of the new substance. The updating process of the substance lists annexed to the relevant by-laws may be initiated ex officio by the General Directorate or upon an application by individual food business operators or an establishment representing them.

88. Biosafety Law [*Biyogüvenlik Kanunu*] (Law N$\underline{\text{o}}$: 5977, Date of Adoption: 18/03/2010, Date of Official Gazette: 26/03/2010, Number of Official Gazette: 27533).

89. By-Law Regarding Genetically Modified Organisms and their Products [*Genetik Yapısı Değiştirilmiş Organizmalar ve Ürünlerine Dair Yönetmelik*] (Date of Official Gazette: 13/08/2010, Number of Official Gazette: 27671). This by-law was prepared on the basis of Regulation (EC) No 1333/2008 of the European Parliament and of the Council of 16 December 2008 on food additives, *Official Journal of the European Union*, L 354, 31/12/2008, pp. 16–33.

90. Turkish Food Codex Food Additives By-Law [*Türk Gıda Kodeksi Gıda Katkı Maddeleri Yönetmeliği*] (Date of Official Gazette: 30/06/2013, Number of Official Gazette: 28693).

91. Turkish Food Codex By-Law on the Common Authorization Procedure Pertaining to Food Additives, Food Enzymes and Flavorings [*Türk Gıda Kodeksi Gıda Katkı Maddeleri, Gıda Enzimleri ve Gıda Aroma Vericilerine İlişkin Ortak İzin Prosedürü Hakkında Yönetmelik*] (Date of Official Gazette: 24/02/2017, Number of Official Gazette: 29989).

3.7.5 Food safety limits

3.7.5.1 Residue limits

The use of plant protection products in Turkey is dependent upon obtaining an authorization in accordance with the *By-Law on the Authorization and Placement on the Market of Plant Protection Products.*[92] The limits of plant protection product residues are fixed in the *Turkish Food Codex Maximum Pesticide Residue Limits By-Law,*[93] which applies to fresh, processed, or composite food that may contain pesticide residue as well as to plant and animal products indicated in a special annex.

The *Turkish Food Codex By-Law on the Classification and Maximum Residue Limits of Pharmacologically Active Substances in Foods of Animal Origin* classifies pharmacologically active substances scientifically and technically detectable in foods of animal origin and determines the maximum residue limits of such substances.[94] This by-law was prepared in conformity with Regulation (EC) No 470/2009[95] and Regulation (EU) No 37/2010.[96] The by-law does not conflict with legislation pertaining to hormones and hormone-like substances, which are prohibited for food producing animals. The use of hormones and hormone-like substances for noncurative purposes is prohibited in Turkey since 1992 (Şevik and Ayaz, 2017). This field is currently regulated by the *Communiqué on Hormones and Hormone-Like Substances Whose Administration to Food Producing Animals is Prohibited or Conditionally Permitted.*[97] However, due to the lack of effective control mechanisms, serious deficiencies exist as regards the prevention of the use of hormones for growth purposes.

3.7.5.2 Contaminant limits

While the *Turkish Food Codex Contaminants By-Law*[98] sets the maximum levels of certain contaminants in foodstuffs, the microbiological criteria applying to foodstuffs are determined by the *Turkish Food Codex Microbiological Criteria By-Law.*[99] The *Turkish Food Codex By-Law on Materials and Articles in Contact with Food*[100] provides special rules pertaining to the specific and total limits of certain components or component groups of materials in contact with food inside or on the surface of food through special legislation. To that effect, various specific communiqués were issued on the basis of relevant EU legislation by the Ministry to set limits pertaining to ceramic articles, materials, and articles made of regenerated cellulose films, certain epoxy derivatives in materials and articles, plastic materials and articles, as well as active and intelligent materials and articles intended to come into contact with food.

---

92. By-Law on the Authorization and Placement on the Market of Plant Protection Products [*Bitki Koruma Ürünlerinin Ruhsatlandırılması ve Piyasaya Arzı Hakkında Yönetmelik*] (Date of Official Gazette: 09/11/2017, Number of Official Gazette: 30235).

93. Turkish Food Codex Maximum Pesticide Residue Limits By-Law [*Türk Gıda Kodeksi Pestisitlerin Maksimum Kalıntı Limitleri Yönetmeliği*] (Date of Official Gazette: 25/11/2016, Number of Official Gazette: 29899 bis). This by-law was prepared on the basis of Regulation (EC) No 396/2005 of the European Parliament and of the Council of 23 February 2005 on maximum residue levels of pesticides in or on food and feed of plant and animal origin and amending Council Directive 91/414/EEC, *Official Journal of the European Union*, L 70, 16/03/2005, pp. 1–16.

94. Turkish Food Codex By-Law on the Classification and Maximum Residue Limits of Pharmacologically Active Substances in Foods of Animal Origin [*Türk Gıda Kodeksi Hayvansal Gıdalarda Bulunabilecek Farmakolojik Aktif Maddelerin Sınıflandırılması ve Maksimum Kalıntı Limitleri Yönetmeliği*] (Date of Official Gazette: 07/03/2017, Number of Official Gazette: 30000).

95. Regulation (EC) No 470/2009 of the European Parliament and of the Council of 6 May 2009 laying down Community procedures for the establishment of residue limits of pharmacologically active substances in foodstuffs of animal origin, repealing Council Regulation (EEC) No 2377/90 and amending Directive 2001/82/EC of the European Parliament and of the Council and Regulation (EC) No 726/2004 of the European Parliament and of the Council, *Official Journal of the European Union*, L 152, 16/06/2009, pp. 11–22.

96. Commission Regulation (EU) No 37/2010 of 22 December 2009 on pharmacologically active substances and their classification regarding maximum residue limits in foodstuffs of animal origin, *Official Journal of the European Union*, L 15, 20/01/2010, pp. 1–72.

97. Communiqué on Hormones and Hormone-Like Substances Whose Administration to Food Producing Animals is Prohibited or Conditionally Permitted [*Gıda Değeri Olan Hayvanlara Uygulanması Yasaklanan ve Belli Şartlara Bağlanan Hormon ve Benzeri Maddeler Hakkında Tebliğ*] (Communiqué No: 2003/18, Date of Official Gazette: 19/11/2016, Number of Official Gazette: 25143).

98. Turkish Food Codex Contaminants By-Law [*Türk Gıda Kodeksi Bulaşanlar Yönetmeliği*] (Date of Official Gazette: 29/12/2011, Number of Official Gazette: 28157 bis 3). This by-law is based on Commission Regulation (EC) No 1881/2006 of 19 December 2006 setting maximum levels for certain contaminants in foodstuffs, *Official Journal of the European Union*, L 364, 20/12/2006, pp. 5–24.

99. Turkish Food Codex Microbiological Criteria By-Law [*Türk Gıda Kodeksi Mikrobiyolojik Kriterler Yönetmeliği*] (Date of Official Gazette: 29/12/2011, Number of Official Gazette: 28157 bis 3). This by-law was prepared on the basis of Commission Regulation (EC) No 2073/2005 of 15 November 2005 on microbiological criteria for foodstuffs, *Official Journal of the European Union*, L 338, 22/12/2005, pp. 1–26.

100. Turkish Food Codex By-Law on Materials and Articles in Contact with Food [*Türk Gıda Kodeksi Gıda ile Temas Eden Madde ve Malzemelere Dair Yönetmelik*] (Date of Official Gazette: 05/04/2018, Number of Official Gazette: 30382). This by-law was drafted by taking into consideration Regulation (EC) No 1935/2004 of the European Parliament and of the Council of 27 October 2004 on materials and articles intended to come into contact with food and repealing Directives 80/590/EEC and 89/109/EEC, *Official Journal of the European Union*, L 338, 13/11/2004, pp. 4–17.

3.7.5.3 Determination of limits

Pursuant to the *Turkish Food Codex Preparation By-Law*,[101] legal regulations relating to the Turkish food codex are to be prepared by the Ministry based primarily on European Union legislation but also taking into account Codex Alimentarius Commission texts, national legislation, international norms, as well as conditions of the country. The Ministry may, if it deems necessary, invite experts from institutions, establishments, universities, and civil society to participate in the drafting process of the food codex. The prepared food codex drafts shall be submitted for the opinion of the said organizations (Article 11).

3.7.6 Process requirements

3.7.6.1 The hygienic regulation of business processes

In the field of hygienic regulation of business processes, Turkish food law is mainly based on Regulation (EC) No 852/2004.[102] The *Food Statute* defines hygiene as "any measure and condition necessary for putting hazards under control and ensuring the suitability of food and feed for human and animal consumption by taking into consideration their intended use" (Article 3 § 1, *35*).

The Ministry of Agriculture and Forestry is empowered to determine general and special hygiene principles, regulations based on hazard analysis and critical control points, health marks indicating that official controls are completed, regulations pertaining to identification marks, and traceability. Primary producers, retail sale businesses, as well as food and feed business operators are bound to comply with hygiene norms determined by the Ministry. Except for primary producers, food and feed business operators are under the obligation of setting up a food and feed safety system based on principles of hazard analysis and critical control points (Article 29). The Ministry is also empowered to determine which food and feed business are subject to registration and approval (Article 30).

The basic framework established by the *Food Statute* as regards food hygiene is supplemented by the detailed provisions of the *Food Hygiene By-Law*[103] and the *Feed Hygiene By-Law*.[104] The former requires the food business operator to be responsible for ensuring the specified hygiene requirements in all phases of production, processing, and distribution under its control. The general principles adopted by the *Food Hygiene By-Law* include the primary responsibility of the food business operator and the concomitant implementation of procedures based on HACCP principles and good hygiene practices.

3.7.6.2 Traceability requirements

The *Food Statute* defines traceability as "the traceability and followability of plant products, food and feed, the animal, or the plant from which food is obtained, a substance intended or expected to be in food and feed during all phases of production, processing, and distribution" (Article 3 § 1, *39*). According to the *Statute*, food and feed business operators are under the obligation to establish a system for tracing—in all phases of production, processing, and distribution food or feed—any substance to be added to food or feed, including the animal from which food is obtained and to provide such information to the Ministry when requested. Also, to ensure traceability, the food and feed to be placed on the market should be appropriately labeled or identified in conformity with the information and documents determined by the Ministry (Article 24).

---

101. Turkish Food Codex Preparation By-Law [*Türk Gıda Kodeksi Hazırlama Yönetmeliği*] (Date of Official Gazette: 29/12/2011, Number of Official Gazette: 28157 bis 3).

102. Regulation (EC) No 852/2004 of the European Parliament and of the Council of 29 April 2004 on the hygiene of foodstuffs, *Official Journal of the European Union*, L 139, 30/04/2004, pp. 1–54.

103. Food Hygiene By-Law [*Gıda Hijyeni Yönetmeliği*] (Date of Official Gazette: 17/12/2011, Number of Official Gazette: 28145).

104. Feed Hygiene By-Law [*Yem Hijyeni Yönetmeliği*] (Date of Official Gazette: 27/12/2011, Number of Official Gazette: 28155).

3.7.7 Labeling

3.7.7.1 Mandatory particulars

Pursuant to the *Turkish Food Codex Food Labeling and Informing Consumers By-Law*,[105] the following particulars are mandatory: (*a*) the name of the food; (*b*) the list of ingredients; (*c*) determined substances or products causing allergies or intolerances; (*d*) the quantity of certain ingredients or categories of ingredients; (*e*) the net quantity of the food; (*f*) the recommended date of consumption or the "use by" date; (*g*) special storage conditions and/or conditions of use; (*h*) the name or business name and address of the food business operator; (*i*) the business registration number or identification mark; (*j*) the country of origin; (*k*) instructions for use where it would be impossible to consume the food appropriately in the absence of such instructions; (*l*) with respect to beverages containing more than 1%–2% by volume of alcohol, the actual alcoholic strength by volume; and (*m*) a nutrition declaration (Article 9 § 1).

3.7.7.2 Labeling of food additives

The *Turkish Food Codex Food Additives By-Law* requires that the packaging of food additives intended for consumption by the final consumer shall contain the following information: (*a*) the name and E-number of each food additive or a sales description which includes the name and E-number of each food additive; (*b*) the statement "for use in food" or the statement "restricted use in food" or a more specific statement indicating the food in which the food additive is intended to be used; (*c*) the name of the source from which the food additive is obtained; (*d*) the species of the animal from which the food additive of animal origin is obtained. The by-law also contains special provisions pertaining to the terms, warnings, and information to be included in sales descriptions and labels of table-top sweeteners. The packaging of food additives should also contain the information determined in specific provisions of the *Turkish Food Codex Food Labeling and Informing Consumers By-Law* and the *By-Law on Genetically Modified Organisms and their Products*.

Since Turkish food law is essentially based on EU standards, the *International Numbering System for Food Additives* (INS) is also in use in Turkey. Although the INS ensures the safety of additives used in food, due to misinformation propagated among the public, Turkish consumers have a negative perception of E-numbers. The consequences of this are observable in the labeling preferences of food manufacturers who sometime refrain from indicating E-numbers and content, indicating only the names of food additives. This practice occasionally causes blockages in international commercial operations.

3.7.7.3 Allergen labeling

The *Turkish Food Codex Food Labeling and Informing Consumers By-Law* provides a list of "certain substances or products causing allergies or intolerances" in its Annex – 1. According to the by-law, information pertaining to any ingredient or processing aid listed in that annex or derived from a substance or product listed in that annex causing allergies or intolerances used in the manufacture or preparation of a food and present in the finished product, even if in an altered form, shall be indicated in the list of ingredients by clearly indicating the name of the substance or product figuring in that annex (Article 24 § 1, *a* and *b*). The names of these substances or products should be emphasized through a typeset that clearly distinguishes them from the rest of the ingredients. The provision of information pertaining to the absence or reduced presence of gluten is facultative. In this respect, only the use of certain statements specifically determined in the by-law is permitted (Article 45).

3.7.7.4 Nutrition labeling

The *Turkish Food Codex Food Labeling and Informing Consumers By-Law* states that the mandatory nutrition declaration is composed of the following information: (*a*) energy value and (*b*) the amounts of fat, saturates, carbohydrate, sugars, protein, and salt. In cases where the salt content is exclusively due to the presence of naturally occurring sodium,

105. Turkish Food Codex Food Labeling and Informing Consumers By-Law [*Türk Gıda Kodeksi Gıda Etiketleme ve Tüketicileri Bilgilendirme Yönetmeliği*] (Date of Official Gazette: 26/01/2017, Number of Official Gazette: 29960 bis). This by-law was prepared by taking account of Regulation (EU) No 1169/2011 of the European Parliament and of the Council of 25 October 2011 on the provision of food information to consumers, amending Regulations (EC) No 1924/2006 and (EC) No 1925/2006 of the European Parliament and of the Council, and repealing Commission Directive 87/250/EEC, Council Directive 90/496/EEC, Commission Directive 1999/10/EC, Directive 2000/13/EC of the European Parliament and of the Council, Commission Directives 2002/67/EC and 2008/5/EC and Commission Regulation (EC) No 608/2004, *Official Journal of the European Union*, L 304, 22/11/2011, pp. 18–63.

a statement relating to this situation may appear in close proximity to the nutrition declaration. It is compulsory to declare the amount of transfat for spreadable fats and margarines, concentrated fats, plant fats, and foods containing these with a transfat content exceeding 2%. The indication in the nutrition declaration of the amounts of mono-unsaturates, polyunsaturates, polyols or sugar alcohol, starch, fiber as well as certain vitamins and minerals present in significant amounts is facultative (Article 35 §§ 1–2).

3.7.7.5 Nutrition and health claims

The *Turkish Food Codex Nutrition and Health Claims By-Law*[106] allows the use of nutrition and health claims in the labeling, presentation, and advertising of foods provided that certain conditions are respected. The use of nutrition and health claims shall not (*a*) be ambiguous, false, or misleading; (*b*) cause doubt about the nutritional sufficiency or the safety of other foods; (*c*) support or promote the excessive consumption of a given food; (*d*) state, allege, or insinuate that a varied and balanced diet cannot provide appropriate quantities of nutrients in general; and (*e*) refer to changes in bodily functions in a manner causing worry in the consumer by means of textual, pictorial, graphic, or symbolic presentations (Article 5 § 2).

3.7.7.6 Medicinal claims

Save for provisions of legislation pertaining to special dietary foods, the *Turkish Food Codex Food Labeling and Informing Consumers By-Law* prohibits preventive, curative, or therapeutic claims. These rules also apply to the advertising and presentation of food, in particular its shape, appearance or packaging, the packaging materials used, the way in which it is arranged, and its form of exhibition (Article 7 §§ 3–4).

3.7.7.7 Modalities of information

The *Turkish Food Codex Food Labeling and Informing Consumers By-Law* stipulates that mandatory food information shall be provided in Turkish. Such information may be given in the official languages of other countries in addition to Turkish. Nonetheless, the information provided in languages other than Turkish does not have to contain all the labeling information foreseen in the food information legislation (Article 18). The mandatory particulars in the labels shall be expressed with words and numbers. These particulars may additionally be expressed by means of pictorial representations or symbols provided that certain conditions are respected (Article 9 § 2).

3.7.8 Conclusion

Despite the development of a substantial food regulation following the establishment of the republic in Turkey, a turning point was marked in 1995 with the acceleration of harmonization efforts with European Union legislation. Since that time, a complex and comprehensive legal corpus was constituted in Turkey through the adoption of successive regulations on various aspects of food safety within a process that has not always followed a coherent and steady course. However, as a consequence of recent constitutional changes, Turkish food law tends to take difficult predictable normative and institutional orientations.

In its present state, Turkish food law appears to be essentially in conformity with international standards. Yet, due to organizational deficiencies, effective tracing and control mechanisms are still not established. This failure manifests itself in relatively frequent scandals related to food services provided in mass consumption places such as hospitals, barracks, schools, dormitories, and prisons. The problem also becomes obvious through "disclosure lists," published in the last 6 years by the Ministry of Agriculture and Forestry, which provide specific information on adulterated food products and indicate the names of firms manufacturing, importing, and selling these products. Moreover, findings of independent researches point out that agricultural products containing pesticide residues exceeding legal maximum limits have reached worrying proportions.

106. Turkish Food Codex Nutrition and Health Claims By-Law [*Türk Gıda Kodeksi Beslenme ve Sağlık Beyanları Yönetmeliği*] (Date of Official Gazette: 26/01/2017, Number of Official Gazette: 29960 bis). This by-law was prepared by taking account of Regulation (EC) No 1924/2006 of the European Parliament and of the Council of 20 December 2006 on nutrition and health claims made on foods, Official Journal of the European Union, *L 404, 30/12/2006*, pp. *9–25*.

Finally, agriculture and livestock policies causing the malfunction of competent public institutions, the insufficiency of agricultural subsidies, the weakening of family farming, and the increase of importations have also negative effects on food safety.

The lack of sound reaction from directly affected groups and public opinion regarding these problems has impeded the development of food safety awareness among both food operators and consumers.

References

I. Literature

Atlı, A., Hayoğlu, İ., Koçak, C., Özer, B., Soyer, A., 2010. Gıda Sanayiinin Hammadde Gereksinimi ve Yeterliliği. In: TMMOB Ziraat Mühendisleri Odası, Türkiye Ziraat Mühendisliği VII. Teknik Kongresi Bildiriler Kitabı, 2, pp. 1097–1112 (Ankara).

Artık, N., Şanlıer, N., Ceyhun Sezgin, A., 2017. Gıda Güvenliği Ve Gıda Mevzuatı. Detay Yayıncılık, Ankara.

Çopuroğlu, G., Kasımoğlu Doğru, A., Ayaz, N.D., 2015. Türk Gıda Mevzuatında Risk Analizi. Etlik Veteriner Mikrobiyoloji Dergisi / Journal of Etlik Veterinary Microbiology 26 (1), 23–28.

Koç, A.A., Bölük, G., Aşçı, S., 2018. Gıda Güvenliği ve Kalite Standartlarının Gıda İmalat Sanayinde Yoğunlaşmaya Etkisi. In: Akdeniz İktisadi ve İdari Bilimler Fakültesi Dergisi, 16, pp. 83–115.

Notermans, S., Powell, S.C., 2005. Introduction. In: Lelieveld, H.L.M., Mostert, M.A., Holah, J. (Eds.), Handbook of Hygiene Control in the Food Industry. Woodhead Publishing, Cambridge, pp. 1–28.

Şevik, S.E., Ayaz, N.D., 2017. Sığır Etlerinde Hormon Kalıntısı Varlığının Araştırılması. In: Veteriner Hekimler Derneği Dergisi/Journal of Turkish Veterinary Medical Society, vol. 88, pp. 13–20 (1).

Tayar, M., 2010. Gıda Güvenliği. İstanbul: Marmara Belediyeler Birliği.

Türkoğlu, H.G., 2011. Roma Cumhuriyet ve İlk İmparatorluk Dönemlerinin İdari Yapısı, 11. Dokuz Eylül Üniversitesi Hukuk Fakültesi Dergisi, pp. 251–289 (2).

Yalçın, H., 2008. AB – Türkiye'de Gıda Mevzuatı ve Tokat İli Gıda Sanayi İşletmelerinin Yapısı. In: Yüksek Lisans Tezi, Gaziosmanpaşa Üniversitesi Fen Bilimleri Enstitüsü Tarım Ekonomisi Anabilim Dalı. Available at: http://earsiv.gop.edu.tr/xmlui/handle/20.500.12881/1684.

Yıldırım, N., 2011–2012. Osmanlı Devleti'nde Gıda Kontrolüne Bakış. In: Sağlık Düşüncesi ve Tıp Kültürü Dergisi, 21, pp. 68–73.

II. Legislation

A. National

1. Laws

Law on Spirit and Alcoholic Beverages Monopoly [*İspirto ve Meşrubat-ı Küuliye İnhisarı Hakkında Kanun*] (Law N°: 790, Date of Adoption: 22/03/1926, Date of Official Gazette: 03/04/1926, Number of Official Gazette: 338).

Law on Animal Health Constabulary [*Hayvan Sağlık Zabıtası Hakkında Kanun*] (Law N° 1234, Date of Adoption: 03/05/1928, Date of Official Gazette: 14/05/1928, Number of Official Gazette: 888).

Municipality Law [*Belediye Kanunu*] (Law N°: 1580, Date of Adoption: 03/04/1930, Date of Official Gazette: 14/04/1930, Number of Official Gazette: 1471).

Public Health Preservation Law [*Umumî Hıfzıssıhha Kanunu*] (Law N° 1593, Date of Adoption: 24/04/1930, Date of Official Gazette: 06/05/1930, Number of Official Gazette: 1489).

Spirit and Spirituous Beverages Monopoly Law [İspirto Ve İspirtolu İçkiler İnhisarı Kanunu] (Law N°: 4250, Date of Adoption: 08/06/1942, Date of Official Gazette: 12/06/1942, Number of Official Gazette: 5130).

Turkish Standards Institute Establishment Law [Türk Standardları Enstitüsü Kuruluş Kanunu] (Law N°: 132, Date of Adoption: 18/11/1960, Date of Official Gazette: 22/11/1960, Number of Official Gazette: 10661).

Law on the Adoption through Modification of the Statutory Decree Regarding the Production, Consumption and Control of Foodstuffs [Gıdaların Üretimi, Tüketimi ve Denetlenmesine Dair Kanun Hükmünde Kararnamenin Değiştirilerek Kabulü Hakkında Kanun] (Law N° 5179, Date of Adoption: 27/05/2004, Date of Official Gazette: 05/06/2004, Number of Official Gazette: 25483).

Biosafety Law [*Biyogüvenlik Kanunu*] (Law N°: 5977, Date of Adoption: 18/03/2010, Date of Official Gazette: 26/03/2010, Number of Official Gazette: 27533).

Veterinary Services, Plant Health, Food and Feed Law [*Veteriner Hizmetleri, Bitki Sağlığı, Gıda ve Yem Kanunu*] (Law N°: 5996, Date of Adoption: 11/06/2010, Date of Official Gazette: 13/06/2010, Number of Official Gazette: 27610).

2. Statutory Decrees

Statutory Decree Regarding the Production, Consumption and Control of Foodstuffs [*Gıdaların Üretimi, Tüketimi ve Denetlenmesine Dair Kanun Hükmünde Kararname*] (Statutory Decree N° 560, Date of Adoption: 24/06/1995, Date of the Enabling Law: 08/06/1995, N° 4113, Date of Official Gazette: 28/06/1995, Number of Official Gazette: 22327).

Statutory Decree on the Organization and Duties of the Ministry of Food, Agriculture and Livestock [*Gıda, Tarım ve Hayvancılık Bakanlığının Teşkilat ve Görevleri Hakkında Kanun Hükmünde Kararname*] (Statutory Decree N° 639, Date of Adoption: 03/06/2011, Date of the Enabling Law: 06/04/2011, N° 6223, Date of Official Gazette: 08/06/2011, Number of Official Gazette: 27958 bis).

Statutory Decree on the Modification of Certain Laws and Statutory Decrees for Accommodating Constitutional Amendments [*Anayasada Yapılan Değişikliklere Uyum Sağlanması Amacıyla Bazı Kanun ve Kanun Hükmünde Kararnamelerde Değişiklik Yapılması Hakkında Kanun Hükmünde Kararname*] (Statutory Decree N° 703, Date of Adoption: 02/07/2018, Date of the Enabling Law: 10/05/2018, N° 7142, Date of Official Gazette: 09/07/2018, Number of Official Gazette: 30473 bis 3).

3. Presidential Decrees

Presidential Decree on the Organization of the Presidency of the Republic [*Cumhurbaşkanlığı Teşkilatı Hakkında Cumhurbaşkanlığı Kararnamesi*] (Presidential Decree N° 1, Date of Official Gazette: 10/07/2018, Number of Official Gazette: 30474).

Presidential Decree on the Organization of Institutions and Establishments Dependent, Related, Connected to Ministries and of other Institutions and Establishments [*Bakanlıklara Bağlı, İlgili, İlişkili Kurum ve Kuruluşlar ile Diğer Kurum ve Kuruluşların Teşkilatı Hakkında Cumhurbaşkanlığı Kararnamesi*] (Presidential Decree N° 4, Date of Official Gazette: 15/07/2018, Number of Official Gazette: 30479).

4. Regulations

Animal Health Constabulary Regulation [*Hayvan Sağlık Zabıtası Nizamnamesi*] (Date of Council of Ministers Decree: 09/08/1931, N° 11656, Date of Official Gazette: 17/09/1931, Number of Official Gazette: 1901).

Regulation Indicating the Particular Qualities of Foodstuffs and of Commodities and Materials Concerning Public Health [*Gıda Maddelerinin ve Umumî Sıhhati İlgilendiren Eşya ve Levazımın Hususî Vasıflarını Gösteren Nizamname*] (Date of Council of Ministers Decree: 11/08/1942, N° 2/18542, Date of Official Gazette: 07/09/1942, Number of Official Gazette: 5204).

Regulation Indicating the Particular Qualities of Foodstuffs and of Commodities and Materials Concerning Public Health [*Gıda Maddelerinin ve Umumî Sağlığı İlgilendiren Eşya ve Levazımın Hususî Vasıflarını Gösteren Tüzük*] (Date of Council of Ministers Decree: 04/08/1952, N° 3/15481, Date of Adoption: 18/10/1952, Number of Official Gazette: 8236).

5. By-Laws

Turkish Food Codex By-Law [*Türk Gıda Kodeksi Yönetmeliği*] (Date of Official Gazette: 16/11/1997, Number of Official Gazette: 23172 bis).

By-Law Regarding Genetically Modified Organisms and their Products [*Genetik Yapısı Değiştirilmiş Organizmalar ve Ürünlerine Dair Yönetmelik*] (Date of Official Gazette: 13/08/2010, Number of Official Gazette: 27671).

Food Hygiene By-Law [*Gıda Hijyeni Yönetmeliği*] (Date of Official Gazette: 17/12/2011, Number of Official Gazette: 28145).

By-Law on the Working Procedures and Principles of Risk Assessment Committees and Commissions [*Risk Değerlendirme Komite ve Komisyonlarının Çalışma Usul ve Esasları Hakkında Yönetmelik*] (Date of Official Gazette: 24/12/2011, Number of Official Gazette: 28152).

Feed Hygiene By-Law [*Yem Hijyeni Yönetmeliği*] (Date of Official Gazette: 27/12/2011, Number of Official Gazette: 28155).

Turkish Food Codex By-Law [*Türk Gıda Kodeksi Yönetmeliği*] (Date of Official Gazette: 29/12/2011, Number of Official Gazette: 28157 bis 3).

Turkish Food Codex Preparation By-Law [*Türk Gıda Kodeksi Hazırlama Yönetmeliği*] (Date of Official Gazette: 29/12/2011, Number of Official Gazette: 28157 bis 3).

Turkish Food Codex Contaminants By-Law [*Türk Gıda Kodeksi Bulaşanlar Yönetmeliği*] (Date of Official Gazette: 29/12/2011, Number of Official Gazette: 28157 bis 3).

Turkish Food Codex Microbiological Criteria By-Law [*Türk Gıda Kodeksi Mikrobiyolojik Kriterler Yönetmeliği*] (Date of Official Gazette: 29/12/2011, Number of Official Gazette: 28157 bis 3).

Turkish Food Codex Food Additives By-Law [*Türk Gıda Kodeksi Gıda Katkı Maddeleri Yönetmeliği*] (Date of Official Gazette: 30/06/2013, Number of Official Gazette: 28693).

Turkish Food Codex Maximum Pesticide Residue Limits By-Law [*Türk Gıda Kodeksi Pestisitlerin Maksimum Kalıntı Limitleri Yönetmeliği*] (Date of Official Gazette: 25/11/2016, Number of Official Gazette: 29899 bis).

Turkish Food Codex Food Labeling and Informing Consumers By-Law [*Türk Gıda Kodeksi Gıda Etiketleme ve Tüketicileri Bilgilendirme Yönetmeliği*] (Date of Official Gazette: 26/01/2017, Number of Official Gazette: 29960 bis).

Turkish Food Codex Nutrition and Health Claims By-Law [*Türk Gıda Kodeksi Beslenme ve Sağlık Beyanları Yönetmeliği*] (Date of Official Gazette: 26/01/2017, Number of Official Gazette: 29960 bis).

Turkish Food Codex By-Law on the Common Authorization Procedure Pertaining to Food Additives, Food Enzymes and Flavorings [*Türk Gıda Kodeksi Gıda Katkı Maddeleri, Gıda Enzimleri ve Gıda Aroma Vericilerine İlişkin Ortak İzin Prosedürü Hakkında Yönetmelik*] (Date of Official Gazette: 24/02/2017, Number of Official Gazette: 29989).

Turkish Food Codex By-Law on the Classification and Maximum Residue Limits of Pharmacologically Active Substances in Foods of Animal Origin [*Türk Gıda Kodeksi Hayvansal Gıdalarda Bulunabilecek Farmakolojik Aktif Maddelerin Sınıflandırılması ve Maksimum Kalıntı Limitleri Yönetmeliği*], (Date of Official Gazette: 07/03/2017, Number of Official Gazette: 30000).

By-Law on the Authorization and Placement on the Market of Plant Protection Products [*Bitki Koruma Ürünlerinin Ruhsatlandırılması ve Piyasaya Arzı Hakkında Yönetmelik*] (Date of Official Gazette: 09/11/2017, Number of Official Gazette: 30235).

Turkish Food Codex By-Law on Materials and Articles in Contact with Food [*Türk Gıda Kodeksi Gıda ile Temas Eden Madde ve Malzemelere Dair Yönetmelik*] (Date of Official Gazette: 05/04/2018, Number of Official Gazette: 30382).

6. Directives

Directive on the Duties of the Central Organization of the Ministry of Food, Agriculture and Livestock [*Gıda, Tarım ve Hayvancılık Bakanlığı Merkez Teşkilatı Görev Yönergesi*] (Date: 18/12/2014, № 13805938/MEV-2011-127/54).

7. Communiqués

Communiqué on Hormones and Hormone-Like Substances Whose Administration to Food Producing Animals is Prohibited or Conditionally Permitted [*Gıda Değeri Olan Hayvanlara Uygulanması Yasaklanan ve Belli Şartlara Bağlanan Hormon ve Benzeri Maddeler Hakkında Tebliğ*] (Communiqué No: 2003/18, Date of Official Gazette: 19/11/2016, Number of Official Gazette: 25143).

B. European

Regulation (EC) No 178/2002 of the European Parliament and of the Council of 28 January 2002 laying down the general principles and requirements of food law, establishing the European Food Safety Authority and laying down procedures in matters of food safety, Official Journal of the European Communities, L 031, 01/02/2002, pp. 1-24.

Regulation (EC) No 852/2004 of the European Parliament and of the Council of 29 April 2004 on the hygiene of foodstuffs, *Official Journal of the European Union*, L 139, 30/04/2004, pp. 1-54.

Regulation (EC) No 1935/2004 of the European Parliament and of the Council of 27 October 2004 on materials and articles intended to come into contact with food and repealing Directives 80/590/EEC and 89/109/EEC, *Official Journal of the European Union*, L 338, 13/11/2004, pp. 4-17.

Regulation (EC) No 396/2005 of the European Parliament and of the Council of 23 February 2005 on maximum residue levels of pesticides in or on food and feed of plant and animal origin and amending Council Directive 91/414/EEC, *Official Journal of the European Union*, L 70, 16/03/2005, pp. 1-16.

Commission Regulation (EC) No 2073/2005 of 15 November 2005 on microbiological criteria for foodstuffs, *Official Journal of the European Union*, L 338, 22/12/2005, pp. 1-26.

Commission Regulation (EC) No 1881/2006 of 19 December 2006 setting maximum levels for certain contaminants in foodstuffs, *Official Journal of the European Union*, L 364, 20/12/2006, pp. 5-24.

Regulation (EC) No 1924/2006 of the European Parliament and of the Council of 20 December 2006 on nutrition and health claims made on foods, *Official Journal of the European Union*, L 404, 30/12/2006, pp. 9-25.

Regulation (EC) No 1333/2008 of the European Parliament and of the Council of 16 December 2008 on food additives, *Official Journal of the European Union*, L 354, 31/12/2008, pp. 16-33.

Regulation (EC) No 470/2009 of the European Parliament and of the Council of 6 May 2009 laying down Community procedures for the establishment of residue limits of pharmacologically active substances in foodstuffs of animal origin, repealing Council Regulation (EEC) No 2377/90 and amending Directive 2001/82/EC of the European Parliament and of the Council and Regulation (EC) No 726/2004 of the European Parliament and of the Council, *Official Journal of the European Union*, L 152, 16/06/2009, pp. 11-22.

Commission Regulation (EU) No 37/2010 of 22 December 2009 on pharmacologically active substances and their classification regarding maximum residue limits in foodstuffs of animal origin, *Official Journal of the European Union*, L 15, 20/01/2010, pp. 1-72.

Regulation (EU) No 1169/2011 of the European Parliament and of the Council of 25 October 2011 on the provision of food information to consumers, amending Regulations (EC) No 1924/2006 and (EC) No 1925/2006 of the European Parliament and of the Council, and repealing Commission Directive 87/250/EEC, Council Directive 90/496/EEC, Commission Directive 1999/10/EC, Directive 2000/13/EC of the European Parliament and of the Council, Commission Directives 2002/67/EC and 2008/5/EC and Commission Regulation (EC) No 608/2004, *Official Journal of the European Union*, L 304, 22/11/2011, pp. 18-63.

III. Documents

Decision No 1/95 of the EC-Turkey Association Council of 22 December 1995 on implementing the final phase of the Customs Union (96/142/EC). Official Journal of the European Communities, 13/02/1996 1−46. L 35.

Commission of the European Communities, SEC (2006) 1390, Commission Staff Working Document, *Turkey 2006 Progress Report*, COM (2006) 649 final, Brussels, 8.11.2006.

Tarım ve Orman Bakanlığı Gıda ve Kontrol Genel Müdürlüğü, 2018. Gıda ve Kontrol Verileri (General Directorate of Food and Control of the Ministry of Agriculture and Forestry, Food and Control Data). Available at: https://www.tarimorman.gov.tr/sgb/Belgeler/SagMenuVeriler/GKGM.pdf.

Codex Alimentarius Commission, Codex Committee on Contaminants in Foods, 2007. Discussion Paper on Aflatoxin in Dried Figs. CX/CF 07/1/20. Available at: http://www.fao.org/tempref/codex/Meetings/CCCF/CCCF1/cf01_20e.pdf.

Codex Alimentarius Commission, Codex Committee on Contaminants in Foods, 2007. Discussion Paper on Maximum Levels for Total Aflatoxins in "ready-to-eat" Almonds, Hazelnuts and Pistachios. CX/CF 07/1/9. Available at: http://www.fao.org/tempref/codex/Meetings/CCCF/CCCF1/cf01_09e.pdf.

Chapter 3.8

The Russian Federation

Altinay Urazbaeva[1] and Yuriy Vasiliev[2]

[1]*Studying Advanced Master Program in European, International Business Law, Leiden University;* [2]*Stavropol Branch, North Caucasus Civil Service Academy, Russia*

3.8.1 Russian food law

Legal requirements on food in the Russian Federation originate at different levels. The Russian Federation is a Member State of the Eurasian Economic Union (EAEU). The EAEU consists of Belarus, Kazakhstan, the Russian Federation, Armenia, and Kyrgyzstan. The EAEU succeeds earlier regional economic organizations such as the Customs Union (CU). Currently applicable food legal requirements originate from these earlier organizations, the EAEU, and from the national level of the Russian Federation.

According to Russian legal doctrine, the legislative system is subdivided into specified branches. "Food law" is not recognized as such a branch. In this sense "food law" does not exist in Russian law. Obviously, this does not prohibit an academic analysis of the law from the perspective of food and the food sector. This makes "Russian food law" a true functional field of academic analysis.

3.8.2 Institutions

Within the government of the Russian Federation, the ministries most involved with food are the Ministry of Agriculture ("Minselkhoz"), the Ministry of Health and Social Development ("Minzdravsotsrazvitiya"), the Ministry of Industry and Trade ("Minpromtorg"), and the Ministry of Economic Development ("Minekonomrazvitiya"). The Federal Agencies responsible for food regulation are "Rospotrebnadzor," "Rosselkhoznadzor," "Rosalkogolregulirovaniye," "Rosstandart," "Rosakkreditatsiya," and "FAS Russia".

The Federal Agency for Supervision of Consumer Rights Protection and Human Well-Being "Rospotrebnadzor" is the federal executive body responsible for developing and implementing state policy and legal regulation in the field of consumer protection, developing and approving state sanitary and epidemiological regulations and hygiene standards, as well as organization and implementation of federal state sanitary and epidemiological control, federal supervision on consumer protection. "Rospotrebnadzor" is directly subordinated to the Government.

The Federal Agency for Veterinary and Phytosanitary Supervision "Rosselkhoznadzor" is a federal executive body, responsible for the control and supervision of veterinary medicine, circulation of medicinal products for veterinary use, quarantine and plant protection, safe handling of pesticides and agrochemicals, soil fertility, quality and safety of grain, cereals, mixed fodders and components for their production, and by-products of grain processing. "Rosselkhoznadzor" is subordinated to the Ministry of Agriculture.

"Rosalkogolregulirovaniye" is the Federal Agency for Regulation of the Alcohol Market. It is subordinated to the Ministry of Finance.

The Federal Agency for Technical Regulation and Metrology "Rosstandart" is the federal executive body responsible for providing state services, managing state property in the field of technical regulation and metrology. "Rosstandart" is subordinated to the Ministry of Industry and Trade.

The Federal Agency for Accreditation Services "Rosakkreditattsiya" is subordinated to the Ministry of Economic Development.

The Federal Antimonopoly Service Agency "FAS Russia" is directly subordinated to the government.

3.8.3 Technical regulation

From 2002 onward, the Russian Federation introduced a system of product safety based in technical regulations. The adoption of the Federal Law of December 27, 2002, No. 184-FZ "On Technical Regulation" marked a new stage in existing technical and legal norms of Russia. It lays the foundation for a radical reform of the entire system. It establishes mandatory requirements for products, production processes, operation, storage, transportation, sale and disposal, performance of works and services, as well as assessment and confirmation of compliance. Active work is being carried out in the EAEU and at the national level to develop and adopt technical regulations.

The Federal Law "on technical regulation" is a comprehensive legislative act of the Russian Federation. It is established at the highest legal level on the basis of the Constitution of the Russian Federation and has a huge socio-economic significance, since it aims at laying down rules for the state regulation on product requirements, including consumer goods, as well as requirements for work and services in the interests of the consumers. The system of technical regulation does not fundamentally distinguish between food and nonfood. In this sense, the Russian Federation has a system of product safety law, rather than food safety law. Nevertheless, in this section we only discuss some of the technical regulations with specific relevance for food and the food sector.

3.8.4 General food safety

The main technical regulation on food safety is the Technical Regulation of the Customs Union TR CU 021/2011 "On safety of food products." It came into force on July 1, 2013, and applies throughout the Russian Federation and the Member States of the Eurasian Economic Union. The Technical Regulation "On safety of food products" is a general horizontal technical regulation. It lays down general framework for all food products, as well as mandatory safety requirements for food and its process.

The objectives of the Technical Regulation "On safety of food products" are as follows:

- to protect human life and health;
- to prevent actions that mislead the purchasers (consumers);
- to protect the environment.

The term "food products" in the Technical Regulation is broad. It includes various types of products. Food products are "*of animal, plant, microbial, mineral, synthetic, or biotechnological origin in natural, treated, or processed form, which are intended for human consumption, including specialized food products, drinking water, packaged in containers, drinking mineral water, alcoholic beverages (including beer and beer-based drinks), soft drinks, biologically active food supplements («БАД»),*[107] *chewing gum, and sourdough starter cultures of microorganisms, yeast, food additives and flavorings, and food raw materials.*"

Safety of food products means "*the state of food products indicating that there is no unacceptable risk associated with adverse effects on humans and future generations.*" Circulated food products throughout the territory of the Customs Union/EAEU must be safe when used as intended within their established period of validity.

The Technical Regulation "On safety of food products" sets out general requirements for food products to be considered as safe, as well as safety requirements for specialized food products.

General requirements for food safety include the following conditions:

- infectious agents, parasitic diseases, and their toxins, which are dangerous for human and animal health, should not be present in food products;
- packaging material and products which are in contact with food must comply with requirements of the Technical Regulation TR CU 005/2011 "On safety of packaging";
- requirements for food additives, flavors, and technological aids used in production of food products are stipulated by the Technical Regulation TR CU 029/2012 "On safety requirements for food additives, flavors, and technological aids";
- GMO lines should pass the state registration of food products produced from food raw materials obtained from GMOs of plant, animal, and microbial origin. If the manufacturer has not used the GMOs in food products and the content of GMOs (0.9% or less) in food products is a random or technically unrecoverable mixture, such food products are not identified as food products containing the GMOs;
- production of baby food products for children of the first life year should be carried out at specialized production facilities, or in specialized workshops, or on specialized processing lines;
- fresh and freshly frozen greenery, vegetables, fruits, and berries must not contain eggs of helminths and cysts of intestinal pathogenic protozoa;

107. Abbreviation «БАД» in Russian means biologically active food supplements.

- content of each food or biologically active substance in enriched food products must be brought to the level of consumption in 100 mL or 100 g, or a single portion of such production of not less than 5% of the level of daily intake.
- shelf life and storage conditions for food products are set by the manufacturer.

Mandatory safety requirements for food products are laid down in 10 Annexes of the Technical Regulation "On safety of food products." The Regulation provides six categories of food safety indicators:

- Microbiological requirements (Annex 1, 2);
- Pathogen-specific requirements (Annex 1, 2);
- Hygiene requirements (Annex 3);
- Permissible levels of radionuclides (Annex 3);
- Requirements for unprocessed raw materials of animal origin (Annex 5);
- Parasitological indicators of safety for fish and crustaceans (Annex 6).

3.8.5 Authorization

3.8.5.1 General

Technical regulations are based on the principle that the quality and safety of industrial food products must be supported by a document. Therefore, an assessment of Conformity is required not only for food products, but also for production processes, storage, transportation, sale, and utilization of food products. In accordance with the requirements of the Technical Regulation "On safety of food products," the assessment of conformity for food products is carried out in the following forms:

1. Declaration of conformity of food products;
2. State registration of specialized food products;
3. State registration of novel food products;
4. Veterinary and Sanitary examination.

The Assessment of Conformity for production processes, storage, transportation, sale, and utilization of food products to the requirements of technical regulations is carried out in the form of state supervision (control), excluding production processes of unprocessed food raw materials of animal origin. The assessment of conformity for production processes of unprocessed food raw materials of animal origin is carried out in the form of state registration.

The Assessment of Conformity for food products of nonindustrial production[108] and food products of catering enterprises (public catering) intended for sale, including sale of these food products, is carried out in the form of state supervision (control) over compliance with the requirements for food products established by the Technical Regulation "On safety of food products" and technical regulations of the Customs Union for certain types of food products.

3.8.5.2 Specialized food products

According to the Article 23(1) of the Technical Regulation "On safety of food products," specialized food products are subject to state registration. State registration is one of the forms confirming the compliance of food products with the requirements of technical regulations.

Specialized food products include the following:

- food products for baby food, including drinking water for baby food;
- food products for dietary curative and dietary preventive nutrition;
- natural mineral water, therapeutic drinking water, therapeutic mineral water with mineralization of more than 1 mg/dm^3 or with less mineralization, containing biologically active substances in an amount not lower than the balneological norms;
- food products for nutrition of athletes, pregnant, and lactating women;
- biologically active food supplements.

108. food products of nonindustrial manufacture are food products produced by citizens at home and (or) in personal part-time farms or citizens engaged in horticulture, truck farming, livestock and other activities (Article 4, TR CU 021/2011).

Production, storage, transportation, and sale of specialized food products subject to state registration is allowed only after its state registration. State registration of specialized food products is carried out at the stage of their preparation for production in the territory of the EAEU. As for the imported specialized food products into the EAEU, they should be registered prior their importation. State registration of specialized food products is permanent; it can be terminated or suspended by the registration authority for specialized food products in cases of noncompliance with the requirements of Technical Regulation "On safety of food products" as a result of state control (supervision) and (or) by decision of the judicial bodies of the Member State.

In order to obtain State Registration, food business operators should prepare a dossier (application documents) and have testing samples done at a laboratory accredited by the EAEU authorities. The State Registration Certificate is issued without expiry date unless the product name, product composition, or the manufacturer's name are changed, and in case the food products are found to be noncompliant with the requirements of technical regulations during inspection. Upon approval, the food products are registered in the Single Register of specialized food products.[109] The applicant has the right to appeal against the decision of the state registration body for specialized food products in court. The state registration of specialized food products is carried out by the authorized body of the EAEU Member State. In the Russian Federation, the Federal Agency for Supervision of Consumer Rights Protections and Human Well-Being "Rospotrebnadzor" is in charge of the state registration of food products.

3.8.5.3 Novel foods

Novel food products (of a new type) are the food products (including food additives and flavors) that were not previously used by man for food in the territory of the Customs Union/EAEU, namely: with a new or intentionally altered primary molecular structure; consisting of or isolated from microorganisms, microscopic fungi, and algae, plants, isolated from animals, obtained from GMOs or with their use, nanomaterials, and nanotechnology products. Novel food products do not include food products using well-known traditional methods and by experience being considered as safe, produced, and already applied technologies, which include ingredients and food additives, already used for human consumption, even if such products and ingredients (components) are produced according to a new recipe.

Novel food products are subject to state registration. The state registration of novel food products is carried out at the stage of its preparation for production for the first time in the territory of the EAEU. Imported novel food products must be registered before their importation for the first time into the EAEU. The state registration of novel food products is carried out by a national authorized body. In the Russian Federation, both the Ministry of Agriculture and the Ministry of Health and Social Development carry out normative and legal regulation in state registration of novel food products of animal origin, including the Federal Agency for Supervision of Consumer Rights Protections and Human Well-Being "Rospotrebnadzor," which is in charge of the state registration of novel food.

After the state registration, the novel food products are no longer considered as novel and are not subject to the state registration by other applicants and under different names. Validity of the state registration for novel food products is unlimited. The Assessment of conformity is required for every item of novel food products according to the procedure of Technical Regulation TR CU 021/2011. The state registration of novel food products may be terminated or suspended by the registration body in cases when harm is revealed by the state control (supervision) and by decision of the judicial bodies of the Member State.

3.8.6 Process requirements

The main safety requirement in production processes of food products is that the manufacturer must develop, implement, and maintain procedures based on the HACCP principles.

HACCP Principles are based on 12 procedures:

1. selection of food production processes to provide food safety;
2. selection of sequence and flow of technological operations for production of food products in order to eliminate contamination of food raw materials and food products;
3. determine controlled stages of technological operations and production stages of food in production control programs;
4. control over food raw materials, technological means, packing materials, products used in food production, as well as facilities for food products to ensure necessary reliability and completeness of control;

109. Collier S, Baldwin N. 2014. All Change for Russia and the Customs Union. The world of Food ingredients. 02.2014/49.

5. control over technological equipment operation to ensure production of food products, meeting the requirements of the Technical Regulation "On safety of food products" and (or) technical regulations of the Customs Union for certain types of food products;
6. documentation of information during controlled stages of technological operations and food control results;
7. compliance with the conditions of storage and transportation of food products;
8. maintain industrial premises, technological equipment, and inventory used in production of food products in order to eliminate contamination of food products;
9. selection of hygiene methods and observance of personal hygiene rules by workers in order to maintain safety of food production;
10. selection of food safety methods, frequency of cleaning, washing, disinfection, deratization, disinsectization of industrial premises, technological equipment, and equipment used in production process of food products;
11. keeping and filing of documentation on paper and (or) electronic media to confirm compliance of food production with the requirements stipulated by the Technical Regulation "On safety of food products" and (or) technical regulations of the Customs Union for certain types of food products;
12. traceability of food products.

Producers must carry out safety procedures in food production processes to ensure that their food products placed on the market comply with the requirements of the Technical Regulation "On safety of food products" and/or technical regulations for certain types of food products. Producers should ensure safety of food production and carry out food control independently and/or with participation of a third party.

To ensure food safety requirements in production process, the producers must determine:

1. a list of hazardous factors that may cause production of food products, which do not meet the requirements of technical regulations;
2. a list of critical control points in the production process;
3. parameter limits, which should be controlled in critical control points;
4. monitoring procedure of critical control points during the production process;
5. operational procedures if indicator values (of para 3) deviate from the established limits;
6. frequency of inspection for conformity of products released into circulation[110] to the requirements of technical regulations;
7. frequency of cleaning, washing, disinfections, deratization, and disinsectization of production premises, as well as cleaning, washing, disinfections of technological equipment and utensils used in food production process;
8. measures to prevent the entry of rodents, insects, synanthropic birds, and animals into production premises.

Producers are obliged to keep documentation regarding the safety measures taken in the production process of food products, including documents confirming safety of unprocessed food raw materials of animal origin, on paper and (or) electronic media. Documents confirming safety of unprocessed food raw materials of animal origin should be stored for 3 years from the date of their issue.

Requirements for employees engaged in food production processes:

1. mandatory medical examinations prior to employment, including periodical medical examinations in accordance with national legislation of the EAEU Member State;
2. Patients with infectious diseases, persons suspected of such diseases, persons who have been in contact with sick infectious diseases, and persons who carry infectious agents should not be allowed to work in production of food products.

3.8.7 Labeling

Labeling of food products must comply with the requirements of Technical Regulation of the Customs Union TR CU 022/2011 "On food products in terms of labeling." This technical regulation applies in the entire territory of the EAEU. It lays down labeling requirements for food products in order to prevent actions that can mislead consumers and ensure consumers' rights to reliable information about food products.

110. Term "food product released into circulation" means purchase and sale including other methods of transfer of food products by manufacturer or importer throughout the Customs Union/EAEU territory" (Article 4, Technical Regulation TR CU 021/2011).

3.8.7.1 Mandatory particulars

The labeling of packaged food products should contain the following information:

1. name of food products (name of food products should be indicated in Russian language and state languages of the CU/EAEU in case of relevant requirements of national legislations of the Member states);
2. composition of food products, except for some food products[111] and unless otherwise provided by technical regulations for certain types of food products;
3. quantity of food products;
4. date of manufacture of food products[112];
5. shelf life of food products;
6. storage conditions of food products established by the manufacturer or provided by technical regulations for certain types of food products. If quality and safety of food products changes after opening of packaging, the storage conditions after opening the package should also be indicated;
7. name and location of the manufacturer of food products or the surname, name, patronymic, and location of the individual entrepreneur, who is the manufacturer of food products, and in the cases established by Technical Regulation "On food products in terms of labeling," the name and location of the person authorized by manufacturer, name, and location of importing organization or the surname, name, patronymic, and location of the individual entrepreneur—importer[113];
8. recommendations and(or) restrictions on use, including preparation of food products when the use of food products is difficult without recommendations or restrictions, or may cause harm to consumers' health, their property, and may lead to a decrease or loss of taste of food products;
9. nutritional value (food value indicators) of food products, taking into account the requirements for labeling of nutrition value;
10. information on presence of components (ingredients) obtained from genetically modified organisms (GMOs), in food products;
11. the single sign (logo) of product circulation in the market of the Member States of the EAEU.

These mandatory particulars bear close resemblance to the labeling requirements suggested by the Codex Alimentarius.

3.8.7.2 Allergens

Ingredients (including food additives, flavors) and biologically active additives which can cause allergic reactions or is contraindicated in certain types of diseases must be indicated in the ingredient list of food products regardless of their quantity.

The most common ingredients, which can cause allergic reactions or are contraindicated in certain types of diseases, are as follows:

1. peanuts and products of its processing;
2. aspartame and aspartame—acesulfame salt;
3. mustard and products of its processing;
4. sulfur dioxide and sulfites, if their total content is more than 10 mg per kilogram or 10 mg per liter in terms of sulfur dioxide;
5. cereals containing gluten, and products of their processing;
6. sesame and its processing products;
7. lupin and products of its processing;
8. mollusks and products of their processing;
9. Milk and products of its processing (including lactose);

111. Composition of food production does not apply to (1) fresh fruits (including berries) and vegetables (including potatoes) that are not peeled, cut, or treated in a similar manner; (2) vinegar obtained from one type of food raw material (without adding other components); (3) food products, consisting of one component, provided that the name of food products allows to determine the presence of this component (Article 4 (4.4./7 of Technical Regulation TR CU 022/2011).

112. Date of manufacture of food products is the end date of technological process of food production (Article 2, TR CU 022/2011).

113. Information on location name of the manufacturer from third countries may be indicated in letters of the Latin alphabet and in Arabic numerals or in the state language (s) of the country at the location of the manufacturer of food products, provided that the name of the country is indicated in Russian (Article 4 (4.8/3) Technical Regulation TR CU 022/2011).

10. nuts and products of their processing;
11. crustaceans and their processing products;
12. Fish and products of its processing (except fish gelatin, used as a basis in preparations containing vitamins and carotenoids);
13. celery and products of its processing;
14. soybean and products of its processing;
15. eggs and products of their processing.

3.8.7.3 Claims

Information about distinctive properties of the food products ("health and nutrition claims") is the information about food products that indicate the characteristics of food products such as food value, origin, composition, and other properties in order to distinguish them from other food products. Such information on distinctive properties of food products is labeled on a voluntary basis. When the information on distinctive properties of food products is used in labeling, including absence of ingredients obtained from GMOs or using GMOs in food products, such information should be confirmed by the producer providing evidence independently or by a third party. The evidence on presence of distinctive properties of food products is subject to filing by the producers or individual entrepreneurs and is presented when required in accordance with the national legislation of the Member State.

Information about distinguishing properties of food products specified in Annex 5[114] to the Technical Regulation "On food products in terms of labeling" can be used only if the conditions specified in this annex are observed, unless otherwise stipulated by the technical regulations for certain types of food products. Example from the Annex 5:

Nutrition indicator or ingredient	Information on the distinctive properties of food products	The condition, the observance of which is mandatory when using information about the distinctive properties of food products in labeling of food products
Energy value (caloric value)	Reduced	The energy value (calorie content) is reduced by at least 30% relative to the energy value (caloric value) of similar food products
Protein	High content	The protein provides at least 20% of the energy value (caloric value) of food products

Table Extract from Annex 5 to Technical Regulation "On food products in terms of labeling."

Information about nutritional properties of the food products should include the amount of relevant nutritional substances that determine the nutritional value of food products in the labeling.

3.8.8 Developments

Currently, in the food market, there is most of all poor-quality dairy products (due to noncompliance with the technical conditions of production). That is why the question of their security in Russia is treated very carefully. On January 11, 2019 in Russia new requirements of technical regulations on the safety of dairy products came into force.

All dairy products according to the regulations are divided into three groups. *Dairy products*, as such, may not contain additives (cereals, garlic, greens, etc.) more than 50% (Group I); milk-containing products may contain more than 50% of additives (sugar, fruit, etc.) - Group II; Group III is products in which part of milk fat (from 0.1% to 50%) is replaced by vegetable oil. Such products should be marked with long names of 10 words: "milk-containing product with milk fat substitute, produced by sour cream technology (cottage cheese, milk, etc.)." Previously, such essentially falsified products in a wide assortment were exhibited in the stores under the name that misled the consumer—"smetanka (sour cream)," "syrok (cheese)," "molochko (milk)," etc.

Now to explain the quality of the product next to them the inscription is added—"contains vegetable oil." However, manufacturers are not in a hurry to switch to the new labeling, and out of a multitude of manufacturers, only two are identified by checks who clearly comply with the requirements of the regulations. In addition, from January 7, 2019, goods without milk fat substitutes will have to be put on store shelves separately from their "less clean" counterparts. Russian Prime Minister D. Medvedev in one of his interviews stated that "such information is needed because the buyer has the right to know what he is getting, and recently vegetable oil is actively used in our industry. This technology allows us to

114. Annex 5. Conditions when using in the labeling of food products information about the distinctive features of food products, TR CU 022/2011.

increase the shelf life of dairy products and reduce its cost price ..., and, frankly speaking, not everyone wants to buy these products." ("Version" newspaper No. 05 (680) from 11 to 17.02.2019). "We welcome any changes that increase consumer awareness of food products. However, they must be combined with informing the public about the benefits and harms of different types of products. The use of vegetable oil not only reduces the cost of production and lengthens the shelf-life, as we mainly position these products, unsaturated fatty acids that are part of vegetable oil, are good for health and vital for people." (O. Medvedev - Doctor of Medical Sciences, Professor of Moscow State University, Head of the National Research Center "Healthy Nutrition") ("Version" newspaper, № 05 (680) from 11 to 17.02.2019).

At the same time, an analysis of judicial practice has shown that the most common violations in this area at present remain:

(1) sale of food products that do not meet the mandatory quality requirements (identification of bacteria of the group of intestinal sticks, nitrates, plant sterols, exceeding the permissible limits of the content of a normalized component in the product, admission to the implementation of expired products[115];

(2) the implementation of food production in violation of the mandatory requirements of the Technical Regulations of the Customs Union TR CU 021/2011 "On food safety"[116] (inclusion in the technological process for the production of food raw materials, the use of which is not provided, the organization of the production process in rooms that do not correspond sanitary standards, etc.).[117]

Rospotrebnadzor (Federal Service on Customers' Rights Protection and Human Well-being Surveillance) has developed a package of bills, tightening administrative, and criminal penalties for falsification of products. We are talking about measures for deceiving the buyer: the law defines falsification as products that have intentionally changed or hidden properties and qualities, information about which is obviously incomplete and unreliable. That is, if the package does not indicate the content of proteins, fats, and carbohydrates, this is counterfeit. The same applies to products for which the figures declared by the manufacturer do not correspond to the actual composition. The use of such products, as a rule, does not threaten health; the logic is that the consumer should be aware of what exactly he buys.

Changes to the Code of Administrative Offenses propose to establish a penalty for the production and importation into the country of falsified food products from 10,000 to 15,000 rubles[118] for officials, from 30,000 to 50,000 rubles for individual entrepreneurs, from 50,000 to 10,0000 rubles for legal entities, it is proposed to establish criminal liability for counterfeit on a large scale and for committing a crime by a group of persons by prior agreement (in both cases, the term of imprisonment is up to 6 months). The package of bills is being approved by the Government for more than a year; however, we hope that it will be submitted to the State Duma for consideration in the near future (Parliamentary newspaper February 2–8, 2018, No. 4).

If the responsibility for low-quality food products will be tightened, this will prevent counterfeiters from entering the retail chains and, especially, to children's and medical institutions.

However, the essence of food security is not only in the quality of food products offered to the buyer. Food security is one of the most important areas of national security. The strategic goal is to provide the population of the country with safe agricultural products, fish, and other products from aquatic bioresources and food. Food security exists when the population of the country is provided with safe agricultural products and other food. Food security exists when all people at any time have physical and economic access to a sufficient amount of safe and nutritious food to satisfy their nutritional needs and preferences for maintaining an active and healthy lifestyle.

On November 18, 2013, Government of the Russian Federation adopted Order No. 2138 "On the Approval of the List of Indicators in the Field of Ensuring Food Security of the Russian Federation." This list is extremely extensive and includes several hundred indicators: 67 targets and 105 monitoring indicators. They are not synchronized with indicators on similar topics with FAO (Food and Agriculture Organization of the United Nations). Much work has been done to create a food safety monitoring system in the Russian Federation; however, it has not yet ended with the creation of a working monitoring system and information about its level and availability comes traditionally from reports and thematic notes at the regional or federal government level.

115. Resolution of the Supreme Court of January 27, 2016 No. 307-АД15-15693; Resolution of the Supreme Court from 10.11.2016 № 310-АД16-10648; Resolution of the Supreme Court from 17.05.2016 № 309-АД16-4234; Resolution of the Supreme Court dated 04.22.2016 № 304-АД165-3291; ATP Guarantor < www.garant.ru >.

116. Decision of the Customs Union Commission from 09.12.2011 No. 880″ On the adoption of technical regulations of the Customs Union "On food safety" < www.tsouz.ru >.

117. Supreme Court Decision from 06.11.2015. № 308-АД15-13715//SPS Garant < www.garant.ru >.

118. one Russian Ruble equals about € 0.013; and $ 0.015.

To form an assessment of the level of food security in the Russian Federation, 10 major food groups have been created, and food security is considered achieved when each person is provided with the possibility of consuming products at rational norms.

Rational consumption rates on average in the country have been achieved in six main groups of products out of 10 (bread, potatoes, meat, eggs, sugar, and vegetable oil). Less rational consumption of fruits (67%), milk and dairy products (75%), vegetables, and fish (84%). The average diet is balanced according to protein and energy content.

The overall level of food independence in Russia is rather stable over the years and in the past 15 years it has changed only in a narrow range of 86%–89%. Although, for certain food products its level is much lower.

Today, there is no shortage of food in the country, but the level of consumption for most food products does not increase as a result of the freezing or reduction of the purchasing power of the majority of the population.

Summing up, it can be noted that, despite some progress in the development of agriculture in the field of production and processing of products, despite the improvement of the food chain and highly productive industrial technologies, the improvement of the regulatory framework and food safety issues cannot be considered resolved. At present, the food problem is shaped by two aspects, first of all: economic affordability and food quality.

Further reading

The European Commission Provides Some Information on Russian Food Law in English Language. https://ec.europa.eu/food/safety/international_affairs/eu_russia/sps_requirements_en.

The US Department of Agriculture Commission Provides Some Information on Russian Food Law in English Language. https://www.fas.usda.gov/regions/russia.

Chapter 3.9

Azerbaijan

Alida Mahmudova

Bona Mente Consulting LLC Law Company, Azerbaijan

3.9.1 Introduction

The Azerbaijani law system is based on civil law system. The Constitution of Azerbaijan Republic was adopted on November 12, 1995 and is the foundation of the legislative system in the Republic. The Constitution created a presidential republic with a separation of powers among the legislative, executive, and judicial branches. The Constitution provides the legal basis for the domestic implementation of international law in general and international human rights law in particular. The international treaties, to which Azerbaijan is a party, are recognized as a constituent part of the internal legal system in accordance with the Article 148 (II) and given a higher hierarchical status in the case of a conflict with a national law in accordance with the Article 151 (Iskandarli, 2017).

The structural elements of the Azerbaijani legal system are based on the former Soviet system, and are divided into the following: (a) rule of law; (b) branch (area) of law; (c) subbranch (subarea) of law; (d) institute of law; (e) subinstitute (Матузов and Малько, 2004).

As far as "food law" is concerned, this is a new branch (area) of law in many countries (especially in post-Soviet countries) emerging only recently due to the integration processes. Therefore, it has not yet been separated into a separate branch or subbranch of law.

"Food law" could be distinguished as a complex subbranch of law that combines the structural elements of several already existing branches of law. These elements can be divided into different types of legal relations that arise at a certain stage of the food chain.

For example:

- agrarian law as a branch of law regulates relations at the first stage of the food chain as primary production of agriculture food products including food products of animal and plant origin as well as agriculture quality issues;
- environmental law regulates the issues of permissible norms for the use of agrochemicals in agriculture and the standards for the maximum permissible residual quantities of chemicals in food products, as well as safety requirements for soil and water resources;
- food import and export issues are regulated by international trade law, which is regarded as including public and private laws;
- with the view that "food law" is normally considered to have two fundamental purposes—the protection of the consumers' health and the prevention of fraud consistent with the first purpose of the Codex ("protecting the health of the consumers and ensuring fair practices in the food trade"), then the relations governing the last stage of the food chain, for example, trading or "retail which means the handling and/or processing of food and its storage at the point of sale or delivery to the final consumer, and includes distribution terminals, catering operations, factory canteens, institutional catering, restaurants, and other similar food service operations, shops, supermarkets, distribution centers, and wholesale outlets"[119] are related to consumer law;
- additionally, there are elements related to intellectual property law.

The dynamic/intensive development of this area of law and its globalization gives us the ground to agree with the opinion that "food law" as a functional area of law also can be considered relevant because as its meaning reaches beyond the closed circle of a group with a special interest (Van der Meulen, 2018).

3.9.2 Most important sources of legislation for food

The laws based on Soviet law are still working in Azerbaijan and they are the source for food legislation—this is the next primary legislation:

- *Law* "On food products" dated November 18, 1999 with last amendments dated February 23, 2018;
- *Law* "On veterinary" dated May 31, 2005 with last amendments dated February 23, 2018 (provisions are regulated the animal origin food products);
- *Law* "On phytosanitary control" dated May 12, 2006 with last amendments dated February 23, 2018 (provisions are regulated the plant origin food products).

As well as the secondary legislation arising from implementation of the primary legislation.

Today, the government is conducting legal and institutional reforms in the country leading to a fundamental change of the entire food safety system.

Thus was adopted the Presidential Decree "On additional measures to improve the food safety system in the Republic of Azerbaijan" *dated 10 February 2017 which stipulates the development of the new law* "On Food Safety"[120] *in accordance with EU food law principals and requirements as well as harmonization of all relative regulatory-legal acts (secondary legislation) in the area of food safety (including standards, sanitary, phytosanitary, and veterinary standards) with international standards and requirements of Codex Alimentarius, IPPC, OIE.*

3.9.3 Developments

Azerbaijan has been investing oil revenues flowing to the country into the infrastructure development, institutional restructuring and development, and capacity building and strengthening legal framework. As a result, rapid economic growth has been observed. Starting in 2008, the Government of Azerbaijan has also adopted various state programs on regional economic development, food security, and food safety development as well as small and medium business development. Unfortunately, these programs ("On Reliable Food Supply of population in the Azerbaijan Republic" dated August 25, 2008 and "On Poverty Reduction and Sustainable Development during the period 2008–15" dated September 15, 2008, etc.) have not been properly implemented and could not provide the required support to the small and medium businesses.

119. Regulation (EC) 178/2002 of the European Parliament and of the Council of 28 January 2002 laying down the general principles and requirements of food law, establishing the European Food Safety Authority and laying down procedures in matters of food safety (Article 3. Other definitions).
120. Draft Law "On Food Safety" drafted with EU lines and expected to adopt during 2019.

And with the decrease of oil revenues in the country due to the fall of oil prices in the world market, Azerbaijan, like many oil-based economies, has had to face stronger economic challenges. It has had to reemphasize its economic reforms aimed at the diversification of the economy with more focus on the development of the nonoil sector including the agriculture sector.

The "Strategic Road Map" covering national economic perspectives and 11 economic sectors consists of 12 documents approved by the Decree of the President of the Republic of Azerbaijan No. 1138 dated December 06, 2016.

The implementation of nine strategic targets for creating a favorable environment to achieve the formation of a production and processing sector of competitive agricultural products is planned based on sustainable development principles as described in the Strategic Roadmap for the years of 2016–20. These strategic targets involve strengthening the sustainability of food safety, increasing production capacity of agricultural products, developing the market of agricultural production means, facilitating the access to relevant resources, as well as finance, upgrading scientific provision and quality of education in agricultural field, developing the system of extension services, facilitating development of market infrastructure and access of producers to market, shaping mechanisms of sustainable use of natural resources, improving business environment in agricultural field, and increasing welfare in rural areas.

For many years, the Government of the Republic of Azerbaijan has recognized the importance of food safety for public health, strengthening export potential, and increase of investments in modern and efficient agricultural production in the country. In order to ensure the implementation of sustainable reforms for the development of the food safety system in the Republic of Azerbaijan, the following have been prepared and adopted:

- "State Program on Reliable Provision of Population with Food Products in the Republic of Azerbaijan for 2008–15" approved by the Decree of the President of the Republic of Azerbaijan dated August 25, 2008;
- "Strategic Road Map for the Production and Processing of Agricultural Products in the Republic of Azerbaijan" approved by Decree No. 1138 of the President of the Republic of Azerbaijan dated December 6, 2016.

The Strategic Road Map "On the Production and Processing of Agricultural Products in the Republic of Azerbaijan" is aimed to improve the food safety system in the country through the following actions:

Action 1.3.1: Improve the existing legislation on the basis of international best practices.

Action 1.3.2: Eliminate overlapping activities of different agencies in the field of food safety control and establish an effective regulatory system.

Action 1.3.3: Adapt food production, processing, storage, transportation, and trading standards to international standards.

Action 1.3.4: Optimize, modernize, and accredit the network of laboratories included in the food safety system and improve the system of certification.

In order to ensure the implementation of the "Strategic Road Map for the Production and Processing of Agricultural Products in the Republic of Azerbaijan", the Decree "On Additional Measures for Improving the Food Safety System in the Republic of Azerbaijan" No 1235 of the President of the Republic of Azerbaijan dated February 10, 2017 was adopted.

In accordance with the above-mentioned Presidential Decree, the Azerbaijan Food Safety Agency (AFSA) has been established and has become fully functional as of January 1, 2018. The establishment of AFSA puts the end to the practice of exercising state food safety control by five different ministries and agencies without any clear roles and responsibilities of the public institutions which resulted in overlap of their activities. The establishment of AFSA has helped scrutinize food safety control mechanisms through the establishment of a single state food safety control system based on risk assessment covering all stages of the food chain and implemented by a single food safety body.

In order to ensure its proper functioning, AFSA was approved by the Decree "On ensuring the activity of the Food Safety Agency of the Republic of Azerbaijan" No. 1681 of the President of the Republic of Azerbaijan dated November 13, 2017. The Decree of the President of the Republic of Azerbaijan No 28 dated May 1, 2018, envisioned introduction of a number of changes and amendments in the AFSA Statute as well as in the existing legal framework to ensure a single state food safety control at all stages of the food chain.

The Azerbaijan Food Safety Institute (AFSI) was established under the Agency with the status of a legal public entity. AFSI, approved by the Order of Cabinet of Ministers of the Republic of Azerbaijan as of May 16, 2018, envisions the conduct of scientific and practical research, risk assessment based on scientific principles, development of technical normative legal acts and minimum quality indicators, laboratory analysis, and expertise, as well as education and public awareness raising in the field of food safety.

In accordance with the Decrees of the President of the Republic of Azerbaijan dated 26 January 2018, the Government of the Republic of Azerbaijan has transferred state-owned property consisting of laboratories, including their administrative buildings and material and technical components from the balance of the Ministry of Economy of the Republic of

Azerbaijan, the Ministry of Agriculture of the Republic of Azerbaijan and the Ministry of Health of the Republic of Azerbaijan to the Food Safety Agency of the Republic. Furthermore, all veterinary and phytosanitary controls have been transferred under the AFSA mandate in accordance with the Decree of the President of the Republic of Azerbaijan dated May 1, 2018 and best international practices.

The Azerbaijan Food Safety Agency is in the process of establishing new food safety systems in accordance with the EU approach: official controls for generic SPS measures (annual control plans, risk management, sampling procedures, corrective measures in case of noncompliance with animal health, plant health, and food safety normative acts); requirements for food hygiene (general and hygiene of products of animal origin), etc.

Therefore, the works carried out in the AFSA are aimed at ensuring the compliance of food safety system of the Republic of Azerbaijan with the requirements of the WTO, European Union, and the relevant international organizations.

Much work is still required to improve the safety and quality of products in the country. The AFSA is undertaking steps and measures to ensure further development of the food safety system in compliance with best international practice.

Thus, the AFSA has drafted the "Food Safety Law" that has been publicly discussed and is currently in the process of approval. The Agency has also drafted the State Program "On ensuring food safety 2019–25." This State Program envisages the full provision of the population with healthy and safe food via significant reduction of the risks of foodborne illnesses, as well as increase of the productivity and competitiveness of agricultural and food products and also strengthening of the export potential especially to the developed countries. The State program contains an action plan guiding the harmonization of the existing national norms and rules on food products with the norms and rules of Codex Alimentarius regarding food safety, International Plant Protection Convention (IPPC) regarding plant products, and International Epizootic Bureau (OIE) regarding animal origin products.

Priority activities aimed to ensure food safety system in the country are as follows:

- Improvement of the food safety–related national regulatory framework with the view of the best international practice which will be aimed at the analysis of sanitary, veterinary, and phytosanitary norms and rules, their compliance with international standards, and development of new technical normative acts.
- Formation of a risk-based food safety system in line with the international standards, particularly improvement of registration and approval system of FBOs and veterinary, phytosanitary, food safety certification systems in accordance with of the international requirements.
- Improvement of infrastructure provision in the area of food safety including optimization of the laboratories' networking and ensuring their international accreditation.
- Formation of risk-based state food safety control system in line with the best international practice.

3.9.4 Role of risk analysis

Priority 1.3 of the State Strategic Roadmap for the agricultural sector aims to "form a food safety system that covers all stages of the value chain and is based on a risk assessment approach."

One of the main duties of the Azerbaijan Food Safety Agency is to conduct risk analysis in the relevant area including risk assessment, management, and communication. This includes developing criteria for the risk-based classification of establishments, collecting data on the actual level of risk of active FBOs, and planning inspection visits on this basis.

The recommendation from the EU Advisor is to ensure that "risk assessment" is separate from "risk management"—in the sense that these terms have in EU legislation and institutions.

Although there is no real "conflict of interest" between scientific risk assessment and regulatory risk management, in order to both offer increased guarantees of independence and professionalism and be able to attract and retain adequately qualified experts, it may be appropriate to house this scientific risk assessment function in the Institute under AFSA, rather than within AFSA "proper."

3.9.5 The addressees of food law

Current Azerbaijan food legislation consists the following definition: *"subjects engaged in food products—state bodies (municipalities), natural, or legal persons operating in all stages of the food chain."*[121]

121. Law "On food products" dated 18 November 1999 (with last amendments dated 23 February 2018).

But the new drafted "food safety law" uses the EU concept of this definition as following: *"food business operators—any individual, including natural or legal persons, state bodies (municipalities), organizations and institutions dealing with food business and responsible for ensuring that the requirements of food safety law are met."*[122]

3.9.6 Codex Alimentarius

As a result of fundamental legal reforms in the country, in accordance with the point 6.1. of the Presidential Decree of the Republic of Azerbaijan "On additional measures to improve food safety system in the Republic of Azerbaijan" dated February 10, 2017 to «*ensure harmonization of regulatory legal acts on food safety standards and regulations with the Codex Alimentarius Commission*».

Today Azerbaijan has already started the harmonization process and adopted the following Codex standards:

- AZS 810–2015 (CODEX STAN 310–2013) «Codex Standard for pomegranate»;
- AZS 837–2015 (CODEX STAN 152–1985) «Codex Standard for wheat flour»;
- AZS 838–2015 (CODEX STAN 33–1981) «Codex Standard for olive oils and pomace oils»;
- AZS 839–2015 (CODEX STAN 87–1981) «Codex Standard for chocolate and chocolate products»;
- AZS 840–2015 (CODEX STAN 279–1981) «Codex Standard for butter»;
- AZS 841–2015 (CAC/MRL-2-2015) «Maximum Residue Limits for Veterinary Drugs in Food»;
- AZS 842–2015 (CAC/RCP1-2015) «Recommended International Code of Practice - General Principles of Food Hygienic»;
- AZS 849–2016 (CODEX STAN 12–1981) «Codex Standard for honey».

3.9.7 Institutional

The new Azerbaijan Food Safety Agency was established by the Presidential Decree of the Republic of Azerbaijan "On additional measures to improve food safety system in the Republic of Azerbaijan" dated February 10, 2017, setting out its Charter and Structure adopted by Decree of the President of the Republic of Azerbaijan "On ensuring activity of the Azerbaijan Food Safety Agency" dated November 13, 2017 implemented on January 2018.

Official control over food safety had been highly fragmented: various issues were controlled by eight agencies within six ministries and committees; their mandates are not clearly defined, which results in overlapping activities as well as gaps in control: State Veterinary Service of the Ministry of Agriculture (veterinary programs, control of zoonotic diseases, and quality and safety of agricultural food products of animal origin, including those presented at numerous farmer markets); State Phytosanitary Service of the Ministry of Agriculture (agricultural products of plant origin, including quarantine issues); State Center for Sanitary and Epidemiological Control of the Ministry of Health (public health and food inspection functions); State Service on Antimonopoly Policy and Consumers' Rights Protection at the Ministry of Economic Development (approvals for EU export, control of foodstuffs at points of sale, as well as certain inspection functions and consumers' rights protection on food products); State Committee for Standardization, Metrology, and Patents (establishment of food standards and assessment of conformity through certification); State Customs Committee (sanitary, veterinary, and phytosanitary border control); Ministry of Environment (control of contaminants in soil and water). Today the functions of all these ministries transferred to Single Authority - Azerbaijan Food Safety Agency since January 1, 2018.

3.9.8 Principles and concepts

Following EU "Food law" the basic principles are risk analysis; precautionary principle; protection of consumers' interests; transparency.

The legal definition of food is based on EU "Food law": "food products—unprocessed, semiprocessed, or fully processed products intended for human nutrition, including special-purpose food products, food additives, biologically active food products, food raw material, drinks (alcoholic and nonalcoholic), drinking water (packaged), gum and water and other substances added to foodstuffs during primary production, production, and processing of food products. This definition shall not include feed products, live animals not intended for human consumption, plants until harvesting, medicines, perfumery and cosmetic products, tobacco and tobacco products, narcotic, and psychotropic substances and contaminants."

122. Draft Law "On Food Safety" drafted with EU lines and expected to adopt during 2019.

The scope of new law shall apply to all stages of production, processing and distribution of food and feed. It shall not apply to primary production for private domestic use or to the domestic preparation, handling or storage of food for private domestic consumption.

The Law shall not apply to the home-made or primary production of the foodstuffs that are intended for personal use.

3.9.9 Standards

Compliance with private food standards in Azerbaijan is currently very low. There is a poor understanding by the industry, as well as by the government, of the differences between HACCP as a possible mandatory preventive instrument and private standards based on HACCP.

In Azerbaijan, of all private standards recognized, only ISO 22000, one of the least popular standards in the EU, is known.

3.9.10 Authorization requirements

In accordance with new legislation, the following products shall be authorized by AFSA: special-purpose food products (baby food, dietary food, therapeutic, and prophylactic food products); novel food products; GMO food products; food and feed additives; biologically active food products; natural mineral waters; materials; and articles in contact with food products.

Work is currently being carried out to harmonize with Codex standards on food additives.

3.9.11 Food safety limits

In Azerbaijan, food safety criteria are fully harmonized with Russia (as major trading partner), but are not harmonized with EU or Codex. It should be noted that the list of products used to develop the tables is based on standardized commodity nomenclature (HS code) of Sanitary Rules and Norms of Azerbaijan "Hygienic Rules on Food Safety and Food Nutrition" dated 2010.

For pesticide residues, while Azerbaijan has detailed regulations and established MRLs, in many cases either substances prohibited in EU are authorized in Azerbaijan or EU MRLs are stricter than Azerbaijani.

Work is currently carried out to transfer to Codex standards on MRLs.

3.9.12 Process requirements

Secondary legislation as Cabinet Ministries Regulation "On General Hygiene Requirements" in accordance with EU Regulation 852/2004 has been drafted and is expected to be adopted during 2019.

Currently, *traceability* is not a regulatory requirement in Azerbaijan, and is practiced by only a few companies, and with deficiencies. Inability of the private sector to ensure and demonstrate traceability can be a significant constraint for export. New legislation envisions the inclusion within the Draft Food Safety Law of provisions which will regulate traceability issues.

New legislation also envisions the inclusion of articles into Draft Food Safety Law which will regulate recall obligations.

Food and feed products which do not comply with the requirements of the legislation in the area of food safety shall be recalled from the consumers, markets, and other sales facilities at the request of the Azerbaijan Food Safety Agency or on the initiative of food business entities.

3.9.13 Labeling

In accordance with national legislation, the labeling of food products shall be accurately, clearly, and fully understandable to consumers and provided with full information on the nutrition and energy value, composition, quantity, shelf life, storage conditions, origin of the product, production method, and terms of use of the food products.

References

Iskandarli, R., 2017. A Guide to the Republic of Azerbaijan Law Research. http://www.nyulawglobal.org/globalex/Azerbaijan1.html.

Матузов, Н.И., Малько, А.В., Юристъ, 2004. Теория государства и права: учебник. С. 150.
van der Meulen, B., 2018. The functional field of food law. The emergence of a functional discipline in the legal science. In: European Institute for Food Law Working Paper 2018/02. https://ssrn.com/abstract=3128103 (Open access).

Further reading

Information and Sources in English Language are Available at: http://www.fao.org/faolex/country-profiles/general-profile/en/?iso3=AZE.

Chapter 3.10

Australia and New Zealand

Joe Lederman
FoodLegal, Australia

3.10.1 Introduction

3.10.1.1 Food law in Australia and New Zealand

Like most developed countries, Australia and New Zealand each have a branch of law that specifically addresses food products. This body of food law has developed over decades from ad-hoc, localized requirements to a binational regulatory framework that addresses everything from the safety and registration requirements for food businesses, to the composition and labeling requirements of food products, to how these products can be sold and marketed.

3.10.1.2 Sources of food legislation

The most prominent source of regulations for food businesses and food products is the *Australia New Zealand Food Standards Code* (Food Standards Code). The Food Standards Code is a binational legislative instrument comprised of four chapters. The first two chapters specify the compositional and labeling requirements for food products sold in both Australia and New Zealand, while chapters three and four, which set requirements for food safety standards and primary production standards respectively, apply only in Australia.

The Food Standards Code is developed and maintained by Food Standards Australia New Zealand (FSANZ), a government agency established under the *Food Standards Australia New Zealand Act 1991*.

Although the Food Standards Code is a binational legislative instrument for Australia and New Zealand, it does not itself contain any enforcement or penalty provisions. Rather, the Food Standards Code is enforced through the Food Acts of each Australian State and Territory and of New Zealand.[123] Each Food Act makes it a criminal offense to sell food that is in breach of the Food Standards Code and prescribes detailed requirements regarding the food safety and registration processes that apply to food businesses.

There are a number of other legislative instruments in effect in Australia and New Zealand that also impose requirements for food products, either directly or indirectly:

- The *Country of Origin (Food Labeling) Information Standard 2016* provides for country of origin labeling requirements for foods for human consumption that are sold in Australia. The *Consumers' Right to Know (Country of Origin of Food) Bill* seeks to introduce limited country of origin labeling requirements for food products sold in New Zealand.
- The *Australian Consumer Law* contained within the *Competition and Consumer Act 2010* applies to all Australian businesses. It contains a broad prohibition on false, misleading, or deceptive representations (including marketing claims) and establishes statutory guarantees and safety requirements for products (including food products) purchased by

123. *Food Act 2001* (ACT), *Food Genetically Modified Organisms Act, 2004* (NT), *Food Food Safety Basic Act, 2003* (NSW), *Food Act 2014* (NZ), *Food Act 2000* (QLD), Food *Act 2002* (SA), *Food Food Safety Basic Act, 2003* (TAS), *Food Act 1984* (VIC), *Food Act 2008* (WA).

consumers. Similar provisions are contained in the New Zealand *Fair Trading Act 1986* and the *Consumer Guarantees Act 1993.*

- The *National Trade Measurement Regulations 2009* (for Australia) and *Weights and Measures Regulations 1999* (for New Zealand) require units of measurement, such as weight or volume, to appear on product packages in a prescribed format.
- The various environmental protection laws of South Australia, the Northern Territory, Queensland, and New South Wales[124] require manufacturers of products in certain recyclable containers sold in each of these jurisdictions to participate in a container deposit scheme.

3.10.1.3 Game-changing events in Australian and New Zealand food law

Arguably the most significant event in the development of food law in Australia and New Zealand was the shift to a binational system from a system of ad-hoc laws across the two countries that had only limited application. To address inconsistencies between different Australian States and Territories, the Australian National Health and Medical Research Council (NHMRC) was tasked with developing uniform nation-wide maintenance of public and individual health standards.[125]

In 1991, the Australian Federal and State and Territory governments formalized an agreement whereby each State and Territory agreed to adopt the federal food standards that had been developed by the NHMRC. In 1996, Australia and New Zealand signed a treaty for the development of uniform food standards. This led to the creation of the Australia New Zealand Food Standards Code, which was adopted by both nations in 2000 and came into full effect in 2002, establishing the current binational framework. The Food Standards Code has had two major revisions, and the current version took effect on March 1, 2016.

3.10.1.4 The role of risk analysis

Risk analysis is a key concept underpinning Australia and New Zealand food laws. In particular, it plays a significant role in the licensing and accreditation of a food business or food premises. The ability to implement different controls depending on level of risk was a key factor in the introduction of food safety programs, as prescribed by Chapter 3 of the Food Standards Code. Food safety program requirements are based on the level of risk presented by a food business, and impose various production and processing standards in addition to the usual end-product standards.

For example, businesses involved in processing and handling seafood or poultry, or businesses supplying food to vulnerable persons, are generally regarded as higher risk businesses requiring licensing, food safety programs, and auditing. Such programs are generally based on *HACCP* principles and commonly address

- Training and knowledge of food handlers;
- Food handling controls in relation to storage, processing, display, packaging, transportation, disposal, and product recalls;
- Health and hygiene of food handlers;
- Cleaning, sanitizing, and maintenance;
- Pest controls;
- Design and construction of premises (e.g., water supply, sewage and waste water disposal, storage of garbage and recyclable matter, ventilation, lighting, floors, walls, ceilings, fixtures, hand washing facilities, storage, toilets, etc.); and
- Food transport vehicles.

Penalties applicable for different types of breach of the requisite Food Acts of the Australian States and Territories will also reflect the different levels of risk to public health posed by any offending conduct.

The labeling requirements of the Food Standards Code also reflect a risk-based approach. Products containing substances known to pose a risk, such as royal jelly, must be labeled with warnings as such. Similarly, product labels must declare the presence of any of a prescribed list of potentially allergenic substances.

124. *Waste Avoidance and Resource Recovery Act 2001* (NSW). *Environment Protection (Beverage Containers and Plastic Bags) Food Sanitation Act, 2017* (NT), *Waste Management and Pollution Control Act 2016* (NT), Environment Protection Act 1993 (SA), *Waste Reduction and Recycling The Punjab food authority act, 2011* (QLD).

125. National Health and Medical Research Council, "*About the NHMRC,*" https://www.nhmrc.gov.au/about.

3.10.1.5 The addressees of food law

In Australia and New Zealand, food laws are primarily addressed to any person or entity in the business of selling food. The definition of "sale" is expressly broadened in the food laws to capture any individual or business involved in supplying food, whether or not at retail and whether or not for a direct financial benefit.

3.10.1.6 Role of the Codex Alimentarius

Australia and New Zealand both participate in the Codex Alimentarius Commission which is responsible for the administration of the Codex Alimentarius (the Codex). Both countries are also highly involved in the review of existing Codex food standards, and the development of new standards.

Importantly, the Codex standards and recommended international codes of practice do not impose direct legal obligations upon Australia or New Zealand.[126] However, one of the functions of FSANZ is to "promote consistency between standards in Australia and New Zealand with those used internationally, based on the best available scientific evidence."[127] As a matter of policy, Australian and New Zealand food standards are designed to be consistent with Codex standards except where necessary.[128] As such, Codex guidelines are highly persuasive in the development of new standards and may be used by applicants seeking to change the Food Standards Code as evidence that amendments may be required.

3.10.2 Institutional framework

3.10.2.1 Food law regulatory bodies

Although the main requirements of the Food Standards Code apply equally in Australia and New Zealand, the enforcement of these and other food laws is the responsibility of local regulatory bodies in each country.

3.10.2.1.1 Australia

In Australia, it is the responsibility of each State and Territory to administer and enforce the provisions of their respective Food Act as it applies to their jurisdiction. State and Territory regulators recognize the "home jurisdiction rule," whereby a food business operating across Australia will be regulated by the food agency of the State or Territory in which its head office is registered.

The relevant State and Territory food regulators are as follows:

New South Wales	New South Wales Food authority
Victoria	Victorian Department of Health
Queensland	Queensland Department of Health
South Australia	South Australian Department of Health
Western Australia	Western Australian Department of Health
Tasmania	Tasmanian Department of Health and Human Services
Australian Capital Territory	Australian Capital Territory Health Protection Service
Northern Territory	Northern Territory Department of Health

The other primary body involved in regulating food businesses in Australia is the Australian Competition and Consumer Commission (ACCC). The ACCC enforces the *Australian Consumer Law* as well as the *Country of Origin (Food Labeling) Information Standard 2016*, and has the power to bring action against food companies in relation to any false, misleading, or deceptive product claims.

Finally, the Federal Department of Agriculture and Water Resources is responsible for assessing the compliance of food products imported into Australia against the Food Standards Code and other relevant laws. The Department distinguishes between "surveillance" and "risk" foods. Surveillance foods are subject to randomized testing of 5% of consignments, while every single consignment of a risk food is inspected at the border.

126. Food Standards Australia New Zealand, *"International Engagement: Codex Alimentarius Commission"* < http://www.foodstandards.gov.au/science/international/codex/pages/default.aspx >.

127. *Food Standards Australia New Zealand Act 1991* s 13.

128. *Food Regulation Agreement 2008,* Cl 3(a) (iii).

3.10.2.1.2 New Zealand

In New Zealand, the Ministry for Primary Industries (MPI) (which subsumed the New Zealand Food Safety Authority in 2012) is responsible for enforcing food laws.[129] The MPI is responsible for inspecting both food products for import at the New Zealand border, as well as food products manufactured in New Zealand.

The New Zealand Commerce Commission is responsible for enforcing the *Fair Trading Act 1986* and bringing actions against food companies for false or misleading representations.

3.10.2.2 The regulatory divide between food and medicine

In both Australia and New Zealand, food and medicines are governed by separate bodies and remain generally distinct. Indeed, each of the Food Acts explicitly asserts that food does not include therapeutic goods (in Australia), or medicines (in New Zealand). This means that medicine regulation is beyond the established functions of FSANZ and each of the food regulators.

In Australia, therapeutic goods are regulated by the Therapeutic Goods Administration (TGA) under the *Therapeutic Goods Act 1989* and include medicines, medical devices, diagnostic tests, vaccines, biologicals, blood and blood products, and other therapeutic goods such as sunscreens, sterilants, and female hygiene products. The TGA may take action where any product or marketing claims present the product as being for therapeutic use. A representation of "therapeutic use,"[130] is defined as a claim of

(a) preventing, diagnosing, curing, or alleviating a disease, ailment, defect, or injury in persons; or
(b) *influencing, inhibiting, or modifying a physiological process in persons;*

Despite this distinction, there remains the potential for considerable overlap between foods and therapeutic goods. In Australia, most dietary supplement products are classified as therapeutic goods; however, products may be classified as foods if their composition is permitted under the Food Standards Code and they are not represented as being for therapeutic use.

New Zealand also distinguishes between food and medicines, but also has separate regulatory categories for dietary supplements (regulated under the *Dietary Supplements Regulations, 1985*) and supplemented foods (regulated under the *Supplemented Food Standard, 2016*) for some products that fall somewhere in-between the food and medicine interface. Supplemented foods are regulated as food products by the MPI, while dietary supplements are regulated alongside medicines by New Zealand medicines regulator Medsafe, although may be manufactured according to food requirements.

It should be noted that the Food Standards Code contains a standard for "*foods for special medical purposes.*"[131] These products are regulated as food products in both Australia and New Zealand and have numerous additional compositional permissions when compared with other foods (for example, they may contain novel foods). On the other hand, a product advertised as a food for special medical purposes must be primarily distributed through pharmacies and medical practitioners and must not be compared to a good that is represented for therapeutic use or make any therapeutic claims.

3.10.2.3 Enforcement powers

Compliance with the Food Standards Code is monitored and enforced by food authorities in the States and Territories in Australia, and the MPI in New Zealand. Enforcement provisions are incorporated into the respective Food Acts within each of these jurisdictions. Depending on the offense in question, the Food Acts provide a range of possible enforcement actions, including:

- Verbal advice
- Written warning
- Statutory orders requiring the business to undertake certain actions to remedy a food safety hazard
- Infringement/penalty notice
- Mandatory recall
- Court proceedings, potentially resulting in:

---

129. New Zealand Ministry of Primary Industries, '*Food Safety*', < https://www.mpi.govt.nz/food-safety/>.
130. *Therapeutic Goods Act 1989* (Cth), s 3.
131. *Australia New Zealand Food Standards Code, Standard 2.9.5.*

- o Substantial fines of up to AU$500,000 as at 2018[132]
- o Imprisonment (for individuals)
- o Publication of details on a public name-and-shame register

Food regulators have a considerable degree of discretion in determining which enforcement action(s) to pursue. Where there are breaches for food safety, or a risk to public health, local authorities are authorized to issue fines and other penalties; however, an educative approach is favored when there is no health risk. Although enforcement is undertaken at a local level, each regulatory body adheres in principle to a uniform compliance strategy.[133]

Competition and consumer law regulators also have significant enforcement powers. In Australia, the ACCC can issue an infringement notice for actions it believes to be in breach of the Australian Consumer Law. It is not uncommon for the ACCC to take court actions against large companies, including food manufacturers, in relation to claims it alleges to be misleading. The maximum fine for any court proceedings can range into the millions of dollars. The New Zealand Commerce Commission has similar powers but does not have a similar history of litigation against New Zealand food companies.

At the border, the Australian Department of Agriculture and Water Resources and New Zealand MPI may refuse entry to noncompliant imported products, and require that either any noncompliance be remedied, or the products in question be reexported or destroyed.

3.10.3 Principles and concepts

3.10.3.1 Principles underpinning food law

The key principle driving food law in Australia and New Zealand is food safety. From the creation of food standards to registration requirements, food safety programs, and recall procedures throughout the supply chain, the regulatory regimes exist to ensure the safety of consumers and prevent and contain food safety incidents. These topics are expanded upon further in this chapter.

3.10.3.2 Important concepts in food law

One of the most important legal concepts in food regulation is the definition of "sale." As described earlier in this chapter, the requirements under the Food Standards Code will only apply where a sale of food takes place.

Other key concepts within the Food Standards Code involve how a particular substance is being used in food, as the type of usage may trigger additional regulatory obligations. These include the following concepts:

- Use as a food additive
- Use as a processing aid
- Use as a nutritive substance

3.10.3.3 Definition of food

It goes without saying that the regulatory framework outlined in this chapter will only apply to products that are food, making the definition of food an important regulatory threshold. Food has a wide definition in each of the relevant Food Acts and includes any substance or thing of a kind used, or represented as being for use, for human consumption (including an ingredient or additive thereof, or a substance that comes into contact with food). Importantly, food is defined to exclude therapeutic goods (in Australia) and medicines (in New Zealand), which are subject to separate regulatory regimes.

3.10.4 Standards

3.10.4.1 What is the role of standards?

The Food Standards Code contains prescriptive Australian and New Zealand legal requirements for composition, labeling, processing, and safety of food. The Food Standards Code represents something of a middle ground between the general requirements of the European Union, and detailed product-specific requirements seen in countries like China.

132. The exchange rate of the Australian Dollar is about 0.73 USD and 0.63 Euro (Early, 2019).

133. *Australia and New Zealand Food Regulation Compliance, Monitoring and Enforcement Strategy*, 2017.

Chapter one of the Food Standards Code sets out general labeling and compositional requirements that apply to all products, while Chapter two prescribes a number of additional requirements for specific product categories such as cheese, bread, formulated caffeinated beverages, and special purpose foods including formulated supplementary sports foods and infant formula products.

3.10.5 Authorization requirements

3.10.5.1 Types of food subject to authorization

In Australia and New Zealand, there is no requirement for any sort of regulatory preapproval or authorization for food products generally. The Food Standards Code provides the starting position that food for sale may contain any food. It then goes on to qualify that such products must not have as an ingredient or component any of the following if not explicitly permitted elsewhere in the Food Standards Code:[134]

- A substance used as a food additive
- A substance used as a nutritive substance
- A substance used as a processing aid
- A novel food

Thus, regulatory authorization must be sought for the use of any of these types of substances, where they are not explicitly permitted for use in that type of food by the Food Standards Code. The definitions of when a substance will fall into each of these categories are set out below. References to Schedules are Schedules to the Food Standards Code.

- **Used as a food additive**
  - A substance added to food to perform a technological purpose listed in schedule 14 (e.g., flavor, thickener, emulsifier), and is[135]:
    - A substance that is identified in Schedule 15 as a substance that may be used as a food additive;
    - An additive permitted at GMP;
    - A coloring permitted at GMP;
    - A coloring permitted to a maximum level; and
    - Any substance that is a nontraditional food and has been concentrated, refined, or synthesized to perform one or more of the technological purposes listed in Schedule 14.
- **Used as a nutritive substance**
  - A substance added to food to achieve a nutritional purpose that is:
    - Permitted by the Food Standards Code to be used as a nutritive substance;
    - A vitamin or mineral; or
    - Any substance (other than an inulin-type fructan, a galacto-oligosaccharide, or a substance normally consumed as a food) concentrated, refined, or synthesized to achieve a nutritional purpose.[136]
- **Used as a processing aid**
  - A substance used in the course of processing to perform a technological purpose in the course of processing and not in the final food for sale, that is listed in schedule 18 or an additive permitted at GMP.[137]
- **Novel food**
  - A nontraditional food (that is, with no history of human consumption in Australia or New Zealand) that requires an assessment of the public health and safety considerations having regard to[138];
    - **(a)** the potential for adverse effects in humans; or
    - **(b)** the composition or structure of the food; or
    - **(c)** the process by which the food has been prepared; or
    - **(d)** the source from which it is derived; or
    - **(e)** patterns and levels of consumption of the food; or
    - **(f)** any other relevant matters.

134. *Australia New Zealand Food Standards Code,* Standard 1.1.1—10 (6).
135. *Australia New Zealand Food Standards Code,* Standard 1.1.2—11(1).
136. *Australia New Zealand Food Standards Code,* Standard 1.1.2—12.
137. *Australia New Zealand Food Standards Code,* Standard 1.1.2—13.
138. *Australia New Zealand Food Standards Code,* Standard 1.5.one to two.

3.10.5.2 Authorization procedure

A substance used in food will require premarket regulatory authorization where it is used as a food additive, processing aid, or nutritive substance, or is a novel food, and is not expressly listed in the Food Standards Code as permitted in food. To obtain this authorization, the party seeking to use the substance must apply to amend the Food Standard Code to permit that substance.

Applications are submitted to FSANZ and must adhere to prescribed content and formatting requirements. The applicant has the burden of establishing the safety of the proposed amendment. FSANZ will undertake a risk assessment based on the scientific data included in the application, and will generally seek submissions from industry. Once FSANZ has determined the safety and viability of a component, it may be added to the list of permitted additives, nutritive substances, processing aids, and novel foods. Applications are free in-principle, but require payment if the applicant seeks to expedite the process or would obtain an exclusive capturable commercial benefit through the application. In this case, application fees payable to FSANZ may exceed AUD$ 100,000. Applicants seeking permission to use a novel food may seek a period of exclusive use.

3.10.5.3 The role of Codex Alimentarius in food additive authorization

Any additive permitted under Codex must be specifically listed in the Food Standards Code in order to be permitted for use in Australia and New Zealand. However, applicants seeking the addition of a new additive may point to inclusion in Codex as evidence of safety, and FSANZ itself may propose an amendment to the Food Standards Code to bring it in line with international standards, including Codex.

3.10.6 Food safety limits

3.10.6.1 Agricultural and veterinary chemical limits

Schedules 20 and 21 of the Food Standards Code establish maximum and extraneous residue limits for agricultural and veterinary chemicals in Australia. Extraneous residue limits apply where the chemical is only indirectly present through environmental sources. Maximum residue limits in New Zealand are set by the *Maximum Residue Levels for Agricultural Compounds Food Notice*.

3.10.6.2 Chemical contaminant and toxicity limits

Schedule 19 of the Food Standards Code establishes maximum levels of metal and nonmetal contaminants, as well as natural toxicants, for both Australia and New Zealand.

3.10.6.3 How limits are set

In Australia, maximum residue limits are registered and approved by the Australian Pesticides and Veterinary Medicines Authority, with levels based on a risk assessment of any health and safety impacts. FSANZ then incorporates these levels in the Food Standards Code. In New Zealand, the MPI assesses and approves maximum residue limits.

Levels for contaminants and toxicants are assessed and set by FSANZ.

3.10.6.4 How do these limits relate to the Codex Alimentarius?

As is the case with other food law aspects, Codex requirements are not binding in Australia or New Zealand unless incorporated into the Food Standards Code. However, both countries seek to maintain international consistency, and look to standards such as Codex when setting maximum limits. The principles used in Australia and New Zealand when setting limits are in line with those established by Codex and are set at such a standard necessary to protect the consumer, but slightly higher than what is typically found in the product in order to avoid disruption to food production and trade.[139]

139. FAO/WHO, CAC, '*CODEX STAN 193–1995 Codex Standard for Contaminants and Toxins in Food and Feed*' 1993–95, last revised in 2015, Annex 1.

3.10.7 Process requirements

3.10.7.1 Business processes to ensure quality and hygiene

As outlined earlier in this chapter, Chapter 3 of the Food Standards Code establishes Food Safety Programs to control how Australian businesses manage the hazards associated with food handling and manufacture. Each relevant Food Act contains further specific requirements as to which businesses must maintain a Food Safety Program and the content of these programs.

Meat and seafood industries are highly regulated, as well as businesses that handle or manufacture food for service to vulnerable persons, which include hospital patients, the elderly, and children. The programs themselves reflect *HACCP* concepts. The program is to be implemented and reviewed by the food business and is subject to periodic audit. The programs themselves must[140]

(a) Systematically identify reasonably expected potential hazards
(b) Identify where each hazard can be controlled and the means of control
(c) Provide for systematic monitoring of these controls
(d) Provide appropriate corrective action for when a hazard is found not to be under control
(e) Provide for regular review of the program by the food business to ensure its adequacy
(f) Provide for appropriate records to be maintained by the business demonstrating action taken under, or compliance with the food safety program

In New Zealand, the Food Act 2014 requires that food businesses either register a food control plan or partake in a national program, depending on the level of food safety risk presented by the business. Businesses such as those that manufacture meat, poultry, or fish must create and register a food control plan addressing issues such as cleaning, waste management, hygiene, and pest control. Lower risk businesses such as horticultural packagers, bread bakeries, and brewers do not need to register a food control program but must comply with the applicable national program for that business.

In both Australia and New Zealand, failure to comply with any of these safety process requirements is a criminal offense under the applicable Food Act.

3.10.7.2 Traceability requirements

In Australia, Standard 3.2.2 of the Food Standards Code creates a "one step back and one step forward" traceability policy. The standard requires a food business to record the name and address of its suppliers. Furthermore, the standard mandates that a business must have an effective recall policy for products it disseminates, which includes holding records of production, batches, distribution, and other relevant information.

Similarly, the New Zealand Food Act 2014 requires any party trading in food to have procedures in place to trace food from the supplier to the food business, while it is under the control of the food business, and from the food business to the next recipient in the supply chain. Different exemptions apply depending on whether the business operates under a food control plan or a national program.

Labeling standards, listed in Standard 1.2.1, require food supplied for retail sale to include a lot number on its packaging, so that defective batches can be identified, tracked, and if necessary, recalled.

3.10.7.3 Recall obligations

In Australia, Standard 3.2.2 of the Food Standards Code requires wholesale suppliers, manufacturers, and importers to have a recall system in place for unsafe food. Similar requirements apply in New Zealand under the Food Act 2014 for any business that trades in food. In addition to these requirements, regulators in both countries have emergency powers to require a food business to undertake a mandatory recall of food that presents a safety risk.

Upon recall, a business must notify its local food enforcement agency. Information on the reasons and details of the recall must then be provided to FSANZ, and the public notified. In Australia, the Australian Consumer Law creates a mandatory reporting requirement for any incident where a product causes death, serious injury, or illness.

140. *Australia New Zealand Food Standards Code*, Standard 3.2.one to five.

3.10.8 Labeling

3.10.8.1 Mandatory labeling particulars

Standard 1.2.1 of the Food Standards Code sets out the information that must be included on a food label or in conjunction with a sale of food.[141] The labeling and information requirements vary depending on whether a product is required to bear a label, as well as the method of sale (for example, by retail or in foodservice).

Information requirement	Retail sales with label	Retail sales with no label	Sales to caterers	Other sales
How must the information be displayed?	On label	Accompanied or displayed with the food	On label	On label
Name or description of the food	✓	✓ (if requested by customer)	✓	✓
Lot identification	✓		✓	✓
Name and address of an Australian or New Zealand supplier	✓		✓	✓
Advisory statements, warning statements and declarations	✓	✓	✓	✓ (if requested by customer)
Statement of ingredients	✓	✓ (if requested by customer)	✓ (can be in documentation provided with sale)	✓ (if requested by customer)
Date marking information	✓		✓	✓ (if requested by customer)
Directions for use and storage	✓	✓	✓	✓ (if requested by customer)
Nutrition information (including information substantiating any nutrition content or health claims)	✓	✓ (if requested by customer)	✓ (can be in documentation provided with sale)	✓ (if requested by customer)
Information relating to any genetic modification or irradiation	✓	✓	✓	✓ (if requested by customer)

Food for retail sale is not required to bear a label, and is subject to fewer information requirements, in the following circumstances:

- If made and packaged on the premises from which it is sold;
- If packaged in the presence of the purchaser;
- If whole or cut fresh fruit and vegetables in transparent packaging;
- If delivered packaged, ready for consumption, at the express order of the purchaser;
- If sold at a fund-raising event; or
- If displayed in an assisted service display cabinet.

Reduced labeling and information requirements also apply to products sold in small packages (less than 100 cm^2) and to inner packages of products.

All retail packaged food products sold in Australia must display a country of origin statement in line with the Country of Origin (Food Labeling) Information Standard 2016. This statement must indicate where the product was grown, made, or produced, and in some cases where it was packaged. Products that are designated as "priority foods" must also include a bar chart specifying the percentage of Australian content in the product, and must display a prescribed kangaroo logo if grown, made, or produced in Australia.

3.10.8.2 Food additive labeling

Standard 1.2.4 of the Food Standards Code sets out how additives must be labeled in an ingredients list. First, the technological function performed by the additive must be declared (taken from a list of additive functions in Schedule 7), followed in brackets by either the additive name or additive INS number. For example:

"Sweeteners (Steviol glycosides, Xylitol)" or "Sweeteners (960, 967)"

141. *Australia New Zealand Food Standards Code*, Standard 1.2.1.

3.10.8.3 Allergen labeling

Standard 1.2.3 sets out labeling requirements regarding the presence of allergenic substances. The Food Standards Code differentiates between declarations, advisory statements, and warnings. The presence of any of the following substances must be declared on pack:

Added sulfites at concentrations of at least 10 mg/kg	Milk
	Peanuts
Cereals containing gluten	Soybeans
Crustacea	Sesame seeds
Egg	Tree nuts
Fish	Lupin

Mandatory advisory statements are contained in Schedule 9 and must be displayed on products containing certain substances, such as bee pollen, which may not be suitable for consumption by all consumers. A prescribed mandatory warning statement is required if a food is or includes as an ingredient royal jelly.

3.10.8.4 Nutrition information labeling

Standard 1.2.8 of the Food Standards Code requires products that must include nutrition information to display a nutrition information panel in the format prescribed in Schedule 12. For the majority of packaged products, this format is as follows:

Nutrition information
Servings per package: (Insert number of servings)
Serving size: g (or mL or other units as appropriate)

	Quantity per Serving	Quantity per 100 g (or 100 mL)
Energy	kJ (cal)	kJ (cal)
Protein, total	g	g
—*	g	g
Fat, total	g	g
—Saturated	g	g
—**	g	g
—Trans	g	g
—**	g	g
—Polyunsaturated	g	g
—**	g	g
—Monounsaturated	g	g
—**	g	g
Cholesterol	Mg	mg
Carbohydrate	g	g
—Sugars	g	g
—**	g	g
—**	g	g
—**	g	g
Dietary fiber, total	g	g
—*	g	g
Sodium	mg (mmol)	mg (mmol)
(insert any other nutrient or biologically active substance to be declared)	g, mg, µg (or other units as appropriate)	g, mg, µg (or other units as appropriate)

* indicates a subgroup nutrient.
** indicates a sub-subgroup nutrient.
The word "total" following "protein" or "dietary fiber" in the first column of the panel need only be included if it is followed immediately by a subgroup.

The exact nutrients that must be listed will vary depending on the categorization of the product (for example, energy drinks have additional labeling requirements), as well as the content of any nutrition content claims.

3.10.8.5 Nutrition content claims and health claims

Standard 1.2.7 of the Food Standards Code sets out the requirements for making nutrition content claims and health claims. A nutrition content claim is a claim about the presence or absence of a bioactive substance, while a health claim is a claim that states or implies a health effect. A health effect includes an effect on biological, physiological, or functional processes; growth and development; physical or mental performance; and a disease, disorder, or condition. A health claim that mentions a serious disease or a biomarker of a serious disease is regulated as a "high-level" health claim, while one that does not is a "general-level" health claim. Most products may only make a health claim if they meet certain nutritional criteria (the Nutrient Profiling Scoring Criterion).

Schedule 4 of the Food Standards Code sets out a number of preapproved nutrition content claims and health claims alongside criteria for use. Products that meet the relevant criteria may make these claims without the need to seek regulatory preapproval or hold supporting scientific evidence. Nutrition content claims may be made about a substance that is not listed in Schedule 4 provided that the claim only states the presence, absence, or quantity of the substance in the product. General-level health claims not in Schedule 4 may be made upon notification to FSANZ, but require the party making the claim to undertake a systematic review and hold scientific evidence substantiating the claim. The Food Standards Code prohibits the use of high-level health claims that are not listed in Schedule 4.

3.10.8.6 Therapeutic and medical claims

The Food Standards Code distinguishes between health claims and therapeutic claims. Specifically, Section 1.2.7-8 prohibits claims that refer to the prevention, diagnosis, cure or alleviation of a disease, disorder, or condition or compare a food with a good that is represented in any way to be for therapeutic use or likely to be taken for therapeutic use. Products that are represented as being for therapeutic use will be regulated as noncompliant therapeutic products rather than as food.

It is therefore important to ensure that any health claims do not cross the line and become therapeutic claims.

3.10.8.7 How do the labeling requirements apply outside of the label?

Section 1.2.1-23 of the Food Standards Code provides that any prohibition relating to labels also applies to an advertisement for that product. In other words, any marketing materials for that product must also not include any statement, information, design, or representation that is prohibited by the Food Standards Code.

Any representation made either on or outside of the label must also not contravene any consumer protection laws, including prohibitions on misleading or deceptive representations.

3.10.9 Human right to food/food security

3.10.9.1 Is food recognized as a human right?

Australia and New Zealand have each ratified the *International Covenant on Economic, Social and Cultural Rights 1966* (*ICESCR*). The ICESCR contains a wide range of rights, including the right of everyone to an adequate standard of living, including adequate food. The ICESCR requires signatories to take steps to progressively achieve these rights, although does not form part of either country's domestic law.[142] A direct right to food is therefore not legally enforceable in either country.[143]

3.10.9.2 Is right to food related to food law?

Currently, a right to food has not been integrated into Australian or New Zealand food law, which primarily focuses on food safety and consumer rights. Given the social security programs of both countries, which seek to provide disadvantaged persons with funds to acquire substance, specific legislation regarding the provision of food does not currently exist.

142. United Nations General Assembly *resolution 2200* (XXI), *"International Covenant on Economic, Social and Cultural Rights,"* Article 11.
143. Australian Human Rights Commission, *"Human Rights Explain: Fact sheet 5: The International Bill of Rights,"* < https://www.humanrights.gov.au/human-rights-explained-fact-sheet-5the-international-bill-rights >.

Chapter 3.11

People's Republic of China

Juanjuan Sun

Food Law, Nantes University of France, Center for Coordination and Innovation of Food Safety Governance, Renmin University, Beijing, China

3.11.1 Concepts, principles, and background

Chinese Food Safety Law[144] recognizes food safety from the scientific perspective that food should be nontoxic and innocuous, satisfies necessary nutritional requirements, and is free of any acute, subacute, or chronic hazards to human health.[145] Ensuring food safety means applying a science based and risk-preventing approach to deal with these hazards and thus protect human health.

However, while emphasizing that there are no such things as "absolute food safety" or "zero risk," it is equally important to point out that risk management for food safety relies not only on scientific assessment for characterizing risk and hazards, but also on risk perception of the public and different stakeholders. For the latter, communication about risk is introduced to provide an interactive exchange of information and opinions concerning risk, risk-related factors, and risk perceptions, among risk assessors, risk managers, consumers, industry, the academic community, and other interested parties. This includes the explanation of risk assessment findings and the basis of risk management decisions.[146] For this reason, a well-structured risk analysis system has been introduced to integrate these three functions, namely risk management, risk assessment, and risk communication. In this way, both through legal principles[147] and institutional arrangements,[148] risk analysis has been integrated into Chinese Food Safety Law.

In addition to risk related legal principles and institutions, the advancements introduced by the revision of the Food Safety Law in 2015 provide for the legal principle of "social governance." As a result, the emphasis put on social governance has become an important feature of China's response to food safety challenges. Briefly, governance is a continuous process of interaction and management involving public and private institutions. As a result, public institutions and private institutions can also become important power centers at different levels thereby coordinating collective actions and sharing interests and responsibilities.

For this reason, emphasis on social cogovernance of food safety is aimed at dealing with the challenge of "everyone's business is no one's business."

By putting social governance as a legal principle, the revised Food Safety Law establishes a system of responsibility including duties and liabilities. For example, in the case of food operators, they assume the primary reasonability for food safety with administrative, civil, and criminal penalties in case of noncompliance. In addition, institutional arrangements such as consumers' complaints and punitive damages, whistle-blowing reports, and media supervision also provide channels and benefits for all society subjects to participate in the governance of food safety.[149]

The harmonization of Chinese food safety governance, including such principles as risk analysis, has occurred in the context of the very quick development of a digital economy and so-called "internet plus" in both food business and food regulation. For the former, the domestic e-commerce in the form of Business to Consumer (B2C) has become increasingly

---

144. Food Safety Law of the People's Republic of China, as revised and adopted at the 14th session of the Standing Committee of the 12th National People's Congress of the People's Republic of China on April 24, 2015, and has been into force on October 1, 2015.

145. The Chinese Food Safety Law, Article 150.

146. Codex Alimentarius Commission, Procedural manual, 21st edition, 2013, p. 114.

147. The Chinese Food Safety Law, Article 4. Accordingly, priority shall be given to prevention, risk management.

148. The Chinese Food Safety Law, Article 23. Accordingly, the food and drug administrative department and other relevant departments of a people's government at or above the county level and the food safety risk assessment expert committee and its technical institution shall, under the principles of "scientific, objective, timely, and open," organize food producers and distributors, food inspection institutions, certification bodies, food industry associations, consumers' associations, and news media to exchange food safety risk assessment information and food safety regulatory information.

149. More introduction of food safety governance in China, see Lepeintre Jerome and Sun Juanjuan (eds.), Building Food Safety Governance in China, Luxembourg Publications Office of the European Union, 2018, open access at: < https://eeas.europa.eu/sites/eeas/files/building_food_safety_governance_in_china_0.pdf >.

popular with the wide use of mobile devices crossing all economic sectors in China. As a type of e-commerce, online food retailing is also a booming market, including, for example, online meal orders.[150]

To ensure the safety of online food, the competent authority for food safety regulation has introduced online food safety regulations. Meanwhile, the "internet plus" regulation not only allows the competent authority to update regulatory tools like a centered monitoring system, but also techniques like "big data" to improve the efficiency of official control. More importantly, internet regulation is characterized by "coregulation,"[151] taking into account state regulation and self-regulation involving the cooperation between the competent authority and online platforms as the role of information intermediaries.

In view of the above, both the legislative framework and regulatory system have been updated to guarantee food safety in a science-based and risk prevention way by involving different stakeholders.

3.11.2 Food safety legislative framework

The role of the basic food law is mainly to ensure the coherence of the whole framework. In China, such a role is played by the Chinese Food Safety Law, which was established in 2009 and revised in 2015.[152] As indicated by its name, this basic law focuses on food safety in order to prioritize public health. Introduced in 2009, this basic food safety law has put food safety regulation on a scientific basis with the establishment of China National Center for Food Safety Risk Assessment (CNCFSRA) as the public organization for carrying out risk assessment. However, due to the separation of official control between agro-food and other food products and the coexistence of the Law on Quality and Safety of Agricultural Product and the Chinese Food Safety Law, the risk assessment for agro-food is not under the CNCFSRA but rather under the framework of agriculture department. In order to keep in line with the Food Safety Law, the Law on Quality and Safety of Agricultural Product which established in 2006 is also under the revision.[153]

The Chinese Food Safety Law, quickly update within 5 years, is historically the strictest to deal with endless food safety issues. In terms of being "strict," compliance is supervised in the most rigorous ways with harsh punishments and stringent accountability and regulatory standards are most precise.

As a result, the current legal requirements provided by the revised Food Safety Law highlight the following critical points:

- Food business operators should assume the primary responsibility for food safety. To this end, they should establish internal rules to reinforce the process-based management. Notably, food distributors like retailors can defend against noncompliance caused by others' fault by providing evidence to show their fulfillment of obligations like verification of suppliers' qualification and product's certification and traceability.[154]
- Official control of special products has been reinforced, including health food, infant formula, and formula food for special medical purpose. The regulatory focus of health food lies in the lists of approved material and health claims that can be used in health food. For infant Formula, both formula registration and quality management during the production are of high importance. When it comes to formula food for special medical purpose, the regulatory focus includes registration and advertisement.
- "Risk governance" has become a key idea to strengthen official controls. In addition to the above-mentioned legal principles of risk prevention and management and risk assessment, the risk-based approach also includes the risk monitoring system under which the negative results are reported to competent authorities who then carry out further investigation. Risk-ranking regulation includes the targets, frequencies, and methods of official control in line with the high or low risks involved in the food categories or business behaviors.
- The regulatory system at the central level has been streamlined by establishing the China Food and Drug Administration (China FDA).[155] Meanwhile, the local governments take overall reasonability for food safety within their

150. Sun Juanjuan and Buijs Jasmin, Online food regulation in China: the role of online platforms as a critical issue, European Food and Feed Law Review, 6|2018, p. 503–513.

151. Christopher T. Marsden, Internet coregulation, European law, regulatory governance and legitimacy in cyberspace, Cambridge University Press, 2011, p. 9.

152. For more information on historical evolution of food legislations in China, see, Sun Juanjuan, Evolution and recent update of food safety governance in China, in, Luigi Costato, Ferdinando Albisinni (ed.), European law and global food law, second edition, Wolters Kluwer, 2016, pp 87–106.

153. For more information on this law and regulation of agro-food, see Review of the "Law of the People's Republic of China on Quality and Safety of Agricultural Products," Journal of Resources and Ecology (2018), 9(1), pp. 106–113.

154. The Chinese Food Safety Law, Article 136.

155. The China Food and Drug Administration was reorganized during the 2018 institutional reform of the State Council, and incorporated into the newly emerged State Administration for Market Regulation (SAMR).

jurisdiction. Local governments at provincial level also introduce local legislation or rules to clarify the legal requirements. An interesting example involves establishing local rules on small food business regulation, since the Food Safety Law provides that the specific measures for the administration of food production or processing workshops and food vendors shall be developed by a province, autonomous region, or municipality directly under the Central Government. In this case, simplification of market access approval is also introduced at local level for a small business or a new business model by taking advantage of "internet plus." For example, only registration of business information rather than approval for market access is necessary for small operators.

The Food Safety Law is supposed to consist of the regulations enacted by the State Council and the rules formulated by the competent authorities in a pyramid structure. For example, to make this Food Safety Law enforceable, the State Council has issued the Regulation on the Implementation of the Food Safety Law of the People's Republic of China.[156] Further, the involved competent authorities also lay down corresponding rules in line with their own jurisdictions.

For example, detailed department rules and relevant normative documents formulated by China FDA are as follows:

- Rules for market access: for example, Measures for the Administration of Food Production Licensing in 2015, the purpose is to regulate the licensing for the production of food and food additives.
- Rules for food recall: Measures for Food Recalls in 2015, which applies to cessation of food production and business operation, recall, and disposal of unsafe foods as well as the supervision and administration.
- Rules for agro-food: Measures for Safety Regulation of Agro-food Distribution at Market in 2016, which provides the inspection and supervision carried out by China FDA on the distribution of agro-food at the markets like wholesale and retail.
- Rules for whistle-blowing, Measures for the Administration of Food and Drug Complaints and Reports in 2016, which realizes the social governance of food safety by encouraging the public participation by means of whistle-blowing.
- Rules for special food: Measures for Administration of Register as to Formula of Special Foods, including the health food, formula food for special medical purpose, and infant formula milk product, respectively.
- Rules for supervision and inspection: for example, Measures for the Investigation and Punishment of Illegal Behavior regarding Online Food Safety in 2016, which typically indicates that online food should be as safe as offline food, and new regulatory instruments should be provided to deal with challenges caused by online business, such as the emphasis put on the legal obligations of online platform.

Additionally, it is important to mention the role of technical rules in law enforcement. As provided by the Food Safety Law, food safety standards involve mandatory execution. No other mandatory food standards other than food safety standards may be developed. Mandatory standards include national and local food safety standards. Food production enterprises are encouraged to develop standards more stringent than the national or local food safety standards for their own application. In China, it is interesting to note that labeling is part of food safety standards, while there is a mandatory standard, GB 7718, on labeling requirements for prepackaged food. However, whether missing information or mistakes on label constitutes a food safety issue and thus entitles a consumer to claim 10 times compensation has become a subject of disputes which is supposed to be clarified in judicial judgments on a case by case basis.

3.11.3 Food safety regulatory system

For a long time, the sector-based regulatory system in the food domain was criticized as a regulatory failure that partly contributed to ongoing food safety issues. To solve problems in food safety regulation, integration of functions or departments and the establishment of a big ministry has become the means of reorganizing the regulatory system for food safety. As mentioned before, one of the purposes in the revision of the Food Safety Law in 2015 was to confirm the establishment of China FDA in 2013. As a ministry-level competent authority for food safety, China FDA had integrated the functions of official control of food safety at the stage of production under the General Administration of Quality Supervision, Inspection and Quarantine (AQSIQ), official control of food safety at the stage of circulation under the State Administration for Industry and Commerce (SAIC), and official control of food safety at the stage of catering under the State Food and Drug Administration (SFDA). The latest reform in early 2018 was to integrate the above-mentioned departments. As a result, SAIC, AQSIQ, CFDA, and the price supervision and antimonopoly law enforcement function of the National Development and Reform Commission, the antimonopoly law enforcement function of the Ministry of

156. Regulation on the Implementation of the Food Safety Law of the People's Republic of China, adopted by State Council, and enter into force on July 20, 2009. Notably, this Regulation is also under revision.

Commerce, and the Anti-Monopoly Committee of the State Council have been integrated into the newly established State Administration for Market Regulation (SAMR), which operates directly under the State Council. The functions under AQSIQ for official control of imported and exported food were not transferred into SAMR but taken over by the General Administration of Customs.

In parallel to the central reform, at the same time, the local governments also reformed organizational arrangements for food safety. The disparities between the so-called single agency as FDA and comprehensive agency for market regulation have given rise to an argument over which model is most appropriate for food safety regulation. Briefly, as a copy of China FDA, local governments have set up a single agency integrating the food safety regulatory *functions* of local food safety office and regulators responsible for food and drug, industry and commerce, and quality inspection. When it comes to a comprehensive agency for market regulation, the local *departments* responsible for food and drug, industry and commerce, quality inspection, and alike were integrated to create a single department/committee for market regulation. Notably, such model has further characterized by "three in one," "four in one," and even "five in one" or "six in one" given the number of departments involved in integration. Nowadays, the establishment of SAMR at the central level may end the dispute since most of the local reforms will adopt the central model to ensure vertical consistence.

3.11.4 Conclusion

With nearly 10 years' effort to reinforce food safety regulation and governance, the status of food safety inside China has been improved. For example, according to the report based on national sampling and testing presented by the China Food and Drug Administration in early 2018, a total of 233,300 batches of samples were sampled across the country in 2017. As a result, the overall average qualified rate was 97.6%, an increase of 0.8 percentage points from 2016 and 2015.[157] Despite such progress and continuing efforts in this aspect, challenges faced by China are twofold.

Firstly, with the new organizational arrangement, it still relies on the internal arrangement, in particular the "chemical" rather than "physical" integration of personnel from different departments to carry out food safety regulation. However, market regulation aims to reduce the regulatory burdens of regulated business operators. That is to say, SAMR needs to figure out how to balance economic regulation and social regulation as to food safety and prioritize the resources of comprehensive law enforcement of food safety regulations. Also, it is questionable whether one big ministry is enough to solve food safety issues, including the continuing separation between official controls of agro food and imported and exported food. Therefore, interdepartment cooperation must ensure food safety regulation by applying the principle of "whole" food supply chain from farm/port to fork. Certainly, as mentioned above, the vertical consistence is another challenge for SAMR with the reforms carried out at local levels.

Secondly, food regulation in China is not only about food safety. For example, food security is always a priority for China given such large population. As indicated in Global Food Security Index in 2017, China is under the pressure to ensure food security given factors such as the strain due to the rapid urbanization and thus demand for more food production, changes in food consumption patterns, the susceptibility to flooding, and where water use is unsustainable. Besides, China also faces the problems of food fraud such as media storm over "fake" labeling of trout as salmon leading to the revelation of serious fraud on fishery products.[158] China has become the second most overweight nation in the world with one of the reasons being increased consumption of more energy-dense, nutrient poor foods with high levels of sugar and saturated fats. As a result, food regulation and governance have also targeted these health issues with official control over the online fraud and voluntary guidance on salt reduction provided by the China Nutrition Society and Nutrition and Health Office under the Chinese Center for Disease Control and Prevention. In view of this, it is better to update the food regulation by taking into account multiple issues including food safety, food quality, and food fraud in a systematic way, for example, through more cooperation among the departments.

Further reading

For a systematic analysis of aspects of Chinese food law, see:

Buijs, J., van der Meulen, B., Jiao, L., n.d.a. China's food safety law. Legal systematic analysis of the 2015 food safety law of the People's Republic of China. European Institute for Food Law Working Paper 2018/01. www.food-law.nl/Working-papers/. (Open access).

157. News, Food safety situation is steady and good according to the national food sampling and testing in 2017, January 24, 2018, available at: http://health.people.com.cn/n1/2018/0124/c14739-29783994.html (last accessed on June 9, 2018).

158. Louis Harkell, China "fake" salmon storm hits country's largest trout farmer, May 25, 2018, available at:< https://www.undercurrentnews.com/2018/05/25/china-fake-salmon-storm-hits-countrys-largest-trout-farmer/> (last accessed on June 9, 2018).

Buijs, J., van der Meulen, B., Jiao, L., n.d.b. Pre-market authorization of food ingredients and products in Chinese food law. Legal systematic analysis of the pre-market authorization requirements of food ingredients and products in the People's Republic of China. European Institute for Food Law Working Paper 2018/06. www.food-law.nl/Working-papers/. (Open access).

Buijs, J., Sun, J., van der Meulen, B., n.d.c. Process requirements in Chinese food law. Legal systematic analysis of process-related requirements for food production and distribution in the People's Republic of China. European Institute for Food Law Working Paper 2018/05. www.food-law.nl/Working-papers/. (Open access).

Buijs, J., van der Meulen, B., Jiao, L., n.d.d. Food information in Chinese food law. Legal systematic analysis of labeling and advertisement requirements of food products in the People's Republic of China This paper discusses the general rules on labeling and advertisement of food products in the People's Republic of China. European Institute for Food Law Working Paper 2018/07. www.food-law.nl/Working-papers/. (Open access).

Jerome, L., Sun, J. (Eds.), 2018. Building Food Safety Governance in China. Luxembourg Publications Office of the European Union (Open access). https://eeas.europa.eu/sites/eeas/files/building_food_safety_governance_in_china_0.pdf.

Chapter 3.12

Republic of Korea

Mungi Sohn

Food Science and Biotechnology, College of Life Sciences, Kyung Hee University, Republic of Korea

3.12.1 Introduction

3.12.1.1 Overall jurisdiction for food safety regulatory system

A single consolidated food safety authority, the Ministry of Food and Drug Safety (MFDS), established in 2013, is responsible for all aspects of food safety of both domestic and imported foods sold in Korea, including both domestic and imported agricultural products, fishery products, meat/poultry products, alcoholic beverages, and processed foods. The Ministry has the power and authority to enact and revise the relevant laws and regulations, develop and implement policies, establish national food standards, and enforce relevant regulations to food business operators. The MFDS is also responsible for all aspects of safety regarding pharmaceutical products, cosmetics, drugs, and medical devices, a combined model of European and North American FDA models. The responsibilities for the primary production of agricultural, livestock, and fishery products at farms, slaughterhouses, and fisheries levels are delegated to the Ministry of Agricultural, Food, and Rural Affairs (MAFRA) and the Ministry of Oceans and Fisheries (MOF).

The two most important food legislations are the "Food Sanitation Act" and the "Sanitary Processing of Livestock Products Act," both enacted in 1962. The Food Sanitation Act defines "food" as "all types of foods ingested excluding those consumed for medicinal purposes." Those food products that are not covered by other individual legislation such as Sanitary Processing of Livestock Products Act or Health Functional Foods Act (Special Act for health claims and dietary supplements) are covered by Food Sanitation Act.

The Ministry of Food and Drug Safety constantly reviews and revises its 10 laws and subordinate statutes directly related to food safety: the Food Sanitation Act, the Sanitary Processing of Livestock Products Act, the Food Safety Framework Act, the Special Act on Safety Control of Children's Diets, the Agricultural and Fishery Products Quality Control Act (cogoverned by Ministry of Agriculture, Food, and Rural Affairs), the Act on Testing and Examination of Food and Drug Products, the Special Act on Imported Food Safety Control, the Act on Food Labeling and Advertisement, and the Pharmaceutical Affairs Act.

3.12.1.2 Food Safety Framework Act: basic concepts, roles, and obligations of government

The Food Safety Framework Act, enacted in 2008, streamlined the concept of food safety, outlined the rights and obligations of the consumers, the duties, and obligations of the industry and the government, and provided the framework for better harmonization of interministerial cooperation among government authorities. The establishment of the National Food Safety Council under the Prime Ministers' Office and the development and implementation of National Food Safety Plans and Strategies for more coordinated and consistent multigovernmental food safety activities were initiated by the Act. The Food Safety Framework Act mandates risk assessment process based on sound scientific evidences and encourages risk communication activities for safer foods and healthier dietary life of the consumers.

The National Food Safety Policy Council oversees national food safety control schemes and plans, reviews, and evaluates the outcome or progress of the scheduled plans assigned to relevant ministries and agencies. The Food Safety Framework Act designated more than 33 food safety−related laws and subordinate statutes managed and implemented by 9 different Ministries and/or 17 provincial and 235 local governments depending on the delegation of responsibility stated in individual laws and its subordinate statutes (enforcement Decrees, enforcement Ordinances, administrative rules and regulations, and local municipal Ordinances, etc.). The consolidation of food safety control responsibility into MFDS enhanced more coordinated approaches and national food safety control capabilities for prompt and efficient implementation of the food safety policy. However, there are still many areas of better cooperation required among ministries for prompt and comprehensive actions.

The MFDS works closely with the Korea Center for Disease Control (KCDC) when a foodborne diseases incident occurs by collecting, testing, and analyzing the food consumed and sold, while KCDC focuses on overall epidemiological surveillance along with the local health officials. When foodborne disease incidents occur in schools, the Ministry of Education (MOE), MFDS, KCDC, and local health officials closely work together as set out in the National Foodborne Disease Surveillance and Prevention Scheme, developed and practiced in accordance with the relevant rules and regulations outlined in relevant laws such as School Meals Act (by MOE), Food Sanitation Act (by MFDS), and Infectious Disease Control and Prevention Act (by KCDC). The food safety−related laws and relevant authorities outlined in the Food Safety Framework Act are listed in Table 3.12.1.

3.12.2 Competent authorities

3.12.2.1 Ministry of Food and Drug Safety

The MFDS has the primary responsibility for regulating all the food safety laws and regulations and coordinates with other agencies to ensure safe food for all domestic and imported food sold in Korea. The MFDS is responsible for enacting and revising major food safety laws, developing national food safety policies and measures, establishing food standards for foods, premarket approvals for food additives and Genetically Modified Organisms for food uses, establishment of maximum residue levels for contaminants, pesticides, and veterinary drugs in foods, establishment of guidelines and mandates for compliance, and facility requirements and rules for HACCP programs.

The MFDS also serves as a food safety control tower by assisting the National Food Safety Council and provides leadership in dealing with multigovernmental tasks such as the National Foodborne Illness Control and Prevention Scheme and the National Antibiotic Resistance Prevention and Control Programs. The MFDS also shares the responsibility of inspecting food facilities, collecting and testing samples, detention, seizure, and/or suspension of the food facility registration and recalls adulterated or misbranded products along with provincial and local governments to ensure nationwide surveillance and monitoring. There are six Regional Korea Food and Drug Administration (KFDA) offices and National Institute of Food and Drug Evaluation (NIFDS) under its direct jurisdiction of MFDS.

3.12.2.2 National Institute of Food and Drug Evaluation

The National Institute of Food and Drug Safety Evaluation (NIFDS) is mainly responsible for scientific risk assessment of food, monitoring, establishing analytical methods, toxicological evaluation, review, and approval of medicinal products including drugs, cosmetics, and medical devices. The NIFDS operates and serves identical roles to those of European Food Safety Authority (EFSA) and European Medicines Agency (EMA) combined.

The NIFDS performs extensive monitoring of chemical and biological contaminants and toxins, risk assessment, and reassessments of contaminants and toxins in every 5 years. The NIFDS monitors national surveillance and monitoring data to provide scientific basis for establishing food standards. The NIFDS works independently for risk assessment and risk communication activities but also works closely with the risk managers in MFDS.

TABLE 3.12.1 List of food safety related acts and relevant authorities in Korea.

Relevant ministries	Relevant laws
Ministry of Food and Drug Safety	Food Sanitation Act, Sanitary Processing of Livestock Products Act Health Functional Food Act, Special Act on Safety Control of Children's Diets, Agricultural/Fishery Products Quality Control Act[159] Act on Testing and Examination of Food and Drug Products, Special Act on Imported Food Safety Control, Food Labeling and Advertisement Act, Pharmaceutical Affairs Act (for veterinary drugs)
Ministry of Health and Welfare (KCDC)	Infectious Disease control and Prevention Act National Health Promotion Act
Ministry of Agriculture, Food, and Rural Affairs	Agricultural/Fishery Products Quality Control Act[160] Food Industry promotion Act, Livestock/Fish Feed control Act, Agrochemicals Control Act, Contagious Animal Diseases Prevention Act, Ginseng Industry Act, Salt Industry Act, Grain Management Act Livestock Industry Act, Fertilizer Control Act, Cattle and Beef Traceability Act, Environment-Friendly Agriculture Promotion Act
Ministry of Justice	Act on Special Measures for the Control of Public Health Crimes[161]
Ministry of Trade, Industry, and Energy	Foreign Trade Act, Industrial Standardization Act Transboundary Movement of Living Modified Organisms Act
Ministry of Education	School Health Act, School Meals Act
Ministry of Strategy and Finance	Liquor Tax Act[162]
Ministry of Environment	Drinking Water Control Act, Water Supply Act
Fair Trade Commission	Product Liability Act

3.12.2.3 Regional Korea Food and Drug Administration

Six Regional Korea Food and Drug Administrations (KFDA) are responsible for enforcement of the food regulations by implementing food safety policies and initiatives nationwide. It has the rights and mandates of inspecting food facilities including manufacturers, collection of samples and regulatory authorities of seizures and recalls, cancellation of registration in accordance with the shared responsibilities empowered to Regional KFDAs along with the regional and local governments to ensure proper and prompt management of food safety control. The KFDA is capable and responsible for independent inspection, monitoring, administrative actions for seizure, recall, and administrative penalties for any adulterated and misbranded food products sold in Korea under its regional jurisdiction.

Inspection of all imported foods is also an important role of KFDA. The Regional KFDAs have 23 imported food inspection posts nationwide and routinely inspect, collect, test samples when food is imported to Korea with a close collaboration with the Korea Customs Service (KCS).

3.12.2.4 Korea Customs Service

The KCS oversees the overall importation of all goods into Korea to ensure that all goods entering and exiting Korea meet the standards of the regulations set out in the Republic of Korea. Electronic Customs Clearing Services developed by KCS are linked with the Import Food Inspection Service Program of KFDA for prompt and transparent imports and clearances of the designated goods in service. When those products such as food, meat and poultry, fisheries, processed foods, food additives, and food packages/contacting materials are imported to Korea, KFDA inspects the product after an import declaration to KCS and KCS issues a permit when KFDA's testing or inspection results are complying with the requirements for imported food of Korea.

159. Cogoverned by Ministry of Agriculture, Food and Rural Affairs.
160. Cogoverned by Ministry of Food and Drug Safety.
161. Cogoverned by Ministry of Health and Welfare (Presidential Ordinance).
162. Safety and Labeling Requirements for Liquors and Alcoholic Beverages are delegated to MFDS.
163. Relevant Regulations and Provisions only.

3.12.2.5 Korea Center for Disease Control

The KCDC, established under the Ministry of Health and Welfare of Korea, oversees all the aspects of infectious disease control and prevention from both waterborne or foodborne diseases, and/or other means of contacts with humans in order to ensure safety and prevention of infectious disease. The MFDS supports KCDC's epidemiological survey by collecting and testing food samples consumed and inspection of the related facilities when foodborne disease occurs. Multigovernmental Nationwide Foodborne Disease Surveillance and Prevention Scheme, consisted of more than 30 agencies and organizations, coordinates for prompt response to foodborne illness incident for prevention and control.

3.12.2.6 Regional and Local governments

Regional and Local governments exercise important roles in the food safety control system by performing routine inspections and surveillance of food-related facilities. However, due to limited resources restricted to their jurisdiction, inspection activities for food manufacturing facilities are limited. In order to enhance local and regional food safety inspection activities, comprehensive food safety information network (CFSIN), operated and maintained by the MFDS, is provided to all relevant competent authorities for regulatory actions information. CFSIN provides all the information and data compiled from the national, provincial, and local governments ranging from identity of the products, manufactures, previous records of test results, recalls, detention, seizure, ingredients used, etc.

3.12.2.7 Other agencies and organizations

The Rural Development Agency (RDA) under the Ministry of Agriculture and Rural Affairs (MOAFRA) is responsible for registration of pesticides use permit for pest control. The Animal and Plant Quarantine Agency (APQA) is responsible for the registration and use of veterinary drugs for animal disease control and sanitary measures in farms and slaughterhouse.

The National Institute of Fisheries Science (NIFS) under the Ministry of Oceans and Fisheries (MOF) is responsible for aquaculture research and fish disease control and monitoring of paralytic shellfish toxin in the near sea.

The National Food Safety Information Service (NFSI) collects and analyses food safety—related information collected from all over the world, and provides the relevant information to consumers, governments, and industry for information. Korea Agency of HACCP Accreditation and Services is taking routine inspection of HACCP program compliance in food manufacturing facilities.

3.12.3 Recent harmonization and modernization efforts

The traditional way of controlling food safety was based on the identity and characteristics of the food itself, namely, meat and poultry products, fishery products, processed foods, and dietary supplements and regulated by Food Sanitation Act, Sanitary Processing of Livestock Products Act, and Health Functional Food Act separately. Product specific legislation has its merit for controlling food safety based on the characteristics of the food products itself. However, some duplication and/or different approaches/concepts caused inefficiency and confusion due to different rules applied by different ministries. Harmonization and modernization efforts were made in some of the areas for overall consistency regardless of the type of the products regulated by different food laws. Recent efforts in harmonizing rules and regulations to improve transparency and efficiency of legal actions were made in three separate areas of government food safety control functions: food import inspection, food labeling approaches, and the testing and examination accreditation and certification areas.

3.12.3.1 Product testing and examination requirements harmonized

The Act on Testing and Examination of Food and Drug Products, enacted in 2013, provided guidance and guidelines required by the food and drug testing laboratories for reliable test results and proper maintenance of equipment, methods of analysis, laboratory accreditation, and performance evaluation for microbial, chemical, and radiological testing of various food and drug products. Different concepts, principles, procedures, and penalties applied and practiced by six different food and drug-related laws were harmonized in the area of product testing, where possible Table 3.12.2.

TABLE 3.12.2 Harmonization of testing and examination requirements.

Items tested	Relevant laws	New law	Requirements
Foods, additives, packaging/containers	Food Sanitation Act	Act on Testing and Examination of Food and Drug Products	Designation/Cancellation of official testing laboratory, requirements and procedures to observe, performance testing and training, education, reporting requirements, penalties and sanctions, etc.
Meat, poultry products	Sanitary Processing of Livestock Products Act		
Drugs, quasidrugs, medicinal herbs	Pharmaceutical Affairs Act		
Medical devices	Medical Device Act		
Cosmetics	Cosmetic Act		
Personal hygiene products	Hygienic Products Act		

3.12.3.2 Food import procedures and requirements harmonized

The Special Act on Imported Food Safety Control, enacted in 2015, incorporated relevant rules and regulations applied by various inspection posts managed by four different food related laws Table 3.12.3. The Act harmonized food import procedures by adopting approaches for preimport point measures, measures at the customs clearance, and market surveillance for continuous follow-up for adulterated foods. Preregistration of overseas food establishment information, delegation of inspection power to six regional Korea Food and Drug Administrations (KFDA) and its inspection posts, sample collection rules and procedures, testing requirements, seizure/shipping/detection rules, penalties, and sanctions for violations are clearly harmonized and streamlined where possible. The Special Act provided sufficient legal basis for the establishment of imported food safety bureau in the Ministry of Food and Drug Safety for proper policy making and consistent import food control actions and by inspection powers divided to various local governments into regional KFDA offices and inspection posts.

3.12.3.3 Food labeling requirements harmonized

Another separate approach to harmonize food labeling standards was made by enacting single Food Labeling and Advertisement Act in 2018 in order to ensure consumers' right-to-know and right-to-choose. Subordinate statutes (Ordinances, Presidential Decrees) are currently developed/prepared for official preannouncement of legislation for enforcement in 2019. Relevant provisions related to labeling standards in three different Acts were incorporated to provide and apply uniform food labeling principles, enhance legal basis of mandates of labeling, more transparent administrative approach and rules to what and how to label, and legal basis for the use of electronic form of information such as QR codes to improve the readability of the food label Table 3.12.4. The Act will provide better transparency and uniform policy guidelines and principles to enhance consumers' right-to-choose and better-informed guidance for the industry.

TABLE 3.12.3 Harmonization efforts for food import procedures and rules.

Items	Relevant acts	New Act	Requirements Harmonized
Foods, additives, packaging/containers	Food Sanitation Act	Special Act on Imported Food Safety control	Delegation of import inspection power to six regional Korea Food and Drug Administrations and other inspection posts, measures, and procedures to take on on-site inspections, overseas' inspections, testing and examination rules, sanctions and penalties for violation, etc.
Meat, poultry products	Sanitary Processing of Livestock Products Act Prevention of Contagious Animal Disease Act[163]		
Foods with health claims, supplements	Health Functional Food Act		

TABLE 3.12.4 Harmonization efforts for labeling requirements for foods in Korea.

Labeling Regulations in Food Sanitation Act,	
Health Functional Food Act,	Food Labeling and Advertisement Act
Sanitary Processing of Livestock Products Act	
● Labeling Standard for Food (nutrition labeling included)	Unified Approach for Labeling Standards:
● Labeling Standard for Genetically Modified Food⟹	- General Rules, labeling Requirements, other labeling details, prohibition of Misleading Indication and Advertisement
[MFDS Notification under Food Sanitation Act]	
● Labeling Standard for Health Functional Food (nutrition Labeling included)	- Consumer education and public relations
[MFDS Notification under Health Functional Food Act]	
● Labeling Standard for Livestock products	- Supplementary rules, penalty, sanction
[MFDS Notification under Sanitary processing of Livestock products Act]	

3.12.4 Food safety regulatory approaches

3.12.4.1 Food safety standards based on sound science and risk analysis principles

Food safety standards and limits are based on the food safety principles of the protection of health of the consumers by ensuring the supply of safe foods that are free from hazardous materials, substances, or contaminants. In order to ensure safe food and gain trust of the consumers, comprehensive efforts and activities are continuously implemented nationwide from information collecting/sharing, on-site inspection, and product testing to surveillance and monitoring. Continuous risk assessment activities of hazardous substances by means of hazard identification, hazard characterization, exposure assessment, and risk characterization process are continuously performed during the food safety control activities.

Food standards are established based on scientific evidence and risk analysis principles are widely applied in compliance with the international norms and food safety principles exercised by Codex Alimentarius Commission. The Ministry of Food and Drug Safety is mainly responsible for risk management activities including risk assessment policy setting, while NIFDS is in charge of the risk assessment. Extensive monitoring of hazardous substances in foods is performed every 5 years and the results are used for revising or establishing food standards for further action if necessary.

Great emphasis has been given to Risk Communication Activities by both MFDS and NIFDS. The Bureau of Consumer Risk Prevention of NIFDS oversees all aspects of consumer education and participation activities, public relations, information and data sharing with consumers and industry, providing guidance, and publishing risk assessment reports to enhance public safety and trust. An emergency alert system and emergency response systems are well established at national level. Efforts to expand and link the national rapid alert system with regional and international levels such as INFOSAN are continued.

3.12.4.2 Food code: product-specific standards and general requirements

In general, mandatory food safety standards for each food product are specified in the Korean Food Code classification and general requirements do apply to all foods simultaneously. Common safety rules and sanitary practices are required on any step of manufacturing, processing, preservation, and cooking for foods sold in Korea.

Product-specific standards consist of specific safety standards related to certain specific products based on their characteristics and the fact whether there is a need to regulate routinely to check chemical or microbiological contamination such as maximum levels for certain heavy metals, toxins, foodborne disease microbes, and/or contaminants during processing, manufacturing, storage, sales, and preparation of raw materials and/or final products.

General requirements such as prohibition of using adulterated food ingredients, chemicals, meats, unapproved pesticides/additives/veterinary drugs, unwholesome or contaminated, and/or disease carrying ingredients apply to all foods.

Other requirements for sanitary facilities and personal hygiene, product testing, preservation, and temperature control requirements, where applicable, are specified in the Food Code.

3.12.4.3 Nationwide emergency alert and response and recall systems upgraded

The National Food Safety Information Service (NFSI), MFDS-affiliated organization, collects, analyses, and disseminates food safety information globally and locally all year round and provides information to governments such as NFDS, industry, and to consumers. Domestic information regarding unsanitary or adulterated foods and unlawful practices is collected by national Adulterated and Misbranded Food Emergency Reporting Call System (1399) and forwarded to MFDS for follow-up actions. After investigation of the report based on on-site inspection and sample collection/testing by regional KFDA and/or local governments, test results are announced, and subsequent legal actions are taken based on the violations identified. The National Food Safety Information Service (NFSI) also develops and provides guidelines and messages by with diverse methods using SNS, pictures, and cartoons to enhance the better understanding of the problem/issue to consumers (children, students, women, elderly groups).

In order to expedite the emergency alert and recall system, a joint government–industry affiliated automatic product recall system is currently used nationwide in all of the convenience stores, departments, medium, and large food stores. The government test results analyzed by the laboratories nationwide are immediately linked and sent to the Emergency Recall System for Hazardous Substances and Products operated by the Korean Chamber of Commerce electronically using the product barcode information, and the system blocks the sales of concerned products at the affiliated stores' cashier for immediate sales blocking. Mobile messages are also sent to store managers to promptly remove recalled products from the market shelf. The system is also linked to national broadcasting TVs to send out warning subtitles or captions in case urgent emergency incidents occur.

The Ministry of Food and Drug also established "Comprehensive Food Safety Information Network (CFSIN)" compiling all the data regarding marketed food products in sale (manufacture, trader, name of the product, ingredients, previous record of violation, hazardous substances identified) for prompt and proper inspection, examination, and recall. The Network information is available to all relevant Ministries and provincial/local governments to use.

3.12.4.4 Premarket approval and authorization

Premarket approval for food additive, pesticides, veterinary drugs, GMO, and new ingredients is required. The standards for proper uses of food additives in most food products are in compliance with the international standards such as Codex and procedure for new uses for a technological purpose is well established for approval. When it comes to the uses of pesticides and veterinary drugs, prior authorization by the Pesticide Control Act and Infectious Animal Disease Control Act, Ordinance for Veterinary Drugs Handling should be obtained by the Ministry of Agriculture, Food, and Rural Affairs (MAFRA) and maximum residual limits should be established by the Ministry of Food and Drug Safety (MFDS) for commodities and meat/poultry products, where applicable. Prior approval of Safety Assessment of GMOs for food use is also required for import. Unapproved or illegal use of those additives, pesticides, veterinary drugs, etc., is subject to recall, seizure, detention, and administrative penalties, where applicable. English versions of Food Code, Food Additive Code, and MRLs for pesticides are available at the Ministry of Food and Drug Safety homepage.

3.12.4.5 Labeling requirements: mandatory information

The labeling standards apply to all packaged foods to be offered as such to the consumers and shall not be described or presented on any label in a manner that is false, misleading, or deceptive. In general, information such as the name of the food, type of food (as classified in Food Code), Expiration Date or Best Before Date (where applicable), ingredients, name and address of the manufacturer, raw materials, list of food additives used, net contents, ingredients known to cause hypersensitivity, country of origin, lot number, quantitative ingredient declaration, nutritional information, and serving size (where applicable) should be included in the label of the foods. Depending on the type of foods classified in Food Code, additional information regarding preservation instructions or "do's and don'ts" should also be included in the label. GMO labeling is also required for those products using GM raw materials. English versions of recent Food labeling Standards and Livestock Food Labeling Standards are available at the Ministry of Food and Drug Safety homepage.

3.12.4.6 Health claims

Medicinal claims are generally prohibited. Health claims can be declared for those products approved by the rules set out in Health Functional Food Act with prior review and approval. English version of recent Labeling Standards for Health Functional Food is available at MFDS homepage.

3.12.5 National surveillance and risk assessment activities

3.12.5.1 Recent surveillance and risk assessment activities

Food Safety Limits are established based on Risk Analysis Principles in accordance with the international norms and standards. Continuous surveillance and risk assessment activities of hazardous substances by means of hazard identification, hazard characterization, exposure assessment, and risk characterization process are performed. The limits for heavy metals (lead, cadmium, mercury, arsenic, inorganic arsenic, tin, etc.) were routinely reviewed by the national intake level of numerous foods consumed by Korean people. Recently, extensive monitoring of 136, 127 cases for 8 heavy metals in 403 varieties of agricultural products (136 varieties), fishery products (107 varieties), meat and poultry products (20 varieties), and processed foods (140 varieties) were carried out Table 3.12.5.

Monitoring of 45,383 cases for 8 mycotoxins (Total Aflatoxin, Aflatoxin B1, M1, Ochratoxin A, Deoxynibalenol, Fumonisin, Zearalenone, Patulin) for above mentioned foods was also carried out. Another 61,296 cases of 50 chemical compounds (benzopyrene, ethylcarbamate, benzene, 3-MCPD, etc.) generated during cooking and processing were monitored in 541 varieties of agricultural products (104 varieties), fishery products (82 varieties), meat and poultry products (30 varieties), and processed foods (325 varieties). Based on the outcome of the exposure assessment study and risk assessment of individual contaminants, the policy decision is made whether or not to lower the intake of certain foods by strengthening or establishing the safety standards.

Recent risk assessment reports for 64 contaminants and hazardous substances compiling all those monitoring results of 243,806 food items (83,340 case of agricultural products, 39,958 cases of fishery products, 45,096 cases of meat/poultry products, 75,412 cases of process foods) for the year 2012–2016 are currently available at www.nifds.go.kr > (NIFDS homepage) for full texts in Korean and a summary in English.

3.12.6 Conclusion

Numerous efforts need to be continued to streamline and harmonize the regulations at the national level when multiple ministries are responsible for food safety control system. Duplication and discrepancy among food safety standards, MRLs for pesticide and veterinary drugs, methods of analysis, and labeling methods were identified during the harmonization process of consolidating related food laws and regulations. Harmonization efforts at the National level need to be exercised and it will also help improving more harmonized approach with the regional and international levels.

TABLE 3.12.5 Extensive monitoring for heavy metals in Foods (2012–15).

Name of heavy metals tested	Agricultural product (# of cases)	Fishery product (# of cases)	Meat/poultry products (# of cases)	Processed foods (# of cases)
Lead (Pb)	11,297	6,630	9,397	5,824
Cadmium (Cd)	11,297	6,630	9,397	5,824
Mercury (Hg), Methyl-Hg	11,297	6,630 788 (Methyl-Hg)	9,397	5,824 78 (Methyl-Hg)
Arsenic (As)	11,297	6,630	9,397	5,824
Inorganic As	–	–	–	–
Tin (Sn)	2,436	–	–	233
Total: 136,127	47,624	27,308	37,588	23,607

References

Websites:

Ministry of Food and Drug Safety, Republic of Korea: http://mfds.go.kr/eng/.

Further reading

Act on Testing and Examination of Food and Drug Products, 2018. Ministry of Food and Drug Safety.

Cosmetic Act, 2018. Ministry of Food and Drug Safety.

Food and Drug Statistical Yearbook, 2017. Ministry of Food and Drug Safety.

Food Code, Ministry of Food and Safety Notification No. 2017-57. http://mfds.go.kr/eng/.

Foods Labeling Standards, Ministry of Food and Safety Notification No. 2016-45. http://mfds.go.kr/eng/. Labeling Standard for Health Functional Food, Ministry of Food and Safety Notification No. 2017-16. http://mfds.go.kr/eng/.

Food Sanitation Act, 2018. Ministry of Food and Drug Safety.

Health Functional Food Act, 2018. Ministry of Food and Drug Safety.

Hygienic Products Control Act, 2018. Ministry of Food and Drug Safety.

Livestock Product Labeling Standards, Ministry of Food and Safety Notification No. 2014-197. http://mfds.go.kr/eng/.

Medical Device Act, 2018. Ministry of Food and Drug Safety.

National Food Safety Information Service. https://www.foodinfo.or.kr/en/.

National Institute of Food and Drug Safety Evaluation, Ministry of Food and Drug Safety http://www.nifds.go.kr/en/.

National Law Information Center, Ministry of Government Legislation, Republic of Korea. http://www.law.go.kr/LSW/eng/.

Pharmaceutical Affairs Act, 2018. Ministry of Health and Welfare.

Sanitary Processing of Livestock Products Act, 2018. Ministry of Food and Drug Safety.

Special Act on Imported Food Safety Control, 2018. Ministry of Food and Drug Safety.

White Paper, 2017. Ministry of Food and Drug Safety.

Chapter 3.13

Japan

Mungi Sohn

Food Science and Biotechnology, College of Life Sciences, Kyung Hee University, Republic of Korea

3.13.1 Introduction

3.13.1.1 Overall jurisdiction for food safety regulatory system

The responsibility for ensuring food safety to protect the health of the public is mainly under the jurisdiction of the Ministry of Health, Labour and Welfare (MHLW). The MHLW regulates all the food safety laws and regulations and coordinates with other agencies to ensure safe food for all domestic and imported foods sold in Japan. The food safety work is carried out on the basis of the Food Sanitation Act, the Abattoir Act, Poultry Slaughtering Business Control, and Poultry Inspection Act. The Health Promotion Act and the Act of Temporary Measures for Enhancing the Control Method of the Food Production Process also play a certain role in the regulation of the manufacture, import, and sale of food, food additives, and apparatus and container/packages. The Ministry shares the responsibility of risk management of food safety with Prefectural and Municipal Governments and local health authorities for implementation.

In an effort to ensure food safety after the occurrence of BSE in 2001, the Food Safety Basic Act was enacted in 2003 and introduced the risk analysis approach as a comprehensive effort to ensure public safety. The approach was to scientifically assess risks in foods and develop necessary measures based on risk assessment. The independent Food Safety Commission (FSC) under Cabinet Office was established for risk assessment and risk communication.

The intergovernmental cooperation and collaboration activities are emphasized among the ministries based on the risk analysis principles, namely, the Ministry of Health, Labor, and Welfare, the Consumer Affairs Agency, and the Ministry of Agriculture, Forestry and Fisheries which are responsible for risk management and risk communication activities, while the FSC is responsible for risk assessment and communication activities.

Recent approaches have been undertaken to enhance consumers' perspectives in consumer affairs and food safety. The Consumer Affairs Agency (CAA), established in 2009, under Cabinet Office took the new important role in consumer affairs and food safety since 2013. The Agency coordinates general maintenance of the environment necessary for ensuring food safety in cooperation with the relevant ministries and agencies related to the emergency responses for food safety, plans, and enforces food labeling regulations specified in the Food Labeling Act. The Food Labeling Act, enacted in 2013, integrated food labeling standards regulated by four different laws, namely, the Food Sanitation Act, the Act on Standardization of Agricultural and Forestry Products (JAS Act), the Health Promotion Act, and the Liquor Tax Act. The labeling standards regulated by different ministries are streamlined in the interest of consumers' perspectives while serving as a tool for applying uniform principles for both consumer protection/safety and the smooth production and distribution of foods. The Consumer Affairs Agency took the responsibility of implementation and risk communication of food labeling issues in Japan while leaving the responsibilities for administrative penalties and sanction actions for business operators to the relevant ministries. Information regarding laws and regulations of Japan is available in the Japanese Law Translation Website < http://www.japaneselawtranslation.go.jp > in English.

3.13.2 Competent authorities

3.13.2.1 Ministry of Health, Labour and Welfare

The Food Sanitation Act, enacted in 1947, is the primary legislation for ensuring food safety by regulating standards and specifications of food, food additives, apparatus, containers/packages, and by prohibiting the sales of harmful and adulterated food and meat products, noncompliant food, additives, pesticides, and veterinary drugs. The Food Sanitation Act formulates guidelines for inspection and guidance for domestic and imported foods, establishes standards for preventive measures against contamination with harmful substances, and provides approval, renewal, and revocation of business licenses and facilities.

The MHLW has the primary responsibility for regulating all the food safety laws and regulations and coordinates with other agencies to ensure safe food for all domestic and imported food sold in Japan. The Department of Food Safety, Pharmaceutical and Food Safety Bureau of the Ministry of Health, Labour and Welfare (MHLW) performs risk management decisions, policy planning, and the establishment of standards and specifications of food, food additives, pesticide residues, animal drug residues in food, and food containers/packaging materials. The Department is also responsible for overall food inspection, health risk management measures for food poisoning and poultry and livestock meat safety, application of HACCP principles, GLP, imported food inspection, labeling for specified uses, nutrition labeling, health claims, dietary supplements, and safety assessment of genetically modified foods. The Ministry closely works with Prefectural and Municipal Governments, responsible for business licenses, inspection of food, and meat-related businesses including restaurants and abattoirs, along with the 7 Regional Bureaus of Health and Welfare, and 31 Quarantine Stations responsible for hygiene inspection for imported foods nationwide.

The Ministry introduced a "positive list system" with respect to pesticides, feed additives, and veterinary drugs for animals that may remain in the food supply in 2003. The positive list system for food additives was introduced in 1995 along with the approval system for the production or processing of food through a comprehensive sanitation management and production process. A recent revision to the Food Sanitation Act in 2018 introduced a positive list system by which only materials that have been assessed for safety are allowed to be used for food apparatus, containers, or food packaging materials to harmonize with the international standards with respect to food apparatus and packing materials. It specifies substances that are permitted for use and those substances not specified are prohibited.

The recent revision of the Food Sanitation Act was based on the consideration for the society with an aging population and growing aim of healthy living, changes in the household structures, and globalization of food trade according to Economic Partnership Agreement and economic globalization. Requirements for food sanitation control to be based on HACCP principles and mandatory requirement for the attachment of a health certificate for imported dairy products and seafood were introduced. In addition to the general sanitation control, all food business operators are required to manage sanitation control using seven principles of HACCP as a general rule.

Some of the strengthening measures against interregional food poisoning incidents were also introduced to enhance mutual cooperation between agencies by forming a council for wide area cooperation. A partnership collaboration system among government bodies and local governments involved should cooperate to prevent the occurrence of food poisoning incidents affecting wide region and for emergency measures. New mandates for collection of information on health-related adverse events associated with foods containing designated ingredients were also introduced.

The Ministry of Health, Labour and Welfare also has the responsibility of meat and poultry product inspection and safety from the slaughter house by the Abattoir Act and Poultry Slaughtering Business Control and Poultry Inspection Act. The Japanese food safety regulatory system resembles the European Union model in terms of the approaches taken to separate the risk management (Ministry of Health, Labour and Welfare vs. European Union DG SANTE) and risk assessment (Food Safety Commission vs. European Food Safety Authority (EFSA)) roles.

3.13.2.2 Food Safety Commission

The FSC is mainly responsible for scientific risk assessment of food and risk communication activities. The Food Safety Basic Act provided basic principles to prioritize national health protection and introduced risk analysis concept (risk assessment, risk management, risk communication) to prevent or minimize the impact of food to human body. The Act outlined the basic principles and the responsibilities for the National Government, Local Governments, and Food-related Business Operators clearly and provided basic direction for policy formulation by risk assessment and implementation to promote exchange of information and cooperation among different agencies and consumer education. The Commission conducts risk assessment, recommends administrative institution appropriate measures based on risk assessment, monitors risk management activities, and collects data and organizes information on domestic and overseas risks.

The FSC assesses risks to human health posed by microorganisms, chemical, and others contained in food and provides guidelines for the risk assessment of food additives, flavoring substances in food. The Commission is also responsible for the Safety Assessment of Genetically Modified Foods and provides Standards for the Safety Assessment of Genetically Modified Foods, Standards for the Safety Assessments of Food Additives Produced Using Genetically Modified Microorganisms, and so forth. Recent risk assessment reports for food additives, pesticides, veterinary medicinal products, chemicals and contaminants, apparatus and containers/packages, prions, natural toxins/mycotoxins, microorganisms and viruses, novel foods, GM foods, feeds and fertilizers, antimicrobial resistant bacteria, and other information are available in FSC Website.

3.13.2.3 Consumer Affairs Agency

The Consumer Affairs Agency, established in 2009, took the new responsibility of risk management activities associated with food labeling in the interest of consumers by implementing and integrating labeling regulations set out in Food Sanitation Law, Japanese Agricultural System (JAS) Law, Health Promotion Law, and Law for Keeping Transaction Record and Relaying Place of Origin Information of Rice and Rice Products. The Agency develops basic plans for consumer policies, collects information concerning consumer accidents form consumers, companies, and governments, analyses consumer accidents, identifies the causes, and promotes the awareness of consumers in an integrated and consistent manner. The Prime Minister should establish the standards for food labeling intended for sale in advance consultation with the relevant ministries related.

The Agency also emphasizes consumer safety and education including food safety and unified approach in food labeling standards that are understandable to consumers and legal systems for companies to implement proper labeling and prevent fraudulent practices and false labeling. The Consumer Affairs Agency provides leadership and acts as a control tower of consumer-related administrative services by gathering, investigating, analyzing, and communicating information and warnings, government-wide efforts for emergency cases, administration and enforcement of laws relevant to consumers, and requests and recommends measures, recommendations, and guidance to relevant ministries and agencies for proper regulatory actions to business operators.[164]

164. Information regarding activities of the Consumer Affairs Agency is available in < http://www.caa.go.jp/en/>.

The Food Labeling Act emphasizes the important role of food labeling in ensuring the safety of ingestion of food and securing the opportunity to make an autonomous and rational choice of food. To ensure proper food label by establishing standards and specifying necessary information for label, the name, allergen, preservation method, expiration date (Best Before Date or Use-By Date), ingredients, additives, nutritional value and caloric value, country of origin, and other information should be displayed in the label when selling food and labeling method should comply with the relevant provisions and regulations. Mandatory allergen labeling is required for those products containing specified ingredients such as shrimp, crab, wheat, buckwheat, egg, dairy products, and peanuts. Foods with Function Claims introduced in 2015 can be used for those foods submitted to the Secretary General of the Consumer Affairs Agency as products whose labels bear function claims based on scientific evidence, under the responsibility of food business operators. Foods with Health Claim system still remain in place, for those products allowed for government-approved Foods for Specified Health Uses and Foods with Nutrient Function Claims.

3.13.2.4 Ministry of Agriculture, Forestry and Fisheries

The Ministry of Agriculture, Forestry and Fisheries (MAFF) is mainly responsible for the sustainable development of agriculture and rural areas, development of basic plan for food, agriculture, and rural areas and measures for securing a stable supply of food by the Food, Agriculture, Rural Area Basic Act enacted in 1999.

3.13.2.5 Prefectural and Municipal Governments

The Public Health Center and Meat Inspection Laboratories in Prefectural and Municipal Governments nationwide routinely inspect and guide food businesses for sanitation and product examination, abattoirs for meat inspections, and exert the approval, renewal, and/or revocation of business licenses based on the result of the inspection. The recent revision of the Food Sanitation Law in 2018 introduced new mandates for local governments to apply food sanitation control based on HACCP principles to businesses, mandatory notification to the local government for voluntary recalls by businesses, and new license notification system for retail businesses.

3.13.3 Conclusion

Numerous efforts have been made to ensure and enhance public health and food safety in Japan along with the efforts for global harmonization of food safety regulations with the careful consideration of societal changes such as aging population, household structure, and economic globalization. Enhanced multigovernmental cooperation and coordination efforts for food poisoning incidents or emergency cases would promote the public health and food safety control in Japan.

Further reading

Cabinet Office, Government of Japan. http://www.cao.go.jp/.
Consumer Affairs Agency, Japan. http://www.caa.go.jp/.
Food Labeling Act, 2013. Consumer Affairs Agency, Japan.
Food Safety Basic Act, 2003. Ministry of Health, Labor and Welfare, Japan.
Food Sanitation Act, 2017. Ministry of Health, Labor and Welfare, Japan.
Food Safety Commission of Japan. http://www.fsc.go.jp/.
Japanese Law Translation, Ministry of Justice, Japan. http://www.japaneselawtranslation.go.jp/.
Ministry of Agriculture, Forestry and Fisheries, Japan. http://www.maff.go.jp/.
Ministry of Health, Labor and Welfare, Japan. http://www.mhlw.go.jp/.
Outlines of partially revised regulations for the Japanese Food Sanitation Act, 2018. Ministry of Health, Labor and Welfare, Japan.
Sanksuk Oh, Understanding the Japanese Food Sanitation Act (I), 2016a. Korea Food Safety Research Institute.
Sanksuk Oh, Understanding the Japanese Food Sanitation Act (II), 2016b. Korea Food Safety Research Institute.
Summary report of meeting for revising the Food Sanitation Act, 2017. Ministry of Health, Labor and Welfare, Japan.

Chapter 3.14

India

V.D. Sattigeri
Food Safety and Analytical Quality Control Laboratory, Central Food Technological Research Institute, Mysuru, Karnataka, India

3.14.1 Introduction

3.14.1.1 Legislation

In India, all food laws and regulations are governed by an Act of Parliament, called the Food Safety and Standards Act, 2006 (FSS act 2006). This act empowers the Government of India to establish an authority called Food safety and Standards authority of India (FSSAI)[165] which is empowered to make rules and regulations to ensure supply of food of safety and quality to the consumers.

Accordingly, FSSAI in 2011 has developed the following regulations:

Food safety and standards (Licensing and Registration of Food business) Regulations, 2011, Food safety and Standards (Packaging and Labeling) Regulations, 2011, Food safety and standards (Food product standards and Food additives) Regulations, 2011, Food safety and standards (Prohibition and Restriction on sales) regulations, 2011.

Food safety and standards (Contaminants, Toxins, and Residues) regulations, 2011, and Food safety and standards (Laboratory and Sample analysis) Regulations, 2011. These regulations are being amended from time to time as food laws are dynamic and have to be kept pace with the changing requirements.

Prior to 2011, India had multiple of rules of regulations, namely the Prevention of Food Adulteration Act, 1954, the Fruit products order, 1955, the Milk and Milk products order, 1992, the Meat Products order, 1993, etc., which were administered by different ministries. The FSS act, 2006, was aimed at consolidating all these laws relating to food and to develop science-based food standards and to regulate and monitor the manufacture, processing, storage, distribution, and import of safe and wholesome food.

With the implementation of new regulations, the emphasis is on production of safe food having contaminants at minimum level. Limits for various contaminants, toxins, and residues are set by risk analysis as required by section 10 and 18 of FSS Act 2006. The risk assessment cell (RAC) established for the purpose will identify the risks that will be assessed, communicated, and managed, based on science.

It is the responsibility of the government to make rules and regulations and the responsibility of the regulator to monitor that they are implemented. However, it is the responsibility of the food business operator (FBO) that all the food laws are complied with. If not, for any lapse, the FBO will be held responsible and will be prosecuted as enunciated in the law.

3.14.1.2 Role of Codex Alimentarius Commission

India is a member of the Joint FAO/WHO Codex Alimentarius Committee. FSSAI represents India on Codex matters of CAC. FSSAI has constituted the National Codex Committee (NCC) in liaison with CAC which has the responsibility of cooperating with the Food Standards Program of the CAC, to formulate the national position on the Codex agenda, to study Codex documents and to collect and review all relevant information, to identify national organizations, and to generate a database for the preparation of base papers on Codex matters. Various shadow committees have been established by the NCC.

165. See: < www.fssai.gov.in >.

3.14.2 Institutional

3.14.2.1 FSSAI

It was the FSS Act, 2006, which paved the way for the internal harmonization of all food laws in which emphasis is more on food safety. FSSAI, an autonomous body established on the 5th of September 2008, is responsible for all food regulations and implementation. It is assigned to the Ministry of Health and Family Welfare (MoHFW). FSSAI has been mandated by the FSS Act, 2006, to perform the following functions:

- Framing regulations and laying down standards and guidelines for food articles; specifying appropriate systems of enforcing standards;
- Laying down mechanisms and guidelines for accreditation of certification of Food Safety Management Systems for food business;
- Laying down procedures and guidelines for accreditation of laboratories and their notification;
- To provide scientific advice and technical support to Central and State governments in framing policy and rules for food safety, quality, nutrition, and advertisement;
- To collect and collate data regarding food consumption, incidence and prevalence of risk, contaminants and hazards in foods, emerging risks, and introduction of rapid alert systems;
- Creating an info network on food safety, manpower development, and capacity building in food laboratories;
- To create general awareness of food safety in all stakeholders.

FSSAI is headed by a nonexecutive chairperson and has 22 members; it has a Chief Executive Officer with headquarters at Delhi. Scientific panels on Nutraceuticals, Pesticides and antibiotic residues, contaminants in the food chain, Biological hazards, Genetically modified organisms and foods, Methods of Analysis and Sampling, Packaging, and labeling, Oils and fats, Cereals and pulses, Food additives, processing aids, flavorings, etc., help the regulator in framing regulations.

At state level, a Commissioner of Food safety is responsible for implementing the regulations. He is assisted by designated officers at district level and food safety officers at lower level.

Section 18 of The FSS Act, 2006 has the provision of risk analysis while determining food standards based on available scientific evidence and in an independent, objective, and transparent manner. Risk management and risk communication have to be conducted to achieve an appropriate level of protection of human life and health, protecting consumers' interests including fair practices in food trade.

3.14.2.2 Risk assessment cell

To improve the food safety framework, under sections 10, 16 and 18 of FSS act 2006, the FSSAI has established an RAC. It will carry out functions of risk analysis to support risk management and risk communication. Risk assessment is done on products, processes, and activities that could result in increased health risk and has direct effect on food safety. The RAC will provide a framework to carry out risk assessment and remedial measures to be taken. Various hazards and risks are identified, prioritized, and managed.

The objectives of RAC are as follows:

- Collection of data, processing of data, preparation of analysis reports;
- Providing this input to relevant scientific panels;
- To involve other scientific experts that it feels necessary
- To coordinate in setting safety limits for contaminants and horizontal/vertical food standards.
- To conduct the risk assessment process: Risk assessment is a resource-intensive and data-driven activity with not just one way to achieve. It varies with the risk and availability of data on scientific evidence.

3.14.2.3 Powers of enforcement

Under the provision of section 29 of the FSS Act, 2006, FSSAI and the State food safety authorities are responsible for the enforcement of the Act. They are responsible for monitoring and verifying the relevant requirements of law at all stages of food business.

Section 32 of the FSS act 2006 empowers the court to issue prohibition orders in cases where an FBO is convicted of any offense; prohibition may be in use of premises and or facilities.

Section 41 of the FSS act 2006 has provisions for Food safety officers of the state to search any premises, seize any article of food or adulterant, or additive to carry out investigation and to launch prosecutions.

Various sections from 48 to 66 in the FSS Act, 2006, provide penalties for offences such as selling food not of desired quality, substandard food, misbranded food, misleading advertisement, foods containing excess of extraneous matter, unhygienic or unsanitary processing of food, adulterated food, unsafe food, and food with false information all call for penalties ranging from Rs.[166] 5000/- to Rs 500,000/-. In case of injury or death of the consumer, apart from the steep penalty, the license may be withdrawn, food recalled from the market, and forfeiture of the establishment and property.

These penalties will also apply to imported food as per section 67 of the FSS Act 2006 in addition to the provisions of the Foreign Trade (Development and Regulation) Act, 1992 and the Customs Act, 1962

Such nonconforming food articles will either be returned to the importer or destroyed.

3.14.3 Principles and concepts

3.14.3.1 Principles

In India, with the principle of one food law and one regulator for one nation, it is now possible to lay down food standards with transparency, consistency, and predictability.

General principles to be followed in administration of the act:

Central government, FSSAI, and State food authorities are guided by the following principles, viz.,

- Endeavor to achieve an appropriate level of protection of human life and health; to follow fair trade practices while applying food safety standards;
- Carry out risk analysis for parameters affecting health;
- Based on any interim studies, in case any harmful effect on health is identified, provisional risk management measures may be adopted to provide protection, pending further scientific confirmation;
- Measures taken shall ensure that they are not trade restrictive at the same time providing protection;
- The measures taken shall be reviewed within reasonable period of time depending on the nature of risk;
- If there is any risk for human health, then the Commissioner of Food Safety of the state or any FSSAI official shall take appropriate steps to communicate the general public;
- When any food fails to comply with food safety requirements, it has to be presumed until the contrary is proven that all the food in that batch, lot, or consignment fails to comply with the requirements;

The Food Authority while framing regulations or specifying standards under the Act shall take into account the following points: Prevalent practices and conditions of agricultural practices, handling, storage and transport conditions, as well as international standards and practices.

The food standards are to be determined on the basis of risk analysis except where there is opinion that such analysis is not appropriate; risk analysis is to be undertaken based on scientific information; protection of the consumers' interests has to be ensured so as to provide informed choices; protection from fraudulent, deceptive, or unfair trade practices which may mislead or harm the consumer and also from unsafe or contaminated food has to be ensured by appropriate legislation.

3.14.3.2 Concepts

There is a shift of focus from "adulteration" to "Food safety." Accordingly, the food law has been changed from "Prevention of Food adulteration Act" to Food safety and Standards Act.

There is a shift from regulatory regime to self-compliance through food safety management systems such as GMP, GHP, or HACCP.

Although uniform national law, the food safety, and standards act and regulations are in place, the implementation is fragmented and inconsistent as the new system is largely built on an old foundation. The same work force is there with only changes in their nomenclature and designation. The FSSAI is now in the process of building a changed mindset to ensure consistent implementation of the law by establishing:

- Food standards of global bench mark,
- Consistent enforcement in all states,
- Facilitating hassle-free imports,

166. The exchange rate of 100 Indian Rupees is about $ 1.4 and € 1.2 (2019).

- Assure credible food testing with NABL-accredited laboratories in government and private areas,
- Food safety practices and codes with global touch, and
- Large scale training and building.

Harmonization: While harmonizing food laws especially the limits for contaminants, the WTO requirements of SPS and TBT have to be considered; as far as possible, they are to be harmonized with Codex regulations as the WTO honors codex limits in case of any trade disputes.

The concepts of food, food additives, labeling, etc., are dealt with in the regulations. These regulations cover definitions, parameters of identity, limits for contaminants, labeling requirements, a positive list of ingredients and food additives, health and nutrition claims, etc.

For food additives, the concept of both vertical and horizontal approach is followed. Tables 1 to 15 of Appendix A of Food safety and standards (food products standards and food additives) Regulations, 2011 cover food additives specific for products, while all food additives cleared by Codex are adopted by FSSAI vide Food safety and Standards (food product standards and food additives) Regulations, 2016.

Section 3 of the FSS Act, 2006 defines food. As per this definition, the term food includes primary, processed, or semiprocessed food, genetically modified food, packaged drinking water, chewing gum, and alcoholic drinks. All these substances are dealt with in the legislation; that is, standards of identity and purity have been prescribed. Tobacco is not considered as food and therefore its presence in foods is prohibited. In fact, Food safety and standards (Prohibition and Restriction on sales) Regulations, 2011 bans some ingredients such as kesari dhal, certain admixtures and foods coated with mineral oil, etc.

3.14.4 Standards

A standard is a document that provides information on requirements, specifications, guidelines, or characteristics that can be used consistently to ensure that the product is fit for the purpose.

In India, we have three types of standards, namely, Process standards that are documented in Food safety and standards (licensing and registration of food business) Regulation, 2011; Product standards described in food safety and standards (Food product standards and Food additives) Regulation 2011 and Information standards that constitute labeling and other communication on ingredients, nutrition and health claims, etc., as stated in Food safety and standards (Packaging and Labeling) Regulations, 2011.

Product standards consist of two aspects, namely quality standards and safety standards. Quality standards include physico-chemical parameters such as moisture, ash, acid insoluble ash, saponification value, iodine value, acid value, fatty acid composition, rancidity tests, etc. Safety standards include limits for contaminants such as toxic metals, pesticides, antibiotics, pharmacologically active substances, aflatoxins, and microbiological contaminants.

Food standards are important as they safeguard the health of the consumer, ensure consumer confidence in the food chain, enable consumers to make informed decisions, and they communicate the safety and quality of the product besides being essential for the domestic and international food trade.

As per the Indian food regulation, a food product is deemed to be adulterated if it does not meet the standards laid down for that particular product in the regulations. A prepackaged food is supposed to meet all labeling requirements of the law.

In India, we have product specific, vertical standards for 377 products which include cereals and pulses, oils and fats, spices and condiments, milk and milk products, meat and meat products, infant foods, fruit and vegetable products, beverages, sweetening agents, fish and fish products, bakery products, etc.

Besides, there is a large group of products which are not standardized and they are covered as proprietary foods. They are defined as products not standardized under the regulations, but they shall contain ingredients that are permitted; the food category under which a proprietary food falls has to be declared and additives permitted for that category may be used in such a food. Further, vitamins and minerals as per the recommended dietary allowances may be used.

Recently, standards which are horizontal are prescribed for health supplements, nutraceuticals, foods for special dietary uses, foods for special medical purposes, and foods with prebiotics and probiotics.

About 9000 provisions for use of about 400 food additives in various categories of foods have been framed during harmonization in alignment of globally acceptable practices.

3.14.5 Role of Codex in standards

For international trade, many a times, if these standards do not meet WTO requirements of SPS and TBT agreements, it may lead to trade barriers and trade disputes. The WTO recognizes the standards laid down by Codex Alimentarius Commission as the final in all trade disputes and that if any country proposes the standards that are stricter than Codex, then it has to be justified by risk analysis. Therefore, India, as far as possible, is adopting the principle of harmonization in the development of its standards.

3.14.6 Authorization requirements

3.14.6.1 Categories

a. Foods for which no authorization is required:
   (i) Foods prepared with ingredients, food additives, nutrients such as vitamins, minerals, amino acids, etc., and plants and botanicals listed in the regulations need not go for authorization requirements.
   (ii) Foods prepared with the above ingredients using conventional methods of processing need not go for prior approval or authorization.
   (iii) Till now, the regulations did not permit the use of GM foods or GM ingredients. But now they are permitted and the label of such products has to mention that it contains GM ingredients if it is more than 5%.[167]

b. Foods for which prior authorization is required:
   (i) Food products prepared from ingredients that are new and not listed in the regulations including plants and botanicals with or without the history of safe use require authorization.
   (ii) Food products with additives not listed in FSSAI regulations shall require authorization.
   (iii) Food products such as novel foods prepared using nonconventional processing method such as nanotechnology require authorization.

3.14.6.2 Procedure for authorization

Prior approval/authorization as given in the regulation, food safety, and standards (approval for nonspecified food and food ingredients) Regulations, 2017[168] has to be taken by the food business operator before such nonspecified food is released into the market.

Procedure for approval:

Form I: for novel food or novel food ingredients or processed with conventional technology:

Foods processed with new, nonconventional technology, new additive, new processing aid including enzymes, and any other nonspecified food.

Information on product category, source of food ingredient, such as animal, chemical, botanical, microbial, etc., functional use, intended use, certificate of analysis, manufacturing process, regulatory issues, details of new technology, safety information/risk analysis/toxicological studies, allergenicity, and history of consumption has to be given by the applicant.

If it is a new additive not listed in Codex, as India has now permitted all food additives listed by Codex, information on chemical name, INS number, proposed usage level, and list of products in which the additive is to be added.

If new processing aid is to be used, for approval, information on specification, enzyme activity, purity, and residue limits has to be submitted.

It is clear from the above that the burden of proof is on the food business operator.

3.14.6.3 Authorization of food additive vis-à-vis Codex

Earlier, India had permitted the food additives as listed in Appendix A, Table 1 to 15 of Food safety and standards (food product standards and food additives) Regulations, 2011. Recently, it has approved all food additives cleared by Codex in Table 1, 2, and 3 of CAC STAN 192 1995 as amended in 2016.

167. GM food labeling requirements; Food Safety and Standards (Labeling and Display) Regulations, 2018, available at < www.fssai.gov.in >.
168. Approval of nonspecified food, new additive and food processed with non-conventional technology; Food Safety and Standards (Approval of nonspecified food and food ingredients) Regulations, 2017; issued on 11 Sept, 2017, < www.fssai.gov.in >.

In Codex, an eight-step procedure is adopted for the approval of new additive wherein all the latest information on its consumption, usage level in foods, exposure studies, toxicological studies, and risk information has to be submitted.

Similar procedure is followed in FSSAI where the latest information on safety, toxicity, and exposure studies has to be submitted by the applicant in a format as given in Section 3.14.6.2. The burden of proof is on the food business operator.

3.14.7 Food safety limits

3.14.7.1 Contaminants

In India, contaminants are dealt with by an exclusive legislation called Food Safety and Standards (Contaminants, Toxins, and Residues) Regulations, 2011. Contaminant is defined as any substance that is not added intentionally to the food, but gets into the food in the process of their cultivation, production, processing, treatment, transport, etc. The contaminants include

i. Metal contaminants such as lead, cadmium, mercury, arsenic, tin, and chromium for which maximum limits in ppm or mg/kg in various foods are prescribed in the regulation;
ii. Aflatoxins, patulin, and ochratoxin in various foods with limits expressed in ppb or mcg/kg;
iii. Naturally occurring toxic substances such as agaric acid, hydrocyanic acid, hypercine, and safrol in foods with limits expressed in ppm or mg/kg;
iv. Pesticides in different types of foods expressed in ppm or mg/kg;
v. Antibiotics and other pharmacologically active substances such as tetracycline, oxytetracycline, trimethoprim, and oxolinic acid in sea foods including shrimps, prawns, fish and fishery products, and honey expressed in ppm or mg/kg; list of prohibited substances is also given.

Recently, these regulations are amended as given in Food Safety and Standards (Contaminants, Toxins, Residues) Amendment Regulations, 2017; The list has 219 insecticides with MRLs. MRLs for antibiotics in meat and meat products are also updated to combat rising antibiotic resistance in humans. Globally, use of antibiotics and pharmacologically active substances is restricted in foods of animal origin and the amendments in the regulations are in tune with this principle.

3.14.7.2 Microorganisms

Food safety is an important concern. The WHO is of the opinion that a good majority of people will experience foodborne disease/food poisoning at some point in their lives; contaminated food can lead to long-term health problems that include cancers and neurological disorders. Therefore, FSSAI has prescribed Microbiological requirements in Appendix B of Food Safety and Standards (food product standards and food additives) Regulations, 2011, for foods such as sea foods, prawns, shrimps, lobster, frozen squid, etc., parameters such as TPC, *Staphylococcus aureus*, *Salmonella* and *Shigella*, *V. cholera*, *V. parahaemolyticus*, and *C. pe*rfringens. Similarly, for milk and milk products, limits for TPC, coliform count, *E. coli*, *Salmonella*, staph. aureus, and *L. monocytogenes* have been prescribed.

Appendix B also has the limits for spices, thermally processed fruits, and vegetables.

3.14.7.3 Pesticides

Pesticides are extensively used in the cultivation of foods which find their way in as residues usually in ppm or ppb in all foods.

The term pesticides may include insecticides, fungicides, rodenticides, and insect repellents that are specifically made to prevent, destroy, repel, or reduce pests.

Each is evaluated for its safety. Tolerance is the term used to denote the amount of a pesticide that is allowed to remain on a food as part of the process of regulating pesticides. In India, tolerances are called maximum residue limits (MRLs).

FSSAI sets MRLs for pesticides by considering the following criteria:

- Residue data obtained from the supervised field trials under Good Agricultural Practice (GAP)
- Acceptable daily intake (ADI) for that particular compound
- Exposure assessment
- Risk analysis

Residue data: Before going for registration/licensing, the manufacturer shall conduct field trials under different agro-environmental conditions and obtain data on the level of residue in the harvested crop.

Acceptable daily intake (ADI): ADI of the chemical is obtained by conducting animal experiments in the laboratory. During the study, the No Observed Adverse Effect Level (NOAEL) of the chemical is obtained. The ADI is obtained by dividing the NOAEL by 100 or in some cases by 1000. It is expressed in mg/kg body weight. In India, mostly ADI published by the Codex Alimentarius Commission is taken for setting MRLs.

Exposure assessment: Exposure of an individual to a particular pesticide through different food sources such as cereals, milk, fruits and vegetables, oils and fats, meat and meat products, etc., is calculated. For this, regional consumption figures of various commodities published by FAO are taken. This figure should not exceed ADI.

Risk analysis: From these data, Theoretical maximum intake and theoretical weekly intake are calculated to find out the short-term or long-term risk involved in the consumption of the pesticide.

Pesticide companies are required to submit a wide variety of scientific studies for review before FSSAI for setting MRLs especially for identifying the possible harmful toxic effect of the chemical on human; the amount of chemical; or its breakdown products likely to remain on food and other possible sources of exposures to the pesticide.

3.14.8 Process requirements

3.14.8.1 Food safety management

All food business operators are required to be registered or licensed in accordance with the procedure laid down in Food Safety and Standards (Licensing and Registration of the Food business) Regulations, 2011. All petty business operators are required to register with FSSAI. A person shall start business only if they have a valid registration or license from Central/ State authority.

To provide assurance of food safety, food business operators must implement an effective Food Safety Management System (FSMS) based on HACCP and suitable prerequisite programs by actively controlling hazards throughout the food chain.

As per the conditions of license under the Food Safety and Standards (Licensing and Registration of food business) Regulations, 2011, every FBO applying for a license shall have an FSMS plan and comply with Schedule 4 of this regulation. Schedule 4 introduces the concept of FSMS based on implementation of GMP and GHP by food business.

Schedule 4	General requirement
Part 1.	General hygienic and sanitary practices to be followed by FBO applying for registration-petty food operation and street food vendors
Part 2.	General hygienic and sanitary practices to be followed by FBO applying for license, manufacturing, processing, packaging, storage, or distribution
Part 3.	General hygienic and sanitary practices to be followed by FBO applying for license for Milk and Milk products
Part 4.	General hygienic and sanitary to be followed by FBO applying for license for slaughter houses and meat processing
Part 5.	General hygienic and sanitary practices for registration of catering

FSSAI has published checklist for food safety inspection. An innovative and extremely effective use of technology—a digital inspection platform—has been created by FSSAI for use by all states. This system, called Food Safety Compliance Regular Inspection and Sampling (FoSCoRIS), has been designed to replace manual inspection being practiced today.

A food regulatory portal has been launched in November 2017 which is unique and comprehensive which caters to domestic and imported food business.[169]

The port hosts multiple IT platforms:

- Fully on-line licensing and registration system (FLRS)
- On-line food import system
- Nationwide integrated network connecting all food testing laboratories on a single technology platform through InfoLNet.
- To demystify standards, Indian Food Standards Quick access (IFSQA) system collates and catalogs vertical and horizontal food standards.

169. < www.foodregulatory.fssai.gov.in >.

3.14.8.2 Traceability

Traceability is the ability to follow the movement of a food article through specified stages of its production, processing, and distribution. Traceability makes it possible to locate any product anywhere in the food chain so that it becomes easy to withdraw.

The labeling regulation for prepackaged foods in India requires the product to be identified with lot number or batch number or code number so that it can be easily identified in the distribution.

3.14.8.3 Food recall

The big and controversial recall by FSSAI was in June 2015 of Nestle's Maggi noodles. FSSAI had ordered its removal following reports that it contained lead content in excess of the permissible limit and that it contained monosodium glutamate, a flavor-enhancer that was not declared on the label. Besides Maggie, several energy drinks, flavored water, syrups, and sauces were recalled in the past as they did not adhere to the prescribed food standards.

Chapter VI, Section 28 of FSS Act, 2006, has provisions relating to food recall procedures and accordingly Food Safety and Standards (food recall procedure) Regulation, 2017 is published as a draft notification on the 18th of January 2017 and this will be effective from the date on which it is notified in the official gazette.

As per the regulation, food recall means the procedure and arrangements that FBOs shall have in place to retrieve food from food chain if problem arises.

The regulation aims to ensure the removal of food under recall from all stages of the food chain. The recall procedure broadly involves the following:

- Initiation of food recall procedure
- Operation of food recall system
- Food recall plan
- Recall communication
- Recall status report
- Food recovery
- Postrecall report
- Termination of recall
- Follow-up action
- Response of FBO/Commissioner of food safety of the state/FSSAI

With reference to the Act provided in Chapter VI , Section 28 of FSS Act 2006, Regulations have been published in the gazette on January 25, 2017 by FSSAI. The following Chapters and Schedules are covered in the regulation.

Chapter I : General; ,
Chapter II : Objectives; ,
Chapter III; Scope; ,
Chapter IV: Food Recall Procedure.
Schedule I : Food Recall Information;
Schedule II : Food Recall Status Format;
Schedule III : Food Recall Termination Request Format.

3.14.9 Labeling

Food labeling is the primary means of communication between the manufacturer/seller on one hand and the purchaser/consumer on the other.

General principles:

- Prepackaged food is not to be labeled with the information that is misleading or deceptive and it shall not convey any impression that is otherwise of its true nature
- The package shall not be described by any words, pictures, or symbols that hide its true nature and conveys the wrong impression.

Nutritional labeling on the food package describes the nutritional properties of that food.

Nutritional claim: It is the label that implies or suggests that the food has a particular nutrient which may include energy value, protein, fat, carbohydrate, fiber, vitamins, minerals, fatty acid profile, etc.

Health claim: Health claim suggests that there exists a relationship between a food or its constituent and health. Health claims include nutrient function claims, other function claims, and reduction disease risk claim.

Health claims should be consistent with the national health policy and shall have the backing of the sound and sufficient body of scientific evidence to substantiate the claim. Health claims consist of two parts: 1. Information on the physiological role of the nutrient or an accepted diet—health relationship and 2. Information of the composition of the product relevant to the physiological role of the nutrient.

The Food Safety and Standards (Packaging and Labeling) Regulations, 2011 prescribe packaging and labeling standards for the food in Indian food legislation. Requirements for packaging of canned products, fruits and vegetables, meat products, drinking water, vegetable oils, milk and milk products, etc., have been laid down in the regulations.

The labeling requirements for prepackaged foods have also been laid down in these regulations. They are as follows:

1. Every package to have a label for displaying information
2. These particulars shall be in English or Hindi compulsorily and in addition in any other language
3. Information not to contain false, misleading, or deceptive or erroneous impression on consumer
4. Label shall not become separated from the container
5. FSSAI logo and license number shall be displayed.

Following labeling particulars shall appear on the label:

1. Name of the product or its trade name and description of food
2. List of ingredients to be listed in descending order of their composition by weight or volume
3. Nutritional information: nutrients such as protein, fat, carbohydrates, and energy value and any other nutrient for which any claim is made on the label; these values are to be given in metric units per 100 g or serving size.

Nutritional information may not be necessary for primary agricultural commodities, spice, tea, coffee, cocoa, sugar, fruits and vegetables, pickles, and foods that are for immediate consumption.

4. Every package to contain veg/nonveg symbol; for nonvegetarian food, the symbol shall consist of brown color filled circle having a diameter as specified in the regulations, and for vegetarian food, it shall be green colored circle.
5. Food additive: Class title together with specific name of the food additive and or INS number to be given on the label.
6. Name and complete address of the manufacturer or packer.
7. Net quantity or volume in metric units.
8. Lot/code/batch identification: A batch number or code number or lot number which is mark of identification that helps in its tracing in the manufacture and distribution.
9. Date of manufacture or packing: The date/month and year in which the commodity is manufactured or packed shall be given on the label.
10. Best before or use by date.
11. Country of origin for the imported food.
12. Instructions for use or any specific storage to be given.

Allergen labeling: In accordance with the guidelines issued by Codex Alimentarius Commission, it is mandatory for the member country to declare the allergens on the food label, and therefore, FSSAI has recently published a draft notification, Food Safety and Standards (Labeling and Display) Regulations, 2018, ingredients known to cause allergy shall be declared on the label separately, as "contains ————(name of the allergen)."

List of allergens:

- Cereals containing gluten such as wheat, rye, barley, oats;
- Crustaceans and their products;
- Fish and fish products;
- Peanuts, tree nuts, and their products;
- Soybeans and their products;
- Sulfite in concentrations of 100 mg/kg or more.

Health claims, nutrition claims, and risk deduction claims and nonaddition claims are permitted by the regulation. However, the following claims are prohibited:

i. No claims shall be made which refer to the suitability of the food for use in the prevention, alleviation, treatment, or cure of a disease, disorder of particular physiological condition
ii. Labels not to use words implying recommendations by medical/nutrition/health professionals

Words such as "recommended by the medical/nutrition/health professionals or any other words which imply or suggest that the food is recommended, prescribed, or approved by medical practitioners or approved for medical purpose" shall not appear on the label.

iii. No product shall claim the term "added nutrients" if such nutrients have been added merely to compensate the nutrients lost or removed during processing of that food.

3.14.10 Apps developed by FSSAI

FSSAI has developed following apps for the stakeholders:

i. On-line platform for food inspection and sampling to bring transparency in food safety inspection;
ii. The web-based FoSCoRIS system will help verify compliances of food safety and hygienic standards;
iii. "Food safety connect" is a consumer centric app that offers a centralized platform to the Indian consumers to raise their concerns related to food safety and hygiene. The mobile app allows the consumers to report any malpractices pertaining to food safety and labeling;
iv. New app by FSSAI allows FBOs to check their compliances regarding food safety.

3.14.11 Human right to food and food security

The National Food Security Act (NFSA), 2013 also known as Right to Food Act is an act of Parliament which aims to provide subsidized food grains to approximately 2/3rd of India's population. It is also expected to provide nutritional security in human life cycle approach by ensuring access to adequate quantity of quality food at affordable prices to people to live a life with dignity. Under the provision, the beneficiaries of public distribution system are entitled to 5 kgs of cereals, rice, wheat, and millets. Lactating and pregnant women will get free cereals. Up to 75% of rural population and 50% of urban population will be covered.

The scheme will provide nutritional support to women and children. The population will receive nutrients required as per recommended dietary allowances as provided in the food regulations. The food supplied to the beneficiaries shall meet the quality and safety requirements set by FSSAI.

The law is enforced by the states with adequate support from the central government and women, children, and down trodden are immensely benefitted by this mega scheme.

3.14.12 Specific issues

i. There is a wide gap between the regulations prescribed by FSSAI and their implementation at state levels by state authorities. This is primarily due to manpower deficit and old mindset. However, these issues are being addressed by FSSAI by capacity building programs.
ii. Harmonization, especially the safety limits as required by WTO agreements, is not happening at the desired pace giving rise to trade disputes.
iii. There is again a wide gap between the draft notification of some regulations and their final gazette notification resulting in the corridor of uncertainty in the minds of food business operators.
iv. Sanitary and hygienic requirements in establishments are many a time not implemented and unfortunately no action is taken by the authorities.
V. FSSAI shall recognize some projects that provide safety to the consumers and fund them so that research organizations are involved in such activities.

Further reading

All legislation mentioned in this section, can be found at: www.fssai.gov.in.

Chapter 3.15

Pakistan

Ahmad Din
National Institute of Food Science & Technology, University of Agriculture, Faisalabad, Pakistan

3.15.1 Food safety standards and regulations

Pakistan is an emerging and developing country in South Asia having strategic and economic importance in the region and at global level. Occupying land crisscrossed by ancient invasion paths, Pakistan was the home of the prehistoric Indus Valley Civilization, which flourished until overrun by Aryans c. 1500 BCE. After being conquered by numerous rulers and powers, it passed to the British as part of India and became a separate Muslim state in 1947. The country originally included the Bengalese territory of East Pakistan, which achieved its separate independence in 1971 as Bangladesh. Pakistan became a republic in 1956. Islamabad is the capital and Karachi the largest city followed by Lahore, Faisalabad, Peshawar, Quetta, Rawalpindi, etc. Pakistan is blessed with many colors of nature having spring, summer, winter and autumn, fertile agriculture lands, green pasturelands, water reservoirs, surplus wheat, rice, maize, corn, cotton and many other cereals, fruits and vegetables, etc., for domestic use and export. The literacy rate of Pakistan is increasing with each passing year.

The food industries have a vital role in economic growth. Trade of Pakistan is improving with the harmonization and compliance of national and international standards. The Federal and Provincial governments of Pakistan are focused on revision and establishment of food standards and legislations to protect consumers and facilitate international trade. In this concern, it is unfortunate that Pakistan does not have a National Food Standards Council. Provincial food authorities are playing their part in the adoption of international best practices. National level conferences, workshops, and seminars are being organized with participation from all stakeholders to discuss and stress upon this important issue of a National Food Standards council. New avenues and ideas have been shared multiple times with scientists, regulators, and investors for harmonization of Food Standard and Safety level to provide the best possible food products in raw as well as processed form at national and international market. Understanding the gravity of this issue, provincial governments have started working on this issue. The Punjab Government has taken the lead by establishing the "Punjab Food authority" that has been extended to all 36 districts of Punjab. In 2018, the Khyber Pakhtunkhawa and Sindh Food Authorities have also started working to enforce the Food Quality and Safety standards across their respective provinces. Baluchistan is working on launching a Food authority in 2019.

Pakistani food laws combine elements from different food legislations which are already established in various industrialized countries. Pakistan Pure Food Rules have added some sections from the Codex Alimentarius Commission. Here, the objectives of the Codex Alimentarius are to protect the health of consumers, to ensure fair practices in the food trade and coordinate all food standards work, whereas the missions of codex Alimentarius are coordination of all food standards, initiation and supervision of draft standards, finalization of food standards, publication of standards worldwide, and to amend standards in face of new technology. Codex Alimentarius ensures that products complying with Codex standards can be bought and sold in the international market without compromising health or interests of consumers. Codex standards ensure that products are safe internationally. Food under the codex is defined as any substance, whether processed, partly processed or raw, which is intended for human consumption and includes drink, chewing gum, and any substance which has been used in the manufacture, preparation, or treatment of food but does not include cosmetics, tobacco, or substances used solely as drugs. The Codex Alimentarius has many components such as intergovernmental body, open to all UN member nations, currently 188 members (99% of world), executive committee overseas commission activities (Chair, three vice chairs, and seven others), secretariat overseas Executive Committee, located at FAO in Rome and corresponds with member states.

3.15.2 Status of food laws and regulations

Pakistan does not have an integrated legal framework but has a set of laws, which deal with various aspects of food safety. These laws, despite the fact that they were enacted a long time ago, have a tremendous capacity to achieve at least a minimum level of food safety. However, these laws remain very poorly enforced. There are four laws that specifically deal with food safety. Three of these laws directly focus on issues related to food safety. While the fourth, the Pakistan Standards and Quality Control Authority Act, is indirectly relevant to food safety.

3.15.2.1 The pure food The Punjab pure food ordinance, 1960

The pure food The Punjab pure food ordinance, 1960 consolidates and amends the law in relation to the preparation and the sale of foods. All provinces and some northern areas have adopted this law with certain amendments. Its aim is to ensure purity of food being supplied to people in the market and, therefore, provides for preventing adulteration.

3.15.2.2 The Cantonment Pure Food Act, 1966

The pure food The Punjab pure food ordinance (1960) does not apply to cantonment areas. There is a separate law for cantonments called "The Cantonment Pure Food Act, 1966." There is no substantial difference between the pure food The Punjab pure food ordinance (1960) and the cantonment pure food act, 1966. Even the rules of operation are very much similar.

3.15.2.3 Pakistan hotels and restaurants act, 1976

Pakistan hotels and restaurant act, 1976 applies to all hotels and restaurants in Pakistan and seeks to control and regulate the rates and standards of service(s) by hotels and restaurants. In addition to other provisions, under section 22(2), the sale of food or beverages that are contaminated, not prepared hygienically, or served in utensils that are not hygienic or clean is an offense. There are express provisions for consumer complaints in the Pakistan restaurants act, 1976, Pakistan penal code, 1860, and Pakistan standards and quality control authority Psqca act (1996). The laws do not prevent citizens from lodging complaints with the concerned government officials; however, the consideration and handling of complaints is a matter of discretion of the officials.

3.15.3 Principles and concepts

3.15.3.1 Food safety principles emerging in food safety standards

Every food establishment uses, processes, and sells food in its own ways. However, the general issues and key principles of food safety remain the same, whatever the style of the operation. All food safety training programs should contain the "big three" factors that could cause food to become unsafe. Food must be kept out of harm's way from human errors, but if one does not train food workers what they are, they will not know why these factors are so important for operating procedures. The basics can make us or break us in one or maybe two food handling mistakes. Those basic three principles that must be trained to all managers and food workers are

- Personal Hygiene for Food Professionals
- Time and Temperature Control
- Cross-contamination Prevention

3.15.3.1.1 Professional personal hygiene

It is not all common sense to everyone. Food workers must observe the highest possible standards of personal hygiene to make certain that food does not become contaminated by pathogenic microorganisms, physical, or chemical hazards. High standards of personal hygiene also play an important part in creating a good public image, as well as protecting food. Handwashing, fingernails, food worker illness policy (including exclusion of ill workers, cuts, burns, bandages, etc.), hair, uniforms, glove use, jewelry, personal cleanliness, or unsanitary habits such as eating, drinking, smoking, or spitting are all part of defining personal hygiene standards. Poor handwashing is one of the leading causes of foodborne illness. "Active Hand Hygiene" is a concept that really helps. There is a benefit to writing down standard operating procedures for the correct handwashing method/safe hands procedure to follow when each crew member is trained about this crucial expectation in their facility (i.e., 20 s handwash, when to wash, if using a nailbrush, type of soap, hand sanitizer, which glove or utensil for which ready-to-eat food task, etc.). Who monitors the process and how does one measure compliance on handwashing? These are questions best answered in writing for each individual operation.

3.15.3.1.2 Time and temperature control of foods

Bacterial growth can be reduced in potentially hazardous foods by limiting the time food in danger zone (140°F—41°F) during any steps of the food flow from receiving through service. The food code recommendation is no more than a cumulative 4 h in the danger zone. Use of a calibrated thermometer to chart time and temperature is based upon one's menu for cold holding (41°F), hot holding (140°F), cooking (based on the food), reheating (165°F), and cooling. Rapid cooling of hot foods (leftovers) or foods cooked several hours advance of service is a special challenge, which allows a six-hour two-stage cooling method (140°F to 70°F in 2 h and 70°F—41°F in 4 h).

3.15.3.1.3 Cross-contamination prevention

This is simply the transfer of harmful microorganisms or substances to food and covers a multitude of potential food mishandling in all stages of food flow. Cross-contamination can occur at any time. The three routes are (1) food to food, (2) hands to food, or (3) equipment to food. Ready-to-eat foods must receive the most care to prevent contamination. Food Safety Policy Food service is frequently dealing with employee turnover, so the job of training staff on professional hygiene, time/temperature, and cross-contamination control is never ending. These three issues contain many separate categories or steps to help keep the foodborne pathogens at bay. An overall "food safety policy" statement is a good idea to start with for all staff that focuses on the group's responsibility to help control these three issues. It is up to each person in charge to help the crew individually understand their responsibility for food safety that is appropriate in their specific food handling tasks. Active managerial control means supervisors must monitor the crew's adherence to your policy, make corrective actions, and set the example. The bottom line is that food safety is not just a matter of making one's facility look clean—you have heard the phrase "so clean you could eat off the floor." Sanitation is important, but do we really want to eat food off the floor anyway? The real mistakes can happen at any step in the flow of food through your facility from receiving, storage, preparation, cooking, holding, cooling, reheating, or serving. Strict attention should be given to training, practicing, and controlling the basics of food safety in one's facility.

3.15.3.2 Standards

Pakistan's food imports are generally regulated by the federal government and food standards are regulated by the provincial governments. Pakistan does not have an integrated legal framework but has a set of laws, which deal with various aspects of food safety. Food standards were established during 1963 in Pakistan Pure Food Laws (PFL). The PFL is the basis for the existing trade-related food quality and safety legislative framework. It covers 104 food items falling under nine broad categories: milk and milk products, edible oils and fat products, beverages, food grains and cereals, starchy food, spices and condiments, sweetening agents, fruits and vegetables, and miscellaneous food products. These regulations address purity issues in raw food and deal with additives, food preservatives, food and synthetic colors, antioxidants, and heavy metals.

Pakistan's Hotels and Restaurant Act, 1976 applies to all hotels and restaurants in Pakistan and seeks to control and regulate the rates and standard of service(s) by hotels and restaurants.

The Pakistan Standards and Quality Control Authority, under the Ministry of Science and Technology, is the national standardization body. In performing its duties and functions, PSQCA is governed by the Psqca act (1996). PSQCA is a member of International Organization for Standardization (ISO) and is the apex body to formulate or adopt international standards. PSQCA also serves as a focal point for national, regional, and international organizations and institutions such as ISO, IEC, Codex Alimentarius and WTO, National Enquiry Point (NEP) for WTO Agreement on Technical Barrier to Trade (TBT), introduce measures through standardization regarding consumer safety and health, and to establish procedure to conformity assessment compliant with national and international standards.

3.15.4 Labeling

Pakistan has now devised a uniform and universal system of imposing labeling and marking requirements on products. However, individual industries or sectors have also implemented some regulations of specific bodies. For example, the Ministry of National Health Services, Regulation and Coordination sets requirements for the pharmaceutical industry. The Ministry of Agriculture sets requirements for pesticides and edible products. Furthermore, in general meeting of 2018 which was held at Directorate of Punjab Food Authority Lahore, Pakistan labeling section was prime consideration and its proper rules have been briefly discussed. In 2019 labeling section will be briefly described.

3.15.5 Conclusion

In millennia, the need of proper knowledge about food laws (regulations and acts, basic concepts, and standards) is increasing day by day for food safety. Pakistan is formulating new sets of regulations, improving and modifying its food laws since 1947 to date. All the stakeholders related to food business play a critical role in increasing awareness of food safety risks. Generally, Food Law may be divided in two parts: (1) a basic food act and (2) regulations. The Act itself sets out broad principles, while regulations contain detailed provisions governing the different categories of products coming under the jurisdiction of each set of regulations. Sometimes food standards, hygienic provisions, lists of food additives, chemical tolerances, and so on are included in basic food control law.

References

Psqca act, 1996. http://www.psqca.com.pk/psqcaact.html.
The Punjab Pure Food Ordinance, 1960.

Further reading

Chaudhary, Z.H., 1974. Manual of Food Laws of Pakistan, vol. 1974. National Law Publications.
S DHHS, https://www.dhhs.nh.gov/dphs/fp/documents/principles.pdf.
https://www.oxfam.org/sites/.../oxfam-dp-the-right-to-adequate-food-20141014.pdf.
Punjab Pure Food Rules, 2018. Published by Punjab Food Authority (PFA), Pakistan.
https://www.slideshare.net/Adrienna/pakistan-status-of-food-laws-and-regulations-2015.
The Punjab Food Authority Act, 2011 (xvi of 2011).
The Punjab Foodstuffs (Control) Act, 1958 (xx of 1958).
The Punjab Pure Food (Amendment) Bill, 2015. Bill no. 45 of 2015.
The Punjab pure food rules, 2007. Framed under the Punjab Pure Food Ordinance, 1960 (w. P. Ordinance vii of 1960).
The Punjab Sugar Factories Control Act, 1950 (xxii of 1950).
West Pakistan Pure Food (Amendment) Act, 1963 (ii of 1963).
https://www.wfp.org/content/food-security-fundamental-human-right.
Wikipedia, https://en.wikipedia.org/wiki/Food_safety.

Chapter 3.16

Eastern Africa

Margherita Paola Poto
K. G. Jebsen Centre for the Law of the Sea, UiT, Tromsø, Norway

3.16.1 Introduction

3.16.1.1 Eastern Africa: defining its jurisdiction

A preliminary observation is due, on the choice to delimit the study area and therefore the jurisdiction object of analysis, since there are at least two main definitions of the Eastern African Region.

The first, based on purely geographical parameters, is provided by the United Nations Statistic Division[170] geo-scheme, and comprises the widest area of 20 Eastern African regions (from Egypt to Mozambique, including all the coastal states and Eastern landlocked countries, and the islands).

The second definition of Eastern Africa is based on an economic integration scheme; it comprises the Member States of East African Community (hereinafter, EAC), signatories of the East African Community Treaty[171] (Tanzania, Kenya, Uganda, Rwanda, Burundi, and recently South Sudan), and has the main purpose to provide an economic integration scheme.

The present contribution will mainly focus on the East African Community as the case study jurisdiction, taking into consideration two opposite and at times contrasting needs:

On the one side, the need to get into the details of national examples, with the consequence of offering a food safety perspective that is not necessarily applicable to all the members of the EAC.

On the other side, the need to identify common principles applicable to the Eastern Africa countries, also beyond the EAC members (as in the cases of the Codex Alimentarius Regional Committee (see Section 3.16.1.3) and of the Common Market for Eastern and Southern Africa, COMESA, both contributing to the principles harmonization beyond the six EAC members).

Although the EAC's objectives have a predominantly economic nature, a broader interpretation of the EAC treaty provisions clarifies that they also cover social, cultural, political, technological, legal, and also peace and security aspects. Such an extended interpretation of the EAC objectives and functions includes food safety at its intersection between legal harmonization, protection of fundamental rights to food, health, and technological development, and protection of local cultures and traditional knowledge. Following such interpretation, a Medicines and Food Safety Unit has been established, to "ensure a high level of consumer protection and restore and maintain confidence in the quality and safety of medicines, food, and health products."[172]

The unit's main objective is "to develop and facilitate implementation of regional policies, regulations strategies, guidelines, and standards to enhance affordability, quality, efficacy, and safety of human and veterinary medicines as well as ensure food safety in the EAC Partner States."

3.16.1.2 Harmonization of food safety principles at regional level

According to a study conducted by GIZ in 2012 (hereinafter, the GIZ Study), as regards food safety and quality regulations, the EAC has reached quite a satisfactory level harmonization in principles and standards[173] and can be considered as one of the leading supranational standard setting organizations (together with the African Union and COMESA).

It is worth anticipating that each of the six countries parties to the EAC is in principle ruled by four different standard levels in food regulation and namely, 1. private standards, 2. national standards, 3. regional standards, and 4. multilateral standards. While the multilateral standards serve as a reference for developing regional and national standards, the regional (in our case, set up but the EAC) guide the development of national standards and regulations.

Though the harmonization process has been quite successful in the EAC, this is true at a very general level, that is to say that the EAC members have agreed in principle to reach some common objectives; such declaration of intents has not always been followed by tangible results, with the consequence that the demand for an effective implementation at national level has still margins of improvement, especially in those country members where the priority is still to achieve the full recognition of the right to food and an equal distribution of safe food to the population (as in the case of South Sudan: see Section 3.16.1.4).

170. < https://unstats.un.org/home/ > visited in April 2018.

171. The Protocol for the Establishment of the EAC Common Market was signed by the five partner states on 20 November 2009 and entered into force on 1 July 2010. South Sudan was admitted into the EAC later, in March 2016. According to the Treaty, the EAC integration is to be achieved through four phases: a customs union, the protocol for which entered into force on 1 January 2005; a common market, which is still under implementation, although the protocol establishing it came into force on 1 July 2010; a monetary union, the protocol establishing which was signed on 30 November 2013; and a political federation. See further: Caroline Nalule, Defining the Scope of Free Movement of Citizens in the East African Community: The East African Court of Justice and its Interpretive Approach, Journal of African Law, 62, 1 (2018), 1–24.

172. < https://www.eac.int/health/medicines-and-food-safety-unit > visited in June 2018.

173. < https://www.giz.de/expertise/downloads/giz2012-en-food-safety.pdf >.

3.16.1.3 Codex Alimentarius, risk analysis, and food safety authority under the EAC principles

The Codex Alimentarius Commission has a regional jurisdiction for the African continent, known as CCAfrica, with a Regional Coordinator (Kenya) and 49 Members, including all the Eastern African States. The activity of the CCAfrica appears to be still at an embryonic stage: yearly technical regional workshops on specific topics are organized by the National Codex Contact Points. For example, in April 2018, a workshop on Risk Assessment and Management of Mycotoxins took place in Kenya; in February 2018, a Regional Technical Workshop for National Codex Contact Points was held in Nairobi, with a specific focus on online web tools, such as the online registration system (ORS) and the Online Commenting System (OCS).

So far, it does not seem that there is a political willingness to identify a Regional Committee for Eastern Africa, nor to approve a common set of principles drawn from the Codex Alimentarius guidelines. Similarly, from the institutional viewpoint, there is no sign of a willingness to establish a Regional Food Safety Authority for Eastern Africa nor for the Eastern Africa Community: some countries within the EAC have autonomously decided to establish a Food Safety Authority (e.g., Tanzania); some others are considering it as a possibility (as for example, Kenya, that in 2009 has approved the National Biosafety Act and established the Biosafety Authority and has been discussing a legislative proposal on the establishment of a Food and Drug Authority); finally, other countries have decided to assign the functions of food safety regulation to the competent Ministries (Rwanda).

3.16.1.4 Food law: from the regional harmonization of food safety standards to the national legal acts on food safety

While at regional level there has been a general—though at traits superficial—attention toward the harmonization of standards, the situation in the six study countries is quite diversified and fragmented with respect to policy statement and regulations on food safety, especially for livestock products.

In *Kenya*, the right to safe food is enshrined in the constitution and consequently assurance of safe food is explicitly stated or implied in public policy documents and these are supported by appropriate legal provisions. These include statutes such as the Food, Drugs and Chemical Substances Act which was revised in 2013, the Public Health Food Labeling Act (2013). A number of national policy documents such as the National Recovery Strategy 2003, Strategy to Revitalize Agriculture 2005, National Livestock Policy 2007, Strategic Plan on Creation of Animal Disease Free Zones 2007, the draft National Food Safety Policy 2010, and the National Dairy Development Policy to support the statutes contain explicit or implied statements about assurance of safe food for the citizens.

Tanzania approved the latest Food Drug and Cosmetics Act in 2011, that contains provisions on food safety, though without explicitly referring to this concept. In 2007, it established the Tanzania Food and Drug Authority (TFDA), an Executive Agency under the Ministry of Health, Community Development, Gender, Elderly and Children (MOHCDGE), with the broad mandate to regulate safety, quality, and effectiveness of food, medicines, cosmetics, medical devices, and diagnostics.

Uganda has not enshrined the right to safe food in the Constitution, nor has approved a food safety law: so far, the actions to provide a food safety legal framework have been taken at government level, by developing and approving a 5-year Uganda Nutrition Action Plan 2011–2016. The objective of the plan is to improve the nutritional status of the Ugandan population, with emphasis on women of reproductive age, children, and infants. The plan is more focused on fighting malnutrition than focusing on safe food. In parallel, the government also produced a short message targeted at district- and lower-level leaders to raise awareness of the nutrition situation in Uganda and introduce the Nutrition Action Plan (Source: FANTA and USAID, 2018).

In *Rwanda*, a joint action of the Ministry of Local Government, Ministry of Health, Ministry of Agriculture, and Animal Resources resulted in the approval of the Rwanda National Food and Nutrition Policy (NFNP) in 2014. As specified in European Commission report on Rwanda, a multisectoral policy framework for nutrition has been established and the NFNP has been recently updated with the National Food and Nutrition Strategic Plan (NFNSP) 2013–18 by providing strategic priorities and implementation guidelines for effective nutrition guidance of multisector stakeholders.[174]

174. < https://ec.europa.eu/europeaid/sites/devco/files/nutrition-fiche-rwanda-2016_en.pdf >.

The implementation of food safety standards, as recommended by the EAC, is still at a preliminary stage in *Burundi.* Only in December 2017, the East African Community Secretariat has initiated the facilitation process of a 4-day benchmarking visit for Burundi Parliamentarians to the TFDA and Ministries responsible for Health, Agriculture, and Trade in the United Republic of Tanzania, in order to support Burundi government in the enactment and implementation of the Burundi National Pharmaceuticals Regulation Law (at the time of writing—Spring, 2018—still pending before the Burundi Parliament). The visit was planned also to enable the Government of Burundi to establish and fully operationalize the Autorité Burundaise de Regulation des Medicaments et des Aliments (ABREMA) as a public autonomous or semi-autonomous agency and legally designated to oversee effective regulation of food and medicinal products in the country in compliance with international and regional food safety principles.[175]

A fully operational food safety regime in *South Sudan* is unfortunately still lacking, due to protracted conflicts and political disputes and at the time of writing the country is facing one of the most severe famines in the last decades.[176] Serge Tissot, FAO representative in South Sudan, declared that "The situation is extremely fragile, and we are close to seeing another famine."[177] The FAO, the UN Children's Fund (UNICEF),[178] and the UN World Food Programme (WFP)[179] warned that, due to the conflicts, progress in preventing hunger-related deaths could be undone, and more people than ever are likely to face starvation-like conditions during May—July unless assistance and access are maintained.

3.16.2 Institutional

3.16.2.1 National and regional food safety authorities: a framework still in progress

As illustrated in Sections 3.16.1.3 and 3.16.1.4, the supervision of the food safety sector in Eastern Africa is still developing and therefore fragmented. Some countries, like Tanzania, adopted the North American FDA model; some others look at the Tanzanian model as the reference for the development of a competent authority but have not come to any significant food safety reform so far (Burundi); and some have decided not to allocate the competence in the food sector to an independent authority (as in the case of Rwanda and Kenya, where however the establishment of a Food Safety Authority has been discussed and it is currently under exam).

As for the case of risk assessment, the FAO raised the issue to devolve the competence to a supranational authority[180]: the food safety challenges related to the lack of a coherent institutional framework were raised in two workshops in Kigali in 2012 and in Addis Ababa in 2013. As an outcome of the workshops, it was planned to establish and develop (1) the African Union (AU) Food Safety Authority and (2) AU-Rapid Alert System for Feed and Food Safety. More specifically on Eastern Africa, the regional collaboration between the African Union-Inter-African Bureau for Animal Resources (AU-IBAR) and FAO EMPRES Food Safety ended up in a Regional Workshop on Enhancing East African's Early Warning Systems for Food Safety (Nairobi, 2014) to help East Africa develop proposals for building or improving existing food safety early warning systems. Nevertheless, such an attempt has not produced any effects on the effective establishment of an independent regional food safety authority with competences on Eastern Africa.

3.16.3 Principles and concepts

3.16.3.1 Food safety principles emerging in the food safety standards

To identify a common set of food safety principles applicable to the East African Community, it is necessary to refer to the GIZ Study mentioned above, as the most complete collection of principles, concepts, and standards applicable to the whole Eastern African Region (pp 97—100). The GIZ Study contains a complete and systematic glossary on the most used terms and principles relevant in food safety regulation, and also the common terms of reference for eventual future regulatory provisions at national or regional level.

175. < https://www.eac.int/press-releases/952-eac-secretariat-facilitates-burundi-parliamentarians-benchmarking-visit-to-tanzania-food-and-drug-authority >.
176. < https://news.un.org/en/story/2018/02/1003552 >.
177. < http://www.fao.org/news/story/en/item/1103429/icode/>.
178. < https://www.unicef.org/>.
179. < https://www1.wfp.org/>.
180. < http://www.fao.org/3/a-i4601e.pdf >.

3.16.4 Standards

3.16.4.1 The GIZ study on the classification of food safety standards for the East African Community and the common market for Eastern and Southern Africa

As mentioned above, standards and their harmonization play a central role both in the East African Community (EAC) and in the Common Market for Eastern and Southern Africa (COMESA). As said above, the GIZ Study on the "Harmonization and Mutual Recognition of Regulations and Standards for Food Safety and Quality in Regional Economic Communities" is probably the most complete and updated reference document, for any analysis on food safety in Eastern Africa. The study focuses on five strategic value chains (cassava, coffee, dairy products, horticulture, and maize) and on six countries (Ethiopia, Kenya, Rwanda, South Sudan, Tanzania, and Uganda). While specifically focusing on vertical standards for strategic products, it also sheds light on the need to group and classify standards, in order to provide a framework for a future common strategy of horizontal regulation in Eastern Africa, by providing basic definitions on food safety, food quality, and related terms.

Moreover, the GIZ Study identifies drawbacks and vulnerabilities in the system that have repercussions on both food security and food safety such as the malfunctioning of the regional trade, with a consequent unbalance between food surplus and food deficit areas.

To overcome the deficiency in relationship and in general the low level of governance of the food sector in Eastern African, the Guidelines' authors foresee the need for the Member States to effectively comply with regulations and standards for food safety in the common attempt to harmonize the national legal provisions for an improved regional cooperation. As a virtuous example of the work conducted so far, the GIZ Study commends the harmonizing activity undertaken both by EAC and COMESA, as well as their advancement in setting up an institutional framework for a regional quality infrastructure.

As said above (Section 3.16.1.2), more specifically, the GIZ Study recognizes that the system of standards is complex and multilayered and it includes 1. private standards, 2. national standards, 3. regional standards, and 4. multilateral standards. While the multilateral standards serve as a reference for developing regional and national standards, the regional guide the development of national standards and regulations.

Such a variety on the one hand shows the great interest and commitment toward the regulation of the food sector and on the other hand mirrors the vulnerability of a system that remains still extremely fragmented, diversified, and poorly implemented at national level.

3.16.4.2 The role of the standards

According to the GIZ Study, there is no unanimity on the binding effects of standards: it is discussed whether they comprise both regulations, also referred to as mandatory standards, and public—private voluntary standards. The GIZ Study does not take a position in this regard, and it rather classifies the standards applicable to Eastern African food safety into two main categories: public regulations and private voluntary standards.

Public regulations and standards:

- Sanitary and Phytosanitary (SPS) regulations: mandatory measures on food safety aspects with the primary objective of protecting human, animal, and plant life and health.
- For members of the WTO, the development of SPS measures is guided by the so called "three sisters": Codex Alimentarius Commission (CAC) of the FAO and the WHO; International Plant Protection Convention (IPPC); and the World Organization for Animal Health (OIE).
- Technical regulations and standards: Technical regulations and standards refer to mandatory or voluntary food quality aspects and conformity assessment requirements (e.g., standards developed by the ISO and/or National Standardization Organizations): Regulations and standards for food quality: partly mandatory (e.g., minimum quality requirements such as size, color, weight, nutritional requirements, label content and formats, etc.), partly voluntary (e.g., East African Organic Standard and Mark), Regulations for conformity assessment (Quality Infrastructure): obligations and guidelines for inspection, methods of sampling, measuring and testing, analysis, accreditation, and certification.

Private voluntary standards:

- Commercial trade and industry standards: Private trade and industry standards are either corporate (when developed by individual firms) or collective standards (when developed by business networks and associations) with the purpose of products harmonization and market transactions' coordination (as in the cases of GlobalGAP/KenyaGAP, Tesco Nature's Choice, Ethiopian Horticulture Producers and Exporters Association (EHPEA) Code of Practice for Sustainable Flower Production)
- Third party standard schemes used by trade and industry: usually developed by nongovernmental organizations; their scope is to tackle sustainability issues or address specific social and ecological issues (as in the cases of Fair Trade, Rainforest Alliance, Utz Certified, organic).

3.16.5 Authorization requirements

3.16.5.1 Is there any?

The Eastern African jurisdictions reveal their vulnerability also with regard to authorization requirements. There is a lack of requirements for both import and export of GMOs and novel food. The criticality is particularly acute in the GMOs sector, where there is still uncertainty and divergence in opinions in the scientific community, on their impact on human health, animal welfare, and biodiversity. What is certain is that the six EAC countries do not seem to have any kind of binding provisions on authorization requirements regarding GMOs foods.[181]

Kenya has raised the issue by approving a contested National Biosafety Act in 2009 (see Section 3.16.1.3) and initiating a process that will hopefully lead to the harmonization of biosafety regulation, in order to create a one-stop GMO approval system at the subregional level.[182] The idea is to establish fast-track GM approval systems that can regulate the introduction of GMOs into Africa. The harmonization approach is supported by the World Bank, USAID and national and regional affiliates of the Consultative Group on International Agricultural Research, and various African academic and research institutions.

3.16.6 Food safety limits

3.16.6.1 Food safety limits for export and for domestic markets: two weights and two measures

The implementation of food safety management systems (FSMS) in Africa has been an ongoing process of the last few decades, since the Codex Alimentarius hygiene code has been accepted as a common term of reference worldwide. Despite the attempts to formally comply with the hygiene requirements, the implementation process in Africa has still shortcomings: 1. As mentioned above, the majority of the Eastern African countries are not yet in line with the Codex Alimentarius Commission requirements, and are neither adequately providing a clear mandate to responsible authorities to prevent food safety problems; 2. The most advanced food law systems, such as Kenya, Tanzania, and Uganda, seem to have two food safety controls, one relatively advanced for the export and one still weak, neglected when not inexistent for the domestic food supply; 3. Last but not least, as observed before, also in the case of food safety limits, the fragmented and overlapping functions of different ministries in food control, the lack of coordination and the limited laboratory capacities hinder the adoption and implementation of limits and quality assurance standards.[183]

181. Some discussion on GMO regulation has been initiated in the other Eastern African countries besides Kenya, but the debate is still at an early stage: < http://www.ofabafrica.org > visited in June 2018.
182. Shenaz Moola and Victor Munnik, GMOs in Africa: food and agriculture, Status report 2007, in < https://www.stopogm.net/sites/stopogm.net/files/GMOAfrica.pdf >.
183. Jamal B Kussaga, Liesbeth Jacxsens, Bendantunguka PM Tiisekwaa and Pieternel A Luning, Food safety management systems performance in African food processing companies: a review of deficiencies and possible improvement strategies, Society of Chemical Industry (2014), J Sci Food Agric 2014; 94: 2154–2169.

3.16.7 Process requirements

As in the other food-related matters, also in the regulation of food processing requirements, the framework is very fragmented and there seems to be very little harmonization in policies among the Eastern African States.

The attempts toward harmonization undertaken by the EAC have not come to a satisfactory scenario.

A recent study on food processing requirements in Eastern Africa highlights such a critical aspect, focusing on the dairy market and on the considerable gap between public and private sectors.[184]

At a policy-sector level, it is noted how national dairy sector policies and institutions differ substantially among Eastern African countries. The study mentions the case of Kenya, where the main policy frameworks are the 2010 Kenya Dairy Master Plan and the Dairy Industry Act; Uganda, the plans date back to the 90s and need to be updated in consideration of the current market dynamics; and Rwanda, with its own National Dairy Strategy for the period 2013–18 to guide the development of the sector and its emphasis on the need to strengthen the public–private partnerships, homogeneity of sectorial institutions, and milk quality assurance. A plan is still missing in Tanzania that aims to develop a National Dairy Master Plan, under the 2004 Dairy Industry Act.

Unfortunately, what is common to all the mentioned countries is the lack of a proper farm policy for the dairy sector. They all still rely on the general agricultural and livestock policy frameworks. Moreover, the dairy industry is at the cross-roads of other sectorial policies in the areas of land, cooperatives, industry, trade, health, and environment. And certainly, such a multitude of directions and policies does not contribute to building up a systematic framework for the dairy sector regulation, so that the absence of specific and clear national dairy policies remains a crucial constraint to the comprehensive development of the industry in the Eastern African Region.

3.16.8 Labeling

A general comprehensive regulatory framework on labeling covering allergens, nutrition, and health claims is still missing for the Eastern Africa region.

As in the cases of the other food safety issues, the framework appears still extremely fragmented and patchy, with some progress limited to certain aspects of labeling, such as the mandatory information.

3.16.8.1 A set of regulations on labeling for the EAC members

Despite the lack of a holistic approach in the labeling regulation, some remarkable attempts to approve general rules have been made in October 2014, when the then five members of the EAC approved a number of regulations regarding food labeling, and namely: 1. EAS 38:2014 Labeling of prepackaged foods - General requirements; EAS 803: 2014 Nutrition labeling - Requirements; EAS 804:2014 Claims - General requirements; EAS 805:2014 Use of nutrition and health claims - Requirements.[185]

3.16.8.2 Mandatory information

Some general common information on mandatory labeling appears as widespread in the Region, thanks to the above mentioned regulations. In particular, there seems to be clarity on what the mandatory information shall include, and namely: 1. The name of the food; 2. The list of ingredients and allergens; 3. The net contents; 4. Name and address of manufacturer, packer, distributor, importer, exporter, or vendor; 5. Country of origin; 6. Lot identification; 7. Date marking and storage instructions; 8. Instructions for use; 9. Quantitative ingredients declaration.

Despite such commendable attempt toward harmonization, as said above the critical aspect is the lack of a general framework in labeling regulation and a very poorly documented, if not inexistent, documentation of its implementation and enforcement.

184. Susan Bingi and Fabien Tondel, Recent developments in the dairy sector in Eastern Africa Towards a regional policy framework for value chain development, Briefing Note, n. 78, September 2015, European Centre for Development Policy Management (ECDPM).
185. < http://resources.selerant.com/food-regulatory-news/east-african-community-labeling-reference-guide-now-available-on-scc >.

3.16.9 Human right to food/food security

3.16.9.1 Food safety as a priority area for the EAC food security plan

In 2011, the EAC has approved a Food Security Action Plan,[186] whose objectives aim to effectively guarantee the diffusion of right to food at large scale. In the priority action areas, the need is mentioned to enhance the efficiency of food utilization, nutrition, food safety, and quality (see priority area n. 3.5 of the Plan). The right to food and the right to safe food are covered under the same regulatory plan.

Nevertheless, it is noteworthy that the real emergency in the EAC is still connected to food security and effective internal policies that do not hinder the African market in this region need to be rethought and developed. The general remark is that such development should start from within, without any external pressure to comply with International standards that ends up in worsening the gap between the Eastern African Region and the rest of the world.

3.16.10 Specific issues

3.16.10.1 The (interrupted) dialogue between EU and EAC

Several attempts toward a dialogue between EU and EAC have been undertaken since the beginning of the year 2000, but the outcomes were not always satisfactory, especially when looking from the Eastern African perspective. The dialogue was formalized on October 16, 2014, when the EAC finalized the negotiations for a region-to-region Economic Partnership Agreement (EPA) with the EU.[187]

The tangible result of such an agreement is an imbalanced protection of the exports, to the detriment of the food quality standards for the products destined to the internal market: if progresses are registered in the improvements of regulations and policies for all EAC exports, the positive trend does not seem to have led to a significant development of the EAC internal market.

Moreover, some perplexities had been raised by single States, worried about the crushing competition force of the EU products on the internal products.

According to Tanzania, such competition is likely to hinder the industrialization and development in East Africa. For other reasons related to alleged human rights violations, Burundi is not part of the agreement, while Uganda has not formally signed the agreement, after having expressed its formal interest to be part of it.

References

https://www.fantaproject.org/.

https://www.usaid.gov/global-health/health-areas/nutrition/countries/uganda-nutrition-profile.

Further reading

Bingi, S., Tondel, F., September 2015. Recent Developments in the Dairy Sector in Eastern Africa towards a Regional Policy Framework for Value Chain Development. Briefing Note, n. 78. European Centre for Development Policy Management (ECDPM).

AC EAC, https://www.eac.int/press-releases/952-eac-secretariat-facilitates-burundi-parliamentarians-benchmarking-visit-to-tanzania-food-and-drug-authority.

http://ec.europa.eu/trade/policy/countries-and-regions/regions/eac/.

https://ec.europa.eu/europeaid/sites/devco/files/nutrition-fiche-rwanda-2016_en.pdf.

http://www.fao.org/3/a-i4601e.pdf.

http://kenya.countrystat.org/fileadmin/user_upload/countrystat_fenix/congo/docs/EAC_Food_Security_Action_Plan.pdf.

Kussaga, J.B., Jacxsens, L., Tiisekwaa, B.P.M., Luning, P.A., 2014. Food safety management systems performance in African food processing companies: a review of deficiencies and possible improvement strategies. Society of Chemical Industry J. Sci. Food Agric. 94, 21.

Moola, S., Munnik, V., 2007. GMOs in Africa: Food and Agriculture, Status Report, pp. 54–2169. https://www.stopogm.net/sites/stopogm.net/files/GMOAfrica.pdf.

Nalule, C., 2018. Defining the scope of free movement of citizens in the East African community: the East African court of justice and its interpretive approach,. J. Afr. Law 62 (1), 1–24. http://resources.selerant.com/food-regulatory-news/east-african-community-labeling-reference-guide-now-available-on-scc.

https://news.un.org/en/story/2018/02/1003552.

https://unstats.un.org/home/.

186. < http://kenya.countrystat.org/fileadmin/user_upload/countrystat_fenix/congo/docs/EAC_Food_Security_Action_Plan.pdf >.

187. < http://ec.europa.eu/trade/policy/countries-and-regions/regions/eac/>.

Chapter 3.17

Republic of South Africa

Bernard Maister
Intellectual Property Unit, University of Cape Town, Cape Town, South Africa

3.17.1 History and background

Following a negotiated process beginning in 1990 and ending in 1994, the Republic of South Africa (RSA) transitioned from apartheid to full democracy. The country's new Constitution, finally enacted in 1997, was intended to "establish a society based on democratic values, social justice, and fundamental human rights" (South African Constitution, Preamble). Included among its socio-economic rights is the right "to have access to ... sufficient food and water" (South African Constitution, Section 27(1) (b)). In a landmark case, *Government of the Republic of South Africa* versus *Grootboom*, the Constitutional Court tied socio-economic rights with political and civil rights stating that the state's "obligation is to provide access to housing, health-care, sufficient food and water, and social security to those unable to support themselves and their dependents" (Grootboom, 93).

RSA has demonstrated that it is capable of responding to food crises. In February 2002, two South African children died from botulism after eating a tin of canned pilchards. After a nationwide food safety alert, including withdrawing all tins with the same batch number, it was determined that this was an isolated event. The family had received a food donation which included a damaged and badly rusted tin of pilchards. Although tragic, the event did prompt a review of the nation's food control system which resulted in several recommendations related to improving coordination between departments involved in food control and developing policy guidelines in regard to food safety and product recalls (DOH, 2004).

16 years later, South African authorities were able to effectively respond to a food safety challenge resulting from a *Listeria* outbreak.[188] Between January 2017 and July 2018, South Africa experienced 1060 confirmed cases, including 216 deaths, of listeriosis ultimately linked to a ready-to-eat processed meat plant. This outbreak, the "largest ever detected *Listeria* outbreak in known history," eventually required the destruction of 5812 tons of affected foods and cost the government the equivalent of $810,000.[189]

A "Listeriosis Emergency Response Plan" was implemented and a "multisectorial incident management team" created under the auspices of the Department of Health (DOH) in April 2018. The goal was to both control the active listeriosis outbreak and improve systems to facilitate prevention and detection of outbreaks.[190] With the support of the WHO, a surveillance system was introduced to improve the detection and identification of affected foods. Food safety laws were updated in regard to ready-to-eat processed meat and poultry.

Despite the improved response to foodborne illness, the food control system remains shared between several government departments, directorates, and authorities with various, sometimes competing and conflicting, components at national, provincial, and local levels.

The lack of an integrated national strategy and duplicative, sometimes contradictory, laws has produced a "fragmentation of legislation, structure, and functions" with multiple agencies from different departments involved in food safety and food control (DOH, 2013). For example, due to different risk management frameworks, inspection methods, compliance verification, and enforcement approaches, "foods of similar risks" may be "inspected at different frequencies and/or in different ways." In some situations, the food industry may be faced with "having to meet multiple and different requirements on the same product" (DOH, 2013).

Food-related legislation in South Africa is often intertwined with other goals such as job creation. In his forward to his department's strategic plan, the head of the Department of Agriculture, Forests, and Fisheries (DAFF) has described

188. Listeriosis is caused by the *Listeria monocytogenes* with infection usually following eating of contaminated food. The disease is most significant in pregnant women, newborns, older adults, and people with weakened immune systems.

189. Food Safety News. September 4, 2018. Available at < https://www.foodsafetynews.com/2018/09/south-africa-declares-end-to-largest-ever-listeria-outbreak/ >.

190. Listeriosis outbreak situation report — 04/05/2018. Available at < http://www.nicd.ac.za/wp-content/uploads/2018/05/Listeriosis-outbreak-situation-report-draft-_07May2018_fordistribution.pdf >.

legislation originating in his department as having to be "catalysts for radical socio-economic transformation," not only a source of "food security for all" but also capable of the "creation of one million decent jobs by 2030" (DAFF, 2015).

Generally, the South African authorities work to keep up with global trends although legislation is only infrequently updated and implemented due to a combination of limited resources and concerns about too frequent changes overwhelming the system.

3.17.2 Food regulatory system

RSA is a constitutional democracy with a three-tier system of government in the form of national, provincial, and local levels constitutionally "distinctive, interdependent, and interrelated." Each of these levels has both legislative and executive authority in their designated spheres.

Food safety and food control systems in RSA are developed and enforced by three government departments: Department of Health (DOH),[191] Department of Agriculture, Forestry and Fisheries (DAFF),[192] and the Department of Trade and Industry (the dti[193]).

3.17.2.1 Complexity of the system—milk as an example

Each level of authority is limited to functions specifically identified in its own legislation. The complexity of the South African system can be illustrated using milk and dairy products as an example.

The control of milk and dairy products is the responsibility of six different authorities at all levels involving four sets of legislation (WHO, 2017). Health-related standards are found in the Foodstuffs, Cosmetics and Disinfectants Act, 1972 and enforced by local and provincial authorities. Milk and dairy quality—related standards are the responsibility of the Directorate: Food Safety and Quality Assurance of DAFF and set by the Agricultural Product Standards Act, 1990. Under the Health Act, 1977, local and provincial authorities also enforce Milking Shed regulations although provincial animal health officers who are responsible for mastitis control. Although mastitis control involves milking shed hygiene, nevertheless the provincial health officers are not authorized to inspect milking sheds which fall under the jurisdiction of local authorities with their own by-laws governing milking sheds (WHO, 2017).

Another example of the complexity of the system is the handling of meat. Abattoirs are under the control of DAFF but once the meat leaves the abattoir control switches to DOH and a separate set of regulations. Food transportation and food handling come under the DOH and are addressed in two separate Acts, National Health Food Safety Basic Act, 2003 (Act of 2003) and Foodstuffs, Cosmetics Disinfectants Act, 1972 (Act 54 of 1972).

In a 2016 briefing to a parliamentary Joint Portfolio Committee on "interdepartmental food safety coordination," it was noted that

"The current situation has resulted in the duplication, overlapping, the inefficient use of scarce resources, multiple decision-making and a lack of coordination between government and nongovernment organizations involved in food control and inspection in South Africa" (Portfolio Committee, 2016).

3.17.2.2 Department of Agriculture, Forestry, and Fisheries

The overriding duty of DAFF, whose legislative mandate is derived from Sections 24 (b) (iii) and 27(1) (b) of the Constitution, is creating consumer confidence through ensuring products of consistent quality based on established standards and both the oversight and support of the agricultural sector thereby ensuring access to and adequate and safe food supply. Imported products are expected to conform to the same standards as those set for locally produced products.

DAFF regulates the sustainability and safety and quality of agricultural and animal products including agricultural processed products, perishables, flowers, and vegetables. DAFF is also responsible for the export of agricultural products, import and export of meat (fresh, frozen, chilled), and registration of pesticides.

Responsibilities for these various activities are distributed among a number of Directorates and Sub-Directorates:

- Food Safety and Quality Assurance
  - This Directorate is responsible for promoting the safety and quality assurance of food of plant and animal origin in terms of Agricultural Product Standards Act, 1990 (Act No. 119 of 1990), and to control the production, sale,

191. < http://www.doh.gov.za/ >.
192. < http://www.daff.gov.za/ >.
193. This is not a mistake. This is the abbreviation used in official documents.

import, and export of certain alcoholic products in terms of the Liquor Products Act, 1989 (Act 60 of 1989). This is also the WTO contact point for SPS notifications.

- Veterinary Public Health
  - This branch facilitates the provision of preventative veterinary activities and regulates the production of safe animal products such as meat, poultry, ostrich, and their by-products. Through local inspectors, DAFF is responsible for the inspection of agricultural products for the presence of insects, pests, and plant and animal diseases.
- Agricultural Inputs Control directorate
  - This Directorate regulates the manufacturing, distribution, importation, sale, use and advertisement of fertilizers, animal feeds, pesticides, stock remedies as well as the operation of sterilizing plants and pest control operators in terms of the Fertilizers, Farm Feeds, Agricultural Remedies, and Stock Remedies Act, 1947 (Act No. 36 of 1947).
- Sub-Directorate Agricultural Product Quality Assurance
  - Establishes criteria for and standardizes quality norms for agricultural and related products such as quality, packaging, marking and labeling, the chemical composition, and microbiological contaminants of the products. These norms are based on the specific needs of the South African market and also harmonized with international standards.
  - It is also the Directorate's responsibility to appoint assignees to undertake inspections at the point of sale, manufacture, packing, or export to ensure that the set standards and requirements are maintained.
- Agricultural Production, Health and Food Safety section (APHFS)
  - responsible for administering legislation applicable to the management of SPS issues in regard to animal diseases, food safety, and plant pests and ensuring compliance to relevant regulatory frameworks;
  - The National Sanitary and Phytosanitary Coordinating Committee (NSPS CC). Serves to coordinate the national sanitary and phytosanitary (SPS) and associated Technical Barriers to Trade (TBT)—related issues.[194]

3.17.2.3 Department of Health (DOH)

The core function of the DOH is ensuring that all foodstuffs shall be safe for human consumption in terms of the Foodstuffs, Cosmetics, and Disinfectant Act, 1972 (FCD Act) and the National Health Food Safety Basic Act, 2003. The activities of DOH are governed by approximately 50 sets of Regulations divided into 13 categories.

"Food control" as defined in the National Health Food Safety Basic Act, 2003 and applicable to the DOH, constitutes "a mandatory, regulatory activity of enforcement by the competent health authority to provide consumer protection and ensure that all food during production, handling, storage, processing, and distribution is safe, wholesome and fit for human consumption; conform to safety requirements; and are honestly and accurately labeled as prescribed by law" (DOH, 2004).

DOH has direct administrative and supervisory functions at all three levels of government—parliamentary, provincial, and municipal—as described below:

- The Directorate: Food Control (the highest level)[195];
  - Administers food legislation including developing and publicizing regulations for food safety, food labeling and developing technical guidelines; regulates the use of food additives, sweeteners, and other ingredients;
  - Informs, educates, and communicates to industry, consumers, the media, government departments, and other stakeholders about food safety matters;
  - Coordinates national activities, such as food product recalls;
  - Establishes national norms and standards;
  - Provides support to provincial and municipal authorities;
  - Serves on national and international bodies that deal with food control matters, e.g., Africa's National Contact Point for the joint FAO/WHO Codex Alimentarius Commission (CAC), International Food Safety Authorities Network (INFOSAN), and the European Union Rapid Alert System for Food and Feed (RASFF);
  - Evaluates risk assessments for DAFF related to agricultural chemicals and food produced through biotechnology;
  - The Food Control Division of DOH is responsible for all ready-to-eat food products;
  - Develops regulations for importers and exporters.
- Provincial level (Environmental Health Services of DOH)

194. Each South African Development Community (SADC) Member State is required to establish a National Committee on SPS Matters to administer the SPS Annex to the SADC Protocol on Trade. Included are the WTO SPS NNA and Enquiry Points (Article 7 of the WTO SPS Agreement).
195. Control at a national level is guided by the broad objectives within the Health Sector Strategic Framework's 10 Point Plan 2000–05 (1999).

    - Coordinates activities within the province, e.g., specific food monitoring programs, food safety alerts;
    - Supports local authorities;
    - Provides specialized services such as import control, e.g., audits and supports Port Health Services.
- Local metro (municipal) level authorities (Environmental Health Services)
    - Provide Environmental Health Practitioners (EHPs) whose duties include food control;
    - Investigates complaints;
    - Enforcement and compliance monitoring;
    - Identifying/controlling health hazards;
    - Employ inspectors to carry out food inspections under the national food safety control program. (DOH does conduct periodic audits of these inspectors.)

As noted, local authorities play a major role in the area of food safety control. Should a local authority not be able to fulfill these functions, provincial authorities are expected to perform them.

3.17.2.4 Department of Trade and Industry (the dti)

In regard to the food industry, the dti and its subsidiary agencies is responsible for

- Accreditation of Calibration and laboratory facilities[196]
- Monitoring and enforcement of compulsory specifications in the food industry including disinfectants
- Legal metrology[197]
- Consumer protection by the National Consumer Commission.[198]

The National Regulator for Compulsory Specification (NRCS) directorate of the dti,[199] which took over the role previously held by the South African Bureau of Standards, is the division of the dti responsible for administering compulsory specifications and other technical regulations in regard to human health, safety, and the environment. In addition to its standards setting role for both local and imported products such as seafood and canned meat and fish products, NRCS has the power to apply sanctions, prevent sales, and seize and destroy products that fail to meet the minimum required standards through the Standards Act, 1993.

3.17.3 Major laws

3.17.3.1 The Constitution

Section 27(1) (b) of the RSA Constitution specifically recognizes a right to food stating that "everyone has the right to have access to … sufficient food and water."

This right is expanded in Section 27(2) which further provides that the "state must take reasonable legislative and other measures, within its available resources, to achieve the progressive realization of each of these rights." Section 28(1) (c) provides for the right to basic nutrition for children and in section 35(2) (e) as a right for detainees and sentenced prisoners.[200]

A number of legislative acts, some overlapping and even contradictory, cover food and food products within the Republic of South Africa (RSA). The following Table 3.17.1 summarizes key legislation in the area of food law.[201]

---

196. < http://www.sanas.org.za >.
197. < http://www.nrcs.org.za/ >.
198. < http://www.thencc.gov.za/ >.
199. < http://www.nrcs.org.za/ >.
200. Constitution of the Republic of South Africa (1996), as amended. Chapter 2, Bill of Rights. A related, and currently controversial constitutional provision, Section 25(4) (a) requires a commitment to land reform and initiatives to bring about equitable access to all South Africa's natural resources.
201. Where relevant, some legislation is discussed in other sections.
202. For the relevant Regulations pertaining to these Acts, e.g., Regulations governing general hygiene requirements for food premises and the transport of food of 12 July 2002; Regulations relating to labeling of alcoholic beverages (No. 109 of 2005), refer to the individual Department websites.
203. The CPA is an overarching piece of legislation that cuts across all industries. It was enacted in order to promote and protect the economic interests of consumers and improve access to, and the quality of information necessary for consumers to make informed choices regarding their well-being and safety.
204. Municipal Health Services are also referred to as Environmental Health Services and rendered by Environmental Health Practitioners (EHP's) (previously known as Health Inspectors).
205. A "Certificate of Acceptability" must be obtained by owners or persons in charge of food-serving facilities from the local authority before food can be handled by any person.

TABLE 3.17.1 Key parliamentary legislation relevant to food law summarized.[202]

ACT	Purpose	Implementation
Agricultural Pests Act No. 36 of 1983.	Provides for measures by which agricultural pests may be prevented and combated; includes restrictions on importation; requires reporting of certain agricultural pests.	DAFF - Directorate: Plant Health (DPH).
Agricultural Products Standards Act No. 119 of 1990 (amended as Agricultural Product Standards Amendment Act 63 of 1998)	Controls and promotes specific product quality and food safety standards for the local market and for export purposes for food of plant origin; provides for control over the import and export of certain agricultural products; controls the sale of certain imported products; appointment and authorization of assignees to do physical inspections under the Act.	DAFF: Directorate: Food Safety and Quality Assurance; perishable products export control Board (PPECB).
Animal Diseases Act No. 35 of 1984.	Regulates on-farm disease control and promotes health of the animals; control of animal diseases and parasites; authorizes permits required for import meat/meat products.	DAFF - Directorate Veterinary Services; Directorate Animal Health)
Consumer Protection Act No. 68 of 2008.[203]	Establishes the National Consumer Commission (NCC), the primary body responsible to ensure consumer protection, e.g., investigations regarding poultry and red meat.	The dti - national Consumer Commission (NCC) - in partnership with NRCS and Municipal environmental Health practitioners.
Fertilizers, Farm Feeds, Agricultural Remedies and Stock Remedies Act, 1947 (Act No 36 of 1947)	Regulates the importation, sale, acquisition, disposal or use of fertilizers, farm feeds, agricultural remedies and stock remedies; provides for the designation of technical advisers and analysts; and to provide for matters incidental thereto.	DAFF
Foodstuff, Cosmetics and Disinfectants Act No. 54 of 1972. (Replaced the Food, Drugs and Disinfectant Act No. 13 of 1929.)	Controls the sale, manufacture and importation and exportation of food stuffs, cosmetics, and disinfectants; protects the consumer from false or misleading food claims; provides sufficient information to make informed choices; provides for labeling of food products (Consumer Protection Act, 2008: regulates labeling of biotech foods and ingredients.); source of the statutory mandate of local authorities to participate in food control activities.	DOH - Directorate: Food control; local enforcement by 52 metro/district municipalities[204]; control of imported foodstuffs by 9 provinces.
International Health Regulations Act, 1974 (Act 28 of 1974)	Provides for the approval of the source of food for consumption at ports, airports, on vessels and on aircraft including the inspection of such premises and the sampling of food by local authorities.	DOH - provincial health departments.
Liquor Products Act, No. 60 of 1989.	Controls the sale and production of liquor and certain liquor products; addresses requirements for all liquor products except beer, sorghum, and medicine; mandates analytical laboratory testing of Liquor products intended for export, import, and national trade for purposes of ensuring safety and quality; tests for compositional characteristics of liquor products including regulated and illegal additives; nonroutine testing such as determination of the authenticity of South African Sauvignon blanc wines.	DAFF - Directorate: Food Safety and Quality Assurance; Additives Laboratory.

Continued

TABLE 3.17.1 Key parliamentary legislation relevant to food law summarized.[202]—cont'd

ACT	Purpose	Implementation
Meat Safety Act No. 40 of 2000	Provides for regulation of meat safety and the safety of animal products; to establish and maintain essential national standards in respect of abattoirs; to regulate the importation and exportation of unprocessed meat; to establish meat safety schemes; manages the national microbial and residue monitoring program.	National (DAFF - Import/Export policy Unit of the Directorate Veterinary Services; Directorate: Food Safety; enforced by the Departments of Agriculture of the nine provinces.)
Medicines and Related Substances Act, 1965 (Act 101 of 1965) as amended, the Medicines and Related Substances Amendment Act, 72 of 2008; Medicines and Related Substances Amendment Act, 14 of 2015.	Regulation of Complementary Medicines including Sport and Health Supplements is considered "Complementary Medicines.".	DOH
National Health Food Safety Basic Act, 2003 (Act of 2003) (replaces Health Act No. 63 of 1977).	In respect of monitoring of *foodstuffs produced, manufactured, and sold locally.* Regulates requirements for transportation and the hygiene of food premises; local foodstuffs; the inspection of premises; stipulates the powers and duties of local authorities; regulations applicable to all food handling situations e.g., restaurants, café's, shabeens, taverns, caterers/suppliers at special events.[205]	DOH – Municipality Health services (8 Metro's and 44 district municipalities)
National Regulator for Compulsory Specifications Act No. 5 of 2008.	Provides for the existence of the NRCS which is responsible for compulsory standards regarding canned meat and frozen/canned fish; exercises import/export control over these products	National (DTI - NRCS)
	National Regulator for compulsory Specifications Act, 2008 (Act No. 5 of 2008) – enforces and sets standard specifications for domestic and imported seafood and canned fish products	
Plant Health Act, No. 35 of 1984.	Phyto-sanitary requirements. (Due deference given to the WTO SPS Agreement)	DAFF

3.17.4 Additional aspects

3.17.4.1 Codex Alimentarius

The only country from Africa at the first session of the Commission, the decision-making body of the Joint FAO/WHO Food Standards Program, in 1963, RSA, formally acceded to the Codex in 1994.

On November 17, 2016, South African Minister of Health approved in Government Gazette No. 1425 the adoption of the Codex General Standard for Food Additives (GSFA). The changes made by the introduction of this gazette will enter into force on November 17, 2017. The intention is to align as far as possible with Codex requirements although at this stage not all regulations are consistent with it. When faced with an absence of local regulations in a particular area, the goal is to follow Codex requirements.

3.17.4.2 Food and associated industries

Part of the NRCS, the FAI is responsible for monitoring and surveillance of factories, processes, and products.[206] FAI is an internationally accredited inspection body and fully complies with *SANS/ISO 17020 General Criteria for the Operation of*

206. < https://www.nrcs.org.za/content.asp?subid=16 >.

Various Types of Bodies Performing Inspection. Recognized by EU authorities as the competent authority for inspection of fish and fishery products for Europe, the FAI assists local and international compliance. FAI inspectors are also recognized by Chinese and Russian authorities.

Its inspectors, some of whom are accredited as expert members of international food inspection bodies are authorized to carry out inspections on behalf of the DOH as well as the NRCS's border enforcement duties. The FAI participates in both national and international food safety activities including those of the Codex Alimentarius Commission as required by the WTO Sanitary and Phytosanitary (SPS) and Technical Barriers to Trade (TBT) agreements.

3.17.4.3 GMO

"The production area of genetically engineered (GE) corn, soybean, and cotton in South Africa is estimated at around 2.7 million hectares. An estimated 94% of corn plantings, more than 95% of soybeans plantings, and all cotton plantings in South Africa are with GE seeds. This confirms South Africa as the ninth largest producer of GE crops in the world and by far the largest in Africa" (GAIN Report: Agricultural Biotechnology Annual Biotechnology in South Africa, 2019).

The primary legislation in RSA dealing with GMOs is the Genetically Modified Organisms Act of 1997 (GMO Act) and its subsidiary legislation.[207] The legislation, which is directed toward limiting potential risks to the environment and human and animal health, provides for regulation of different GMO classes such as laboratory or glasshouse development ("contained" use), as food and feed ("commodity clearance"), environmental release ("field" trials), commercial release ("general" release), and transboundary movement.

Under the South African Committee on Genetic Experimentation (SAGENE), research into genetically modified (GM) crops has occurred in RSA since 1979 with the first field trials being conducted in 1989. GM insect-resistant cotton and maize was approved for commercial release in 1997. Since the beginning of 2008, the South African government has granted over 1000 permits for GM maize experimentation, cultivation, import, and export (Jaffer, 2014).

The GMO Act, amended in 2006, fulfills all the obligations under the Cartagena Protocol on Biosafety, which was included as an annex to the Amendment Act (Act No. 23 of 2006). As shown in the following Table 3.17.2, there are also a number of other laws imposing additional rules on GMO-related activities.

The GMO Act established three regulatory bodies with specific functions: the Executive Council (EC), the Registrar, and the Advisory Council (AC).

The EC, comprising experts appointed by the Minister (DAFF) from those departments specified in the GMO Act, has various functions impacting GMO-related activities and their monitoring. The Registrar, appointed by the Minister in consultation with the EC, is in charge of administering the GMO Act exercising those powers and duties specifically provided in the GMO Act, e.g., examining applications for GMO-related activity, issuing permits or extensions, and amending or withdrawing permits; ensuring that all necessary measures are taken to protect the environment as well as human and animal health; arranges for inspection of GMO facilities by inspectors appointed by the Registrar. These inspectors have the power to conduct routine, unannounced, and warrantless inspections of facilities registered for conducting GMO-related activities, and take samples of GMOs.

The AC is a national advisory body on all matters having to do with GMO-related activities including drafting of GMO-related laws and guidelines. Of its 10 members appointed by the Minister, 8 must be knowledgeable in the field of GMO-science and two must be from the public sector. Of the latter, one must have expertise in ecological issues and the other familiar with the effects of GMOs on human and animal health. This Committee evaluates risk assessments, comprising scientific data relating to food, feed, and environmental impact, submitted with every application. Based on the findings of the Committee, the application is recommended to the Executive Council for a decision. Comments from the public may also be considered in evaluating an application.

The process of risk assessment suggested by DAFF includes various "scenarios" such as the "likelihood that the competitive abilities of wild relatives occurring in undisturbed wild-lands will be altered by hybridization with transgenic crops," the "likelihood that a new type of arable weed will be produced by gene flow between the transgenic crop and its relatives," and the "likelihood that the transgenic crop will become a volunteer problem on arable land or wild areas" (DAFF Guideline, 2004). The recommendation report, outlining the conclusions of the review, should include information about the following safety issues: (i) Food and feed (toxicological studies, allergenicity studies, compositional analysis,

207. The 1997 GMO Act is available on the Department of Agriculture, Forestry and Fisheries (DAFF) website, at < http://www.daff.gov.za/doaDev/sideMenu/acts/15%20 GMOs%20No15%20% 281997%29.pdf >. The Genetically Modified Organisms Amendment Act No. 23 of 2006 (Apr. 17, 2007) is available at < http://www.info.gov.za/view/DownloadFileAction?id=67850 >.

TABLE 3.17.2 Key GMO legislation in South Africa.

ACT	Purpose	Implementation
Genetically Modified Organisms Amendment Act (No. 23 of 2006), and its subsequent amendments and regulations. (amended the genetically Modified Organisms Act No. 15 of 1997)	Research and development, i.e., contained and field trial use, general release activities, import and export, transport, use; revisions ensure compliance with the Cartagena protocol on Biosafety in 2003 (included as an annex).	DAFF: Directorate Genetic Resource: Biosafety; Administered by the Registrar of the GMO Act; two regulatory bodies (i) the executive council and (ii) the Advisory committee evaluate and decide on applications.
National Environmental Management Biodiversity Act (Act no. 10 of 2004; NEMBA)	Monitors the environmental impacts of released GMOs; provides a mechanism whereby the Minister of environmental Affairs may request a GMO environmental impact assessment (EIA).	Department of Environmental Affairs (DEA); GMO Research and Monitoring Unit of South African National Biodiversity Institute (SANBI).
National Environmental Management Act Amendment Act (Act no. 8 of 2004) (amended National Environmental Management Act (Act no. 107 of 1998)).	Provides general principles for decision making with regards to activities affecting the environment;	DEA (which has published guidance regarding objectives of an EIA and the administrative procedure to follow once criteria triggering an EIA have been met).
Foodstuffs, Cosmetic and Disinfectants Act (FCD Act; Act No. 54 of 1972)	Mandatory GM food labeling regulations requiring that GMO foodstuffs must be labeled—where they differ significantly from existing foodstuffs, i.e., composition, nutritional value, mode of storage, preparation, allergenicity, or genes with human or animal origin—must be labeled).	DOH.
Consumer Protection Act (CPA, Act No. 68 of 2008); Regulations (R. 293 of 2011).	Requires labeling for all GM goods; Regulations require food producers, importers, and packagers to choose one of three mandatory labels for GM foods (see text)	The dti.

nutritional analysis pathogenicity, feeding trials) and (ii) Environmental (weediness/invasiveness, gene flow, altered plant pest potential, nontarget organisms, impact on biodiversity) (DAFF Guideline).

Foodstuffs obtained through genetic modification must be labeled as such.[208] Regulations (R. 293 of 2011) were published requiring the use of one of three mandatory labels for GM foods: (i) "containing GMOs" where the GM content is at least 5%; (ii) "produced using genetic modification" for food produced directly from GMO sources; or (iii) "may contain GMOs" when argued that it is scientifically impractical and not feasible to test food for GM content.

Voluntary labels include (i) "does not contain GMOs" where the GM content is less than 1%; (ii) "GM content is less than 5%" where GM content is between 1% and 5%; and (iii) "may contain genetically modified ingredients" if it cannot be detected.

In 2012, draft amendments were published changing the wording from "labeling genetically modified organisms" to "labeling genetically modified ingredients or components." The importance of this change is that *individual ingredients* will have to be labeled as "containing GM" in the ingredients table and not the whole product.

The law imposes this requirement if the composition, nutritional value, mode of storage, preparation, or cooking "of the foodstuff differs significantly from the characteristic composition of the corresponding existing foodstuff …"[209] The Consumer Protection Act, which imposes labeling requirements on food items containing a certain level of GMOs, also criminalizes certain acts. It makes it an offense for anyone to "alter, obscure, falsify, remove, or omit… labeling… without authority."[210]

208. Regulations Relating to the Labeling of Foodstuffs Obtained Through Certain Techniques of Genetic Modification, GG No. 25908 (Feb. 26, 2010).
209. *Id.* § 2.
210. Consumer Protection Act No. 68 of 2008, § 110, 526 GG No. 467 (Apr. 29, 2009).

In a 2013 review of the South African GMO governance noted that "... the robustness of the governance of GMOs is reflected in the track record of safety since 1990, as no significant impact from accidents or adverse effects on humans or animals have been recorded" (Jansen van Rijssen 2013).

3.17.4.4 Pesticide screening

To protect both consumer health and the environment, most countries have established Maximum Residue Limits (MRLs) for pesticides used in agriculture. In order to satisfy these goals and ensure that exports comply with these standards,

South African legislation requires that only chemical remedies registered in terms of The Fertilizers, Farm Feeds, Agricultural Remedies and Stock Remedies Act (Act No. 36 of 1947) are used and agricultural products for export should not exceed the maximum MRLs for the importing countries.[211]

The DAFF Directorate: South African Agricultural Food, Quarantine and Inspection Services (SAAFQIS) laboratories have the responsibility of auditing agricultural products for compliance with maximum residue limits (MRLs) as prescribed by both internationally agreed standards and individual importing countries. Laboratory analysis includes searching for banned or illegal substances.

The Perishable Products Exporting Control Board (PPECB) provides fruit, vegetable, and tea samples mainly from consignments intended for export. Wine samples are screened if requested by exporters but are also screened to monitor the product from different wine-growing areas.

Commodities are analyzed for pesticide residues using gas- and liquid chromatography with each product having its own range of pesticide specifications. The Chemical Residue Laboratory only reports results and does not decide if a consignment is rejected or accepted.

3.17.5 Labeling

Food labeling is primarily the responsibility of the DOH, Directorate: Food Control.[212] Relevant legislation in this regard includes the Foodstuffs, Cosmetics, and Disinfectants Act, no 54/1972, the Agricultural Product Standards Act and the National Health Act (Act 61 of 2003), and the Regulations that fall under each.

In March 2010, the "New Regulations Relating to the Labeling and Advertising of Foodstuffs" (No. R. 146), which had not been updated since 1993, were published by the DOH, coming into effect in March 2012.[213] These were intended to fulfill an interim role with more comprehensive labeling planned at a later time. As part of its goal to encourage a healthier lifestyle, the DOH wanted to ensure that labels should be legible, accurate, and informative; misleading information was removed and consumers educated (Koen, 2016).

In 2014, the Department of Health proposed new Regulations (R. 429) which are intended to replace the Regulation 146.[214] Updated requirements include mandatory nutritional information and regulations regarding health claims, reduction of disease risk claims, and slimming/weight loss claims. Another aspect of the proposed regulations involves marketing to children. Not only is advertising unhealthy food to schoolchildren not permitted but even allowed ads may not contain such manipulative devices as references to celebrities, sports stars, or even cartoon characters.

Foodstuffs are required to be described in such a manner that the information related to the contents and/or composition of a product is indicated in close proximity to the product's name on the main panel of the packaging in letter sizes stipulated in the regulations. Ingredients must be listed in order of descending mass. To include the logos of endorsement entities, such as Weigh Less, Diabetes SA, etc., in order to promote the food's nutritional or diet-related characteristics, the approval of the Director General of DOH is required. The relevant organization must provide proof that it is involved in health promotion supported by evidence-based nutrition. However, any representation by words, logos, marks, etc., that suggest the food product has been endorsed by a medical entity such as a health practitioner or a professional organization is prohibited.[215]

---

211. South African MRL Lists are available on the DAFF website at < https://www.daff.gov.za/daffweb3/Branches/Agricultural-Production-Health-Food-Safety/Food-Safety-Quality-Assurance/Maximum-Residue-Limits >.
212. Department Of Health, Republic of South Africa. Regulations relating to the labeling and advertising of foodstuffs: (R146/2010) [homepage on the Internet]. c2016. Available from: < http://www.health.gov.za/index.php/2014-03-17-09-38/legislation/joomla-split-menu/category/96-2010r >.
213. Foodstuffs, Cosmetics and Disinfectants Act, 1972 (Act 54 of 1972). Regulations relating to the labeling and advertising of foodstuffs. No. R 146. Government Gazette, No 32975, 1 March 2010. Available at < http://www.kwanalu.co.za/upload/files/reg0146.pdf >.
214. Department of Health, Republic of South Africa. Regulations relating to the labeling and advertising of food: amendment (R429/2014). Pretoria: Department of Health, 2014).
215. 13(a).

Mandatory information required by the new regulations includes the name and address of the food manufacturer, importer, or distributor, the country of origin ("product of … ," "produced in … ," "packed in … ," etc.), the net content expressed in the International System of Units, and a specific date by when the product should be used. Words such as "nutritious," "healthy," "wholesome," etc., are no longer permitted. The regulations include a list of permitted health and nutrient claims and identify statements and descriptions that are considered misleading and therefore prohibited.

Following comments by industry and the scientific community, a nutrient profiling system applicable to all categories of food was proposed. Following studies and validation evaluations, a modified version of the *Food Standards Australia New Zealand* model (released in 2012) was recommended for South African use to determine whether a food product is eligible to carry a nutrient and/or health claim[216] (Koen, 2016).

In order to add nutritional claims, e.g., high in fiber, low fat, to the label, the manufacturer must have had the food tested in an accredited laboratory and a nutritional table to support the claim must be shown. Terms such as "Sugar free" or "Fat free" are permitted provided that they conform to the specific criteria stipulated in the food labeling regulations. For example, foods can only be labeled as "Low fat" if they contain no more than 3 g of total fat per 100 g (solids) or 1.5 g of total fat per 100 mL (liquids).

Common allergens (gluten, milk, eggs, soya, peanuts and tree nuts, shellfish or crustaceans, or significant cereals wheat, rye, barley, oats, and titricale which is a cross between wheat and rye) must be indicated on food labels.

Uncommon allergens (i.e., an allergen which is not classified as a common allergen as described above) must be disclosed by manufacturers upon request by a consumer, inspector, or the DOH.

Regarding Health Claims, "function" claims may be made for specifically identified nutrients listed in the legislation provided the appropriate wording is used (57(1)). So, for example, regarding beta-carotene, the label may state "Beta-carotene functions as a tissue antioxidant and so keeps cells healthy" or, as an example of the various vitamins referenced, "Vitamin A is necessary for normal vision/for the maintenance of good vision" (R. 429, Table 4). Also permitted are "Reduction of Disease Claims" (e.g., regarding calcium and osteoporosis, "Regular exercise and a healthy diet high in calcium and an adequate Vitamin D status may assist to maintain good bone health and may reduce the risk of osteoporosis or osteoporotic fractures later in life") (R. 429, Table 5). Other categories of approved claims include "Approved health claims for oral health," "100% intact whole grain" health claim, and "Approved health claims for physical performance/exercise."

Only claims specifically identified in the legislation are permitted. The regulations fail to provide for "the potential approval or assessment of any new health and nutrition claims in the event that valid and scientific claims are discovered in future." (GAIN, 2014) Where health claims or nutrient claims form part of the trade or brand name, the use of the trade or brand name must eventually be phased out. An interesting, and perhaps uniquely South African, requirement is that foods displaying a religious symbol (e.g., Kosher, Halaal) are required to be made available without such a symbol (16(1) (b)).[217]

Enforcement of the Regulations is the responsibility of the Environmental Health Practitioners (EHPs), employed by the Municipal Health Services of the metro and district municipalities. The local municipalities and EHPs are supported by the Food Control Directorate of the DOH in regard to the interpretation and implementation of the labeling provisions. The Food Control Directorate also provides support to the Port Health Services of the nine provinces, who are responsible for the control of imported foodstuffs.

During the Parliamentary debate on these regulations, Dr. Ntsuabele (Director of Food Safety and Quality Assurance, DAFF) raised concerns about the government's capacity to implement these regulations resulting in variable compliance among international and local food companies with only the larger and mainly international companies being able to comply with the regulations (GAIN Report, 2014).

216. The FSANZ model is based on the nutritional value per 100 g of food, and three categories are used. A nutrient profile calculator is available on the website of the Department of Health, South Africa at < http://www.health.gov.za/phocadownload/FoodInfor/NPC_NWU.html >.

217. "The following information or declarations shall not be reflected on a label or advertisement of a food: … endorsements by specific religious entities, unless food business operators give consumers their constitutional right off freedom of choice, by making such foods without any particular religious endorsement available on the shelf at all times." (R. 429. 16(1) (b)).

References

Briefing to The Joint Portfolio Committees, February 2–3, 2016. Progress Report on the Inter-departmental Food Safety Coordination Committee and the Establishment of Food Control Agency in South Africa. Available at: https://www.google.com/search?q=PROGRESS+REPORT+ON+THE+INTER-DEPARTMENTAL+FOOD+SAFETY+COORDINATION+COMMITTEE+AND+THE+ESTABLISHMENT+OF+FOOD+CONTROL+AGENCY+IN+SOUTH+AFRICA&rlz=1C5CHFA_enUS512ZA521&oq=PROGRESS+REPORT+ON+THE+INTER-DEPARTMENTAL+FOOD+SAFETY+COORDINATION+COMMITTEE+AND+THE+ESTABLISHMENT+OF+FOOD+CONTROL+AGENCY+IN+SOUTH-+AFRICA&aqs=chrome.69i57.1432j0j8&sourceid=chrome&ie=UTF-8.

Consumer Protection Act: Regulations (G.N. No. R. 293 of 2011).

Department of Agriculture, Forestry and Fisheries, March 2015. 2015/16 to 2019/20 Strategic Plan. Available at: https://www.daff.gov.za/doaDev/topMenu/DAFF_SP_%20complete.pdf.

DOH, June 2013. Report: food safety and food control in South Africa: specific reference to meat labelling. Report from the Department of Health (DOH); Department of Trade and Industry (DTI) and Department of Agriculture. Forest. & Fish. (DAFF) 21. Available at: http://pmg-assets.s3-website-eu-west-1.amazonaws.com/130621food.pdf.

Department of Health, June 2004. Policy Guidelines: National Food Safety Alerts and Official Food Product Recalls in South Africa. Available at: https://www.ehrn.co.za/download/reg_meattrans.pdf.

Food Safety and Nutrition, 2017. Food Law Guidelines. Regional Office for Africa. WHO. Available at: https://www.afro.who.int/sites/default/files/2017-06/Food%20Safety%20and%20Nutrition%20Food%20Law%20Guidelines.pdf.

Genetically Modified Organisms Act, 1997 (Act No. 15 of 1997), May 2004. DAFF. Available at: https://www.nda.agric.za/docs/geneticresources/applicant%20-%20%20May%202004.pdf.

GAIN Report: "Amendments to Regulation Relating to Food Labelling and advertising.", August 22, 2014. USDA Foreign Agricultural Service. https://gain.fas.usda.gov/Recent%20GAIN%20Publications/Amendments%20to%20regulation%20relating%20to%20food%20labelling%20and%20advertising_Pretoria_South%20Africa%20-%20Republic%20of_8-26-2014.pdf.

GAIN Report: Agricultural Biotechnology Annual Biotechnology in South Africa, February 5, 2019. Available at: https://gain.fas.usda.gov/Recent%20GAIN%20Publications/Agricultural%20Biotechnology%20Annual_Pretoria_South%20Africa%20-%20Republic%20of_2-5-2019.pdf (Gain: Annual Biotechnology in South Africa.).

Jaffer, Z., August 26, 2014. SA Only Country Allowing GM Staples: We Have No Choice. Journalist. Available at: http://www.thejournalist.org.za/kau-kauru/gm-staples.

Jansen van Rijssen, F.W., Jane Morris, E., Eloff, J.N., August 7, 2013. A critical scientific review on South African governance of genetically modified organisms (GMOs). Afr. J. Biotechnol. 12 (32), 5010–5021.

Koen, N., Blaauw, R., Wentzel-Viljoen, E., 2016. Food and nutrition labelling: the past, present and the way forward. S. Afr. J. Clin. Nutr. 29 (1), 13–21.

Further reading

Annual Report, 2017/2018. Department of Agriculture, Forestry and Fisheries. Available at: https://www.daff.gov.za/Daffweb3/Portals/0/Annual%20Report/AR_Final_28%20September.pdf.

Department of Health, May 29, 2014. Foodstuffs, Cosmetics and Disinfectants Act, 1972 (Act no.54 of 1972): Regulations Relating to the Labeling and Advertising of Foods: Amendment. No. R. 429 (R. 429).

Labadarios, D., Mchiza, Z.J., Steyn, N.P., Gericke, G., Maunder, E.M., Davidsa, Y.D., Parkera, W., 2011. Food security in South Africa: a review of national surveys. Bull. World Health Organ. 89, 891–899.

Government of the Republic of South Africa and Others V Grootboom and Others (CCT11/00) [2000] ZACC 19; 2001 (1) SA 46; 2000 (11) BCLR 1169, October 4, 2000. Available at: http://www.saflii.org/za/cases/ZACC/2000/19.html.

Guideline Document for Use by the Advisory Committee when Considering Proposals/Applications for Activities with Genetically Modified Organisms.

Regional Stakeholders Workshop on Sanitary and Phytosanitary Awareness Creation, September 12–13, 2013. Pretoria, South Africa, Available at: https://extranet.sadc.int/files/6213/8190/6085/National_SPS_Committee_-_Jeremiah_Manyuwa.pdf.

Restrictions on Genetically Modified Organisms: South Africa. Library of Congress, 2015. Available at: https://www.loc.gov/law/help/restrictions-on-gmos/south-africa.php.

Shisana, O., et al., 2013. South African National Health and Nutrition Examination Survey (SANHANES-1). HSRC Press, Cape Town. Available at: http://www.hsrc.ac.za/uploads/pageNews/72/SANHANES-launch%20edition%20(online%20version).pdf.

Standard Setting Process of the Codex Alimentarius Commission (CAC) A Handbook for Guidance of Participation of African Countries, 2012. African Union - Interafrican Bureau for Animal Resources.

The Right to Access to Nutritious Food in South Africa. 2016-2017. South African Human Rights Commission. Brief Prepared by Yuri Ramkissoon, Senior Researcher. Available at: https://www.sahrc.org.za/home/21/files/Research%20Brief%20on%20The%20Right%20to%20Food%202016-2017.pdf.

Chapter 3.18

Private food law

Bernd van der Meulen

GHI, Prof. Comparative Food Law, Renmin University of China School of Law, University of Copenhagen, European Institute for Food Law, Amsterdam, The Netherlands

3.18.1 Introduction

States and inter/supranational organizations are not the only actors regulating the food chain. From the 1990s onward, the business sector is involved in creating a parallel legal universe in national and international trade relations. For many producers, these market requirements have an impact just as big or even bigger than legal obligations. A field of "private food law" (Van der Meulen, 2011) is emerging. Private standards are often referred to as "voluntary.". They are voluntary in so far as legal obligations only occur for businesses that contractually undertake to comply with these standards. In situations where gate keepers to the market such as retail chains and other distributers make certification according to a certain standard a condition for market access, economically businesses have little alternative but to comply. In such situation the label "voluntary" has limited significance.

3.18.2 Triangular structure

Private food law consists of three structure elements. First of all, there is the standard. The standard expresses the requirements businesses promise their products meet. Second, an independent third party performs audits to check if businesses live up to their promises. If they do, they provide certification. The certificate is the third element. It is required and accepted as evidence that products from the certified supplier possess certain quality aspects.

3.18.3 Standards

Usually standards address credence aspects of foods. These aspects range from sustainability (such as marine stewardship council) to social fairness (Fair-Trade, UTZ), animal welfare (Beter leven), and food safety.

Food safety standards address primary production including fresh products (GlobalG.A.P.), industry brand products (FSSC 22000 - Food Safety System Certification based on ISO 22000), and retail brand products. The leading retail standards are in Europe the British Retail Consortium (BRC) and the International Featured Standard (IFS) and in United States of America, Australia, and New Zealand the Safe Quality Food standard (SQF). Private food safety standards are all based on HACCP as elaborated in the Codex Alimentarius. In this way, they bring an obligation to comply with Codex-HACCP requirements into the private contractual relation of (internationally) trading businesses.

3.18.4 Standard setting organizations

Standards can be placed in the market from any side in the food chain. Retail brands derive their position from the purchasing power of the big international retail chains. GlobalG.A.P. started as a retail standard. In a bid for democratization, producers have been given a position in the system as well. FSSC 22000 was the initiative of certification organizations. The Fair-Trade scheme has been created and is maintained by an NGO.

Sometimes public authorities are involved in setting standards, for example, in ISO (the International Standardization Organization), giving them a semipublic flavor.

3.18.5 Harmonization

The proliferation of private standards is a reason for concern. Producers may face the need of multiple certifications for the same issue. To deal with the situation, the globally leading retail chains have set up the Global Food Safety Initiative (GFSI). GFSI benchmarks food safety standards. Participating retailers have committed themselves to accepting products that are certified against any GFSI accepted standard: "certified once, accepted everywhere."

3.18.6 Enforcement

Compliance with private standards is enforced via certification. Auditors check compliance. In case of noncompliance, certification is denied or withdrawn. Customers may delist suppliers who are not or no longer certified.

3.18.7 Accreditation

Most countries provide systems of accreditation for certifying bodies. To acquire and keep accreditation, certifying bodies must provide evidence of the quality of their work.

References

van der Meulen, B., 2011. In: Private Food Law: Governing Food Chains Through Contracts Law, Self-Regulation, Private Standards, Audits and Certification Schemes. Wageningen Academic Publishers (Open access). http://www.wageningenacademic.com/doi/book/10.3920/978-90-8686-730-1.

Further reading

On Animal Welfare Standards, see: Carolina Toschi Maciel 'Public Morals in Private Hands? A Study into the Evolving Path of Farm Animal Welfare Governance' (Ph.D. thesis). http://edepot.wur.nl/343363. (Open access).

Chapter 3.19

Conclusions

Bernard Maister[1] **and Bernd van der Meulen**[2]

[1]*Intellectual Property Unit, University of Cape Town, Cape Town, South Africa;* [2]*GHI, Prof. Comparative Food Law, Renmin University of China School of Law, University of Copenhagen, European Institute for Food Law, Amsterdam, The Netherlands*

From this chapter, food law emerges as a two-sided concept. On the one hand, in many countries, it is that part of a legal system which labels itself as such. In EU law, there is even a definition. On the other hand, it is the field of interest of scholars and commentators who look at the law from the perspective of food and/or the food sector. Obviously, the delineation of this field of interest is open to those who have this interest. As became apparent in the section on the United States of America, opinions differ as to what should be included and what should be excluded. It seems generally accepted that the regulation of food safety and consumer protection should be included. The same consensus cannot be found regarding the question whether the human right to food should be included or rather approached as a separate subject.

It appears that around the globe, food legal systems were created or fundamentally altered in response to crises and incidents. Mostly, these are food safety crises, but also radical changes in political structures (for example in postsoviet countries) may spark restructuring of the food legal system.

This chapter demonstrates that food law has become an increasingly important national and international legal system impacting many countries.

The reasons for this are varied—outbreaks of foodborne disease have prompted government intervention, concerns regarding population health-related issues such as obesity, diabetes, etc., scientific advances in food-related science, concerns about global weather changes, …

This can be seen in both the range of countries covered in this chapter and the extent to which their legislation is involved in the process of bringing food to the consumer regardless of whether it originates "naturally" in the ground or the animal kingdom, is produced "artificially" from natural or scientific sources, or is processed or raw.

The recognition of the importance of food law is also demonstrated by the wide acceptance, and incorporation into national legislation, of the international standards originating from organizations such as the FAO and WTO.

The range and extent of food law discussed in the various country sections of this chapter demonstrate that thinking of food law as merely a limited part of a country's legal system would be doing the field a disservice. In fact, as described in all countries, the "specialty" of food law encompasses a wide range of activities ranging from genetic modification of seeds, administering antibiotics or hormones to farm animals, stipulating limits on advertising food products, border controls in the form of managing exports and imports ... the list goes on.

While the wide-ranging reach of food law may be reassuring to its practitioners, some concerns about its boundaries remain. In the United States section, for example, the significance of banning *foie gras* production is used an example of what may, or may not, be reasonably considered a "food law issue."

International Standards are important.

Whether successfully or not, whether to provide high quality food to its citizens, or just enhance its export numbers, each country has tried to incorporate international standards into its food system.

Yet despite the availability, and apparent clarity, of international standards, national regulations are not always well delineated. Even in well-established systems such as that of the United States, for example, issues exist in the area of food and color additives and what constitutes generally accepted "safe" substances.

Food law impacts many levels of government.

In most of the countries reviewed, the point is made that the implementation of food legislation involves more than one level of government.

This has implications for the enforcement of legislation. In some countries, South Africa is an example, while the legislation may satisfy current international standards, enforcement depends on local municipal inspectors.

Food labeling has matched growing consumer sophistication.

Consumers, concerned about health, are demanding more information about the products they consume. Public health campaigns have increased awareness of the risks of too much salt (hypertension), sugars/carbohydrates (obesity, diabetes), dairy products (allergies, milk intolerance), aluminum (Alzheimer's), calorie content (obesity, diabetes), etc.

A related area involves health/medicinal claims. Most countries accept that unsubstantiated health claims are unacceptable.

It is apparent that while great strides have been made in the area of food law, a great deal of work in this area still has to be done.

Consumers, educated or otherwise, will continue to be a force in directing food law. Demands for safer and healthier food will have an impact on many areas ranging from GMO research to the contents of labels to how animals are fed, raised, and processed.

Historically, it appears that food law evolution has been driven by consumers. Concerns about producing unsafe food, whether as a result of a progressive social conscience or fear of litigation, will force food producers, whether fast food purveyors or GMO-developing corporations, to lobby for new or modified laws.

National food laws will be impacted by developments in related societal areas such as socially responsible food harvesting whether from land (e.g., farming methods, environmental impacts) or sea (e.g., overfishing, pollution).

All countries (governments) recognize that it is their responsibility to feed their citizens.

But merely providing sufficient food for their population is not enough. Consumers throughout the world are now more educated, demanding, and discerning, particularly in regard to their health and wellbeing, both of which are directly linked to the food they eat.

While the countries reviewed in this chapter are different in many ways (political systems, geography, population size, level of infrastructure, etc.), all have accepted the importance of there being globally harmonized and scientifically based regulations.

One means of achieving this goal is through incorporating the principles and standards of the Codex Alimentarius. Azerbaijan, for example, in its "Strategic Road Map for 2016–20," targets upgrading its scientific resources to meet this challenge.

While the importance of global harmonization is well accepted, recognition is given to the fact that a country's regulations may not always be in accordance with Codex standards. This is well illustrated by an example from the United States where Congressional legislation regarding certain food colorants contradicts the Codex standard, thereby preventing the FDA from adopting the latter's standards regarding these colorants (see Section on the United States).

Another important area subscribed to by these countries involves "risk analysis," a term covering the procedures related to risk assessment (a scientific process), risk management (the work of regulators and legislators), and risk communication. Again, the Codex provides the global standard although the relationship of these activities can be complex.

Turkey, for example, has established a "risk assessment unit" within the Ministry of Agriculture and Rural Affairs whose function is to oversee a number of risk assessment "commissions" consisting of representatives from various disciplines such as research facilities, university faculty, etc. These commissions act in an advisory capacity regarding risk analysis formulations.

Finally, achieving universally harmonized legislation is meaningless without the power of enforcement. Again, the countries discussed in this chapter all recognize the importance of this but are at different levels of enforcement.

The reports of the individual countries in this chapter suggest that, while hurdles remain to be overcome, there is some hope that, via satisfaction of humankind's most basic need and desire, an adequate fulfilling diet, the linked goals of a healthy population, and universal cooperation may, with the appropriate scientific standards as guidance, be achieved.

Chapter 4

The global harmonization initiative

Huub Lelieveld and Veslemøy Andersen
Global Harmonization Initiative (GHI), Vienna, Austria

4.1 Introduction

It is 2020 and although the world population is still growing, in principle there is no need for malnutrition in the world. Earth has the capacity to feed everyone properly. There are many reasons for the still > 800 million undernourished people in the world (FAO et al., 2018). Some of them are regrettably beyond the control of most of us, they have to do with wars and corruption. There are, however, also other reasons that can be addressed by the many of us who are motivated to do so. The initiators of the Global Harmonization Initiative (GHI) singled out one of them when in 2004 it became obvious that too often food is destroyed by authorities for legally correct but scientifically and morally wrong reasons. In most cases when food is destroyed by authorities, this is because according to the local laws and regulations the food is deemed unfit for consumption, because it may harm the consumer. It was felt that scientists should be able to define when food can be considered safe or should be considered unsafe. Many of the food safety regulations have been made many decades ago, when detection limits of chemicals were such that when a potentially toxic substance was detected, the concentration indeed was such that it would do harm. Today, the limit of detection for most chemicals is at least a million (10^6) lower and that has changed the meaning of "absence" When regulations demand the absence of a substance, 50 years ago may have meant some milligrams per kilogram. Today the same regulation may stipulate less than a picogram or femtogram for the same substance. With perhaps few exceptions, politicians and judges are not food scientists or toxicologists and they follow the law. For them absence means absence and therefore the requirement of absence has to be replaced by a concentration figure based on good science. At the time most of the food safety regulations were developed, many people, including scientists and even many if not all toxicologists, maintained the idea that if a substance is harmful, it may do harm, in any quantity, although already in the 16th century Paracelsus has observed and written down that "Alle Dinge sind Gift und nichts ist ohne Gift; allein die dosis machts, daß ein Ding kein Gift sei" (Paracelsus, 1538). In English: "All things are poison and nothing is without poison; only the dose makes that a thing is no poison." On the same page Paracelsus writes that admission of a poison in the right amount can be healing. Because this is such a basic truth in food safety, GHI has produced a document, available in 20 languages, that explains the matter clearly, at least for scientists. It is the intention to produce other versions of the document that are also suitable for nonscientists, politicians, and legal professionals, and in even more languages (GHI, 2016).

As everybody has witnessed in the past decades, food is traveling over the world as never before. As a consequence, severe safety incidents that have the source in one country may easily be spread to many other countries. After the shocks caused by the mad cow disease (bovine spongiform encephalopathy, BSE) in the United Kingdom in 1987; various *Escherichia coli* O157:H7 outbreaks, starting in the US in 1993; adulteration of milk with melamine in China in 2008; and more recently the *Listeria monocytogenes* outbreaks in South Africa in 2017 and Spain in 2019, it is clear that good and enforceable food safety laws and regulations are needed. This is particularly so because of the cases where the responsible companies knew about the safety issue but failed to warn the authorities (examples are *Salmonella* in peanut butter in the US in 2009 and *Listeria* in pork in Spain in 2019).

Despite reactions from national governments to tighten legislation to restore diminishing consumer confidence, such food safety incidents illustrate the critical and urgent need for global harmonization of food safety laws and regulations. As the world grows more tightly connected, so does the impetus for harmonizing food safety regulations and reaching global consensus on their founding principles (Motarjemi et al., 2001). Globally, annually foodborne diseases still make 600 million people ill and cause 420,000 premature deaths. Most of the diseases are of microbiological nature (WHO, 2015).

Ensuring Global Food Safety. https://doi.org/10.1016/B978-0-12-816011-4.00004-5

4.2 Food and nutrient security

In the past decade, it has become very clear again that food security is still an immense problem. In 2018, the Food and Agriculture Organization of the United Nations (FAO) published "The State of Food Security and Nutrition in the World 2018″ that was jointly prepared by the FAO, the International Fund for Agricultural Development (IFAD), the United Nations Children's Fund (UNICEF), the World Health Organization (WHO), and the World Food Programme (WFP). The very extensive report describes what can be done and must be done to end food insecurity (hunger and malnutrition) by 2030, addressing climate changes in addition to the regrettably more common and perpetuating issue of wars. The report is freely downloadable (FAO et al., 2018), and therefore there is no point in reproducing any part of it. Instead, this chapter will focus on what food scientists together can contribute to diminishing food insecurity and how to do so without compromising food safety.

While one could argue that standards aimed at improving food safety will also reduce losses due to food spoilage, and consequently enhance food availability, it should still be noted that harmonization of food safety legislation will not single-handedly ensure all individuals access to safe and wholesome foods (Motarjemi et al., 2001). Proper consumer protection requires additional critical components, including food control management, food inspection, enforcement, laboratory services, foodborne disease surveillance systems and public information, education, and communication. This responsibility lies with all stakeholders—governments, food producers, food processors, retailers, consumers, as well as academic and scientific institutions. For a more detailed discussion of capacity and need for capacity building in food safety, refer to chapter 30: Capacity Building: Harmonization and Achieving Food Safety and chapter 31: Capacity Building: Building Analytical Capacity for Microbial Food Safety.

4.3 International standards

In 1963 the FAO and WHO jointly established the Codex Alimentarius Commission (CAC or Codex), mandated by the United Nations to produce standards to be adopted by its (currently 188) member countries. The purpose is protecting consumers' health and ensuring fair practices in the food trade. Codex has done a tremendous amount of work since its inception. For that purpose, Working Parties (WP) had been and continue to be formed that work on specific tasks. The intention is to obtain consensus about food safety requirements, particularly with respect to chemicals and microorganisms in food and food products. A huge task, not easy and never ending. The work is supported by governments who have delegates in the WP. These delegates are not necessarily scientists and are always representing their country and thus have to follow instructions from the governments they represent. Due to the involvement of the governments of 188 countries, the procedures to come to consensus for standards (WHO/FAO, 2018) are complex and time consuming. Nevertheless, the work of Codex is essential, very important, and of high quality. The standards, however, are on chemical and microbiological food safety; they do not describe how to meet the standards. "Understanding Codex" (WHO/FAO, 2016) provides a detailed description of what Codex Alimentarius is, what they do and how. Many other organizations provide standards and guidance on how to meet food safety requirements. These include the International Organization for Standardization (ISO, 2020), the International Dairy Federation (IDF, 2020), the European Hygienic Engineering and Design Group (EHEDG, 2020), 3-A Sanitary Standards, Inc. (3-A SSI, 2020), and NSF International (NSF International, 2020). There are many more, in particularly national and regional ones. Those mentioned operate internationally and have representatives all over the world.

4.4 The global harmonization initiative

GHI is not a standardization organization and does in no way compete with CAC or any of the standardization and guidance organizations. GHI addresses differences in food safety laws and regulations between countries when this hampers movement of food from one country to another, leading to spoilage of the food at the border or confiscation and destruction of safe and healthy food, while at the time more than a billion (10^9) people in the world starve. Today that number is still about 820 million (WHO, 2019). GHI is of the opinion that incidents based on differences in such laws and regulations may be juridically correct but morally wrong. Incidents resulting from such differences would not happen if the food safety laws and regulations would be globally harmonized.

This subject was discussed during a meeting between a number of food scientists at the Annual Meeting of the Institute of Food Technologists (IFT) in 2004. Representatives of the International Division of IFT and the European Federation of Food Science and Technology (EFFoST) discussed that there is no scientific justification for the differences and agreed to develop activities to try to harmonize the relevant laws and regulations (Lelieveld et al., 2006).

Differences in regulations also hamper the development of technologies aimed at preserving food with minimal loss of nutrients, because authorities in every country demand proof of the safety and reliability of new technologies, according to their own legal specifications. This means that safety tests need to be done in many countries and that may make it so costly that companies decide to stop such developments.

The same applies to the introduction of new ingredients, maybe only new to some parts of the world. The usual demand is that the safety of such ingredients is tested using animals and that this again must be done in nationally recognized and accredited laboratories. If tests are not mutually recognized, again this leads to huge costs and because of the use of animals, make companies hesitate to use new ingredients.

Much food is lost due to crops being affected by insects and microorganisms, including molds that produce toxins that in high concentrations, many times present on crops, may kill humans in a short period of time. In low concentrations many of these toxins cause cancer in humans. In some regions, almost half of the yield is lost for those reasons. Other losses are caused by viruses and bacteria that become increasing more resistant to common pesticides. A significant part of these problems can be and are being successfully addressed by genetic modification (GM). Regrettably, however, the world is still far from coming to an agreement about the safety of GM-based food. While in some countries most of the staple food is GM, in other countries it is strictly forbidden. This is largely not because of differences in scientific opinions but because of emotional arguments that, however, are still the cause of famines killing millions of people. A technology developed in the past decade, known as CRISPR-Cas9 (Lemay and Moineau, 2020) and similar more recently developed technologies, makes it possible to very accurately modify genes such that the results are predictable and quickly available for testing. The technology can be used to make plants resistant to certain diseases and may make plants more resistant to changes in the climate, resulting in too dry or too wet soil for many crops. In the EU, the technology is forbidden until it has been demonstrated to be safe in the same way as required for more traditional GM, a time-consuming and expensive procedure. In most of the rest of the world, the technology can be applied without restrictions. Again, having globally harmonized science-based regulations about the use of GM for the development of plant-based food would stimulate the development of crops that may improve food security.

GHI cannot and will not attempt to try to recommend changes of all food safety laws and regulations, that would be an impossible task and as discussed above, it is in the domain of other organizations, particularly the Codex. It is the intention of GHI, however, to address those differences that are important now because they are an impediment to food security in many countries with many people and low yields of crops and need to be addressed sooner rather than later and should not wait for agreements between government delegations.

This concept is written in the Charter of GHI that in 2007 was converted to a constitution, needed to be officially registered. It is important to note that GHI will not attempt to effect direct change in any food safety legislation or regulation. Instead, GHI intends to work with individual food scientists and technologists worldwide to obtain consensuses on the science that should be used to improve food safety laws and regulations. Once published, such scientific consensus documents may be used by any stakeholder to demonstrate to authorities that changes are needed and beneficial for everybody.

GHI does not aim at consensuses between organizations and governments, but between individual scientists and technologists, globally, regardless of their affiliations. Nobody in GHI is expected to represent his or her employer in any way, but use their own scientific conscience. GHI's philosophy is that once global consensus is obtained and published, stakeholders will use such information to achieve the desired changes.

4.5 GHI association

In October 2007, the initiative gained formal legal status as a nongovernmental, nonprofit association, and was registered in Vienna, Austria, as "GHI Association - Globale Harmonisierungs Initiative für Gesetze und Verordnungen im Bereich Lebensmittel." The GHI goal of "Achieving consensus on the science of food regulations and legislation to ensure the global availability of safe and wholesome food products for all consumers" was carefully incorporated in the Constitution (Lelieveld, 2009), which can be downloaded from the GHI website in both the original German version and the English translation. GHI Board carefully ensured that the German translation and other changes necessary to comply with Austrian law did by no means alter the objectives as described in the Charter.

The Constitution stipulates that a General Assembly (GA) be formed that meets at least once per 3 years to evaluate the performance of the Board and elects its members.

From its inception, GHI has recognized that impartiality of the scientific consensus process would be an essential requirement to enable cooperation with scientists from all over the world. This has led to the establishment of the Supervisory Board, which is also firmly embedded in the Constitution. Consisting of both competent individual scientists and representatives of independent scientific member organizations across the globe, its main mission is to safeguard the impartiality, integrity, and overall transparency of the consensus process (Lelieveld et al., 2006). The Supervisory Board reports to the GA. The composition of both the Board and Supervisory Board can be found on the relevant webpage.

Initiated and officially founded by five individuals, from Austria, India, Korea, The Netherlands, and the United States of America, GHI now has members in most countries in the world and has even appointed ambassadors in many countries and that number is continually increasing.

4.6 GHI ambassador programme

To facilitate communication with local scientists around the world, a GHI Ambassadorship programme was initiated in 2010. GHI Ambassadors can be scientists who are fluent not only in their own language, but also in English. GHI Ambassadors may invite scientists in their country or region to join GHI, foster new opportunities for information sharing, and serve as front-line representatives for GHI. The ambassadors play an important role in the organization's goal in particular because they can have the contacts needed to propagate the messages of GHI and can do so effectively in the local language. At the time of writing this chapter, GHI has approx. 90 ambassadors but has got members in many more countries, and therefore, it is expected that the number will continue to increase, also because many ambassadors want a fellow ambassador to have a sparring partner and several countries do already have more ambassadors. For details, see the ambassador page on the GHI website.

The Ambassador Programme Guidelines contain explicit actions to be performed by Ambassadors, serving as a guidance document for their endeavors. GHI Ambassadors are the association's proactive representatives and advocates at the local and regional level.

4.7 GHI working groups

To address issues that are seen as most important to make safe and wholesome (nutritious) food available for everybody in the world, GHI meanwhile has established many working groups (WGs). A WG provides the starting point for a consensus-building forum in which individual scientists share their expertise and come to initial agreement on the scientific principles that may support informed and objective global regulatory and legislative decision-making with respect to food safety, security, and nutrition issues.

Any food scientist may propose a WG by submitting a proposal describing the issue to be addressed and the benefits from obtaining global consensus about solving the issue by suggestion changes in food laws and/or regulations. Although GHI can help with identifying experts that may join a proposed WG, in principle, the proposer should also suggest a Chair and members for the WG. Proposers should realize that the GHI consensus process is independent of companies, governments, or any pressure group. Members of a WG, like all GHI members, do not represent their employers. Membership is individual and contributions to GHI are based on the scientific conscience of motivated members. The complete proposal should be submitted to the Board of GHI who will make a decision within 2 months. Approval of a proposal that is supporting the goal of GHI and is not conflicting with the GHI constitution cannot be withheld without strong scientific arguments that then must be clearly explained to the proposer to allow him to edit the text to meet valid arguments. The protocol to develop a global scientific consensus document can be found on the GHI website. Developing a consensus document may take up to 70 weeks, in which time the result may also be that consensus is not obtained, in which case the WG may instead write a discussion document addressing the differences and proposing activities to resolve them.

4.7.1 Working group nomenclature of food safety and quality

When attempting to harmonize regulations, like when attempting to make a deal between parties, it must be ascertained that the parties understand each other. That requires that the words or expressions that are used have the same meaning in the minds of everybody involved. Here, however, there is a huge problem because in many situations those communicating with each other *think* that they understand the words or expressions in the same way, while they are not aware of that often *this is not so*. This is more worrisome if parties use different languages and thus need translations to understand each other. An example on the food side is that it is globally seen as important that anything added to food is properly regulated. The definitions of food additives, however, differ between countries and organizations. The definition of food additives in English is in Canada 42 words long, that of neighboring United States of America 28 words and Japan needs just 23 words (in the English translation). The definition of the European Union is 94 words long and that of the CAC 103. Another example is that in the United States of America a billion means 1,000,000,000 and in Europe, a billion means 1,000,000,000,000 or 1000 times more. Imagine what this means if regulations prescribe the acceptable concentration of potentially toxic substances in parts per billion. It becomes worse if we talk about parts per trillion. All reasons to try to get the world to understand each other when talking about food safety. This is why the Working Group (WG) Nomenclature of Food Safety and Quality was established that WG aims to harmonize definitions in the above areas, starting with Russian and UK English.

4.7.2 Working group chemical food safety

Chemical food safety always has the attention from the public because regrettably most people seem to link a chemical name to danger. That is also why the public demands from governments tight regulations on chemicals in general and in particular on chemicals in food. That medicines are chemicals too seems not to bother most people. The public pressure results in regulations that protect them but may also lead to unrealistic legal demands that affect food security. Food that is safe to eat may not meet all regulatory requirements. When a chemical is found to exceed the allowed concentration, the food will be confiscated and destroyed, while sometimes such requirements are exaggerated. This is particularly so if the regulations require absence, which means that even natural concentration can make food not meeting the legislative requirements. Such regulations may and do lead to judging that healthy food is seen as unsafe and consequently destroyed.

The bottleneck to correcting such regulations is that most people believe that if something is poisonous it must not be allowed, also not in small amounts. The thought is toxic = toxic and toxic means dangerous. In an early stage therefore, the WG developed a document attempting to explain the relation between the amount of a chemical and toxicity. It basically repeats what Paracelsus described in the 16th century also referred to in the introduction: "Alle Dinge sind Gift, und nichts ist ohne Gift; allein die Dosis machts, daß ein Ding kein Gift sei" (All things are poison, and nothing is without poison; it is only the dose that stops a thing from being a poison; Paracelsus, 1538). Because of the ever-increasing sensitivity of methods of analysis, when the legal requirement is "absence" it is the progress in methods of analysis that determine what is legally allowed. If regulations remain what they are, with time all food will be illegal, because all substances occur in nature, sometimes in high but sometimes in tiny amounts that could not be detected in the past but today they can. Missing these tiny amounts may actually be unhealthy. The WG will investigate and formulate chemical safety issues in a broader sense, not only toxicological, since safety issues and standardization of chemical exposures have multifaceted consequences for the global economy, management of chemical safety data, and the ultimate goal of legislation—namely, the abatement of risks to human health through reducing exposure to chemicals via food to safe levels.

4.7.3 Working group education and training of food handlers

Although in the past decades literacy in the world has improved significantly, there are still many illiterate people. In 2018, literacy in Sub-Saharan Africa was only 62%, up 8% from 2013. Everywhere in the developing world the literacy of women is even 5%–10% lower (UNESCO Institute of Statistics, 2020). Further, often illiteracy is highest in rural areas; an example is India where in 2011 literacy of men in urban areas was 84%, 16% higher than in rural areas. For women again it is worse: 79% in urban and 58% in rural areas (MHRD, 2016). In some parts of Africa, literacy is even below 30% for men and 15% for women (UNESCO Institute of Statistics, 2020). Hence, to reduce the incidences of foodborne diseases from 600 million and the 420,000 premature deaths (WHO, 2015), there is a serious need of material for safe handling of food that can be understood by illiterate people who indeed handle food for so many. The WG will develop material and courses about food safety and hygiene and actively promote effective training of food handlers with limited or no reading skills.

4.7.4 Working group ethics in food safety practices

The purpose of companies usually is to make profit, and whatever the perceived or real disadvantages, leaving a significant part of running the society to companies seems to work reasonably well, although also depending on where in the world and in what period of time. Companies employ people and so do governments and companies and laborers pay taxes. Consequently, companies and governments are for a large deal driven by maximizing financial resources. Regrettably, sometimes those responsible find income more important than the health of people. This then may lead to erroneous, unethical, and also criminal behavior. Such activities include misinformation about the ingredients in food, when cheap and unhealthy substitutes are used or hiding the presence of unhealthy contaminants, be it chemical, physical, or microbiological. This can have severe consequences like consumers suffering from illnesses or even dying. The WG Ethics in Food Safety Practices is working on measures intended to make it more difficult to hide misbehavior. Working against the background of a series of serious food safety incidents in the past decades, e.g., those with melamine in milk products and *Salmonella* in peanut butter, the WG will investigate, formulate, and propose specific approaches to ethical food safety practices. The WG will then also propose additions or modifications to regulations and legislation to enhance ethics in the food chain.

4.7.5 Working group food microbiology

Although there are food safety cases with chemicals that result in a large number of victims, most food safety incidents are caused by microorganisms. While the chemical incidents often are by criminal intend, microbiological incidents are not. Nevertheless there are many more microbiological incidents that chemical ones. Both chemicals and microorganisms are everywhere and settle on food and are inhaled continually. For both, a normal concentration of the usual types in the environment is harmless. The difference is that an acceptable concentration of chemicals remains acceptable. Microorganisms, however, given the chance, multiply and as a consequence an acceptable concentration of microorganisms can easily become unacceptable, and if the microorganisms are pathogenic, they may cause serious harm and even death. Food is perfect for microorganisms to multiply and do so very fast if the temperature is optimal, which is often the same temperature humans like too.

Consumers at home should take care that food once exposed to the environment is not kept for a too long time at too high a temperature. The foods they buy need to be microbially safe and therefore regulations should be clear and correct. To make certain that this is the case, the food chain, from the source (farm, sea, lake, or fishery) to the retailer needs to be under control to avoid microorganisms from multiplying and in particular to ensure that pathogenic microorganisms are not present or, if presence is unavoidable, that storage and cooking instructions are such that when consumed the food will not cause harm. As with chemicals, demanding total absence in not needed. Giving in by governments, after an outbreak, to demands for legislation that (certain types of) microorganisms must be absent is unnecessary and may make food unaffordable because of the costs of measures to comply with such legislation and the high losses if a product does not meet such requirements. Moreover, differences in requirements between two sides of a border may cause food that is considered safe on one side being considered unsafe on the other side and would then be confiscated and destroyed.

The WG will strive for evidence-based and rational harmonization of criteria, regulations, and legislation relevant to the microbiology of foods. The WG will investigate and build scientific consensus around the food safety and spoilage challenges associated with microorganisms in foods to reduce morbidity and mortality. The WG will also address pathogen reduction by available technologies and instrumentation to measure and detect pathogenic microorganisms.

4.7.6 Working group food packaging materials

Unless food is consumed fresh, immediately after picking or catching, food needs to be transported to the place of consumption or may have to be stored. Usually that requires packaging of the food, to protect it from contamination by dirt, insects, and microorganisms. If food is sterilized or pasteurized, the packaging also serves to protect it against recontamination. In the past decades, it became clear that packing materials may also have adverse effects on the food because constituents of packing material may migrate into the product and these substances may be harmful. This has become a challenge to the food industry, the manufacturers of packaging material, and the manufacturers of packaging machines. Countries and regional blocks have different types of regulations, standards, and recommendations with respect to food contact materials, or none at all. Countries that are developing food packaging regulations find that regulations across the world differ and are confusing. The WG will address the differences and will develop proposals to harmonize the regulations.

4.7.7 Working group food preservation technologies

To feed people living in areas where no food is harvested and to ensure that there is also food in times that food does not grow, food needs to be preserved. During the ages, many technologies have been developed, more often than not by try and error. The early preservation technologies—mostly still used for many products—are drying, salting, and fermentation. In the 19th century, thanks to Nicolas Appert, preservation by heat was introduced. This involved packing the product in jars or cans, sealing them, and subjecting them to a heat treatment, initially using boiling water. When it was discovered that the methods were not always successful, the temperature was increased by using water under pressure, using autoclaves, pressure vessels. It was Louis Pasteur who discovered that heating worked because it killed the microbes that caused the spoilage. Still today, many and in particular particulate foods are still preserved by this method. Technologies to do this on a larger scale were developed in the 20th century and today it is common to find thermally pasteurized and sterilized products everywhere in the world, often using very advanced automatic systems. A drawback of using heat is that the taste of the product is affected. For some products, this was good, but for many products consumers preferred the taste of fresh food. In the second half of the 20th century, much research was done to limit the negative consequence of heat treatments while retaining the effect on microorganisms. This led to methods to improve heat transfer to the product as well as, for liquid products, continuous heat treatments, to be followed by aseptic packaging, to prevent recontamination. Although there have been attempts earlier, in the 1970s people started to find nonthermal ways of preserving food (other than drying and salting), inactivating microorganisms without affecting taste and at the same time also preserving more of the valuable nutrients, of which it had become clear that a part of these got lost by heating. These methods include the use of gamma radiation that changes the DNA (deoxyribonucleic acid) of the microorganism so that they are no longer capable of multiplying, subjecting the product to pressures exceeding 500 MPa (5000 bars) that most microbes do not survive and using pulsed electric fields, that perforate the membranes of microorganisms resulting in lysis of the microbial cell and death. Although these methods inactivate or kill microorganisms, none of them affect the chemistry of flavors and nutrients, with few exceptions, depending on product. As with everything that is new, people were concerned that the food preserved by the new methods was not healthy and demanded regulations to prevent or restrict the application of the new methods, ignoring the evidence that the new methods cause significantly less changes to the product than the thermal methods. The WG Food Preservation Technologies addresses this and many other issues. In 2018, the WG produced the first GHI global scientific consensus document, in which it recommends that international regulatory bodies recognize and accept the finding of the Joint FAO/IAEA/WHO Expert Committee on Food Irradiation (JECFI) first published in 1981 and, subsequently, adopted by FAO/IAEA/WHO Joint Study Group High-Dose Irradiation in 1999. These organizations concluded that food irradiated to any dose appropriate to inactivate harmful microorganisms is both safe to consume and nutritionally adequate and does not result in any toxicological hazard (Koutchma et al., 2018).

The WG will collect and make available information about existing international regulations on novel thermal and nonthermal technologies, exchange and summarize knowledge, and set up a database of applications in industry world-wide. The intention is to produce recommendations to accelerate and clarify the path for the validation and approval of new processes, eventually leading to harmonization of relevant international requirements.

4.7.8 Working group food safety in relation to religious dietary laws

Differences in opinions about the way food must be prepared to meet the religious requirements of the many groups within the various religions are a very big hurdle to trade, in particular with respect to neighboring countries. The opinions about what is kosher or halal are huge and the various groups have difficulty in compromising. In addition, there is the problem that in many countries the food and processing of food has also to comply with local regulations. This WG was founded at the request of GHI members living in the Middle East. Members from countries in Sub-Saharan Africa, Asia, America, and Europe joined too. One of the problems is the large number of certificates issued by the equally large number of organizations that claim that they have the correct interpretation of the Qur'an (Koran) or the Kashrut (Jewish dietary law) and that is confusing. The WG Food Safety in Relation to Religious Dietary Laws has undertaken to try to harmonize certification and symbols.

The WG will strive for evidence-based and rational harmonization of food safety regulations and legislation relevant to religious food and drink production and will develop proposals to harmonize certification and certification symbols. Globally harmonized food safety regulations will make it possible to better ensure that foods produced according to religious dietary laws meet the necessary criteria and are truthfully marketed to consumers.

4.7.9 Working group genetic toxicology and genomics

In most countries there is a legal requirement to prove the safety of new food ingredients and of contamination of food (such as mycotoxins) using animals. The animals used to prove the safety, however, are given much more of the chemical under investigation than will ever be consumed by humans and using the results obtained may have little value for determining the toxicity for humans. Moreover, the relation between the sensitivity of animals and humans for developing cancer from food constituents is at least doubtful. The methods take a long time, many animals, and are costly. More than 2 decades ago, it has been shown that in vitro methods using competent **human** liver cells are significantly more relevant for estimating genetic toxicity. The hepatoma cells (HepG2) used in these methods also dramatically reduce the number of false positives that are obtained using rodent cell lines. They give faster, more accurate and to humans very relevant results. This is a tremendous simplification of the contents of the chapter by Firouz Darroudi in this book (Testing for food safety using human competent liver cells (HepG2) : A review). Regulators all over the world should consider replacing the current demands involving animals and rodent cell lines by the methods described in that chapter.

4.7.10 Working group global incident alert network

Food crimes take place all over the world and cause illnesses and deaths. These crimes often remain hidden until an alarming number of people get ill and eventually it turns out that it has been the food that has been consumed. One might think that authorities should watch the safety of food that retailers offer, but in reality that is not easily done. Firstly, in case of adulterations, inspectors do not know what to look for. Even when there is a case of a consumer getting ill, it may take quite some time to identify the cause. Secondly, the number of products on the market is huge and inspection services do not have the capacity for regularly inspecting food, unless they have a suspicion and know what to look for. The number of victims can be high. In the case of peanut butter contaminated with *Salmonella*, there were 9 fatal cases and 714 reported illnesses. With melamine in milk powder, almost 300,000 children had been diagnosed with melamine-related kidney stones of which 52,000 were hospitalized and at least 6 children died (WHO, 2009). The incident happened despite the fact that already in 2004 it was clear that melamine in feed caused kidney failure in pet dogs, while in 2007 the US Food and Drug Administration found that wheat protein powder from China, used in pet food, was the cause of deaths of cats and dogs and ordered the recall of pet foods likely to be contaminated with melamine (Li et al., 2019). It is worrying that nobody took measures after revealing that contaminated feed and food contained melamine in high concentrations already in 1979 and 1980 (Cattaneo and Ntoni, 1979; Cattaneo and Ceriani, 1988). It took many victims before actions were taken. It is very likely that several professionals knew about the adulteration but did not report it, because of the potentially severe consequences for their positions and their families. It is for that reason that the WG Global Incident Alert Network was established, with the main objective to make it possible to report food crimes before there are victims, in such a way that the reporter cannot be identified.

4.7.11 Working group mycotoxins

In 2015 the International Agency for Research on Cancer mentioned in the report "Mycotoxin Control in Low- and Middle-Income Countries" that an estimated 500 million of the poorest people in Sub-Saharan Africa, Latin America, and Asia are exposed to mycotoxins at levels that substantially increase mortality and morbidity (Pitt et al., 2012). Bath et al. (2010) reported that poor harvesting practices, improper drying, handling, packaging, storage, and transport conditions contribute to fungal growth and increase the risk of mycotoxin production. The chronic incidence of aflatoxin in diets is evident from the presence of aflatoxin M1 in human breast milk in Ghana, Nigeria, Sierra Leone, Sudan, Thailand, and the United Arab Emirates, and in umbilical cord blood samples in Ghana, Kenya, Nigeria, and Sierra Leone (Bhat and Vasanthi, 2003). There is a vast amount of literature about the presence of mycotoxins in staple food and it varies between countries and weather conditions. In certain years, the deoxynivalenol content in wheat was found to be more than 500 mg/kg in some countries (Jelinek et al., 1989). Sometimes people have no choice than to eat food heavily contaminated with mycotoxins (Ferrão et al., 2017).

The interest in mycotoxins is increasing with time, but most of the published literature is about the occurrence, health effects, and analysis. The WG Mycotoxins is collecting methods to reduce the presence of mycotoxins that have been shown to be practical because they are successfully applied in some regions, but the experience is not shared with farmers in other parts of the world. When ready, the WG Education and Training of Food Handlers (see section 4.7.3 Working group education and training of food handlers.) may produce training material that is understandable by smallholder farmers.

4.7.12 Working group nanotechnology and food

Particles with the size of 100 nm and smaller have always been present in food (Rogers, 2016) and our body is full of nanoparticles—we would die without them, because enzymes and parts of blood cells and many other constituents of the human body are in the nano-size scale. Nature is full of nano-size particles, in the soil, in the air, and everywhere. The reason why mankind still exists is that our body during evolution has learned to use and protect against the environment, including nanoparticles. Enzymes, without which life is impossible, are in the order of 10 nm. Apart from scientists nobody other than scientists knew the term "nano" and the same applies to other expressions that indicate very small size (pico, femto). When technologists found ways to measure and control the size of particles and published about them, like with everything that is new in the eyes of many, concerns about "nano" became a new hype and thus also the demand to regulate the technology. Scientists are studying how nanoparticles can be used to improve food, e.g., by making some essential food components easily accessible to the digestive system, and yes, self-evidently like everything else, also applications need legislation, but based on sound science and the same everywhere. With the exception of the European Union, most nations have not finalized or have minimal regulations regarding the use of nanoparticles in foods and food packaging. The WG intends to produce proposals for nanotechnology-related food safety legislation with global harmonization in mind.

4.7.13 Working group nutrition

GHI is working toward sufficient safe and healthy food for everybody in the world. This means that the food should be sufficient in nutrients. The meaning about "sufficient nutrients" differs around the world. Genetic differences between populations complicate harmonization of regulations with respect to nutrition. There is also much confusion about informing the consumer about the nutrients present in or absent from food, what should be on the label, and what will not help the consumer but only cause confusion. What makes life even more difficult is that there are many self-proclaimed experts who give advice lacking any scientific foundation, but because of the way presented, for many consumers such advices are very believable, while sometimes they can just be unhealthy. There is little evidence that popular diets do what they promise and are indeed healthy (Ruden et al., 2007). The WG will strive for evidence-based global harmonization of nutrition regulations and legislation. The goal is to publish a nutrition white paper that discusses global, consistent nutrition policy with an action plan on nutrition and recommendations for harmonization of dietary guidance.

4.7.14 Working group reducing postharvest losses

In a large part of the world there is a tremendous loss of food during transport from the place of harvesting to the place of selling. Reasons can be the way of harvesting, the way of intermediate storage, and the way of transport. In many cases smallholder farmers in less developed regions have no idea how to do this in such a way that the harvest is not consumed by insect or animals, not damaged by squeezing it in to small crates, or not to have it destroyed during transport on a bumpy road. The WG will identify technologies and procedures to reduce food loss and food waste around the globe, and to identify those technologies and procedures that are appropriate to be proposed for harmonized regulations.

4.7.15 Working group science communication

Writing recommendations for those who need them is not easy for most scientists and as a consequence much of the available knowledge is not understood and therefore also not implemented. The way to write recommendations for scientists of the same discipline can be difficult enough if one wants to be understood correctly. Writing for scientists in other disciplines is certainly more difficult, but the target audiences of GHI generally are not scientists, but politicians, governments, journalists, managers, farmers, and the electorate. To avoid that the work done is basically lost because it is not understood by the target group, the WG Science Communication was established, composed of professional science communicators.

The current Chairs of the WG are aware that scientific activities continually influence the lives of everybody. Therefore, there is a strong need to focus on the transparency and high integrity of research that is essential for public health, wellbeing, and safety, as well as for environmental protection and diversity preservation. In order for scientists to work in the public interest, a continuous dialogue with both the publics and the regulators is required. The WG will establish effective methods and strategies for science communication, suitable for different target audiences (Bogueva et al., 2019).

4.7.16 Working group food law and regulations

Most of the GHI WGs work is on specific topics, but in most cases, neither the Chair(s) nor the members know much about the laws and regulations that have to do with the subjects they study. When they need to find out about the regulatory aspects, they need much time to firstly find the relevant documents and often even more time to understand the meaning of the language used. For that reason the WG Food Laws and Regulations was initiated. Meanwhile, the WG has expanded the scope, which also now includes advising the European Food Safety Authority using information obtained from the other WGs.

Regulations may also differ where the laws do not differ, that is a matter of interpretation of the laws by the regulators. The WG will try to understand the causes of such differences, because they usually are not based on differences of opinions between subject-matter experts. Then the WG will propose possible instruments for food law implementation, having scientific evidence as a main pillar.

4.8 GHI library

To make it easy for GHI members to have access to GHI documents and to documents that have been made available by GHI members, an electronic GHI Library has been set up. Many GHI documents can be downloaded from that library, such as articles written by GHI members, GHI consensus documents, and information sheets (sometimes in many languages). In addition slides for presentations or complete presentations can be downloaded for use by GHI members who have been invited to give presentations about GHI or about some aspects of activities of GHI. Members who have material available that they are willing to share are encouraged to send such information to librarian@globalharmonization.net. This applies also to slides translated into other languages that may be useful in countries where they speak the same language.

4.9 Conclusion

Harmonized food safety regulations no doubt would reduce the unnecessary destruction of healthy food and thereby improve food security. To achieve harmonization, however, publishing scientific consensus statements is not enough on its own. To promote acceptance of scientific consensuses, the target publics need to understand the documents and understand them correctly. Therefore, it is essential to educate people in what food safety is, and that at all levels, because eventually it is the electorate who decides what they want, science or not. The contributions of the WG Nomenclature of Food Safety and Quality, the WG Education and Training of Food Handlers, and the WG Science Communication will play an essential role in this respect as do the GHI Ambassadors, the ones who can ensure that their countrymen correctly understand important messages that can easily be misinterpreted and then may have an effect opposite to what has been intended.

Regrettably, food safety not only depends on laws and regulations, because there are those who have no scruples to adulterate food or sell contaminated food for increasing profit at the expense of the health of consumers. It is hoped that the activities of the WG Ethics in Food Safety Practices and the WG Global Incident Alert Network will succeed in reducing food safety incidents by providing governments with information they can use to stop food crimes.

With the activities of the global membership of GHI, led by the increasing number of WGs, we expect that with time GHI will be successful in realizing its goal: Achieving consensus on the science of food regulations and legislation to ensure the global availability of safe and wholesome food products for all consumers.

References

3-A SSI, 2020. 3-A Sanitary Standards, Inc. Accessed 10 February 2020. http://www.3-a.org/About/History.

Bhat, R.V., Vasanthi, S., 2003. Mycotoxin Food Safety Risk in Developing Countries, Vision 2020 for Food. International Food Policy Research Institute. Accessed 19 January 2020. https://www.researchgate.net/publication/285475730_Mycotoxin_food_safety_risks_in_developing_countries_Food_safety_in_food_security_and_food_Trade_Vision_2020_for_Food.

Bhat, R., Rai, R.V., Ravishankar and, A.A., Karim, A.A., 2010. Mycotoxins in food and feed: present status and future concerns. Compr. Rev. Food Sci. Food Saf. 9, 57–81.

Bogueva, D., Duca, E., Hristozova, N., 2019. GHI WG Science Communication. Personal communication.

Cattaneo, P.C., Ca ntoni, C., 1979. Identification and determination of melamine in animal meals. Tec. Molit. 30 (5), 371–374.

Cattaneo, P., Ceriani, L., 1988. Melamine in animal meals. Tec. Molit. 39 (1), 28–32. Downloadable from. https://eurekamag.com/research/001/883/001883365.php. Accessed 13 January 2020.

EHEDG, 2020. European Hygienic Engineering and Design Group. https://www.ehedg.org. Accessed 10 February 2020.

FAO, IFAD, UNICEF, WFP, WHO, 2018. The State of Food Security and Nutrition in the World 2018. Building Climate Resilience for Food Security and Nutrition. FAO, Rome. Licence: CC BY-NC-SA 3.0 IGO. http://www.fao.org/3/i9553en/i9553en.pdf. Accessed 9 January 2020.

Ferrão, J., Bell, V., Fernandes, T.H., 2017. Mycotoxins, food safety and security in sub-saharan Africa. SM J. Food Nutri. Disord. 3 (2), 1021.

GHI, 2016. Harmonisation: feeding people, fueling innovation. In English: https://www.globalharmonization.net/sites/default/files/pdf/final_ghi_ShareSheet1_FeedPeople_Feb2016.pdf. In other languages: https://www.globalharmonization.net/share-sheets-and-consensus-documents. Accessed 5 January 2020.

IDF, 2020. International Dairy Federation Standards. https://www.fil-idf.org/working-areas-strategic/standards/. Accessed 10 February 2020.

ISO, 2020. International Organization for Standardization - about Us. https://www.iso.org/about-us.html. Accessed 10 February 2020.

Jelinek, C.F., Pohland, A.E., Wood, G.E., 1989. Worldwide occurrence of mycotoxins in foods and feeds - an update. J. Assoc. Off. Anal. Chem. 72 (2), 223–230.

Koutchma, T., Keener, L., Kotilainen, H., 2018. Global Harmonization Initiative (GHI) Consensus Document on Food Irradiation. GHI, Vienna, 2018. Downloadable from. https://www.globalharmonization.net/sites/default/files/pdf/GHI-Food-Irradiation_October-2018_revised_04-2019.pdf. Accessed 18 January 2020.

Lelieveld, H., Keener, L., Boisrobert, C., 2006. Global harmonisation of food regulations and legislation—the global harmonization initiative. New Food 4, 58–59.

Lelieveld, H., 2009. Progress with the global harmonization initiative. Trends Food Sci. Technol. 20, S82–S84.

Lemay, M.-L., Moineau, S., 2020. How are genes modified? Cross breeding, mutagenesis and CRISPR-Cas9. In: Andersen, V. (Ed.), Genetically Modified and Irradiated Food - Controversial Issues: Facts versus Perceptions. Elsevier/Academic Press, pp. 39–54.

Li, Q., Song, P., Wen, J., 2019. Melamine and food safety: a 10-year review. Curr. Opin. Food Sci. 30, 79–84.

MHRD, 2016. Adult Education. Department of School Education & Literacy. Ministry of Human Resource Development, Government of India. https://mhrd.gov.in/adult-education#main-content. Accessed 16 January 2020.

NSF International, 2020. NSF the Public Health and Safety Organization. https://www.nsf.org/about-nsf/. Accessed 10 February 2020.

Paracelsus, 1538. Bombastus Ab Hohenheim, A. P. T. - Septem Defensiones, Page 509. First German Edition Published by Perna in Basel, in 1574. Text Cited Is from "Theophrast Paracelsus: Werke", Published by Will-Erich Peukert. Wissenschaftliche Buchgesellschaft, Darmstadt, 1965.

Pitt, J.I., Wild, C.P., Baan, R.A., Gelderblom, W.C.A., Miller, J.D., Riley, R.T., et al., 2012. In: Wild, C.P., Miller, J.D., Groopman, J.D. (Eds.), Improving Public Health through Mycotoxin Control. Lyon, France: International Agency for Research on Cancer (IARC Scientific Publications Series, No. 158), 2015. Mycotoxin Control in Low- and Middle-Income Countries. WHO International Agency for Research on Cancer, Lyon (France), 2015.

Rogers, M.A., 2016. Naturally occurring nanoparticles in food. Curr. Opin. Food Sci. 7, 14–19.

Ruden, D.M., Rasouli, P., Lu, X., 2007. Potential long-term consequences of fad diets on health, cancer, and longevity: lessons learned from model organism studies. Technol. Cancer Res. Treat. 6 (2), 1–8.

UNESCO Institute of Statistics, 2020. Data for the Sustainable Development Goals - Education. http://data.uis.unesco.org/Index.aspx?DataSetCode=EDULIT_DS&popupcustomise=true&lang=en#. Accessed 16 January 2020.

WHO, 2009. Toxicological and Health Aspects of Melamine and Cyanuric Acid: Report of a WHO Expert Meeting in Collaboration with FAO, Supported by Health Canada, Ottawa, Canada, 1–4 December 2008. WHO, Geneva, Zwitzerland.

WHO, 2015. WHO Estimates of the Global Burden of Foodborne Diseases: Foodborne Disease Burden Epidemiology Reference Group 2007-2015. World Health Organization, ISBN 978 92 4 156516 5 downloadable from. https://apps.who.int/iris/bitstream/handle/10665/199350/9789241565165_eng.pdf. Accessed 11 January 2020.

WHO/FAO, 2016. Codex Alimentarius - Understanding Codex. World Health Organization, Rome, ISBN 978-92-5-109236-1. Downloadable from. http://www.fao.org/3/a-i5667e.pdf. Accessed 9 February 2020.

WHO/FAO, 2018. Codex Alimentarius Commission Procedural Manual, twenty-sixth ed. World Health Organization Food and Agricultural Organization of the United Nations Rome, Italy. 2018 Downloadable from. http://www.fao.org/3/i8608en/I8608EN.pdf. Accessed 11 January 2020.

WHO, 2019. World Hunger Is Still Not Going Down after Three Years and Obesity Is Still Growing – UN Report. https://www.who.int/news-room/detail/15-07-2019-world-hunger-is-still-not-going-down-after-three-years-and-obesity-is-still-growing-un-report. Accessed 10 February 2020.

Chapter 5

Food safety regulations within countries of increasing global supplier impact

Odel Yun LI[1] and Xian-Ming Shi[2]

[1]*Shanghai Jiao Tong University, Shanghai Legislative Research Institute, Shanghai, China;* [2]*MOST-USDA Joint Research Center for Food Safety, School of Agriculture and Biology, State Key Lab of Microbial Metabolism, Shanghai Jiao Tong University, Shanghai, China*

5.1 Introduction

Food safety is vitally important to the development of the human world from ancient to modern times. Although the specific legislation of food in different countries is not completely consistent (due to the differences in the level of scientific and technological development, the level of economic development and religious belief), the goal is the same—legislation to ensure food safety. In order to ensure the world's food safety, all countries have formulated strict food safety laws. In addition, relevant international organizations have formulated corresponding international recommendations on food safety. Therefore, food safety is not the goal of a group or an organization, but depends on the joint efforts of various countries and relevant international food safety organizations.

Food Safety Regulations in our increasingly globalized world especially within countries of growing global supply impact is important and challenging for governments and commercial organizations. The regulations complexly link food safety, nutrition, and food security[1] and could minimize the risk of any foodborne hazards that can be caused by factors including large-scale production practices, complex food supply chains, and globalized distribution networks.[2] This chapter will mainly focus on the introduction of important food suppliers in the current global scope, analyze the geographical distribution of suppliers in the current global field, the impact of e-commerce platform on food safety, and discuss what factors affect the building of food laws in various countries and the formulation of the international food laws.

5.1.1 International food suppliers

According to the top 500 Forbes rankings in the past 5 years, and also based on companies' revenue, profit, assets, and market value, at present, the world's largest food suppliers include:

5.1.1.1 Nestlé[3]

The Swiss food giant Nestlé has an annual turnover of $92.285 billion and is the world's largest food manufacturer and one of the top multinational corporations. The company which is based in Switzerland (Vevey) was founded in 1866. It is a leading nutritious, health, and wellness company and it is worldwide well known for producing chocolate bars and instant coffee. At present, it offers a rich product line suitable for the local market and different cultures. The company is very

1. * Dr. Odel Yun Li is a PhD from KoGuan Law School, Shanghai Jiao Tong University. She is also an assistant professor at Shanghai Legislative Research Institute. This chapter is a phased research achievement of Shanghai philosophy and Social Sciences Planning Project (2020BFX010) and Chinese national major project "Research on the legal system construction of Chinese genetically modified food" (18ZDA147). ** Dr. Xian-Ming Shi is a Professor and the Director in the MOST-USDA Joint Research Center for Food Safety, Shanghai Jiao Tong University. WHO, https://www.who.int/news-room/fact-sheets/detail/food-safety.
2. Nyachuba, D.G., 2010. Food borne illness: Is it on the rise? Nutr. Rev. 66, 257−269. https://doi.org/10.1111/j.1753-4887.2010.00286.x.
3. https://expandedramblings.com/index.php/nestle-statistics-and-facts/.

Ensuring Global Food Safety. https://doi.org/10.1016/B978-0-12-816011-4.00026-4

diversified in its range of products combining over 2000 brands.[4] Currently, Nestlé has more than 400 factories in more than 80 countries on five continents. The production and sales of all products are completed by more than 200 departments under the leadership of its headquarters. Nestlé is known as the "most international multinational group" because 98% of Nestlé's sales come from the company operation abroad.

5.1.1.2 Cargill[5]

The Minnesota-based Cargill Company was founded in 1865 by Mr. William Wallace Cargill. After 141 years of operation, Cargill has become a multinational privately held professional company in commodity trade, processing, transportation, and risk management.[6] At present, Cargill is one of the largest global trading, processing and sales company, covering agricultural products, food, financial, and industrial products and services. With about 160,000 employees in 70 countries, Cargill is the world's largest private holding company, the largest manufacturer of animal nutrition and agricultural products.

5.1.1.3 Unilever[7]

Headquartered in Rotterdam, the Netherlands, and London, the United Kingdom, respectively, Unilever is a transnational leader in the food and beauty and personal care business. Following its goal to make sustainable living a natural thing Unilever built an empire based on beloved iconic products and brands. Unilever has a huge company network in more than 70 countries, 400 brands, and nearly 161,000 employees. It is the second-largest consumer goods manufacturer in the world. Its annual revenue exceeds US$55 billion, and it is one of the most profitable companies in the world.

5.1.1.4 PepsiCo[8]

PepsiCo Inc. is a beverage and leisure food company. Pepsi, originally introduced as Brad's Drink in 1893, has since grown into a multibillion dollar brand.[9] With 263,000 employees in more than 200 countries and regions around the world, and sales revenue of $64.661 billion in 2018, it is one of the largest food and beverage company in the world. The company is currently headquartered in New York City. There are 22 different brands in the Pepsi portfolio that generate more than 1 billion in annual retail sales.[10]

5.1.1.5 Kraft Heinz foods[11]

Kraft Heinz Foods Inc. was formed in 2015. It's the third largest food company in North America and world fifth largest.[12] There are more than 200 brands presenting the Kraft Heinz portfolio, and the number of countries Kraft Heinz sells their products in are 200. There are 39,000 employees in Kraft Heinz. The annual net sales number is $26.268 billion in 2018.

5.1.1.6 InBev[13]

Anheuser-Busch InBev is a listed company with headquarters in Leuven, Belgium. Budweiser InBev is the world's leading beermaker and one of the world's top five consumer goods companies. As a large consumer-centric and sales-driven company, InBev operates more than 500 brands, including global flagship brands such as Budweiser, Stella Artois and Baker Beer; fast-growing multinational best-selling brands such as Leffe, Hoegaarden, Bud Light, Skol, and Brahma. The company's goal is to maintain its name as the world best beer company bringing people together for a better world.[14]

---

4. www.nestle.com.
5. https://expandedramblings.com/index.php/cargill-statistics-and-facts/.
6. https://www.cargill.com/about/cargill-history.
7. https://expandedramblings.com/index.php/unilever-statistics-and-facts/.
8. https://expandedramblings.com/index.php/pepsi-statistics-and-facts/.
9. http://www.pepsistore.com/history.asp.
10. https://www.unilever.com/brands/?category=408118.
11. https://expandedramblings.com/index.php/kraft-heinz-statistics-facts/.
12. www.kraftheinzcompany.com.
13. https://expandedramblings.com/index.php/anheuser-busch-inbev-statistics-facts/.
14. https://www.ab-inbev.com/.

5.1.1.7 *Mars*[15]

Mars is a privately held internationally renowned multinational fast-moving consumer goods company. It is a global manufacturer of confectionery and pet food and animal care services and other food product. The company produces and sells many famous brands of chocolate, candy, and pet food all over the world. It has many well-known brands at home and abroad, including Dove Chocolate, Snickers, M&M's Chocolate, Big Red, etc. Brand products with its high quality and good reputation have won the support and recognition of the vast number of consumers around the world.[16]

5.1.1.8 *Coca-cola*[17]

Coca-Cola company, which made the famous soft drink Coca-Cola or Coke, was created in 1886 by John Stith Pemberton in a pharmacy in Atlanta City, Georgia, the USA[18] and brought to the world by businessman Asa Griggs Candler, sold as a beverage with medicinal application for temperance and intended as a patent medicine (Eschner, Kat (March 29, 2017)). Coca-Cola's Creator Said the Drink Would Make You Smarter.[19] The name Coca-Cola refers to the kola tree grown in the tropical rainforest of Africa. The nuts the tree produces contain caffeine used as one of the main ingredients to flavor the famous drink. The fruit is the main material for making coke drinks. On March 12, 1894, the first bottled Coca-Cola began to be sold. The current Coca-Cola formula remains a trade secret despite the many experimental recreations and recipes have been published. There are more than 200 countries that Coke is sold in, and there are 62,600 employees. The annual revenue is $31.8 billion in 2018.

5.1.1.9 *Tyson*[20]

TYSON-Foods, Inc. is an American multinational food industry specialized corporation. It is the first American chicken brand, and the world's largest processor, marketer and supplier of chicken, beef, pork, and many brands of deep processing and convenience food. At present, it ranks the second-largest food processing company among the Fortune 500 in the world with a revenue of more than 40 billion US dollars a year. There are 121,000 employees and sells products in 117 countries.

5.1.1.10 *Danone*

The leading food company Danone has its headquarters in Paris, France. The company was founded in Barcelona, Spain, in 1919 as a small yogurt maker company. At present, Danone has four different product divisions: Fresh Dairy Products, Waters, Early Life Nutrition, and Medical Nutrition, and employs over 99,000 people worldwide. The company market value is $42 billion. Its products range from dairy, aqua drinks, infant formula, to food with medical purposes.[21]

Based on the above presented data of the ranking of the major international food companies, it is evidentiary that the main food suppliers in the world are still concentrated in European countries and the United States. However, with the economic globalization and free trade, BRICS countries (Brazil, Russia, India, China, and South Africa) and some other countries have gradually increased the share of global food trade through low-cost and other quality and price advantages.

Due to the increased attention to food safety regulations including supply chain, coordination, and international trade, the emergence of new and more strict food safety standards are important because of factors like the growth of trade in nonperishable and valuable products, advances in hazard detection and epidemiology, high health scandals, scientific and regulatory consensus on the best approaches to risk management and recognition of global standards, and approaches under the World Trade Organization (WTO).[22]

---

15. https://expandedramblings.com/index.php/mars-statistics-and-facts/.
16. https://www.mars.com/made-by-mars/food.
17. https://expandedramblings.com/index.php/coca-cola-statistics/.
18. https://www.coca-colacompany.com/history.
19. http://www.smithsonianmag.com/smart-news/coca-colas-creator-said-drink0would-make-you-smarter-180962665/.
20. https://expandedramblings.com/index.php/tyson-foods-statistics-and-facts/.
21. www.danone.com.
22. more can be found here: International Food Policy Research Institute (IFPRI), 2014. Food Safety and Developing Markets. https://pdfs.semanticscholar.org/7ea1/5791f8f515fee2b585f101f7eca39d56b32f.pdf.

5.1.2 Global food supply chain

The adoption of third-party certification, public standards and private standards in the global food industry, the increasing control of food retailers and brand processors over the global food value chain, and the improvement of food supply chain safety requirements based on sustainable development goals may be achieved through the promotion of food safety standards and certification requirements. Governance has an impact. In addition to the complex relationship of related enterprises in the supply chain caused by transactions, the improvement of food safety standards and certification caused by negative reports of global food safety incidents and the requirements of food safety certification of some nongovernmental organizations have brought far-reaching impact on the governance of the aquatic product value chain.

5.1.3 The impact of E-commerce platform on global food supply

In the process of the food industry developing with the Internet, frequent incidents about food safety have aroused widespread concern on social responsibility in the food supply chain at home and abroad. The government and relevant departments have attached great importance to and encouraged food supply chain enterprises to fulfill their social responsibility, so it is of great significance to further explore and study the social responsibility of food supply chain enterprises from the perspective of government encouragement.

In recent years, with the rapid development of the Internet and the emergence of e-commerce, the supply chain presents a new development model, which makes the supply chain enterprises gradually return to customer centered, and the operation efficiency of the supply chain has been greatly improved. From the perspective of the Internet, the food supply chain has its development characteristics and changes because of the particularity of its industry. Its main characteristics are as follows:

5.1.3.1 The number of node enterprises in the food supply chain has decreased

The rise of the Internet has reduced the circulation link in the food supply chain to a certain extent. The level of the food supply chain has been reduced, and the distance between the food supplier and the end consumer has been shortened, which enables the food enterprises to approach the users at a lower cost. To some extent, consumers can also understand food enterprises.

5.1.3.2 The role of third-party e-commerce platforms is becoming increasingly prominent

With the popularity of e-commerce and the growing express industry, food suppliers choose to cooperate more with third-party e-commerce platforms, such as Jingdong Mall, Tianmao, and other e-commerce giants. With the help of the Internet integrated platform, while establishing brand image and visibility, they can achieve more economic benefits.

5.1.3.3 Social responsibility of third-party e-commerce platforms

The accession of third-party e-commerce platforms has changed the role of CSR (Corporate Social Responsibility) in the food supply chain under the background of the Internet. Third-party Internet platforms need to fulfill their social responsibilities and be responsible to the businesses and consumers of their platforms. Once food quality and safety problems occur, the impact is not only on the food enterprises involved but also on the credibility of third-party e-commerce platform and all businesses on the platform.

5.2 Regulations of global food suppliers by international law and standards

5.2.1 The recommendations of the codex alimentarius commission[23]

The Code of International Food Code stipulates the corresponding food safety standards, including technical standards, safety standards, and related certification standards.[24] These standards and guidelines also play a corresponding role in restricting international food suppliers, but also have a certain degree of soft law.

Soft law in international legal regulation of food safety is mainly embodied in international soft law in cross-government governance network and international soft law formulated by relevant international organizations. The former refers to the nontreaty agreements reached by the government departments of food safety in different countries in

23. Also refer to Chapter 2.
24. http://www.fao.org/fao-who-codexalimentarius/en/.

the form of bilateral memorandums. The latter refers to international documents such as guidelines, action plans, declarations of principles, and standards related to food safety formulated by professional international organizations, regional international organizations, and international nongovernmental organizations. These nonlegally binding but effective international soft laws play a unique complementary and auxiliary role in the international legal regulation of food safety, which deserves our attention.

5.2.2 Sanitary and phytosanitary standards of the World Trade Organization

The agreement on the implementation of sanitary and phytosanitary measures (SPS Agreement) is a new agreement reached in the Uruguay round, which belongs to the WTO multilateral agreement on trade in goods. SPS agreement takes the regulation of sanitary and phytosanitary measures (SPS measures) that affect international trade as its duty. In the Uruguay Round negotiations, many countries proposed to formulate SPS agreement, which put forward specific and strict requirements for animal and plant quarantine in international trade. It is the product of animal and plant quarantine work infiltrated by WTO agreement principles.

The WTO promotes free trade by regulating the behavior of governments, but its premise is to ensure food safety. Countries have considerable autonomy in food safety standards, but importing countries make judgments on the setting of high food safety standards on exporting countries that do not always conform to the principle of fairness and may discriminate against the setting of ultrahigh food safety standards by international soft law cited in WTO. The case judgments in WTO also have considerable judgment power, but the defects still need to be improved.

5.3 Regulations of global food suppliers by domestic laws

5.3.1 USA

There are four major agencies involved in food standards management in the United States, including the Food and Drug Administration (FDA), the Environmental Protection Agency (EPA), Agricultural Market Authority (AMA), and the United States Department of Agriculture (USDA). Taking FDA as an example, the FDA regulatory standards mainly include three items: Animal Feed Regulatory Program Standards (AFRPS), Manufactured Food Regulatory Program Standards (MFRPS), and Voluntary National Retail Food Regulatory Program Standards, VNRFRPS. Taking MFRPS as an example, MFRPS is an important part of the Food Safety System in the United States. The goal is to establish a complete regulatory system based on risk to help the federal and state governments to regulate their behavior to reduce the incidence of food borne diseases. Its characteristics mainly include two aspects: one is unity. MFRPS establishes a unified criterion to measure and improve the supervision behavior of food production and processing; second, it constantly improves itself. Through continuous exchanges with regulatory authorities and partners in public health cooperation, regulatory procedures are constantly improved, food safety supervision is strengthened, and federal food safety is maintained. In the process of these standards, indirect governance of suppliers will be involved. However, if the unsafe food provided by these suppliers is due to their misconduct, the damage caused will be compensated through the tort law of the United States, thus resolving the dispute.[25]

5.3.2 EU

On May 6, 2013, the European Commission adopted a package of proposals[26] related to food safety.[27] In terms of the legal system, a supranational centralized supervision system should be established, and in terms of legal content, the EU food safety standards should be further improved to realize the fundamental adjustment of EU food safety law. The EU Food Safety Administration is endowed with the power of centralized transnational supervision, and the EU Food Safety Import and Export Access Mechanism is unified at abroad. At the same time, the package of proposals also drastically reformed the existing food safety laws in the EU, and raised the requirements of food safety, mainly in two aspects: expanding the scope of food safety laws and improving standards: (1) Expanding the scope of food safety laws; (2) Emphatically expanding the safety of imported and exported goods and food. A full scope of supervision; (3) Improving food safety standards.

25. *See* Lemos, M.H., Schneider, S., Roberts, M., et al., 2015. U.S. Food & Law Policy Panel.
26. This new official controls regulation has been published in the official journal as regulation 625/2017. It comes into force at the end of 2019.
27. http://www.fao.org/3/y5871e/y5871e0l.htm.

This is mainly reflected in the further refinement of the existing risk prevention and control mechanism and stricter regulations on international trade related to food safety. In the details of the internal risk prevention and control mechanism, the first part and second chapter of the package of proposals for animal products listed in detail the possible animal diseases and risks and marked the risk level. For some high-risk levels, the EU is required to intervene and control directly, which further strengthens the control of animal food safety at the EU level. In the import of foreign goods, the proposal made stricter provisions. The proposal makes more detailed and stringent provisions on the destruction of import and export goods with food safety risks. Once possible risks arise, the relevant goods will be destroyed immediately.

Finally, the proposal places particular emphasis on strengthening food safety regulation of goods from third world countries. The package proposed that goods from third world countries should be subject to higher levels of food safety risks, and amended the terms of collaboration and cooperation between relevant departments to make it more effective to strengthen the collaboration among relevant departments of food safety. The European Commission believes that such coordinated regulation is particularly important to prevent possible food safety risks in goods from third world countries and to ensure that regulators respond quickly and accurately.[28]

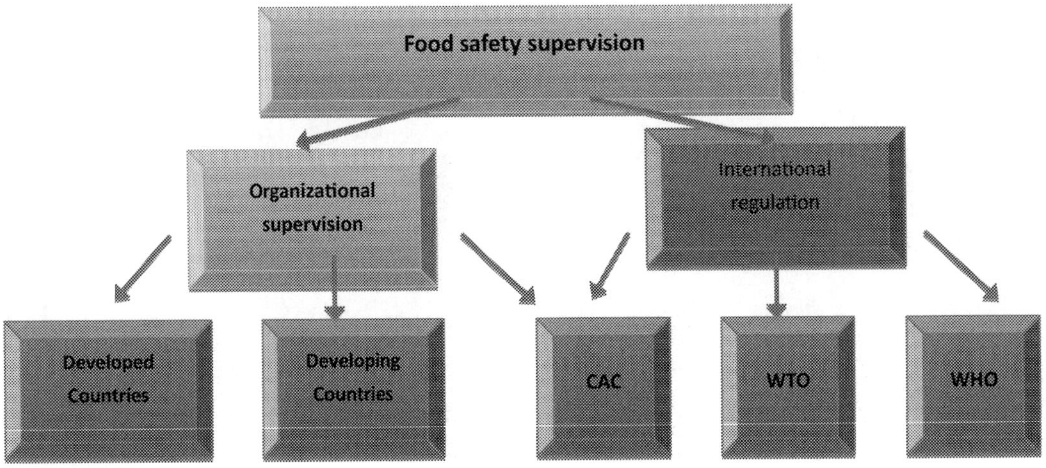

5.4 Conclusion: supplier change and global food safety regulation

At present, the global food supply chain has undergone two profound changes: first, from a vertical perspective, with the continuous refinement of the social division of labor, the food supply chain is usually divided into product research and development, agricultural production, processing and manufacturing, warehousing and transportation, marketing and sales, and other different links; second, from a horizontal perspective, with the development of trade globalization and logistics industry, food raw material supply, sales, and production have changed from localization to regionalization and globalization. Food supply chains are distributed in different countries and regions. The internationalization and complexity of the food supply chain lead to the corresponding extension of the regulatory chain, which increases the difficulty of government regulation. Problems in each link may lead to global food safety incidents. Therefore, in today's economic globalization, it is difficult for any country or region to stand alone in the face of food safety problems. All countries in the world, including China,[29] must strengthen international governance of food safety.

Countries and international organizations should strengthen cooperation to further strengthen the management of the supply chain of imported food. Firstly, food safety representative offices should be set up abroad to act as food safety foreign-related agencies of exporting countries and importing countries to communicate with other countries on food safety standards, food safety supervision system, and food safety laws and regulations system, to enhance mutual understanding and trust. Secondly, the establishment of food safety integrity files. Establish food safety integrity archives with major food exporting countries. Enterprises need to submit integrity archives in international food trade. The integrity level of food

28. *See* Bernd Van Der Meulen. EU Food Law Handbook 2014.

29. Also See Chapter 2, Ines Härtel (ed.) Handbook of Agri-Food Law in China, Germany, European Union, Springer 2018; JéromeLepeintre and Juanjuan Sun, Building Food Safety Governance in China, Luxembourg Publications Office of the European Union, 2018 and several working papers at: www.food-law.nl.

exporting enterprises is regarded as an important criterion for deciding whether to grant imports. Finally, a global food safety integrity archives database is gradually established and a global food safety integrity system is established. Thirdly, we should jointly combat illegal food import and export. Establish regular cooperation mechanism, like INFOSAN[30] with major countries to jointly combat illegal food import and export, and jointly combat illegal smuggling, illegal transit, illegal entrepot, and other criminal acts by means of technical exchanges, personnel exchanges, information exchanges, and special joint action, so as to build a green and safe food trade environment for countries.

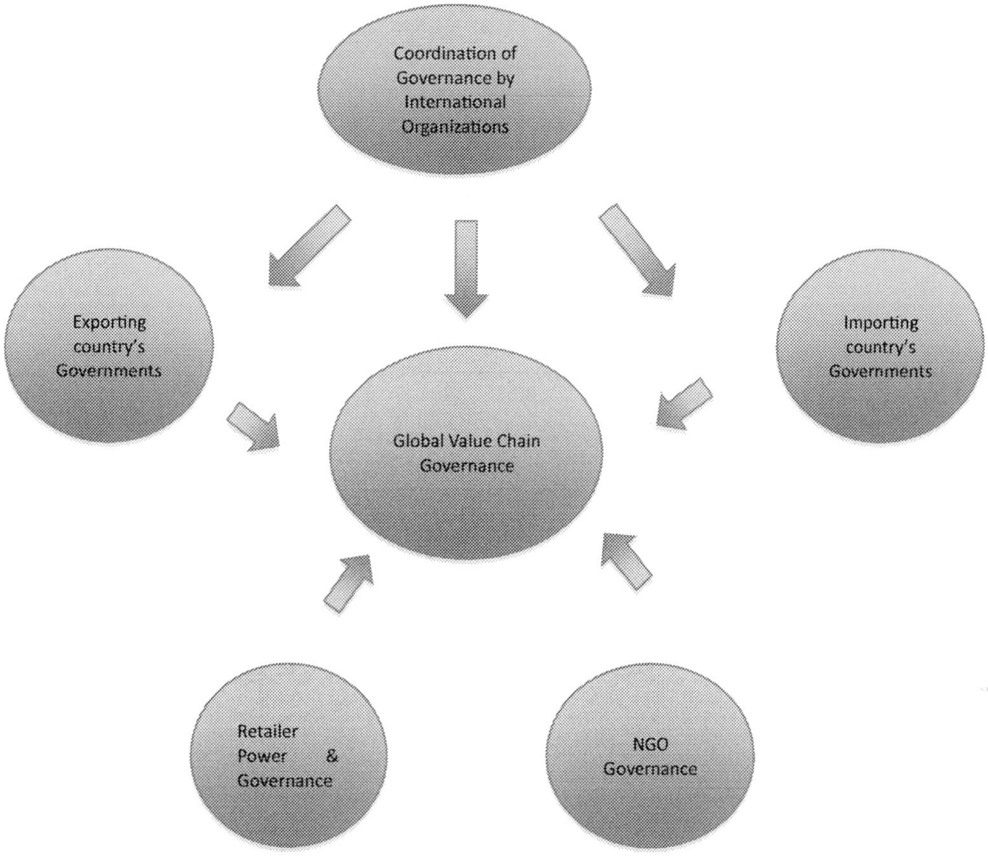

The above figure reveals the governing bodies and their relationships of the global food value chain: the coordination of importing countries, exporting countries, suppliers, nongovernmental organizations, and international organizations can have an impact on the global food value chain, and international law can have an impact on the governments of exporting and importing countries, but the influence on suppliers is mainly achieved through national legislation, which is reflected in each case. Differentiation of domestic legislation. Despite the restrictions of international law, countries still set up a high level of food safety supervision model within a certain range, and mainly concentrated in the regulatory areas mentioned above. Overall, however, the governing bodies have played a positive role in promoting the global food value chain.

Further reading

n.d. Analysis of Food Safety Management System in the United States. [Online]. http://www.mofcom.gov.cn/article/i/dxfw/nbgz/201701/20170102497593.shtml.Analysisabout American food safety administration system (Accessed 6 January 2019).

n.d. Animal Feed Regulatory Program Standards (AFRPS). [Online], Available: https://www.fda.gov/federal-state-local-tribal-and-territorial-officials/regulatory-program-standards/animal-feed-regulatory-program-st andards-afrps. (Accessed 6 January 2019).

Awaysheh, A., Klassen, R.D., 2010. The impact of supply chain structure on the use of supplier socially responsible practices. Int. J. Oper. Prod. Manag. 30 (12), 1246–1268.

30. INFOSAN is a voluntary technical network managed by the Food and Agriculture Organization (FAO) and WHO that brings together national authorities playing a role in food safety. https://www.who.int/foodsafety/areas_work/infosan/infosan_archives/en/.

Eugenia, L., 2013. The EU's new regulatory framework on Official Controls, Animal Health, Plant Health and Seeds. Eur. J. Risk Regul. (3), 381.

n.d. FDA Provides $21.8 Million to States for FSMA Produce Safety Rule Implementation. [Online], Available: https://www.fda.gov/NewsEvents/Newsroom/PressAnnouncements/ucm519817.htm (Accessed 9 September 2019).

Gilles, B., 2013. Horsemeat crisis about to tighten French Food Law. Eur. Food Feed Law Rev. 8.

Ji, P., Ma, X., Li, G., 2015. Developing green purchasing relationships for the manufacturing industry:an evolutionary game theory perspective. Int. J. Prod. Econ. 166, 155–162.

Jones, P., Hillier, D., Comfort, D., 2014. Assurance of the leading UK food retailers'corporate social responsibility sustainability reports. Corp. Govern. 14 (1), 130–138.

Luis González, V., 2013. The European Commission proposal to simplify rationaliseand standardise food controls. Eur. Food Feed Law Rev. 308.

Ma, X., 2010. Analysis on quality control in food supply chain based on dynamics evolutionary game model [C]. Optoelectron. Image Process. 1, 259–262.

Manning, L., 2013. Corporate and consumer social responsibility in the food supply chain. Br. Food J. 115 (1), 9–29.

n.d. Manufactured Food Regulatory Program Standards 2016 Updates. [Online], Available: https://www.fda.gov/Food/NewsEvents/ConstituentUpdates/ucm523145.htm. (Accessed 5 October 2018).

n.d. Manufactured Food Regulatory Program Standards. [Online], Available: https://www.fda.gov/downloads/ForFederalStateandLocal Officials/ProgramsInitiatives/RegulatoryPrgmStnds/UCM606483.pdf. (Accessed 30 September 2018).

n.d. Manufactured Food Regulatory Program Standards (MFRPS) [Online], Available: https://www.fda.gov/ForFederalStateandLocalOfficials/ProgramsInitiatives/RegulatoryPrgmStnds/ucm475064.htm. (Accessed 6 December 2018).

MFRPS fact sheet, 2019-8-06 [Online], Available. https://www.fda.gov/downloads/ForFederal StateandLocalOfficials/ProgramsInitiatives/RegulatoryPrgmStnds/UCM52 3942.pdf.

Spence, L., Bourlakis, M., 2009. The evolution from corporate social responsibility to supply chain responsibility: the case of Waitrose. Supply Chain Manag. Int. J. 14 (4), 291–302.

Sweet & Maxwell, 2013. Commission Proposes to Modernise, Simplify and Strengthen EU's Agri-Food chain[J]. EU Focus, p. 308.

n.d. Voluntary National Retail Food Regulatory Program Standards [Online], Available: https://www.fda.gov/food/retail-food-protection/voluntary-national-retail-food-regulatory-program-standards. (Accessed 30 July 2019).

Wieke Willemijn Huizing Edinger, 2004. A legal perspective on EU competence to regulate the 'healthiness' of food. Eur. Food & Feed Law Rev.

Xiao, T., Chen, G., 2009. Wholesale pricing and evolutionarily stable strategies of retailers with imperfectly observable objective. Eur. J. Oper. Res. 196 (3), 1190–1201.

Xiao, T., Yu, G., 2006. Supply chain disruption management and evolutionarily stable strategies of retailers in the quantity-setting duopoly situation with homogeneous goods. Eur. J. Oper. Res. 173 (2), 648–668.

Yu, H., Zeng, A.Z., Zhao, L., 2009. Analyzing the evolutionary stability of the vendor-managed inventory supply chains. Comput. Ind. Eng. 56 (1), 274–282.

Chapter 6

A simplified guide to understanding and using food safety objectives and performance objectives

L.G.M. Gorris[1], M.B. Cole[2] and The International Commission on Microbiological Specifications for Foods[3]
[1]Food Safety Expert, Food Safety Futures, Nijmegen, The Netherlands; [2]Head, School of Agriculture Food and Wine. University of Adelaide, Urrbrae, SA, Australia; [3]www.icmsf.org

6.1 Introduction

Diseases caused by foodborne pathogens constitute a major worldwide public health problem and preventing or mitigating them is a priority for many societies that requires close collaboration between public and private sectors. Microbiological foodborne diseases are typically caused by bacteria or their metabolites, toxigenic fungi, parasites, and viruses. The importance of different foodborne diseases varies between countries depending on foods consumed, food processing, preparation, handling, storage techniques employed, and sensitivity of the population or of subpopulations. While the total elimination of foodborne disease remains an unattainable goal, much work is done by food safety professionals in government, academia, and industry to continuously reduce the burden of illness due to contaminated food. However, reducing foodborne illnesses will always have a cost to society. "Cost" includes money as well as considerations of culture, eating habits, etc. For example, banning a certain food commodity, such as unpasteurized milk or products made thereof, may be acceptable to some countries, but not to others. While virtually all countries aim to reduce foodborne illnesses, hardly any country has explicitly stated explicitly to what degree they accept current levels of risk or to what lower level they aim to reduce the number of foodborne illnesses in their country. Also, between different governments or even within governments there may be different opinions about how a society wishes to balance cost with the reduction in foodborne illnesses aimed for.

Countries have traditionally attempted to establish benchmarks for the accepted level of food safety by setting microbiological criteria (MCs) for raw or for finished processed products (ICMSF, 2002, 2011, 2018; CAC, 2013; Zwietering et al., 2015; Gorris and Cordier, 2019). However, in isolation, the frequency and extent of sampling used in traditional food testing programs may not provide a high degree of consumer protection, which primarily depends on the adherence to Good Practices (Good Agricultural Practice [GAP], Good Manufacturing Practice [GMP], Good Hygiene Practice [GHP]) and the principles of Hazard Analysis Critical Control Point (HACCP). In many cases, when no in most cases, MCs are typically decided on by governments without a clear view of the level of risk a food poses to consumers and without estimating the impact of setting MCs in terms of maintaining status quo or reducing the risk of foodborne diseases. Sometimes, MCs established by national governments for different foods have been viewed by other countries as barriers to international trade, in particular where a country would impose a stricter level than the level that international food safety bodies such as Codex Alimentarius might have advised. Over 100 countries have signed an important food safety–related agreement, generally referred to as the "Sanitary and Phytosanitary (SPS) Agreement" of the World Trade Organization (WTO, 1994). This agreement states that "while a country has the sovereign right to decide on the degree of protection it wishes for its citizens, it must provide, if required, the scientific evidence on which this level of protection rests." It follows that if a country sets an MC—or any other form of limit—for a particular microorganism or group of microorganisms in a particular food product, it must be able to explain the rationale and justification for the criterion on the basis of scientific

Ensuring Global Food Safety. https://doi.org/10.1016/B978-0-12-816011-4.00018-5

data, public health risk, and other legitimate societal considerations. Another WTO agreement, the "Technical Barriers to Trade (TBT) Agreement" (WTO, 1995), also requires that a country must not ask for a higher degree of safety for imported goods than it does for goods produced in its own country.

6.2 Good practices and hazard analysis critical control point

Realizing the shortcomings and lack of food safety assurance provided by traditional food control approaches relying on inspection and limited sampling and testing of lots, the concept of HACCP was developed in the early 1970s and by now, through Codex Alimentarius and national governments, advocated or even mandated almost worldwide. The HACCP concept has provided major improvements in the production of safe foods. The HACCP principles help food business operators to focus on the specific hazards possibly associated to a particular food commodity that are reasonably likely to affect public health when left uncontrolled, and to design food products, processing, commercialization, preparation, and use conditions such that they in concert consistently control those hazards. To be successful, HACCP needs to build on good practices such as GAPs, GMPs, and GHPs, which are prerequisite programs to prevent and/or minimize the occurrence of microbiological, chemical, and physical hazards in food products as they pass through the food supply chain up to the point of consumption. HACCP then involves an assessment of significant hazards not dealt with in a specific production step by good practices alone and defines additional control measures needed at that step that are critical for the safety of the end product. HACCP plans will typically state limits to hazards as well as parameters associated to control measures, monitoring procedures, and corrective actions. HACCP plans are plant/factory and line specific and do not directly link the effectiveness of operational control over hazards to the expected level of health protection (e.g., a reduction in the number of foodborne illnesses occurring in a country) or the acceptable level of a hazard in a food (e.g., a current or future level of acceptable risk articulated by a government). Notably, in recent years, governments and other food safety professionals have updated the Codex Alimentarius guidelines on the use of Good Practices and HACCP (CAC, 2019).

6.3 Setting public health goals—the concept of appropriate level of protection

During the past few decades, there has been increased interest and effort in developing tools to more effectively link the requirements of food safety management systems implemented in the various steps throughout food supply chains with currently expected or planned levels of public health protection or acceptable risk to consumers. This chapter introduces two such tools that essentially are risk-based metrics functioning as benchmarks for food safety: the "Food Safety Objective" (FSO) and the "Performance Objective" (PO) (ICMSF, 2002, 2005, 2018; CAC, 2007a; Gkogka et al., 2013a,b). These metrics can be used to communicate food safety requirements to industry, trade partners, consumers, and other countries. Good practices and the HACCP system remain essential food safety management tools for achieving FSOs or POs.

Setting goals for public health or benchmarks for food safety is the right and responsibility of governments. These goals or benchmarks should specify the maximum level of harmful bacteria that may be present in a food. The term "level" in the case of FSOs or POs can signify a concentration of a microorganism, a frequency of its occurrence (i.e., prevalence), or both (CAC, 2007a; CAC, 2007b). Where possible, the determination of this level should be based on scientific and legitimate societal factors. Costs may include industry costs for reformulation and changes in processing, consumer costs due to increased prices, loss of quality, or reduced availability of certain products, and regulatory costs in terms of surveillance and enforcement.

In many countries, governments rely on disease and food surveillance data in combination with expert advice in the areas of epidemiology, food microbiology, and food technology to evaluate which types and levels of harmful microorganisms in foods may cause disease, including the severity and likelihood of disease, i.e., the consumer risk level. The level of consumer risk can be expressed in a qualitative way (e.g., high, medium, or low risk), or where possible, as the number of cases of foodborne disease per number of people per year. Particularly in developing countries, disease surveillance data are limited or not available at all. In such instances, estimates of the risk level have to be based on clinical information available (e.g., how many stool samples have been found to contain salmonellae) in combination with results from microbiological surveys of foods, evaluations of the types of foods that are produced, how they are produced, and how they are stored, prepared, and used. Increasingly more countries use scientific techniques such as Microbiological Risk Assessment (MRA) to estimate the risk of illnesses using detailed knowledge of the relationship between the number of microorganisms in foods and the occurrence of foodborne diseases. The Joint FAO/WHO Expert Meetings on Microbiological Risk Assessment (JEMRA) by now have produced a large number of authoritative MRA studies, guidance on different aspects of risk assessment, and associated training, which are all available online (JEMRA, 2021).

Whatever method is used to estimate the risk of foodborne illness associated to certain hazards associated to food, the next step is to decide whether this risk can be tolerated or needs to be reduced. The level of risk that a society is willing to accept (as articulated by the relevant competent authority) is referred to as the "Appropriate Level of Protection" (ALOP). Importing countries with stricter requirements for a specific hazard (e.g., some type of harmful bacteria or chemical) may be asked to determine a value for the ALOP according to the SPS agreement (WTO, 1994). When a country is willing to accept the current risk of illnesses, that level may be considered as the default or status quo ALOP. Locally produced as well as imported foods should not compromise the ALOP, and a government may take action accordingly when it expects an undue impact on ALOP. However, most countries will wish to lower the incidence of foodborne disease and may set targets for future ALOPs (FAO/WHO, 2000; EFSA, 2007, 2017). For instance, the United States of America launched the "Healthy People 2010" program with, in the food safety area, a stated governmental ambition to half the burden of illness caused by four key pathogens by the year 2010 as compared to the 1997 baseline (FDA-FSIS, 2010). In this case, the new reduced levels of illness cannot be considered as an ALOP in WTO terms (FAO/WHO, 2000), but rather as a Public Health Target (PHT) or ambition that may be stipulated by an authority to the local industries but cannot actually be considered mandatory for trading partners importing into the country.

6.4 Food safety objectives

When a government expresses PHTs or ALOPs relative to the incidence of disease, this does not provide food processors, producers, handlers, retailers, or trade partners with tangible information about what they need to do to contribute to delivering toward an ALOP or a PHT. To be meaningful, the benchmarks for acceptable food safety levels set by governments need to be translated into parameters that can be used by food industries to establish the correct stringency of the food safety management system that they operate and that can be assessed by government agencies for surveillance and enforcement. The concepts of FSOs and POs have been proposed to serve this specific purpose: to inform private and public stakeholders of the acceptable food safety levels in a tangible and quantitative way. When FSO and/or POs values have been established, these can be operationalized to the level of MCs that can be used for verification of the correct functioning of the food safety management system (Zwietering et al., 2015). ICMSF (2018) contains several case studies (i.e., aflatoxins, *Listeria monocytogenes*, *Salmonella* spp., pathogenic *E. coli*, *Vibrio* spp., *Campylobacter* spp. in certain food commodities) on how either ALOPs or PHTs may be used to derive FSO/PO and ultimate MC values. The position of these various concepts in the course of a very much simplified food chain is illustrated in Fig. 6.1.

According to Codex Alimentarius (CAC, 2007b), an FSO is "the maximum frequency and/or concentration of a hazard in a food at the time of consumption that provides or contributes to the appropriate level of (health) protection (ALOP)." Obviously, FSOs should only be established when they do contribute to meeting an ALOP or PHT, i.e., when it is anticipated that an FSO will deliver the required level of food safety. FSOs transform an ALOP or PHT into a concentration and/or frequency (i.e., a level) of a hazard in a certain food product at the point where that food is consumed. An extensive treatise of the role of FSOs, POs, and related metrics is provided in ICMSF (2018).

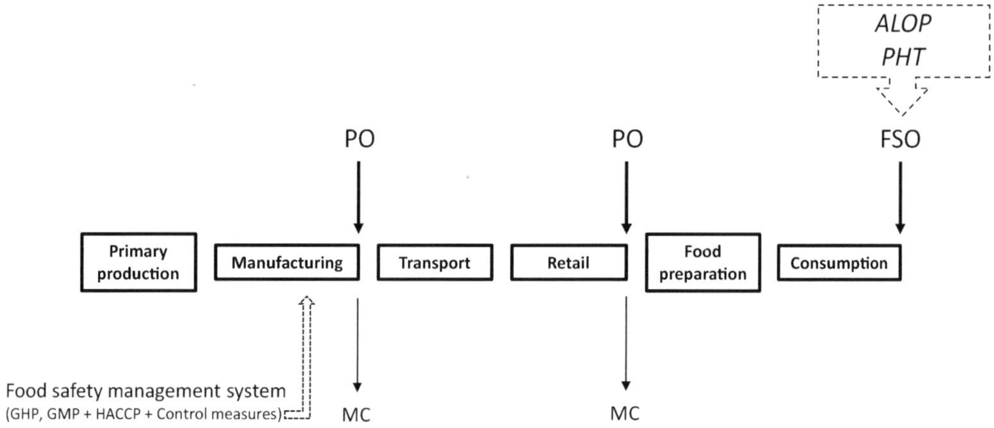

FIGURE 6.1 Simplified food supply chain indicating the position of a food safety objective (FSO) and a performance objective (PO) and a microbiological criterion (MC) derived from it at the specific steps of retail and manufacturing. The FSO has been set against an Appropriate Level of Protection (ALOP) or Public Health Target (PHT).

The FSO thus specifies an ultimate food safety benchmark for the whole food chain to achieve, but it does not specify how this benchmark is to be achieved (Gorris, 2005; Gorris et al., 2006). Hence, the FSO gives flexibility to the food business operators in the food chain to use production and processing techniques that best suit their situation but may well be significantly different across different comparable value chains as long as the maximum hazard level specified at consumption is not exceeded. For instance, milk is typically rendered safe by heat processing; however, in the future this may also be achieved by other technologies that rely less or not at all on heating and may offer quality benefits to consumers. This is also important in international trade since different techniques may be used in different countries. The "equivalence" of these techniques in delivering an articulated level of safety may then be the key point of compliance that ensures adequate consumer protection without imposing an unjustified barrier to trade.

6.5 Performance objectives

For some food hazards, the FSO is likely to be very low, sometimes referred to as "absent in a serving of food at the time of consumption." This, for instance, would likely be the case for *Salmonella* spp. or pathogenic *E. coli* in prepared food. For some other pathogens, such as *L. monocytogenes* in ready-to-eat foods that do not support its growth or of *Bacillus cereus* in any food, a particular low level of the pathogen may be tolerated in the food at the point of consumption when the risk to consumers is deemed acceptable by the relevant competent authority. For a processor that makes ingredients or foods that require cooking prior to consumption, an FSO level at consumption may be very difficult to use as a guideline in the factory, and also for governmental surveillance and enforcement, the point of consumption is not a practical reference point (Gorris, 2005; Gorris et al., 2006). Therefore, a level that must be met at earlier steps in the food chain would be much more practical to have as a food safety benchmark. This level is called a "Performance Objective" (ICMSF, 2002, 2018; Zwietering et al., 2015). A PO may be obtained from an FSO, as will be explained below, but this is not necessarily always the case. A PO is defined by Codex Alimentarius as "the maximum frequency and/or concentration of a hazard in a food at a specified step in the food chain before consumption that provides or contributes to an FSO or ALOP, as applicable (CAC, 2007b). Foods that need to be cooked before consumption may contain harmful bacteria that can contaminate other foods in a kitchen. Reducing the likelihood of cross-contamination from these products could be important in achieving an ALOP or public health target. The level of contamination that should not be exceeded in such a situation at a step earlier than consumption is a PO. For example, raw chicken may be contaminated with *Salmonella*. Although thorough cooking will make the chicken safe (i.e., will deliver the absence of *Salmonella* in a serving of food), raw chicken may contaminate other foods during the preparation of a meal. A PO of "no more than a specified percentage of raw chicken carcasses may contain *Salmonella*" may in that case reduce the likelihood that *Salmonella* will contaminate other foods and it will need to be delivered by good hygienic practices and avoidance of cross-contamination where the chicken is prepared for consumption. POs can be calculated from the FSO by considering expected bacterial contamination and/or growth as well as reduction between the points of the FSO backward to the point where the PO is required. In ICMSF (2002, 2018), a so-called conceptual equation is used to explain this in more detail in terms of principles and case examples.

6.6 The difference between food safety objectives, performance objectives, and microbiological criteria

MCs ideally are accompanied by information that makes it very clear to stakeholders at what point in a value chain an MC applies and what its target performance is, such as food product, target microorganism, sampling plan parameters and limits, analytical units and methods, and typical corrective actions for noncompliance lots (CAC, 2013). Traditionally, MCs have been designed to be used for testing a shipment or lot of food for acceptance or rejection, as a means to verify that the food safety management system used for producing the food was operating well. However, also in situations where no prior knowledge of the processing conditions is available, MCs can be relevant indicators of lot acceptability with the assumption that when lots comply to well-designed MCs, they need to be produced with a sound food safety management system that has assured adequate operational control. MCs have also been found very useful for verification of process control and management of environmental contamination (ICMSF, 2011; Zwietering et al., 2014).

In contrast to MCs, FSO or PO values are maximum levels and do not directly relate to verification sampling and testing (Gorris et al., 2006). However, an MC can be derived from an FSO or a PO (Van Schothorst, 2009; Zwietering et al., 2010, 2015) such that it is a risk-based benchmark allowing testing of foods for specific microorganisms as an effective means for validation of control measures and verification of operational control. There are several approaches to sampling (e.g., lot testing, process control testing), but they all compare the results obtained against a predetermined limit (i.e., a level of microorganisms in terms of concentration, frequency, or both). Approaches and tools have now been developed to relate the performance of attribute, or presence/absence testing, and sampling plans to the level of a microbiological hazard that could be detected with a certain probability (Legan et al., 2001; Van Schothorst et al., 2009; Zwietering et al., 2015). It is expected that these approaches help to quantify and document more clearly and consistently that food safety management systems and their key control measures do, in fact, achieve the desired level of food safety stipulated by ALOPs or PHTs.

6.7 Responsibility for setting a food safety objective

Deciding if and when to use an FSO is the responsibility of governments (CAC, 2007a); the decision on what is or is not considered acceptable in terms of food safety is the traditional role of government, but the actual expression of a concentration and/or frequency of a hazard (e.g., related to infectious bacteria or to bacterial toxins) in a food at the time of consumption (the FSO) was introduced in the last decade or so as a very new concept based on sound science and risk appraisal by governmental competent authorities. Governments have the risk management responsibility of deciding on acceptable risk levels of required food safety levels. To evaluate risks and safety, they typically consult with risk assessors mostly residing in public organizations, who typically are experts in foodborne disease epidemiology, food microbiology and food processing, statistics, etc., as well as other stakeholders (e.g., industry, consumers, trading partners, etc.) to decide what the FSO value should be. Occasionally, a very quick reaction is needed, and use is then, for instance, made of expert panels consulted on short notice, and a risk management decision made. The SPS agreement requires that, in such instances, these values are considered as interim measures.

As noted earlier, FSOs should only be developed in situations where they will have an impact on public health in terms of contributing to food safety. It is, therefore, not necessary to establish FSOs for all foods. Understanding which hazards are important in what foods, predicting future food safety concerns and, importantly, designing food processing, and preparation procedures that will prevent or minimize foodborne diseases from occurring are major goals of food microbiological research conducted by many food professionals in academia, government, and industry around the world. Experts in these areas can assist governments in the development of realistic FSOs.

6.8 Setting a performance objective

When an FSO value has been set, POs upstream in the food supply chain may be derived from this value by taking into account the changes that will occur in the level and/or frequency of the hazard (e.g., the relevant harmful bacteria) between the points where POs are set and the point of consumption. These POs may be stricter than the FSO to account for contamination or growth of harmful bacteria downstream of the point of the PO, e.g., during distribution, preparation, storage, and use of a particular food. On the other hand, where appropriate, PO values may be more lenient than the FSO value, for instance, if the product is adequately cooked just before consumption. POs may be set by both government and industry. Considering the diversity of industry, governments may decide to set POs as a means to achieve FSOs at the point of consumption (CAC, 2007a). Governments may also set POs in the absence of FSOs or, for instance, in cases where raw foods are seen as a source of cross-contamination as was explained previously. POs can be set at one or more steps along the food chain where control measures can and should be applied to prevent foodborne diseases, for example, at points where it is important that all products remain below a particular level. POs, like any other microbiological limit for finished products, should take into consideration the initial level of the hazard before any treatment, as well as the decreases and possible increases of that hazard level, if any, prior to consumption (ICMSF, 2002, 2018). This approach has been fundamental to safe food processing for decades and will not change with the introduction and implementation of an FSO or PO. In fact, the FSO and PO are additional tools that the food industry can use to build food safety into their products (Fig. 6.2).

6.9 Responsibility for compliance with the food safety objective

Marketing of food that is not harmful to consumers when used in the intended way is the responsibility of the various food business operators engaged along the food production chain. This is a fundamental responsibility and the expertise and guidance to discharge has been continuously improved at the international level over the last 50–60 years. The introduction of the concepts of FSO and PO and linking these to MCs is expected to further enhance good food safety management, making food professionals involved in the various parts of a food value chain even more aware of the fact that they share this responsibility. Governments or third parties can assess programs, such as good practices and HACCP,

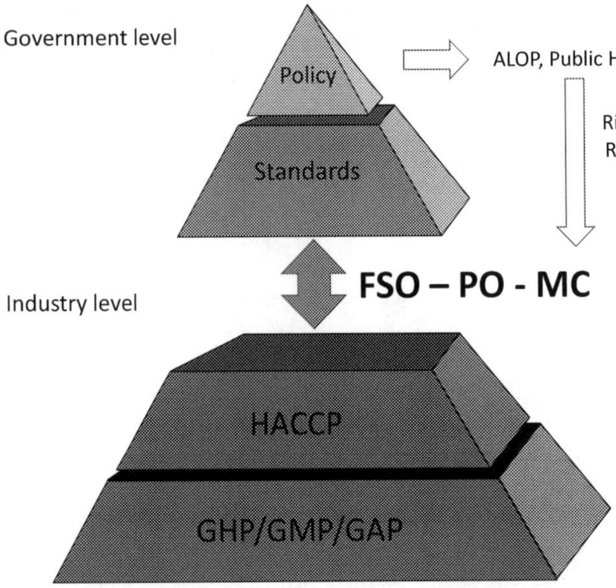

FIGURE 6.2 Food Safety Objective (FSO) and Performance Objective (PO) are risk-based metrics through which governments may communicate food safety levels to be met by certain food industries in order to contribute to a policy level ALOP (Appropriate Level of Protection) or public health target. Governments risk management and risk assessment to establish risk-based metrics. Industry will be using their Good Practices and HACCP to meet the specified metrics, which can be verified by a Microbiological Criterion (MC) derived from an FSO or PO.

to confirm the likelihood that the products will meet the FSOs/POs. This may be extended across national boundaries, as some countries will ask that imported products are produced under food safety management programs based on GHP and HACCP, using not only MCs as benchmarks for safety, but possibly also FSOs or POs.

6.10 Meeting the food safety objective

Since the FSO is the maximum level of a hazard at the point of consumption, this level will frequently be very low. Because of this, measuring this level is impossible in most cases. Compliance with POs set at earlier steps in the food chain can sometimes be checked by microbiological testing. However, in most cases, validation of control measures, verification of the results of monitoring critical control points, as well as auditing good practices and HACCP systems, will provide the reliable evidence that POs and thus the FSO will be met. MCs can be derived from FSOs and POs, if such levels are available. If such levels are not stated, MCs can be developed, if appropriate. The ICMSF (2011, 2018) has provided extensive guidance on the establishment of suitable MCs.

6.11 Not all food safety objectives are feasible

When establishing FSOs, governments should determine through discussions with relevant experts and stakeholders what feasible FSO values should be. In some cases, it may turn out that it is not possible to comply with a set FSO level in practice, and a government may decide to set a less stringent FSO. Such an FSO may be set temporarily until improvements in processing technology make it possible to set a lower (more stringent) FSO. An alternative would be to keep the more stringent FSO and to provide a period during which processing procedures can be changed to meet the FSO. In the first case, it may be appropriate to communicate to consumers the particular risk associated with consuming the product. An alternative approach is the banning of a product; e.g., banning of high-risk tissues (spinal cord, root ganglia, tonsils) of beef to be sold for human consumption due to the inability to detect and/or eliminate bovine spongiform encephalopathy (BSE).

6.12 Concluding remarks

FSOs and POs are relatively new concepts in food safety management that have been introduced to further assist government and industry in communicating and complying with PHTs or Appropriate Levels of Protection. These tools may function as outcome-based benchmarks for designing appropriate food safety management systems in and along food supply chains, leaving flexibility to the industries involved regarding the technologies and systems deployed to deliver against the benchmarks. Hence FSOs and POs build on, rather than replace, existing food safety practices and concepts. Using the new risk management tools of FSO and PO may well allow for food safety control measures and regulations to be developed and implemented more rapidly and more proportional to actual consumer risk levels. Many of the food safety issues that the world faces today are very complex in nature, frequently requiring a through-value-chain approach and often requiring more than one control measure to effectively manage a consumer risk. The continuously improvement and sharpening of science- and risk-based food safety management practices offers a robust framework that facilitates clear communication between stakeholders on the most effective food safety management options, thereby speeding up the development of effective and proportional risk management.

A good example of a Codex Alimentarius code that benefited from these developments is the "Code of Hygienic Practice for Powdered Formulae for Infants and Young Children" (CAC, 2008). This code addressed the emerging public health threat of *Enterobacter sakazakii* (*Cronobacter* spp.) and was developed in a rather short period of time thanks to several dedicated MRAs that, at the time, were compiled by FAO and WHO through JEMRA, and which are available online (JEMRA, 2021). These risk assessments considered a range of control strategies during both manufacture and subsequent use of powdered infant formula that could be implemented to reduce risk. Importantly, the approach facilitated risk managers to not only compile the global advisory code, but also helped to establish important and urgent advice to several different stakeholders, including caregivers of infants, consumers, and industry.

Conceivably, the new risk-based approaches and tools may also be used to assess whether novel processes that utilize combinations of control measures provide a level of protection equivalent to traditional processing methods. For industry, this risk-based process development approach provides a roadmap for safe innovation and will encourage the development of innovative technologies. For academia, many opportunities exist to make great contributions to the success of risk-based process development through the development of innovative process technologies, mathematical modeling, and for fundamental research in preservative based multiple-hurdle preservation (Stewart et al., 2002; Membré et al., 2007; Zwietering et al., 2010; Anderson et al., 2011; Gkogka et al., 2013a,b).

6.13 About the ICMSF

The International Commission on Microbiological Specifications for Foods (ICMSF) was formed in 1962 under the International Union of Microbiological Societies (IUMS). Through the IUMS, the ICMSF has been linked to the International Union of Biological Societies (IUBS) and to the World Health Organization (WHO) of the United Nations. The Commission's primary goal is to provide timely, science-based advice and guidance to governments and industries on appraising and controlling the microbiological safety of foods in support of public health protection and facilitation of fair trade. Guidance and advice of the Commission is meant to be practical for various stakeholders to use within their scope of responsibility in food safety. With around 20 members from around the globe, with day jobs in public as well as private organizations relevant to food safety, the Commission is a small but very active volunteer organization. Further information about ICMSF can be found at www.ICMSF.org.

Acknowledgments

This chapter is based on an original paper from the International Commission on Microbiological Specifications for Food (http://www.icmsf.org/publications/guide/). ICMSF members and consultants that worked on the original paper and its translation are gratefully acknowledged.

References

Anderson, N.M., Larkin, J.W., Cole, M.B., Skinner, G.E., Whiting, R.C., Gorris, L.G.M., Rodriquez, A., Buchanan, R., Stewart, C.M., Hanlin, J.H., Keener, L., Hall, P.A., 2011. Food safety objective approach for controlling *Clostridium botulinum* growth and toxin production in commercially sterile foods. J. Food Protect. 74, 1956–1989.

CAC, Codex Alimentarius Commission, 2007a. Principles and Guidelines for the Conduct of Microbiological Risk Management (MRM). Codex Alimentarius Commission, CAC/GL-63. Food and Agriculture Organization, Rome. https://www.fao.org/input/download/standards/11776/CXG_077e.pdf.

CAC, Codex Alimentarius Commission, 2007b. Joint FAO/WHO food standards programme. In: Procedural Manual, seventeenth ed. Codex Alimentarius Commission. ISSN 1020-8070.

CAC, Codex Alimentarius Commission, 2008. Code of Hygienic Practice for Powdered Formulae for Infants and Young Children. Codex Alimentarius Commission. CAC/RCP 66-2008. https://www.fao.org/input/download/standards/11026/CXP_066e.pdf.

CAC, Codex Alimentarius Commission, 2013. Principles and Guidelines for the Establishment and Application of Microbiological Criteria Related to Foods. CAC/GL 21-1997. Food and Agriculture Organization. World Health Organization, Rome, Italy. https://www.fao.org/input/download/standards/394/CXG_021e.pdf.

CAC, Codex Alimentarius Commission, 2019. General Principles of Food Hygiene: Good Hygiene Practices (GHPs) and the Hazard Analysis and Critical Control Point (HACCP) System. Annex IV. Report REP20/FH of the 51st Session of CCFH (Cleveland, USA. Available at: link).

EFSA, European Food Safety Authority, Panel on Biological Hazards, 2007. Opinion of the Scientific Panel on biological hazards (BIOHAZ) on microbiological criteria and targets based on risk analysis. EFSA J. 462, 1–29 (link).

EFSA, European Food Safety Authority, Panel on Biological Hazards, 2017. Scientific Opinion on the guidance on the requirements for the development of microbiological criteria. EFSA J. 15 (11), 5052–5110 link).

FAO/WHO, 2000. The Interaction between Assessors and Managers of Microbiological Hazards in Food. Report of a FAO/WHO Expert Consultation, Kiel, Germany Food and Agricultural Organization of the United Nations/World Health Organization: Rome/Geneva. Available from: https://www.fao.org/3/ae586e/ae586e.pdf.

FDA-FSIS, 2010. U.S. Food and Drug Administration and the U.S. Department of Agriculture Food Safety and Inspection Service. https://www.cdc.gov/nchs/data/hpdata2010/hp2010_final_review_focus_area_10.pdf.

Gkogka, E., Reij, M.W., Gorris, L.G.M., Zwietering, M.H., 2013a. Risk assessment strategies as a tool in the application of the appropriate level of protection (ALOP) and food safety objective (FSO) by risk managers. Int. J. Food Microbiol. 167, 8–28.

Gkogka, E., Reij, M.W., Gorris, L.G.M., Zwietering, M.H., 2013b. The Application of the Appropriate Level of Protection (ALOP) and Food Safety Objective (FSO) concepts in food safety management, using Listeria monocytogenes in deli meats as a case study. Food Control 29, 382–393.

Gorris, L.G.M., 2005. Food safety objective: an integral part of food chain management. Food Control 16, 801–809, 2005.

Gorris, L.G.M., Bassett, J., Membré, J.-M., 2006. Food Safety Objectives and related concepts: the role of the food industry. In: Motarjemi, Y., Adams, M. (Eds.), Emerging Foodborne Pathogens. Woodhead Publ, ISBN 1-85573963-1.

Gorris, L.G.M., Cordier, J.-L., 2019. Microbiological criteria and indicator microorganisms. In: Doyle, M.P., Diez-Gonzalez, F., Hill, C. (Eds.), Food Microbiology: Fundamentals and Frontiers, fifth ed. ASM Press, Washington, DC.

ICMSF, International Commission on Microbiological Specifications for Foods, 2002. Microorganisms in Foods 7. Microbiological Testing in Food Safety Management. Kluwer Academic/Plenum Publishers, New York, ISBN 0-306-47262-7.

ICMSF, International Commission on Microbiological Specifications for Foods, 2005. Impact of Food Safety Objectives on microbiological food safety management. Proceedings of a workshop held on 9–11 April 2003 Marseille, France. Food Control 16 (9), 775–832.

ICMSF, International Commission on Microbiological Specifications for Foods, 2011. Microorganisms in Foods 8. Use of Data for Assessing Process Control and Product Acceptance. Springer, New York, NY, ISBN 978-1-4419-9373-1.

ICMSF, International Commission on Microbiological Specifications for Foods, 2018. Microorganisms in Foods 7. Microbiological Testing in Food Safety Management, second ed. Springer, Cham, Switzerland, ISBN 978-3-319-68458-1.

JEMRA, 2021. Joint FAO/WHO Expert Meetings on Microbiological Risk Assessment. Microbiological Risks and JEMRA. Food and Agricultural Organization of the United Nations (link).

Legan, J.D., Vandeven, M.H., Dahms, S., Cole, M.B., 2001. Determining the concentration of microorganisms controlled by attributes sampling plans. Food Control 12 (3), 137–147.

Membré, J.-M., Bassett, J., Gorris, L.G.M., 2007. Applying the Food Safety Objective and related concepts to thermal inactivation of Salmonella in poultry meat. J. Food Protect. 70 (9), 2036–2044.

Stewart, C.M., Tompkin, R.B., Cole, M.B., 2002. Food safety: new concepts for the new millennium. Innovat. Food Sci. Emerg. Technol. 3, 105–112.

Van Schothorst, M., Zwietering, M.H., Ross, T., Buchanan, R.L., M. B. Cole & the International Commission on Microbiological Specifications for Foods, 2009. Relating microbiological criteria to food safety objectives and performance objectives. Food Control 20, 967–979.

WTO, 1994. The WTO Agreement on the Application of Sanitary and Phytosanitary Measures (SPS Agreement). World Trade Organization, Geneva. Available at: https://www.wto.org/english/tratop_e/sps_e/spsagr_e.htm.

WTO, 1995. Uruguay Round Agreement, Agreement on Technical Barriers to Trade. World Trade Organization, Geneva. Available at: http://www.wto.org/english/docs_e/legal_e/17-tbt.pdf.

Zwietering, M.H., Stewart, C.M., Whiting, R.C., ICMSF, 2010. Validation of control measures in a food chain using the FSO concept. Food Control 21, 1716–1722.

Zwietering, M.H., Ross, T., Gorris, L.G.M., 2014. Food safety assurance systems: microbiological testing, sampling plans, and microbiological criteria. In: Motarjemi, Y. (Ed.), Encyclopedia of Food Safety, vol. 4. Academic Press, Waltham, MA, pp. 244–253.

Zwietering, M.H., Gorris, L.G.M., J.M. Farber and the Example 5A Codex Working Group, 2015. Operationalising a performance objective with a microbiological criterion using a risk-based approach. Food Control 58, 33–42.

Further reading

Cole, M.B., Tompkin, R.B., 2005. Microbiological performance objectives and criteria. In: Sofos, J. (Ed.), Improving the Safety of Fresh Meat. Woodhead Publishing Ltd, Cambridge.

Caipo, M., Cahill, S., Kojima, M., Carolissen, V., Bruno, A., 2015. Development of microbiological criteria for food. Special issue with 8 papers. Food Control 58, 1–50.

Gkogka, E., 2019. Food Safety Management Strategies Based on Acceptable Risk and Risk Acceptance. PhD thesis. Wageningen University, Wageningen, The Netherlands, ISBN 978-94-6395-036-7. https://doi.org/10.18174/496132. (link).

Chapter 7

Regulating emerging food trends: a case study in insects as food for humans

Adina Alexandra Baicu
University of Agronomic Sciences and Veterinary Medicine of Bucharest, Romania

7.1 Introduction

Eating insects is not only a challenge during entertaining shows, but turns out to be a rather normal, ancient source of food worldwide. If Asian and African populations consider insects as a potential meal, most of the Westerners are still confronted with the disgust factor, which appears to have deep cultural roots, grown together with the delimitation of religions.

Insects can be the contributing protein-rich ingredients to meet future global demand for a planet that is expected to shelter and feed nine billion inhabitants in the first half of this millennium, when the food supply should go up by 70% (United Nations, 2015). Entomophagy (the term for "eating insects") can be part of the answer not only for the population growth but also for undernourishment—it causes 3 million deaths among children every year (Black et al., 2013; Fasolato et al., 2018; Nadeau et al., 2015). Lack of micronutrients causes 2 billion people to suffer from the so-called "hidden hunger" (IFPRI, 2016). Including insects as food can lead to a wholesome diet; it can be a solution to address malnutrition in developing countries (Nadeau et al., 2015).

In addition, the slow global warming warns us to reduce gas emissions when harvesting our food. Farming insects for human (and animal) consumption can be a viable alternative for traditional agriculture.

Despite the proven benefits of entomophagy, possible associated risks have not been consistently inventoried. Moreover, no uniform legislative system is in place neither clear standards for breeding, farming, and commercialization of insect food have been developed so far. National or international legislation acknowledging the possibility of using insects as food or food ingredients is sparse, as for many human communities entomophagy was not an option. Lack of legal certainty inhibits innovation and sets barriers for international trade.

7.2 Where and what?

At the beginning of time, insects used to be a large source of food for humans; collecting insects intended for food was a seasonal activity mainly done by women, as a corresponding skill to hunting, done by our male ancestors (van Huis, 2017). Anthropologists have discovered that among their pharmaceutical, artistic, or alchemistic use, insects were recognized as food both in a raw form (uncooked) and sun-dried (Sutton, 1995).

Nowadays, the Western cultures have developed and handed down to generations the perception of insects as a source of pest, dirt, and disgust, while some Asian and African populations continue to include insects in their normal diet. The determining factor of such segregation was religion. It parted the population according to the corresponding religious practices and taboos; under these circumstances, new groups of people have been formed: those that kept the same eating habits, groups only eating vegetables, and groups that stopped eating red meat or avoided certain types of protein (including insects) (Sillow, 1983).

Similar to selecting other items in categories in terms of what is edible and what not, what needs to be cooked and what can be eaten raw, people have accumulated over time a sum of 2300 types of edible insects that are consumed by two billions of humans (Zielińska et al., 2018). They observed that, depending on the insect type, the larvae, pupae, and/or the adult stage are proper for consumption. In tropical areas, the availability and size of the insect are higher which resulted in

Ensuring Global Food Safety. https://doi.org/10.1016/B978-0-12-816011-4.00014-8

gathering food with fewer risks than hunting vertebrates (van Huis, 2017). The actual number of insects considered to be edible has been mapped (Jongema, 2017) and it can be observed that despite the popular belief that insects are consumed in tropical areas, they are recognized as food in other geographic zones too, like Japan, China, or Mexico.

As of species, the most consumed are beetles, caterpillars, followed by ants, bees, and wasps. The least populated groups of edible species are those of dragonflies, termites, flies, cockroaches, spiders, and others (Jongema, 2017). It is estimated that about 2000 species are consumed in the world (Jongema, 2017) and that approximately 30% of the population consumes insects (Kim et al., 2019; Zielińska et al., 2018).

Even if entomophagy enjoys of a smaller acceptance among those parts of the population, in Europe and the United States of America (Zielińska et al., 2018) insects can be found in restaurants and supermarkets and can take Western forms: snacks, pizza, pasta, burgers, sprinkles, or chips (Ulrich et al., 2017). The products are also marketed online and can be ordered as ready-to-eat or ready-to-cook-variants (Fasolato et al., 2018). This shows that, if the psychological barrier is overcome, Westerners could transform eating insects in a trend that can slowly become a normal eating habit. In this case, education can help to build knowledge and customs, combined with slowly introducing insects in already preferred and familiar shapes (Vanhonacker et al., 2013), or in the form of convenience food (De Boer, Schösler and Boersema, 2013). However, to be able to substitute animal protein successfully, whole and not processed insects should be consumed. Processed insects may also not meet the desired sustainability anymore (Hartmann and Siegrist, 2017).

7.3 Why eating insects?

Scientific findings underline the nutritional value of insects. They are a rich source of protein, minerals, vitamins, and omega-3 fatty acids (Payne et al., 2016; Roos and van Huis, 2017; Arnold van Huis, 2013; Zielińska et al., 2018).

Insect agriculture implies lower greenhouse emissions, so a lower impact on the environment. Farming insects require less extensive land use and resources than the known energy and agricultural land requirements for obtaining conventional animal protein (Oonincx and de Boer, 2012). Moreover, the use of pesticides can be reduced (DeFoliart, 1997)—some insects that are harmful to crop might be fit for human consumption; for example, grasshopper (*Sphenarium purpurascens*) is consumed by Central Mexicans and can be manually harvested, thus chemicals are avoided and a second food option is on the table (Ghosh et al., 2017).

Due to the novelty in research also, insects should be tested for possible toxic compounds and associated foodborne pathogens. Preliminary testing is already ongoing: a study performed in Belgium (Poma et al., 2017) compared the levels of risky chemicals found in farmed insects with those found in traditional protein sources (fish, eggs, meat). The results suggest that there was no significant difference between the two groups in that respect.

Another safety concern is that insects can possibly carry allergens. Detection methods are being developed (Belluco et al., 2013; Kim et al., 2019; Ribeiro et al., 2018; Ulrich et al., 2017) and further investigations are needed to establish the responsible molecular mechanism for insect food allergies. Studies for other possible risks are limited, but the subject starts to gain interest. Controlling the way insects are multiplied, fed, and depending on their development stage and how they are processed can limit the potential hazards (Gallo, 2019).

Last but not least, for some parts of the population, entomophagy is trendy (Fasolato et al., 2018) and it can open new markets, boost innovation, and change attitudes toward food.

7.4 The consumers are having a say

Despite the proven socio-economic benefits of including insects as a source of protein, consumer readiness can be a serious barrier for businesses in terms of product development and entering the market with insect food. It is important to identify the type of people that would consume insects, because they can become the trendsetters, while their incipient preferences can serve as basis for future product design, mass production, and quality expectations to be met.

After the request of the scientists, policymakers, and of the public sector to channel the attention to other possible food sources due to population dynamics and climate change, consumer studies regarding entomophagy have emerged.

In countries where people are comfortable with eating insects, the preference is to consume them as a whole, rather than hidden in the form of powder or other shapes. A study applied to the Kenyan consumers revealed that more than three-quarters of the respondents would consume insects. They originated from areas where entomophagy is a traditional, familiar aspect and their positive reaction was mainly driven by past experience, availability of the food source, and accessibility (Pambo et al., 2016).

In Europe, the consumer acceptance for entomophagy is slowly increasing in the last years, directly proportional to the level of understanding regarding the nutritional and environmental advantages (Schouteten et al., 2016).

A consumer study has shown that Swiss consumers would perceive the people eating insects as innovative, creative, health aware, and respectful of the environment. In addition, the insect consumer and a vegetarian would be considered more similar than a meat consumer (Hartmann et al., 2018).

Dutch consumers, self-determined to eat insects, agreed that at the beginning it is the curiosity and perception of an insect-based meal being as environmentally friendly and healthy that drive them to consume it. For developing a regular habit, the price, availability, and taste of edible insects should be similar to traditional Western protein sources (House, 2016). The sustainability factor is not sufficient for a new food product to enter and last on the market, but affordability, sensory qualities, and possibility to easily include it into the actual eating behavior as well as availability are key attributes for consumers to integrate entomophagy in the long run (House, 2016; Tan et al., 2016).

Sex is also a factor to be taken into account when addressing the consumers' willingness to adopt entomophagy. In 2015, Verbeke (Verbeke, 2015) worked on profiling the Western consumer that would potentially switch from meat to insects. Men are 2.17 times more likely than women to get over the "yuck factor" and consider insects as a possible source of protein. Similarly, a more recent study performed on Italian consumers concluded that men would be more open to eating insects and the most tempting would be crickets, while the least inviting species is the giant water bug (Tuccillo et al., 2020).

It is important to give people the opportunity to try insects as food since those who already tasted before and are familiar with the concept are more willing to embrace entomophagy than those who never ate insects. A survey carried out on Dutch and Australian consumers has shown that the majority of the participants were not aware of the benefits of consuming insects, neither thought of possible risks associated. However, because during the study they were exposed to insect food and were able to taste, they did enjoy the flavor of two items containing whole insects and insect flour. Moreover, the participants considered the insect product they tasted as natural food (Ulrich et al., 2017).

Food aesthetics, as well as texture and flavor play a large role in consumer acceptance when it comes to edible insects (Mishyna et al., 2019). The young Italian population is not ready to consume insects, but if the insect is hidden, taking different shapes (i.e., flour), the respondents were not completely reluctant anymore (Cavallo and Materia, 2018). Germans had comparable reaction, their disgust was triggered when they were able to see the whole insect rather than when the insect ingredient is not visible (processed in pasta, granola or protein bars) (Orsi et al., 2019). The Hungarian consumers that contemplate to reduce meat intake in the near future, would be willing to substitute that protein with protein of insect sources. Similar to other Europeans, Hungarians are generally not ready to adopt entomophagy, but if the insects are integrated into a socially acceptable dish and with familiar flavors, they might reflect on the idea. The consumers that are generally open to experience new, unfamiliar food products are the ones that would be open to trying edible insects (Gere et al., 2017). The attitude of the Romanian consumers regarding entomophagy was also explored. According to the study, about three-quarters of the participants are not inspired to consume insects; they associate insects with disgust and see them as unsafe for consumption. The Romanian respondents that would try insects would prefer them to be part of a salad or in a meat dish. The most inviting species would be crickets and locusts, while the most appealing cooking technique for insects would be frying (Andronoiu et al., 2018).

Exploratory studies to establish the presently preferred types of food recipes could help in adapting insects to the Western sensory profile (Hartmann et al., 2015). Building a generally positive image of consuming insects can lead to one barrier less in selling and marketing insects in areas where such eating behavior is not a social norm. The first step to increase the probability that people experiment and are curious about edible insects is displaying the familiar shapes and Westernize insect dishes. However, to determine habitual consumption, pleasant taste is an important factor, together with an acceptable price and availability of the products (Tan et al., 2017).

7.5 Regulatory aspects regarding insects for human consumption

7.5.1 Codex Alimentarius

Up until now, the Codex Alimentarius guidelines do not include anything about breeding, rearing, quality and safety assurance, and commercialization regarding edible insects. Codex Alimentarius only refers to insects as "impurities." The lack of Codex Alimentarius recommendations has determined concerned nations to install their own rules to regulate edible insects according to their own understanding and local needs. Such an approach does not necessarily assure a smooth collaboration between the countries and trade is not facilitated if the regulations differ.

7.5.2 Regulating edible insects in the European Union

If before 2018, when Regulation (EC) 2015/2283 came into force, European legislation regarding the status of edible insects was not clear; the actual legislative framework clarifies that insects are a novel food in the EU.

The old novel food regulation, Regulation (EC) 258/97, deemed as "novel" any food or food ingredient that was not consumed to a significant degree in one of the Member States before the May 15, 1997. In addition, a novel food would fall under one of four listed categories. One of them was "food ingredients isolated from animals." The definition of "novel food" did not clarify whether insects or insect parts were novel food in the EU or not. Later, in 2015, the European Parliament made clear that insects are within the scope of the novel food regulation, thus authorization is required to enter the European market. The premarket approval under the old novel food regulation was lengthy and cumbersome, determining the EU market to be less competitive and attractive for international companies and product development. Because some Member States (exempli gratia Belgium, Netherlands, Denmark) were against the decision that insects are novel food, they installed own regulations for edible insects (Lähteenmäki-Uutela et al., 2017).

Starting with January 2018, the new novel food regulation has introduced a simplified and centralized procedure for authorization of novel foods and aims to clarify the data requirements. The expected consequences are to stimulate innovation and to reduce costs and time spent by companies when dealing with the hurdle of an authorization process. Small and medium companies, start-ups, could be, therefore, encouraged to enter the market with a novel food.

Regulation (EC) 2015/2283 already introduces in the Recitals (the eighth) that edible insects are within its scope. The updated novel food definition includes "food consisting of, isolated from or produced from animals or their parts."

Having clarified that insects are considered novel food in the EU, food business operators will submit a request to the European Commission and the dossier is directed to the European Food Safety Authority (EFSA) for evaluation. The new novel food regulation opens two possible paths for placing a novel food or novel food ingredient on the European market: via the standard procedure or via the notification procedure (for foods that are novel for the EU but have been traditionally consumed in third countries). The second procedure, that of notification, is the fast-track option and requires the food business operator to demonstrate that the food product in question has been uninterruptedly consumed for at least 25 years in at least one country outside the EU. For four months, EFSA or the Member States can raise safety objections and if they are clarified, the novel food can be introduced on the market based on notification. The simplified procedure can be the case for many insects, as various species were traditionally consumed in other regions of the world. Once an insect is approved under the notification procedure, it enters the Union List of Authorized Novel Foods. On the flip side, the notification procedure does not guarantee data protection to the applicant, opposed to following the standard procedure. The first half of 2021 marks the first authorisation given to an edible insect in the EU. It is the dried yellow mealworm that successfully passed the risk assessment procedures performed by EFSA and, subsequently, received the greenlight from all 27 EU Member States.

No hygiene rules are yet in place in the EU regarding edible insects. However, a new section is in preparation and expected to be added in Annex III of the Regulation (EC) 853/2004 to establish minimum hygiene requirements for insects as food.

7.5.3 Regulating edible insects in the USA

Insects as food are sold in the United States and they fall under the attention of Food and Drug Administration (FDA). No specific legislation points to edible insects, but companies involved in such business should be registered to the FDA and receive facility inspections regarding the safety of the products obtained. Under the Food, Drug, and Cosmetics Act, food business operators must follow the Good Manufacturing Practices and the factory has to be approved by the FDA, so a firm should prepare a dossier that results from following the regulations and best practices, similar to the portfolio used for any other type of food product.

A highlight about the insect sourcing was made by the FDA, suggesting that insects that can fall under the definition of food should be reared and not captured from the wild. Edible insects should be documented as wholesome food, and when farmed they should benefit from clean water and fit feed sources (Van Huis and Dunkel, 2016).

Clear federal rules on insect food are missing in the United States of America. However, edible insects are on the market despite the legal uncertainty.

7.5.4 Regulating edible insects in Canada

History of use is a concept that applies in many jurisdictions when it comes to food safety. According to the definition in the Food and Drugs Act (Codification, 2010), Canada allows edible insects on the market and considers them as food. However, if a new insect is to be introduced on the market without prior evidence of being consumed anywhere else in the world, it will need to be approved. Thus, it will fall under the novel food status and will require authorization by the Novel Foods Section Bureau of Microbial Hazards together with the Bureau of Chemical Safety and the Bureau of Nutritional Sciences of the Food Directorate of Health Canada. The novelty of a food product on the Canadian territory entails, consequently, both safety and nutritional sufficiency checks (Codification, 2010). Food industry representatives are the ones that have to demonstrate whether a certain insect has a history of use. The dossier must compile a set of different data: from scientific and nonscientific publications, evidence, and information about presumable adverse effects, details about possible processing methods, explanations about the type of harvest (controlled or from the wild), quantity and frequency for normal consumption, and nutritional profile (Directorate, 2006).

Canadian food labeling requirements expect food producers to highlight if allergens are present in the product. Insects are not on the list of allergens, but food–insect business operators have self-adopted the idea to label the warning (Lähteenmäki-Uutela et al., 2017).

7.5.5 Regulating edible insects in Australia and New Zealand

Import and consumption of insects are allowed in Australia and New Zealand. The Food Standards Code is the legal act for both states and it is released by a common agency called Fsanz (Food Standards Australia New Zealand). The code does not make any reference about the safety or marketing requirements for edible insects, but it introduces the notions of traditional food, nontraditional food, and novel food. Insects can fall under any of the three listed categories. A novel food can be introduced on the Australian or New Zealand market after it has filled an inquiry to Fzans. Then, the Advisory Committee on Novel Foods will comment whether the product is indeed novel or traditional. If deemed as novel, it means the aliment raises a health or safety doubt and just after it is clarified, it can enter the positive list on the Food Standards Code (Australia New Zealand Food Standards Code, n.d.). Three types of insects have come to the attention of the Advisory Committee on Novel Foods and they were all classified as nontraditional (Fzans, 2017). This means that the insects in question are not traditionally consumed in Australia or New Zealand, but they neither imply a safety risk.

7.5.6 Regulating edible insects in Africa and Asia

Even though Africa and Asia have a history in entomophagy and more inhabitants of the two continents are accustomed to it, no clear, harmonized legal frameworks are in place. The situation might be like that, since insects are generally considered as food by significant parts of the population, so no different regulations are needed.

Both in Asia and Africa insects are generally harvested from the wild, so safety control is difficult to be obtained—the approach might work for local populations or markets, but for international trade, a set of food safety rules should be implemented.

Pieces of legislation regarding insects as food are only a minority in Asia. In China, some local food safety standards are in place for edible insects, and for putting a new insect on the market, a novel food authorization is needed (Measures, n.d.). Edible insects are also used as medicines or as health foods in China (Lähteenmäki-Uutela et al., 2017). In Thailand, a set of Good Agricultural Practices is in place for farming crickets (Standard, 2010) and South Korea is updating legislation and adds insects (mealworms and crickets, for now) to be considered as food.

7.6 Conclusions

The popularity of entomophagy is increasing among the general population, and studies to find acceptable solutions both for curious consumers that are not yet ready to accept eating whole insects and for those that would are gaining weight.

In order to support industry and researchers to develop studies on insect-based food, as well as to strengthen consumers' willingness to consider entomophagy, policymakers should update legislation accordingly and keep pace with the trend timely. The actual status of global legislation on insects as food for humans is not consistent among the different jurisdictions or it is completely absent. The explanation for this situation is that some nations view insects as a common source of food; thus, no specific legislation is required other than the general standards for food. In areas where entomophagy was not a habit or is only trending up now, policymakers have not anticipated the proportion, so the set of rules are lagging behind. Food legislation is a dynamic system and should mingle together and keep up with global trends, innovations, and needs.

Nonetheless, despite a proven history of safe use of some insect species, more in-depth research on safety and quality is needed to have solid proofs to stand behind laws and regulations. In this case, a close collaboration among scientists, industry, legislators, and the general public is vital for advancing entomophagy as a standing solution for the upcoming global challenges.

The publication of a set of standards on edible insects within Codex Alimentarius can serve for the governments as a basis for national legislation that follows trade requirements, so complying with the Sanitary and Phytosanitary Agreement. A harmonized regulatory framework in the field of edible insects would not only facilitate trade, but it would also guide food business operators to perform under uniform quality and safety standards.

References

Andronoiu, D., Ilie, C.T., Botez, E., 2018. Study upon Romanian consumers' attitude about entomophagy. Agricultura 105 (1–2), 96–100. https://doi.org/10.15835/agrisp.v105i1-2.13047.

Australia New Zealand Food Standards Code. (n.d.). No Title. Retrieved from Standard 1.5.1.-Novel Food website: https://www.legislation.gov.au/Details/F2017C00324.

Belluco, S., Losasso, C., Maggioletti, M., Alonzi, C.C., Paoletti, M.G., Ricci, A., 2013. Edible insects in a food safety and nutritional perspective: a critical review. Compr. Rev. Food Sci. Food Saf. 12 (3), 296–313. https://doi.org/10.1111/1541-4337.12014.

Black, R.E., Victora, C.G., Walker, S.P., Bhutta, Z.A., Christian, P., De Onis, M., Uauy, R., 2013. Maternal and child undernutrition and overweight in low-income and middle-income countries. Lancet 382 (9890), 427–451. https://doi.org/10.1016/S0140-6736(13)60937-X.

Cavallo, C., Materia, V.C., 2018. Insects or not insects? Dilemmas or attraction for young generations: a case in Italy. Int. J. Food Syst. Dynam. 9 (3), 226–239. https://doi.org/10.18461/ijfsd.v9i3.932.

Codification, C., 2010. Food and Drug Regulations Règlement sur les aliments et drogues. Drugs.

De Boer, J., Schösler, H., Boersema, J.J., 2013. Motivational differences in food orientation and the choice of snacks made from lentils, locusts, seaweed or "hybrid" meat. Food Qual. Prefer. 28 (1), 32–35. https://doi.org/10.1016/j.foodqual.2012.07.008.

DeFoliart, G.R., 1997. An overview of the role of edible insects in preserving biodiversity. Ecol. Food Nutr. 36 (2–4), 109–132. https://doi.org/10.1080/03670244.1997.9991510.

Directorate, F., Products, H., Branch, F., Canada, H., Procedure, N., Procedure, S.O., Process, N., 2006. Guidelines for the Safety Assessment of Novel Foods.

Fasolato, L., Cardazzo, B., Carraro, L., Fontana, F., Novelli, E., Balzan, S., 2018. Edible processed insects from e-commerce: food safety with a focus on the *Bacillus cereus* group. Food Microbiol. 76, 296–303. https://doi.org/10.1016/j.fm.2018.06.008.

Fzans, 2017. Novel Food. Retrieved from http://www.foodstandards.gov.au/industry/novel/Pages/default.aspx.

Gallo, M., 2019. Novel foods: insects - safety issues. In: Encyclopedia of Food Security and Sustainability. https://doi.org/10.1016/b978-0-08-100596-5.22134-3.

Gere, A., Székely, G., Kovács, S., Kókai, Z., Sipos, L., 2017. Readiness to adopt insects in Hungary: a case study. Food Qual. Prefer. 59 (February), 81–86. https://doi.org/10.1016/j.foodqual.2017.02.005.

Ghosh, S., Lee, S.M., Jung, C., Meyer-Rochow, V.B., 2017. Nutritional composition of five commercial edible insects in South Korea. J. Asia Pac. Entomol. 20 (2), 686–694. https://doi.org/10.1016/j.aspen.2017.04.003.

Hartmann, C., Ruby, M.B., Schmidt, P., Siegrist, M., 2018. Brave, health-conscious, and environmentally friendly: positive impressions of insect food product consumers. Food Qual. Prefer. 68, 64–71. https://doi.org/10.1016/j.foodqual.2018.02.001.

Hartmann, C., Shi, J., Giusto, A., Siegrist, M., 2015. The psychology of eating insects: a cross-cultural comparison between Germany and China. Food Qual. Prefer. 44, 148–156. https://doi.org/10.1016/j.foodqual.2015.04.013.

Hartmann, C., Siegrist, M., 2017. Insekten als Lebensmittel: wahrnehmung und Akzeptanz – erkenntnisse aus der aktuellen Studienlage. Ernährungs Umschau 64 (3), 44–50. https://doi.org/10.4455/eu.2017.010.

House, J., 2016. Consumer acceptance of insect-based foods in The Netherlands: academic and commercial implications. Appetite 107 (September 2015), 47–58. https://doi.org/10.1016/j.appet.2016.07.023.

IFPRI, 2016. Global nutrition report: from promise to impact. In: Global Nutrition Report - from Promise to Impact: Ending Malnutrition by 2030. Retrieved from globalnutritionreport.org/the-report/.

Jongema, 2017. Lista De Insetos Comestíbles, pp. 1–100.

Kim, M.J., Kim, S.Y., Jung, S.K., Kim, M.Y., Kim, H.Y., 2019. Development and validation of ultrafast PCR assays to detect six species of edible insects. Food Control 103 (December 2018), 21–26. https://doi.org/10.1016/j.foodcont.2019.03.039.

Lähteenmäki-Uutela, A., Grmelová, N., Hénault-Ethier, L., Deschamps, M.H., Vandenberg, G.W., Zhao, A., Nemane, V., 2017. Insects as food and feed: laws of the European Union, United States, Canada, Mexico, Australia, and China. European Food Feed Law Rev. 12 (1), 22–36. https://doi.org/10.3920/JIFF2015.x002.2.

Measures, N.R.F.M. n.d. No Title. Retrieved from https://baike.so.com/doc/6128631-6341791.html.

Mishyna, M., Chen, J., Benjamin, O., 2019. Sensory attributes of edible insects and insect-based foods – future outlooks for enhancing consumer appeal. Trends Food Sci. Technol.

Nadeau, L., Nadeau, I., Franklin, F., Dunkel, F., 2015. The potential for entomophagy to address undernutrition. Ecol. Food Nutr. 54 (3), 200–208. https://doi.org/10.1080/03670244.2014.930032.

Oonincx, D.G.A.B., de Boer, I.J.M., 2012. Environmental impact of the production of mealworms as a protein source for humans - a life cycle assessment. PLoS One 7 (12), 1–5. https://doi.org/10.1371/journal.pone.0051145.

Orsi, L., Voege, L.L., Stranieri, S., 2019. Eating edible insects as sustainable food? exploring the determinants of consumer acceptance in Germany. Food Res. Int. 125.

Pambo, K., Okello, J., Mbeche, R., Kinyuru, J., 2016. Consumer acceptance of edible insects for non-meat protein in western Kenya. African Assoc. Agric. Econ. (November 2017), 1–20. Retrieved from https://ageconsearch.umn.edu/record/246317/.

Payne, C.L.R., Scarborough, P., Rayner, M., Nonaka, K., 2016. A systematic review of nutrient composition data available for twelve commercially available edible insects, and comparison with reference values. Trends Food Sci. Technol. 47, 69–77. https://doi.org/10.1016/j.tifs.2015.10.012.

Poma, G., Cuykx, M., Amato, E., Calaprice, C., Focant, J.F., Covaci, A., 2017. Evaluation of hazardous chemicals in edible insects and insect-based food intended for human consumption. Food Chem. Toxicol. 100, 70–79. https://doi.org/10.1016/j.fct.2016.12.006.

Ribeiro, J.C., Cunha, L.M., Sousa-Pinto, B., Fonseca, J., 2018. Allergic risks of consuming edible insects: a systematic review. Mol. Nutr. Food Res. 62 (1), 1–31. https://doi.org/10.1002/mnfr.201700030.

Roos, N., van Huis, A., 2017. Consuming insects: are there health benefits? J. Insects Food Feed 3 (4), 225–229. https://doi.org/10.3920/JIFF2017.x007.

Schouteten, J.J., De Steur, H., De Pelsmaeker, S., Lagast, S., Juvinal, J.G., De Bourdeaudhuij, I., Gellynck, X., 2016. Emotional and sensory profiling of insect-, plant- and meat-based burgers under blind, expected and informed conditions. Food Qual. Prefer. 52, 27–31. https://doi.org/10.1016/j.foodqual.2016.03.011.

Sillow, C.-A., 1983. Notes on Ngangela and Nkoya Ethnozoology: Ants and Termites (Etnologiska Studier). Goteborgs Etnografiska Museum.

Standard, T.A., 2010. GOOD AGRICULTURAL PRACTICES for National Bureau of Agricultural Commodity and Food Standards National Bureau of Agricultural Commodity and Food Standards.

Sutton, M.Q., 1995. Archaeological aspects of insect use. J. Archaeol. Method Theor 2 (3), 253–298. https://doi.org/10.1007/BF02229009.

Tan, H.S.G., Fischer, A.R.H., van Trijp, H.C.M., Stieger, M., 2016. Tasty but nasty? Exploring the role of sensory-liking and food appropriateness in the willingness to eat unusual novel foods like insects. Food Qual. Prefer. 48, 293–302. https://doi.org/10.1016/j.foodqual.2015.11.001.

Tan, H.S.G., Verbaan, Y.T., Stieger, M., 2017. How will better products improve the sensory-liking and willingness to buy insect-based foods? Food Res. Int. 92, 95–105. https://doi.org/10.1016/j.foodres.2016.12.021.

Tuccillo, F., Marino, M.G., Torri, L., 2020. Italian consumers' attitudes towards entomophagy: influence of human factors and properties of insects and insect-based food. Food Res. Int. 137.

Ulrich, S., Kühn, U., Biermaier, B., Piacenza, N., Schwaiger, K., Gottschalk, C., Gareis, M., 2017. Direct identification of edible insects by MALDI-TOF mass spectrometry. Food Control 76, 96–101. https://doi.org/10.1016/j.foodcont.2017.01.010.

United Nations, 2015. World Population Prospects: The 2015 Revision, Methodology of the United Nations Population Estimates and Projections (242). Retrieved from http://www.ghbook.ir/index.php?name=&فرهنگ و رسانه های نوین&option=com_dbook&task=readonline&book_id=13650&page=73&chkhashk=ED9C9491B4&Itemid=218&lang=fa&tmpl=component.

van Huis, A., 2017. Did early humans consume insects? J. Insects Food Feed 3 (3), 161–163. https://doi.org/10.3920/JIFF2017.x006.

Van Huis, A., Dunkel, F.V., 2016. Edible insects: a neglected and promising food source. In: Sustainable Protein Sources. https://doi.org/10.1016/B978-0-12-802778-3.00021-4.

van Huis, Arnold, 2013. Potential of insects as food and feed in assuring food security. Annu. Rev. Entomol. 58 (1), 563–583. https://doi.org/10.1146/annurev-ento-120811-153704.

Vanhonacker, F., Van Loo, E.J., Gellynck, X., Verbeke, W., 2013. Flemish consumer attitudes towards more sustainable food choices. Appetite 62, 7–16. https://doi.org/10.1016/j.appet.2012.11.003.

Verbeke, W., 2015. Profiling consumers who are ready to adopt insects as a meat substitute in a Western society. Food Qual. Prefer. 39, 147–155. https://doi.org/10.1016/j.foodqual.2014.07.008.

Zielińska, E., Karaś, M., Baraniak, B., 2018. Comparison of functional properties of edible insects and protein preparations thereof. LWT - Food Sci. Technol. (Lebensmittel-Wissenschaft -Technol.) 91 (October 2017), 168–174. https://doi.org/10.1016/j.lwt.2018.01.058.

Chapter 8

Some thoughts on the potential of global harmonization of antimicrobials regulation with a focus on chemical foodsafety

Jaap C. Hanekamp[1,2,3]

[1]*University College Roosevelt, Middelburg, the Netherlands;* [2]*Environmental Health Sciences, University of Massachusetts Amherst, Amherst, MA, United States;* [3]*HAN-Research, Zoetermeer, the Netherlands*

8.1 Introduction

Setting scientific and policy standards that benchmark the risks and benefits of products intended for human consumption is of great consequence for industry, policymakers, and consumers. The safety of food products consumed is more often than not defined as *chemical* product safety, meaning that the food product is regarded as "safe" when man-made chemicals such as antibiotics and/or pesticides are absent or only present at very low levels. Food safety is usually defined as such.

In this chapter, antimicrobials (antibiotics; the terms are used interchangeably henceforth) as a tool in the rearing of food animals are discussed from the perspective of the consumer. As antimicrobials create benefits for the producers, the risks to consumers that simultaneously benefit from antimicrobials use in terms of food abundance (food security) and concomitant lower prices need to be understood and balanced effectively. We will thus focus on the topics of chemical food safety, relevant policies, and the benefits and risks when dealing with global food safety and food security. As the introductory statement to this book makes clear: "Today's reality is that there are differences between countries that too often lead to severe measures such as the destruction of huge quantities of food to protect consumers, lacking any scientific justification, while a large part of the human population in the world suffers from undernourishment."

This will be our leading perspective as to address the many-facetted aspects of antimicrobials in the world of food production. Varied questions could be addressed here, but first let's take a closer look at the actual use of antimicrobials. Subsequently we will raise the issues that will be tackled in this chapter.

8.2 Global estimates of antimicrobials in food animals—the wrong and the right trousers

Demand for food products containing animal is rising globally. That has many implications, and focusing on the use of antimicrobials, it seems certain that total global use will be on the rise. Van Boeckel et al. (2015) made an estimation of the current use of antimicrobials in food animals with an outlook on future use. These authors estimate that the global average annual consumption of antimicrobials per kilogram of animals produced was 45 mg/kg for cattle, 148 mg/kg chicken, and 172 mg/kg for pigs in 2010. Starting from this baseline (2010), the authors estimate that between 2010 and 2030, the global consumption of antimicrobials will increase by some 67%, from 63,151 ± 1560 to 105,596 ± 3605 tons. China, for instance, is the largest consumer of livestock antimicrobials, some 14,500 tons in 2010 (Fig. 8.1A), according to van Boeckel et al. In the United States, however, Domestic sales and distribution of medically important antimicrobials approved for use in food-producing animals: decreased by 33% from 2016 through 2017, decreased by 43% from 2015

Ensuring Global Food Safety. https://doi.org/10.1016/B978-0-12-816011-4.00027-6

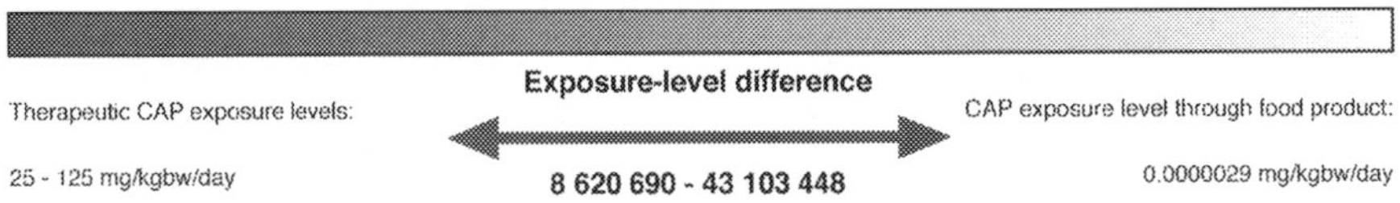

FIGURE 8.1 Cap exposure level differences between medication and food.

(the year of peak sales) through 2017, decreased by 28% from 2009 (the first year of reported sales) through 2017. Tetracyclines, which represent the largest volume of these domestic sales (3,535,701 kg in 2017), decreased by 40% from 2016 through 2017 (FDA, 2018).

Bu et al. (2016) estimate the human use of antibiotics at the Chinese national level at some 89,700 tons for 2011. If the numbers of van Boeckel et al. and Bu et al. can be compared, human use outstrips veterinary use by roughly a factor of 6. In Germany, Van Boeckel et al. estimate the veterinary consumption, in 2010, at some 1900 tons. This is close to the estimate made by Meyer et al. (2013), which came to a veterinary use of 1700 tons, while human use was estimated to be 300 tons. Here, veterinary use of antimicrobials outstrips human use by roughly the same factor of 6. More countries could be scrutinized here, but the general gist seems clear: antimicrobial use is both veterinarily and medically important, and the one might outstrip the other, but they both add to the total use of antimicrobials sizeably.

Thus, antimicrobial use is widespread in both the veterinary and human world. That we will observe an increase in veterinary use the coming years seems inevitable. The likelihood of presence of antimicrobials in human food products of animal origin will thereby increase as well, especially considering growing international trade. That requires a global approach of regulation that could be implemented across the board.

This global approach, however, can have two characteristics or "trousers," the right and the wrong ones. The trousers metaphor comes from the famous stop-motion Wallace & Gromit animated film "*The Wrong Trousers*" by Aardman Animations. In that film, the woeful hero, Wallace, becomes trapped in a pair of remote-controlled "Techno Trousers," which he gave to his dog Gromit as a birthday present. Instead of making Wallace's life easier, they literally carry him off in directions he does not wish to go, commandeered by the penguin criminal Feathers McGraw, who is lodging at Wallace & Gromit's house. Prins and Rayner (2007) referred to this animated film as a means to explain the wrongful approach of climate change policies.

The metaphor invoked is that policies that have a global outreach could well march us off into unwanted and counterproductive directions for all sorts of reasons, not excluding political powerplay by and between countries. The "wrong trousers" approach predicates upon changing the world *first* in order to meet the goal of, in this case, reducing antimicrobials use, however laudable that may be. The "right trousers" seeks ways to build on possibilities and dynamics already present in the world of food production, which needs to grow the coming years as to feed the world population. The former top-down approach might be called, for lack of better terminology, *utopian* with all its ramifications (see Hanekamp, 2015). The latter might be called, in Popper's terminology (see e.g., Popper 1986), *piecemeal engineering*, the slow and careful tinkering with the present state of the world toward a future that is more sustainable.

We will use the chloramphenicol (henceforth CAP) case, which rose to prominence in 2001 in Europe, as an example of the "wrong trousers." Subsequently, we will propose a way forward that engages the likely future growth of antimicrobial uses in the food chain head on, and formats it into a system of transparency and (non)necessity. We will wrap up our discussions with some notes on antimicrobial resistance. First, we will spend a few words on the nature of antimicrobials.

8.3 The "nature" of antimicrobials

With the discovery of penicillin in 1928 by Alexander Fleming, the human potential to tackle bacterial infections in both humans and animals grew immeasurably. Penicillin is made by a fungus (*Penicillium notatum*), yet most antibiotics we now know today are derived from actinomycetes, nature's topmost antibiotic producers. Not only do they produce antibiotics in a huge variety, they also produce chemicals that kill fungi, parasitic worms, and even insects (Hopwood, 2007). Many other pharmaceuticals such as antitumor agents and immuno-suppressants are also derived from the actinomycetes (Walsh, 2003).

The streptomycetes, belonging to the actinomycetes, account for well over two thirds of these commercially and therapeutically significant antibiotics, which are produced by means of complex "secondary metabolic" pathways (Bibb, 2005). Streptomycetes therefore are the most important source of antibiotics for medical, veterinary, and agricultural use.

Streptomycetes are a group of gram-positive filamentous bacteria, ubiquitous soil bacteria found worldwide. They are among the most numerous and ubiquitous soil bacteria adapted to the utilization of plant remains and are key in this environment because of their broad potential of metabolic processes and biotransformations. These include degradation of the insoluble remains of other organisms making streptomycetes imperative organisms in carbon recycling (Bentley et al., 2002). Streptomycetes are members of the same taxonomic order as the causative agents of tuberculosis (TB) and leprosy (*Mycobacterium tuberculosis* and *M. leprae*).

While secondary metabolites must in all likelihood confer an adaptive advantage, the roles of most are not fully understood. This is true even for antibiotics. Antibiotics give a competitive advantage compared to other ground dwelling microorganisms that need to tap into the same biomass as nutrition (Shi and Zusman, 1993). An ensnarement strategy was observed as well, where competing organisms are attracted and subsequently killed by excreted antibiotics and subsequently consumed supplying additional nutrition (Chater, 2006). Streptomycin was the first antibiotic to be discovered to be effective against tuberculosis TB. CAP (Ehrlich et al., 1947) was the first antibiotic to be synthetically produced and was shown to be effective against typhoid (Patel and Banker, 1949).

8.4 A precautionary tale and chloramphenicol

In the European Union, the core regulatory framework for food is Regulation 178/2002/EC on general principles of food law (the "General Food Law"). Under this regulation, "food" (or "foodstuff") denotes "any substance or product, whether processed, partially processed, or unprocessed, intended to be, or reasonably expected to be ingested by humans." The regulation applies broadly to "all stages of the production, processing, and distribution of food." Its general objective is to provide "a high level of protection of human life and health and the protection of consumers interests." It tries to accomplish this objective, by setting general rules for all products, including dietary supplements and fortified foods that are placed on the EU market. General requirements deal with food safety, presentation, traceability, and related responsibilities of food business operators. Importantly, the regulation also establishes the European Food Safety Authority ("EFSA") and defines its tasks and powers (Hanekamp et al., 2003; Hanekamp and Bast, 2015; Bergkamp and Hanekamp, 2018).

EU food law also pursues several specific objectives. As Article 5(1) of the General Food Law states, EU food law pursues one or more of the general objectives of a "high level of protection of human life and health and the protection of consumers' interests, including fair practices in food trade, taking account of, where appropriate, the protection of animal health and welfare, plant health and the environment." Whenever the EU pursues a "high level of protection," the precautionary principle comes into play. The area of food law is a prime example. Over the last several decades, the EU has taken a series of precautionary measures in relation to food-related hazards. For instance, precautionary measures were taken in response to the BSE crisis, dioxin in poultry meat, and several other food scares. Interestingly, however, the EU has not been precautionary with respect to all possible hazards, or even all food-related hazards. Unlike the United States, the EU has been unperturbed by certain known and obvious risks arising from the consumption of foods. For instance, the use of raw milk for the production of cheese is permitted, even though there have been incidents in Europe involving contamination of raw milk cheese with listeria and other bacteria, which resulted in morbidity and mortality. In the case of raw milk cheese, however, the EU did not impose any such measure, while the US did regulate the risk (see e.g., Carrique-Mas et al., 2003; Quaglia et al., 2008; Oliver et al., 2009).

In relation to antibiotics used in animal rearing, regulation is pervasive and precautionary. This is partly related to reduce the chronic exposure through food as much as possible, but is also in part due to precautionary risk averseness. Both these aspects are well illustrated by the CAP case. Article 7 thereof describes the precautionary principle as follows (178/2002/EC):

1. In specific circumstances where, following an assessment of available information, the possibility of harmful effects on health is identified but scientific uncertainty persists, provisional risk management measures necessary to ensure the high level of health protection chosen in the Community may be adopted, pending further scientific information for a more comprehensive risk assessment.
2. Measures adopted on the basis of paragraph 1 shall be proportionate and no more restrictive of trade than is required to achieve the high level of health protection chosen in the Community, regard being had to technical and economic feasibility and other factors regarded as legitimate in the matter under consideration. The measures shall be reviewed within a reasonable period of time, depending on the nature of the risk to life or health identified and the type of scientific information needed to clarify the scientific uncertainty and to conduct a more comprehensive risk assessment."

In 2001, the detection of CAP, a broad-spectrum antibiotic, in shrimp imported into Europe from Asian countries was branded as yet another food scandal. The initial European response was to close European borders to fish products, mainly shrimp, from these countries and make laboratories work overtime to analyze numerous batches of imported goods for the presence of this antibiotic. Some European countries went so far as to have food products containing the antibiotic destroyed as public health was deemed to be at stake. This regulatory response spilt over to other major seafood-importing countries such as the United States.

The legislative background to this mainly European response was found in the now defunct Council Regulation EEC No. 2377/90 (1990), which was implemented to establish maximum residue limits of veterinary medicinal products in foodstuffs of animal origin. This so-called "MRL Regulation" (maximum residue limit) introduced Community procedures to evaluate the safety of residues of pharmacologically active substances according to human food safety requirements. Commission Regulation (EU) No 37/2010 on pharmacologically active substances and their classification regarding maximum residue limits in foodstuffs of animal origin is now in force in the EU.

A pharmacologically active substance may be used in food-producing animals, only if it receives a favorable evaluation. If it is considered necessary for the protection of human health, maximum residue limits ("MRLs") are established. They are the points of reference for setting withdrawal periods in marketing authorizations as well as for the control of residues in the Member States and at border inspection posts. Directive 96/23/EC ("the Residue Control Directive") contains specific requirements, in particular for the control of pharmacologically active substances that may be used as veterinary medicinal products in food-producing animals (Council Directive, 1996). This includes primarily sampling and investigation procedures, requirements as to the documentation for their use, indication for sanctions in case of noncompliance, requirements for targeted investigations, and for the setting up and reporting of monitoring programs.

Commission Regulation (EU) No 37/2010 contains a Table (2) of prohibited substances for which no maximum toxicological levels (Tolerable Daily Intake—TDI, also known as Acceptable Daily Intake—ADI) can be fixed, either from lack of toxicological or pharmacological data, e.g., the absence of a definable NOAEL (No Observed Adverse Effect Level) or LOAEL (Lowest Observed Adverse Effect Level) or because of purported genotoxic characteristics of the compound in question.[1] These substances are consequently not allowed in the animal food production chain. So-called zero-tolerance levels are in force for the prohibited antimicrobials as found:

- Lack of scientific data de facto makes the establishment of a TDI not feasible;
- The absence of a TDI and the subsequent impossibility to establish a maximum residue limit (MRL), in regulatory terms is understood as "dangerous at any dose" requiring zero-tolerance regulation;
- With the introduction of zero tolerance, a veterinary ban on Table 2 compounds (such as CAP) is in place, whereby the listed compounds, when producers compliance is achieved, would disappear from the food chain;
- When zero tolerance was implemented, analytical equipment was only capable to detect at the Limit of Detection (LOD) of ppm (parts per million; mg per kg); nowadays, LODs are at least ppb (parts per billion; μg per kg), obviously depending on analyzed chemicals.

No TDI could be established for CAP due to the lack of scientific information to assess its carcinogenicity and effects on reproduction, and because the compound showed some genotoxic activity (IPCS-INCHEM). Overall, CAP—and the other Table 2 substances—should not be detected in food products at all, regardless of concentrations. The presence of CAP in food products, which can be detected by any type of analytical apparatus, is a violation of European law and moreover deemed to be a threat to public health. In consequence, food containing the smallest amount of these residues is considered unfit for human consumption. For all intents and purposes, zero tolerance is best understood as zero concentration, a molecular prohibition. Only when Table 2 substances are completely absent from food the risks are deemed completely absent. The presence of CAP in food products is solely related to illicit veterinary use; other sources are not taken into account, or indeed considered, as they are not included in the legislation. Chloroform, chlorpromazine, colchicine, dapsone, dimetridazole, metronidazole, nitrofurans (including furazolidone), and ronidazole are the other compounds in Table 2 of Commission Regulation No 37/2010.

In 2009, the failure of this policy was, to some extent, corrected by the designation of Minimum Required Performance Limit (MRPL) as targets for regulatory levels of concern for banned antibiotics (Regulation 470/2009), which serve as so-called Reference Points for Action (EFSA, 2013). However, this MRPL approach has not eliminated the fallacious drivers for zero tolerance. Fundamentally, the policy remains based on the misconception that risks arise from exposure to low-level concentrations of chemicals, such as CAP (Hanekamp and Bast, 2015) It also assumes that an unambiguous causal

1. Genotoxic agents (chemicals, ionizing radiation) are those capable of causing damage to DNA. Such damage can potentially lead to the formation of a malignant tumor.

link can be established between the detection of some banned compound and an illegal activity in food production. However, some 16 years ago our own findings show already (Hanekamp et al., 2003) CAP to be a natural component in different food products. In 2010, unambiguous evidence was published on the ecological background of CAP in herbs and grass (Berendsen et al., 2010). In the authors' words: "... Among other plant materials, samples of the Artemisia family retrieved from Mongolia and from Utah, USA, and a therapeutic herb mixture obtained from local stores in the Netherlands proved to contain CAP at levels ranging from 0.1 to 450 μg/kg." This research was followed up in 2013, in which other plant material was found to contain CAP of natural origin (Berendsen et al., 2013).

The discussion on the presence of CAP in plant material has come to the fore because of the increasing capabilities of analytical equipment. As said, these capabilities have greatly improved over the past decades. We have now entered the realm of atto- (part per quintillion; 10^{-18}) and zeptomoles (part per sextillion; 10^{-21}) of detectable analytes (Pagnotti et al., 2011). Basically, this means that the zero-tolerance level is shifting to ever lower exposure levels. Advances in "cleaner" consumable goods production are thus offset by increased detection capacities, whereby the ecological threshold for compounds such as CAP can be crossed, and in fact has been crossed as shown above.

8.5 Risk profile of foods containing CAP—of exposure levels and toxicological models

Despite a ban in animal food production, CAP is still used as human medication. It has a wide spectrum of activity against gram-positive and gram-negative bacteria. CAP therapy is usually restricted to serious infections when other drugs are not as effective. Internal infections are only rarely treated with CAP. Ophthalmic infections, however, are still treated with CAP. A number of registered pharmaceutical products containing CAP that are used to treat eye infections are on the market in the Western World. In Asian countries it is still widely used against, e.g., typhoid.

It is important to first show the gap between the medicinal exposure to CAP and the exposure as a result of CAP presence in food products. We will take 1 μg of CAP per kg food product (1 ppb; 0.001 mg/kg product) as an arbitrary but realistic starting point. As said, in the CAP episode referred to above, imported shrimp was found to contain between 1 and 10 ppb of CAP. Assuming that the exposed consumer weighs 70 kg and consumes 200 g of said undefined food product on a daily basis, the intake is 0.0000029 mg/kgbw/day (0.0029 μg/kgbw/day). Compared to the therapeutic exposure to CAP (which of course cannot be considered a life-time event) the exposure difference amounts to the following (Fig. 8.1).

The Dutch National Institute for Public Health and Environment (RIVM; Janssen et al., 2001) calculated the risk of exposure to CAP found in shrimp, applying the conservative linear nonthreshold model (LNT) by default used for genotoxic carcinogens (CAP purportedly being genotoxic). The RIVM came to the conclusion that exposure to 1−5 μg/kgbw/day of CAP through food containing CAP posed a risk to humans at the level of a $1{:}10^6$ added cancer risk, the so-called Maximum Tolerable Risk (MTR-) level (see on definition and background of the MTR: USFDA, 1973; Mantel and Bryan, 1961). This means that the regulatory standard for acceptable risk from a carcinogen is one additional case of cancer in a population of one million people. Thus, applying the conservative LNT dose−response model for the calculated exposure—that is 0.0029 μg/kgbw/day—the risk of daily exposure would be a factor 345−1725 lower than the MTR. That would amount to a vanishingly small risk, if the LNT model in fact is empirically and theoretically valid for carcinogens.

Reiterating, the above calculation is based on the conjecture that an LNT dose−response correlation is valid. This regulatory model asserts that for genotoxic carcinogenic substances and ionizing radiation, any level of exposure—except for zero—implies a health risk (Kathren, 2003). Only zero exposure is ultimately deemed to be safe. Conversely, any nonzero exposure has a nonzero probability of causing cancer. This so-called "one hit" dose−response model maintains that exposure to even one molecule or ionizing photon may result in irreversible health damage. The potential effects of CAP, or any other genotoxic carcinogen, at very low-level exposures are derived from this model as, of course, actually observing those effects in human populations would be out of the question, as the effects are too small.

This conservative toxicological model is increasingly coming under fire (see e.g., Moustacchi, 2000; Calabrese and Baldwin, 2003; Ricci et al., 2012, Bast and Hanekamp, 2014). The French Academy of Sciences and the French National Academy of Medicine, for instance, have raised serious doubts as to the validity of the LNT model, effectively abandoning it (Tubiana and Aurengo, 2005; Tubiana et al., 2006). Already in 1996, Goldman noted the palpable incongruity of the LNT model when he linearly calculated the increased risk of cancer, because of increased cosmic radiation, if the entire world population would add a one-inch lift to their shoes (sic):

"As an extreme extrapolation, consider that everyone on Earth adds a 1-inch lift to their shoes for just 1 year. The resultant very small increase in cosmic ray dose (it doubles for every 2000 m in altitude), multiplied by the very large population of the Earth, would yield a collective dose large enough to kill about 1500 people with cancer over the next 50 years. Of course no epidemiological confirmation of this increment could ever be made, and although the math is approximately correct, the underlying assumptions should be questioned. Most of the environmental risks we now face from present or proposed activities probably are of this magnitude, and many of our policies say that prudence requires us to reduce these small values even further. We do not seem to have a realistic process whereby we can uniformly both protect the public health and avoid seemingly frivolous prevention schemes" (Goldman, 1996).

Goldman, despite his flippant exemplar, does describe the basic scientific and regulatory assumptions of the LNT model appropriately, even when people are actually exposed, under normal conditions, to doses several thousand fold or even several hundred 1000 fold lower than tested animals (Calabrese, 2009), say, for example, through foods.

What is more, ingestion of varying carcinogens associated with a *de minimis* risk of cancer (<1 cancer (disease)/million/lifetime exposure; the MTR referred to above and used in the calculation) would still imply a human exposure to many trillions of carcinogenic molecules every day of a human lifetime. This value approaches and at times exceeds some 18 orders of magnitude greater than a single carcinogenic molecule (Calabrese et al., 2012).

Taking aflatoxin B1 as an example of a well-known genotoxic carcinogen (group 1; IARC, 2012), the maximum regulatory level in certain foods (such as almonds, pistachios, and apricot kernels), intended for direct human consumption or use as an ingredient in foodstuffs, is between 2 and 12 ppb (Commission Regulation No 165/2010). Choosing 8 ppb would roughly amount to 1.54×10^{16} aflatoxin B1 molecules per kilogram of a certain contaminated food product. Exposure to aflatoxin B1, even with a limited consumption of contaminated foods, by far exceeds the LNT exposure restriction of a single molecule. The same goes for CAP: 1.5 μg would amount to some 2.80×10^{15} molecules. Overall, the LNT model is ill suited for determining risk of low level chemicals exposure through foods and will impede effective global harmonization of food chemicals exposure regulation (see further Hanekamp and Calabrese, 2021a,b).

Also, considering the Second Law of Thermodynamics, each individual interaction between molecules does not produce a chemical reaction and, therefore, a biological effect (see Koch, 1983). Molecules must have a minimum activation energy to react with receptor molecules, between 16 and 42 kJ/mol for biochemical reactions in vivo. So, only a percentage of the total amount of molecules present in an organism has the required activation energy to react. Obviously, for a reaction to occur at all, the molecules must collide with the receptor molecules at a sufficient energy level and the proper orientation.

8.6 Toward a straightforward resolution—Intended Normal Use

Existing regulations list CAP and other antibiotics as substances prohibited in animal food production as no acceptable daily intake could scientifically be derived. Through zero tolerance, regulatory bodies propagate the view that residues of nonallowed compounds in food constitute a hazard to consumer health, a view at odds with scientific knowledge. Efforts to enforce zero tolerance for antibiotics have evoked concerns for reliable analytical methods able to generate transparent and reproducible results (Schröder, 2002), regulatory harmony, practical modes of prevention, and useful risk assessments. Sensitivity of analytical methods in effect determines the operational definitions for "zero" (LODs), and as the analytical sensitivities reach ppb and ppt levels, the costs for equipment and tests limit surveillance and furthermore increase the probability of detection. Detection as such says very little, if anything, of the toxicological relevance thereof.

Additionally, when considering increasing analytical capabilities, CAP and many other antibiotics have been detected in different aquatic environments at different locations such as hospital effluents, sewage, wastewater treatment plants, river water, and drinking water (Hirsch et al., 1999; Lindberg et al., 2004; Loraine and Pettigrove, 2006; Papa et al., 2007; Watkinson et al., 2009; Zhou et al., 2013). Thereby another, much more diffuse source has come to the fore that might potentially contaminate food. These contaminants are sometimes referred to as PPCPs (Pharmaceuticals and Personal Care Products). PPCPs comprise all drugs, diagnostic agents (such as X-ray contrast media), "nutraceuticals" (bioactive food supplements), and other consumer chemicals, such as fragrances and sunscreen agents (Kummerer, 2008). From the perspective of the point of entry into the aquatic environment, namely wastewater and wastewater treatment plant discharge, PPCPs are referred to as organic wastewater contaminants (Kolpin et al., 2002; Fatta-Kassinos et al., 2010). The human and veterinary use of pharmaceuticals results in a diffuse dispersion in the environment, comparable to for instance pesticides. These can be traced in food and food products.

Detection per se of forbidden veterinary products and their metabolites in food or feed should not immediately be seen as a concern for human or animal health. Indeed, the MRPL is now functioning as a tolerance limit by which the problems that surfaced in 2001 with the detection of CAP in shrimp will be reduced. However, ambiguity remains, as the *toxicological relevance* of these concentrations are not addressed as such. With analytical technology advancing, MRPLs will be lowered whereby problems may develop anew. Therefore, in order to fundamentally address food safety in relation to veterinary medicinal products irrespective of source, we propose the following schemes based on "Intended Normal Use." By that we mean that clinical substances applied in the animal-producing field are authorized products used with an intentional normal purpose. This means that food safety regulation is not meant to tackle intended misuse. Again, its focus is on safeguarding public health. Below (Fig. 8.2) we have depicted a decision tree by which future regulation could best be organized:

The buttress of the decision tree is the risk approach. A number of effective risk tools are on hand namely: the MTR level, the Tolerable Daily Intake (TDI); the related MRL and the Toxicologically Insignificant Exposure level (TIE). With the latter, the concept of hormesis is embraced as a means to advance a rational approach of low-level exposures of chemicals in food, which need not be zero (Hanekamp and Calabrese, 2007). Hormesis is the physiological adaptive response to low levels of stress or damage, resulting in enhanced robustness of some physiological systems for a finite period. More specifically, hormesis is defined as a moderate overcompensation to a perturbation in the homeostasis of an organism. The fundamental conceptual facets of hormesis are respectively (1) the disruption of homeostasis; (2) the moderate overcompensation, (3) the reestablishment of homeostasis; and (4) the adaptive nature of the overall process (see e.g., Kaiser, 2003).

Hormesis redefines our concept of pollution and contamination; questions the premise that pollutants are unconditionally bad; and therefore, acknowledges that the human organism does have adaptive capabilities. This is innovative because modern environmental and public health legislation is built in large part on the moral dichotomies of good versus evil, clean versus dirty, natural versus unnatural, but also safety versus health. Chemical substances—be it natural or synthetic—are not either bad or good; they are both, depending on exposure levels and adaptive responses from the exposed organisms (Hanekamp, 2008). Policies on chemicals safety, as proposed here in relation to antibiotics in food, need to abandon the simplistic moral dichotomy of good and evil in order to be able to mature into regulation that truly addresses the safety and health of citizens in relation to food consumption Fig. 8.3).

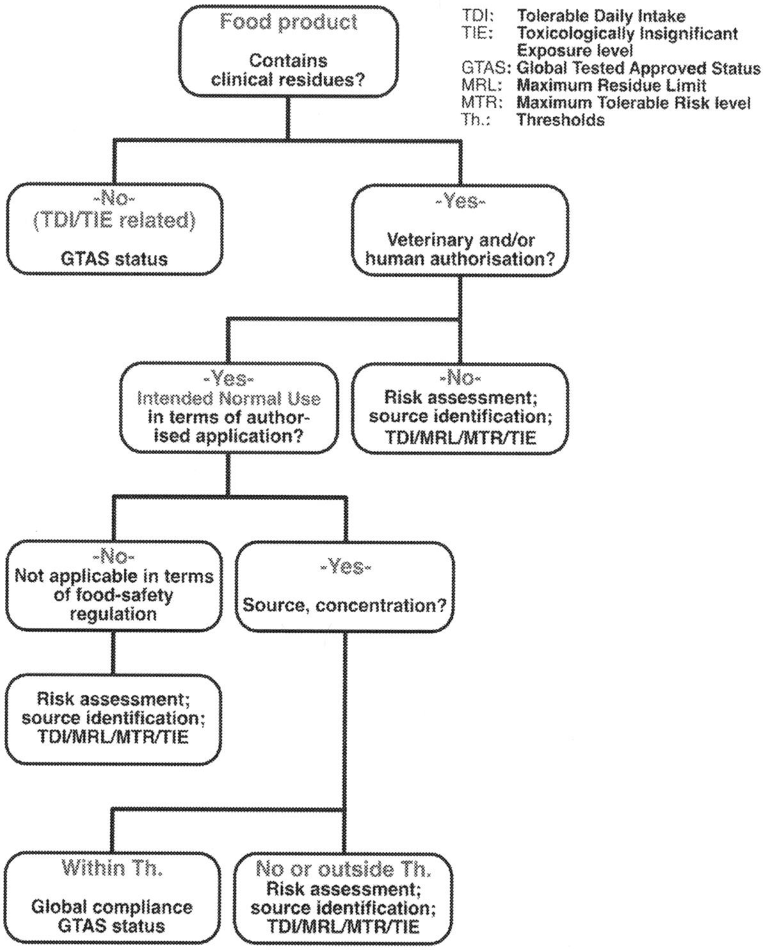

FIGURE 8.2 Food safety regulatory decision tree for clinical residues.

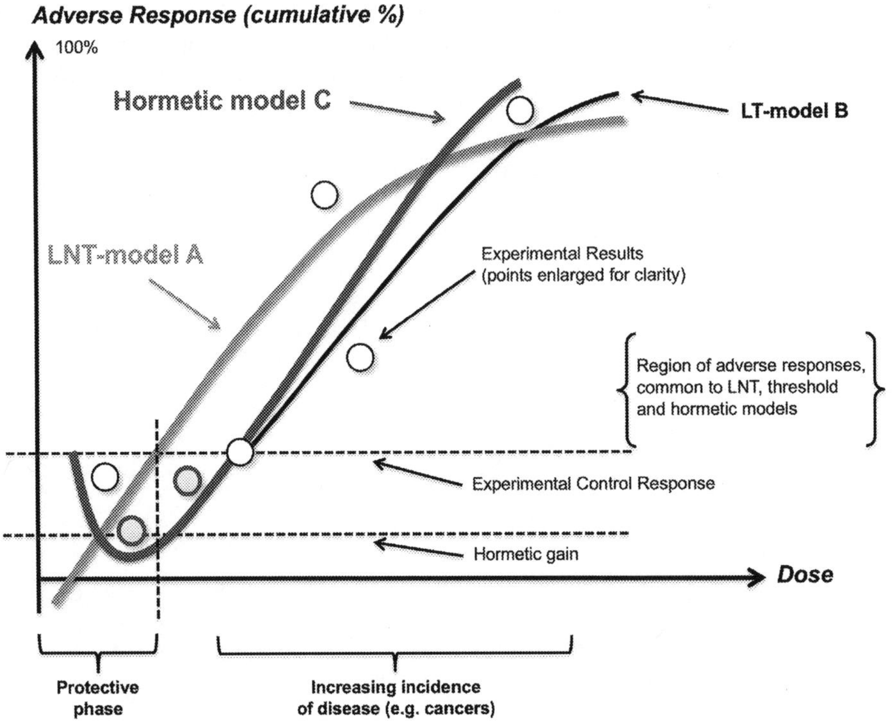

FIGURE 8.3 Three models of dose-response: (A) linear nonthreshold model, (B) linear threshold model, and (C) hormetic model.

With the concept of a TIE, the hormetic part of the dose—response curve is "translated" into a toxicological threshold. Insignificance is understood **not** as an evaluatory outcome based on an MTR (Maximum Tolerable Risk) level of 1:1 000 000 as is done with the threshold of toxicological concern concept (Kroes et al., 2004), but as a direct toxicological dose—response matter (Calabrese and Cook, 2005). Exposure at some (low) level is therein not regarded as potentially hazardous but understood within the framework of a hormetic adaptive response.

As shown in the above depicted decision tree (Fig. 8.2), INU is not limited to authorized veterinary use but also includes human clinical use as the aquatic environment can potentially be a source of excreted clinical residues for the food production chain (see below). Obviously, concentrations present in food as a result of exposure of clinical products through the aquatic environment are expected to be lower than normal intended veterinary use. Drinking water derived from either surface- or groundwater contains numerous excreted clinical compounds at low levels not associated with any measurable risk for human health, whereby a strict risk-avoiding approach in terms of zero tolerance within the food sector is disproportionate and unnecessary.

The inclusion of the INU of human medication as a result of which the food chain is exposed to these substances (through the aquatic environment) is a logical consequence of the risk-based approach proposed here and resolutely widens the window of risk-based food safety regulation. It is essential that proof-of-harm of low-dose toxicity supersede the current regulatory approach. When proof-of-harm as a result of food residue exposure surfaces, then the substance needs to be listed on a Universally Banned Substances List. This is related to the fact that proof of a negative is difficult to do and potentially will generate a legal *probatio diabolica* (Hanekamp et al., 2003). Lack of data to establish an MRL as such is not sufficient ground to ban certain veterinary products, even more so when those products are authorized as human medication and damaging effects surface only as a result of human therapeutic use. Indeed, with any authorized human and veterinary medication, a balance is struck between toxicity and beneficial effects at the biological active dosage. Risks materializing at the human therapeutic level cannot be indicative for a veterinary ban.

The risk approach should have a global jurisdiction in order to avoid trade barrier issues based on analytical noncompliance. Therefore a "Global Tested Approved Status" is introduced in the decision tree. Tools of analysis need to be horizontally unified in order to generate global compliance and a level playing field as to preclude trade barriers, always keeping in mind the local circumstances of food production. Trade between nations will benefit from international cross-compliance, in which properly analyzed goods will be accepted unreservedly by importing nations. In cases where nonauthorized substances are detected, a risk analysis of observed concentrations needs to be undertaken with food safety as the sole objective.

Obviously, the proposed global approach rightly raises the "wrong trousers issue" with which we started this conversation. What needs to be taken into account is the fact that the INU approach understands the world as it is: antibiotics, rightly or wrongly, are used in the food chain, environmental presence of a myriad of pharmacons (including antimicrobials) is a reality and impinges on the food chain, food production needs to be augmented to feed the world population, which might include the use of antimicrobials, and so on. These real-world issues cannot be solved "mechanically and idealistically"; what is required is a clear perspective on the situatedness of food production in terms of biogeochemical and weather environment, pest and microorganisms control, animal and human health, transport, storage, retail, etc. What we have tried to do here is to give food producers, national governmental competent authorities, risk assessors, and managers a tool to assess food products within the confines of agreed upon parameters; nothing more, nothing less. And these parameters should be understood both on the local and global level and never be understood as static ex cathedra.

This leaves us with the use of veterinary antibiotics and antimicrobial resistance as a potential global threat to human (and animal) health. The choice usually offered seems clear enough and proves another "mechanistic" example: reduce or eliminate the use of veterinary antimicrobials and antimicrobial resistance, if present, will be reduced. As everything else in the world, things are not that simple (see e.g., Cox, 2006; Cox and Singer, 2012; see also Phillips et al., 2004). A number of trade-offs are part of the antimicrobial equation: antimicrobial resistance (in animals and humans), resistance traits flow (animal to humans and vice versa) animal welfare (containment of animal diseases), control of zoonoses (see e.g., Cox, 2005), food availability (security), food safety (as discussed in this chapter), feed use, etc. These trade-offs, next to the toxicological issues at play we have focused on in this chapter, need to be taken into account in making well-informed choices as to the use or abandonment of specific antimicrobials in food animals (see e.g., Chattopadhyay, 2014; Chen et al., 2015).

References

Bast, A., Hanekamp, J.C., 2014. You can't always get what you want—linearity as the golden ratio of toxicology. Dose-Response 12 (4), 664–672.

Bentley, S.D., Chater, K.F., Cerdeño-Tárraga, A.-M., Challis, G.L., Thomson, N.R., James, K.D., Harris, D.E., Quail, M.A., Kieser, H., Harper, D., Bateman, A., Brown, S., Chandra, G., Chen, C.W., Collins, M., Cronin, A., Fraser, A., Goble, A., Hidalgo, J., Hornsby, T., Howarth, S., Huang, C.-H., Kieser, T., Larke, L., Murphy, L., Oliver, K., O'Neil, S., Rabbinowitsch, E., Rajandream, M.-A., Rutherford, K., Rutter, S., Seeger, K., Saunders, D., Sharp, S., Squares, R., Squares, S., Taylor, K., Warren, T., Wietzorrek, A., Woodward, J., Barrell, B.G., Parkhill, J., Hopwood, D.A., 2002. Complete genome sequence of the model actinomycete *Streptomyces coelicolor* A3(2). Nature 417, 141–147.

Berendsen, B., Stolker, L., de Jong, J., Nielen, M., Tserendorj, E., Sodnomdarjaa, R., Cannavan, A., Elliott, C., 2010. Evidence of natural occurrence of the banned antibiotic chloramphenicol in herbs and grass. Anal. Bioanal. Chem. 397, 1955–1963.

Berendsen, B., Pikkemaat, M., Römkens, P., Wegh, R., van Sisseren, M., Stolker, L., Nielen, M., 2013. Occurrence of chloramphenicol in crops through natural production by bacteria in soil. J. Agric. Food Chem. 61, 4004–4010.

Bergkamp, L., Hanekamp, J.C., 2018. European food law and the precautionary principle: paradoxical effects of the EU's precautionary food policies. In: Bremers, H., Purnhagen, K. (Eds.), Regulating and Managing Food Safety in the EU. A Legal-Economic Perspective, vol. 6. Springer, Switzerland, pp. 217–244.

Bibb, M.J., 2005. Regulation of secondary metabolism in streptomycetes. Curr. Opin. Microbiol. 8, 208–215.

Bu, Q., Wang, B., Huang, J., Liu, K., Deng, S., Wang, Y., Yu, G., 2016. Estimating the use of antibiotics for humans across China. Chemosphere 144, 1384–1390.

Calabrese, E.J., Baldwin, L.A., 2003. Toxicology rethinks its central belief. Hormesis demands a repraisal of the way risks are assessed. Nature 421, 691–692.

Calabrese, E.J., Cook, R.R., 2005. Hormesis: how it could affect the risk assessment process. Hum. Exp. Toxicol. 24, 265–270.

Calabrese, E.J., 2009. The road to linearity: why linearity at low doses became the basis for carcinogen risk assessment. Arch. Toxicol. 83, 203–225.

Calabrese, E.J., Cook, R.R., Hanekamp, J.C., 2012. Linear no threshold (LNT)—the new homeopathy. Environ. Toxicol. Chem. 31, 2723.

Carrique-Mas, J.J., Hökeberg, I., Andersson, Y., Arneborn, M., Tham, W., Danielsson-Tham, M.L., Osterman, B., Leffler, M., Steen, M., Eriksson, E., Hedin, G., Giesecke, J., 2003. Febrile gastroenteritis after eating on-farm manufactured fresh cheese – an outbreak of listeriosis? Epidemiol. Infect. 130, 79–86.

Chater, K.F., 2006. Streptomyces inside-out: a new perspective on the bacteria that provide us with antibiotics. Philos. Trans. R. Soc. B 361, 761–768.

Chattopadhyay, M.K., 2014. Use of antibiotics as feed additives: a burning question. Front. Microbiol. 5. https://doi.org/10.3389/fmicb.2014.00334.

Chen, C.-Y., Yan, X., Jackson, C.R., 2015. Antimicrobial Resistance and Food Safety. Methods and Techniques. Elsevier, Academic Press, Amsterdam.

Commission Regulation (EU) No 165/2010 of 26 February 2010 amending Regulation (EC) No 1881/2006 setting maximum levels for certain contaminants in foodstuffs as regards aflatoxins. Off. J. Eur. Union L50, 8–12.

Commission Regulation (EU) No 37/2010 of 22 December 2009 on pharmacologically active substances and their classification regarding maximum residue limits in foodstuffs of animal origin. Off. J. Eur. Union L15, 2–72.

Council Directive 96/23/EC of 29 April 1996 on measures to monitor certain substances and residues thereof in live animals and animal products and repealing Directives 85/358/EEC and 86/469/EEC and Decisions 89/187/EEC and 91/664/EEC. Off. J. Eur. Communities L125, 10–32.

Cox, L.A., 2005. Potential human health benefits of antibiotics used in food animals: a case study of virginiamycin. Environ. Int. 31, 549–563.

Cox, L.A., 2006. Quantitative Health Risk Analysis Methods. Modeling the Human Health Impacts of Antibiotics Used in Food Animals. Springer, NYC.

Cox, L.A., Singer, R.S., 2012. Confusion over antibiotic resistance: ecological correlation is not evidence of causation. Foodborne Pathog Dis 9 (8), 776.

EFSA Panel on Contaminants in the Food Chain (CONTAM), 2013. Guidance on methodological principles and scientific methods to be taken into account when establishing Reference Points for Action (RPAs) for non- allowed pharmacologically active substances present in food of animal origin. EFSA J. 11 (4), 3195.

Ehrlich, J., Bartz, Q.R., Smith, R.M., Joslyn, D.A., 1947. Chloromycetin, a new antibiotic from a soil actinomycete. Science 106, 417.

Fatta-Kassinos, D., Bester, K., Kümmerer, K. (Eds.), 2010. Xenobiotics in the Urban Water Cycle. Mass Flows, Environmental Processes, Mitigation and Treatment Strategies. Springer, Dordrecht.

FDA, 2018. The 2017 Summary Report on Antimicrobials Sold or Distributed for Use in Food-Producing Animals.

Goldman, M., 1996. Cancer risk of low-level exposure. Science 271, 1821–1822.

Hanekamp, J.C., 2008. Micronutrients, hormesis and the aptitude for the maturation of regulation. Am. J. Pharmacol. Toxicol. 3 (1), 141–148.

Hanekamp, J.C., 2015. Utopia and Gospel: Unearthing the Good News in Precautionary Culture. Dissertation. Tilburg University, The Netherlands.

Hanekamp, J.C., Bast, A., 2015. Antibiotics exposure and health risks: chloramphenicol. Environ. Toxicol. Pharmacol. 39 (1), 213–220.

Hanekamp, J.C., Calabrese, E.J., 2007. Chloramphenicol, European legislation and hormesis. Dose-Response 5, 91–93.

Hanekamp, J.C., Calabrese, E.J., 2021a. Reflections on chemical risk assessment or how (not) to serve society with science. Sci. Total Environ. 792, 148511.

Hanekamp, J.C., Calabrese, E.J., 2021b. Tradeoffs of chemicals regulation – The science and tacit knowledge of decisions. Sci. Total Environ. 794, 148566.

Hanekamp, J.C., Frapporti, G., Olieman, K., 2003. Chloramphenicol, food safety and precautionary thinking in Europe. Environ. Liabil. 6, 209–221.

Hirsch, R., Ternes, T., Haberer, K., Kratz, K.-L., 1999. Occurrence of antibiotics in the aquatic environment. Sci. Total Environ. 225, 109–118.

Hopwood, D.A., 2007. Streptomyces in Nature and Medicine. The Antibiotic Makers. Oxford University Press, Oxford.

IARC, 2012. IARC monographs on the evaluation of carcinogenic risks to humans; 100(F), pp. 225–248.

Janssen, P.A.H., Baars, A.J., Pieters, M.N., 2001. Advies Met Betrekking Tot Chlooramfenicol in Garnalen. RIVM/CSR, Bilthoven, The Netherlands [Recommendations on chloramphenicol in shrimp.] The food exposure was calculated from the 'reasonable worst-case' scenario of a weekly shrimp consumption of 8.4 g contaminated with 10μg CAP per kg shrimp. Kgbw stands for kilogram bodyweight. Toxicological data are usually related to this unit.

Kaiser, J., 2003. Sipping from a poisoned chalice. Science 320, 376–378.

Kathren, R.L., 2003. Historical development of the linear nonthreshold dose-response model as applied to radiation. Pierce Law Rev. 1, 5–30.

Koch, R., 1983. A threshold concept of environmental Pollutants. Chemosphere 12 (1), 17–21.

Kolpin, D.W., Furlong, E.T., Meyer, M.T., Thurman, M.E., Zaugg, S.D., Barber, L.B., Buxton, H.T., 2002. Pharmaceutical, hormones, and other organic waste water contaminants in U.S. streams, 1999-2000: a national reconnaissance. Environ. Sci. Technol. 36, 1202–1211.

Kroes, R., Renwick, A.G., Cheeseman, M., Kleiner, J., Mangelsdort, I., Piersma, A., Schilter, B., Schlatter, J., Van Schothorst, F., Vos, J.G., Wurtzen, G., 2004. Structure-based thresholds of toxicological concern (TTC): guidance for application to substances present at low levels in the diet. Food Chem. Toxicol. 42, 65–83.

Kummerer, K. (Ed.), 2008. Pharmaceuticals in the Environment. Sources, Fate, Effect and Risks. Springer Verlag, Berlin.

Lindberg, R., Jarnheimer, P.-A., Olsen, B., Johansson, M., Tyskind, M., 2004. Determination of antibiotic substances in hospital sewage water using solid phase extraction and liquid chromatography/mass spectrometry and group analogue internal standards. Chemosphere 57, 1479–1488.

Loraine, G.A., Pettigrove, M.E., 2006. Seasonal variations in concentrations of pharmaceuticals and personal Care products in drinking water and reclaimed wastewater in Southern California. Environ. Sci. Technol. 40, 687–695.

Mantel, N., Bryan, W.R., 1961. Safety testing of carcinogenic agents. J. Natl. Cancer Inst. 27, 455–470.

Meyer, E., Gastmeier, P., Deja, M., Schwab, F., 2013. Antibiotic consumption and resistance: Data from Europe and Germany. Int. J. Med. Microbiol. 303, 388–395.

Moustacchi, E., 2000. DNA damage and repair: consequences on dose-responses. Mutat. Res. 464, 35–40.

Oliver, S.P., Boor, K.J., Murphy, S.C., Murinda, S.E., 2009. Food safety hazards associated with consumption of raw milk. Foodborne Pathog. Dis. 6 (7), 793–806.

Pagnotti, V.S., Chubatyi, N.D., McEwen, C.N., 2011. Solvent assisted inlet ionization: an ultrasensitive new liquid introduction ionization method for mass spectrometry. Anal. Chem. 83, 3981–3985.

Papa, E., Fick, J., Lindberg, R., Johansson, M., Gramatica, P., Andersson, P.L., 2007. Multivariate chemical mapping of antibiotics and identification of structurally representative substances. Environ. Sci. Technol. 41, 1653–1661.

Patel, J.C., Banker, D.D., 1949. Chloramphenicol in typhoid fever. A preliminary report of clinical trial in 6 cases. Br. Med. J. 22 (2), 908–909.

Phillips, I., Casewell, M., Cox, T., De Groot, B., Friis, C., Jones, R., Nightingale, C., Preston, R., Waddell, J., 2004. Does the use of antibiotics in food animals pose a risk to human health? A critical review of published data. J. Antimicrob. Chemother. 53, 28–52.

Popper, K.R., 1986. Utopia and violence. World Aff. 149 (1), 3–9.

Quaglia, N.C., Dambrosio, A., Normanno, G., Parisi, A., Patrono, R., Ranieri, G., Rella, A., Celano, G.V., 2008. High occurrence of *Helicobacter pylori* in raw goat, sheep and cow milk inferred by glmM gene: a risk of food-borne infection? Int. J. Food Microbiol. 124, 43–47.

Rayner, S., Prins, G., 2007. The Wrong Trousers: Radically Rethinking Climate Policy. Institute for Science, Innovation and Society, Oxford, UK.

Regulation (Ec) No 470/2009 of the European Parliament and of the Council of 6 May 2009 laying down Community procedures for the establishment of residue limits of pharmacologically active substances in foodstuffs of animal origin, repealing Council Regulation (EEC) No 2377/90 and amending Directive 2001/82/EC of the European Parliament and of the Council and Regulation (EC) No 726/2004 of the European Parliament and of the Council. Off. J. Eur. Union L152, 11−22.

Ricci, P.F., Straja, S.R., Cox Jr., A.L., 2012. Changing the risk paradigms can be good for our health: J-shaped, linear and threshold dose-response models. Dose-Response 10, 177−189.

Schröder, U., 2002. Final Report on Chloramphenicol − Laboratory Comparison Study. Bundesforschungsanstalt für Fischerei. Institut für Fischereitechnik und Fischqualität, GFR, Hamburg (This report can be obtained through the authors.).

Shi, W., Zusman, D., 1993. Fatal attraction. Nature 366, 414−415.

Tubiana, M., Aurengo, A., 2005. Dose−effect relationship and estimation of the carcinogenic effects of low doses of ionising radiation: the Joint Report of the Académie des Sciences (Paris) and of the Académie Nationale de Médecine. Int. J. Low Radiat. 2 (3/4), 1−19.

Tubiana, M., Aurengo, A., Averbeck, D., Masse, R., 2006. Recent reports on the effect of low doses of ionizing radiation and its dose−effect relationship. Radiat. Environ. Biophys. 44, 245−251.

U.S. Food and Drug Administration (USFDA), 1973. Compounds used in food-producing animals. Procedures for determining acceptability of assay methods used for assuring the absence of residues in edible products of such animals. Proposed rule. Federal Register 19226−19230. July 19.

Van Boeckel, T.P., Brower, C., Gilbert, M., Grenfell, B.T., Levin, S.A., Robinson, T.P., Teillant, A., Laxminarayan, R., 2015. Global trends in antimicrobial use in food animals. Proc. Natl. Acad. Sci. Unit. States Am. 112 (18), 5649−5654.

Walsh, C., 2003. Antibiotics. Actions, Origins, Resistance. ASM Press, Washington D.C.

Watkinson, A.J., Murby, E.J., Kolpin, D.W., Costanzo, S.D., 2009. The occurrence of antibiotics in an urban watershed: from wastewater to drinking water. Sci. Total Environ. 407, 2711−2723.

Zhou, L.-J., Ying, G.-G., Liu, S., Zhao, J.-L., Yang, B., Chen, Z.-F., Lai, H.-J., 2013. Occurrence and fate of eleven classes of antibiotics in two typical wastewater treatment plants in South China. Sci. Total Environ. 452−453, 365−376.

Chapter 9

Substantiating regular, qualified, and traditional health claims

Bert Schwitters[1,a] and Jaap C. Hanekamp[2,3,4]
[1]Independent Researcher; [2]University College Roosevelt, Middelburg, the Netherlands; [3]Environmental Health Sciences, University of Massachusetts Amherst, Amherst, MA, United States; [4]HAN-Research, Zoetermeer, the Netherlands

9.1 Introduction and background

Around the world, governments are subjecting commercial communication concerning the relationship between foods/foodstuffs and human health to various types of evaluation, scrutiny, and control. In general, subjecting commercial communication to pre- or postmarket entry evaluation is primarily based on the political notion that government should protect consumers against misleading information. However, instead of determining whether commercial communication is or is not misleading on the strict basis of legal definitions, concepts, and principles that determine the truthfulness of information, governments seek to substitute legal scrutiny of commercial speech by courts of law for scrutiny by scientific bodies or panels with expertise in the field of assessing research performed in the field of nutrition and human health.

In such a regulatory approach, which, in many instances, also seeks to address public health via regulatory measures that have consumer protection as their stated legal basis, a predominant role is attributed to "science." Not only must "science" assess the relevant commercial communication, but, doing so, it must also provide the foundations for regulations by way of which governments seek to "organize" markets and protect consumers. At the same, "science" is called upon to deliver data in support of public health programs and initiatives fostering positive behavioral change related to diet, physical activity, and other lifestyle factors. This makes that the criteria applied in the scientific assessment of health claims used in commercial communication are no longer chosen for the purpose of assessing its truthfulness and no more than its truthfulness. The scientific criteria applied in assessing health claims have become legal instruments in designing and addressing public health issues and "dietary choice" policy frameworks to facilitate consumers' choices.

At first sight, "science" might very well enjoy its new influential position. But the fact that it is caught in a hybrid scientific-political/legislative setting that seeks to combine the categorically different legal assessment of the truthfulness of information with the scientific assessment of the certainty of information based on research outcomes puts "science" in the difficult if not impossible position of "producer of truth." In order to fulfill the concomitant tasks, truth is equated with certainty, and both terms are used interchangeably. However, even when "science" would produce a 95% or higher probability as required in science as confirmation of a hypothesis as the probability that comes really close to certainty, up to being undistinguishable from it, such a level of probability still does not produce truth.

Put differently, science is usually understood as empirical in nature (although mathematics and (scientific) reasoning cannot be reduced thereto). Through experimentation one tries to establish basic regularities of the world. What the empirical sciences produce are contingent propositions, that are not necessarily true or false: "chemical A interacts with protein X resulting in effect Y"; "thallium has the atomic weight of w"; "the lethal dose of X for mice is Y"; "the consumption of this food adds to our health and longevity."

a Bert Schwitters is an independent author/researcher who published Health Claims Censored, a fundamental critique on the European Nutrition and Health Claims Regulation 1924/2006/EC. www.healthclaimscensored.com – b.schwitters@defactopublications.com.

Ensuring Global Food Safety. https://doi.org/10.1016/B978-0-12-816011-4.00019-7

9.2 When truth and certainty must compete

These and many other propositions generated by the empirical sciences are all conditionally true, given various facts and evidence. None of these propositions are logically necessary. It is logically possible for these statements to be false, say, due to measurement errors, mistakes in experimental set-ups, incorrect starting materials, the limitations of available facts, and so on. Thus, scientific arguments start from empirical premises and draw only probabilistic conclusions, prone to correction. To be sure, we do not doubt the measurements of the atomic weight of thallium, for instance. The premise of trust is ever present, and quite rightly so. But, as the business of science expands, this premise can be undermined.

Thus, no scientific results will give us definitive answers to our many questions (Briggs, 2016). Many scientists developed the risky habit of insisting that their conditional truths are necessary truths. Some have gone further downhill by insisting fallaciously that their probable truths are universally true. The compelling statement "science has shown that ..." should be taken with a grain of salt, and sometimes perhaps even more than that, say, a truckload. Wholesome skepticism thus is a balancing act, as Michael Polanyi showed, between orthodoxy and dissent, between the quietist "everybody knows that ..." and the twitchy "forget everything you know about ..."

Evidently, when truth and certainty must compete for "pole position," truthful yet uncertain information is lost. In terms of health claims, examples of uncertain information are claims that truthfully reflect probable or possible, yet inconclusive effects caused by a well-characterized food or foodstuff. In terms of health claims, examples of uncertain and misleading information and claims that do not truthfully reflect products' effects have raised concerns among authorities, politicians, and consumer advocates. Therefore, the issues of inconclusive and traditional scientific evidence used to explain the claimed beneficial effect of a well-characterized food or food component have to be clearly explicated in the context of the claims platform.

9.3 Qualifying the certainty of information

When truthfully "qualifying" a claim concerning the beneficial effects that may be expected from consuming a food or foodstuff, a business operator informs consumers about the scientifically assessed possibility or probability—not the absolute certainty, which is impossible—that such and such benefits may result from consuming the food or foodstuff. Pending the emergence of conclusive scientific evidence, such "qualified claims" expose consumers to outcomes of emerging scientific evidence, suggesting that using the researched products or ingredients will probably or may possibly benefit them. In legal terms, a qualified claim claims probability and disclaims certainty.

Such claims truthfully express the fact that the relevant scientific community has not yet reached the highest scientifically possible level of certainty regarding the claim, but that there exists emerging evidence and agreement sufficiently supporting the *possibility or probability* of the claimed effect. For a statement to be truthful, it does not need to express unequivocally conclusive scientific evidence. A well-formulated "qualified health claim," correctly reflecting the strength (level/grade of scientific certainty) of the underlying evidence as well as the totality of the available scientific data, may be characterized as a truthful, objectively correct representation.

This is especially relevant when the use of a food/foodstuff is based on long-standing and often "tacit" traditional knowledge and experience. This kind of knowledge is not based on scientific evidence as we understand it today, but on nutritional and/or medicinal practice. In the European Union, traditional medicinal claims are accepted on the basis of sufficient "bibliographical or expert evidence to the effect that the medicinal product in question, or a corresponding product has been in medicinal use throughout a period of at least 30 years preceding the date of the application, including at least 15 years within the Community." Such evidence shall be accompanied by "a bibliographic review of safety data together with an expert report, and where required by the competent authority, upon additional request, data necessary for assessing the safety of the medicinal product." Moreover, the data on traditional use shall be deemed "insufficient, especially if pharmacological effects or efficacy are not plausible on the basis of long-standing use and experience" (Article 16).

9.4 RCT's and plausibility

To connect dietary patterns (including supplementation and fortification) with human health and thereby assess benefits and risks, methods other than establishing such connections on the basis of describing and examining the traditional use and that use's biological plausibility have been developed especially in Western science. In the domain of nutrition and health, observational epidemiologic studies are particularly relevant when it comes to long and well-established—traditional—use. In addition, intervention trials (especially Randomized Controlled Trials—RCTs), all sorts of models and simulations, *in* and ex vivo animal and human studies, in vitro research, and the like are some of the well-known examples. Accordingly, methods might be mechanistic in nature— e.g., elucidating metabolic pathways in *in* or ex vivo animal/human studies—or methods might be phenomenological in nature—e.g., an RCT giving some insight in the efficacy or effectiveness of a certain treatment. Specific endpoints might comprise the number of healthy life years and life expectancy, motor-, cognitive-, neurologic-, and metabolic function, wellbeing, satiety and hunger, specific biomarkers, and the like. In the relevant fields of science, most of these methods have been exhaustively described and standardized (Bast et al., 2013).

In the context of substantiating health claims, it is important that all these methods are used appropriately and uniformly, i.e., in accordance with the commonly accepted standardized methodologies. It is equally important that outcomes shall be classified or "graded" in accordance with the chosen methodology and the strength of the evidence attributed to the applied methodology. This will provide clarity concerning the probability or likelihood of the health claim. Much work on the grading and classifying of evidence has been performed by the ILSI organization (Aggett et al., 2005; ILSI, 2010).

From a political, regulatory and (Western/mainstream) scientific point of view, the RCT is overarchingly regarded as the "gold standard" for establishing a relationship between nutrition and health. Indeed, the European Union identifies in the "organization of pertinent scientific data" that must be provided in support of a health claim a "hierarchy of study design" where RCTs rank at the top of this ostensible scientific pyramid. RCTs thus are given legal sanction and preference with respect to the approval or rejection of certain health claims for certain foods or food products.

Answering the question of how and why things work is the quintessence of science. RCTs themselves, however, do not further mechanistic knowledge. Obviously, that knowledge is usually available before any RCT is attempted. After all, one cannot do an RCT "in the blind." Moreover, if the results of RCTs cannot be readily extrapolated, then the lack or absence of mechanistic knowledge exacerbates ignorance: put in the agent—food/foodstuff—and outcome(s) the effect(s) without the RCT as such being able to elucidate the underlying mechanism(s). In fact, the concrete regulatory focus on RCTs as a primary means to substantiate some health claim will hamper that search. Making RCTs a requirement that must be met in the substantiation of health claims, truthful claims based on plausibility will not pass this bar.

9.5 Traditional medicinal products in the EU

While the European legislature insists on RCT-type substantiation in the field of health claims made for foods/foodstuffs to the exclusion of plausibility based on traditional use, it accepts, in the field of medicinal claims/indications, bibliographical or expert evidence. Article 16c.1 of the European Medicinal Products Directive provides that "expert evidence to the effect that the medicinal product in question, or a corresponding product has been in medicinal use throughout a period of at least 30 years preceding the date of the application, including at least 15 years within the Community" must form part of applications for traditional herbal medicinal products ("THMPs"):

[1. The application shall be accompanied by:]

16c.1.c bibliographical or expert evidence to the effect that the medicinal product in question, or a corresponding product has been in medicinal use throughout a period of at least 30 years preceding the date of the application, including at least 15 years within the Community. At the request of the Member State where the application for traditional-use registration has been submitted, the Committee for Herbal Medicinal Products shall draw up an opinion on the adequacy of the evidence of the long-standing use of the product, or of the corresponding product. The Member State shall submit relevant documentation supporting the referral;

(d) a bibliographic review of safety data together with an expert report, and where required by the competent authority, upon additional request, data necessary for assessing the safety of the medicinal product."

In addition to the "medicinal use" requirement, the expert shall also demonstrate that "the data on the traditional use of the medicinal product are sufficient":

Article 16a.1.(e)

[1. A simplified registration procedure (hereinafter "traditional-use registration") is hereby established for herbal medicinal products which fulfill all of the following criteria:]

(e) the data on the traditional use of the medicinal product are sufficient; in particular the product proves not to be harmful in the specified conditions of use and the pharmacological effects or efficacy of the medicinal product are plausible on the basis of long-standing use and experience.

Plausibility is a key criterion, because Article 16e.1.d provides that "the data on traditional use are insufficient, especially if pharmacological effects or efficacy are not plausible on the basis of long-standing use and experience [...]."

9.6 Health claims based on traditional use

Five years after the European Nutrition and Health Claims Regulation entered into force, the European Commission sent a *Discussion Paper* to the Member States regarding the serious legal problem that had existed all these years. The Commission formulated the problem as:

> *... the current different legal treatment of botanicals in foods and medicines legislation with respect to health claims/therapeutic indications.*

Precisely put, the "different treatment" concerns the different treatment with respect to "traditional health claims/ traditional medical claims." The different treatment enforces a discriminatory constraint in the field of health claims, by unconditionally prohibiting the use of *truthful* traditional health claims in the market for foods, while strictly reserving entitlement to practically unhindered market entry for *technically and legally equivalent* traditional medical claims in the adjacent and competing market for medicinal products. In legal terms, *dissimilar conditions* are applied to equivalent situations.

Although the European Commission described the problem as one existing within the European Union, the issue is of global importance, especially in terms of the concept of Global Harmonization and/or Mutual Recognition. Traditional use of foods/foodstuffs in the maintenance of health and the prevention and treatment of disease in non-Western populations is mainly based on traditional knowledge. In this regard, the notion that traditional health claims and traditional medicinal claims are technically and legally equivalent is highly relevant.

First of all, equivalence exists when the claims concern the traditional character of the use. Tradition is generally defined as "the passing down from generation to generation of the same customs, beliefs, etc., especially by word of mouth." In human populations, the "generation time" typically ranges from 20 to 30 years. Therefore, that traditional use must go back more than 20—30 years.

The second equivalence criterion concerns the establishment of the truthfulness of commercial information that is based on or refers to a tradition. In the case of information concerning traditional use, the demonstration of truthfulness requires, e.g., bibliographical evidence, describing or documenting the tradition. It may also require evidence written by recognized experts in the field of a particular tradition. Scientific evidence of the "Gold Standard" (randomized/placebo controlled clinical trials—RCTs) type is not required to establish truthfulness of the traditional use. In fact, traditional use cannot be substantiated by way of generally accepted scientific evidence, because the evidence of the use is not of a "scientific" nature.

The third equivalence criterion is the sufficiency of the data on the traditional use presented in the bibliographical or expert evidence, especially the plausibility on the basis of the long-standing use and experience. Plausibility is the equivalence criterion shared by traditional medicinal and traditional health claims.

9.7 Basic evidential requirements

By allowing therapeutic claims based on traditional use for *medicinal* products, the European legislature has accepted and acknowledged that nonscientific ("traditional") evidence can produce information that is not misleading or false, unambiguous, objectively correct, and representing "*toute la réalité*." Consumers are sufficiently protected when the evidence is unbiased and based on adequate and accurate facts. "Informative equivalence" between traditional *health* claims and traditional *therapeutic* claims exists when information meets the basic evidential requirement that establishes the truthfulness of a description of a tradition.

In the United States, issues concerning the truthfulness of "representations" have led to numerous court cases. In that context, the term "reasonable basis" is often used when it comes to substantiating the credibility of representations such as health claims. In the context of this chapter, which was written for the Global Harmonization Initiative (GHI) as a Working Document, it is noteworthy that US Courts have declared that the "reasonable basis" standard is met when there is a *causal* connection between the evidence proffered as support and the representation at issue at the time it is made, and the representation is supported by competent and reliable scientific supporting evidence, including tests, analysis, research, and studies that:

(i) are based on the expertise of professionals in the relevant area;
(ii) conducted and evaluated in an objective manner;
(iii) by a person qualified to do so;
(iv) use procedures accepted in the profession to yield accurate and reliable results (U.S. District Court).

In European legislation concerning traditional therapeutic claims, the term "plausible" is used as a sufficient cause-and-effect criterion. However, there is no legal definition of the term "plausible." With regards to traditional therapeutic claims, the plausibility requirement is fulfilled when "long-standing use and experience" is documented in the form of "bibliographical or expert evidence, to the effect that the medicinal product in question, or a corresponding product, has been in medicinal use throughout a period of at least 30 years preceding the date of the application, including at least 15 years within the Community" (Article 16c).

Evidently, an intelligent and reliable evaluation of facts is often difficult or impossible without the application of some scientific, technical, or other specialized knowledge. The most common source of this knowledge is "the expert." Experts with the necessary technical or professional qualifications need not necessarily be scientists. Skilled gardeners, millers, mechanics, cooks, sculptors, housewives, business operators, nutritionists, etc., are experts whose expertise does not necessarily depend on scientific training and/or university degrees.

In view of the provision that all European regulations must provide a "high level of protection," especially for consumers, "expert evidence" regarding information based on traditional (therapeutic) use is supposed to reach levels of certainty and protection that are equal to the level of certainty supposedly reached by "scientific substantiation." Scientific substantiation shall meet the level of certainty attributed to "generally accepted scientific evidence." Likewise, bibliographical or expert evidence concerning traditional use shall meet this level of certainty. Apparently, at least in the European Union, the plausibility criterion meets the standard of generally accepted scientific evidence.

In addition to the plausibility criterion, the food or foodstuff shall be well characterized. In the case of traditional foods, this is of particular importance given that some traditional foods have changed names during long periods, sometimes thousands of years, of use. Also, some traditional foods have different names in different regions or cultures. Moreover, methods of production and preparation may have evolved over time. This requires meticulous research of a food's principal components and composition requirements.

9.8 Qualifying the expert

In line with US standards, GHI proposes that experts shall meet technical or professional qualifications that permit the collecting, production, and evaluation of evidence in a particular field of knowledge or activity. A person who is qualified as an expert by knowledge, skill, experience, training, or education should be allowed to provide evidence in the form of an Opinion if:

(a) the expert's scientific, technical, or other specialized knowledge will help the "trier of fact" understand the relevant facts;
(b) the Opinion is based on sufficient facts or verifiable data;
(c) the Opinion is the product of the application of reliable principles and methods;
(d) the expert has demonstrated that he/she has reliably applied the principles and methods to the relevant facts.

The fields of knowledge and methodology which may be drawn upon are not limited merely to the "scientific" or "technical" ones, but extend to all "specialized" knowledge, in casu to the traditional use of foods to achieve health benefits. Therefore, the expert is viewed, not in a narrow sense, but as a person qualified by "knowledge, skill, experience, training, or education." Thus, within the scope of GHI's proposition, experts are not only experts in the strictest sense of the word, e.g., physician and nutritionists, but also members of the large group of "skilled" experts.

9.9 Reliability of the expert's opinion

To assess the reliability of an Expert Opinion, the following specific factors are selected from the Federal Rules of Evidence, applied by the courts in the United States:

(1) whether the theory, technique, principle(s), and method(s) applied by the expert can be or have been tested—that is, whether the expert's theory, technique, etc., can be challenged in some objective sense, or whether the expert's approach is instead subjective or conclusory, and cannot reasonably be assessed for reliability;
(2) whether the technique, principle, method, or theory has/have been subject to peer review and publication;
(3) the known or potential rate of error of the technique, principle, method, or theory when applied;
(4) the existence and maintenance of any standards and controls;
(5) whether the technique, principle, method, or theory has/have been generally accepted in the relevant community of experts.

These factors shall also be applicable in assessing the reliability of nonscientific expert opinion, depending upon the particular circumstances of the particular case at issue. The factors are neither exclusive nor dispositive. Not all of the specific factors can apply to every type of expert opinion. Lack of peer review or publication shall not be dispositive where the expert's opinion was supported by "widely accepted scientific knowledge." In any case, the expert shall demonstrate that his/her Opinion is based on data that can be accessed and verified by other experts and/or the trier of fact.

There may exist other factors relevant in determining whether an expert Opinion is sufficiently reliable to be considered by the "trier of fact." These factors include:

(1) Whether experts are proposing to testify about matters growing naturally and directly out of research they have conducted independent of the specific request made by the ultimate trier of fact, or whether they have developed their opinions expressly for purposes of producing an Opinion.
(2) Whether the expert has unjustifiably extrapolated from a generally accepted premise to an unfounded conclusion.
(3) Whether the expert has adequately accounted for obvious alternative explanations.
(4) Whether the expert is being as careful as he would be in his regular professional work outside his paid consulting and whether the expert employs in writing his Opinion the same level of intellectual rigor that characterizes the practice of an expert in the relevant field.
(5) Whether the field of expertise claimed by the expert is known to reach reliable results for the type of Opinion the expert would give.

9.10 Principles and methodology

Expert Opinion cannot be rejected or opposed simply because the expert uses one test rather than another, when both tests are accepted in the field and both reach reliable results. The focus, of course, must be solely on principles and methodology, not on the conclusions they generate. Yet, conclusions and methodology are not entirely distinct from one another.

When an expert purports to apply principles and methods in accordance with professional standards, and yet reaches a conclusion that other experts in the field would not reach, the ultimate trier of fact may fairly suspect that the principles and methods may not have been faithfully or correctly applied. In such cases, peer review of the expert Opinion must take place. In this respect, vigorous examination of the expert Opinion and presentation of contrary evidence are the traditional and appropriate means of assessing the correctness and validity of the Opinion.

When an Expert Opinion is found to be reliable, this does not necessarily mean that contradictory expert Opinion is unreliable. It is suggested that Opinions that are the product of competing principles or methods in the same field of expertise shall be permitted.

9.11 Degree of scrutiny

The ultimate "trier of fact" should accept Opinions written by any expert, not only Opinions based on "scientific" knowledge, but also Opinions based on "technical" and "other specialized" knowledge." While the relevant factors for determining reliability will vary from expertise to expertise, the authors of this chapter reject the premise that an expert Opinion should be treated more permissively simply because it is outside the realm of science. An opinion from an expert who is not a scientist should receive the same degree of scrutiny for reliability as an opinion from an expert who purports to be a scientist.

Some types of expert Opinion will be more objectively verifiable, and subject to the expectations of falsifiability, peer review, and publication, than others. Some types of expert Opinion will not rely on anything like a scientific method, and so will have to be evaluated by reference to other standard principles attendant to the particular area of expertise. The ultimate "trier of fact" in all cases of proffered expert Opinion must find that it is properly grounded, well reasoned, based on data proven factual and not speculative before it can be accepted as the basis for policies and/or regulations.

The expert Opinion must be grounded in an accepted body of learning or experience in the expert's field, and the expert must explain how the conclusion is so grounded. The Opinion must be the product of reliable principles and methods that are reliably applied to the facts of the case. While the terms "principles" and "methods" may convey a certain impression when applied to scientific knowledge, they remain relevant when applied to Opinion based on technical, historical, or other specialized knowledge.

9.12 Extrapolating results obtained in diseased subjects

Experts in the field of traditional use who are asked to present Opinions that demonstrate a relationship between a food/foodstuff and *health* often have to solve the problem that traditional use is mostly medicinal/therapeutic and not strictly nutritional/physiological. The difference between medicinal and nutritional use has been framed by legislators to a degree that escapes consumers and therapists. In the highly theoretical regulatory settings of most "Western" countries, medicinal use turns a food/foodstuff into a medicinal substance. This often precludes or hampers the use of such a substance as a nutritional agent (food/foodstuff) in the maintenance of health.

In the framework of the European Nutrition and Health Claims Regulation, the European Food Safety Authority (EFSA) has proposed that this problem must be tackled as follows:

When a particular study submitted for the scientific substantiation of the claim has been conducted in a study group (e.g., subjects with a disease) which is different from the target population for a claim (e.g., the general population or subgroups thereof), the NDA Panel [EFSA] considers whether the results from that study can be extrapolated to the target population for the claim. In principle:

[...]

(ii) results from studies performed in subjects with a disease (i.e. type 2 diabetic patients) that affects the function mentioned in the claim (e.g. reduction of postprandial blood glucose responses) can be extrapolated to the target population for a claim (e.g., the general population) as long as the effect of the food/constituent on the beneficial physiological effect which is mentioned in the claim is also reasonably expected to occur in subjects without the disease (e.g. if it can be established that the mechanism by which the food/constituent exerts a beneficial effect on the disease is the same by which it could reduce the risk of a disease in the target population). If subjects with a disease are under pharmacological treatment, the Panel considers whether the effect of the food/constituent is also reasonably expected to occur in subjects without medication.

Experts in the field of traditional use could follow this approach when the traditional use concerns diseased subjects. They must demonstrate that the relevant beneficial physiological effect is also reasonably expected to occur in subjects without the disease. Prior to this, they must, of course, elucidate the physiological mechanism of action. This harks back to the problem that RCTs are not fit to elucidate the mechanism by which the food/foodstuff exerts a beneficial effect. Other study types are required to establish the physiological mechanism. In this regard, the plausibility criterion comes into play.

9.13 Plausibility

According to the renowned British epidemiologist and statistician Austin Bradford Hill, when "our observations reveal an association between two variables [a food/foodstuff and a particular aspect of human health], perfectly clear-cut and beyond what we would care to attribute to the play of chance, [w]hat aspects of that association should we especially consider before deciding that the most likely interpretation of it is causation?" (Hill, 1987) Among the aspects strength, consistency, specificity, temporality, biological gradient, coherence, experiment, and analogy, Hill also placed plausibility. However, with regard to that aspect, Hill noted:

It will be helpful if the causation we suspect is biologically plausible. But this is a feature I am convinced we cannot demand. What is biologically plausible depends upon the biological knowledge of the day. To quote again from my Alfred Watson Memorial Lecture (Hill, 1962), there was '... no biological knowledge to support (or to refute) Pott's observation in the 18th century of the excess of cancer in chimney sweeps. It was lack of biological knowledge in the 19th that led to a prize essayist writing on the value and the fallacy of statistics to conclude, amongst other 'absurd' associations, that 'it could be no more ridiculous for the strange who passed the night in the steerage of an emigrant ship to ascribe the typhus, which he there contracted, to the vermin with which bodies of the sick might be infected.' And coming to nearer times, in the 20th century there was no biological knowledge to support the evidence against rubella.' In short, the association we observe may be one new to science or medicine and we must not dismiss it too light-heartedly as just too odd. As Sherlock Holmes advised Dr. Watson, 'when you have eliminated the impossible, whatever remains, however improbable, *must be the truth.'*

Considering that each of the scientific evidential criteria applied by the World Health Organization (convincing, probable, possible) requires a plausibility assessment, and that a demonstration of plausibility is required in European regulations concerning traditional (therapeutic) use, it is not only "helpful" but necessary that the "suspected causation" is biologically plausible to sustain a traditional health claim.

Given the fact that we lack generally accepted—harmonized—criteria for plausibility, GHI proposes that, along the lines proposed in this Working Document, a group of experts in the field of traditional use should answer the following question:

In case, beyond what one would care to attribute to the play of chance, a well-documented traditional use of a food/foodstuff reveals a perfectly clear-cut association between that food/foodstuff and a specific, well-described, effect on human health, what criteria of plausibility should we especially consider before deciding that the most likely interpretation of the association is causation?

GHI's focus on "plausibility" originates from the fact that this is the main *scientific* aspect that is still undefined and, as a consequence, unharmonized in the field of substantiating health claims. In addition, when we would be capable of defining criteria for plausibility, this will have a "harmonizing" influence on all the different regulations existing in various jurisdictions without necessarily effacing these—incompatible—legal and constitutional differences. Put differently, instead of trying to harmonize the different regulatory systems into one WW system, GHI sees this "plausibility initiative" as the only approach that might be successful in "symphonizing" these different systems so that they may coexist "in harmony."

9.14 The way forward

- We recommend that "triers" of health claims in the United States, Europe, and elsewhere favor allowance of nutrient—health and nutrient—disease relationship claims backed by credible scientific and/or expert evidence, regardless of the inconclusiveness of that evidence, provided that the evidence meets the criterion of plausibility.
- Regarding the particular characteristics of traditional health claims, the long tradition of the use should make it possible to eliminate the need for clinical trials, in so far as the efficacy of the nutrient or other substance, the food category, the food, or one of its constituents is plausible on the basis of established and long-standing use and experience.
- A traditional health claim is defined as a claim that states, suggests, or implies that a traditional relationship exists between the consumption of a nutrient or other substance, a food category, a food or one of its constituents and health.
- A traditional reduction of disease risk claim is defined as a traditional health claim that states, suggests, or implies that a traditional relationship exists between the consumption of a nutrient or other substance, a food category, a food, or one of its constituents and the significant reduction of a risk factor in the development of a human disease.
- Traditional health claims shall be based on bibliographical or expert evidence to the effect that the effect on health is plausible and that the traditional health claim and the nutrient or other substance, the food category or the food, or one of its constituents, in respect of which the traditional health claim is to be made, have been in use throughout a period of at least 30 years.
- Any trier of fact shall:
  - **(a)** verify that the relevant nutrient or other substance, the food category, the food, or one of its constituents has been sufficiently characterized;
  - **(b)** verify that the proposed wording of the traditional health claim is supported by adequate bibliographical or expert evidence;
  - **(c)** verify that the wording of the traditional health claim expressly confirms the nutritional ("food") character of the relevant nutrient of other substance and that the claim cannot be construed or interpreted as implying a medicinal effect;
  - **(d)** verify that the traditional health claim and the nutrient or other substance, the food category, the food, or one of its constituents, in respect of which the traditional health claim is to be made, have been in use throughout a period of at least 30 years;
  - **(d)** verify the plausibility of the effect.

References

Aggett, P.J., Antoine, J.-M., Asp, N.-G., Bellisle, F., Contor, L., Cummings, J.H., Howlett, J., Müller, D.J.G., Persin, C., Pijls, L.T.J., Rechkemmer, G., Tuijtelaars, S., Verhagen, H., 2005. PASSCLAIM process for the assessment of scientific support for claims on foods - consensus on criteria. Eur. J. Nutr. 44 (Suppl. 1), I/1—I/2.

Article 16 of the European Medicinal Products Directive of 6 November 2001 on the Community Code Relating to Medicinal Products for Human Use as Amended by Directive 2004/24/EC of 31 March 2004.

Article 16c.1.(C) of Directive 2001/83/EC of 6 November 2001 on the Community Code Relating to Medicinal Products for Human Use.

Bast, Briggs, W.M., Calabrese, E.J., Fenech, M.F., Hanekamp, J.C., Heaney, R., Rijkers, G., Schwitters, B., Verhoeven, P., 2013. Scientism, legalism and precaution — contending with regulating nutrition and health claims in Europe. Eur. Food Feed Law 6.

Briggs, W., 2016. Uncertainty. The Soul of Modeling, Probability & Statistics. Springer, Switzerland.

Commission Regulation (EC) No 353/2008 of 18 April 2008 Establishing Implementing Rules for Applications for Authorisation of Health Claims as provided for in Article 15 of Regulation (EC) No 1924/2006/EC.

Discussion Paper on Health Claims on Botanicals Used in Foods.

General scientific guidance for stakeholders on health claim applications. Nutrition and Allergies (NDA). EFSA J., January 18, 2016. https://doi.org/10.2903/j.efsa.2016.4367.

Federal Rules of Evidence. Article VII. Opinions and Expert Testimony. Rule 702. Testimony by Expert Witness.
Hill, A.B., 1987. The environment and disease: association or causation. In: Evolution of Epidemiologic Ideas: Annotated Readings on Concepts in Methods - Sander Greenland. Epidemiology Resources Inc., Massachusetts, pp. 7–12.
Beyond Passclaim; Summary Report of a Workshop Held in December 2009, May 2010. ILSI.
Regulation (EC) No 1924/2006 of 20 December 2006 on Nutrition and Health Claims Made on Foods.
U.S. District Court for the District of Utah in the Case Federal Trade Commission ("FTC") against Basic Research, LLC ("Basic Research").
Diet, Nutrition and the Prevention of Chronic Diseases; WHO Technical Report Series – 916, 2003. World Health Organization, Geneva.

Chapter 10

Benefits and risks of organic food

H.K.S. De Zoysa[1,2] and Viduranga Y. Waisundara[3]

[1]*Department of Bioprocess Technology, Faculty of Technology, Rajarata University of Sri Lanka, Anuradhapura, North Central Province, Sri Lanka;* [2]*Department of Biology, University of Naples Federico II, Naples, Italy;* [3]*Australian College of Business & Technology - Kandy Campus, Peradeniya Road, Kandy, Central Province, Sri Lanka*

10.1 The modern food market

According to the data from the FAO (2014), it is estimated that about 805 million people, or one out of nine, around the world are malnourished. This statistic in sub-Saharan Africa, in particular, is as high as one out of four. In general, 98% of those suffering from hunger live in developing countries, while 37 million hungry people reside in Asia, Africa, and Latin America (Jouzi et al., 2017). Although these numbers have shown a remarkable decline as compared to the past (specifically in Latin America), there is still a long way to go on the road of eradicating hunger. As the population and subsequent consumption around the world is growing, the demand for food, feed, and fuel will also increase in the future. Moreover, in the developing world, diets are changing and people are putting extra pressure on natural resources as they consume more dairy products and meat (Godfray et al., 2010; Seufert et al., 2012). It is estimated that by 2050, the demand for agricultural products will grow by 1.1% annually as the world's population reaches around 9 billion (Alexandratos and Bruinsma, 2012).

In the past decade, the global food market has seen a trend toward consumer preference geared toward combating health issues (Menrad, 2003; Siró et al., 2008). This development has primarily been driven by concerns over rising obesity rates (WHO, 2007) and appeals to the food industry to provide healthier products (Seiders and Petty, 2004). Also, consumers are increasingly aware on taking necessary actions themselves to favor foods that might help them maintain or improve their health and wellness (Siró et al., 2008; Verbeke, 2005). Parallel to the trend toward healthier food, a trend toward more environmentally friendly or "green" food products has emerged at present. This is often disguised under "ethical consumption" (Harrison et al., 2005) or the trend toward "sustainability" in marketing (Belz and Peattie, 2009). This encompasses diverse issues such as toxic emissions, climate change, biodiversity, and animal welfare, to name but a few.

Organic agriculture involves many practices that emphasize farming based on ecosystem management, integrated cropping and livestock systems, diversity of products, reliance on natural pest, and disease control without conventional chemical treatments. The main objective of organic agriculture is to produce healthy and sustainable food using exclusively biological and ecological processes (Azadi et al., 2011). Establishing that consuming organic food enhances perceived well-being would have significant theoretical implications related to the explanation of variables and processes underlying consumer well-being (Apaolaza et al., 2018).

10.2 Why organic food?

The concept of "organic food" has becomes popular among the consumers at present. Over the last 2 decades due to issues arisen with food safety, quality attention has been directed more toward organic foods leading to a high demand (Adebayo et al., 2020; Chekima et al., 2019; Mesnage et al., 2020). In addition to this, overexploitation, environmental destruction, and natural resource utilization have encouraged people to engage in organic agriculture and consume organic foods (Chekima et al., 2019). Thus, many farmers started organic farming in a large scale to meet this requirement. Nowadays, one can see many kinds of products that are available everywhere with the name "organic" on the packaging, and many people are consuming these products without proper awareness (Mesnage et al., 2020). However, most of the organic

Ensuring Global Food Safety. https://doi.org/10.1016/B978-0-12-816011-4.00008-2

farming practices and productions of organic food do not really meet the concept of the holistic approach of organic agriculture. Therefore, organic foods need to fulfill its holistic means ensuring sustainable organic farming including maintenance of health, agro-ecosystem, soil, biological cycles, and the environment (Chekima et al., 2019; Hurtado-Barroso et al., 2017; Tsion and Steven, 2019). Nonetheless, the claim "organic" has resulted in the need, or rather the belief, to ensure that those products are safer than nonorganic foods by reducing or avoiding consumers' exposure to synthetic chemicals and pesticides during the phases of farming, processing, and storage (Mesnage et al., 2020). It has to be borne in mind nevertheless that "organic" pesticides are not necessarily safer than synthetic pesticides; the latter are tested for safety, whereas the organic ones, being natural, often have not been subjected to safety assessment. This is one void which has not been considered by present scientists and even regulatory bodies. Organic foods supposedly do not include any growth hormones or chemicals and genetically modified materials to enhance the growth and production (Basha et al., 2015). Therefore, many countries have an increased consumption in organic food as they do not comprise pesticides, herbicides, chemical fertilizers, and other synthetic substances at each of these phases.

10.2.1 Consumer attitude, behavioral intentions, and preference toward organic and nonorganic food products

Consumer preference toward organic food has been increasing primarily because of the belief that organic food is free from the synthetic chemicals. Organic food production practices are conducted to the best of the abilities, in a sustainable way concerning the safety and economical aspects as well (Adebayo et al., 2020; Tsion and Steven, 2019). Typically, organic food has a higher market price and is one of reasons why many farmers are focusing their attention toward this practice (Adebayo et al., 2020). It is typically the attitude which determines the evaluation of the purpose of organic food purchasing—mostly in the forms of being tangible, intangible, or abstract. These can affect consumers' favorability or unfavorability toward organic foods. Likewise, attitude can be considered as vital to guess, and it explains human behavior in any direction toward organic foods (Dangi et al., 2020). Also, consumer attitude toward organic food consumption can be categorized as functional and constructive. A stable and structured consumer mindset comes under the functional attitudes that have been present at length, whereas temporal and transient belong to a constructive attitude. However, consumers may have both these attitudes together at the same time instead of separately at the point of deciding to purchase organic food (Dangi et al., 2020).

Identification of the consumers' attitudes toward organic products is an important factor to analyze, since organic foods have become and already have been recognized in the domestic marketplaces of many countries (Analuiza et al., 2020). However, the theory of planned behavior has revealed the attitude to be the most important of all factors and that it can be considered as a significant predictor to understand the organic food purchasing preference by the consumers. This is so, because it evaluates both positive and negative feelings of individuals and its direct effects on purchasing behavior of organic food (Asif et al., 2018). Nevertheless, while many studies have found several concerns and awareness in developed countries related to this, less information is available from the developing countries.

When it comes to the perspective of purchasing and consumption of organic foods, there are several main influences that consumers desire to purchase based on their personal satisfaction, namely, health benefits, environmental concerns, and price. These also correlate with consumers beliefs, psychological effects, subjective knowledge and norms, life style, social consciousness, and lack of confidence in conventional foods (Adebayo et al., 2020; Analuiza et al., 2020; Apaolaza et al., 2018; Asif et al., 2018; Basha et al., 2015; Hansen et al., 2018; Janssen, 2018). Nevertheless, based on consumer self-satisfaction, those attitudes have become a mediator to purchase organic foods based on the health aspects, food safety, and environmental concerns (Fig. 10.1) (Butlera and Stergiadis, 2020; Çabuk et al., 2014; Hansen et al., 2018). Furthermore, it is important that the practical implication for consumer policy is aimed to make sure to enhance consumers' wellbeing and promotion of organic foods. This can be also used by marketers and producers of organic foods (Apaolaza et al., 2018). In addition, there are some other interconnected factors that belong to these influences which can be listed as follows (Adebayo et al., 2020; Analuiza et al., 2020; Apaolaza et al., 2018; Ashaolu and Ashaolu, 2020; Asif et al., 2018; Basha et al., 2015; Britwum et al., 2021; Butlera and Stergiadis, 2020; Chekima et al., 2019; Dangi et al., 2020; Denver and Christensen, 2015; Gomiero, 2018; Hansen et al., 2018; Hwang and Chung, 2019; Janssen, 2018; Kamenidou et al., 2020):

- Brand
- Packing
- Availability
- Labeling
- Taste and quality

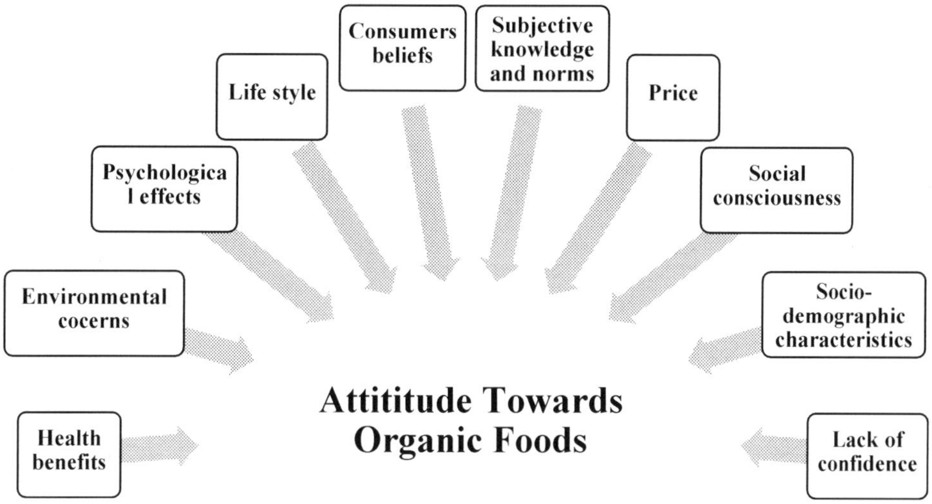

FIGURE 10.1 Main factors which influence consumers towards selecting organic foods.

- Locality
- Retail store reputation
- Urbanization
- Nostalgia
- Certification of origin
- Monthly income
- Cultural background
- Consumers' emotional value
- Gender-based preference
- Social norms
- Family influence
- Animal welfare
- Support local farmers

Some studies have already identified gender-based behavior when it comes to purchase of organic foods, since mainly women showed a higher interest (Analuiza et al., 2020; Ashaolu and Ashaolu, 2020; Denver and Christensen, 2015; Hansen et al., 2018). Also, consumer behavior with their psychological processes influences the purchasing decisions and postpurchasing decisions. This is also related with the confidence of the consumers organic foods based on the labeling and certification process (Britwum et al., 2021; Dangi et al., 2020). However, consumers in the United States have moderate confidence in the United States Department of Agriculture (USDA) and they rely more on unbiased sources of information such as researchers and farming groups. Nevertheless, many other countries have revealed their trust on the USDA as a standard organic certification, although it does not distinguish between organic and conventional food (Britwum et al., 2021; Hansen et al., 2018) and it has been found that USDA labeling is stronger than the usual organic food labeling (Britwum et al., 2021; Gomiero, 2018). Further, consumers from other countries tend to not trust the organic certification process, which has led to the decrease of purchasing organic foods—this has been the case in Thailand.

Consumer behavior is a key factor which impacts the environment, as nowadays consumers tend to purchase organic food based on increased awareness on the environmental degradation (Basha et al., 2015). Although, some studies have reported that environmental concerns are not a valid attitude all the time for the usage of organic foods, it can also be argued entirely as an unrealistic approach if it does not comply with the sustainable consumption of organic foods. Most of the time consumers tend to be satisfied with their personal benefits rather than environmental concerns (Chekima et al., 2019; Hansen et al., 2018). There are some product attributes that have surfaced recently as being responsible actors contributing as barriers for organic food consumption. Those factors could be categorized into intrinsic and extrinsic factors. The intrinsic factors mainly depend on appearance, smell, color, texture, and test, whereas extrinsic factors depend on price, brand, certification, and packaging (Dangi et al., 2020). However, contradictory reports have been observed due to the lack of studies on socio-demographic characteristics, because some studies have reported these characteristics not being considered as a factor to predict consumer purchasing behavior and vice versa (Analuiza et al., 2020; Denver and

Christensen, 2015; Janssen, 2018). Also, young children in the household can encourage their parents and elders to buy organic foods due to their influences, but when they get old, this preference changes (Janssen, 2018). A generational cohort behavioral study would provide its necessity for future policymaking and implementation. Generational cohort marketing as a whole has been recognized as a very effective approach (Kamenidou et al., 2020). By considering all the above aspects of consumer attitudes, this understanding would provide essential information that can be useful for policymakers, aid marketers, producers to promote their products by awareness of ethical consumer habits, credibly certified labeling, well-designed product packaging, and digital and promotional advertisement (Chekima et al., 2019; Hansen et al., 2018).

Further, other than the facts mentioned above, there are other points that need to be addressed. These are difficulties or barriers linked with the consumers' preference which can change the consumers' intention (Asif et al., 2018; Hansen et al., 2018). Further, some studies have reported that perceived behavior control is not valid for purchasing Halal and organic food (Asif et al., 2018). Altogether, we can list the difficulties and limitations as follows (Analuiza et al., 2020; Ashaolu and Ashaolu, 2020; Asif et al., 2018; Basha et al., 2015; Hansen et al., 2018; Janssen, 2018):

- High price
- Irregular supply and availability
- Trust
- Lack of communication with retails outlets and producers
- Genuineness of organic certification
- Socio-demographic characteristics

These indicate consumers' willingness to purchase the organic foods rather than the conventional and local foods. As per these facts, consumers' main preference that has been demonstrated is that foods should not be dangerous to their health. Thus, consumers' quality expectation tends to prefer organic foods over conventional foods because the awareness of pollutants, contamination, and harmful effects of conventional foods has increased and has led consumers to purchase organic foods as an alternative (Basha et al., 2015). However, a study conducted by Chandrashekar (2014) reported that not all organic consumers are following and having the same approach or methods toward organic foods. This also depends on socio-demographic characteristics and geographical agricultural practices. Besides, positive consumer attitude does not imply purchasing organic foods all the times. This ideology has revealed other factors interfering consumer attitude and behavior. The intermediate factors have been identified as seasonal availability, price, and socio-demographic characteristics. These factors can also act as positive or negative influences of purchasing organic foods and conventional foods (Analuiza et al., 2020; Chandrashekar, 2014).

One of the main reasons for not purchasing organic foods when compared with conventional foods is the price, because organic food is sold at a higher price than the conventional foods. Therefore, price and quality relationship has been identified as the main biasness of purchase of conventional foods; it makes sense as price is the main sensitive factor for consumers of conventional food (Analuiza et al., 2020; Chekima et al., 2019; Hwang and Chung, 2019). However, when focusing on the price of organic foods and promoting them, it has already caused some controversial attitudes from consumers even though some retailers try to sell organic foods at a lower price. With that, consumers' impression is getting negative and casting doubts as selling low-priced organic foods. This comes mainly because of the strong affinity with higher price and higher product quality of the organic foods than conventional foods (Hwang and Chung, 2019). In addition, based on the freshness of products, consumers tend to buy local varieties as regional and local foods networks have been introduced and are on the rise. Therefore, many consumers prefer to select their food through specialized stores and direct sales channels such as farmers' market and their farm shops (Janssen, 2018). Overall, the present tendency of the organic food consumers' purchasing behavior has been shown to be socially motivated factors; those factors facilitate more purchasing and advertising of the organic foods than the conventional ones because those factors support the rural community, improve animal welfare, support small family farms and the rural community, and reduce gasoline consumption, thereby, benefiting the environment (Britwum et al., 2021; Hwang and Chung, 2019).

10.3 Organic food production and market

Organic food production implies that the agriculture practices used are void of fertilizers, pesticides, herbicides, chemical fertilizers, synthetics chemical substances, growth hormones, antibiotics, and genetically modified materials (Basha et al., 2015; David et al., 2020) where they represent significant environmental concerns. There are supposedly 2.8 million organic food producers in the world (David et al., 2020; Willer et al., 2020). However, from a historical point of view, at a global scale, the green revolution has been increasing the global level production of organic foods. Also, there are some barriers that have arisen related to these productions such as land availability, water, and lack of access to capital and labor availability (Jouzi et al., 2017).

The worldwide production of organic food has extended up to 186 countries by the end of 2018, but only 103 countries already have organic food-based regulations (Willer et al., 2020). However, when compared with the latest information about organic agricultural land (71.5 million hectares) with last available information (69.8 million hectares), it is a 2.9% increment of use of lands with a lack of available information from countries such as Brazil and India as they have very large organic farming areas. Australia holds the highest reported organic agricultural lands (35.7 million hectares) followed by Argentina (3.6 million hectares) and China (3.1 million hectares). Regionally, Oceania (36 million hectares) and Europe (15.6 million hectares) hold the largest areas for the organic agricultural lands—it is around 72% altogether (50% and 22%, respectively), whereas Latin America (8 million hectares), Asia (6.5 million hectares), North America (3.3 million hectares), and Africa (2 million hectares) share 28% together. United States of America, Germany, and France were reported as the highest market for Organic foods, and India, Uganda, and Ethiopia were the countries with the top three highest producers. Also, among the producers, 47% contribute from Asia, followed by 28% from Africa, 15% from Europe, 8% from Latin America, and 1% each from North America and Oceania (Willer et al., 2020).

According to the survey about retail sales of the Research Institute of Organic Agriculture (FiBL), the United States of America is the highest single market with 40.6 billion Euros followed by Germany with 10.9 billion Euros, France with 9.1 billion Euros, and China 8.1 billion Euros for organic foods. When considering per capita consumption, Denmark and Switzerland each shared 312 Euros and followed by Sweden (231 Euros), Luxembourg (221 Euros), and Austria (205 Euros). Also, the highest share of the total organic food market is from Denmark (11.5%), followed by 9.9% from Switzerland, 9.6% from Sweden, 8.9% from Austria, and 8% from Luxembourg (Willer et al., 2020).

10.3.1 Farming types

Practicing of organic agriculture farming has traditionally existed, where modern practices began in the 1920s as a result of pesticide use that was popularized in 1960s. In the 1970s this conventional farming has been increasing with greater knowledge and developments of technology (Hurtado-Barroso et al., 2017; Kuchler et al., 2020). However, organic farming always depends on its holistic approach including the aspects discussed above by Chekima et al. (2019), Tsion and Steven (2019). Its holistic production management system always promoted and enhanced all its aspects. It emphasized the use of best management practices with adopting local conditions (Hurtado-Barroso et al., 2017). However, up to date, most of the countries are struggling with their soil fertility as it has been deteriorating due to heavy use of chemically synthesized compounds in conventional farm practices. Always, conventional farming focuses on a higher yield with higher productivity and there are several types of agricultural uses in conventional farming such as tilling and mono-cropping with use of chemical pesticides, herbicides, fertilizers and other synthetic chemical substances, etc (Basha et al., 2015; Hurtado-Barroso et al., 2017; Mesnage et al., 2020; Tsion and Steven, 2019). Nevertheless, as conventional agriculture farming has developed well and is still famous, many issues could be seen related to those with increasing global human population along with climate change, food security, social instability, health and environmental aspects, etc. Most of these issues were reported from the developing nations rather than developed nations (Jouzi et al., 2017; Tsion and Steven, 2019). Therefore, in order to address all these issues, people tend to turn toward sustainable agriculture farming before starting of conventional farming (Jouzi et al., 2017).

In this viewpoint, integrated organic farming is able to facilitate consumers' requirements that arise due to conventional farming. Nowadays, the organic food production belongs not only to organic agriculture, but it includes the wild collection, area for beekeeping, and as a nonagricultural area, it consists of forests, aquaculture, and grazing areas as well. From those available areas, 67% is shared by agricultural land and crops, 32% for wild collection, and 0.6% for other organic farming activities (Willer et al., 2020). Thus, agro-ecological, integrated, pest management including organic farming practices have evolved as suggested by many researchers (Jouzi et al., 2017; Tsion and Steven, 2019). Based on regional and climate differences, organic farming might be practiced differently based on local conditions. Therefore, based on these aspects, many studies have been conducted and researchers have proposed organic farming as a solution to avoid negative impacts caused by conventional farming. These holistic management practices would improve long-term environmental sustainability by ensuring the production of organic foods in an environmentally friendly way. This includes organic livestock farming with the awareness of farm animal health, welfare, and ethical standards (Jouzi et al., 2017). Large-scale and small-scale farming practices in organic food production are followed by developed countries, although large scale framing is practiced more frequently. Small-scale farmers' contribution is more crucial for food security. However, still, those farmers also face many serious challenges to adopt completely into organic production (Jouzi et al., 2017).

Available productions in organic farming up to now are listed in Table 10.1. There are many large-scale and small-scale farming practices in organic food productions; developed countries most of the time practice large-scale farming, whereas developing countries, practices small-scale. It has also been revealed as to why nearly 75% of organic farmers are from

TABLE 10.1 Organic farming products in the world (Willer et al., 2020).

Farming and land use types with product group	Products
Arable crops	Cereals, dry pulses, flowers and ornamentals plants, fresh vegetables and melons, hops, industrial crops, medicinal and aromatic plants, mushrooms, oil seeds, root crops, seeds and seedlings, strawberries, sugarcane, tobacco, and textile crops
Permanent crops	Citrus fruits, cocoa, coconut, coffee, berries, fruits (tropical, subtropical and temperate), grapes, nuts, nurseries and aromatic plants, olives, tea, and other permanent crops
Wild collection and beekeeping areas	Apiculture, wild berries, wild cereals, forest honey, wild fruit, wild medicinal and aromatic farms, wild mushrooms, wild nuts, wild oil plants, palm sugar, wild palm trees, wild rosehips, wild bamboo, wild rice, wild palmito, wild baobab, wild garlic, wild aloe vera, natural gums, beehives, seaweeds, and other wild collections
Aquaculture	Mussels, salmons, carps, sturgeon, rainbow trouts, trouts, aquatic plants, seabream, shrimps, sea bass, oysters, bream, and bass
Organic live stock	Bovine animals, sheep, pigs, and poultry

developing countries. Though organic farming conducted by small-scale farmers are a majority, those farmers still face many serious challenges to totally adapt to organic production. Also, small-scale farmers' contribution is more crucial in food security aspects (Asian et al., 2019; Jouzi et al., 2017). Furthermore, other challenges for farmers in developing countries are their lack of awareness of benefits of organic production, less sense on the pricing of the products, inadequate storages and infra-structure facilities, and nature of the product pertaining to its perishability (Asian et al., 2019).

In addition to using land in organic farming, there is another method called hydroponic production. This technique utilizes the growth of terrestrial plants with their roots in nutrient-dissolved water or water with dissolved fertilizers without soil. This is considered a solution for organic farming where the places or regions have less water and land availability (Gomiero, 2018). This can also be considered as small-scale practicing of organic farming although these areas are still under study and development. Therefore, this method is worthy and suitable for urbanized areas. However, some debates are there on conventional and organic hydroponics mainly in the certification process, as it is not a holistic approach satisfying the definition of organic farming with typical greenhouses. Also, the practice is not responsible for biological conservation, ecological balance, etc. Thus, based on those aspects, some countries are not allowed to sell organic hydroponic products, especially in the EU (Gomiero, 2018).

When considering farm animals, they are expected to be fed with feed that are organically certified, organically farmed and those without any animal by-products or treated with any hormones or antibiotics, stocking density should be followed-up according to the recommended standards for each animal group along with access to open air (Ashaolu and Ashaolu, 2020; Gomiero, 2018). However, under certain conditions, some regulation bodies and countries, such as the United States of America, allow the use of antibiotics as growth promoters; synthetic somatotropin as a growth hormone is allowed in the United States, whereas other countries do not allow its use. In addition, some regulations depend on the region. For instance, the EU recommendations slightly differ from those in the United States (Gomiero, 2018).

10.3.2 Retail marketing aspects of organic food

The consumer attitudes including increased health benefits, awareness, and government encouragements boost the higher demand for organic foods. More importantly, the whole organic system is based on the triangle which includes demand, production, and the market (Kuchler et al., 2020; Peng, 2019). Therefore, focus on marketing aspects has been increasing because of an accelerated growing trend in organic farming with higher purchase demand (Peng, 2019). Globally, the United States of America has the largest market for organic production (Peng, 2019; Willer et al., 2020). Also, to support the organic food market, USDA National Organic Program (NOP) was established in 2002 and it comprises specific requirements based on the method of organic production (Britwum et al., 2021). Though organic food labels are the consumers' main attention toward the purchase of organic foods, more often than none, retail consumers are confused by labeling foods labeled as *natural*. In addition, food suppliers do not need to meet any of this if they sell food without specificity and only they need it, if they claim any organic foods under the USDA claim. USDA-certified organic foods are grown and processed according to federal guidelines addressing, among many factors, soil quality, animal raising practices, pest and weed control, and use of additives. Organic producers rely on natural substances and physical, mechanical, or biologically based farming methods to the fullest extent possible. By USDA definitions, a produce can be called organic if it is certified to have grown on soil that had no prohibited substances (synthetic fertilizers and pesticides) applied for 3 years prior to harvest. In instances when a grower has to use a synthetic substance to achieve a specific purpose, the substance must first be approved according to criteria that examine its effects on human health and the environment (Kuchler et al., 2020).

Consumers' response is important on organic food marketing and it is tightly connected with the shape of the world economy (Hwang and Chung, 2019; Peng, 2019). Currently, the outbreak of COVID-19 may directly affect the organic food market. Nevertheless, the United States of America and the European financial crisis have significantly affected the global economy and it creates uncertainty (Peng, 2019). Since November 2019, COVID-19 has increased this uncertainty worldwide, affecting all the sectors including the organic market. Hence, such extreme conditions may slowdown the organic market. Therefore, this kind of downtrend would bring many challenges and it is important that balance between consumer and organic food industry is maintained.

To avoid the organic food downtrend, frozen foods, catering services, candies, baby food products, raw materials for processed food, and convenient foods would provide more attraction, rapid boost, and encouragement in order to have a higher market share. Besides, these approaches would be more attractive toward organic food for the next generation of consumers with a highly educated and a high-income community. Based on those aspects and consumers' demand, specialized supermarket chains have already established including discounts stores. Those would provide a pathway to continue the expansion of the organic food industry by overcoming challenges in the organic food market even under the downtrend of economic conditions (Peng, 2019). Nevertheless, searching on the web by consumers has been increasing including the online purchase. It does not mean online purchase is the only influence but it has also increased purchasing of natural food. Therefore, online platforms and web searching methods have affected organic foods selling in retails markets (Kuchler et al., 2020). However, modern media approaches by using documentary films and media images toward marketing aspects of organic foods revealed that modern media approach has influenced consumer behavior providing economic benefits for the marketers and their firms (Ma et al., 2020).

Some studies have suggested a different set of recommendations such as less costly production to give maximum value to the purchasing and marketing of the green concept; these would effectively enable sales, in order to attract more consumers for a lengthier period of time. In addition, these implementations would also be helpful for the policymakers making consumers rights, safety, and health including the environmental concerns of paramount importance (Nguyen et al., 2019). In addition to that, there are many aspects that are available especially in organic food production based on marketing aspects such as food inflation, standards, and certification (Peng, 2019).

10.4 Impact and benefits of organic food

Existing evidence to date cannot support the view that organic products have a higher nutritional quality than conventional ones, because studies concerning the health effects of organic foods in humans are still scarce and sometimes inconclusive. Apart from the direct effects attributed to organic products, it is noteworthy that other factors may play a role in defining the beneficial effect of consuming organic products. Interestingly, there is evidence that people who purchase organic foods generally have a healthier lifestyle, including a healthier diet, than those who do not buy them, thus potentially reducing the risk of several major diseases, independently of the potential additional health effects brought by organic foods. Indeed, comparisons of diets containing organic and nonorganic foods have been challenged because consumers of organic food most frequently maintain a healthy lifestyle. For these reasons, there is still insufficient evidence to recommend organic over conventional products, at least from a nutritional point of view (Dall'Asta et al., 2020). There have been no significant differences in most nutrients, with the exception of higher nitrogen content in conventional produce, and a higher titratable acidity and phosphorus in organic produce. Better quality research that accounts for the many confounding variables is needed to elucidate potential differences in nutrients and the clinical importance of nutrients that may be different (Forman and Silverstein, 2012).

10.4.1 Nutritional composition

Here we are discussing nutritional composition and aspects of organic foods over conventional foods. There is plenty of research conducted and therefore more attention has arisen based on those aspects. Nutrition composition is one main aspect of consumers and this reflects organic standards and it enhances their belief toward organic foods (Lairon, 2010; Popa et al., 2019; Zheng et al., 2019). Organic farming does not use any chemical fertilizers and only uses natural fertilizers, which comprise animal and green manure that are created in overall less nutrient content compared to conventional foods. Nonetheless, when compared with organic husbandry with the conventional ones, animals in organic husbandry receive less concentrated feeds, and often, they are allowed to graze outdoors (Popa et al., 2019). Also, organic products have lower calories with high palatability including products such as fruits and vegetable, milk and dairy foods, and meat (Dall'Asta et al., 2020). Moreover, organic foods are rich with macro- and micronutrients, vitamins, antioxidants, and other phyto-micronutrients. This is due to the high nutrition value of the organic foods (Lairon, 2010; Mesnage et al., 2020; Popa et al., 2019; Zheng et al., 2019). Overall, since they are rich in beneficial nutrients with less insecticides or chemically synthesized compounds, people have a lower risk of obesity, allergies, and noncommunicable diseases. Also, most of the organic vegetables are less in nitrogen content and rich with the dry matter, vitamins, and secondary metabolites (Lairon, 2010; Popa et al., 2019). Further, no significant difference could be seen majorly between organic and conventional foods except some controlled conditions (Dall'Asta et al., 2020; Lairon, 2010; Popa et al., 2019).

Besides, there are some other factors responsible for the change of nutritional compositions, especially for the vegetables, where the composition of nutrition can be changed according to the maturity levels at the harvesting stages (Mesnage et al., 2020). Mainly, among the fruits, it directly depends on the ripening and it also affects the change of production of antioxidants and phenolic compositions (Mesnage et al., 2020; Popa et al., 2019). In addition to that, seasonal changes, genetic variation, environmental factors (light, temperature, altitude, soil type, soil pH, and air pollution), and nature of raw materials of feeds affect the nutritional values (Mesnage et al., 2020; Popa et al., 2019).

10.4.1.1 Macronutrients

10.4.1.1.1 Carbohydrates

Carbohydrates are the main energy source of humans (Popa et al., 2019). However, when compared with the composition of the carbohydrates in organic foods, it has not reported a higher level than conventional products except for a few products such as potatoes (Zheng et al., 2019). Organic pasta has been reported with a low level of carbohydrates (Dall'Asta et al., 2020). Nevertheless, the composition of carbohydrates for organic products is limited and we need further data to come into a convincing comparison with conventional foods.

10.4.1.1.2 Proteins

Most of the studies reported a significant difference between organic and conventional products including fruits, vegetable, and animals. Also, some studies reported no significant effect on the protein and amino acid contents based on the farming methods or husbandry methods (Lairon, 2010; Zheng et al., 2019), but the grain has been recorded with a higher amount of presence of essential amino acids than conventional products (Popa et al., 2019). In addition, some conventionally produced pasta, rice, and other cereals have been reported with higher values than the organically produced ones. Other results have also shown that the organically produced jams, chocolate spreads, and honey have enhanced protein content than conventional products (Dall'Asta et al., 2020). Nevertheless, many studies have recorded that the highest amount of proteins are present in conventional products compared to organic products and conventionally produced pasta, rice, and other cereals have been reported with higher values than the organically produced ones (Hurtado-Barroso et al., 2017).

10.4.1.1.3 Fats, fatty acids

When considering the fats and fatty acid composition, organically produced animals have shown that their products have a higher presence of polyunsaturated fatty acids (PUFAs) (Lairon, 2010). In addition, some studies have reported differences in meat products such as chicken consisting of different fats or fatty acid profile. Egg products only show a significant difference of fats for egg yolk, whereas white and albumen fats were similar for both organically produced and conventionally produced eggs. However, the composition of PUFA or monounsaturated fatty acids (MUFAs) did not show any significant difference. Also, the presences of PUFA and MUFA reported at a higher level in organically produced milk and fruits also contained higher fat contents than the conventionally produced ones (Forman and Silverstein, 2012; Gomiero, 2018; Hurtado-Barroso et al., 2017; Popa et al., 2019; Zheng et al., 2019). The reason behind the different fatty acid compositions of organically produced products is due to the influence of animal feeds directly affecting the composition of fats in milk, eggs, and meat. Also, the presence of PUFA including essential fatty acids provides better health benefits to human health. Also, presence of omega-3 fatty acids provides similar health benefits as plants oil and fish oils that are rich with fatty acids like linoleic, palmitoleic, γ-linolenic, and docosapentaenoic acids. They are also present at a higher level as other beneficial fatty acids in organically produced products (Bavec et al., 2019; Forman and Silverstein, 2012; Popa et al., 2019; Vigar et al., 2019).

10.4.1.2 Micronutrients

10.4.1.2.1 Minerals

Minerals are the most essential nutrients required for maintaining human health. That is the one of the reasons why people are buying more organic foods because they are rich with minerals (Adebayo et al., 2020). Most of the studies have revealed the presence of higher levels of trace elements or minerals in organically produced food than conventionally produced foods (Adebayo et al., 2020; Dall'Asta et al., 2020; Lairon, 2010; Zheng et al., 2019). Also, some studies have reported that there is no effect on the mineral contents, and other nutrients are linked with agricultural management systems (Mesnage et al., 2020; Popa et al., 2019; Zheng et al., 2019).

10.4.1.2.2 Vitamins

Vitamins also play a major role in the health aspects of the human body. Most of the studies have been conducted on ascorbic acid (Vitamin C) and that is one of main key water-soluble vitamins among other vitamins (Lairon, 2010; Popa et al., 2019; Zheng et al., 2019). Some studies revealed that organically grown fruits and vegetables (especially leafy vegetables) had a higher amount of vitamins than conventionally grown products, mainly, Vitamins C and E, whereas some studies have revealed no differences (Forman and Silverstein, 2012; Gomiero, 2018; Hurtado-Barroso et al., 2017). Usually, the vitamins we can typically find in organic foods are E, B1, and B2 (Gomiero, 2018; Lairon, 2010). In addition, for the composition of vitamins, it is not affected by the method of agricultural management, albeit organically produced eggs are to be considered an important source, including milk and meat (Popa et al., 2019). Moreover, apart from the farming management system, composition of vitamin and content may be varied based on yield, maturity, and by the time of year (Zheng et al., 2019).

10.4.1.2.3 Antioxidants

The antioxidant composition can be improved and reached to a high level with organic farming practices (Popa et al., 2019). Mainly, antioxidants are rich in fruits and vegetable as those with higher phenolic, carotenoids, anthocyanins, and flavonoid compounds (Bavec et al., 2019; Forman and Silverstein, 2012; Hurtado-Barroso et al., 2017; Popa et al., 2019; Zheng et al., 2019). In addition, there are some other antioxidants categories such as carotenes, carotenoids, lutein, lycopene, resveratrol, procyanidins (OPC), anthocyanins, ellagic acid, chlorogenic acid, caffeoylquinic acid, hydroxycinnamic acids, tannins, kaempferol, naringenin, punicalagin, isoflavone, myricetin, quercetin, catechin, and epicatechin that are found in both organic foods as well as in conventional products (Zheng et al., 2019). Some studies have also reported organic foods and their products to be richer with antioxidants than conventional foods, whereas some studies have not reported anything as such (Gomiero, 2018; Gustavsen and Hegnes, 2020; Hurtado-Barroso et al., 2017; Zheng et al., 2019). In addition, as antioxidants are sensitive to some factors such as environmental temperatures, the change of composition of antioxidants such as flavonoids and polyphenols depends on the cultivation or farming, harvesting stages, processing methods, and storage conditions (Mesnage et al., 2020; Zheng et al., 2019).

10.4.1.2.4 Phyto-micronutrients

These compounds are abundant mainly in fruits and vegetables, where most of them are secondary metabolites such as resveratrol, polyphenols, and nonprovitaminic carotenoids (Lairon, 2010). These compounds interfere with plant cellular level mechanisms in order to prevent diseases, while some are produced due to external stress (phytoalexins). Same as other vitamins and antioxidants, these also depend on many factors such as maturity of harvest, cultivar, environmental temperature or light intensity, and cropping systems (Lairon, 2010). Basically, these compounds are higher in organic products than conventional products (Dall'Asta et al., 2020; Lairon, 2010; Popa et al., 2019). Some compounds are used in organic farming for pest and disease control purposes using the natural extracts rich with pyrethrins, rotenone, copper salts, and sulfur (Lairon, 2010).

10.4.2 Health benefits

This section mainly focuses on human health and the benefits of organic foods as we discussed in Section 4.1 about nutritional composition. One of the main reasons for getting health benefits from organic foods is the lower level of pesticides and insecticides (Besson et al., 2019; Mesnage et al., 2020). Also, organic food consumers have adapted to a healthier diet plan with reduced processed food, less additives, and plasticizers; additionally, they have a physically active lifestyle (Mesnage et al., 2020; Vigar et al., 2019). Also, consumption of organic food has shown less overweight/obesity among adults, cardiovascular disease, different types of cancer (postmenopausal breast cancer, non-Hodgkin lymphoma, and all lymphomas), atherosclerosis, and diabetes mellitus type 2 (Bavec et al., 2019; Forman and Silverstein, 2012; Gomiero, 2018; Hurtado-Barroso et al., 2017; Mesnage et al., 2020; Mie et al., 2017). During organic animal husbandry, nontherapeutic antibiotics were not used and it reduced the development of drug-resistant organisms in the animals while preventing the colonization of those microbes in the human gut (Forman and Silverstein, 2012; Gomiero, 2018). Also, reduction and prohibition of synthetic chemicals are helpful for the prevention of diseases that were caused by chronic exposure like dermatologic conditions, respiratory problems, depression, endocrine issues, developmental, reproductive issues, hepatotoxicity, neurotoxicity, reproductive toxicity, genotoxic effects, memory disorders, neurologic deficits (Parkinson's disease), miscarriages, improper functioning of the immune system, childhood leukemia or lymphomas [due to occupational exposure during pregnancy or residential use of pesticides during pregnancy of human (autism)], or childhood and birth defects (Bavec et al., 2019; Forman and Silverstein, 2012; Gomiero, 2018; Hurtado-Barroso et al., 2017; Mesnage et al., 2020; Mie et al., 2017). Moreover, organic foods would provide positive health benefits on atopic, allergic, eczema, and asthma symptoms as well as other hypersensitivity diseases (Hurtado-Barroso et al., 2017). These kinds of health benefits are also beneficial to organic food farmers as they are less exposed to synthetics chemical herbicides, pesticides, fungicide, and fertilizers. However, conventional practices have been reported to cause many severe damages caused to farmers including chronic diseases, gene and mitochondrial damages, effects on some signaling pathways, for instance, insulin pathway, and its enhance insulin resistance inside the body and obesity (Bavec et al., 2019; Forman and Silverstein, 2012; Gomiero, 2018; Mie et al., 2017). In addition, conventionally farmed animals using hormone treatments have clearly reported adverse effects on human hormone metabolism. For instance, some hormones like estrogen-treated animal meat and milk have been reported as the main reason for the increasing of early breast cancers and early development of puberty (Forman and Silverstein, 2012; Hurtado-Barroso et al., 2017). However, some studies have demonstrated that estrogen-treated cow's milk was not bad for children and not significantly different concentration was found in both organically and conventionally farmed animals (Forman and Silverstein, 2012; Hurtado-Barroso et al., 2017). Furthermore, it is important that further long-term studies are carried out on children fed with food with hormone-treated animals and how it impacts endocrine disruptions. Moreover, according to Mesnage et al., (2020), as organic dietary patterns remain unproven, its health benefits and other linked lifestyle factors remain largely doubtful.

10.4.3 Environmental concerns

The environment is one of the main concerns of organic farming worldwide. Therefore, consumer concerns, together with an environmentally friendly way of food production, have brought about an organic food concept. Compared with conventional agriculture, organic agriculture provides more benefits such as less damage to soil and groundwater, ecosystem balance, and conservation of biodiversity (Analuiza et al., 2020; Çabuk et al., 2014; Jäger and Weber, 2020; Rao and Ravishankar, 2019). However, in the last few decades, expansion of global agriculture has caused a huge damage to the environment and biodiversity which has led to climate change and global warming events (Hole et al., 2005). The recent findings of organic farming have been reported that more than 30% of biodiversity has been saved by organic farms and its concept has reduced greenhouse gas emissions and increased energy efficiency (Forman and Silverstein, 2012; Jäger and Weber, 2020; Jouzi et al., 2017). Therefore, we can consider that organic farming is the best solution to overcome usual challenges belonging to the agricultural sector where secure food quality and meeting the consumer demand, biodiversity conservation and sustainable use of natural resources, concerns of human health, and wellbeing are all taken care of (Lorenz and Lal, 2016). Also, giving priority to safety including sustainable utilization of an environmentally friendly organic food production promotes consumers' attraction and their willingness to pay more for organic foods (Ashaolu and Ashaolu, 2020; Çabuk et al., 2014). In addition, the organic system would simplify the conventional system as a traditional procedure, which can be used as a cost-effective cultivation technique, enhance nutrient recycling and retention process including water retention capacity, and sustain diverse ecosystems (plants, animals, and insects), saving local terrestrial and aquatic wildlife, easy pest, and disease control and making sure of pollination and balance of ecosystems (Forman and Silverstein, 2012; Lorenz and Lal, 2016).

While it is believed that organic farming provides only positive outputs to the environment, there are some negative impacts that have not been considered by organic food consumers. Therefore, attention should be given to these in order to prevent an imbalance between the ecosystem and human health. Organic farming may cause diseases due to the use of the natural fertilizers and pesticides, animal manures, tillage practices, postharvest residues, and irrigation (Lorenz and Lal,

2016; Reganold and Wachter, 2016). These can change the environmental balance of organic farming and the surrounding environment. However, when compared with conventional farming practices, this is negligible with developing modern organic practices. In conventional farming systems, the addition of chemical fertilizers and relevant synthetic compounds such as pesticides, herbicides, weedicides, fungicides, etc., has reportedly caused damages to environment and consumers also (Rao and Ravishankar, 2019). These are further discussed in Section 10.4.4.

10.4.4 Safety aspects

One of the main purposes of organic farming is health and safety aspects of humans because it is claimed that it promotes less usage of toxins, toxic metabolites, pesticide residues, toxic heavy metals, and chemically synthesized fertilizers. Therefore, as those usages are limited and prevented, some studies showed these factors are also acting as motivating factors among consumers for their organic food preference. Therefore, it leads to low residues of chemically synthesized compounds and less exposure of farmers to those toxic substances as well (Çabuk et al., 2014; Hurtado-Barroso et al., 2017; Mie et al., 2017; Rao and Ravishankar, 2019; Vigar et al., 2019). Many consumers think food being organic is always safe and provides only benefits to them (Çabuk et al., 2014; Ferelli and Micallef, 2019). In this sense, many food scientists believe and disagree organic foods always provide safe and health aspects to consumers (Ashaolu and Ashaolu, 2020). However, many consumers consider many different aspects as important to be considered organic foods as we discussed in Section 10.2.1. These main factors are considered by organic consumers based on their safety and environmental concerns (Britwum et al., 2021).

Absence of pesticide residues is the main reason consumers tend to purchase organic foods. There are many studies that have also been conducted based on this perspective and these investigations are associated with consumer preference toward organic foods (Ferelli and Micallef, 2019). Many toxic pesticides have been reported, such as dichlorodiphenyltrichloroethane (DDT), 2,4-Dichlorophenoxyacetic acid (2,4-D), β-Hexachlorocyclohexane (BHC), hexachlorobenzene, dieldrin, lindane, polychlorinated dibenzofurans (PCDD/FS), polychlorinated biphenyls (PCBs), or polychlorinated-p-dioxins, pyrethroid, and azole (Bavec et al., 2019; Gomiero, 2018; Rao and Ravishankar, 2019). Also, glyphosate and chlormequat were reported as widely used chemical pesticides in conventional farming practices. Nevertheless, some of these compounds can be found in both organic and conventional foods until they get degraded (Bavec et al., 2019; Gomiero, 2018). In addition, most of the pesticide residues are also been reported from the natural environment at a higher level for a longer time and some are also at undetectable levels (Gomiero, 2018; Mie et al., 2017). These residues and substances are identified by the human immune system as xenobiotics; long-term accumulation of nondegradable residues may cause cancer risk as those interacting with other molecules are also responsible for the change in the human gut microbiome (Hurtado-Barroso et al., 2017; Rao and Ravishankar, 2019). Nonetheless, there are some residues often found in organic food such as natural toxins, bromide ion, copper, and spinosad that are of lesser concern (Gomiero, 2018). The USDA - NOP, the European Food Safety Authority (ESFA), EU, Food Safety Modernization Act (FSMA), and Food and Drug Administration (FDA) have recommended and approved the permitted pesticides to use in organic farming in order to minimize and reduce the use of chemically synthetized fertilizers including giving priorities on other safety aspects (Gomiero, 2018; Mie et al., 2017; Rao and Ravishankar, 2019).

Presence of heavy metals is also one of the main concerns of farming and food production industries, as conventional farming uses heavily mineralized fertilizers which lead to concentration and accumulation of heavy metals in soil and food products. Nevertheless, some studies have revealed that organic farming practices showed less accumulation of Cd (cadmium) compared with conventional practices. Normally, Cd comes with contaminated phosphate fertilizers (Ferelli and Micallef, 2019; Gomiero, 2018). The accumulation of heavy metals also depends on the environmental condition in a particular area (contamination level of the environment), physiological pathways, soil pH and oxidation reduction processes, etc (Bavec et al., 2019).

Some toxic byproducts have also been reported to have serious concerns when it comes to safety aspects (Gomiero, 2018; Hurtado-Barroso et al., 2017). Most of those toxic byproducts appear in foods due to exposure to different environmental conditions during transport and storage (Gomiero, 2018; Mie et al., 2017). For instance, mycotoxins are one of the byproducts that are produced by molds and are found mainly in crops, resistant to decomposed or digestion and continue to persist through the food chain (Ferelli and Micallef, 2019; Gomiero, 2018). However, as fungicides are prohibited in organic farming, some arguments on farming management practices lead to an increased risk, while some studies showed there is no difference of contamination between organic and conventional farming practices (Gomiero, 2018; Hurtado-Barroso et al., 2017). Nevertheless, some other toxins such as phytotoxins which plant endogenous toxin higher accumulation, deoxynivalenol (DON), and ochratoxin A (OTA) can be seen (Ferelli and Micallef, 2019; Mie et al., 2017). Altogether, these toxins present in organic food depend on the location, land legacy, use of disease, and infection, they depend on the materials, humidity, plant species or farming animals, harvesting time/stage and year, transport and warehouse condition, and processing methods (Ferelli and Micallef, 2019; Gomiero, 2018; Mie et al., 2017). Studies have also been reported the association of some microbes especially bacteria with plants where they enhance the production of chemotoxins (Ferelli and Micallef, 2019).

Microbial contamination is another concern about organic food farming management on both consumers' and farmers' ends, because organic foods are more susceptible to microbial contamination compared with conventional products (Gomiero, 2018). One of the main microbial contaminations occurs with *E. coli* O157:H7, which is mostly detected from leafy and fresh organic foods, and is also an indicator of fecal bacteria contamination (Ferelli and Micallef, 2019; Gomiero, 2018; Hurtado-Barroso et al., 2017). A lower level of contamination of coliform, total aerobic mesophilic bacteria, yeast, and mold counts was reported much less in organically produced products such as raw milk, while somatic cell and *Staphylococcus aureus* counts were reported as high in some foods (Hurtado-Barroso et al., 2017). In addition, some studies have reported the presence of foodborne pathogens like *E. coli* O157:H7, *Listeria monocytogenes*, *Salmonella enterica* subsp. *enterica* and other *Salmonella* spp., *Campylobacter* spp, *Clostridium botulinum*, and Hepatitis A in both conventional and organic foods (Table 10.2). However, a few studies have reported there is no difference of contamination of microbes in both organic and conventional foods (Ferelli and Micallef, 2019; Hurtado-Barroso et al., 2017; Rao and Ravishankar, 2019). Nevertheless, due to the higher-level microbial contamination, the presence of antibiotic residues and associated antibiotic-resistant microorganisms, and probiotics in animal products also have proven a serious risk for organic food consumers as they are associated with the human gut microbes. It also causes many immunological disorders, hepatotoxicity, and neurotoxicity effects. Only a few studies have been conducted on these aspects and less information is available about direct effects of organic foods on the human gut microbiome. In addition, environmental pollutants and exposures to them are also responsible for changing of gut microbiomes of humans (Ferelli and Micallef, 2019; Forman and Silverstein, 2012; Gomiero, 2018; Hurtado-Barroso et al., 2017; Mesnage et al., 2020). It is important to make sure that pathogenic and other harmful contaminants are less in organic foods during farming, postharvest handling, and at the processing methods including the storage, transport, and selling (Ferelli and Micallef, 2019).

Other common issues related to safety aspects of organic foods occur in the presence of transgenic compounds, hormones (growth hormones, sex steroids, periodic fertilization of animals), and additives (Bavec et al., 2019; Çabuk et al., 2014; Dangi et al., 2020; Hwang and Chung, 2019; Janssen, 2018; Mie et al., 2017; Rao and Ravishankar, 2019). Organic food that is free from artificial additives is also one of the leading concerns among organic food consumers to purchase organic foods (Dangi et al., 2020; Janssen, 2018). The regulations about conventional foods and its production also mentioned the recommended levels including preservatives and coloring agents. However, when it comes to organic production, it is only recommended and implemented as less use of these compounds (Britwum et al., 2021; Forman and Silverstein, 2012; Gomiero, 2018; Mesnage et al., 2020). Also, both animal- and plant-based organic food production have already banned the use of hormones (growth hormones, sex steroids, and periodic fertilization of animals), as well as transgenics compounds like GMO. However, use of these compounds causes severe effects that were not reported and a few pieces of information are available; hence, more long-term effects and negative evidence (endocrine and genetics effects) need to be investigated further as no any clear conclusions are available (Bavec et al., 2019).

10.5 Limitations, gaps, and future research

As emphasized and highlighted many times before in this chapter, organic food is often considered more environmentally friendly than conventionally produced food, since the list of chemical pesticides and synthetic fertilizers allowed for organic farming is certainly much more limited (EFSA, 2016; Gomiero, 2018). Treu et al. (2017) studied the carbon food prints and the land used of the conventional and organic diets in Germany. They observed that the carbon footprints of the average conventional and organic diets are essentially equal, but the average land used in organic diet is approximately 40% more than the average land used in conventional diet. The average conventional diet contains 45% more meat than the average organic diet and both carbon footprints and land use can be reduced by eating less animal-based foods. However, they highlighted the significant uncertainties associated with their study. Therefore, the authors suggested more systematic studies that compare the environmental impacts of conventional and organic food products, in order to support a more robust comparison between conventional and organic diets.

One major problem of contemporary agricultural production is the intensification of cultivation practices, resulting in an inferior quality product in terms of microelements or functional element-based constituents. Another difficulty in assessing the contribution of organic food consumption to a health outcome is the lack of biomarkers, although biomonitoring of the exposure to synthetic pesticides could be used (Mesnage et al., 2020). Although many food policy makers and scientists believe that the total food production in organic farming could be enough to feed the global population (Tscharntke et al.,

TABLE 10.2 Foodborne pathogens found in animal and plant products and contamination sources.

Foodborne pathogens	Food type	Contamination sources	References
Escherichia coli O157:H7	Leafy green vegetables and meat	Pathogen associated with foods, wildlife and livestock, contaminated irrigation water and surface water, use of raw or composted manure, meteorological factors (temperature and elevated precipitation)	Clements and Bihn (2019), Ferelli and Micallef (2019), Hurtado-Barroso et al. (2017), Pradhan et al. (2019)
Salmonella spp.	Eggs, poultry, meat, unpasteurized milk or milk products, fruits and vegetables, and nuts	Wildlife and livestock, contaminated irrigation water and surface water, use of raw, or composted manure	Clements and Bihn (2019), Ferelli and Micallef (2019), Gomiero (2018), Pradhan et al. (2019)
Campylobacter spp.	Poultry, meat, unpasteurized milk or milk products, fruits and vegetables, and raw or undercooked foods	Wildlife and livestock, contaminated irrigation water and surface water, use of raw, or composted manure	Clements and Bihn (2019)
Listeria monocytogenes	Ready-to-eat deli meats, unpasteurized milk and dairy products, fresh produce, and smoked seafood	Wildlife and livestock, contaminated irrigation water and surface water, use of raw or composted manure, meteorological factors (elevated precipitation)	Ferelli and Micallef (2019), Pradhan et al. (2019)
Noroviruses	Ready-to-eat foods (salad and other)	Touched by infected person or vomit or feces from an infected person, contaminated irrigation water and surface water	Ferelli and Micallef (2019), Pradhan et al. (2019)
Hepatitis a virus	Raw or undercooked shellfish from contaminated water, raw produce, contaminated drinking water, uncooked foods	Food or water contaminated by stool from an infected person, contaminated irrigation water and surface water, meteorological factors (elevated precipitation)	Ferelli and Micallef (2019), Pradhan et al. (2019)
Toxoplasma gondii	Meat products (pork, lamb, and goat meat)	Contaminated raw or undercooked meat products, contaminated irrigation water, and surface water	Pradhan et al. (2019)

2012; Badgley et al., 2007), low yield in organic farming is one of the most important issues regarding the ability of organic farming to improve food security. Therefore, a higher yield is not the absolute solution to the problem of food insecurity and there are multiple social, political, and economic contributing factors in this regard (Ponisio et al., 2015; Vasilikiotis, 2000).

10.6 Conclusions

It is obvious that organic consumers choose organic foods with purported claims over those without claims. As an implication for marketing and consumer communication, it is known that organic consumers can be a target groups seeking for functional food characteristics in food as well. The communication of functional characteristics and any worries attached to these does not seem to pose a "threat" to favorable buying intentions by organic consumers, not even intensive buyers of organic food with their rather different ideology. Occasional organic buyers, however, might appear more receptive to organic food over conventional food. Policy wise, when the public shifts toward a greener planet, the food industry should implement green policies. The uprising of the green movement is inevitable due to food contamination that is happening around the world. With a growing improvement in people's awareness, green food will be the mainstream and will improve the future health of human populations. Cities around the world are also engaging in food and agriculture practice. Organic food advocates the claim that it is better for the environment than the conventional foods, but there is no clear evidence or justification for this claim. Yet, the organic food market has experienced unprecedented growth over recent years. The reasons for this is hypothesized to be consumers submitting for choosing organics include healthfulness, taste, environment friendliness, safety, and local agriculture support.

References

Adebayo, S.A., Omotesho, K.F., Ajibade, T.B., Adetayo, G.B., 2020. Assessment of consumers' behaviour towards organic foods in Ilorin Metropolis, Kwara State, Nigeria. Ann. West Univ. Timişoara – Ser. Biol. 23, 3–10.

Alexandratos, N., Bruinsma, J., 2012. World Agriculture towards 2030/2050: The 2012 Revision. ESA Work. Pap, vol. 3.

Analuiza, J.C.C., Checa Morales, C., Perea, J., 2020. Consumer's perceptions of organic foods in Ambato, Ecuador. Esic Mark. Econ. Bus. J. 51, 263–279. https://doi.org/10.2139/ssrn.3609943.

Apaolaza, V., Hartmann, P., D'Souza, C., López, C.M., 2018. Eat organic – feel good? The relationship between organic food consumption, health concern and subjective wellbeing. Food Qual. Prefer. 63, 51–62. https://doi.org/10.1016/j.foodqual.2017.07.011.

Ashaolu, T.J., Ashaolu, J.O., 2020. Perspectives on the trends, challenges and benefits of green, smart and organic (GSO) foods. Int. J. Gastron. Food Sci. 22, 1–7. https://doi.org/10.1016/j.ijgfs.2020.100273.

Asian, S., Hafezalkotob, A., John, J.J., 2019. Sharing economy in organic food supply chains: a pathway to sustainable development. Int. J. Prod. Econ. 218, 322–338. https://doi.org/10.1016/j.ijpe.2019.06.010.

Asif, M., Xuhui, W., Nasiri, A., Ayyub, S., 2018. Determinant factors influencing organic food purchase intention and the moderating role of awareness: a comparative analysis. Food Qual. Prefer. 63, 144–150. https://doi.org/10.1016/j.foodqual.2017.08.006.

Azadi, H., Schoonbeek, S., Mahmoudi, H., Derudder, B., De Maeyer, P., Witlox, F., 2011. Organic agriculture and sustainable food production system: main potentials. Agric. Ecosyst. Environ. 144, 92.

Badgley, C., Moghtader, J., Quintero, E., Zakem, E., Chappell, M.J., Aviles-Vazquez, K., Perfecto, I., 2007. Organic agriculture and the global food supply. Renew. Agric. Food Syst. 22 (2), 86–108.

Basha, M.B., Mason, C., Shamsudin, M.F., Hussain, H.I., Salem, M.A., 2015. Consumers attitude towards organic food. Procedia Econ. Financ. 31, 444–452. https://doi.org/10.1016/s2212-5671(15)01219-8.

Bavec, M., Bavec, F., Bavec, A., Robačer, M., 2019. Healthy facts of organic. Food. Biomed. J. Sci. Tech. Res. 20, 14802–14805. https://doi.org/10.26717/bjstr.2019.20.003403.

Belz, F.-M., Peattie, K., 2009. Sustainability Marketing. A Global Perspective. Wiley, Chichester.

Besson, T., Lalot, F., Bochard, N., Flaudias, V., Zerhouni, O., 2019. The calories underestimation of "organic" food: exploring the impact of implicit evaluations. Appetite 137, 134–144. https://doi.org/10.1016/j.appet.2019.02.019.

Britwum, K., Bernard, J.C., Albrecht, S.E., 2021. Does importance influence confidence in organic food attributes? Food Qual. Prefer. 87. https://doi.org/10.1016/j.foodqual.2020.104056.

Butlera, G., Stergiadis, S., 2020. Organic milk: does it confer health benefits? | Elsevier enhanced reader. In: Givens, D.I. (Ed.), Milk and Dairy Foods: Their Functionality in Human Health and Disease. Academic Press, pp. 121–143. https://doi.org/10.1016/B978-0-12-815603-2.00005-X.

Çabuk, S., Tanrikulu, C., Gelibolu, L., 2014. Understanding organic food consumption: attitude as a mediator. Int. J. Consum. Stud. 38, 337–345. https://doi.org/10.1111/ijcs.12094.

Chandrashekar, H.M., 2014. Consumers perception towards organic products-a study in Mysore City. Int. J. Res. Bus. Stud. Manag. 1, 52–67.

Chekima, B., Chekima, K., Chekima, K., 2019. Understanding factors underlying actual consumption of organic food: the moderating effect of future orientation. Food Qual. Prefer. 74, 49–58. https://doi.org/10.1016/j.foodqual.2018.12.010.

Clements, D.P., Bihn, E.A., 2019. The impact of food safety training on the adoption of good agricultural practices on farms. In: Biswas, D., Micallef, S.A. (Eds.), Safety and Practice for Organic Food. Elsevier, pp. 321–344. https://doi.org/10.1016/B978-0-12-812060-6.00016-7.

Dall'Asta, M., Angelino, D., Pellegrini, N., Martini, D., 2020. The nutritional quality of organic and conventional food products sold in Italy: results from the food labelling of Italian products (FLIP) study. Nutrients 12, 1–13. https://doi.org/10.3390/nu12051273.

Dangi, N., Narula, S.A., Gupta, S.K., 2020. Influences on purchase intentions of organic food consumers in an emerging economy. J. Asia Bus. Stud. 1–22. https://doi.org/10.1108/JABS-12-2019-0364.

David, A., Ahmed, R.R., Ganeshkumar, C., Sankar, J.G., 2020. Consumer purchasing process of organic food product: an empirical analysis. Food Saf. Manag. 21, 128–132.

Denver, S., Christensen, T., 2015. Organic food and health concerns: a dietary approach using observed data. NJAS Wagening J. Life Sci. 74–75, 9–15. https://doi.org/10.1016/j.njas.2015.05.001.

EFSA, 2016. The 2014 European Union Report on Pesticide Residues in Food European Food Safety Authority. Parma, Italy. https://www.efsa.europa.eu/en/efsajournal/pub/4611.

FAO, 2014. The State of Food Insecurity in the World. FAO, IFAD and WFP, Rome.

Ferelli, A.M.C., Micallef, S.A., 2019. Food safety risks and issues associated with farming and handling practices for organic certified fresh produce. In: Biswas, D., Micallef, S.A. (Eds.), Safety and Practice for Organic Food. Elsevier, pp. 151–180. https://doi.org/10.1016/B978-0-12-812060-6.00007-6.

Forman, J., Silverstein, J., 2012. Organic foods: health and environmental advantages and disadvantages. Pediatrics 130, 1406–1415. https://doi.org/10.1542/peds.2012-2579.

Godfray, H.C.J., Beddington, J.R., Crute, I.R., Haddad, L., Lawrence, D., Muir, J.F., et al., 2010. Food security: the challenge of feeding 9 billion people. Science 327 (5967), 812–818.

Gomiero, T., 2018. Food quality assessment in organic vs. conventional agricultural produce: findings and issues. Appl. Soil Ecol. 123, 714–728. https://doi.org/10.1016/j.apsoil.2017.10.014.

Gustavsen, G.W., Hegnes, A.W., 2020. Individuals' personality and consumption of organic food. J. Clean. Prod. 245, 1–9. https://doi.org/10.1016/j.jclepro.2019.118772.

Hansen, T., Sørensen, M.I., Eriksen, M.L.R., 2018. How the interplay between consumer motivations and values influences organic food identity and behavior. Food Pol. 74, 39–52. https://doi.org/10.1016/j.foodpol.2017.11.003.

Harrison, R., Newholm, T., Shaw, D. (Eds.), 2005. The Ethical Consumer. Sage, Los Angeles.

Hole, D.G., Perkins, A.J., Wilson, J.D., Alexander, I.H., Grice, P.V., Evans, A.D., 2005. Does organic farming benefit biodiversity? Biol. Conserv. 122, 113–130. https://doi.org/10.1016/j.biocon.2004.07.018.

Hurtado-Barroso, S., Tresserra-Rimbau, A., Vallverdú-Queralt, A., Lamuela-Raventós, R.M., 2017. Organic food and the impact on human health. Crit. Rev. Food Sci. Nutr. 59, 704–714. https://doi.org/10.1080/10408398.2017.1394815.

Hwang, J., Chung, J.E., 2019. What drives consumers to certain retailers for organic food purchase: the role of fit for consumers' retail store preference. J. Retailing Consum. Serv. 47, 293–306. https://doi.org/10.1016/j.jretconser.2018.12.005.

Jäger, A.K., Weber, A., 2020. Can you believe it? The effects of benefit type versus construal level on advertisement credibility and purchase intention for organic food. J. Clean. Prod. 257, 1–12. https://doi.org/10.1016/j.jclepro.2020.120543.

Janssen, M., 2018. Determinants of organic food purchases: evidence from household panel data. Food Qual. Prefer. 68, 19–28. https://doi.org/10.1016/j.foodqual.2018.02.002.

Jouzi, Z., Azadi, H., Taheri, F., Zarafshani, K., Gebrehiwot, K., Van Passel, S., Lebailly, P., 2017. Organic farming and small-scale farmers: main opportunities and challenges. Ecol. Econ. 132, 144–154. https://doi.org/10.1016/j.ecolecon.2016.10.016.

Kamenidou, I., Stavrianea, A., Bara, E.-Z., 2020. Generational differences toward organic food behavior: insights from five generational cohorts. Sustainability 12, 1–25. https://doi.org/10.3390/su12062299.

Kuchler, F., Bowman, M., Sweitzer, M., Greene, C., 2020. Evidence from retail food markets that consumers are confused by natural and organic food labels. J. Consum. Pol. 43, 379–395. https://doi.org/10.1007/s10603-018-9396-x.

Lairon, D., 2010. Nutritional quality and safety of organic food. A review. Agron. Sustain. Dev. 30, 33–41. https://doi.org/10.1051/agro/2009019.

Lorenz, K., Lal, R., 2016. Environmental impact of organic agriculture. In: Sparks, D.L. (Ed.), Advances in Agronomy. Academic Press Inc., pp. 99–152. https://doi.org/10.1016/bs.agron.2016.05.003

Ma, J., Seenivasan, S., Yan, B., 2020. Media influences on consumption trends: effects of the film Food, Inc. on organic food sales in the U.S. Int. J. Res. Market. 37, 320–335. https://doi.org/10.1016/j.ijresmar.2019.08.004.

Menrad, K., 2003. Market and marketing of functional food in Europe. J. Food Eng. 56 (2–3), 181–188.

Mesnage, R., Tsakiris, I.N., Antoniou, M.N., Tsatsakis, A., 2020. Limitations in the evidential basis supporting health benefits from a decreased exposure to pesticides through organic food consumption. Curr. Opin. Toxicol. 19, 50–55. https://doi.org/10.1016/j.cotox.2019.11.003.

Mie, A., Andersen, H.R., Gunnarsson, S., Kahl, J., Kesse-Guyot, E., Rembiałkowska, E., Quaglio, G., Grandjean, P., 2017. Human health implications of organic food and organic agriculture: a comprehensive review. Environ. Heal. A Glob. Access Sci. Source 16, 1–22. https://doi.org/10.1186/s12940-017-0315-4.

Nguyen, H., Nguyen, N., Nguyen, B., Lobo, A., Vu, P., 2019. Organic food purchases in an emerging market: the influence of consumers' personal factors and green marketing practices of food stores. Int. J. Environ. Res. Publ. Health 16, 1–17. https://doi.org/10.3390/ijerph16061037.

Peng, M., 2019. The growing market of organic foods: impact on the us and global economy. In: Biswas, D., Micallef, S.A. (Eds.), Safety and Practice for Organic Food. Elsevier, pp. 3–22. https://doi.org/10.1016/B978-0-12-812060-6.00001-5.

Ponisio, L.C., M'Gonigle, L.K., Mace, K.C., Palomino, J., de Valpine, P., Kremen, C., 2015. Diversification practices reduce organic to conventional yield gap. Proc. R. Soc. Lond. B Biol. Sci. 282 (1799), 20141396.

Popa, M.E., Mitelut, A.C., Popa, E.E., Stan, A., Popa, V.I., 2019. Organic foods contribution to nutritional quality and value. Trends Food Sci. Technol. 84, 15–18. https://doi.org/10.1016/j.tifs.2018.01.003.

Pradhan, A.K., Pang, H., Mishra, A., 2019. Foodborne disease outbreaks associated with organic foods: animal and plant products. In: Biswas, D., Micallef, S.A. (Eds.), Safety and Practice for Organic Food. Elsevier, pp. 135–150. https://doi.org/10.1016/B978-0-12-812060-6.00006-4.

Rao, A.P., Ravishankar, S., 2019. Alternatives to pest and disease control in preharvest, and washing and processing in postharvest levels for organic produce. In: Biswas, D., Micallef, S.A. (Eds.), Safety and Practice for Organic Food. Elsevier, pp. 213–226. https://doi.org/10.1016/B978-0-12-812060-6.00010-6.

Reganold, J.P., Wachter, J.M., 2016. Organic agriculture in the twenty-first century. Nat. Plants 2, 1–8.

Seiders, K., Petty, R.D., 2004. Obesity and the role of food marketing: a policy analysis of issues and remedies. J. Publ. Pol. Market. 23, 153–169.

Seufert, V., 2012. Organic Agriculture as an Opportunity for Sustainable Agricultural Development. Available at: http://www.mcgill.ca/isid/files/isid/pb_2012_13_seufert.pdf.

Siró, I., Kápolna, E., Kápolna, B., Lugasi, A., 2008. Functional food. Product development, marketing and consumer acceptance – a review. Appetite 51 (3), 456–467.

Treu, H., Nordborg, M., Cederberg, C., Heuer, T., Claupein, E., Hoffmann, H., Berndes, G., 2017. Carbon footprints and land use of conventional and organic diets in Germany. J. Clean. Prod. 161, 127–141.

Tscharntke, T., Clough, Y., Wanger, T.C., Jackson, L., Motzke, I., Perfecto, I., et al., 2012. Global food security, biodiversity conservation and the future of agricultural intensification. Biol. Conserv. 151 (1), 53–59.

Tsion, K., Steven, W., 2019. An overview use and impact of organic and synthetic farm inputs in developed and developing countries. Afr. J. Food Nutr. Sci. 19, 14517–14540. https://doi.org/10.18697/ajfand.86.15825.

Vasilikiotis, C., 2000. Can Organic Farming "Feed the World". ESPM-Division of Insect Biology 201. University of California, Berkeley.

Verbeke, W., 2005. Consumer acceptance of functional foods: socio-demographic, cognitive and attitudinal determinants. Food Qual. Prefer. 16 (1), 45–57.

Vigar, M., Oliver, A., Robinson, L., 2019. A systematic review of organic versus conventional food consumption: is there a measurable benefit on human health? Nutrients 12, 1–32. https://doi.org/10.3390/nu12010007.

WHO, 2007. The Challenge of Obesity in the WHO European Region and the Strategies for Response. World Health Organisation, Copenhagen.

Willer, H., Schlatter, B., Trávníček, J., Kemper, L., Lernoud, J., 2020. The World of Organic Agriculture. Statistics and Emerging Trends 2020., The World of Organic Agriculture. Statistics and Emerging Trends 2020. Research Institute of Organic Agriculture (FiBL), Switzerland.

Zheng, Y., Yu, X., Yang, H., Wang, S., 2019. From a perspective of nutrition: importance of organic foods over conventional counterparts. In: Biswas, D., Micallef, S.A. (Eds.), Safety and Practice for Organic Food. Elsevier, pp. 75–134. https://doi.org/10.1016/B978-0-12-812060-6.00005-2.

Chapter 11

Mycotoxin management: an international challenge

Rebeca López-García
Logre International Food Science Consulting, Mexico

11.1 Introduction

Mycotoxins are a group of diverse toxic secondary fungal metabolites that contaminate cereal grains as well as other agricultural products and are of concern to human and animal health. The total number of mycotoxins is not known, but according to the Council for Agricultural Science and Technology (CAST) on mycotoxins (CAST, 2003; Gruber-Dorninger et al., 2017), toxic fungi metabolites could potentially be in the thousands. However, the number of mycotoxins known to be involved in disease is considerably less. The major classes of mycotoxins of concern are aflatoxins, tricothecenes, fumonisins, zearalenone, ochratoxin A, and ergot alkaloids. However, new mycotoxins are coming into the spotlight. Emerging compounds that include enniantins, beauvericin, sterigmatocystin, penitrem A, cyclopiazonic acid, mycophenolic acid, roquefortine (A,B), emodine, penicillic acid, fusarenon, and citrinin among others may not be currently considered for regulation, but study of their impact on public health may identify new compounds of interest (Gruber-Dorninger, 2017). Consuming foods contaminated with high levels of certain mycotoxins can cause the rapid onset of mycotoxicosis, a severe illness characterized by vomiting, abdominal pain, pulmonary edema, convulsions, coma, and in some rare cases death (Dohlman, 2003). However, lethal cases are not common and in many cases are associated with lack of food security since the food affected would be considered inedible under conditions where food supply is adequate. Among mycotoxins, the most widely recognized risk comes from aflatoxins and particularly aflatoxin B_1 (AFB_1) since it is listed as a known human carcinogen by the International Agency for Research on Cancer (IARC, 1993; IARC, 2012). In addition, aflatoxins are of particular toxicological concern for populations with high incidence of hepatitis B because the rate of liver cancer is up to 60 times greater in people with hepatitis B than in healthy people who are exposed to aflatoxins (Miller, 1996). Other mycotoxins have also been considered as suspected carcinogens (IARC, 1993) and human exposure may be direct through the consumption of contaminated agricultural commodities or indirect through the consumption of animal products from animals that have been fed with contaminated feed. Mycotoxins can be detected in meat, milk, and eggs and can also result in economic losses from animal health and productivity problems (CAST, 2003). The identification of the role of microbiota in health and disease has led to the study of the impact of mycotoxins in the gut. It has been identified that mycotoxins cause perturbation in the gut, particularly the intestinal epithelium. It has been suggested that there is a bidirectional relationship between mycotoxins and gut microbiota, thus involving gut microbiota in the development of mycotoxicosis (Liew and Mohd-Redzwan, 2018).

Climate change is expected to increase the incidence and severity of pests and diseases in crops and might have a significant effect on mycotoxins and their regulations (van Egmond, 2013). In addition, changes in dietary patterns such as the adoption of plant-based diets may pose a significant shift in exposure, and thus, create the need for additional evaluation. Mycotoxin contamination of food and feedstuffs poses food safety and economic concerns and control has great impact on grain and agricultural product trade.

Ensuring Global Food Safety. https://doi.org/10.1016/B978-0-12-816011-4.00024-0

11.2 Mycotoxin regulations

Countries have the legitimate right to protect their consumers from the toxic effects of these naturally occurring compounds. However, setting limits for unavoidable contaminants is not an easy task since it is not always straightforward and minimizing the risk may not be economically feasible. Socio-economic factors such as cost–benefit considerations, trade issues, and sufficiency of food supply are equally important in the decision-making process to come to meaningful regulations (van Egmond, 2013). Therefore, regulatory bodies must continually assess the levels of allowable exposure to humans by using a sound risk assessment process (CAST, 2003). Risk assessment is defined as the scientific evaluation of the probability of occurrence of known or potential adverse health effects resulting from human exposure to foodborne hazards and is the primary scientific basis for the establishment of regulations (FAO, 2004). Its main components include hazard identification, a qualitative indication that a contaminant can cause adverse effects on health that depends on the availability of toxicological data; and hazard characterization, a qualitative and quantitative evaluation of the nature of the adverse effects that depends on the availability of exposure data. In addition to risk assessment, other factors such as availability of adequate sampling and analysis procedures also play an important role in the establishment of regulations (van Egmond, 2013).

In addition to the data used for the risk assessment process, risk management of mycotoxins is affected by other scientific and socio-economic factors such as the distribution of mycotoxin concentrations within a lot, the availability of analytical methods, legislation in other countries that are trading partners, and availability of a sufficient food supply (FAO, 2004). Although the World Trade Organization's Sanitary and Phytosanitary agreement states that standards must be based on sound risk assessments, economic and sociological implications of regulations are also very important since there are diverging perceptions of tolerable health risks and these are largely associated with the level of economic development and the susceptibility of a nation's crops to contamination. The availability of sampling and analytical methods also plays an important role in establishing limits since tolerance levels that do not have a reasonable expectation of being measured may waste resources and condemn products that are perfectly fit for consumption (Smith et al., 1994). The regulatory philosophy may also change in different areas of the World, since it should not jeopardize the availability of basic commodities at reasonable prices. So, in developing countries, the adequate level of protection must take into consideration the amount of food available. If food supplies are already limited, drastic legal measures may cause food shortages and excessive consequences Thus, there is a wide range of varying standards among different national or multilateral agencies (Dohlman, 2003). According to a survey of worldwide mycotoxin regulations by the Food and Agriculture Organization of the United Nations (FAO), the number of countries that have mycotoxin regulations has been steadily increasing (FAO, 2004). In 2003, approximately 100 countries were known to have established specific limits for various combinations of mycotoxins and commodities, often accompanied by prescribed or recommended procedures for sampling and analysis.

The United Nations Environment Programme and the World Health Organization International Program have both declared that humans have a right to food free from mycotoxins that could cause significant risk to human health, animal health, and market access (Logrieco et al., 2018). The European Commission and the Chinese government have paid much attention to mycotoxin contamination due to the increased trade of agricultural products between European and Chinese partners (Logrieco, 2018). The MycoKey Project (integrated and innovative key actions for mycotoxin management in the food and feed chains) is an initiative funded by the European Commission under the Horizon 2020 programme, Societal challenge 2 "Food security, sustainable agriculture and forestry, marine, maritime, and inland water research and bioeconomy challenge" topic "Biological contamination of crops and the food chain." Under this project, 32 partners from Europe, China, Nigeria, and Argentina, including research institutions, industries, and associations, will work together for 4 years, focusing on aflatoxins, deoxynivalenol, zearalenone, ochratoxin A, and fumonisins. The MyToolBox project, another initiative under the Horizon 2020 programme, is a project that facilitates a multiactor partnership to develop novel intervention strategies aimed at achieving up to 90% reduction in crop losses due to mycotoxin contamination for major food and feed crops. Both projects, MycoKey and MyToolBox are expected to provide innovative integrated solutions for sustainable mycotoxin management along numerous food and feed chains in both developed and developing countries (Logrieco et al., 2018).

11.3 Harmonized regulations

Until around the late 1990s, the setting of regulatory limits was mainly a national affair. However, gradually, several economic communities have harmonized their regulations. According to the 2004 FAO report, the following trading blocks have harmonized regulations: Australia/New Zealand, the European Union (EU), MERCOSUR (*Mercado Común del Sur*—Common Market of the South), and the Association of Southeast Asian Nations (ASEAN).

11.3.1 Australia/New Zealand

In 2002, New Zealand and Australia initiated a joint regulatory approach to be codified in the Australia New Zealand Food Standards Code. This transition occurred parallel to an evolution from hazard to risk-based standards. Common limits are now applied for total aflatoxins in peanuts and tree nuts and ergot (the sclerotium of *Claviceps purpurea*, not actually a mycotoxin but a dormant winter form of the fungus containing the ergot alkaloids mycotoxins). The harmonized standards also include limits for phomopsins in lupin seeds and products thereof and for agaric acid in food containing mushrooms and alcoholic beverages. These limits are unique to Australia and New Zealand and have not been reported anywhere else in the World. Although some analytical methods are recommended in import regulations, mycotoxin testing in New Zealand and Australia is based on a performance approach. For this, laboratories must be accredited and use appropriate, suitably validated methods. Ochratoxin A (OTA) and the *Fusarium* toxins (T-2, nivalenol, acetodeoxynivalenol, zearalenone, and fumonisins) were also included in the original risk assessment, but it was concluded that it was premature to establish maximum permitted concentrations in food for these mycotoxins. There have been no subsequent amendments up to date to include these toxins in the harmonized regulations (Cressey, 2008).

11.3.2 European Union

The EU has harmonized regulations for AFB_1 in various feeds since 1976 including official protocols for sampling and analysis. In 1998, harmonized regulations for mycotoxins in foods that included sampling protocols and methods of analysis came into force and have been gradually expanding. In 2004, a significant expansion was initiated for several mycotoxin/food combinations. For foods, harmonized regulations include patulin, AFB_1, aflatoxin M_1, ochratoxin A, and DON (deoxynivalenol) in infant and follow-up formulae; ochratoxin A in coffee, wine, beer, spices, grape juice, cocoa, and cocoa products; several *Fusarium* mycotoxins, i.e., tricothecenes (T-2 and HT-2 toxins in addition to DON), fumonisins and zearalenone, and ochratoxin A.

The EU has increased awareness of the impact of climate change and its impact on mycotoxin production. Thus, the European Food Safety Authority (EFSA) is working to establish appropriate controls to minimize the impact of the presence of mycotoxins. Currently, EFSA is working on comprehensive risk assessments for aflatoxins and ochratoxins in foods. In 2018, EFSA published the last opinion in a series of four evaluating whether it is appropriate to set a group health-based guidance value for mycotoxins and their modified forms. Based on the ongoing risk assessment activities, the results of the MycoKey and MyToolBox projects, the reports from the Rapid Alert System for Food and Feed (RASFF), the EU will continue to add on and improve current controls to enhance public health protection.

11.3.3 MERCOSUR

MERCOSUR includes Argentina, Brazil, Paraguay, and Uruguay. Bolivia, Chile, Colombia, Ecuador, and Peru have associate member status. Venezuela signed a membership agreement on June 17, 2006, but before becoming a full member, its entry must be ratified by the Paraguayan and Brazilian parliaments. A formal process of harmonization of national regulations took place in 1994. Through this process, common maximum limits were established for aflatoxins B_1, B_2, G_1, and G_2 for peanuts and corn and for aflatoxin M_1 in dairy products. These limits included sampling plans and analytical methods for each mycotoxin/commodity combination. The general criteria and guidelines followed for the harmonization of regional mycotoxin regulations include: the definition of priority products in relevance to trade and health relevance; the comparison of national regulations with international standards or regulations; the use of risk assessment data as reference; and the application of equivalency principles and risk analysis. After all the specific steps were taken to develop the final approval of the common regulations for mycotoxin maximum limits, the approved regulations were incorporated into the four members' legal bodies and became mandatory. However, implementation of the regulations may be challenging due to a lack of data that represent all member countries. Up to date, full member countries apply common limits for total aflatoxins in peanuts, maize, and products thereof, and for aflatoxin M_1 in fluid and powdered milk. In addition, each member country can publish its own legislation for other non-harmonized products. Such is the case of Uruguay where maximum levels for other mycotoxins have been established to address specific mycotoxins of concern for the country. Brazil has a proposal on maximum limits for other relevant mycotoxins as well. In special cases, international references can be used (Lindner Schreiner, 2008).

11.3.4 ASEAN

Current member countries include Brunei Darussalam, Cambodia, Indonesia, the Lao People's Democratic Republic, Malaysia, Myanmar, the Philippines, Singapore, Thailand, and Vietnam. Most of these countries have specific regulations for mycotoxins. Harmonized regulations are not yet established. However, an ASEAN Task Force on Codex Alimentarius has taken a common position to support the 0.5 μg/kg level for aflatoxin M_1 in milk. ASEAN Reference Testing Laboratories have been established for mycotoxins as well as pesticide residues, veterinary drugs, microbiology, heavy metals, and genetically modified organisms. In addition, other documents such as the ASEAN Common Food Control Requirements, the ASEAN Common Principles for Food Control Systems, the ASEAN Common Principles and Requirements for the Labeling of Prepackaged Food, and the ASEAN Common Principles and Requirements for Food Hygiene have been finalized and will serve as guiding principles for the ASEAN relevant food bodies (Le Chau, 2006; Anukul et al., 2013). Even when the mycotoxin contamination in food, feed, and foodstuffs in ASEAN has been monitored more closely in the past 10 years, relevant information remains insufficient. Therefore, it is difficult to assess the exposure of the ASEAN population to mycotoxins. An indirect evaluation of the systems is through the study of detentions in export markets using systems such as the European rapid alert system (RASFF). ASEAN countries are still looking for continuing support from national and provincial governments as well as local authorities to encourage and fund activities that contribute to mycotoxin management.

11.3.5 Codex Alimentarius

The Codex Alimentarius Commission (CAC), supported by FAO and the World Health Organization (WHO), aims to facilitate world trade and protect the health of consumers through the development of international standards for foods and feeds. Within the CAC, the Codex Committee on Food Additives and Contaminants derives maximum limits (standards) for additives and contaminants in food, which are decisive in trade conflicts. The Joint FAO/WHO Expert Committee on Food Additives and Contaminants (JECFA) is the scientific body that develops advisory documents on food additives and contaminants for the CAC. JECFA uses the formal risk assessment approach to evaluate contaminants. Information presented to JECFA includes the hazard identification and characterization data. JECFA then evaluates all the toxicological information and gives a Provisional Tolerable Weekly Intake or a Provisional Tolerable Daily Intake. The term provisional is used to express the tentative nature of the evaluation in view of the paucity of reliable data on the consequences of human exposure at levels approaching those with which JECFA is concerned. The evaluation is based on the determination of a No-Observed-Adverse-Effect-Level (NOAEL) in toxicological studies and the application of an uncertainty factor that is calculated using the lowest NOAEL in animal studies and divided by a factor of 100 (10 for the extrapolation from animal studies or interspecies differences and 10 for the variation among different individuals or intraspecies differences). If the availability of toxicological data is inadequate, then a higher safety factor is used. This calculation derives a tolerable intake level. This approach is not considered appropriate for toxins where carcinogenicity is the basis for concern (i.e., aflatoxins). A no-effect concentration limit cannot be established for genotoxic compounds since any small dose will have a proportionally small effect. The imposition of a complete elimination of the carcinogen would be appropriate. However, natural contaminants cannot be eliminated from a food or feed without outlawing the commodity they contaminate. For these cases, JECFA has established the ALARA (As Low as Reasonably Achievable) level. This may be considered as the level where the contaminant has been minimized to an irreducible level. ALARA is formally defined as the concentration of a substance that cannot be eliminated from a food without involving the discard of the food altogether or seriously compromising the food supply.

The Codex Committee on Contaminants (and Toxins) in foods (CCCF) has continued work to establish Maximum Limits (ML) and Codes of Practice (CP) to address mycotoxin contamination, management, and potential impact on public health and international trade. Codex MLs and CPs are essential tools for building a shared global view of acceptable practice. Ideally addressing mycotoxin contamination requires that countries adopt Codex MLs into national legislation and adopt CPs to the local context to facilitate uptake of good practices by value-chain operators. Countries may adopt standards that differ from Codex recommendations if such action can be justified by a risk assessment and if the same level of protection applies to imports as well as local production. Effective regulatory oversight to ensure that foods reaching the market are within established regulatory limits depends on the political will both to develop technical capacities and facilities in the country and to provide the financial resources necessary to run monitoring and surveillance programs (Clarke and Fattori, 2013).

An issue observed during the latest CCCF meetings is the need for data that are more geographically representative in order to establish MLs since data presented for the establishment of new limits have relied heavily on data from a few countries and regions. Delegates noted that careful data analysis was required to prevent the erroneous inclusion of outliers, which may result in overestimation of percentiles, which in turn would result in overly conservative MLs (CCCF, 2019).

11.4 Trade impact of regulations

It is difficult to establish the overall cost of mycotoxin contamination in a consistent and uniform manner since the lack of information on the health costs and other economic losses from mycotoxin-related human illness is not easy to establish due to the complication in exactly determining cause-and-effect relationships between mycotoxins and the chronic diseases they are suspected of causing (Dohlman, 2003). The most information on economic impact is available for the United States but, according to Logrieco et al. (2018), it is reasonable to anticipate that the information is broadly similar to agricultural economies in other developed countries. The cost of mycotoxin contamination to the U.S. economy was estimated to be between $2 and $3 billion per year depending on the year. Testing is a considerable expense with 10%–20% of the projected loss spent to help ensure food safety. Most of the costs of mycotoxin contamination are borne directly by farmers and are only indirectly passed along to consumers (Logrieco et al., 2018). Mycotoxin contamination is associated with economic losses due to their impact on productivity and world trade. Globalization of trade has complicated the regulatory control of mycotoxins since regulatory standards may become bargaining chips in trade negotiations. While developed countries have well-developed infrastructures for monitoring food safety and quality, people in developing countries may not be as protected due to the lack of food safety, quality monitoring, and available resources for an effective enforcement infrastructure in developing countries (Cardwell et al., 2001). In some cases, it has been clear that developing countries have experienced market losses due to persistent mycotoxin problems or the imposition of new, stricter regulations by importing countries (Dohlman, 2003). One study has estimated that crop losses (corn, wheat, and peanuts) from mycotoxin contamination in the United States amount to 932 million dollars annually, in addition to losses averaging 466 million dollars annually from regulatory enforcement, testing, and other quality control measures (CAST, 2003). However, large economic losses due to stricter limits may have greater impact on the major exporters of commodities susceptible to contamination. Trade disputes from mycotoxin contamination arise because it is recognized as an unavoidable risk and there are several factors that influence the level of contamination in cereals and grains that are purely environmental (weather and insect infestation) and, thus, are difficult or impossible to control. In addition, perception of tolerable risks is based largely on the level of economic development and the susceptibility of the nation's crops to contamination (Dohlman, 2003). A study by Wilson and Otsuki (2001) estimated that for a group of 46 countries, including the United States, the adoption of a harmonized standard based on the Codex guidelines would increase trade of cereals and nuts by more than six billion dollars or more than 50% compared with the divergent standards in effect during 1998. However, the EU regulation is among the strictest in the world (4 ng/g total aflatoxins in foods other than peanuts and 15 ng/g in peanuts) with these concentrations well under the harmonized Codex standards (Wu, 2008). When these European harmonized limits were communicated, the international trade community was afraid of the impact of this regulation from an importing commercial block on the economies of the exporting countries. In fact, several studies indicated that the standards could cause severe economic losses in the United States (US), Argentina, and Africa, without any considerable gain in health benefits for European consumers (Otsuki et al., 2001; Wu, 2004). A more recent World Bank study (2005) indicated that Otsuki et al. had overestimated the impact of the EU aflatoxin standard on Africa due to the lack of productivity and other commercial factors that influenced African exports. This trade study showed, however, that the major losses were incurred by the more successful exporters of these commodities, Turkey, Brazil, and Iran. Wu (2008) estimated that these studies do not take into account the multiple stakeholders and price fluctuations that are inherent in adjusting to the EU standard and do not account for the fact that in certain circumstances, a stricter food standard can actually result in economic benefits to high-quality export markets.

On the other hand, there is a well-recognized fear that the imposition of stricter standards in higher value markets could have a more negative impact on the health of the population in the exporting countries that have, in fact, more vulnerable populations. This is because the best commodities are selected for the demanding export market and the lower quality, more contaminated products may stay in the domestic market and eaten by consumers with low income that may be more susceptible due to higher incidence of hepatitis B. Wu estimated (2004) that, in the EU, a change in aflatoxin standards from 20 to 10 ng/g, much less from 10 to 2 ng/g, would likely reduce the risk of mortality by an amount so small that it would not be detected. However, areas with high incidence of hepatitis B and C (China and Sub-Saharan Africa) could be at greater health risk.

Most countries recognize that placing standards on the level of mycotoxin entering the food chain is prudent, but there are still diverging perceptions on how to balance economic costs and health benefits. This issue is especially relevant for countries that lack the means to implement stronger quality control measures. One generalized conclusion of all these studies is that regardless of the level of economic impact calculated using different models, it is important for policymakers to consider the implications of both health and economic outcomes when developing harmonized standards for mycotoxins. The creation of these standards will continue since regulating mycotoxin contamination is a clear goal for the 21st century.

11.5 Technical assistance

The setting of stricter regulatory limits should always be paired with the technical assistance required to minimize the initial risk of mycotoxin contamination and, thus, lessen the likelihood of exceeding the regulatory limits. A World Bank trade analysis (Diaz Rios and Jaffee, 2008) reported that after 6 years of harmonized standards for mycotoxins in Europe, the "lost" trade for the Sub-Saharan countries attributed to the EU standards is very low in contrast with the original estimates. For most of these countries, the EU standards were significant neither as a barrier to trade nor as a catalyst for proactive action since Sub-Saharan Africa's edible groundnut export sector had been gradually losing competitiveness for decades. The stricter EU standards were counteracted by competitive countries with the implementation of control systems. Since there is now better knowledge of the factors associated with the prevention and control of aflatoxin contamination, upgrades to production of susceptible commodities can gradually incorporate Good Agricultural Practices (GAP) and the Hazard Analysis and Critical Control Points (HACCP) systems. These improvements are combined with end-product inspections at different stages of the supply chain, implemented by officials before exporting. The World Bank study (2008) shows that even if the EU had adopted the less stringent Codex standards, Africa's groundnut exports to the EU would not have been higher. Analysis of the notifications reported in the RASFF shows that nearly 80% of African consignments intercepted by EU authorities over 2004–06 would have failed even the Codex standards. If the EU had adopted the Codex standards, however, the major beneficiaries would have been the more competitive groundnut exporters, Argentina, the US, Brazil, China, and Egypt. All these countries have made investments in upgrading their production systems to not only boost quality and productivity, but also achieve better control with subsequent beneficial impact on food safety. However, some of these exporting countries may not have the technical resources to upgrade their systems accordingly. Such has been the case of coffee exporting countries that had to face the more stringent European requirement for Ochratoxin A. In this case, FAO acknowledged the need to support these countries with technical assistance to develop systems that were appropriate to improve the quality of coffee sent to the export market and control the contamination with OTA. The implementation of an Integrated Mycotoxin Management System in Ecuador (Lopez-Garcia et al., 2008) showed that technical assistance is important, but the effectiveness of such systems relies on the ability to reach producers and demonstrate the benefits of applying proper controls. This may be difficult because other economic factors may prevent the producer from getting a higher price for a seemingly higher value product. However, it is important to show producers that the use of proper controls does, in fact, have a beneficial economic impact in the long term, and for them, the benefit may only be future market access.

International cooperation models such as the previously described MycoKey and MyToolBox can set better platforms for understanding mycotoxin contamination, generating scientific data and helping exporting countries to not only access high value markets but also protect local population through the adoption of harmonized standards with upgraded codes of practice.

11.6 Conclusion

The impact of mycotoxin contamination around the world is well-recognized and it is only natural that countries will continue to set regulatory limits to protect public health as well as their access to different markets. However, the setting of these regulations must always be based on sound risk assessment processes combined with the development of adequate sampling and analysis methods. In addition, policymakers should always consider the economic impact of setting different standards. In harmonization processes, it is important to find a delicate equilibrium between the economic impact and an adequate level of protection at all levels. Risk perception and the desired level of protection are most definitely different in countries with different resources and food availability. However, it is the duty of international organizations to protect public health overall, including that of the most vulnerable populations that may not even be aware of the risks associated with mycotoxin contamination. The development and implementation of proper control systems is also important not only to decrease mycotoxin contamination and maintain access to higher value markets but also to achieve higher quality products that will hopefully fetch better value in world trade with the subsequent benefit to producers and health protection of the most vulnerable population.

References

Anukul, N., Vangnai, K., Mahakarnchanakul, W., 2013. Significance of regulation limits in mycotoxin contamination in Asia and risk management programs at the national level. J. Food Drug Anal. 12 (13), 227–241.

Cardwell, K.F., Desjardins, A., Henry, S.H., Munkvold, Robens, J., August 2001. Mycotoxins: The Costs of Achieving Food Security and Food Quality. ASPSnet. American Phytopathological Society. www.apsnet.org/online/feature/mycotoxin/top.html. (Accessed 10 March 2009).

CAST, 2003. Mycotoxin: risks in plant, animal, and human systems. In: Richard, J.L., Payne, G.A. (Eds.), Council for Agricultural Science and Technology Task Force Report No. 139. Iowa, Ames.

CCCF, 2019. Codex Alimentarius Commission. Report of the 13rd Session of the Codex Committee on Contaminants in Foods. Ygyakarta, Indonesia, 29 April–3 May 2019.

Clarke, R., Fattori, V., 2013. Codex standards: a global tool for aflatoxin management. In: Unnevehr, L., Grace, D. (Eds.), Aflatoxins: Finding Solutions for Improved Food Safety, 2020 Vision Focus 20(13). International Food Policy Research Institute (IFPRI), Washington, D.C.

Cressey, P., 2008. Fungal downunder: mycotoxin risk management in New Zealand and Australia. In: Presented at the Fifth World Mycotoxin Forum. 17–18 November 2008. Noordwijk, the Netherlands.

Diaz Rios, L.B., Jaffee, S., 2008. Barrier, Catalyst or Distraction? Standards, Competitiveness, and Africa's Groundnut Exports to Europe. Agriculture and Rural Development Discussion Paper 39 World Bank. http://siteresources.worldbank.org/INTARD/Resources/AflatoxinPaperWEB.pdf. (Accessed 10 March 2009).

Dohlman, E., 2003. Mycotoxin hazards and regulations: impacts on food and animal feed crop trade. In: Buzby, J. (Ed.), International Trade and Food Safety: Economic Theory and Case Studies, Agricultural Economic Report 828. USDA, ERS.

FAO, 2004. Worldwide Regulations for Mycotoxins in Food and Feed in 2003. FAO Food and Nutrition Paper 81. Food and Agriculture Organization of the United Nations, Rome, Italy.

Gruber-Dorninger, C., Novak, B., Nagl, V., Berthiller, F., 2017. Emerging mycotoxins: beyond traditionally determined food contaminants. J. Agric. Food Chem. 65 (33), 7052–7070. https://doi.org/10.1021/acs.jafc.6b03413.

International Agency for Research on Cancer, 1993. Some naturally occurring substances: food items and constituents, heterocyclic aromatic amines and mycotoxins. Summary of data reported and evaluation. IARC (Int. Agency Res. Cancer) Monogr. Eval. Carcinog. Risks Hum. 56. Last updated 08/21/1997. http://monographs.iarc.fr/ENG/Monographs/vol56/volume56.pdf (Accessed March 16, 2009), Lyon, France.

International Agency for Research on Cancer, 2012. Chemical agents and related occupations. A review of human carcinogens. IARC (Int. Agency Res. Cancer) 100. Lyon, France. file:///C:/Users/rebec/Downloads/mono100F-23_new.pdf. (Accessed 28 October 2021).

Le Chau, G., 2006. ASEAN approaches to standardization and conformity assessment procedures and their impact on trade. In: Presented at: Regional Workshop on the Importance of Rules of Origin and Standards in Regional Integration, 26–27 June 2006, Hainan, China.

Liew, W.-P.-P., Mohd-Redzwan, S., 2018. Mycotoxin: its impact on gut health and microbiota. Front. Cell. Infect. Microbiol. 8, 60. https://doi.org/10.3389/fcimb.2018.00060.

Lindner Schreiner, L., 2008. Mycotoxins: regulatory measures in MERCOSUR, the common market of the Southern Cone. In: Presented at the Fifth World Mycotoxin Forum. 1–18 November 2008. Noordwijk, the Netherlands.

Logrieco, A.F., Miller, J.D., Eskola, M., Krska, R., Ayalew, A., Bandyopadhyay, R., Battilani, P., Bhatnagar, D., Chulze, S., De Saeger, S., Li, P., Perrone, G., Poapolathep, A., Rahayu, E.S., Shephard, G.S., Stepman, F., Zhang, H., Leslie, J.F., 2018. The mycotox charter: increasing awareness of, and concerted action for, minimizing mycotoxin exposure worldwide. Toxins 10 (4), 149. https://doi.org/10.3390/toxins10040149. Published online 2018 April 4.

Lopez-Garcia, R., Mallmann, C., Augusto, Pineiro, M., 2008. Design and implementation of an integrated management system for ochratoxin A in the coffee production chain. Food Addit. Contam. 25 (2), 231–240.

Miller, J.D., 1996. Food-borne natural carcinogens: issues and priorities. Afr. Newsl. 6 (Suppl. 1). http://www.ttl.fi/Internet/English/Information/Electronic+journals/African+Newsletter/1996-01+Supplement/06.htm. (Accessed 16 March 2009).

Otsuki, T., Wilson, J.S., Sewadeh, M., 2001. What price precaution? European harmonization of aflatoxin regulations and African groundnut exports. Eur. Rev. Agric. Econ. 28, 263–283.

Smith, J.W., Lewis, C.W., Anderson, H.G., Solomons, G.L., 1994. Mycotoxins in Human and Animal Health. Technical Report. European Commission, Directorate XII. Science, Research and Development, Agro-Industrial Research Division, Brussels, Belgium.

Van Egmond, H., 2013. Mycotoxins: risks, regulations, and European cooperation. Zbornik Matice Srpske za Prirodne Nauke 125, 7–20.

Wilson, J., Otsuki, T., October 2001. Global Trade and Food Safety: Winners and Losers in a Fragmented System. The World Bank. http://www-wds.worldbank.org/external/default/WDSContentServer/IW3P/IB/2001/12/11/000094946_01110204024949/Rendered/PDF/multi0page.pdf. (Accessed 27 February 2009).

World Bank, 2005. Food Safety and Agricultural Health Standards. Challenges and Opportunities for Developing Country Exports. Report No. 31207 (Washington, D.C., USA).

Wu, F., 2004. Mycotoxin risk assessment for the purpose of setting international standards. Environ. Sci. Technol. 38, 4049–4055.

Wu, F., 2008. A tale of two commodities: how EU mycotoxin regulations have affected U.S. tree nut industries. World Mycotoxin J. 1 (1), 95–101.

Chapter 12

Novel food processing technologies and regulatory hurdles

Gustavo V. Barbosa-Cánovas[1], Daniela Bermúdez-Aguirre[1], Beatriz Gonçalves Franco[1], Kezban Candoğan[2] and Ga Young Shin[1]

[1]*Center for Nonthermal Processing of Food, Washington State University, Pullman, WA, United States;* [2]*Faculty of Engineering, Department of Food Engineering, Ankara University, Ankara, Turkey*

12.1 Introduction

The search for new food processing and preservation technologies is more than 100 years old (Lelieveld and Keener, 2007), but in the last 30 years food scientists have accelerated the development of newer technologies capable of producing and maintaining a final food product with fresh-like characteristics. Moreover, the goal has been to achieve higher quality products while also ensuring the microbiological safety of the food.

A number of novel food processing technologies are still under development, some of which have been recognized and approved by regulatory agencies for their use in food processing facilities. These technologies have numerous advantages and have been shown to be very efficient in food processing as they better maintain the sensory and nutritional characteristics of the food, resulting in a final product similar to fresh food.

One promising and innovative nonthermal technology is High Pressure Processing (HPP) that is being used worldwide by industry for the pasteurization and decontamination of specific products such as cold cuts, seafood, fruit, and vegetable juices. HHP in combination with thermal processing received approval by Food and Drug Administration (FDA) in the United States of America for use as a processing alternative to the sterilization of food (NCFST: National Center for Food Safety and Technology, 2009). This process is known as Pressure-Assisted Thermal Sterilization (PATS).

The same year (2009) microwave technology was also approved by the FDA as a sterilization process. In this case, this technology is known as Microwave-Assisted Thermal Sterilization (MATS) (Tang, 2009).

Pulsed electric fields (PEF), a very successful nonthermal technology, is also used by the industry to pasteurize pumpable food like fruit and vegetable juices and milk to name a few. This technology is becoming quite popular in the potato industry to soften the tissue prior to cutting bringing significant energy savings, while reducing oil intake at the time of frying. Other applications of PEF include its use as a predrying technique and as an extraction aide.

Most of the requirements for validation and acceptance of these alternative technologies for food processing have now been set up; however, more work needs to be done to identify optimum processing conditions. It is worth mentioning to validate a technology for sterilization that is necessary to identify mathematical models to accurately predict microbial inactivation, and at the same time, microbiological studies should be conducted that will confirm the required level of inactivation. A series of additional validation experiments must be conducted as well, like toxicological ones, interaction between food and packaging, chemical stability, and eventual harmful effects (Koutchma and Keener, 2015).

Because fast development of new technologies, including those at the lab or pilot plan levels, there is a need (lack) of valuable information related to the processes (for example, standardization of process variables, how to report results from different processes) and all of them should be addressed worldwide.

Global harmonization of food regulations and legislation on novel technologies is very relevant and should be done at the earliest; this will facilitate future actions in this domain by food scientists and regulatory agencies. This harmonization will have important drivers, where the most important one is to offer safe food regardless of the processing technology used. Initial attempts have been made to address the above mentioned issues with the establishment of the Global Harmonization Initiative (GHI) (Lelieveld and Keener, 2007).

Ensuring Global Food Safety. https://doi.org/10.1016/B978-0-12-816011-4.00006-9

It is the case, because the advancement of new technologies, FDA changed the definition of pasteurization where other variables besides time and temperature could be used to reach equivalent levels of microbial inactivation. The new definition states "Any process, treatment, or combination thereof, that is applied to food to reduce the most resistant microorganism(s) of public health significance to a level that is not likely to present a public health risk under normal conditions of distribution and storage" (NACMCF, 2006).

This chapter focuses on existing and upcoming regulations and legislation of some novel food processing technologies such as HPP, PEF, irradiation, and microwave. Future actions around these technologies should be pursued with a global perspective to facilitate standardization of the new processes variables, recommended/allowable doses, how to report results, and how to effectively regulate them for the benefit of food manufacturers, regulatory agencies, and the consumer.

12.2 Novel technologies

The search for alternative methods for food pasteurization and sterilization which would generate a safe product with higher quality nutrient content and sensorial properties prompted food scientists to explore other stress factors for microbial inactivation—those other than heat or heat applied in a different way compared to traditional thermal sterilization methods by conduction and convection. Two very broad fields of food processing technologies are currently under research (Table 12.1), nonthermal technologies, in which the stress factor is different from heat using physical hurdles such as pressure, electromagnetic fields, and sound waves, among others, and novel thermal processing technologies, which mainly use energy generated by microwave and radio frequency. A safe product should also be free of toxic substances and contact of food with certain materials during processing should be avoided (Lelieveld and Keener, 2007).

Thus, careful evaluation and validation of the overall quality (including safety) of food products processed by emerging technologies is an essential requirement before a product can be commercialized.

It is worth mentioning that most novel technologies were first studied as prospective microbial inactivation technologies to improve the safety of food. However, important results in the final characteristics of many food items were also observed: such as intact nutrient content in most of the novel food products; unique sensorial properties like color, texture, and appearance; and formation of new aroma compounds. Thus, the search for microbial inactivation technologies not only yielded the possibility of a safer product, but also improved overall product quality, and provided new ingredients for the development of other novel food products.

Another point worthy of note relates to the connection that novel technologies have with environmental concerns; most of the developed emerging technologies are environmentally friendly and may even offer significant energy savings. These savings are due, in part, to much shorter processing times.

TABLE 12.1 Examples of novel food processing and preservation technologies from around the world.

Nonthermal technologies	Thermal technologies
High pressure processing[a]	Microwave[a]
Pulsed electric fields[a]	Radio frequency[a]
Irradiation[a]	Ohmic heating[a]
Ultraviolet[a]	Inductive heating
Ultrasound	
Cold plasma	
Ozone[a]	
Dense phase carbon dioxide[a]	
Supercritical water	
High pressure homogenization[a]	
Electrolyzed oxidizing water	
Pulsed light	

[a]*In use by the Food Industry.*

Sustainability in an industrial setting is very much associated with Corporate Social Responsibility which is based on social, environmental, and economical pillars, often referred as the triple bottom line or people, planet, and profit (Elkington, 2004). Novel processing techniques are being targeted to achieve these key elements of sustainability, reducing the environmental impact of food processing by reducing waste, minimizing the use of natural resources (e.g., energy and water), and providing safe, nutritious and high-quality products for the consumer.

12.3 Nonthermal technologies

The list of nonthermal technologies under research is quite long (see Table 12.1), the most popular being HPP, which 30 years ago was under scientific investigation in a few research centers worldwide. Today, HPP applications are all around the world not only for use in research, but in industry for a number of purposes, including the processing of a number of food products with high acceptance by the consumer.

HPP was initially used to pasteurize high-acid foods such as jams and jellies, with successful results. Another use that followed was microbial inactivation in packaged foods, as demonstrated in ready-to-eat and deli meat products, where their shelf life was significantly extended. Today, the list of commercially pressurized products includes meat products, dairy products, guacamole, and other dips, sauces, oysters, juices, fruit and vegetable products, smoothies, and more. HPP as a nonthermal pasteurization technology has been approved in the United States by the FDA as well as in Europe under Regulation (EU) 2015/2283 (2015).

The big challenge for food scientists in the past was to use HPP for processing shelf-stable, low-acid foods (i.e., for spore inactivation); many attempts were thus conducted using extremely elevated high pressures (up to 1 GPa). Instead of increasing pressure to nonpractical levels, a different approach was taken, to combine high pressure with a thermal treatment. The heat is coming from two sources, the one generated while compressing the media, the so-called compression heating, and the other by an external source such as hot water. The food product and the compressing media, typically water, are preheated to temperatures in the range of 70−90°C and then, the packaged food products are placed in the pressurizing vessel and treated by high pressure for a few minutes. This combined process is known as PATS (Pressure-Assisted Thermal Sterilization) and indeed sterilization is reached by properly selecting processing conditions. This approach allows to inactivate bacterial spores, viruses, fungal ascospores, prions, as well as to reduce enzyme activity in low-acid foods. PATS allows to process foods at lower temperature and processing times than conventional thermal treatments, and thus, better quality products are obtained.

A PATS Consortium composed of NCFST (now Institute for Food Safety and Health, IFSH); International Product Safety Consultants, Inc.; Avure Technologies, Inc.; Dual Use Science and Technology (DUST); selected industry members; and scientists collaborated in the evaluation of this technology (Stewart et al., 2016). The goal of the PATS Consortium was to develop validation protocols and demonstrate efficacy of PATS for the production of a commercially sterile, low-acid mashed potato product. Studies include qualification of the equipment, product and package, and process performance. Mashed potatoes incorporated with *Clostridium botulinum* inoculated packs were PATS treated, and as a result, sterilization levels were reached. Based on this result, FDA approved PATS as a valid sterilization process (NCFST, 2009; Stewart et al., 2016), a significant, remarkable achievement. Even though PATS offers a number of appealing characteristics including being environmentally friendly, it has not yet been adopted by the food industry due to the high equipment and operational costs. In the future, this technology might be used for high value-added food products.

Over the last few decades, HPP and PEF are probably the two most intensively researched nonthermal technologies. This fact is based on their potential use in various industrial activities such as pasteurization, sterilization, tissue softening, or extraction. Available information for food processors on the benefits of HPP and PEF for food processing has been published and delivered in many venues (Ramaswamy et al., 2004, 2005), informing both the industrial community and the consumer.

Irradiation in the form of gamma energy, high-energy electrons (electron beam irradiation), or X-rays is another nonthermal technology that has been approved for various foods. More than 50 countries used this technology for commercial purposes (NACMCF, 2006). Since 1980, many institutions such as the World Health Organization (WHO), Food and Agriculture (FAO), International Atomic Energy Agency (IAEA), Codex Alimentarius Commission, European Food Safety Authority (EFSA), the US FDA, and Department of Agriculture (USDA) have supported irradiation on foods as a safe technology (Koutchma et al., 2018).

The first use of irradiation dates back to 1958 when it was recognized by the FDA as an additive more than a process, even though irradiation is really just energy applied to food (HPS, 2009). Currently, irradiation facilities are regulated by federal and/or state licensing agencies, and according to the actual dose applied, its effect on inactivating microorganisms, extending product shelf life, or inhibiting postharvest changes in certain food products can vary quite significantly. Another important advantage of food irradiation processing is that it is a cold process which, in many cases, does not significantly alter physico-chemical characters of the treated product. It can be applied to food after its final packaging, without leaving residues to the food (Loaharanu and Ahmed, 1991). The list of irradiated food products is very comprehensive; a few examples are flour, potatoes, spices, pork, fruit, fresh vegetables, poultry, meat, pet food, and specific food items for NASA Space Programs and immune compromised patients. Several "fact sheets" have been published on irradiation (FDA, 1997, 2012; HPS, 2009; Keener, 2009) to show the advantages of this technology and to prove the erroneous belief about radioactivity in foods following irradiation. The use of other terms instead of irradiation, such as cold pasteurization, has been suggested (Ehlermann, 2009). In 2012, the FDA allowed the use of 4.5 kilogray (kGy) as a maximum absorbed dose of ionizing radiation on foods to eliminate foodborne pathogens and extend shelf life.

Other nonthermal technologies under research for microbial inactivation include PEF, ultrasound, cold plasma, ozone, dense phase carbon dioxide, pulsed light, and filtration. PEF technology is probably the most explored on this list as evidenced by the number of products and microorganisms tested with this technology, and the important advances and encouraging results with PEF as an alternative for liquid food pasteurization. This technology has fundamental parameters such as the electric field intensity (10−80 kV/cm), pulse duration (1−100 μs), frequency (up to 10 kHz), to inactivate microorganism. Further nonthermal examples include the use of ultraviolet and ozone for water disinfection; however, they are still under research for opaque liquids (e.g., milk) and liquids with small particulates, representing new challenges for food scientists. Cold plasma has been researched for the effect on the surface of foods such as fruits, vegetables, meats, and powders. A better understanding on how the promising cold plasma process works is in progress, as well as the identification of pros and cons.

12.4 Thermal technologies

In the search for new alternatives to conventional thermal processing of food, the application of heat generated inside the food seems to be a viable option for preserving the quality of products and inactivating microorganisms. These days, microwave is one of the alternative thermal technologies which is receiving significant attention for pasteurization and sterilization.

Microwaves rapidly generate heat throughout the volume of food because of the complete interaction between microwave, polar water molecules, and charged ions in food. Microwaves cause polar water molecules in food to constantly rotate and couple with the electromagnetic field. Molecular friction of water molecules resulting from dipolar rotation can generate heat. Because water constitutes a major portion of most food products, it is the primary component that interacts with microwaves due to its strong dipole rotation. Furthermore, heat can be generated through ionic migration that positive and negative ions of dissolved salts in food interact with the electric field by moving toward the oppositely charged regions of the electrical field and disrupt the hydrogen bonds with water (Tang, 2015).

Washington State University (WSU) developed a technology known as MATS which properly addresses the nonuniform electromagnetic heating by using water as an intermediate heat source. Because heat transfer time from the heating source to the food is short, food quality is excellent after this process. WSU participated and led an important Academic-Industry-Government Consortium in the development and implementation of this technology where the other partners were the National Food Processors Association (NFPA, a consultant on regulatory filing) and scientists of the FDA Low Moisture Food Group. This Consortium submitted in October 2008 a formal petition to FDA to use microwave as a sterilization technology after collecting relevant information to demonstrate the merits of the petition.

Microbial validation was conducted using *Clostridium sporogenes* strain PA 3679 spores as surrogates for *C. botulinum* spores type A and B. The first product selected to validate the technology was mashed potato as a homogeneous, low acid food that was prepacked in trays. The prepacked trays were treated in the WSU semicontinuous microwave system at 45 kW, 915 MHz.

Based on the information provided, FDA accepted in October of 2009 the petition submitted by the Consortium related to the sterilization of mashed potato. A second prepacked food, salmon fillet with sauce as a nonhomogenous food in pouches, was accepted in December 2010 (Tang, 2015).

Other nonconventional thermal technologies under research, development, and implementation are radio frequency, inductive heating, and ohmic heating. All of them are viable options to process foods based on the reports coming from many different sources. All of them, like microwaves, have advantages and disadvantages. One of the challenges facing these technologies is to identify which food products will benefit by using them.

There is no question that the approval of PATS and MATS by FDA to sterilize foods is a turning point in the world of food processing and preservation.

12.5 Legislative issues concerning novel technologies

There is a significant number of worldwide regulations concerning farming and agricultural activities, food formulation and labeling, and of course food production, processing, and preservation technologies. The pressing need for a global regulation of all food processing activities has been ongoing for some time. A new process or product must meet a series of standards and regulations before it can be commercialized.

Depending on the type of product, there is a list of safety regulations that a food item must meet; these are related to the guidelines established in the HACCP (Hazard Analysis Critical Control Point) and GMPs (Good Manufacturing Practices) documents and must be implemented during processing (FDA, 2020). A number of international regulations and guidelines must be followed, such as the recommendations of the World Trade Organization (WTO) to ensure the safety of novel products or processes.

With the creation of WTO in 1995, the search for global harmonization in food safety regulations became an important issue needing to be addressed by many countries. Agreements, such as the Agreement on the Application of Sanitary and Phytosanitary Measures (SPS), together with the Codex Alimentarius Commission, encouraged countries to work toward international legislation on food standards, regulations, and guidelines (Motarjemi et al., 2001).

The WTO is not in charge of developing food safety standards, but it does have the authority to limit and control some food safety actions by using the SPS Agreement and Technical Barriers to Trade (TBT) Agreement, which covers other food safety and food quality issues (Mansour, 2004). The Codex Alimentarius definitely plays a significant role in global harmonization, together with other nongovernmental institutions, not only for novel technologies, but also for other food science issues that require globalization (Newsome, 2007).

12.6 Global harmonization concerning novel technologies

There are many regulations concerning food safety around the world; in most cases, each country has established their own standards for specific food products. These regulations are established to show that a specific food product or ingredient is safe for human consumption or food formulation. However, showing that the same product is also safe in other countries usually requires compliance with different legislations. This is often time consuming and costly, not to mention delays in the food processing chain (Lelieveld and Keener, 2007; Sawyer et al., 2008).

On the other hand, many foodborne outbreaks have been generated because of imported food that has served as a vehicle for pathogens from one country to another, even in countries with strict and modern control systems at borders (Motarjemi et al., 2001). Nevertheless, without international trade of foodstuffs, many products would not be commercialized in specific and remote markets. In addition, travelers are also a source of widespread pathogenic bacteria, as shown by Notermans and Lelieveld (2001); the probability of contracting a foodborne disease is very high in specific countries because of poor general hygiene practices when handling food (e.g., 63% in Egypt).

The main priority at present is to establish regulations and legislation for food safety that would declare that a particular food product is safe regardless of where it is processed (Lelieveld and Keener, 2007; Motarjemi et al., 2001). Global harmonization of these regulations and protocols should be addressed, including a system that monitors and assures that this legislation is followed (Lelieveld and Keener, 2007).

Harmonization, as defined by Horton (1997) "*exists when two or more countries have a common set of requirements in place*." Where according to him, global harmonization of food issues is important but difficult. With harmonization, however, the consumer will benefit greatly in the assurance that a product is of equal quality worldwide; competition between countries in terms of trade will also be more fairly balanced (Motarjemi et al., 2001).

From an economic point of view, food that is regulated under international harmonization protocols would be more economical because of the costs avoided in having to comply with standards that differ between countries (Horton, 1997).

There is a number of internationally recognized institutions that advise on regulation and legislation of foods, such as the *Codex Alimentarius* (a joint FAO/WHO Commission), WTO, EFSA, and the US FDA.

Nevertheless, some countries either do not participate in these organizations (Lelieveld and Keener, 2007) or do not apply the international agreements even if they have been officially accepted by that country. In the United States, the FDA is the main regulatory agency established for the approval of novel food processing technologies and novel products. The FDA is going through the most comprehensive reform in the history of the United States after the adoption of the Food Safety Modernization Act. This law has given the FDA new authorities to regulate the way foods are grown, harvested, and processed. The law grants the FDA a number of new powers, including mandatory recall authority, which the agency has sought for many years. The law was prompted after many reported incidents of foodborne illnesses during the first decade of the 2000s. The objective of these changes is ensuring food safety through prevention of microbial contamination rather than just reacting to the problem after it has already occurred. With new changes in food safety regulations also come new compliance challenges for the food industry (Rasco, 2017).

In the European Union, Regulation EC 258/97 was similarly used to regulate novel foods and novel ingredients for a number of years (Regulation (EC) No 258/97, 1997). According to this regulation, Novel Food is defined as a type of food that does not have a significant history of consumption or is produced by a method that has not previously been used for food. As of January 1, 2018, the new Regulation (EU) 2015/2283 on novel foods is applicable. It repeals and replaces Regulation (EC) No 258/97 and Regulation (EC) No 1852/2001 (an addendum to clarify and expand the scope of Regulation (EC) No 258/97) which were in force until December 31, 2017.

The new regulation improves conditions so that food businesses can easily bring new and innovative foods to the EU market, while maintaining a high level of food safety for European consumers. Some features and improvements of the new regulation include expanded categories of novel foods; establishment of a Union list of authorized novel foods; centralized safety evaluation of the novel foods by EFSA; facilitate the marketing of traditional foods from countries outside the EU; establishment of deadlines for the safety evaluation and authorization procedure approvals; promotion of innovation by granting an individual authorization for 5 years to place on the market a novel food; and any food business operator can place an authorized novel food on the European Union market, provided the authorized conditions of use, labeling requirements, and specifications are respected (Regulation (EU) 2015/2283, 2015).

There are many reported attempts to regulate food in other nations by various organizations with specific standards for food products (Motarjemi et al., 2001), but these are in small geographical zones related to specific foodstuffs. Some of the hurdles in global harmonization encountered in the past by the United States and European Union involve economic factors, mainly due to small producers who are afraid of losses in competing with multinational brands, or who fear that the process would lead to Americanization (Horton, 1997).

A GHI was launched in 2004 by the International Division of the Institute of Food Technologists and the European Federation of Food Science and Technology to eliminate differences in regulations and legislation (Lelieveld and Keener, 2007) and set up a basis for food safety regulations. The GHI has shown important progress as portrayed by Lelieveld (2009). During a workshop organized by GHI in 2007 in Lisbon, Portugal, four working groups discussed global harmonization. One of them, the food preservation group, focused on HPP of foods (Lelieveld, 2009) because of the impact that this novel technology has on food processing around the world. However, without regulations and standards, producing a safe food product using this method would be difficult. Many industries are interested in novel food processing technologies, but with the lack of global regulations to monitor them, investors have doubts about their acceptance by consumers. There is no question that global harmonization on regulations will ensure global food safety and food safety for the consumer (Koutchma et al., 2018).

In 2006, a group of scientists in academia and industry from 32 organizations (mainly in Europe) met together to establish a research consortium, NovelQ, based on the use of novel technologies (De Vries et al., 2007). Their research was mainly focused on HPP and PEF for food pasteurization and sterilization, and the use of cold plasma for surface disinfection, but they also addressed issues related to other novel thermal technologies such as microwave, ohmic heating, and radio frequency. The consortium planned a series of interconnected projects in stages for each technology, from basic food science and kinetics, packaging, consumer perception, development of techniques, to technology transfer and management, which would allow a comprehensive overview of each technology. This consortium developed an ambitious plan to align the development of new food processing technologies following the GHI expectations.

Progress in HPP research in the last 30 years has been remarkable. Hundreds of references can be found on HPP pertaining to not only microbial inactivation, but to food components within a large number of food products, and studies of packaging materials for these products (Health Canada, 2014; USDA, 2012). However, the lack of uniformity in reporting processing conditions continues to be a problem. Reported come-up times in HPP technology, for example, is something noticeably absent in many references, which makes the comparison between treatments difficult. Moreover, the thermal effects during high pressure were not reported by many researchers in early studies (Balasubramaniam et al., 2004) because of the unknown effects of compression heating at the beginning of the technology. Most studies conducted in the first years of HPP research showed important results, but without knowledge of the thermal effects.

Other obstacles in comparing past and present HPP treatments are differences in equipment design, configuration, and operation, effect of natural differences in food composition, process variables, lack of definitions, and ambiguousness in reporting data (Balasubramaniam et al., 2004). Some of the process engineering aspects that should be reported in HPP studies are process time (including come-up time, holding time, decompression time), process pressure, initial temperature, product temperature at process pressure (during compression heating), pressure-transmitting fluid, and packaging material.

Recently, a "Cold Pressure Council" was established in the United States with the mission statement "to lead, facilitate, and promote industry standardization, user education, and consumer awareness of high pressure processing." Among other tasks, this council grants a "High Pressure Certified Seal" that could be incorporated on the product packaging, indicating that the process for the product has been validated (Tonello, 2018).

All of these efforts will establish the basis of global legislation, making safe food products a reality in every aspect, and taking advantage of all the unique features gained from novel food processing technologies.

12.7 Final remarks

Global harmonization of all novel technologies is indeed a requirement that must be addressed at all times. It has been shown that there are still many barriers between countries in commercializing novel products. Regulatory agencies must work diligently with research centers and nongovernmental agencies in following the GHI and help develop the basis of food safety regulations that are the main concerns for emerging technologies. A standardization process from equipment manufacturers is an important starting point which can be addressed according to the type of product and major ingredients in product formulation. The identification of benchmarks for each product according to key characteristics helps to set up the basis for process standardization, which could lead to a barrier-free international commercialization of novel food products and technologies.

References

Balasubramaniam, B., Ting, E.Y., Stewart, C.M., Robbins, J.A., 2004. Recommended laboratory practices for conducting high-pressure microbial inactivation experiments. Innovat. Food Sci. Emerg. Technol. 5, 299–306.

De Vries, H.S.M., Lelieveld, H., Knorr, D., 2007. Consortium researches novel processing methods. Food Technol. 61 (11), 34–39.

Elkington, J., 2004. Enter the triple bottom line. In: Henriques, A., Richardson, J. (Eds.), The Triple Bottom Line, Does It All Add Up? Assessing the Sustainability of Business and CSR. Earthscan Publications Ltd., London, pp. 1–16.

Ehlermann, D.A.E., 2009. The RADURA-terminology and food irradiation. Food Control 20, 526–528.

FDA, 1997. Irradiation in the production, processing and handling of food, Final rule. Food and Drug Administration. Fed. Regist. 62 (232), 64107–64121.

FDA, 2012. Irradiation in the production, processing and handling of food. Food and Drug Administration. Fed. Regist. 77 (231), 71312–71316.

FDA, 2020. Title 21 Food and Drugs. Part 120 Hazard Analysis and Critical Control Point (HACCP) Systems, Subpart 24 Process Controls. e-Code of Federal Regulations. U.S. Government Printing Office, Washington, D.C.

HPS, 2009. Food Irradiation, Health Physics Society Fact Sheet. Retrieved from: https://hps.org/documents/Food_Irradiation_Fact_Sheet.pdf.

Health Canada, 2014. Guidance for Industry on Novelty Determination of High Pressure Processing (HPP)-treated Food Products. Division 28 of Part B of the Food and Drug Regulations (Ottawa, Ontario, Canada).

Horton, L., 1997. The United States Food and Drug Administration: its role, authority, history harmonization activities, and cooperation with the European Union. In: European Union Studies Association (EUSA), Biennial Conference, 5th, 29 May–June 1, Seattle, WA, p. 36.

Keener, K.M., 2009. Food Irradiation. Fact Sheet (FSR 98- 13). North Carolina State University, Department of Food Science, Raleigh, NC.

Koutchma, T., Keener, L., 2015. Novel food safety technologies emerge in food production. Food Safety Magazine.

Koutchma, T., Keener, L., Kotilainen, H., 2018. GHI consensus document on food irradiation. Global Harmonization Initiative. Retrieved from. https://www.globalharmonization.net/sites/default/files/pdf/GHI-Food-Irradiation_October-2018.pdf.

Lelieveld, H., 2009. Progress with the global harmonization initiative. Trends Food Sci. Technol. 20, S82–S84.

Lelieveld, H., Keener, L., 2007. Global harmonization of food regulations and legislation—the global harmonization initiative. Trends Food Sci. Technol. 18, S15–S19.

Loaharanu, P., Ahmed, M., 1991. Advantages and disadvantages of the use of irradiation for food preservation. J. Agric. Environ. Ethics 4, 14–30.

Mansour, M., 2004. One world for all: international harmonization of food regulations. J. Food Sci. 69 (4), 127–129.

Motarjemi, Y., van Schothorst, M., Käferstein, F., 2001. Future challenges in global harmonization of food safety legislation. Food Control 12, 339–346.

NACMCF, 2006. National Advisory Committee on Microbiological Criteria for Foods Requisite scientific parameters for establishing the equivalence of alternative methods of pasteurization. J. Food Protect. 69 (5), 1190–1216.

NCFST, 27 February 2009. National Center for Food Safety and Technology Receives Regulatory Acceptance of Novel Food Sterilization Process. Press release, Summit-Argo, IL.

Newsome, R., 2007. Codex vital in global harmonization. Food Technol. 61 (10).

Notermans, S., Lelieveld, H., 2001. Food Safety: a burning issue in the past, present and future. Food Eng. Ingredients (May), 33–38.

Ramaswamy, R., Balasubramaniam, V.M., Kaletunc, G., 2004. High Pressure Processing. Fact Sheet for Food Processors. FSE-1-04. Ohio State University Extension Fact Sheet, Columbus, OH.

Ramaswamy, R., Jin, T., Balasubramaniam, V.M., Zhang, H., 2005. Pulsed Electric Fields Processing. Fact Sheet for Food Processors. FSE 2-05. Ohio State University Extension Fact Sheet, Columbus, OH.

Rasco, B., 2017. Laws and regulations for novel food processing technology. Ultrasound: Adv. Food Process. Preserv. 499–524.
Regulation (EC) No 258/97 (1997) of the European Parliament and of the Council of 27 January 1997 Concerning Novel Foods and Novel Food Ingredients.
Regulation (EU) 2015/2283 (2015). Retrieved from European Union official website: https://ec.europa.eu/food/safety/novel_food/legislation_en.
Sawyer, E.N., Kerr, W.A., Hobbs, J.E., 2008. Consumer preferences and international harmonization of organic standards. Food Pol. 33, 607–615.
Stewart, C.M., Dunne, C.P., Keener, L., 2016. Pressure-assisted thermal sterilization validation. In: High Pressure Processing of Food. Springer, New York, pp. 687–716.
Tang, J., 2009. Personal Communication. Washington State University, Pullman, WA.
Tang, J., 2015. Unlocking potentials of microwaves for food safety and quality. J. Food Sci. 80 (8), E1776–E1793.
Tonello, C., 2018. Personal Communication. Institute of Food Technologists Annual Meeting, Chicago, IL.
United States Department of Agriculture (USDA), 2012. High Pressure Processing (HPP) and Inspection Program Personnel (IPP) Verification Responsibilities. FSIS Directive 6120.2. Retrieved from: https://www.fsis.usda.gov/wps/wcm/connect/a64961fa-ed6f-44d1-b637-62232a18f998/6120.2.pdf?MOD=AJPERES.

Chapter 13

Processing issues: acrylamide, furan, and *trans* fatty acids

Lauren S. Jackson[1] and Fadwa Al-Taher[2]
[1]*U.S. Food and Drug Administration, Division of Food Processing Science & Technology, Bedford Park, IL, United States;* [2]*VDF FutureCeuticals, Inc., Momence, IL, United States*

13.1 Introduction

Processed foods have become a way of life in the modern world (Rupp, 2003). According to recent estimates, over 60% of foods purchased in the United States, Canada, and the United Kingdom are processed (Moubarac et al., 2013a,b; Poti et al., 2015). Food processing is defined as any procedure that alters food from its natural state, such as heating, canning, freezing, drying, milling, and fermenting (Poti et al., 2015). Processing increases consumer convenience, allows a variety of food choices, and can improve the taste, texture, and flavor of food (Rupp, 2003). In addition, as many ingredients and foods are globally sourced (fruits and vegetables, spices, seafood products, etc.), processing allows for a more consistent food supply.

One of the most important beneficial aspects of some food processes is its ability to inactivate pathogenic microorganisms, enzymes, and toxins that may be present in food (van Boekel et al., 2010). Processing unit operations such as washing, trimming, milling, leaching, and mechanical separation may decrease the natural toxicity of some raw materials by eliminating specific undesirable components (Sikorski, 2005). However, processing can decrease nutrient levels and bioavailability and produce chemical hazards (Rupp, 2003). Industrial thermal processes and conventional cooking operations may induce the formation of harmful compounds such as mutagenic compounds, carcinogenic heterocyclic aromatic amines, acrolein, polycyclic aromatic hydrocarbons, furan, acrylamide, and *N*-nitrosamines. Other detrimental chemical changes in food that can occur as the result of processing include the formation of *trans* fatty acids (TFAs) during the hydrogenation of fats and the creation of chloropropanols and chloroesters during the production of hydrolyzed vegetable protein and some soy sauce products (Hunter, 2005; Hamlet and Sadd, 2009).

Processing-induced food toxicants such as acrylamide, furan, and TFAs have gained widespread attention. The precursors and mechanisms of formation of these compounds are different, but they all are formed during processing of food. Acrylamide and furan are primarily formed during thermal processing (frying, baking, roasting, toasting, etc.) of carbohydrate-rich foods via Maillard-type reactions. TFAs are formed when liquid oils are partially hydrogenated to improve their plasticity or oxidative stability. There have been concerns about the potential health issues associated with the dietary intake of acrylamide, furan, and TFAs. Thus, much attention has been focused on finding ways to reduce or prevent the formation of these three process-induced hazards without compromising food safety, sensory properties (e.g., taste, texture, and color), or nutritional quality. This chapter will provide an overview of the factors affecting the formation of acrylamide, furan, and TFAs in processed food and will give a perspective on the current tools that are being used to manage these chemical hazards.

13.2 Acrylamide

13.2.1 Introduction

Acrylamide (2-propenamide; $H_2C{=}CH{-}CO{-}NH_2$) is a colorless and odorless water-soluble solid in its pure form. The compound is an industrial chemical primarily used in the preparation of polyacrylamide, a flocculant used in the treatment of the municipal water supply and for removal of suspended solids from industrial waste-water before discharge. Other applications of polyacrylamide are in paper and pulp processing, in dyes and adhesives, as cosmetic additives, in the

Ensuring Global Food Safety. https://doi.org/10.1016/B978-0-12-816011-4.00021-5

formulation of grouting agents, and as gels used for electrophoretic separations (Koszucka and Nowak, 2018). Acrylamide is known to be a neurotoxin, a genotoxin and a carcinogen in animals, and a possible carcinogen in humans as defined by the International Agency for Research on Cancer (IARC) (IARC, 1994).

In 2002, researchers at the Swedish National Food Administration and Stockholm University reported finding acrylamide at levels over 1000 μg/kg in a wide range of heated, carbohydrate-rich foods such as fried, baked and roasted potato products, breakfast cereals, breads, and crackers (Tareke et al., 2002). This resulted in heightened worldwide interest in determining the mechanisms by which acrylamide is formed, its occurrence and levels in food, and its effects on human health. Acrylamide is generated in foods that are subjected to high-temperature (>120°C) processes such as frying, baking, roasting, and extrusion and can be found in industrially processed foods as well as those prepared in restaurants or food service operations or at home by the consumer. Acrylamide is generated during heat treatment as a result mainly of the Maillard reaction between certain amino acids and reducing sugars in food, although other mechanisms have been proposed (Mottram et al., 2002; Tareke et al., 2002; Ryberg et al., 2003; Lineback et al., 2012).

Since the discovery of acrylamide in food, a wealth of information has been gathered on the levels of acrylamide in food, toxicological properties of the compound, mechanisms by which acrylamide is formed in food, and methods for reducing the acrylamide content of food. Several excellent resources are available for obtaining up-to-date information on the status of acrylamide. For example, the FAO/WHO Acrylamide in Food Network (http://www.acrylamide-food.org/) was established as a result of the June 2002 FAO/WHO Consultation on the Health Risks of Acrylamide in Food. This network functions as a global resource and inventory of ongoing research on acrylamide in food and includes formal research, surveillance/monitoring, and industry investigations. The HEATOX project, an international multidisciplinary program funded by the European Commission (EC) from 2002 to 2007, focused on the health risks associated with hazardous compounds such as acrylamide and other similar compounds formed in heat-treated carbohydrate-rich foods. Research topics explored by the HEATOX project included mechanisms of formation, impact of raw material composition, inhibiting factors, and the effects of cooking and processing methods on formation of hazardous compounds such as acrylamide. A final report on findings from the study was published (HEATOX, 2007a).

Since this chapter serves only to summarize the current status of the acrylamide issue, the reader is referred to several comprehensive volumes (Friedman and Mottram, 2005; Skog and Alexander, 2006; Stadler and Lineback, 2009; Gökmen, 2015; Halford and Curtis, 2019) on acrylamide and other heat-produced food toxicants, status reports on acrylamide compiled by the Council for Agricultural Science and Technology (CAST, 2006) and the European Food Safety Authority (EFSA) Panel on Contaminants in the Food Chain (CONTAM) (EFSA, 2015), a number of comprehensive reviews (Friedman, 2003; Lineback et al., 2012; Arvanitoyannis and Dionisopoulou, 2012; Pedreschi et al., 2014; Krishnakumar and Visvanathan, 2014), and proceedings from a symposium on "The Chemistry and Toxicology of Acrylamide" published in the *Journal of Agricultural and Food Chemistry* (2008, volume 56, number 15) an issue of *Food* Additives and Contaminants (2007, volume 24, supplement 1) devoted to the topic of acrylamide in food.

13.2.2 Occurrence and levels of acrylamide in food

Since the discovery of acrylamide in food in 2002, there have been considerable efforts from national regulatory agencies such as the U.S. Food and Drug Administration (FDA) and multinational organizations such as the World Health Organization (WHO) and the EC's Directorate General Joint Research Center to gather data on the acrylamide content of foods. The FDA developed a database of acrylamide levels of individual food products purchased in the United States in 2002–04 (FDA, 2019a) and for composite food samples obtained from FDA's Total Diet Studies conducted in 2002–06 (FDA, 2019b). Initially, food groups were chosen for analysis if they were previously reported to contain acrylamide or if they contributed significantly to the diet of infants or young adults. Most products were analyzed as received, while others were examined before and after cooking. More recently (in 2011 and 2015), FDA surveyed food products obtained at retail markets or restaurants throughout the United States for acrylamide (FDA, 2019a; Abt et al., 2019). The products and brands chosen for this survey were known to contain high or variable levels of acrylamide and were not identical to those included in FDA's 2002–06 surveys (FDA, 2019a).

FDA's exploratory data on acrylamide levels in different categories of foods surveyed in 2011 and 2015 are summarized in Table 13.1. The survey indicates wide variations in levels of acrylamide in different food categories and in different brands of the same food category. The variations result from different levels of acrylamide precursors present and cooking or processing conditions such as temperature or time. Comparison of the data from the earlier FDA surveys (2002–06) to those from the more recent ones (2011; 2015) indicates that there were significant decreases in acrylamide levels in several food categories such as potato chips and crackers, suggesting that mitigation procedures may be responsible for the reductions in acrylamide content (FDA, 2019a; Abt et al., 2019). In contrast, concentrations of acrylamide in other food categories did not appreciably decrease.

TABLE 13.1 Summary of acrylamide levels in various food categories based on FDA survey (2011–15).[a]

Food category	Acrylamide concentration (μg/kg)[b]	Food product with highest concentration of acrylamide
Beans	<10–160	Vegetarian baked beans
Beverages (other than coffee)	<10–100	Prune juice
Breads and bakery products	<10–102	Whole grain bread
Breakfast foods (other than cereal)	<10–353	Wheat gluten-free blueberry muffins
Candy and sweeteners	<10–2160	Blackstrap molasses
Cereals	<10–1210	Corn-based ready-to-eat breakfast cereal
Coffee (brewed)	<10	
Coffee (ground, not brewed)	70–1080	Instant coffee
Cookies and granola bars	<10–1450	Ginger snap cookies
Crackers	<10–2110	Graham crackers
Entrees and snacks (restaurant)	<10–30	Beef taco
Entrees and snacks (frozen)	<10–408	Buffalo-style chicken sandwich
French fries and other potato products (consumer)	<10–2090	Hashbrowns (fried)
French fries and other potato products (restaurant and take-out)	<10–1999	Hashbrowns (fried)
Infant foods	<10–1774	Graham honey sticks
Nut butters	<10–570	Almond butter
Nuts and fruits	<10–450	Ripe pitted olives
Potato chips	<10–8440	Sweet potato chips
Pretzels	20–496	Mini pretzels
Snack foods (miscellaneous)	<10–3244	Veggie chips
Soup	<10–260	Chili with beans
Tortilla chips	<10–610	Sweet potato tortilla chips

[a]*Data from FDA, 2019a. Survey data on acrylamide in food. https://www.fda.gov/food/chemicals/survey-data-acrylamide-food.*
[b]*Analytical method limit of quantitation (LOQ) was <10 μg/kg.*

Since 2002, member states within the European Union (EU) have been surveying foods for acrylamide content (EFSA, 2019). In 2007, the EC adopted a recommendation on the monitoring of acrylamide levels in foods available in the EU; this recommendation was extended by EC Recommendation 2010/307/EU (EC, 2010). As a result, surveys on the acrylamide content of food obtained throughout the EU were compiled by the EFSA. The monitoring program was intended to target foods known to contain high acrylamide levels and/or contribute significantly to the human dietary intake. Reports for acrylamide monitoring efforts for 2007–10 are available for review (EFSA, 2012). Based on the results of the surveys, there were downward trends in acrylamide levels in the category "processed cereal-based foods for infants and young children" and the subcategories "nonpotato-based savory snacks" and "biscuits and rusks for infants and young children." In contrast, acrylamide levels increased in the "coffee and coffee substitutes" category and in the subcategories "crisp bread," "instant coffee," and "French fries from fresh potatoes" though for the latter category this trend was not consistent across Europe. In general, acrylamide levels in surveyed foods in the EU were similar to those reported in FDA's exploratory surveys.

Table 13.2 lists the top 20 foods that contribute the most acrylamide to the U.S. diet for populations >2 years of age based on FDA's most recent survey data (2011 and 2015). Foods that contribute most to acrylamide exposure include breakfast cereal, cookies, crackers, corn snacks, brewed coffee, and fried potato products. An earlier exposure assessment for U.S. consumers using exploratory data gathered in 2002–06 shows that the same food categories contributed most to acrylamide intake (Doerge et al., 2008).

TABLE 13.2 Top 20 foods contributing acrylamide to the U.S. diet.[a]

Food category	Mean acrylamide intake (μg/kg body weight/day)	Cumulative percentile
Breakfast cereal	0.050	0.17
French fries (RF)[b]	0.047	0.33
Potato chips	0.038	0.46
Cookies	0.030	0.56
Crackers	0.022	0.64
Corn snacks	0.019	0.70
Brewed coffee	0.018	0.77
Fried potatoes (other than French fries)	0.010	0.80
French fries (CF)[c]	0.010	0.84
Pizza	0.010	0.87
Soft bread	0.008	0.90
Granola bars	0.008	0.92
Pretzels	0.006	0.94
Pancakes	0.004	0.96
Pies and cakes	0.003	0.97
Toast	0.003	0.98
Breaded chicken	0.002	0.99
Potato skins	0.001	0.99
Doughnuts	0.001	0.99
Brownies	0.000	0.99

[a]*Data from Abt, E., Robin, L.P., McGrath, S., Srinivasan, J., DiNovi, M., Adachi, Y., et al., 2019. Acrylamide levels and dietary exposure from foods in the United States, an update based on 2011–2015 data. Food Addit. Contam. 36, 1475–1490.*
[b]*RF = restaurant fries.*
[c]*CF = consumer fries (includes fried and baked).*

The top contributors of acrylamide to the diet differ from country to country due to differences in the national diet and the manner in which foods are prepared (CAST, 2006; Friedman and Levin, 2008; Mills et al., 2009). In general, foods that contribute most to dietary exposure include processed foods such as potato products (French fries, potato chips, baked potatoes), bakery and cereal products (breakfast cereals, bread, cereal, crackers, cookies, cakes), and coffee. Although cereal products are a major dietary source of acrylamide, the percentage of total acrylamide that cereals contribute varies for different populations, ranging from ~24% in the diet of Swedish adults to ~44% in the diet of Belgian adolescents, with the American diet at 40%. The contribution of potato products to total acrylamide intake ranges from ~29% for adult Norwegian women to ~69% for Dutch children/adolescents with ~38% for the U.S. population (Bagdonaite et al., 2008; Dybing et al., 2005; FDA, 2006).

The contribution of coffee to the dietary intake of acrylamide varies widely demographically and can be high in countries with a high coffee consumption. It ranges from ~8% for the United States, to 13% for the Netherlands, to ~28% for Norwegian adults, and to ~39% for Swedish adults (Bagdonaite et al., 2008; Friedman and Levin, 2008; FDA, 2006; Dybing et al., 2005). These data indicate that the amount of acrylamide that coffee contributes to the diet may be important in some populations.

The average dietary intake of acrylamide has been estimated by the Joint FAO/WHO Expert Committee on Food Additives (JECFA) (JECFA, 2011) for the general population as well as by different national organizations for the inhabitants of their respective countries (Wenzl et al., 2007; Wenzl and Anklam, 2007). A mean daily intake of 0.2–1.0 μg/kg body weight/day was estimated by JECFA for adult consumers, and 0.6–1.8 μg/kg body weight/day for adult consumers of food items with higher concentrations of acrylamide (JECFA, 2011). In 2015, EFSA estimated mean exposure to dietary acrylamide across survey and age groups to be 0.4−1.9 μg/kg body weight/day, and 0.6−3.4 μg/kg body weight/day for 95th percentile consumers (EFSA, 2015). These exposure estimates are similar to those estimated by the FDA for U.S. consumers: 0.44 μg/kg body weight/day for average consumers >2 years of age, and 0.95 μg/kg body weight/day for high consumers >2 years of age (FDA, 2006). More recent calculations of U.S. exposure based on FDA's 2011 and 2015 survey data are similar to earlier exposure estimates (Abt et al., 2019). For all countries, dietary acrylamide exposure is not evenly distributed in the general population. Children and teenagers have a higher acrylamide intake than adults as they have a higher food consumption per kg of body weight and since they tend to consume more acrylamide-rich foods than adults (Brisson et al., 2014). Intake estimation for different countries may deviate from the above levels due to dissimilarities in the eating habits and food items consumed, and differences in the methods used to model dietary intake (Boon et al., 2005; Fohgelberg et al., 2005).

Although biomarkers (urinary metabolites and hemoglobin adducts) were not used to estimate exposure to acrylamide in the exposure assessments described above, several studies have evaluated whether biomarkers could be used to estimate short- and long-term dietary exposure to acrylamide (Dybing et al., 2005; Bjellaas et al., 2007; Brantsaeter et al., 2008; Brisson et al., 2014). In a study conducted using pregnant women in Norway, Brantsaeter et al. (2008) found that dietary exposure to acrylamide calculated with a food frequency questionnaire correlated highly with the levels of acrylamide urinary metabolites in study participants. More recently, Brisson et al. (2014) published the result of a study on the relationship between dietary acrylamide exposure and biomarkers (urinary metabolites and hemoglobin adducts) of internal dose in Canadian teenagers. The results of the study found that acrylamide intake during the 2 days prior to biomarker measurement was a significant predictor to urinary metabolite concentrations, while acrylamide intake over the past month had a high degree of correlation with hemoglobin adduct levels.

13.2.3 Mechanism of formation

Research suggests that acrylamide forms in foods mainly through the Maillard reaction between reducing sugars or a source of carbonyl compounds and certain amino acids. Model systems demonstrated that asparagine is the major amino acid precursor (Mottram et al., 2002; Zyzak et al., 2003; Stadler et al., 2002; Becalski et al., 2004). This explains the occurrence of acrylamide in potato- and grain-based foods, which are particularly rich in free asparagine and typically contain high levels of carbonyl compounds such as reducing sugars (Mottram et al., 2002). Maillard reaction products are also responsible for the desirable flavors and colors of heat-processed foods. Consequently, methods for preventing acrylamide formation by limiting the Maillard reaction frequently compromise the flavor and color of cooked or processed food.

Fig. 13.1 shows the major pathways generating acrylamide from asparagine and a source of carbonyl compounds. Acrylamide in food is derived mainly from heat-induced reactions (temp. >120°C) between the amino group of free asparagine and the carbonyl group of reducing sugars such as glucose during baking, frying, and other thermal treatments. Studies have shown that the major mechanistic pathway in the formation of acrylamide in foods involves a Schiff's base where a decarboxylation is necessary, followed by further degradation of the decarboxylated Schiff's base. This degradation step involves cleavage of a nitrogen−carbon bond, which can occur by two different mechanisms. One involves direct degradation of the decarboxylated Schiff's base to form acrylamide via elimination of an imine and the other involves hydrolysis of the decarboxylated Schiff base to yield aminopropionamide and a carbonyl compound (Zyzak et al., 2003; Stadler et al., 2004; Blank et al., 2005; Granvogl and Schieberle, 2006; Claus et al., 2006a; Friedman and Levin, 2008). It is important to note that the yield of acrylamide in simple model systems is low, whereby less than 1% of asparagine is converted to acrylamide (Stadler et al., 2002; Becalski et al., 2003; Surdyk et al., 2004; Stadler, 2006). These yields are even further reduced in more complex systems such as food.

A number of acrylamide-forming pathways without the involvement of asparagine have been proposed. One involves the reaction of acrylic acid, generated from Maillard reactions or from lipid degradation reactions, with ammonia, released during heating of amino acids or proteins, and/or proteins, to form acrylamide (Yaylayan and Stadler, 2005). Another possible pathway involves release of acrylamide from wheat gluten heated under pyrolysis conditions (Claus et al., 2006a). Overall, the pathways not involving asparagine are not believed to be major ones involved in acrylamide formation in cereal- and potato-based products (Lineback et al., 2012).

The detailed mechanism at each step in the formation of acrylamide may depend on the species involved (type of carbonyl compound), the chemical environment (water content, pH), and processing or cooking temperature (Mills et al., 2009). Stadler et al. (2002) reported equal reactivities of fructose, glucose, and sucrose in model systems with asparagine at 180°C. In contrast, Claeys et al. (2005) found sucrose to have roughly half the reactivity of glucose to react with asparagine to form acrylamide in the temperature range of 140 to 200°C. Although acrylamide can form in the presence of water,

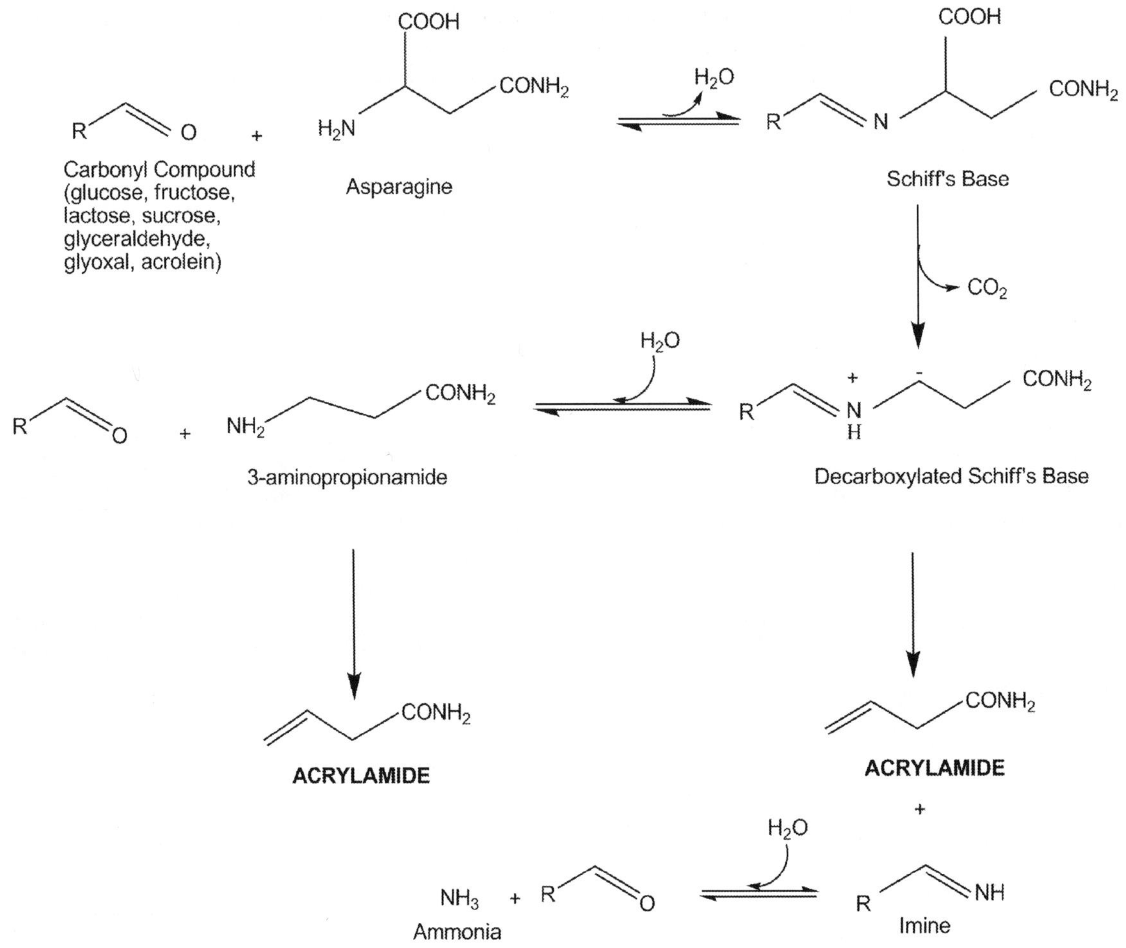

FIGURE 13.1 Mechanisms of formation of acrylamide in thermally processed/cooked food. *Adapted from Zyzak, D.V., Sanders, R.A., Stojanovic, M., Tallmadge, D.H., Eberhart, B.L., Ewald, D.K., 2003. Acrylamide formation mechanism in heated foods. J. Agric. Food Chem. 51, 4782–4787, Blank, I., Robert, F., Goldmann, T., Pollien, P., Varga, N., Devaud, S., 2005. Mechanism of Acrylamide Formation: Maillard-induced transformations of asparagine. In: Friedman, M., Mottram, D. (Eds.), Chemistry and Safety of Acrylamide in Food. Springer, New York, NY, pp. 171–189, Yaylayan, V.A., Locas, C.P., Wronowski, A., O'Brien, J., 2005. Mechanistic pathways of formation of acrylamide from different amino acids. In: Friedman, M., Mottram, D. (Eds.), Chemistry and Safety of Acrylamide in Food. Springer, New York, pp. 191–204, Granvogl, M., Schieberle, P., 2006. Thermally generated 3-aminopropionamide as a transient intermediate in the formation of acrylamide. J. Agric. Food Chem. 54, 5933–5938, Claus, A., Weisz, G.M., Scieber, A., Carle, R., 2006a. Pyrolytic acrylamide formation from purified wheat gluten and gluten-supplemented wheat bread rolls. Mol. Nutr. Food Res., 50, 87–93, Claus, A., Schreiter, P., Weber, A., Graeff, S., Herrmann, W., Claupein, W., Schieber, A., Carle, R., 2006b. Influence of agronomic factors and extraction rate on the acrylamide contents in yeast-leavened breads. J. Agric. Food Chem., 54, 8968–8976 and Friedman, M., Levin, C.E., 2008. Review of methods for the reduction of dietary content and toxicity of acrylamide. J. Agric. Food Chem. 56, 6113–6140.*

acrylamide tends to form in dry environments such as the crust of breads and the outer surface of fried potatoes (Surdyk et al., 2004; Friedman and Levin, 2008). Examples of foods that contain high levels of water yet contain relatively high amounts of acrylamide include prune juice and canned black olives (Amrein et al., 2007; FDA, 2019a).

13.2.4 Factors affecting formation

13.2.4.1 Processing conditions

Processing and cooking conditions such as temperature and time are important factors affecting the levels of acrylamide in model systems and in foods (potato products, cereal-based foods, roasted almonds, coffee, etc.) (Mottram et al., 2002; Tareke et al., 2002; Stadler et al., 2002; Taubert et al., 2004; Jackson and Al-Taher, 2005; Amrein et al., 2007; Friedman and Levin, 2008; Lineback et al., 2012). However, the manner in which heat is transferred to a food (e.g., frying, baking, roasting, microwave-heating) does not impact the rate of acrylamide formation (Stadler et al., 2004). In general, acrylamide content increases in the temperature range of 120–185°C, then decreases when the food is heated at higher temperatures

(Mottram et al., 2002; Tareke et al., 2002; Ryberg et al., 2003). The mechanism(s) by which acrylamide degrades at higher temperatures have not been identified, but it has been hypothesized that acrylamide reacts with other food components or that another reaction mechanism occurs that bypasses acrylamide (Rydberg et al., 2005; CAST, 2006).

Acrylamide levels for deep-fat fried French fries ranged from 265 μg/kg for potatoes fried at 150°C for 6 min to 2130 μg/kg for French fries prepared at 190°C for 5 min (Jackson and Al-Taher, 2005). At frying temperatures of 180–190°C, acrylamide levels in French fries increased exponentially at the end of the frying process (Jackson and Al-Taher, 2005). Similar results were reported by Rydberg et al. (2003), Matthaus et al. (2004) and Pedresci et al. (2005). In potato slices with low surface-to-volume ratios (SVRs), acrylamide levels increased with increasing frying temperatures as well as with frying time, reaching maximum levels of 2500 μg/kg. However, in samples with higher SVRs, acrylamide levels were the greatest at 160–180°C with maximal acrylamide formation of 18,000 μg/kg, then decreased with higher frying temperatures and more prolonged frying times (Taubert et al., 2004).

A study assessing the effects of time and temperature on acrylamide formation in wheat bread (Surdyk et al., 2004) showed that the majority (99%) of acrylamide was formed in the crust layer and that levels increased with baking time and temperature. Studies performed on the effects of baking times and temperature (180 and 200°C) on acrylamide formation in gingerbread showed that acrylamide formation occurred linearly over a 20 min baking period (Amrein et al., 2004).

Coffee is a complex matrix in terms of acrylamide formation and reduction. It has been shown that roasting time and temperature had an impact on acrylamide formation in coffee beans (Taeymans et al., 2004). Although coffee beans are roasted at very high temperatures (240–300°C), significant amounts of acrylamide are formed during the first few minutes of roasting, then levels decrease exponentially toward the end of the roasting cycle (Taeymans et al., 2004; Stadler, 2006; Banchero et al., 2013). Guenther et al. (2007) found that maximum acrylamide levels in Robusta (3800 μg/kg) and Arabica (500 μg/kg) coffee beans were generated in the first minutes of the roasting process. Kinetic models and spiking experiments with isotope-labeled acrylamide have shown that >95% of acrylamide formed in coffee beans is degraded during roasting (Bagdonaite et al., 2008). These findings may explain why light roasted coffees contain higher amounts of acrylamide than dark roasted coffees (Guenther et al., 2007). The higher levels of acrylamide found in roasted Robusta rather than Arabica coffee beans may be attributed to higher asparagine content in Robusta green coffee (Lantz et al., 2006).

In most cases, acrylamide concentrations are highly correlated to the degree of browning on the surface (crust) of cooked or processed potato- and cereal-based foods (Surdyk et al., 2004; Amrein et al., 2004; Jackson and Al-Taher, 2005; CAST, 2006). Since acrylamide and the brown color of cooked foods are formed during the Maillard reaction, it is likely that acrylamide is formed parallel with browning. Thus, the degree of surface browning could be used as a visual indicator of acrylamide formation during cooking. However, the extent of browning may not necessarily indicate the acrylamide content of a food when additives such as ammonium bicarbonate (Amrein et al., 2004) or ingredients that contain asparagine are added (Surdyk et al., 2004).

13.2.4.2 Raw material composition

The presence and concentration of precursors affect acrylamide formation, but presence of compounds that compete with reducing sugars and amino acids in the Maillard reaction are also important compositional factors. In potato products, acrylamide levels are highly correlated with glucose and fructose concentrations in the uncooked potato tubers, whereas asparagine levels do not predict acrylamide levels in cooked potato products (Amrein et al., 2003; Becalski et al., 2004; CAST, 2006; Haase, 2006). When prepared under similar conditions, acrylamide contents of thermally processed potato products can vary depending on potato cultivar mainly due to the variability in the reducing sugar content of the different cultivars (Amrein et al., 2003; Haase, 2006). It has been recommended that potatoes used for roasting or frying should be mature and contain less than 1 g/kg reducing sugars on a wet weight basis (Biedermann-Brem et al., 2003; De Wilde et al., 2005; FoodDrinkEurope, 2019).

The high variability in reducing sugar content among potatoes of the same cultivar suggests that storage conditions may have a stronger influence on sugar content of raw potato tubers than cultivar (Noti et al., 2003). Short-term storage of potatoes at 4°C (e.g., in the refrigerator of a supermarket) significantly increases the potential for acrylamide formation (Biedermann et al., 2002; Noti et al., 2003). Cooling potatoes to temperatures less than 6°C causes the reducing sugars to increase through the phenomenon known as "cold sweeting," thus increasing the potential for acrylamide formation (Biedermann et al., 2002; Noti et al., 2003; Jackson and Al-Taher, 2005). Reconditioning of potatoes at room temperature following cold storage results in significant reductions in the content of reducing sugars and the acrylamide forming potential (Haase, 2006). Controlling storage conditions so that potatoes are kept at temperatures between 6 and 10°C is essential for minimizing reducing sugar accumulation in potatoes (FoodDrinkEurope, 2019; FDA, 2016a, 2017a).

Agronomic factors influence the composition of potatoes and thus influence acrylamide formation during cooking or thermal processing. Lowering the amount of nitrogen fertilizer used during potato cultivation resulted in a 30%–65% increase in acrylamide levels in fried potato products. In addition to fertilizer application rate, dry and hot weather seems to increase acrylamide formation by increasing the content of reducing sugars in potato cultivars (De Meulenaer et al., 2008). These studies indicate that acrylamide content of potato products can be reduced through agricultural practices and by carefully selecting potato cultivars with low levels of reducing sugars. Although reducing sugar content is a major determinant of acrylamide forming potential in potatoes, a study by Tran et al. (2017) found that reducing the free asparagine content of conventional potato varieties by transforming them with a plasmid (pSIM1278) was able to reduce their acrylamide-forming potential by 62%–67%.

In cereal-based foods, acrylamide levels are more highly correlated with levels of asparagine rather than reducing sugars (Konings et al., 2007). No correlation was reported between reducing sugar and acrylamide contents of heated flour or breads (Claus et al., 2006b). In contrast, asparagine levels of the flour significantly affected acrylamide contents with levels differing by a factor of five due to marked differences in free asparagine and crude protein contents.

Use of nitrogen fertilizers during the growth of cereal crops caused an increase in amino acid and protein contents of the grain, thus increasing acrylamide levels in bread products (Claus et al., 2006b). Therefore, to minimize acrylamide levels, nitrogen fertilization should be adjusted to the minimum requirement of the crops. In addition, providing adequate sulfur fertilization for wheat and rye has been shown to reduce asparagine levels in these crops (Halford et al., 2007; FoodDrinkEurope, 2019).

13.2.5 Prevention and mitigation

International food monitoring agencies in collaboration with the food industry and academia have put forth strategies for reducing acrylamide levels in food. Food manufacturers in Europe have worked with researchers through Food-DrinkEurope (formerly known as the CIAA) to produce a series of guidelines known as the "Food Drink Europe Acrylamide Toolbox" for reducing acrylamide levels in different food classes (potato products, cereal-based foods, coffee, and coffee mixtures) (FoodDrinkEurope, 2019). In 2009, the FAO/WHO Codex Alimentarius Commission developed a "Code of Practice for the Reduction of Acrylamide in Foods" which was written to provide national and local authorities, manufacturers, and other relevant bodies with guidance to prevent and reduce formation of acrylamide in potato products and cereal products (CODEX Alimentarius Commission, 2009). A guidance document that provides information to help growers, manufacturers, and food service operators reduce acrylamide levels in potato-based foods, cereal-based foods, coffee, and foods prepared in food service operations was issued by the FDA in 2016 (FDA, 2016a). The FDA also provides information to consumers on ways to reduce acrylamide exposure through recommendations on diet choices, potato storage, and food preparation (FDA, 2016b, 2017a).

It is important to recognize that there is no common mitigation strategy that is effective for all food categories, and that acrylamide-lowering approaches can adversely affect food quality and safety. Considerations have to be made for scale-up of mitigation procedures, as approaches that are effective in laboratory- or pilot-scale trials may not be feasible or effective when applied at the factory or industrial scale (Medeiros Vinci et al., 2012; Lineback et al., 2012).

Acrylamide mitigation strategies fall into five major categories: methods that interrupt reactions leading to acrylamide formation, treatments that reduce the levels of acrylamide precursors, approaches that reduce the cooking or processing time and/or temperature, procedures that change the product recipe or formulation, and agronomic methods (Jung et al., 2003; Taubert et al., 2004; CAST, 2006; Amrein et al., 2007; Muttucumaru et al., 2008; Mestdagh et al., 2008a; FoodDrinkEurope, 2019). These mitigation strategies are summarized below and described in detail by Konings et al. (2007), Guenther et al. (2007), Foot et al. (2007), Friedman and Levin (2008), Capuano and Fogliano (2011), Pedreschi et al. (2014), Medeiros Vinci et al. (2012), and Baskar and Aiswarya (2018).

13.2.5.1 Methods that interrupt reactions leading to acrylamide formation

Several approaches have been somewhat successful at inhibiting acrylamide formation by preventing the key reactions responsible for generating the compound. Lowering the pH of foods by adding or using dips containing citric, acetic, ascorbic, lactic, or phosphoric acids blocks the nucleophilic addition of asparagine with a carbonyl compound, preventing the formation of the Schiff's base, a critical intermediate in the formation of acrylamide (Mestdagh et al., 2008a). Although this approach was successful at lowering acrylamide levels by up to 90% in fried potato products (Jung et al., 2003; Pedresci et al., 2004; Kita et al., 2005; Jackson and Al-Taher, 2005; Mestdagh et al., 2008a), treatments that lower the pH of foods may cause foods to have an undesirable taste (Pedreschi et al., 2014).

Another approach that blocks acrylamide formation by preventing Schiff's base formation includes the addition of mono- and divalent cations (Na^+ or Ca^{2+}) to foods. Modest to high (>90%) reduction in acrylamide levels in fried potato products was reported using this approach (Lindsay and Jang, 2005; Park et al., 2005; Gökmen and Senyuva, 2007; Mestdagh et al., 2008a; Sadd et al., 2008). Use of calcium salts (calcium chloride and calcium lactate) was successful at reducing acrylamide content of cookies, biscuits, and other baked goods (Levine and Ryan, 2009; Açar et al., 2012). Calcium salt concentrations must be kept as low as possible since they tend to lead to off-flavors and changes in product texture (Pedreschi et al., 2014).

The addition of proteins or free amino acids other than asparagine (lysine, glutamine, glycine, cysteine) was studied as a method for reducing acrylamide formation by causing competitive reactions and/or covalently binding acrylamide via Michael addition reactions (Hanley et al., 2005; Claeys et al., 2005; Mestdagh et al., 2008a; Sadd et al., 2008). One of the more efficacious treatments used glycine to reduce acrylamide concentrations in cereal products by 30%–70% (Bråthen et al., 2005; Kita et al., 2005; Fink et al., 2006). As with the use of other additives, careful consideration must be made to ensure that the levels of amino acids have minimal impact on the sensory properties of the foods to which they are added.

Antioxidants have been examined as an acrylamide mitigation strategy by scavenging free radicals produced during the Maillard reaction. Laboratory-scale experiments have shown mixed results on the use of plant extracts, spices, and purified antioxidants on acrylamide formation in potato chips, French fries, and baked cereal products (Morales et al., 2008; Jin et al., 2013). Studies by Zhang (Zhang et al., 2007; Zhang and Zhang, 2007) demonstrated >70% reduction in acrylamide content of potato chips, French Fries, and fried bread sticks that were either treated with, or contained, bamboo leaf antioxidants or green tea extract. Similarly, a liquid flavonoid-rich spice mix was found to reduce acrylamide levels in fried potatoes by up to 50% (Fernandez et al., 2003). Hedegaard et al. (2008) added aqueous rosemary extract, rosemary oil, and dried rosemary leaves to a bread model system and demonstrated reductions of 62%, 67%, and 57% of acrylamide content, respectively, compared to bread without treatments. In contrast, studies by Vattem and Shetty (2002) and Rydberg et al. (2003) found that use of phenolic antioxidants from cranberry and oregano and ascorbic acid increased or had negligible effects on acrylamide content of potato- and/or cereal-based foods. Further research is needed to evaluate the effects of antioxidants on acrylamide formation in different types of products, and the impact effective treatments have on the sensory properties of food.

13.2.5.2 Treatments that reduce the levels of acrylamide precursors

Since reducing sugars and free asparagine are the major acrylamide precursors in food, removing either of these substrates is a strategy for reducing acrylamide formation (CAST, 2006; Morales et al., 2008; FoodDrinkEurope, 2019). Procedures for reducing free asparagine and sugar levels in foods include rinsing and blanching treatments, use of asparaginase or microorganisms (fermentation), and control of storage conditions (for potato products).

Use of rinsing, washing, and soaking treatments for fresh-cut potatoes has been shown to have a low to moderate degree of effectiveness in reducing acrylamide formation during frying. Soaking and blanching treatments reduce acrylamide formation by leaching out sugars and asparagine from the surface of the potato slice. Washing and blanching treatments (time, temperature, pH, etc.) can be manipulated to optimize leaching of acrylamide precursors (Medeiros Vinci et al., 2012).

Soaking potato slices in room temperature water for at least 15 min before frying resulted in over 50% reduction in acrylamide in lab-scale experiments (Grob et al., 2003; Jackson and Al-Taher, 2005). Since water or steam blanching is an important unit operation in the production of frozen French fries, there has been interest in optimizing this treatment for possible acrylamide reduction in these products (Pedreschi et al., 2006, 2011; Mestdaugh et al., 2008b; Medeiros Vinci et al., 2010). Blanching slices in warm or hot water removed more glucose and asparagine from sliced potatoes than immersing them in water at ambient temperatures (Pedreschi et al., 2006, 2014). However, the studies to date indicate that use of a low blanching water temperature (70°C) and longer time (15 min) results in effective mitigation of acrylamide in fried potato products (French fries; potato chips) without sacrificing desirable flavor and color attributes (Pedreschi et al., 2004, 2007; Mestdagh et al., 2008b; Zhang et al., 2018). Soaking and blanching treatments reduce acrylamide formation by leaching out sugars and asparagine from the surface of the potato slice.

Use of asparaginase, an enzyme that hydrolyzes asparagine into aspartic acid and ammonia, has been successful at reducing acrylamide levels in potato products, wheat-based bakery products, and coffee (Zyzak et al., 2003; Vass et al., 2004; Amrein et al., 2004; Ciesarova et al., 2006; Pedreschi et al., 2008, 2011; Hendriksen et al., 2009, 2013; Kumar et al., 2014; Xu et al., 2015, 2016). Since asparaginase targets asparagine and not other amino acids involved in the Maillard reaction, browning, flavor development, and overall food quality are not significantly impacted by treatment with the enzyme. Asparaginase treatments are theoretically more useful for reducing acrylamide levels in cereal products since asparagine is known to be the limiting factor in acrylamide formation. Amrein et al. (2004) reported that asparaginase treatment of gingerbread dough resulted in a 75% decrease in free asparagine and a 55% reduction in acrylamide levels in the baked product. Similarly, Hendriksen et al. (2009) and Kumar et al. (2014) obtained 34%–90% reduction in final acrylamide content of semisweet biscuits, ginger biscuits, crisp breads, and other bakery goods with asparaginase treatment of the dough. Critical parameters that influenced the effectiveness of the treatment included enzyme dose, dough resting time, and water content.

Immersion of potato slices (chips and French fries) with asparaginase prior to frying has resulted in low (15%–30%) reductions in acrylamide levels, mainly due to the limited ability of the enzyme to interact with asparagine present in the surface of potato slices (Pedreschi et al., 2008, 2011). However, by treating potato pieces with asparaginase after a blanching step, acrylamide levels in French fries could be lowered by 60%–90% (Hendriksen et al., 2009; Pedreschi et al., 2011). The greater effectiveness of the asparaginase treatment was believed to be due to the blanching step which increased permeability of the cells located at the surface of the potato slices and enhanced release of asparagine.

Mitigation of acrylamide formation in roasted coffee has been a challenge since treatments that reduce acrylamide precursors also affect the flavor and aroma of coffee. However, some studies have shown that acrylamide mitigation could be achieved by treating wetted green coffee beans with asparaginase. In laboratory-scale experiments, use of asparaginase alone and with a steam to treat green coffee beans prior to roasting resulted in 55%–74% and 69%–86% reduction in acrylamide levels, respectively (Hendriksen et al., 2013; Xu et al., 2015). Clearly, more work is needed to determine if use of asparaginase is a viable method for reducing acrylamide formation in roasted coffee products.

At present, two commercial asparaginase preparations are available: Acrylaway derived from *Aspergillus oryzae* (Novozymes, Denmark) and PreventASe produced by *Aspergillus niger* (DSM Food Specialties, Denmark). Asparaginase has received generally recognized as safe (GRAS) status from the FDA, and the enzyme has been explored for use in the United States, Australia, New Zealand, China, Russia, Mexico, and several European countries (Xu et al., 2016). A thorough review on the use of asparaginase for acrylamide mitigation in potato, cereal, and coffee products was written by Xu et al. (2016).

Extensive fermentation with yeast is a way to reduce acrylamide content of bread by eliminating free asparagine, the main determinant of acrylamide formation in cereal products (Fredriksson et al., 2004; Konings et al., 2007). Prolonged (2 h) yeast fermentation of whole wheat and rye dough caused an 87% and 77% reduction in acrylamide concentrations in whole grain and rye breads, respectively (Fredriksson et al., 2004). Sourdough fermentation was less effective than yeast fermentation in reducing the asparagine content of the dough, and thus not effective in acrylamide reduction (Fredriksson et al., 2004). Use of yeast fermentation was examined for reducing the amounts of reducing sugar, in the surface of potato slices prior to frying (Zhou et al., 2015). Under optimized conditions, fermented potato slices contained 70% less acrylamide than unfermented potato slices after deep-frying. Several yeast strains have been commercially developed to reduce acrylamide levels in bakery and fried potato products (Gorton, 2017; Campbell, 2017).

Several ingredients may increase acrylamide formation during baking of cereal-based products. A study by Amrein et al. (2004) showed that the baking agent, ammonium hydrogen carbonate, enhanced acrylamide formation in bakery products. The enhancing effect of ammonium hydrogen carbonate may be due to (1) the action of NH_3 as a base in the retro-aldol reactions leading to sugar fragments, (2) facilitated retro-aldol-type reactions of imines in their protonated forms leading to sugar fragments, and (3) oxidation of the enaminols whereby glyoxal and other reactive sugar fragments are formed (Amrein et al., 2006). In contrast, use of sodium bicarbonate as an alternative baking agent to ammonium hydrogen carbonate reduced acrylamide content by more than 60% (Amrein et al., 2004). Similarly, use of sucrose rather than honey or inverted sugar syrup can reduce acrylamide formation in cookies, breads, and cakes (Amrein et al., 2004; Gökmen et al., 2007; FoodDrinkEurope, 2019).

13.2.5.3 Changing processing/cooking conditions

Acrylamide formation in foods can be controlled by carefully regulating processing or cooking conditions (Medeiros Vinci et al., 2012; Lineback et al., 2012). However, since the Maillard reaction is responsible for the production of desirable flavor and color compounds in heated food, reducing the cooking time and temperature to minimize acrylamide formation may compromise food color, flavor, and texture. High temperature cooking and processing treatments (i.e., >150°C) which result in surface browning are most responsible for acrylamide formation in food. In contrast, boiling and microwave cooking that typically result in food temperatures <120°C do not result in significant acrylamide formation (FoodDrinkEurope, 2019).

Numerous studies have evaluated the effects of frying time and temperature on acrylamide formation in potato chips or French fries, and all reported that acrylamide rapidly formed toward the end of the frying process when potato surfaces developed a brown crust (Taemans et al., 2004; Jackson and Al-Taher, 2005; Pedreschi et al., 2005; Gökmen et al., 2006; Pedreschi et al., 2007). Conditions that minimize acrylamide in French fries involve frying or baking potato pieces as long as necessary to get the surface golden in color and the texture crispy (Grob et al., 2003; Jackson and Al-Taher, 2005; FoodDrinkEurope, 2019) and using cuts of potatoes that have lower surface area (Medeiras Vinci et al., 2012). Type of frying oil has negligible impact on acrylamide formation in fried potato products, and acrylamide does not form in appreciable amounts in potatoes that are microwaved or boiled (Jackson and Al-Taher, 2005; Mestdagh et al., 2007).

Overall, the available information suggests that prolonged baking or excessive browning should be avoided to minimize acrylamide formation in baked cereal-based foods (Koning et al., 2007; Ahrné et al., 2007; FoodDrinkEurope, 2019). Steam-assisted baking at 165°C resulted in lower acrylamide concentrations in cookies than those baked at the same temperature in natural or forced convection ovens, mainly due to reduced surface browning that occurs in this type of oven (Isleroglu et al., 2012). Since acrylamide formation increases linearly in the baking process, an important factor for minimizing acrylamide formation is to determine the proper cooking end-point. In bakery products, the degree of surface browning could be used as a visual indicator of acrylamide formation (FoodDrinkEurope, 2019).

As mentioned previously, acrylamide concentrations in roasted coffee differ widely depending on roasting conditions. Acrylamide levels in lighter roasted coffees tend to be higher than in darker roasted products since acrylamide levels tend to decrease during the roasting process (Taeymans et al., 2004; Stadler, 2006). Although increasing roasting time and/or temperatures may appear to be a viable approach for controlling acrylamide levels in coffee, the more extreme roasting conditions can result in undesirable taste or aroma. Other mitigations strategies effective in potato- and cereal-based foods such as use of additives have limited applicability to coffee since they alter coffee flavor and overall quality. Use of supercritical CO_2 extraction processing to reduce acrylamide levels in roasted coffee beans was examined by Banchero et al. (2013). The treatment did not alter the caffeine content and resulted in 79% removal of acrylamide in roasted coffee beans. Further work is needed to determine whether supercritical fluid extract treatments affect sensory properties of coffee.

13.2.5.4 Agronomic factors

Selective breeding of potatoes or cereal grains is a potential method for controlling acrylamide levels by decreasing levels of acrylamide precursors (reducing sugars and asparagine, respectively) (CAST, 2006; Lineback et al., 2012; Baskar and Aiswarya, 2018; FoodDrinkEurope, 2019). Since 2002, several groups of researchers demonstrated the importance of variety selection on formation of acrylamide in potato products (Ryberg et al., 2003; Grob et al., 2003; Foot et al., 2007). Amrein et al. (2003) found a 50-fold variation in total reducing sugars in the different potato cultivars they studied. Control of sugar content of potato tubers and processing conditions are believed to be the major mitigation methods for reducing acrylamide content of potato products (Lineback et al., 2012). Cultivars with low reducing sugar content were found to be more suitable for potato products (French fries, potato chips) cooked or processed at high temperatures.

In addition to cultivar selection, an inverse correlation between the amount of nitrogen fertilizer applied in potato cultivation and acrylamide content in foods was established as decreased fertilizer enhanced reducing sugar expression (De Wilde et al., 2006; Lineback et al., 2012). Climatic conditions such as the amount of rainfall can also influence the acrylamide-forming potential of potatoes. For example, potatoes grown in hot dry summers were found to result in lower acrylamide levels when used to produce fried potato products (De Meulenaer et al., 2008).

Asparagine is the major determinant of the acrylamide forming potential in products made from cereal grain. Studies by Claus et al. (2006b) revealed a significant impact of cultivar and fertilizer application rate on acrylamide levels in bakery products mainly due to differences in the asparagine and crude protein contents of the raw material (Konings et al., 2007). A study showed that nitrogen fertilization resulted in high amounts of amino acid and protein contents, causing increased acrylamide levels in breads, ranging from 10.6 to 55.6 μg/kg (Claus et al., 2006b). When wheat was grown with sulfate depletion, asparagine levels were 3600−5200 μg/kg compared to 600−900 μg/kg in wheat grown in soils with proper amounts of sulfate fertilizer (Muttucumaru et al., 2006; Friedman and Levin, 2008; Baskar and Aiswarya, 2018).

More research is needed to understand the factors affecting acrylamide formation in food. A better understanding of these factors will enable food manufacturers to minimize the acrylamide content of food without compromising food safety and quality.

13.2.6 Health effects of dietary acrylamide

Experimental studies in animals have shown acrylamide to have neurotoxic, genotoxic, and carcinogenic properties (Rice, 2005; CAST, 2006; Klaunig, 2008; Doerge et al., 2008; LoPachin and Gavin, 2008). Chronic acrylamide exposure was shown to be carcinogenic in rats and mice, causing tumors at many organ sites when given in drinking water or by other means (Bull et al., 1984; Johnson et al., 1986; Friedman et al., 1995; Klaunig, 2008). In mice, acrylamide increased the incidence of lung and skin tumors (Bull et al., 1984). In two bioassays in rats, acrylamide administered in drinking water consistently produced mesotheliomas of the testes, thyroid tumors, and mammary gland tumors (Friedman et al., 1995; Johnson et al., 1986). In one of the rat bioassays, acrylamide induced formation of pituitary tumors, pheochromocytomas, and uterine and brain tumors (Johnson et al., 1986).

In 2012, FDA National Center for Toxicological Research completed a 2-year chronic rodent carcinogenicity bioassay in rats and mice under the auspices of the National Toxicology Program (NTP, 2012). This assay included study of dose–response relationships, histopathology, and correlation of adduct levels in target tissues with tumor incidence. Female and male rats and mice (50 in each category) were exposed to acrylamide from drinking water [0–50 parts per million (ppm)] for 2 years. The results indicated that male and female rats and mice receiving acrylamide had increased rates of types of cancers: male rats had increased rates of cancer in the pancreatic islets and of malignant mesotheliomas, and female rats also had increased rates of cancers in the clitoral gland, liver, mammary gland, skin, and mouth or tongue. Male and female mice had increased rates of cancer in the Harderian gland, lung, and stomach; female mice also had increased rates of cancer in the mammary gland, skin, and ovary (NTP, 2012).

Although animal studies have shown that exposure to acrylamide causes a higher incidence of tumors, intake levels at which it causes cancer are much greater than those typically consumed by humans (Burley et al., 2010). The mechanism by which acrylamide causes cancer in laboratory animals is not clear. Genotoxic and nongenotoxic mechanisms have been suggested (Hogervorst et al., 2008). In animals, acrylamide is metabolized by conjugation with glutathione or is oxidized to the epoxide glycidamide (2,3-epoxypropionamide) by an enzymatic reaction involving cytochrome P450 2E1. Both acrylamide and glycidamide can form adducts with DNA (Besaratinia and Pfeifer, 2004). Possible nongenotoxic mechanisms by which acrylamide can induce cancer are that acrylamide reacts with glutathione and may influence the redox status of cells, or it may interfere with DNA repair (Hogervorst et al., 2008). Glycidamide has been found to induce DNA damage, primarily by formation of N7-glycidamide-guanine adducts (Watzek et al., 2012). Based on the results of rodent carcinogenicity studies, IARC deemed acrylamide to be a likely human carcinogen through the formation of glycidamide (IARC, 1994).

More than one third of the calories that is consumed by the U.S. and European populations are derived from food that contains acrylamide (Friedman and Levin, 2008). Human dietary exposure to acrylamide is not negligible and can amount to 100 μg/day. Studies have been done to determine whether the amount of acrylamide in the human diet is an important cancer risk factor. Several reviews have been written on the relationship between dietary acrylamide and human cancer risk (Lipworth et al., 2012; Riboldi et al., 2014; Virk-Baker et al., 2014).

The vast majority of the epidemiological studies examining the relationship between dietary intake of acrylamide and cancers of the colon, rectum, kidney, bladder, and breast have found no association between intake of acrylamide-containing foods and the risk of these cancers (Mucci and Wilson, 2008; Mucci et al., 2003, 2004, 2006; Burley et al., 2010; Lipworth et al., 2012). However, epidemiological studies conducted by Hogervorst et al. (2008) and Olesen et al. (2008) have shown that dietary exposure to acrylamide increased the risks of renal cancer and ovarian cancer, respectively. Virk-Baker et al. (2014) pointed out that the majority of the epidemiology studies used a single time point to estimate exposure to acrylamide-containing foods rather than multiple time points or long-term exposure. Since food consumption choices can vary due to seasonality, social factors, and other factors, exposure estimates may not have been accurately captured in previous epidemiology studies. Clearly, more work is needed to determine the health consequence of acrylamide from dietary sources. Future studies that use improved dietary exposure estimates using biomarkers and improved dietary assessment tools are needed (Virk-Baker et al., 2014).

In 2015, the EFSA Scientific Panel on CONTAM adopted an opinion on acrylamide in food (EFSA, 2015). Based on animal studies, EFSA confirmed previous evaluations that acrylamide in food potentially increases the risk of developing cancer for consumers in all age groups. Since acrylamide is present in a wide range of everyday foods, this concern applies to all consumers, but children are the most exposed age group on a body weight basis. Possible harmful effects of acrylamide on the nervous system, pre- and postnatal development, and male reproduction were not considered to be a concern, based on current levels of dietary exposure. Although epidemiological studies have not demonstrated acrylamide to be a human carcinogen, the margin of exposure (MoE) indicates a concern for neoplastic effects based on animal studies (EFSA, 2015).

13.2.7 Regulatory status/risk management

Owing to its carcinogenic potential in animals and its MoE, acrylamide is considered a potential health hazard (JECFA, 2005; EFSA, 2015). Therefore, food authorities in many countries have recommended or requested food manufacturers to take measures to limit acrylamide formation in their products (Amrein et al., 2007; FDA, 2016a; EC, 2017).

In 2002, the German minimization concept for acrylamide was established by the Federal Office of Consumer Protection and Food Safety (BVL). The concept was based on a voluntary agreement among the BVL, the Federal Ministry of Food, Agriculture and Consumer Protection (BMELV), the German federal state authorities, and the stakeholders of the affected industry. The aim of the minimization concept was to achieve gradual reduction of acrylamide content of foods by avoiding formation (Göbel and Kliemant, 2007).

In 2017, the EU published a food safety regulation (2017/2158) that aims to reduce the levels of acrylamide in specific food. The regulation, which went into effect in April 2018, requires that all food business operators (FBOs) put into place steps to control acrylamide levels within their food safety management systems, so that levels of acrylamide are as low as reasonably achievable. The legislation applies to all FBOs that produce or place on the market French fries and other cut potato products; potato chips, snacks, and other products made from potato dough; bread, breakfast cereals; fine bakery wares such as cookies, biscuits, rusks, cereal bars, etc.; coffee, coffee substitutes; and baby food and processed cereal-based foods intended for infants and young children (EC, 2017). Firms could use the suggested mitigation procedures described in the Annexes to the regulation (EC, 2017), the Food Drink Europe Acrylamide Toolbox (FoodDrinkEurope, 2019) or the FAO/WHO Codex "Code of Practice for the Reduction of Acrylamide in Foods" (CODEX Alimentarius Commission, 2009). The benchmark levels set out in the regulation Annex are generic performance indicators for the various food categories covered by the regulation (EC, 2017).

In 2016, the FDA issued a guidance document that describes ways for growers, food manufacturers, and food service operators to reduce acrylamide levels in potato- and cereal-based foods (FDA, 2016a). Although the guidance document does not require firms to reduce acrylamide levels, it recommends that manufacturers consider approaches for mitigating acrylamide formation in their products.

Several regulatory agencies have issued advice to consumers for reducing exposure to dietary acrylamide (Health Canada, 2009; FDA, 2016a, b, 2017a; Food Standards Agency, 2018). Additional advice on ways to reduce cancer risk from acrylamide is available from the HEATOX Project (HEATOX, 2007b), the American Cancer Society (2019), and the National Institutes of Health National Cancer Institute (National Cancer Institute, 2019). Overall, these groups suggest consumers not to make major dietary changes, but to eat a balanced diet, choosing a variety of foods that are low in *trans* fat and saturated fat, and rich in high-fiber grains, fruits, and vegetables.

13.3 Furan

13.3.1 Introduction

Furan and methyl furans, volatile (boiling point = 31.4°C; furan) cyclic ether compounds, are found in cigarette smoke and are used in the production of resins and lacquers, agrochemicals, and pharmaceuticals (Goldmann et al., 2005). Furan and furan derivatives have long been known to occur in heated foods and contribute to the sensory properties of food (Maga, 1979; Zoller et al., 2007). However, attention has been brought to the presence of furan in a wide variety of heated processed foods (coffee, juices, soups, and canned and jarred fruits and vegetables, including baby foods) by the FDA following the posting of data on the occurrence of the contaminant in food on its website in 2004 (FDA, 2017b). The concerns over furan stem from its classification as a "possible carcinogen to humans" (Group 2B) IARC (1995) and the finding that the compound causes cancer in rodents (NTP, 1993).

Shortly after the FDA published data on furan levels in food, there was interest by the scientific and regulatory community to determine occurrence of the contaminant in different food categories, estimate exposure to furan, explore the mechanisms of formation of furan, and access methods for reducing furan levels in foods.

The following sections summarize the current state of information on occurrence and levels of furan in food, mechanisms of formation, processing factors, health effects, and regulatory status. More complete information regarding these topics can be obtained from Vranová and Ciesarová (2009), CODEX Alimentarius Commission (2011), Seok et al. (2015), EFSA (2017), and Kettlitz et al. (2019).

13.3.2 Occurrence and levels of furan in food

Since furan has become a potential food safety issue, several international food agencies such as the FDA and the EFSA have launched monitoring programs to survey the furan content of selected foods and beverages (Altaki et al., 2009). Information on furan levels in thermally treated foods in the United States can be obtained from surveys conducted by the FDA in 2004–08 (FDA, 2017b). Table 13.3 summarizes the data from FDA's exploratory survey on furan levels in different food categories. FDA's exploratory survey included analysis of baby and infant foods (jarred foods, infant

TABLE 13.3 Summary of furan levels in various food categories.[a]

Food category	Furan concentration (μg/kg)	Food product with highest level of acrylamide (furan concentration)
Infant and toddler foods	N.D.−111	Jarred garden vegetables
Infant formula	ND.−26.9	Milk-based infant formula concentrate
Mixtures (soups, sauces, broths, chili)	<5.0−125	Vegetable beef soup
Fish (canned)	<5.0−8.1	Canned sardines in tomato sauce
Canned fruit and vegetables	N.D.−122	Baked beans with pork
Fruit and vegetable juices, punches and drinks	N.D.−30.5	Prune juice
Bread	N.D.−<2.0	Whole wheat bread
Breakfast cereals	9.2−47.5	Corn flakes
Crackers and crispbreads	4.2−18.6	Whole grain crispbread
Bakery products	N.D.−30.1	Ginger snaps
Meats	ND.−39.2	Sausage
Nuts and nut butters	2.1−7.5	Peanut butter
Beer	N.D.−4.4	−
Nutrition drinks	N.D.−174	Strawberry shake
Dairy and eggs	N.D.−15.3	Evaporated milk
Fats and oils	N.D.−5.4	Vegetable shortening
Jams, jellies and preserves	1.3−37.4	Apple butter spread
Gravies	13.3−173.6	Roast Turkey gravy
Desserts	<0.8−13	Rice pudding
Snacks	<3.2−64.7	Pretzels
Candy	<0.8−5.5	Red licorice
Coffee (brewed)	N.D.−84.2	−
Dried fruit	N.D.−2.2	Dried cranberries
Chocolate drink mixes, cocoa, chocolate syrups, and chocolate drinks	0.5−10.3	Chocolate drink
Miscellaneous	N.D.−88.3	Maple syrup

N.D. = Not Detected.
[a]*Data from FDA, 2017b. Exploratory Data on Furan in Food. U.S. Food and Drug Administration, Center for Food Safety and Applied Nutrition. https://www.fda.gov/food/chemicals/exploratory-data-furan-food. (Accessed 20 October 2019).*

formulas), canned mixed foods (soups, sauces, broths, chili), canned fish and meat products, canned fruit, fruit juices and vegetables, dried fruit, breads and other bakery goods, breakfast cereals, snack foods, nut butters, beers, and other foods. Of the foods analyzed, the highest furan levels were found in jarred infant/toddler foods (sweet potatoes, 73−108 μg/kg), canned baked beans (56−122 μg/kg), brewed coffee (34−84 μg/kg), and canned soups containing meat and vegetables (50−125 μg/kg). Not included in FDA's survey were analyses of the food items for methyl furan (i.e., 2-methyl furan and 3-methyl furan).

Becalski et al. (2009) published a survey of 176 food items purchased in retail stores in Canada for furan, 2-methyl furan and 3-methyl furan. Most of the sampled products were packaged in cans or jars. Furan was detected at levels of above 1 μg/kg in all nonbaby foods with concentrations ranging from 1.1 to 1230 μg/kg and a median concentration of 28 μg/kg. Furan was detected above 1 μg/kg in baby foods, with concentrations ranging from 8.5 to 331 μg/kg and a median concentration of 66.2 μg/kg. Foods that tended to have higher concentrations of furan included coffee, canned beans, canned pasta, and canned meat-containing products (Becalski et al., 2009). In a more recent survey, Becalski et al. (2016) evaluated furan, 2-methylfuran and 3-methylfuran contents of coffees obtained in Canada and found concentrations similar to results of his earlier survey (Becalski et al., 2009).

Data are available on furan levels in food commodities obtained in Europe, Africa, and Asia. In a survey of food items purchased in Spain, Altaki et al. (2009) reported the highest levels of furan (820 and 1100 μg/kg or ppb) in powdered instant coffees and in a jarred baby food entrée containing spaghetti and beef (40 μg/kg). A wide variety of food items purchased in Switzerland were analyzed for furan (Reinhard et al., 2004; Zoller et al., 2007). Furan levels were higher in jarred baby foods containing both meat and vegetables (21−153 μg/kg) than in baby foods containing solely meat, vegetables, or fruit. Minimal amounts of furan were found in jarred baby foods containing solely meat (3−6 μg/kg) or fruit (1−16 μg/kg). Similar to FDA's survey, Zoller et al. (2007) found moderate amounts (17−80 μg/kg) of furan in jarred vegetable foods. Breads contained up to 30 μg/kg furan, and the majority of the furan was present in the crust. Ground roasted coffee beans contained up to 5900 μg/kg furan, while brews of the coffees contained from 13 to 151 μg/kg. Shen et al. (2016) analyzed 126 thermally processed food items obtained in China for the presence of furan, 2-methyl furan, 2,5-dimethylfuran, and other alkylfurans. The survey found that furan and 2-alkylfurans were detected in 70.6% of food samples with total furan concentrations ranging from 0.5 to 789.4 μg/kg.

The EC requested that EU member states to collect data on furan levels in heat-treated commercial food products. In response to the request, a total of 20 countries have so far submitted analytical results for furan content in food to EFSA. A total of 5050 complete results were reported for foods sampled between 2004 and 2010 (EFSA, 2011). Of the 21 different food categories (5 coffee and 16 noncoffee categories), the highest concentrations of furan were found in coffee products, baby foods, and canned soups. Mean furan levels in brewed coffee, instant coffee powder, roasted ground coffee, unspecified coffee, and roasted coffee beans were 45 μg/kg, 394 μg/kg, 1936 μg/kg, 2016 μg/kg for unspecified coffee product, and 3660 μg/kg for roasted coffee beans. Overall, the data submitted to EFSA are similar to FDA's exploratory survey results.

Dietary exposure assessments to furan were developed by FDA in 2004 and revised in 2007 (FDA, 2007). The FDA 2007 assessment reported a mean and 90th percentile of 0.26 and 0.61 μg furan/kg body weight/day for 2+-year-olds, respectively, and 0.41 and 0.99 μg furan/kg body weight/day for 0 to 1 year-olds from the consumption of adult and infant foods, respectively (Bolger et al., 2009). Based on the 2007 exposure assessment conducted by the FDA (FDA, 2007), brewed coffee contributes the greatest proportion of furan to the adult diet followed by chili, cereals, snack foods, and soups containing meat (Table 13.4).

In 2011, EFSA published estimates exposure to furan for different populations by combining pooled furan occurrence values obtained through 2004−10 monitoring efforts (EFSA, 2011). Mean exposure was estimated to range between 0.03 and 0.59 μg furan/kg body weight/day for adults, between 0.02 and 0.13 μg furan/kg body weight/day for adolescents, between 0.04 and 0.22 μg furan/kg body weight/day for other children, between 0.05 and 0.31 μg furan/kg body weight/day for toddlers, and between 0.09 and 0.22 μg furan/kg body weight/day for infants. Brewed coffee was the major contributor to exposure for adults with an average of 85% of total furan exposure, while major contributors to furan

TABLE 13.4 Exposure to furan from adult food categories.[a]

Food category	Amount of furan contributed to diet (μg/kg body weight/day)
Brewed coffee	0.15
Chili	0.04
Cereals	0.01
Salty snacks	0.01
Soups containing meat	0.01
Pork and beans	0.004
Canned pasta	0.004
Canned pasta	0.004
Canned string beans	0.004
Pasta sauces	0.001
Juices	0.001
Canned tuna (water packed)	0.000008

[a]From FDA (2007).

exposure in toddlers and other children were fruit juice, milk-based products, cereal-based products, and jarred baby foods (for toddlers, only) (EFSA, 2011). More work is needed to expand the database on furan and methyl furan levels in food, to determine variability in furan and methyl furan levels between and within brands, and to evaluate the effects of home cooking on furan and methyl furan levels.

13.3.3 Mechanisms of formation

Furan forms in a wide variety of foods of different composition suggesting several pathways of formation. Precursors that have been found to form furan during thermal treatment include ascorbic acid and its derivatives, Maillard reaction systems containing reducing sugars and amino acids, lipids containing unsaturated fatty acids, carotenoids, and organic acids (Maga, 1979; Frankel et al., 1987; Mottram, 1991; Locas and Yaylayan, 2004; Becalski and Seaman, 2005; Fan, 2005; Limacher et al., 2007, 2008; Fan et al., 2008; Van Lancker et al., 2011). An overview of the major routes of formation of furan from amino acids, carbohydrates and ascorbic acid, and polyunsaturated fatty acid precursors is shown in Fig. 13.2.

Mechanistic studies on furan formation were carried out by Locas and Yaylayan (2004) using model systems containing $^{13}C_3$-labelled sugars, amino acids, and ascorbic acid heated at 250°C (pyrolysis conditions). Among the precursors studied, ascorbic acid had the highest potential to form furan, followed by glycolaldehyde/alanine > erythrose > ribose/serine > sucrose/serine > fructose/serine > glucose/cysteine. In similar experiments, Mark et al. (2006) found that ascorbic acid had the greatest potential to form furan in simple model systems than other precursors (glyceryl trilinolenate, linolenic acid, trilinolein, and erythrose).

Ascorbic acid and its derivatives (isoascorbic acid, dehydroascorbic acid) have been shown to be a major precursor of furan formed during different processing conditions (pyrolysis, thermal decomposition, roasting, retorting, γ-irradiation) of aqueous solutions and in food (Fan, 2005; Becalski and Seaman, 2005; Fan and Geveke, 2007; Limacher et al., 2007). In experiments using simple model systems, levels of furan formed ranged from 0.1 μmol furan/mol ascorbic acid to about 1 mmol furan/mol ascorbic acid (Limacher et al., 2008). In more complex systems such as food, Limacher et al. (2007, 2008) found that formation of furan from ascorbic acid was significantly reduced due to competing reactions. These results illustrate the difficulty in extrapolating results from simple model systems to foods.

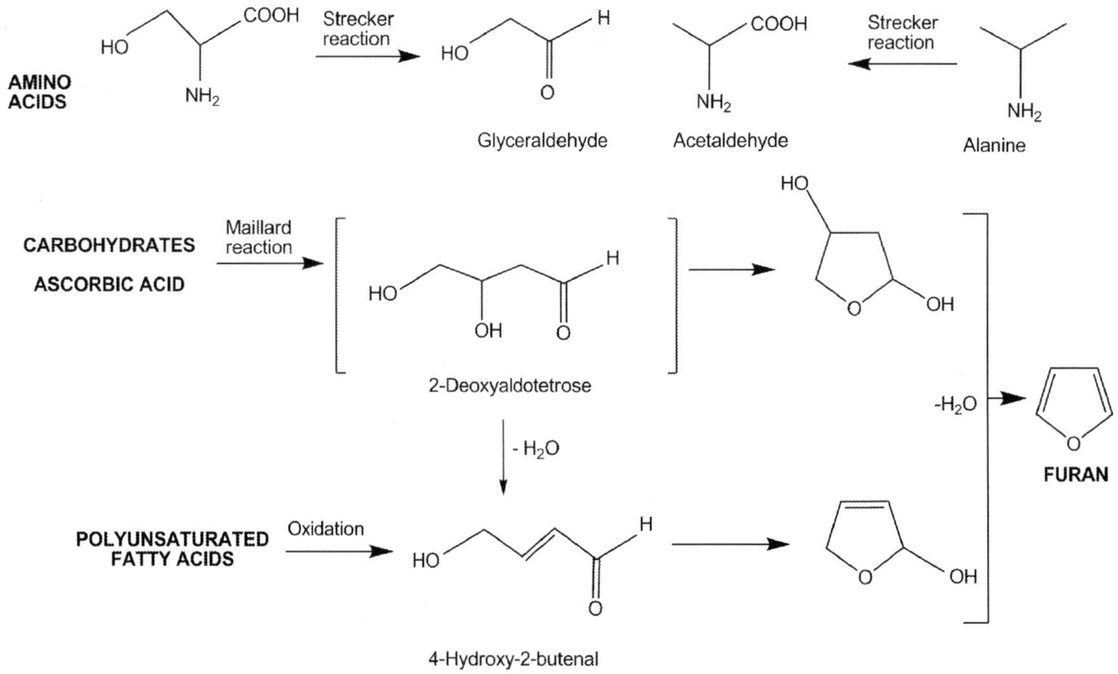

FIGURE 13.2 Mechanisms of formation of furan in heated foods from amino acids, carbohydrates, ascorbic acid and polyunsaturated fatty acid precursors. *Adapted from Locas, C.P., Yaylayan, V.A., 2004. Origin and mechanistic pathways of formation of the parent furan- a food toxicant. J. Agric. Food Chem. 42, 6830–6836, Mark, J., Pollien, P., Lindinger, C., Blank, I., Mark, T., 2006. Quantitation of furan and methylfuran formed in different precursor systems by proton transfer reaction mass spectrometry. J. Agric. Food Chem. 54, 2786–2793 and Crews, C., Castle, L., 2007. A review of the occurrence, formation and analysis of furan in heat-processed foods. Trends Food Sci. Technol. 18, 365–372.*

Pyrolysis of sugars at extreme temperatures resulted in formation of furan and alkylated furan derivatives. Hexoses, pentoses, tetroses, and polysaccharides generated furan and its derivatives, while glyceraldehyde, a triose, was less efficient (Locas and Yaylayan, 2004). Overall, sugars or amino acids when heated alone were not efficient at producing furans, but when heated in combination, are significant precursors of furan in heated food (Locas and Yaylayan, 2004; Fan et al., 2008). Of the sugars studied, erythrose was efficient at producing furan, but only when heated at temperatures (i.e., >250°C) that do not represent typical food-processing conditions (Locas and Yaylayan, 2004; Mark et al., 2006).

In experiments using labeled sugars, Limacher et al. (2008) identified two major pathways of furan formation from sugars: (1) from the intact sugar skeleton, and (2) by recombination of reactive C2 and C3 fragments. Experiments by Van Lancker et al. (2011) found that under roasting and pressure-cooking conditions, furan was formed from glucose via its intact skeleton, and that its formation pathways from glucose were not amino acid dependent. However, presence of some amino acids, such as serine and alanine, promoted furan production by providing an additional formation pathway. Serine, threonine, and alanine promote furan formation by recombination of C2 fragments such as acetaldehyde and glycolaldehyde (Vranova and Ciesarova, 2009).

Several amino acids alone, such as serine or cysteine, can undergo thermal degradation to produce furan, while others such as alanine, threonine, and aspartic acid require the presence of reducing sugars to produce furan (Locas and Yaylayan, 2004). Reducing sugars in the presence of amino acids undergo the Maillard reaction and generate reactive intermediates that can ultimately form furan. In simple model systems simulating roasting conditions, furan and methyl furan were formed at levels of up to 330 μmol/mol precursor from mixtures of sugars and selected amino acids (Limacher et al., 2008).

Furan can be formed during thermal treatment of unsaturated fatty acids and the yield of furan increases with the degree of unsaturation (Becalski and Seaman, 2005; Hasnip et al., 2006; Crews and Castle, 2007). Becalski and Seaman (2005) found that furan can be formed by oxidation of polyunsaturated fatty acids at elevated temperatures and that formation was reduced by addition of antioxidants such as tocopherol acetate. Overall, these results suggest that the formation of furan is linked with the process of free radical autoxidation (Locas and Yalyayan, 2004; Crews and Castle, 2007).

A considerable amount of information has been generated on formation of furan and its derivatives (2-methyl furan and further alkylated derivatives) in model systems containing a variety of precursors. Although these studies provide important information about possible mechanisms of formation of furan, they do not necessarily mimic furan formation in food since food systems tend to be more complex in nature and processing conditions are typically less severe.

13.3.4 Factors affecting furan formation and mitigation in food

Many factors can affect formation furan, but those thought to have the greatest effect include food composition and processing or cooking conditions. However, when the available information is examined, it is difficult to clearly define the effects of these factors on furan levels in food. This may be due to the inherent volatility of furan, making quantitation difficult and the fact that furan is formed by at least several mechanisms and with different classes of food precursors. If survey (Zoller et al., 2007; Becalski et al., 2009; EFSA, 2011; FDA, 2017b) and experimental (Limacher et al., 2008) data are examined, it becomes apparent that furan levels tend to be greater in foods containing complex mixtures of carbohydrates, fatty acids, and proteins than those of simple composition. As mentioned previously, furan can form from thermal processing of a variety of different food components. In a simple food system such as apple cider (containing mainly sugars with small amounts of ascorbic acid and fatty acids), very low amounts of furan were formed during heating at 120°C for 10 min. In contrast, more furan was produced by heating pumpkin puree, a food matrix that contained a mixture of starch, simple sugars, fatty acids, carotenoids, and ascorbic acid (Limacher et al., 2008).

The presence of phosphate and the pH of the food may influence furan formation. In a model system consisting of ascorbic acid and fructose, phosphate increased furan formation (Fan et al., 2008). It was speculated that phosphate may increase furan production from sugars through the formation of reactive intermediate compounds from the Maillard reaction which can recombine to form furan. More furan was formed from heated ascorbic acid or sucrose solutions as the pH was reduced from 8 to 3 (Fan, 2005; Fan et al., 2008). Palmers et al. (2016) investigated the effects of pH and oxygen availability on furan formation in potato purees during pasteurization or sterilization thermal treatments. The results indicate that lowering the oxygen concentration or pH prior to thermal processing decreased furan production.

At present, there are few systematic studies on the effects of cooking/processing temperature and time on furan formation in food. Fan et al. (2008) reported that significant amounts of furan were produced in apple cider only at higher temperatures (>100°C) and prolonged times (>4 min). Similarly, Hasnip et al. (2006) reported greater levels of furan in bread toasted at 17–200°C than at 140–160°C. Use of high-pressure processing of vegetable purees has been shown to have reduced furan levels by 81%–96% compared retorted purees (Palmers et al., 2014).

Furan levels tend to be greater in foods that are heat processed in cans or jars since furan that is formed cannot escape from the closed container (Roberts et al., 2008; Kettlitz et al., 2019). Typical home cooking does not appreciably contribute to furan formation, since furan is likely released to the atmosphere from open cooking vessels (Kettlitz et al., 2019). Several reports (Roberts et al., 2008; Hasnip and Castle, 2006; Goldmann et al., 2005) have shown that furan in jarred or canned foods is lost during opening, during cooking (at lower temperatures), and during stirring/mixing. Of the cooking methods studied, heating canned and jarred foods in an open saucepan resulted in the greatest losses of furan than microwave cooked foods (Roberts et al., 2008). Furan levels in cooked foods and beverages decreased upon sitting and stirring foods before consumption also reduced furan levels (Goldmann et al., 2005; Roberts et al., 2008; Al-Taher et al., 2008). Becalski et al. (2016) reported that brewing coffee samples in laboratory as per manufacturers' instructions resulted in 27%–85% loss of furans, and that allowing brewed coffee to stand for up to 30 min resulted in 3%–47% additional losses. A study on the effects of preparation methods on furan levels in brewed coffee showed that furan levels increased as a function of brew strength and decreased significantly when coffee brews were heated in an open carafe for 1 h (Al-Taher et al., 2008). Heating jars of baby food by immersion in a hot water bath followed by stirring the contents of the jar resulted in decreases in furan levels (Goldmann et al., 2005; Roberts et al., 2008).

Overall, the work to date indicates that furan in retail foods persists during the normal heating practices that precede consumption (Hasnip et al., 2006). However, treatments that may reduce consumer exposure to furan include cooking in open containers and stirring food before consuming. More work is needed on determining the effects of food preparation/processing procedures on furan and methyl furan levels in food as they are consumed. Research is also needed to obtain a better understanding of the mechanisms by which they are formed in actual food systems. This may enable the development of mitigation approaches for furan and methyl furan in food.

13.3.5 Health effects of dietary furan

Furan is a hepatocarcinogen in rats and mice and induces hepatocellular carcinomas and cholangiocarcinomas (NTP, 1993). Furan has also been shown to cause a dose-dependent increase in mononuclear leukemia in male and female rats (NTP, 1993). The mechanism of carcinogenic activity of furan in rodents has not been elucidated but may be due to the activation of furan to the toxic metabolite, cis-2-butene-1,4-dial, by cytochrome P450 in the liver (Fransson-Steen et al., 1997; Hamadeh et al., 2004). There is some evidence that cis-2-butene-1,4-dial can react with DNA in target cells and induce the production of tumors (Wilson et al., 1992; Chen et al., 1995).

There is currently little information on the reproductive and developmental toxicities of furan and the toxicological effects of furan in humans from dietary sources (Heppner and Schlatter, 2007). Jun et al. (2008) measured urinary furan levels in healthy individuals consuming a normal diet and detected up to 3.14 ppb in over half (56%) of the subjects. In individuals with detectable urinary furan, the level of γ-glutamyltranspeptidase, a marker for liver disease, was strongly correlated with urinary furan concentration. This study points to the need to evaluate the metabolic fate and potential toxicity of dietary furan in humans.

13.3.6 Regulatory status

To date, FDA and other regulatory agencies have not initiated any regulatory action on furan. This is mainly due to a need to better understand mechanisms of formation of furan and furan derivatives, and a lack of information on effective approaches to significantly reduce levels of furan in food.

In 2011, the CODEX Committee on Contaminants in Food (CCCF) wrote a discussion paper on furan, which summarized information on such topics as toxicology and epidemiology, analytical methods, occurrence and exposure, formation, and mitigation approaches (CODEX Alimentarius Commission, 2011). Based on the need for more information on approaches for effectively reducing furan levels in food, CODEX CCCF concluded that it is premature to develop a Code of Practice document for furan (CODEX Alimentarius Commission, 2011).

Based on its ability to cause cancer in experimental animals, furan was classified as a group 2B carcinogen by IARC (IARC, 1995). In 2011, JECFA concluded that human exposure to furan from food constitutes a human health concern. This conclusion was based on the use of the MoE approach applied to benchmark dose modeling of rodent liver tumor incidence data (JECFA, 2011; Kettlitz et al., 2019). JECFA's calculated MoEs indicate a human health concern for a carcinogenic compound that might act via a DNA-reactive genotoxic metabolite.

Recently, the EC requested that EFSA provide an assessment on the risk to human health from the presence of furan and methyl furans (2-methylfuran, 3-methylfuran, and 2,5-dimethylfuran) in food (EFSA, 2017). Based on this request, EFSA evaluated survey data on furan levels in food, exposure estimates, and toxicological assessments, and published an opinion on the human health risks associated with consumption of furan and furans derivatives in food. The EFSA Panel on CONTAM used an MoE approach for the risk characterization using as a reference point a benchmark dose lower confidence limit for a benchmark response of 10% of 0.064 mg/kg body weight (bw) per day for the incidence of cholangiofibrosis in the rat. The calculated MOEs indicate that exposure to furan and methyl furans in food could lead to possible long-term liver damage. Gaps in the data used for the 2017 EFSA assessment are a lack of information on the levels of alkyl furans in food and factors affecting formation and dissipation of alkyl furans (Kettlitz et al., 2019). This information is critical for development of regulations regarding furan in food, and potential approaches for the food industry and consumers to reduce exposure to furan and furan derivatives.

13.4 Trans fatty acids

13.4.1 Introduction

TFAs have received great attention since the 1990s when studies began to emerge about their adverse effects on human health (Brownell and Pomeranz, 2014). TFAs, like saturated fatty acids, raise blood levels of low-density lipoprotein cholesterol, thereby increasing the risk of coronary heart disease (Mozaffarian et al., 2006; Teegala et al., 2009). At equivalent dietary levels, TFA may also increase the risk of coronary heart disease more than saturated fats. This is because, unlike saturated fats, TFAs also reduce blood levels of high-density lipoprotein cholesterol and increase blood levels of triglycerides (Judd et al., 1994; Ascherio and Willett, 1997; Hu et al., 1997; Mensink et al., 2003; Lopez-Garcia, 2005; Teegala et al., 2009; Remig et al., 2010). A study conducted in 2006 estimated that between 30,000 and 100,000 cardiac deaths per year in the United States were due to the consumption of *trans* fats (Mozaffarian et al., 2006). In light of these findings, consumers are advised to limit their intake of *trans* fats.

Trans fat is the common name for a type of unsaturated fat with *trans*-isomer fatty acid (s). *Trans* fats may be monounsaturated or polyunsaturated. *Trans* fats occur naturally in trace amounts in some animal-based foods (i.e., dairy and beef fat) due to bacterial transformation of unsaturated fatty acids in the rumen of ruminant animals (Dijkstra et al., 2008; Aldai et al., 2013). TFAs can be present at trace levels in most vegetable oils and in variable amounts in a wide range of foods, including most foods made with partially hydrogenated oils (PHOs), such as baked goods and fried foods, and some margarine products (Food Standards Australia New Zealand, 2005). Typically, ordinary vegetable oils, such as soybean, corn, cottonseed, sunflower, peanut, and olive, are moderately low in saturated acids, and the double bonds within unsaturated acids are in the *cis* configuration. Vegetable oils are sometimes hydrogenated in processed foods to improve their oxidative stability, extend their shelf-life, and give a more desirable texture (List, 2004a,b). For example, by using partially hydrogenated vegetable oil to make some margarine products, manufacturers can produce a spreadable topping that is lower in saturated fat than butter. Similarly, manufacturers can produce shortenings to manufacture French fries, flaky piecrusts, and crispy crackers.

At present, the TFAs formed by hydrogenation are considered nutritionally undesirable. Hydrogenation is a process whereby hydrogen atoms are added to cis-unsaturated fats in the presence of a catalyst, breaking the double bond and making them more saturated. These saturated fats have a higher melting point, which makes them desirable for baking and extends their shelf-life. However, the process frequently converts some cis-isomers into *trans*-unsaturated fats instead of hydrogenating them completely (Dijkstra et al., 2008; Aldai et al., 2013). Concerns exist about the potential health effects of TFA, especially those that are derived from partial hydrogenation of vegetable oils (Mozaffarian et al., 2006; Teegala et al., 2009). As a result, many countries including the United States, Canada, and some European countries have either placed restrictions on the use of TFA in processed foods or mandated labeling requirements for TFA in foods.

13.4.2 Regulatory status/risk management

Many countries have acted to reduce consumption of *trans* fats. Denmark was the first country, in March 2003, to issue an order (Order no. 160), strictly regulating the content of *trans* fats in foods. The order did not apply to TFAs in animal fats, but banned PHOs (Dijkstra et al., 2008; List, 2004b). A limit of 2% TFA of fats and oils was allowed for human consumption and a *trans*-free claim could be made if the product had less than 1% TFA.

Switzerland followed Denmark's *trans* fats ban, and implemented the same regulation beginning in April 2008 (Anonymous, 2008). In 2015, the EC adopted a report to the European Parliament and the Council regarding *trans* fats in foods and in the overall diet of the EU population (EC, 2015). In April 2019, the EC adopted a Commission Regulation amending Annex III to Regulation (EC) No 1925/2006 of the European Parliament and of the Council in regards to trans fat, other than trans fat naturally occurring in fat of animal origin. The regulation set a maximum limit of *trans* fat, other than *trans* fat naturally occurring in fat of animal origin, in food which is intended for the final consumer and food intended for supply to retail, of 2 g per 100 g of fat.

On July 11, 2003, the FDA announced final rules requiring that *trans* fat be stated in the nutrition label of conventional foods and dietary supplements (Dijkstra et al., 2008; FDA, 2003). As of January 1, 2006, the *trans* fat content of foods must be labeled as a separate line on the Nutrition Facts label. Products containing less than 0.5 g/serving of *trans* fat can be listed as 0 g *trans* fat on the food label (Dijkstra et al., 2008; FDA, 2003). In 2015, the FDA determined that PHOs, a major source of artificial *trans* fat in the food supply, are no longer "Generally Recognized as Safe," or GRAS. Since June 18, 2018, food manufacturers are not permitted to add PHOs to food. However, a compliance date was extended to January 1, 2020 to allow for foods that were manufactured prior to June 18, 2018 to work through their distribution (FDA, 2018).

Since December 2005, Health Canada has required that food labels list the amount of *trans* fat for most foods. Products containing less than 0.049 g of *trans* fat or less per serving may be labeled as free of *trans* fats (Canadian Food Inspection Agency, 2019). Similar to FDA efforts, on September 17, 2018, Health Canada banned the use of PHOs in foods. The ban came into effect with the addition of PHOs to the "List of Contaminants and other Adulterating Substances in Foods." Food manufacturers are now prohibited to add PHOs to foods sold in Canada. The ban on PHOs pertains to both Canadian and imported foods, as well as those prepared in all food service establishments (Health Canada, 2019).

13.4.3 Hydrogenation

Hydrogenation is important for two reasons in the fats and oils industry. It converts the liquid oils into semisolid or plastic fats for special applications, such as in shortenings and margarine, and it improves the oxidative stability of the oil (Nawar, 1996; Dijkstra et al., 2008). Hydrogenation involves the reaction between unsaturated liquid oil and hydrogen adsorbed on a metal catalyst. The oil is mixed with a suitable catalyst (usually nickel), heated (140−225°C), then exposed, while stirred, to hydrogen at pressures up to 60 psi (Nawar, 1996). First, a carbon−metal complex is formed at either end of the olefinic bond. This intermediate complex then reacts with an atom of catalyst-adsorbed hydrogen to form an unstable half-hydrogenated state in which the olefin is attached to the catalyst by only one link and is thus free to rotate. The half-hydrogenated compound may either (a) react with another hydrogen atom and separate from the catalyst to generate the saturated product or (b) lose a hydrogen atom to the nickel catalyst to restore the double bond. The regenerated double bond can be either in the same position (cis) as in the unhydrogenated compound, or a positional and/or geometric isomer (*trans*) of the original double bond (Nawar, 1996; Dijkstra et al., 2008). The hydrogenation process is usually monitored by determining the change in refractive index, which is related to the degree of saturation of the oil. The hydrogenated oil is then cooled and the catalyst is removed by filtration.

The rate of hydrogenation and the formation of *trans* acids depend on the processing conditions (hydrogen pressure, intensity of agitation, temperature, and kind and concentration of catalyst). For many years, manufacturers of fats have been trying to devise hydrogenation processes that minimize isomerization while avoiding the formation of high amounts of fully saturated material (Nawar, 1996; Patterson, 1994).

Temperature is the most important variable in *trans* isomer control. Lowering the temperature below the traditional levels of 140°C causes a significant *trans* isomer reduction, but an increase in saturate levels (Patterson, 1994). A TFA level of 6% can be achieved with a hydrogenation temperature of 40−60°C with a nickel catalyst, but this decreases the catalytic activity. It is, therefore, important to use a catalyst that can saturate the double bonds at these lower temperatures. Conventional hydrogenation plants are configured for typical initiation reaction temperatures in the 110−150°C range (Patterson, 1994).

13.4.4 Decreasing *trans* fatty acids in fats and oils

The food industry has reduced the *trans* fat content in foods through the use of several techniques. These include use of lauric oils (coconut, palm kernel), palm fractions, completely hardened vegetable oils (stearines), interesterification, lower-temperature deodorization to minimize thermal *trans* formation, and chemical refining rather than physical refining (List, 2004b; Menaa et al., 2013). Methods that can be used to provide an oil blend with the required physical and chemical properties include the blending of different oils and fats, single- or multistage fractionation, interesterification, and combination of these processes. A *trans*-free margarine fat blend can, for example, be produced by blending a lauric oil and a nonlauric oil, fully hydrogenating the mixture, randomizing the fully hydrogenated fat by interesterification, fractionating the interesterified product to eliminate high-melting triglycerides and/or low-melting triglycerides, and blending the olein or mid-fraction with a liquid oil. This oil can also be subjected to a direct interesterification process to reduce the saturated fatty acid content of the fat blend (Dijkstra et al., 2008). Plant breeding and use of transgenic techniques have been used to create oilseeds with altered fatty acid composition.

13.4.4.1 Interesterification

The process of random interesterification was used in Europe during the 1950s and 1960s. This process can change the original order of distribution of the fatty acids in the triglyceride-producing products. Chemical and enzymatic interesterification processes can affect physical properties by changing the melting properties and in some cases the crystal behavior of the original oil or fat. Commercially, interesterification is used for processing edible fats and oils to produce confectionery or coating fats, margarine oils, cooking oils, frying fats, shortenings, and other special application products (O'Brien, 2004). Unlike hydrogenation, interesterification does not affect the degree of saturation or cause isomerization of the fatty acid double bond. In this process, fatty acids are removed from the glyceride molecules, these acids are shuffled, and replaced on the glyceride in random positions with the aid of a catalyst (O'Brien, 2004; List, 2004a; Menaa et al., 2013). Chemical interesterification has disadvantages. It forms by-products such as soaps, methyl esters, and partial glycerides, thus requiring postprocessing procedures to remove them (List, 2004b).

Enzymatic interesterification is currently being used commercially in North America and Europe to formulate margarine and shortening oils (Mensink et al., 2016). The major advantages of the enzymatic interesterification over the chemical processes are that the lipase catalysts are specific, and the reaction can be better controlled (O'Brien, 2004). Oil modification by lipases is performed under anhydrous conditions at temperatures up to 160°F (70°C). Catalysts/enzymes are expensive, but there are no side-products and no postprocessing (List, 2004b; O'Brien, 2004).

Interesterification can be used to formulate products with less saturated or isomerized fatty acids for the production of products with low or no *trans*-acids (O'Brien, 2004). For example, interesterifying palm stearine with fully hydrogenated high-erucic acid rapeseed oil can provide a hardstock that allows *trans*-free margarines to be made that also contain low saturated fatty acids (Patterson, 1994).

13.4.4.2 Fractionation

A *trans*-free margarine fat blend can also be produced by fractionating the interesterified product to remove high-melting triglycerides and/or low melting triglycerides and blending the olein or mid-fraction with a liquid oil (Patterson, 1994). Dry fractionation is the standard industry method (List, 2004a). Palm oil is the most popular oil used for fractionation. Palm oil (IV 51–53) is fractionated into olein (IV 56–59) and stearine (IV 32–36) fractions (List, 2004a). Fractionation produces products with different properties. Molten fat forms crystals when chilled, and these crystals can be separated from the mother liquor by filtration. The crystals are isolated as a filter cake and are called the stearin fraction and the filtrate is referred to as the olein fraction (O'Brien, 2004).

Single-stage fractionation generates an olein fraction with a cloud point of <10°C. It is used as a substitute for soft oils in frying or cooking or it can be fractionated further (O'Brien, 2004). The olein (more liquid) fraction can then be fractionated into mid-fractions, super oleins, and top oleins. The palm mid-fractions (IV 42–48) are further processed into harder fractions (IV 32–36). Fractionation of the stearine (more solid) (IV 32–36) gives softer and super stearines (IV 40–42 and 17–21, respectively). These fractions can be incorporated into margarines/shortenings and confectionery fats. However, replacing *trans* fat with highly saturated palm fractions will increase saturated acid content (List, 2004a).

Multistage fractionation of palm oil is applied to produce high-IV superoleins (IV > 65) and soft-palm mid-fractions (S-PMF with IV 45–47). High-IV superoleins combine a low cloud point with good oxidative stability and are therefore used as frying oil and salad oil. S-PMF is increasingly used as *trans*-free hardstock in margarines and shortenings (O'Brien, 2004). Dry fractionation is also used for the modification of other vegetable oils (cottonseed oil, partially hydrogenated soybean oil, etc.) and animal fat (lard, fish, oil, etc.) (Dijkstra et al., 2008).

13.4.4.3 Modified fatty acid composition

Several transgenic and plant breeding techniques have been used to create oilseeds with altered fatty acid composition. These include low-linolenic soybean, high-oleic corn (maize) and soybean, high-oleic and mid-oleic sunflower, and high-saturate lines (List, 2004b). The low-linolenic soybean and the high-oleic sunflower, soybean, and corn oils can be used as frying oils and can reduce TFAs in fried snack foods. High-oleic oils (i.e., palmitic acid or stearic acid) are used without hydrogenation for frying and in high-stability applications. High-saturate oils are used in margarines, spreads, and shortenings. They may require blending with higher melting components (i.e., palm oil, cottonseed/soybean stearines) or interesterification of their glyceride structures for functionality (List, 2004b).

13.5 Conclusions

Processing plays an essential role in the modern world in providing a safe, palatable, nutritious, and consistent food supply. Despite these benefits, processing can result in the formation of chemical hazards such as the heat-produced toxicants, acrylamide and furan, and TFAs, a product of hydrogenation of oils. All three compounds have been found to have adverse physiological effects in laboratory animals and possible humans. Acrylamide and furan are produced by chemical reactions responsible for the desirable flavor, aroma, and color of cooked foods. TFAs are generated during processes that improve the texture of oils and render them more stable to oxidative stresses. Changing processing conditions to minimize formation of acrylamide, furan, and TFAs can result in undesirable effects on food safety and quality. Regulatory agencies in collaboration with the food industry and academia have placed an emphasis on gathering more information on the mechanisms by which they are formed and the health consequences resulting from dietary exposure to these compounds. This information will provide ways for managing these chemical hazards in food.

References

Abt, E., Robin, L.P., McGrath, S., Srinivasan, J., DiNovi, M., Adachi, Y., Chirtel, S., 2019. Acrylamide levels and dietary exposure from foods in the United States, an update based on 2011–2015 data. Food Addit. Contam. 36, 1475–1490.

Açar, Ö.Ç., Pollio, M., Di Monaco, R., Fogliano, V., Gökmen, V., 2012. Effect of calcium on acrylamide level and sensory properties of cookies. Food Bioprocess Technol. 5, 519–526.

Ahrné, L., Andersson, C.-G., Flosberg, Rosén, J., Lingnert, H., 2007. Effect of crust temperature and water content on acrylamide formation during baking of white bread: steam and falling temperature baking. LWT-Food Sci. Technol. 40, 1708–1715.

Aldai, N., de Renobales, M., Barron, L.J.R., Kramer, J.K.G., 2013. What are the trans fatty acids issues in foods after discontinuation of industrially produced trans fats? Ruminant products, vegetable oils and synthetic supplements. Eur. J. Lipid Sci. Technol. 115, 1378–1401.

Al-Taher, F., Didsatha-Amnarj, Y., Jackson, L., Varelis, P., Coulson, M., 2008. Development of a Headspace Solid Phase Micro Extraction GC/MS Method to Measure Furan in Foods and Beverages and its Application to Survey Work. Abstract. American Chemical Society Meeting, April 6–10, 2008, New Orleans, LA.

Altaki, M.S., Santos, F.J., Galceran, M.T., 2009. Automated headspace solid-phase microextraction versus headspace for the analysis of furan in foods by gas chromatography-mass spectrometry. Talanta 78, 1315–1320.

American Cancer Society, 2019. Acrylamide and Cancer Risk. https://www.cancer.org/cancer/cancer-causes/acrylamide.html (Accessed 20 October 2019).

Amrein, T.M., Bachmann, S., Notti, A., Biedermann, M., Barbosa, M.F., Biedermann, B.S., Grob, K., Keiser, A., Realini, P., Escher, F., Amado, R., 2003. Potential of acrylamide formation, sugars, and free asparagine in potatoes: a comparison of cultivars and farming systems. J. Agric. Food Chem. 51, 5556–5560.

Amrein, T.M., Schonbachler, B., Escher, F., Amado, R., 2004. Acrylamide in gingerbread: critical factors for formation and possible ways for reduction. J. Agric. Food Chem. 52, 4282–4288.

Amrein, T.M., Andres, L., Manzardo, G.G., Amado, R., 2006. Investigations on the promoting effect of ammonium hydrogencarbonate on the formation of acrylamide in model systems. J. Agric. Food Chem. 54, 10253–10261.

Amrein, T.M., Andres, L., Escher, F., Amado, R., 2007. Occurrence of acrylamide in selected foods and mitigation options. Food Addit. Contam. 24 (S1), 13–25.

Anonymous, 2008. Deadly Fats: Why Are We Still Eating Them? the Independent. http://www.independent.co.uk/life-style/health-and-wellbeing/healthy-living/deadly-fats-why-are-we-still-eating-them-843400.html (Accessed 20 October 2019).

Arvanitoyannis, I.S., Dionisopoulou, N., 2012. Acrylamide: formation, occurrence, in food products, detection methods, and legislation. Crit. Rev. Food Sci. Nutr. 54, 708–733.

Ascherio, A., Willett, W.C., 1997. Health effects of trans fatty acids. Am. J. Clin. Nutr. 66, 1006S–10S.

Bagdonaite, K., Derler, K., Murkovic, M., 2008. Determination of acrylamide during roasting of coffee. J. Agric. Food Chem. 56, 6081–6086.

Banchero, M., Pellegrino, G., Manna, L., 2013. Supercritical fluid extraction as a potential mitigation strategy for the reduction of acrylamide level in coffee. J. Food Eng. 115, 292–297.

Baskar, G., Aiswarya, E., 2018. Overview on mitigation of acrylamide in starchy fried and baked foods. J. Sci. Food Agric. 98, 4385–4394.

Becalski, A., Lau, B.P., Lewis, D., Seaman, S.W., 2003. Acrylamide in foods: occurrence, sources, and modeling. J. Agric. Food Chem. 51, 802–808.

Becalski, A., Lau, B.P.-Y., Lewis, D., Seaman, S.W., Hayward, S., Sahagian, M., Ramesh, M., Leclerc, Y., 2004. Acrylamide in French fries: influence of free amino acids and sugars. J. Agric. Food Chem. 52, 3801–3806.

Becalski, A., Seaman, S., 2005. Furan precursors in food: a model study and development of a simple headspace method for determination of furan. J. AOAC Int. 88, 102–106.

Becalski, A., Hayward, S., Krakalovich, T., Pelletier, L., Roscoe, V., Vavasour, E., 2009. Development of an analytical method and survey of foods for furan, 2-methylfuran and 3-methylfuran with estimated exposure. Food Addit. Contam. 27, 764–775.

Becalski, A., Halldorson, T., Hayward, S., Roscoe, V., 2016. Furan, 2-methylfuran and 3-methylfuran in coffee on the Canadian market. J. Food Compos. Anal. 47, 113–119.

Besaratinia, A., Pfeifer, G.P., 2004. Genotoxicity of acrylamide and glycidamide. J. Natl. Cancer Inst. 96, 1023–1029.

Biedermann, M., Noti, A., Biedermann-Brem, S., Mozzarti, V., Grob, K., 2002. Experiments on acrylamide formation and possibilities to decrease the potential of acrylamide formation in potatoes. Mitt. Lebensm. Hyg. 93, 668–687.

Biedermann-Brem, S., Noti, A., Grob, K., Imhof, D., Bazzocco, D., Pfefferle, A., 2003. How much reducing sugar may potatoes contain to avoid excessive acrylamide formation during roasting and baking? Eur. Food Res. Technol. 217, 369–373.
Bjellaas, T., Stølen, L.H., Haugen, M., Paulsen, J.E., Alexander, A., Lundanes, E., Becher, G., 2007. Urinary acrylamide metabolites as biomarkers for short-term dietary exposure to acrylamide. Food Chem. Toxicol. 45, 1020–1026.
Blank, I., Robert, F., Goldmann, T., Pollien, P., Varga, N., Devaud, S., Saucy, F., Huynh-Ba, T., Stadler, R.H., 2005. Mechanism of acrylamide formation: maillard-induced transformations of asparagine. In: Friedman, M., Mottram, D. (Eds.), Chemistry and Safety of Acrylamide in Food. Springer, New York, NY, pp. 171–189.
Bolger, P.M., Tao, S.S.-H., Dinovi, M., 2009. Hazards of dietary furan. In: Stadler, R.H., Lineback, D.R. (Eds.), Induced Food Toxicants: Occurrence, Formation, Mitigation and Health Risks. John Wiley & Sons, Hoboken, NJ, pp. 117–134.
Boon, P.E., de Mul, A., van der Voet, H., van Donkersgoed, G., Brette, M., Van Klaveren, J.D., 2005. Calculations of dietary exposure to acrylamide. Mutat. Res. 580, 143–155.
Brantsaeter, A.L., Haugen, M., de Mul, A., Bjellaas, T., Becher, G., Van Klaveren, J., Alexander, J., Meltzer, H.M., 2008. Exploration of different methods to assess dietary acrylamide exposure in pregnant women participating in the Norwegian Mother and Child Cohort Study (MoBa). Food Chem. Toxicol. 46, 2808–2814.
Bråthen, E., Kita, A., Knutsen, S., Wicklund, T., 2005. Addition of glycine reduces the content of acrylamide in cereal and potato products. J. Agric. Food Chem. 53, 3259–3264.
Brisson, B., Ayotte, P., Normandin, L., Gaudreau, E., Bienvenu, J.-F., Fennell, T.R., Blanchet, C., Paneuf, D., Lapointe, C., Bonvalot, Y., Gagne, M., Courteau, M., Snyder, R.W., Bouchard, M., 2014. Relation between dietary acrylamide exposure and biomarkers of internal dose in Canadian teenagers. J. Expo. Sci. Environ. Epidemiol. 24, 215–221.
Brownell, K.D., Pomeranz, J.L., 2014. The trans-fat ban — food regulation and long-term health. N. Engl. J. Med. 370, 1773–1775.
Bull, R.J., Robinson, M., Laurie, R.d, Stoner, G.D., Greisiger, E., Meier, J.R., Stober, J., 1984. Carcinogenic activity of acrylamide in the skin and lung of Swiss-ICR mice. Cancer Lett. 24, 209–212.
Burley, V.J., Greenwood, D.C., Hepworth, S.J., Fraser, L.K., de Kok, T.M., van Breda, S.G., Kyrtopoulos, S.A., Botsivali, M., Kleinjans, J., McKinney, P.A., Cade, J.E., 2010. Dietary acrylamide intake and risk of breast cancer in the UK women's cohort. Br. J. Cancer 103, 1749–1754.
Campbell, S., December 14, 2017. The Bane of Snack Foods: Acrylamide. Food Safety Magazine. https://www.foodsafetymagazine.com/signature-series/the-bane-of-snack-foods-acrylamide/ (Accessed 20 October 2019).
Canadian Food Inspection Agency, 2019. Trans Fatty Acid Claims. https://www.inspection.gc.ca/food/requirements-and-guidance/labelling/industry/nutrient-content/specific-claim-requirements/eng/1389907770176/1389907817577?chap=6 (Accessed 20 October 2019).
Capuano, E., Fogliano, V., 2011. Acrylamide and 5-hydroxymethylfurfural (HMF): a review on metabolism, toxicity, occurrence in food and mitigation strategies. LWT Food Sci. Technol. 44, 793–810.
CAST, June 2006. Acrylamide in Food. Number 32, Available at: https://www.cast-science.org/wp-content/uploads/2018/12/acrylamide_ip.pdf (Accessed 20 October, 2019).
Chen, L.J., Hecht, S.S., Peterson, L.A., 1995. Identification of cis-2-butene-1,4-dial as a microsomal metabolite of furan. Chem. Res. Toxicol. 8, 903–906.
Ciesarova, Z., Kiss, E., Boegl, P., 2006. Impact of L-asparaginase on acrylamide content in potato products. J. Food Nutr. Res. 45, 141–146.
Claeys, W.L., DeVleeschouwer, K., Hendrickx, M.E., 2005. Kinetics of acrylamide formation and elimination during heating of an asparagine-sugar model system. J. Agric. Food Chem. 53, 9999–10005.
Claus, A., Weisz, G.M., Scieber, A., Carle, R., 2006a. Pyrolytic acrylamide formation from purified wheat gluten and gluten-supplemented wheat bread rolls. Mol. Nutr. Food Res. 50, 87–93.
Claus, A., Schreiter, P., Weber, A., Graeff, S., Herrmann, W., Claupein, W., Schieber, A., Carle, R., 2006b. Influence of agronomic factors and extraction rate on the acrylamide contents in yeast-leavened breads. J. Agric. Food Chem. 54, 8968–8976.
CODEX Alimentarius Commission, 2009. Code of practice for the reduction of acrylamide in foods. CAC/RCP 67–2009.
CODEX Alimentarius Commission, 2011. Discussion Paper on Furan. http://www.fao.org/tempref/codex/Meetings/CCCF/cccf5/cf05_13e.pdf (Accessed 20 October 2019).
Crews, C., Castle, L., 2007. A review of the occurrence, formation and analysis of furan in heat-processed foods. Trends Food Sci. Technol. 18, 365–372.
De Meulenaer, B., De Wilde, T., Mestdagh, F., Govaert, Y., Ooghe, W., Fraselle, S., et al., 2008. Comparison of potato varieties between seasons and their potential for acrylamide formation. J. Sci. Food Agric. 88, 313–318.
De Wilde, T., De Meulenaer, B., Mestagh, F., Govaert, Y., Vandeburie, S., Ooghe, W., Fraselle, S., Demeulemeester, K., Van Peteghem, C., Calus, A., Degroot, J.-M., Verhé, R., 2005. Influence of storage practices on acrylamide formation during potato frying. J. Agric. Food Chem. 53, 6550–6557.
De Wilde, T., De Meulenaer, B., Mestdagh, F., Govaert, Y., Ooghe, W., Fraselle, S., Demeulemeester, K., Van Peteghem, C., Calus, A., Degroodt, J.-M., Verhé, R., 2006. Influence of fertilization on acrylamide formation during frying of potatoes harvested in 2003. J. Agric. Food Chem. 54, 404–408.
Dijkstra, A.J., Hamilton, R.J., Hamm, W. (Eds.), 2008. Trans Fatty Acids. Wiley-Blackwell, Hoboken, NJ.
Doerge, D.R., Young, J.F., Chen, J.J., DiNovi, M.J., Henry, S.H., 2008. Using dietary exposure and physiologically based pharmacokinetic/pharmacodynamic modeling in human risk extrapolations for acrylamide toxicity. J. Agric. Food Chem. 56, 6031–6038.
Dybing, E., Farmer, P.B., Andersen, M., Fennell, T.R., Lalljie, S.P.D., Müller, D.J.G., Olin, S., Petersen, B.J., Schlatter, J., Scholz, G., Scimeca, J.A., Slimani, N., Törnqvist, M., Tuijtelaars, S., Verger, P., 2005. Human exposure and internal dose assessments of acrylamide in food. Food Chem. Toxicol. 43, 271–278.

EC, 2010. Commission recommendations of 2 June 2010 on the monitoring of acrylamide levels in food. Official Journal of the European Union 137 (3.6.2010), 4–10.

EC, 2015. Report from the Commission to the European Parliament and the Council Regarding Trans Fats in Foods and in the Overall Diet of the Union Population. https://ec.europa.eu/food/sites/food/files/safety/docs/fs_labelling-nutrition_trans-fats-report_en.pdf (Accessed 20 October 2019).

EC, 2017. Establishing mitigation measures and benchmark levels for the reduction of the presence of acrylamide in food. Off. J. Eur. Union, 21.11.2017. https://eur-lex.europa.eu/legal-content/EN/TXT/PDF/?uri=CELEX:32017R2158&from=EN (Accessed 20 October 2019).

EFSA, 2011. Update on furan levels in food from monitoring years 2004-2010 and exposure assessment. EFSA J. 9 (9), 2347, 1–33. https://efsa.onlinelibrary.wiley.com/doi/epdf/10.2903/j.efsa.2011.2347 (Accessed 20 October 2019).

EFSA, 2012. Update on acrylamide levels in food from monitoring years 2007 to 2010. EFSA J. 10 (2938). https://doi.org/10.2903/j.efsa.2012.2938.

EFSA, 2015. Scientific opinion on acrylamide in food. EFSA Panel on contaminants in the food Chain (CONTAM). EFSA J. 13 (6), 4104.

EFSA, 2017. Scientific opinion on the risks for public health related to the presence of furan and methylfurans in food. EFSA J. 15, 5005.

EFSA, 2019. Acrylamide Database. Acrylamide Levels in Food. https://ec.europa.eu/food/sites/food/files/safety/docs/cs_contaminants_catalogue_acrylamide_db_study_area1_en.pdf (Accessed 20 October 2019).

Fan, X., 2005. Formation of furan from carbohydrates and ascorbic acid following exposure to ionizing radiation and thermal processing. J. Agric. Food Chem. 53, 7826–7831.

Fan, X., Geveke, D.J., 2007. Furan formation in sugar solution and apple cider upon ultraviolet treatment. J. Agric. Food Chem. 55, 7816–7821.

Fan, X., Huang, L., Sokorai, K.J.B., 2008. Factors affecting thermally induced furan formation. J. Agric. Food Chem. 56, 9490–9494.

FDA, 2003. Food labeling: trans fatty acids in nutrition labeling. Fed. Regist. 68, 41433–41506.

FDA, 2006. The 2006 Exposure Assessment for Acrylamide. https://wayback.archive-it.org/7993/20170111181833/http://www.fda.gov/downloads/Food/FoodborneIllnessContaminants/UCM197239.pdf (Accessed 20 October 2019).

FDA, 2007. An Updated Exposure Assessment for Furan from Consumption of Adult and Baby Foods. U.S. Food and Drug Administration, Center for Food Safety and Applied Nutrition. https://wayback.archive-it.org/7993/20170112012203/http://www.fda.gov/Food/FoodborneIllnessContaminants/ChemicalContaminants/ucm110770.htm (Accessed 20 October 2019).

FDA, March 2016a. Guidance for Industry- Acrylamide in Foods. U.S. Food and Drug Administration, Center for Food Safety and Applied Nutrition. https://www.fda.gov/food/cfsan-constituent-updates/fda-issues-final-guidance-industry-how-reduce-acrylamide-certain-foods (Accessed 20 October 2019).

FDA, March 2016b. You Can Help Cut Acrylamide in Your Diet. https://www.fda.gov/consumers/consumer-updates/you-can-help-cut-acrylamide-your-diet (Accessed 20 October 2019).

FDA, 2017a. Acrylamide and Diet, Food Storage, and Food Preparation. https://www.fda.gov/food/chemicals/acrylamide-and-diet-food-storage-and-food-preparation (Accessed 20 October 2019).

FDA, 2017b. Exploratory Data on Furan in Food. U.S. Food and Drug Administration, Center for Food Safety and Applied Nutrition. https://www.fda.gov/food/chemicals/exploratory-data-furan-food (Accessed 20 October 2019).

FDA, 2018. Trans Fat. U.S. Food and Drug Administration, Center for Food Safety & Applied Nutrition. https://www.fda.gov/food/food-additives-petitions/trans-fat (Accessed 20 October 2019).

FDA, 2019a. Survey Data on Acrylamide in Food. https://www.fda.gov/food/chemicals/survey-data-acrylamide-food (Accessed 20 October 2019.

FDA, 2019b. Survey Data on Acrylamide in Food from Total Diet Study Results. https://www.fda.gov/food/chemicals/survey-data-acrylamide-food-total-diet-study-results (Accessed 20 October 2019).

Fernandez, S., Kurppa, L., Hyvönen, L., 2003. Content of acrylamide decreased in potato chips with addition of a proprietary flavonoid spice mix (Flavomare) in frying. Innov. Food Technol. 18, 24–26.

Fink, M., Andersson, R., Rosen, J., Aman, P., 2006. Effect of added asparagine and glycine on acrylamide content in yeast-leavened bread. Cereal Chem. 83, 218–222.

Fohgelberg, P., Rosen, J., Hellenäs, K.-E., Abramsson-Zetterberg, L., 2005. The acrylamide intake via some common baby food for children in Sweden during their first year of life - an improved method for analysis of acrylamide. Food Chem. Toxicol. 43, 951–959.

FoodDrinkEurope, 2019. The Food Drink Europe Acrylamide Toolbox. https://www.fooddrinkeurope.eu/publication/fooddrinkeurope-updates-industry-wide-acrylamide-toolbox/ (Accessed 18 October 2019).

Food Standards Agency, January 2018. Acrylamide, Information on the Risks of Acrylamide and How You Can Reduce the Chances of Being Harmed by it. https://www.food.gov.uk/safety-hygiene/acrylamide#how-to-reduce-acrylamide-at-home (Accessed 20 October 2019).

Food Standards Australia New Zealand, April 12, 2005. Trans Fatty Acids. Fact Sheets. http://www.foodstandards.gov.au/newsroom/factsheets/factsheets2005/ (Accessed 1/9/09).

Foot, R.J., Haase, N.U., Grob, K., Gondé, P., 2007. Acrylamide in fried and roasted potato products: a review on progress in mitigation. Food Addit. Contam. 24 (S1), 37–46.

Frankel, E.N., Neff, W.E., Selke, E., Brooks, D.D., 1987. Thermal and metal-catalyzed decomposition of methyl linolenate hydroperoxides. Lipids 22, 322–327.

Fransson-Steen, R., Goldsworthy, T.L., Kedderis, G.L., Maronpot, R.R., 1997. Furan-induced liver cell proliferation and apoptosis in female B6C3F1 mice. Toxicology 118, 195–204.

Fredricksson, H., Tallving, J., Rosen, J., Aman, P., 2004. Fermentation reduces free asparagine in dough and acrylamide content in bread. Cereal Chem. 81, 650–653.

Friedman, M.A., Dulak, L.H., Stedham, M.A., 1995. Lifetime oncogenicity study in rats with acrylamide. Fund. Appl. Toxicol. 27, 95–105.

Friedman, M., 2003. Chemistry, biochemistry, and safety of acrylamide. J. Agric. Food Chem. 51, 4504–4526.

Friedman, M., Mottram, D. (Eds.), 2005. Chemistry and Safety of Acrylamide in Food, Advances in Experimental Medicine and Biology, vol. 561. Springer, NY.

Friedman, M., Levin, C.E., 2008. Review of methods for the reduction of dietary content and toxicity of acrylamide. J. Agric. Food Chem. 56, 6113–6140.

Göbel, A., Kliemant, A., 2007. The German minimization concept for acrylamide. Food Addit. Contam. 24 (S1), 82–90.

Gökmen, V. (Ed.), 2015. Acrylamide in Food. Analysis, Content and Potential Health Effects. Academic Press, London.

Gökmen, V., Senyuva, H.Z., 2007. Acrylamide formation is prevented by divalent cations during the Maillard reaction. Food Chem. 103, 196–203.

Gökmen, V., Palazoglu, T.K., Senyuva, H.Z., 2006. Relation between the acrylamide formation and time-temperature history of surface and core regions of French fries. J. Food Eng. 77, 972–976.

Gökmen, V., Acar, O., Köksel, H., Acar, J., 2007. Effects of dough formula and baking conditions on acrylamide and hydroxymethylfurfural formation in cookies. Food Chem. 104, 1136–1142.

Goldmann, T., Périsset, A., Scanlan, F., Stadler, R.H., 2005. Rapid determination of furan in heated foodstuffs by isotope dilution solid phase micro-extraction-gas chromatography – mass spectrometry (SPME-GC-MS). Analyst 130, 878–883.

Gorton, L., 2017. Special Bakers Yeast Strain Cuts Acrylamide Risk. https://www.bakingbusiness.com/articles/45155-special-bakers-yeast-strain-cuts-acrylamide-risk (Accessed 20 October 2019).

Granvogl, M., Schieberle, P., 2006. Thermally generated 3-aminopropionamide as a transient intermediate in the formation of acrylamide. J. Agric. Food Chem. 54, 5933–5938.

Grob, K., Biedermann, M., Biedermann-Brem, S., Noti, A., Imhof, D., Amrein, T., Pfefferle, A., Bazzocco, D., 2003. French fries with less than 10 mg/kg acrylamide. A collaboration between cooks and analysts. Eur. Food Res. Technol. 217, 185–194.

Guenther, H., Anklam, E., Wenzl, T., Stadler, R.H., 2007. Acrylamide in coffee: review of progress in analysis, formation and level reduction. Food Addit. Contam. 24 (S1), 60–70.

Haase, N.U., 2006. The formation of acrylamide in potato products. In: Skog, K., Alexander, J. (Eds.), Acrylamide and Other Hazardous Compounds in Heat-Treated Foods. England: Woodhead Publishing, Cambridge, pp. 41–59.

Halford, N.G., Curtis, N.Y., 2019. Acrylamide in Food. UK: World Scientific Publishing, London.

Halford, N.G., Muttucumaru, N., Curtis, T.Y., Parry, M.A., 2007. Genetic and agronomic approaches to decreasing acrylamide precursors in crop plants. Food Addit. Contam. 24 (Suppl. 1), 26–36.

Hamadeh, H.K., Jayadev, S., Gaillard, E.T., Huang, Q., Stoll, R., Blanchard, K., Chou, J., Tucker, C.J., Collins, J., Maronpot, R., Bushel, P., Afshari, C.A., 2004. Integration of clinical and gene expression endpoints to explore furan-mediated hepatotoxicity. Mutat. Res. 549, 169–183.

Hamlet, C.G., Sadd, P.A., 2009. Chloropropanols and chloroesters. In: Stadler, R.H., Lineback, D.R. (Eds.), Process-induced Food Toxicants. NJ: Wiley, Hoboken, pp. 175–214.

Hanley, A.B., Offen, C., Clarke, M., Ing, B., Roberts, M., Burch, R., 2005. Acrylamide reduction in processed foods. In: Friedman, M., Mottram, D. (Eds.), Chemistry and Safety of Acrylamide in Food. Advances in Experimental Medicine and Biology, 561. Springer, Boston, MA, pp. 387–392.

Hasnip, S., Crews, C., Castle, L., 2006. Some factors affecting the formation of furan in heated foods. Food Addit. Contam. 23, 219–227.

Health Canada, February 2009. Acrylamide. What You Can Do to Reduce Exposure. In: https://www.canada.ca/en/health-canada/services/food-nutrition/food-safety/chemical-contaminants/food-processing-induced-chemicals/acrylamide/acrylamide-what-you-reduce-exposure-food-processing-induced-chemicals.html (Accessed 20 October 2019).

Health Canada, 2019. Fats. https://www.canada.ca/en/health-canada/services/nutrients/fats.html (Accessed 20 October 2019).

HEATOX, 2007a. HEATOX, Heat-Generated Food Toxicants: Identification, Characterisation and Risk Minimization. Final Report. https://www.health-petfood.nl/wp-content/uploads/2019/06/Heatox_InforFINAL07-1.pdf (Accessed 20 October 2019).

HEATOX, 2007b. Guidelines to Authorities and Consumer Organisations on Home Cooking and Consumption. http://www.ub.edu/cecem/Index/documents_index/D59_guidelines_to_authorities_and_consumer_organisations_on_home_cooking_and_consumption.pdf (Accessed 20 October 2019).

Hedegaard, R.V., Granby, K., Frandsen, H., Thygesen, J., Skibsted, L.H., 2008. Acrylamide in bread. Effect of prooxidants and antioxidants. Eur. Food Res. Technol. 227, 519–525.

Hendriksen, H.V., Kornbrust, B.A., Ostergaard, P.R., Stringer, M.A., 2009. Evaluating the potential for enzymatic acrylamide mitigation in a range of food products using an asparaginase from *Aspergillus oryzae*. J. Agric. Food Chem. 57 (10), 4168–4176.

Hendriksen, H.V., Bodolfsen, G., Baumann, M.J., 2013. Asparaginase for acrylamide mitigation in food. Aspect Appl. Biol. 116, 41–40.

Heppner, C.W., Schlatter, J.R., 2007. Data requirements for risk assessment of furan in food. Food Addit. Contam. 24 (S1), 114–121.

Hogervorst, J.G., Schouten, L.J., Konings, E.J., Goldbohm, R.A., Van den Brandt, P.A., 2008. Dietary acrylamide and the risk of renal cell, bladder and prostate cancer. Am. J. Clin. Nutr. 87, 1428–1438.

Hu, F.B., Stampfer, M.J., Manson, J.E., Rimm, E., Colditz, G.A., Rosner, B.A., Hennekens, C.H., Willett, W.C., 1997. Dietary fat intake and the risk of coronary heart disease in women. N. Engl. J. Med. 337, 1491–1499.

Hunter, J.E., 2005. Dietary levels of *trans*-fatty acids: basis for health concerns and industry efforts to limit their use. Nutr. Res. 25, 49–513.

IARC, 1994. Acrylamide. Some Industrial Chemicals; IARC Monographs on the Evaluation of Carcinogenic Risks to Humans. International Agency for Research on Cancer, Lyon, France.

IARC, 1995. Furan. IARC Monographs on the Evaluation of Carcinogenic Risks to Humans, vol. 63. International Agency for Research on Cancer, Lyon, France, pp. 3194–3407.

Isleroglu, H., Kemerli, T., Sakin-Yilmazer, M., Guyen, G., Ozdenstan, O., Uren, A., Kaymak-Ertekin, F., 2012. Effect of steam baking on acrylamide formation and browning kinetics of cookies. J. Food Sci. 7, E257–E263.

Jackson, L., Al-Taher, F., 2005. Effects of consumer food preparation on acrylamide formation. In: Friedman, M., Mottram, D. (Eds.), Chemistry and Safety of Acrylamide in Food. Advances in Experimental Medicine and Biology, 562. Springer, Boston, MA, pp. 447–465.

JECFA, 2005. Summary and Conclusions of the Sixty-Fourth Meeting of the Joint FAO/WHO Expert Committee on Food Additives (JECFA). JECFA/65/SC.

JECFA, 2011. Safety Evaluation of Certain Contaminants in Food. Seventy-Second Meeting of the Joint FAO/WHO Expert Committee on Food Additives (JECFA). http://www.inchem.org/documents/jecfa/jecmono/v63je01.pdf (Accessed 20 October 2019).

Jin, C., Wu, X., Zhang, Y., 2013. Relationship between antioxidants and acrylamide formation: a review. Food Res. Int. 51, 611–620.

Johnson, K.A., Gorzinski, S.J., Bodner, K.M., Campbell, R.A., Wolf, C.H., Friedman, M.A., Mast, R.W., 1986. Chronic toxicity and oncogenicity study on acrylamide incorporated in the drinking water of Fischer 344 rats. Toxicol. Appl. Pharmacol. 85, 154–168.

Judd, J.T., Clevidence, B.A., Muesing, R.A., Wittes, J., Sunkin, M.E., Podczasy, J.J., 1994. Dietary *trans* fatty acids: effects on plasma lipids and lipoproteins of healthy men and women. Am. J. Clin. Nutr. 59, 861–868.

Jun, H.-J., Lee, K.-G., Lee, Y.-K., Woo, G.-J., Park, Y.S., Lee, S.-J., 2008. Correlation of urinary furan with plasma γ-glutamyltranspeptidase levels in healthy men and women. Food Chem. Toxicol. 46, 1753–1759.

Jung, M.Y., Choi, D.S., Ju, J.W., 2003. A novel technique for limitation of acrylamide formation in fried and baked corn chips and in French fries. J. Food Sci. 68, 1287–1290.

Kettlitz, B., Scholz, G., Theurillat, V., Czelovszky, J., Buck, N.R., O'Hagan, S., Mavromichali, E., Ahrens, K., Kraehenbuehl, K., Scozzi, G., Weck, M., Vinci, C., Sobieraj, M., Stadler, R.H., 2019. Furan and methylfurans in foods: an update on occurrence, mitigation, and risk assessment. Compr. Rev. Food Sci. Food Saf. 18, 738–752.

Kita, A., Bråthen, E., Knutsen, S., Wicklund, T., 2005. Effective ways of decreasing acrylamide content in potato crisps during processing. J. Agric. Food Chem. 52, 1287–1290.

Klaunig, J.E., 2008. Acrylamide toxicity. J. Agric. Food Chem. 56, 5984–5988.

Konings, E.J.M., Ashby, P., Hamlet, C.G., Thompson, G.A.K., 2007. Acrylamide in cereal and cereal products: a review on the progress in level reduction. Food Addit. Contam. 24 (S1), 47–59.

Koszucka, A., Nowak, A., 2018. Thermal processing food-related toxicants: a review. Crit. Rev. Food Sci. Nutr. 12, 1–18.

Krishnakumar, T., Visvanathan, R., 2014. Acrylamide in food products: a review. J. Food Process. Technol. 5, 1–10.

Kumar, N.S.M., Shimray, C.A., Indrani, D., Manoonmani, H.K., 2014. Reduction of acrylamide formation in sweet bread with L-asparaginase treatment. Food Bioprocess Technol. 7 (3), 741–748.

Lantz, I., Ternité, R., Wilkens, J., Hoenicke, K., Guenther, H., van der Stegen, G.H.D., 2006. Studies on acrylamide levels in roasting, storage and brewing of coffee. Mol. Nutr. Food Res. 50, 1039–1046.

Levine, R.A., Ryan, S.M., 2009. Determining the effect of calcium cations on acrylamide formation in cooked wheat products using a model system. J. Agric. Food Chem. 57, 6823–6829.

Limacher, A., Kerler, J., Conde-Petit, B., Blank, I., 2007. Formation of furan and methylfuran from ascorbic acid in model systems and food. Food Addit. Contam. 24 (S1), 122–135.

Limacher, A., Kerler, J., Davidek, T., Schmalzried, F., Blank, I., 2008. Formation of furan and methylfuran by Maillard-type reactions in model systems and food. J. Agric. Food Chem. 56, 3639–3647.

Lindsay, R.C., Jang, S.J., 2005. Chemical intervention strategies for substantial suppression of acrylamide formation in fried potato products. In: Friedman, M., Mottram, D. (Eds.), Chemistry and Safety of Acrylamide in Food. Advances in Experimental Medicine and Biology, 561. Springer, Boston, MA, pp. 393–404.

Lineback, D.R., Coughlin, J.R., Stadler, R.H., 2012. Acrylamide in foods: a review of the science and future considerations. Annu. Rev. Food Sci. Technol. 3, 15–35.

Lipworth, L., Sonderman, J.S., Tarone, McLaughlin, J.K., 2012. Review of epidemiologic studies of dietary acrylamide intake and the risk of cancer. Eur. J. Cancer Prev. 21, 375–385.

List, G.R., 2004a. Decreasing *trans* and saturated fatty acid content in food oils. Food Technol. 58, 23–31.

List, G.R., 2004b. Processing and reformulation for nutrition labeling of *trans* fatty acids. Lipid Technol. 16, 173–177.

Locas, C.P., Yaylayan, V.A., 2004. Origin and mechanistic pathways of formation of the parent furan- A food toxicant. J. Agric. Food Chem. 42, 6830–6836.

LoPachin, R.M., Gavin, T., 2008. Acrylamide-induced nerve terminal damage: relevance to neurotoxic and neurodegenerative mechanisms. J. Agric. Food Chem. 56, 5994–6003.

Lopez-Garcia, E., 2005. Consumption of trans fatty acids is related to plasma biomarkers of inflammation and endothelial dysfunction. J. Nutr. 35, 562–566.

Maga, J.A., 1979. Furans in foods. CRC Crit. Rev. Food Sci. Nutr. 11, 355–400.

Mark, J., Pollien, P., Lindinger, C., Blank, I., Mark, T., 2006. Quantitation of furan and methylfuran formed in different precursor systems by proton transfer reaction mass spectrometry. J. Agric. Food Chem. 54, 2786–2793.

Matthäus, B., Haase, N.U., Vosmann, K., 2004. Factors affecting the concentration of acrylamide during deep-fat frying of potatoes. Eur. J. Lipid Sci. Technol. 106, 793–801.

Medeiros Vinci, R., Mestdagh, F., De Muer, N., Van Peteghem, C., De Meulenaer, B., 2010. Effective quality control of incoming potatoes as an acrylamide mitigation strategy for the French fries industry. Food Addit. Contam. 27, 417–425.

Medeiros Vinci, R., Mestdagh, F., De Meulenaer, B., 2012. Acrylamide formation in fried potato products- Present and future, a critic review on mitigation strategies. Food Chem. 133, 1138–1154.

Menaa, F., Menaa, A., Menaa, B., Tréton, J., 2013. Trans-fatty acids, dangerous bonds for health? A background review paper of their use, consumption, health implications and regulation in France. Eur. J. Nutr. 52, 1289–1302.

Mensink, R.P., Zock, P.L., Kester, A.D., Katan, M.D., 2003. Effects of dietary fatty acids and carbohydrates on the ratio of serum total to HDL cholesterol and o serum lipids and apolipoproteins: a meta-analysis of 60 controlled trials. Am. J. Clin. Nutr. 77, 1146–1155.

Mensink, R.P., Sanders, T.A., Baer, D.J., Hayes, K.C., Howles, P.N., Marangoni, A., 2016. The increasing use of interesterified lipids in the food supply and their effects on health parameters. Adv. Nutr. 7, 719–729.

Mestdagh, F., De Meulenaer, B., Van Peteghem, C., 2007. Influence of oil degradation on the amounts of acrylamide generated in a model system and in French fries. Food Chem. 100, 1153–1159.

Mestdagh, F., Maertens, J., Cucu, T., Delporte, K., van Peteghem, C., De Meulenaer, B., 2008a. Impact of additives to lower the formation of acrylamide in a potato model system through pH reduction and other mechanisms. Food Chem. 107, 26–31.

Mestdagh, F., de Wilder, T., Fraselle, S., Govaert, Y., Ooghe, W., Degroodt, J.M., Verhe, R., van Peteghem, C., De Meulenaer, B., 2008b. Optimization of the blanching process to reduce acrylamide in fried potatoes. LWT Food Sci. Technol. 41, 1648–1654.

Mills, C., Mottram, D.S., Wedzicha, B.L., 2009. Acrylamide. In: Stadler, R.H., Lineback, D.R. (Eds.), Process-induced Food Toxicants. Wiley, Hoboken, NJ, pp. 23–50.

Morales, F., Capuano, E., Fogliano, V., 2008. Mitigation strategies to reduce acrylamide formation in fried potato products. Ann. N. Y. Acad. Sci. 1126, 89–100.

Mottram, D.S., 1991. Meat. In: Maarse, H. (Ed.), Volatile Compounds in Foods and Beverages (pp. 207–177). Marcel Dekker, New York.

Mottram, D.S., Wedzicha, B.L., Dodson, A.T., 2002. Acrylamide is formed in the Maillard reaction. Nature 419, 448–449.

Moubarac, J.C., Martins, A.P., Claro, R.M., Levy, R.B., Cannon, G., Monteiro, C.A., 2013a. Consumption of ultra-processed foods and likely impact on human health: evidence from Canada. Publ. Health Nutr. 16, 2240–2248.

Moubarac, J.C., Claro, R.M., Baraldi, L.G., Levy, R.B., Martins, A.P., Cannon, G., Monteiro, C.A., 2013b. International differences in cost and consumption of ready-to-consume food and drink products: United Kingdom and Brazil, 2008–2009. Global Publ. Health 8, 845–856.

Mozaffarian, D., Katan, M.B., Ascherio, A., Stamfer, M.J., Willett, W.C., 2006. Trans fatty acids and cardiovascular disease. N. Engl. J. Med. 354, 1601–1613.

Mucci, L.A., Wilson, K.M., 2008. Acrylamide intake through diet and human cancer risk. J. Agric. Food Chem. 56, 6013–6019.

Mucci, L.A., Dickman, P.W., Steineck, G., Adami, H.O., Augustsson, K., 2003. Dietary acrylamide and cancer of the large bowel, kidney and bladder: absence of an association in a population study in Sweden. Br. J. Cancer 88, 84–89.

Mucci, L.A., Lindblad, P., Steineck, G., Adami, H.O., 2004. Dietary acrylamide and risk of renal cell cancer. Int. J. Cancer 109, 774–776.

Mucci, L.A., Adami, H.O., Wolk, A., 2006. Prospective study of dietary acrylamide and risk of colorectal cancer among women. Int. J. Cancer 118, 169–173.

Muttucumaru, N., Halford, N.G., Elmore, J.S., Dodson, A.T., Parry, M., Shewry, P.R., 2006. Formation of high levels of acrylamide during the processing of flour derived from sulfate-deprived wheat. J. Agric. Food Chem. 54, 8951–8955.

Muttucumaru, N., Elmore, J.S., Curtis, T., Mottram, D.S., Parry, M.A.J., Halford, N.G., 2008. Reducing acrylamide precursors in raw materials derived from wheat and potato. J. Agric. Food Chem. 56, 6167–6172.

National Cancer Institute, 2019. National Institute of Health (NIH). Acrylamide and Cancer Risk. https://www.cancer.gov/about-cancer/causes-prevention/risk/diet/acrylamide-fact-sheet (Accessed 20 October 2019).

NTP, 1993. Toxicology and Carcinogenesis Studies of Furan in F-344/N Rats and B6C3F1 Mice. NTP Technical Report 402. U.S. Department of Health and Human Services, Public Health Service. National Institute of Health, Research Triangle Park, NC, USA.

NTP, 2012 July. Toxicology and carcinogenesis studies of acrylamide (CASRN 79-06-1) in F344/N rats and B6C3F1 mice (feed and drinking water studies). Natl. Toxicol. Progr. Tech. Rep. 575, 1–234.

Nawar, W., 1996. Lipids. In: Fennema, O.R. (Ed.), Food Chemistry, third ed. Marcel Dekker, New York, pp. 225–319.

Noti, A., Biedermann-Brem, S., Biedermann, M., Grob, K., Albisser, P., Realini, P., 2003. Storage of potatoes at low temperatures should be avoided to prevent increased acrylamide formation during frying or roasting. Mitt. Lebensm. Hyg. 94, 167–180.

O'Brien, R.D., 2004. Fats and Oils: Formulating and Processing for Applications, second ed. CRC Press, Boca Raton, FL.

Olessen, P.T., Olsen, A., Frandsen, H., Frederiksen, K., Overvad, K., Tjonneland, A., 2008. Acrylamide exposure and incidence of breast cancer among postmenopausal women in the Danish Diet, Cancer and Health Study. Int. J. Cancer 122, 2094–2100.

Palmers, S., Grauwet, T., Kebede, B.T., Hendrickx, M.E., Van Loey, A., 2014. Reduction of furan formation by high-pressure high-temperature treatment of individual vegetable purees. Food Bioprocess Technol. 7 (9), 2679–2693.

Palmers, S., Grauwet, T., Vanden Avenne, L., Verhaeghe, T., Kebede, B.T., Hendrickx, M.E., Van Loey, A., 2016. Effect of oxygen availability and pH on the furan concentration formed during thermal preservation of plant-based foods. Food Addit. Contam. 33, 612–622.

Park, Y.W., Yang, H.W., Storkson, J.M., Albright, K.J., Liu, W., Lindsay, R.C., 2005. Controlling acrylamide in French fry and potato chip models and a mathematical model of acrylamide formation — acrylamide: acidulants, phytate and calcium. In: Friedman, M., Mottram, D. (Eds.), Chemistry and Safety of Acrylamide in Food. Advances in Experimental Medicine and Biology, 561. Springer, Boston, MA, pp. 343–356.

Patterson, H.B.W., 1994. Hydrogenation of Fats and Oils: Theory and Practice. AOAC Press, Champaign, IL.

Pedreschi, F., Kaack, K., Granby, K., 2004. Reduction of acrylamide formation in potato slices during frying. LWT Food Sci. Technol. 37, 679–685.

Pedreschi, F., Moyano, P., Kaack, K., Granby, K., 2005. Color changes and acrylamide formation in fried potato slices. Food Res. Int. 38, 1–9.

Pedreschi, F., Kaack, K., Granby, K., 2006. Acrylamide content and color development in fried potato strips. Food Res. Int. 39, 40–46.

Pedreschi, F., Kaack, K., Granby, K., Troncoso, E., 2007. Acrylamide reduction under different pre-treatments in French fries. J. Food Eng. 79, 1287–1294.

Pedreschi, F., Kaack, K., Granby, K., 2008. The effect of asparaginase on acrylamide formation in French fries. Food Chem. 109, 386–392.

Pedreschi, F., Mariotti, S., Granby, K., Risum, J., 2011. Acrylamide reduction in potato chips by using commercial asparaginase in combination with conventional blanching. Food Sci. Technol. 44, 1473–1476.

Pedreschi, F., Mariotti, M.S., Granby, K., 2014. Current issues in dietary acrylamide: formation, mitigation and risk assessment. J. Sci. Food Agric. 94, 9–20.

Poti, J.M., Mendez, M.A., Ng, S.W., Popkin, B.M., 2015. Is the degree of food processing and convenience linked with the nutritional quality of foods purchased by US households? Am. J. Clin. Nutr. 101, 1251–1262.

Reinhard, H., Sager, F., Zimmermann, H., Zoller, O., 2004. Furan in foods on the Swiss market- method and results. Mittl. Lebnsm. Hyg. 95, 532–535.

Remig, V., Franklin, B., Margolis, S., Kostas, G., Nece, T., Street, J.C., 2010. Trans fats in America: a review of their use, consumption, health implications, and regulation. J. Am. Diet Assoc. 110, 585–592.

Riboldi, B.P., Vinhas, A.M., Moreira, J.D., 2014. Risks of dietary acrylamide exposure: a system review. Food Chem. 157, 310–322.

Rice, J.M., 2005. The carcinogenicity of acrylamide. Mutat. Res. 580, 3–20.

Roberts, D., Crews, C., Grundy, H., Mills, C., Matthews, W., 2008. Effect of consumer cooking on furan in convenience foods. Food Addit. Contam. 25, 25–31.

Rupp, H., 2003. Chemical and physical hazards produced during food processing, storage and preparation. In: Schmidt, R.H., Rodrick, G.E. (Eds.), Food Safety Handbook. John Wiley & Sons, Hoboken, NJ, pp. 233–263.

Rydberg, P., Erikson, S., Tareke, E., Karlsson, P., Ehrenberg, L., Törnqvist, M., 2003. Investigations of factors that influence the acrylamide content of heated foodstuffs. J. Agric. Food Chem. 51, 7012–7018.

Rydberg, P., Eriksson, S., Tareke, E., Karlsson, P., Ehrenberg, L., Törnqvist, M., 2005. Factors that influence the acrylamide content of heated foods. Chemistry and Safety of Acrylamide in Food. Advances in Experimental Medicine and Biology. In: Friedman, M., Mottram, D. (Eds.), 561. Springer, Boston, MA, pp. 317–328.

Sadd, P.A., Hamlet, C.G., Liang, L., 2008. Effectiveness of methods for reducing acrylamide in bakery products. J. Agric. Food Chem. 56, 6154–6161.

Seok, Y.-J., Her, J.-Y., Kim, Y.-G., Kim, M.Y., Jeong, S.Y., Kim, M.K., Lee, J., Kim, C.-I., Yoon, H.-J., Lee, K.-G., 2015. Furan in thermally processed foods: a review. Toxicol. Res. 31, 241–253.

Shen, M.Y., Liu, Q., Jia, H., Jiang, Y., Nie, S., Xie, J.H., Li, C., Xie, M.Y., 2016. Simultaneous determination of furan and 2-alkylfurans in heat-processed foods by automated static headspace gas chromatography-mass spectrometry. LWT Food Sci. Technol. 72, 44–54.

Sikorski, Z.E., 2005. The effect of processing on the nutritional value and toxicity of foods. In: Dabrowski, W.M., Sikorski, Z.E. (Eds.), Toxins in Food. CRC Press, Boca Raton, FL, pp. 285–312.

Skog, K., Alexander, J. (Eds.), 2006. Acrylamide and Other Hazardous Compounds in Heat-Treated Foods. England: Woodhead Publishing, Cambridge,.

Stadler, R.H., 2006. The formation of acrylamide in cereal products and coffee. In: Skog, K., Alexander, J. (Eds.), Acrylamide and Other Hazardous Compounds in Heat-Treated Foods. Woodhead Publishing, Cambridge, England, pp. 23–40.

Stadler, R.H., Lineback, D.R. (Eds.), 2009. Process-induced Food Toxicants. Wiley & Sons, Hoboken, NJ.

Stadler, R.H., Blank, I., Varga, N., Robert, F., Hau, J., Guy, A.P., Robert, M.-C., Rieliker, S., 2002. Acrylamide formed in the Maillard reaction. Nature 419, 449.

Stadler, R.H., Robert, F., Riediker, S., Varga, N., Davidek, T., Devaud, S., Goldmann, T., Hau, J., Blank, I., 2004. In-depth mechanistic study on the formation of acrylamide and other vinylogous compounds by the Maillard reaction. J. Agric. Food Chem. 52, 5550–5558.

Surdyk, N., Rosén, J., Andersson, R., Aman, P., 2004. Effects of asparagine, fructose, and baking conditions on acrylamide content in yeast-leavened wheat bread. J. Agric. Food Chem. 52, 2047–2051.

Taeymans, D., Wood, J., Ashby, P., Blank, I., Studer, A., Stadler, R.H., Gonde, P., Van Eijck, P., Lalljie, S., Lingnert, H., Lindblom, M., Matissek, R., Muller, D., Tallmadge, D., O'Brien, J., Thompson, S., Silvani, D., Whitmore, T., 2004. A review of acrylamide: an industry perspective on research, analysis, formation, and control,. Crit. Rev. Food Sci. Nutr. 44, 323–347.

Tareke, E., Rydberg, P., Karlsson, P., Eriksson, S., Törnqvist, M., 2002. Analysis of acrylamide, a carcinogen formed in heated foodstuffs. J. Agric. Food Chem. 50, 4998–5006.

Taubert, D., Harlfinger, S., Henkes, L., Berkels, R., Schömig, E., 2004. Influence of processing parameters on acrylamide formation during frying of potatoes. J. Agric. Food Chem. 52, 2735–2739.

Teegala, S.M., Willett, W.C., Mozaffarian, D., 2009. Consumption and health effects of trans fatty acids: a review. J. AOAC Int. 92, 1250–1257.

Tran, N.L., Barraj, L.M., Collinge, S., 2017. Reduction in dietary acrylamide exposure- Impact of potatoes with low acrylamide potential. Risk Anal. 37, 1754–1767.

Van Boekel, M., Fogliano, V., Pellegrini, N., Stanton, C., Scholz, G., Lalljie, S., Somoza, V., Knorr, D., Jastri, P.R., Eisenbrand, G., 2010. A review on the beneficial aspects of food processing. Mol. Nutr. Food Res. 54, 1215–1247.

Van Lancker, F., Adams, A., Owczarek-Fendor, A., De Meulenaer, B., De Kimpe, N., 2011. Mechanistic insights into furan formation in Maillard model systems. J. Agric. Food Chem. 59, 229–235.

Vass, M., Amrein, T.M., Schönbächler, B., Esher, F., Amadom, R., 2004. Ways to reduce the acrylamide formation in cracker products. Czech J. Food Sci. 22, 19–21.

Vattem, D.A., Shetty, K., 2003. Acrylamide: a model for mechanism of formation and its reduction. Innovat. Food Sci. Emerg. Technol. 4, 331–338.

Virk-Baker, M.K., Nagy, T.R., Barnes, S., Groopman, J., 2014. Dietary acrylamide and human cancer: a systemic review of literature. Nutr. Cancer 66, 744–790.

Vranová, J., Ciesarová, Z., 2009. Furan in food- A review. Czech J. Food Sci. 27, 1–10.

Watzek, N., Böhm, N., Feld, J., Scherbl, D., Berger, F., Merz, K.H., Lampen, A., Reemtsma, T., Tannenbaum, S.R., Skipper, P.L., Baum, M., Richling, E., Eisenbrand, G., 2012. N7-glycidamide-guanine DNA adduct formation by orally ingested acrylamide in rats: a dose-response study encompassing human diet-related exposure levels. Chem. Res. Toxicol. 25, 381–390.

Wenzl, T., Anklam, E., 2007. European Union database of acrylamide levels in food: update and critical review of data collection. Food Addit. Contam. 24 (S1), 5–12.

Wenzl, T., Lachenmeier, D.W., Gökmen, V., 2007. Analysis of heat-induced contaminants (acrylamide, chloropropanols and furan) in carbohydrate-rich foods. Anal. Bioanal. Chem. 389, 119–137.

Wilson, D.M., Goldsworthy, T.L., Popp, J.A., Butterworth, B.E., 1992. Evaluation of genotoxicity, pathological lesions, and cell proliferation in livers of rats and mice treated with furan. Environ. Mol. Mutagen. 19, 209–222.

Xu, F., Khalid, P., Oruna-Concha, M.-J., Elmore, J.S., 2015. Effect of asparaginase on flavor formation in roasted coffee. In: Taylor, A.J., Mottram, D.S. (Eds.), Flavour Science, Proceedings of the XIV Weurman Flavour Research Symposium. Context Products, Packington, UK, pp. 563–566.

Xu, F.,M., Oruna-Concha, M.-J., Elmore, J.S., 2016. The use of asparaginase to reduce acrylamide levels in cooked food. Food Chem. 210, 163–171.

Yaylayan, V.A., Stadler, R.H., 2005. Acrylamide formation in food: a mechanistic perspective. J. AOAC Int. 88, 262–267.

Yaylayan, V.A., Locas, C.P., Wronowski, A., O'Brien, J., 2005. Mechanistic pathways of formation of acrylamide from different amino acids. In: Friedman, M., Mottram, D. (Eds.), Chemistry and Safety of Acrylamide in Food. Springer, New York, pp. 191–204.

Zhang, Y., Zhang, Y., 2007. Study on reduction of acrylamide in fried bread sticks by addition of antioxidant of bamboo leaves and extract of green tea. Asia Pac. J. Clin. Nutr. 16 (Suppl. 1), 131–136.

Zhang, Y., Chen, J., Zhang, X., 2007. Addition of antioxidant of bamboo leaves (AOB) effectively reduces acrylamide formation in potato crisps and French fries. J. Agric. Food Chem. 55, 523–528.

Zhang, Y., Kahl, D.H.W., Bizimungu, B., Lu, Z.-X., 2018. Effects of blanching treatments on acrylamide, asparagine, reducing sugars and colour in potato chips. J. Food Sci. Technol. 55, 4028–4041.

Zoller, O., Sager, F., Reinhard, H., 2007. Furan in food: headspace method and product survey. Food Addit. Contam. 24 (S1), 91–107.

Zhou, W., Wang, M., Chen, J., Zhang, R., 2015. The effect of biological (yeast) treatment conditions on acrylamide formation in deep-fried potatoes. Food Sci. Biotechnol. 24, 561–566.

Zyzak, D.V., Sanders, R.A., Stojanovic, M., Tallmadge, D.H., Eberhart, B.L., Ewald, D.K., Gruber, D.C., Morsch, T.R., Strothers, M.A., Rizzi, G.P., Villagran, M.D., 2003. Acrylamide formation mechanism in heated foods. J. Agric. Food Chem. 51, 4782–4787.

Chapter 14

Food safety and regulatory survey of food additives and other substances in human food

Larry Keener
International Product Safety Consultants, Seattle, WA, United States

14.1 Introduction

According to a joint Extension Services report from Iowa, Kansas, and Nebraska, by Redlinger and Nelson (1983), there are presently upwards of 2800 additives approved for use in the United States (US). In the US, there are greater than 400 million pounds of additives used annually in processed meats and meat bearing products (Food Product Development, 1980). The average US citizen is reported to consume between 140 and 150 pounds of additives per year (Redlinger and Nelson, 1993). During the period from 1978 to 2008, the US food additives business grew from about $1 billion to nearly $13.0 billion in annual sales (Fredonia, 2004). The market size of food ingredients worldwide for 2015 was reported at $67 billion US dollars and the projections for 2020 are at $85.4 billion dollars representing nearly 30% increase in market value over the 5 -year interval (www.statista.com/statistics/627706/market-size-of-food-ingredients-worldwide/) (see Fig. 14.1).

Worldwide, approximately 98% of the additive compounds consumed include sugar; corn sweeteners; salt; citric acid; pepper; vegetable color; mustard; yeast; and baking powder (Redlinger and Nelson, 1993).

14.1.1 Food additive

A food additive is a substance or mixture of substances, other than basic food stuff, present in food as a result of any aspect of production, processing, storage, or packaging. Food processors have in their armamentaria many thousands of chemicals they may add to foods intended for human consumption. Some are deleterious, some are harmless, and some are beneficial. According to the Food Protection Committee of the National Academy of Sciences, which evaluates the safety of additives in the United States, every one of the chemicals in this class must serve one or more of the following purposes or functions: improve nutritional value; enhance food quality or consumer acceptability; improve keeping characteristic; make food more readily accessible; and facilitate its preparation. The reality of the matter, when viewed globally, is that these fundamental precepts do not have universal acceptance. There is an abundance of discord and disagreement internationally as to what exactly is, and what is not, a food additive.

14.1.1.1 Codex Alimentarius

A "food additive" according to Codex is "any substance not normally consumed as a food by itself and not normally used as a typical ingredient of the food, whether or not it has nutritive value, the intentional addition of which to food for a technological (including organoleptic) purpose in the manufacture, processing, preparation, treatment, packing, packaging, transport or holding of such food results, or may be reasonably expected to result, (directly or indirectly) in it or its by-products becoming a component of or otherwise affecting the characteristics of such foods." The definition goes on to state that the term *does not include* contaminants or substances added to food for maintaining or improving nutritional qualities

Ensuring Global Food Safety. https://doi.org/10.1016/B978-0-12-816011-4.00005-7

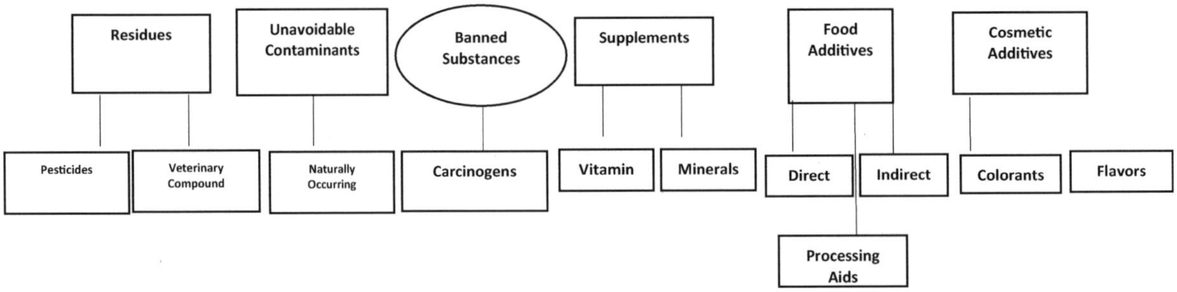

FIGURE 14.1 Inventory of substances in human food. *After Schultz, H.W., 1981, Food Law Handbook, AVI Publishing Westport CT, USA13. US Code of Federal Regulations 21CFR101.100 (A)(3), US Food and Drug Administration 1986.*

or sodium chloride (Codex, 1991). It is easily conceived that confusion might rapidly ensue using this definition as to what is and what is not a food additive. The issue is further complicated when one considers the implications associated with the multitude of national definitions and norms related to food additives. Salt (sodium chloride), for example, is used for both flavor development and preservation. Which begs the question is salt a food additive or not? Codex says not. The very explicit statement in Codex regulations to exempt sodium chloride from this classification is very confusing. In the US, salt is classified as a GRAS substance (generally recognized as safe) and here too it is exempt from the requirements of the food additive regulations.

The **International Numbering System for Food Additives (INS)** is a European-based naming system for food additives, aimed at providing a short designation of what may be a lengthy actual name (Centre for Food Safety, 2006). It is defined by Codex Alimentarius, the international food standards organization of the World Health Organization (WHO) and Food and Agriculture Organization (FAO) of the United Nations (UN). The information is published in the document *Class Names and the International Numbering System for Food Additives*, first published in 1989, with revisions in 2008 and 2011. The INS is an open list, and therefore subject to the inclusion of additional additives or removal of existing ones on an ongoing basis. The EU approved additives are listed with an "E," and those approved for Australia and New Zealand with an "A" and for the US with a U, even though the US does not use the INS numbering system. For example, the vegetable gum, carrageenan, is designated by the INS register as 407(A) (E) (U). Similarly, the thickening agent hydroxypropyl methylcellulose is designated 464 (A) (E) by this method. The INS registry currently contains on the order of 1400 entries.

14.1.1.2 United States food additive regulation

In the US, a Food Additive is defined at Title 21cfr170.3 (e) (1); a Food *additive* includes all substances not exempted by section 201(s) of the US Food Drug and Cosmetic Act, the intended use of which results or may reasonably be expected to result, directly or indirectly, either in their becoming a component of food or otherwise affecting the characteristics of food. A material used in the production of containers and packages is subject to the definition if it may reasonably be expected to become a component, or to affect the characteristics, directly or indirectly, of food packed in the container. "Affecting the characteristics of food" does not include such physical effects, as protecting contents of packages, preserving shape, and preventing moisture loss. If there is no migration of a packaging component from the package to the food, it does not become a component of the food and thus is not a food additive. A substance that does not become a component of food, but that is used, for example, in preparing an ingredient of the food to give a different flavor, texture, or other characteristic in the food, may be a food additive (FDA, 2017). Also in contrast to Codex, the US Food and Drug Administration, for example, accept that food additives "promote health and wellness," "improve nutritional value," and "flavor development and stabilization." Clearly, these functional characteristics are in conflict with Codex's food additive definition.

In the United States the matter of food additives is convoluted and a complicated state of affairs. There are direct food additives or substances added deliberately to cause a specific outcome or effect in the food. For example, the low-calorie sweetener aspartame, which is used in beverages, puddings, yogurt, chewing gum, and other foods, is considered a direct additive. By and large, one would expect that all direct food additives would be included in the food ingredient declaration on the package label. There are also indirect additives. These are substances, permitted by regulation for inclusion in foods, which find their way into foods unintentionally and are not expected to have an impact on the food, favorable or otherwise. For instance, minute amounts of packaging substances may find their way into foods during storage. Or in the course of food production very small quantities of machine oil or lubricants come in contact with the food. These too would be considered indirect food additives. Indirect additive are not likely to be declared on food labels.

The Food Additives Amendment to the FD&C Act, passed in 1958, requires FDA approval for the use of an additive prior to its inclusion in food. It also requires the manufacturer to prove an additive's safety for the ways it will be used.

The Food Additives Amendment exempted two groups of substances from the food additive regulation process. All substances that FDA or the US Department of Agriculture (USDA) had determined were safe for use in specific food prior to the 1958 amendment were designated as *prior-sanctioned substances (21cfr181)*. Examples of prior-sanctioned substances are sodium nitrite and potassium nitrite used to preserve luncheon meats.

A second category of substances excluded from the food additive regulation process are generally recognized as safe or *GRAS substances*. GRAS substances are those whose use is generally recognized by experts as safe, based on their extensive history of use in food before 1958 or based on published scientific evidence. Sodium chloride, sugar, spices, vitamins, and monosodium glutamate are classified as GRAS substances, along with several hundred other food ingredients. Manufacturers may also petition FDA to review the use of a substance to determine if it is GRAS.

Since 1958, FDA and USDA have continued to monitor all prior sanctioned and GRAS substances in light of new scientific information. If new evidence suggests that a GRAS or prior sanctioned substance may be unsafe, federal authorities can prohibit its use or require further studies to determine its safety. For example, Titanium Dioxide, a GRAS substance, is a common additive in many **foods**, personal care, and other consumer products. It is sometimes used as a whitener and sometimes as an anticaking agent (to prevent the product from clumping). In recent years, concerns have been raised about the health effects of this compound in food. There are reports that it may promote type 2 diabetes and or that it is a carcinogen. Research linking the use of titanium dioxide as a food additive to gut inflammation and bowel cancer in rats has prompted the French government to initiate further analysis into the substance. In May, 2021, The European Food Safety Authority ("EFSA") published its updated safety assessment of titanium dioxide (E171) as a food additive. Based on that review the agency reported that it could not calculate an acceptable daily intake (ADI) value for titanium dioxide in food. Ultimately, the agency concluded that it could not confirm the safety of E171 in food that is intended for human consumption (Van Vooren, B., et al, Aug. 2021) (Eu Plans Ban on Titanium Dioxide in Food). It is likely that these findings, questioning the safety of this additive, in the EU will cause the FDA to review the GRAS standing of Titanium Dioxide.

The Food Additives Amendments of the FDC&A include a provision which prohibits the approval of an additive if it is found to cause cancer in humans or animals. This clause is often referred to as the Delaney Clause, named for its Congressional sponsor, Rep. James Delaney (D-N.Y.). Sections 402(a) (3) and 402(a) (4) of the FDC&A also stipulate that a food is adulterated, not fit for human consumption, if it contains in part any decomposed or other substance that may be injurious to the public health. Thus, when a substance is added to food, directly or indirectly, in a manner inconsistent with the regulations, it may be deemed adulterated.

14.1.1.3 China's National Food Safety Standard for food additives

China's Food Safety National Standards for the Usage of Food Additives GB 2760-2011 was published on the April 20, 2011 and entered into force on the 20th of June of that same year. This standard specifies the basic principle for the usage of food additives, food additive types, application scope, and maximum allowable dose level. It is interesting to note that this legislation was written on the heels of the 2008 and 2010 incidents involving dairy product and infant formula contaminated with melamine. BBC News Asia—Pacific reported in a June 2010 article that upwards 60 tons of melamine tainted dairy materials had been seized by the government. Test samples were reported to have contained more than 500 times the allowed levels established for melamine in dairy powders. The use of melamine in milk in 2008 was reported by government officials to have killed six infants and made more than 300,000 people ill. The food safety failures involving substances added to human food attracted global attention and raised significant concerns about the safety of foods produced in China. Moreover, these spectacular food safety failures threatened China's nascent flirtations with free market economics and potentially scuttling its burgeoning international trade in food. Its trading partners, across the globe, grew increasingly apprehensive about the safety products, especially food products, exported by China. The United States, for example, issued an import alert for Chinese-made food products, calling for foods to be stopped at the border unless importers could certify that they were either free of dairy or free of melamine (Reuters, 2008). Many other countries were reported to also have increased scrutiny of all Chinese exports of milk and egg products.

According to the Chinese National Food Safety Standard for Uses of Food Additives (GB 2760-2014), a food additive refers to an artificially chemosynthetic or natural substance to be added to foods in order to improve food quality, color, fragrance, and taste, and for the purpose of preservation and processing technology. The scope of food additives in China also includes flavorings, gum-based substances in chewing gum, and processing aids. Permitted food additives are classified into 22 varieties according to their functions. More than 2400 permitted food additives are listed in GB 2760-2014 regulation.

In December 2017, China's National Center for Food Safety Risk Assessment released the first draft National Food Safety Standard for the Use of Food Additives (GB 2760) to revise and update its current version issued in 2014 (Centre for Food Safety, 2006). The draft Standard amends the use of some food additives to reflect the latest regulatory developments at home and abroad.

The draft Standard made a number of substantive changes relating to the use of food additives. Listed here are several examples of the changes to the regulations of 2014:

- Names of 22 food additives are updated. For example, the current Chinese name of "benzoic acid and its sodium salt" is revised to "benzoic acid and its sodium salt (benzoic acid, sodium benzoate)." Since a food manufacturer or operator may elect to declare the Chinese name of a food additive on the food label (Codex, 1991), change to a food additive's Chinese name in the standard may impact the corresponding declaration on the food label.
- The draft updates the Chinese Numbering System (CNS numbers) for 14 food additives, and incorporates 22 food additives approved by the National Health and Family Planning Commission since the publication of the 2014 version. 68 English names and International Numbering System (INS numbers) of food additives are revised to be in line with the Codex General Standard of Food Additives. The draft states that, in case of any inconsistency, the most recent INS numbers shall prevail.
- "Nutrition enhancer" is now included in the scope of "food additives" so as to be consistent with its definition under China's Food Safety Law. Accordingly, this functional class of food additives is now added to the draft GB 2760s Appendix D—Functional Classes of Food Additives.

It is clear that China has rushed to update its food additive regulations to be more in line with those of the EU and Codex. It is also interesting to note the evolution of the GB 2060 regulations since their first publication in 2011 and further to contemplate the impact of the melamine scandal on the country's food additive and evolving food safety regulations.

14.1.2 Processing aids

Processing aids and carry-over compounds, subclassifications of food additives, are also an exceedingly confusing subject. Carry-over compounds are components of food or food ingredients that are added indirectly (carried-over) to another food but have no technological effect in that final food. According to Codex, processing aids are "*A 'chaotic subject' where member states have disparate views due to their own individual experiences and long histories of regulatory development*" (CAC, 2001). The subject may even be chaotic *within* national regulatory frameworks. Consider that in the US, regulations of the Federal Food, Drug and Cosmetic Act allow for the definition and use of Processing Aids (FDA, 1986) in the production of food that are intended for human use. By contrast, however, the US Department of Agriculture's Food Safety and Inspection Service regulations are mute on the subject. Likewise, the Canadian Food Inspection Agency does not have specific regulations governing this class of food additive compounds called processing aids (Salminen, 2005).

Processing aids are an important but complex class of food additives. They play a role in facilitating the stabilization and preservation of many food products. Hydrogen peroxide, for example, is used in processing liquid eggs and allows the egg to be pasteurized using mild thermal processing conditions that will not result in protein denaturation. Utilizing this heat and hydrogen peroxide process requires the use of yet another processing aid, the enzyme Catalase. Catalase is added to the process for the removal of residual hydrogen peroxide and thereby enables compliance with the aspect of the US regulation requiring that the processing aid is removed and not present in the final food. Chemicals in this grouping are used to promote separation, clarification, mixing, blending, and foam suppression, as well as for management of material flow characteristics. Sand and silica are excellent examples of flow control agents that are processing aids. Included in this classification is a very broad array of both chemical and biological agents. Processing aids may be derived from biotechnology, from other natural foods sources, or they may result from synthesis in a chemical manufacturing facility. Organic acids (lactic, Citric, Acetic, or Tartaric) are approved food additives in the US but under certain circumstances of use they are also processing aids. The global increase in awareness of food allergens, intolerances, and other food sensitivities has called in to question the practice of not labeling all substances added to human food.

In terms of substances that are allowed for addition to human foods, processing aids hold a unique status in that many countries have elected to exempt them from labeling requirements. This exemption frequently raises questions and concerns for both regulatory officials and consumers about the public health status and safety of foods to which these substances have been added. The vagaries and nuances in this area are enormous and also frequently an impediment to international trade.

14.1.2.1 Japanese legislation and regulations

Japan defines processing aids as "substances that are added to a food during the processing of such food but are removed from the food before it is prepared in its finished form,<substances> that are added to a food during processing that are converted into components ubiquitously present in the food, and do not significantly increase the level of the constituents naturally found in food, or <substances> that are added to a food for their technical or functional effect in the processing but are present in the finished food at an insignificant level and do not have any technical or functional effect in the food" (CAC, 2007).

14.1.2.2 Codex Alimentarius

Codex's definition takes a somewhat different view of processing aids and it has defined this class of additives accordingly: "*Processing aid* - any substance or material not including apparatus or utensils, and not consumed as a food ingredient by itself, intentionally used in the processing of raw materials, foods, or its ingredients to fulfill a certain technological purpose during treatment or processing and which may result in the nonintentional but unavoidable presence of residues or derivatives in the final product"(CAC, 1991).

14.1.2.3 Australia and New Zealand

According to the Food Standards Australia New Zealand Act 1991 and the Australia New Zealand Food Standards Code 1.3.3 (March 2016), a processing aid is any substance used in the processing of raw materials, foods, or ingredients, to fulfill a technological purpose relating to treatment or processing, but does not perform a technological function in the final food; and the substance is used in the course of manufacture of a food at the lowest level necessary to achieve a function in the processing of that food, irrespective of any maximum permitted level specified (Food Standards Gazette No. FSC96; April 10, 2015). Standard 1.3.3 is further divided into two main sections: Division 2; processing aids that may be used with any food generally permitted processing aids for all foods (1) A substance listed in subsection (2) may be *used as a processing aid in any food if it is used at a level necessary to achieve a technological purpose in the processing of that food. Division 3; processing aids that can be used with specified foods. In this class are listed very specific compounds for use in a limited few categories of food and water. There are four main subdivisions of this section including Processing Aids for Water; Bleaching, Washing, and Peeling Agents; Extraction Solvents; Processing aids that perform various technological purposes; and a very curious provision for the Microbial control agent—dimethyl dicarbonate. It is also very noteworthy that this standard, given its great detail on the use and application of processing aids, is mute on the subject of labeling requirements for processing aids.

14.1.2.4 United States

According to the US Food Drug and Cosmetic Act (FDC&A) at 21 CFR 101.100 (a) (3) (ii) a processing aid is (a) Substances added to food during processing but are removed before the food is packaged in the finished form which does not have a technical or functional effect in the finished food, (b) Substances added to a food during processing are converted into constituents normally present in the food, and do not significantly increase the amount of the constituents naturally found in the food, and (c) Substances that are added to a food for their technical or functional effect in the processing but are present in the finished food at insignificant levels and so not have any technical or functional effect in that food. US regulations exempt processing aids declaration on food packaging.

14.1.2.5 Canada

"There is currently no formal regulatory definition for the term 'processing aid' in the Food and Drug Regulations." There is an administrative definition to differentiate between processing aids and food additives. Accordingly, "A processing aid is a substance that is used to aid in the processing of a food but is not intended to have a functional effect in the final food."

Also, the use of a processing aid should not result in any residues of the substance in the final food or, if traces are present, they should be negligible and every reasonable effort should be made to remove and minimize residues (Salminen, 2009).

14.1.2.6 China

Processing aids shall be used in the course of food processing with necessity, and shall reduce the dosage as far as possible under the precondition of reaching the desired effect. C.1.2 The processing aid shall be generally removed before the finalized products, if impossible to remove it completely; the residue quantity shall be minimized, where the residue limits shall have an adverse effect on health and shall not play the functional role in final products. C.1.3 The processing aid shall meet relevant requirements on quality and specification. Further the Chinese regulations for processing aids are divided in to two broad sections: Section 1 pertains to "processing aids that can be used in all kinds of food processing and the residue quantity needs no restriction with the names of the processing aids ranking in Chinese Phonetic Alphabet (excluding Enzyme preparation)." This section contains listing of some 38 substances that meet this standard. A lengthier list is included for Section 2 compounds. In this class are the Processing Aids that Require Clarification of the Functions and Scope of Use. There are nearly 80 compounds listed as meeting this requirement for clarification of use and function. Like the standards of Australia and New Zealand, the Chinese Standards are also mute on the subject of labeling (Chinese Standards for Food Additives - GB2760-2015 appendix C).

14.1.3 Cosmetic additives—comparison of EU and US color additive regulations

Following the Industrial Revolution, the transition to new manufacturing processes in the period from about 1760 to sometime between 1820 and 1840, foods were increasingly mass processed on a large scale and new technologies including preservation frequently altered the natural appearance of foods (Downham and Collins, 2000). To overcome these cosmetic changes, inexpensive and stable synthetic and mineral dyes with high tinctorial strengths and bright shades were excessively applied to a wide variety of foods. However, with advances in science and especially toxicology, some of the color additives such as copper sulfate, mercury sulfide, lead chromate, and indigo were shown to have toxic properties. As a public health measure to restrict the use of certain food color compounds in the US, a list of approved food colors was published in 1906 (US Pure Food and Drug Act) (Burrows, 2009). The Pure Food and Drug Act was a centerpiece of progressive reforms in the United States during early 20th century. In 1960, the Color Additive Amendment incorporated the Delaney Clause prohibiting the addition of substances to human food that were known to induce cancer in humans or animals (Sanchez, 2015). In the United Kingdom, several colorants were prohibited in 1923 and, in 1957, a legally binding list of permitted colors was established (Burrows, 2009). A joint expert committee on food additives (JECFA), administered jointly by the FAO and WHO, was established in 1956, and has since provided an extensive review of 1500 substances, including food colors, setting the standards for safety assessment globally. Also, the International Program on Chemical Safety of the WHO assesses the health impact of chemicals in food.

Both the EU and United States of America take a risk-based approach to managing and controlling color additives in the food supply, and while the approaches are similar, the outcomes are frequently different. At the present there are color compounds approved by the US FDA that are not allowed or accepted for food use in the European Union. This discord in risk assessment outcomes and regulation frequently causes concern for public health, food safety, and is also an impediment to trade between the European Union and the United States. EU exports arriving in the US must adjust for the differences between the various regulations, otherwise the food could be deemed adulterated, or misbranded and therefore subject to regulatory enforcement action by the US.

The overlap between the US and EU food color authorizations is limited. Currently, there are approximately 30 color compounds or groups of colors approved for use in food in both the EU and US. Only six colors of synthetic origin are authorized by both EU and US legislation. Four color additives approved in the US are not permitted in the EU: the three synthetic colors, namely Orange B, Citrus Red No. 2, and FD&C Green No. 3 (Fast Green FCF) and toasted partially defatted cooked cottonseed flour. In addition, there are 16 color additives authorized in the EU that are prohibited from use in the US, including nine colors of synthetic origin and lutein, vegetable carbon, aluminum, silver and gold, chlorophylls and chlorophyllins, and calcium carbonate (Lehto et al., 2017).

14.1.3.1 US color additive regulations

In the USA, the statutes pertaining to food colors are primarily enforced by the USFDA, an agency of the Department of Health and Human Services, under Title 21 of the Code of Federal Regulations (CFR, 2016). Parts 70 through 82 and part

101 of the code contain rules on petitions and labeling and list the specifications and rules for use of approved color additives. Color additives are regulated apart from other food additives as a specific class of substances added to food and animal feed with the purpose or capable of imparting color to food.

According to the US Food and Drug Administration (1993), "Technically, a color additive is any dye, pigment, or substance that can impart color when added or applied to a food, drug, and cosmetic or to the human body." The Food and Drug Administration (FDA) is responsible for regulating all color additives used in the United States. All color additives permitted for use in foods are classified as "certifiable" or "exempt from certification." Certifiable color additives (see Table 14.1) are manmade, with each batch being tested by manufacturer and FDA. This "premarket approval" process, known as color additive certification, assures the safety, quality, consistency, and strength of the color additive prior to its use in foods. There are nine certified colors approved for use in food in the United States. One example is FD&C Yellow No. 6, which is used in cereals, bakery goods, snack foods, and other foods.

Color additives that are exempt from certification (see Table 14.2) include pigments derived from natural sources such as vegetables, minerals, or animals, and man-made counterparts of natural derivatives. For example, caramel color is produced commercially by heating sugar and other carbohydrates under strictly controlled conditions for use in sauces, gravies, soft drinks, baked goods, and other foods. Beet juice extract is another example of an exempt color additive. It is derived from the vegetable root and the pigment extracted for use in a myriad of food products. Certifiable color additives generally do not impart undesirable flavors to foods, while color derived from foods such as beets and cranberries can produce such unintended effects. Certifiable color additives are further classified as either "dyes" or "lakes." Dyes are water-soluble compounds. By contrast, Lakes are water insoluble.

There are currently nine certifiable colors approved for use in the United States. Whether a color additive is certifiable or exempt from certification has no bearing on its overall safety. Both types of color additives are subject to rigorous standards of safety prior to their approval for use in foods (FDA, 1993). Colored foods or food ingredients, which contribute their own color when mixed with other foods, such as chocolate in chocolate milk or cherries in cherry yoghurt, are not considered as color additives (21 CFR §70.3). Applying cherry juice to color cherry yoghurt, however, is defined as color addition (Matulka and Tardy, 2014). Similarly, ingredients such as riboflavin and beta-carotene are only considered to be color additives when they are intended to impart color, not when they are added for nutritive value as additives with GRAS status; the coloring effect must then remain unimportant (21 CFR §70.3(g)).

14.1.3.2 EU color additives regulation

The European Union (EU) is a political and economic union, consisting of 28 member states that are subject to the obligations and the privileges of the membership. Every member state is part of the founding treaties of the union and is subjected to binding laws within the common legislative and judicial institutions. In order for the EU to adopt policies that concern defense and foreign affairs, all member states must agree unanimously. The union is threatened by many factors

TABLE 14.1 US FDA certified colors.

Certifiable colors 21cfr74(a)
FD&C Blue No. 1 (Dye and Lake)
FD&C Blue No. 2 (Dye and Lake)
FD&C Green No. 3 (Dye and Lake)
FD&C Red No. 3 (Dye)
FD&C Red No. 40 (Dye and Lake)
FD&C Yellow No. 5 (Dye and Lake)
FD&C Yellow No. 6 (Dye and Lake)
Orange B[a]
Citrus Red No. 2[a]

[a]*These food color additives are restricted to specific uses.*

TABLE 14.2 US FDA colors exempt from certification.

Colors exempt from certification 21Cfr73(a)
Annatto extract
B-Apo-8′-carotenal[a]
Beta-carotene
Beet powder
Canthaxanthin
Caramel color
Carrot oil
Cochineal extract (carmine)
Cottonseed flour, toasted partially defatted, cooked
Ferrous gluconate[a]
Fruit juice, Grape color extract[a]
Grape skin extract[a] (enocianina)
Paprika
Paprika oleoresin
Riboflavin
Saffron
Titanium dioxide[a]
Turmeric
Turmeric oleoresin
Vegetable juice

[a]*These food color additives are restricted to specific uses.*

including the great economic disparity within the community among its member states. To that end, the British government has elected to secede from the union as of June 2016. The impact of "Brexit" (a portmanteau of the words "Britain" and "exit," the nickname for a British exit of the European Union) is not fully understood. But many pundits have suggested that the impact on trade between EU members and Britain will likely suffer and so too many of the agreed standards of practice relating specifically to the trade in food.

The EU has spent many years arriving at its current structure in its regulation of food and feed. The first directive for food additives agreed to was for food colors in 1962. They used the E-number classification system. This was followed by other directives for preservatives, antioxidants, and emulsifiers. This system still allowed the member states to specify which foods could contain the substances and the maximum levels permitted.

Harmonization of food additives was advocated throughout the community, and was achieved to some extent in the Framework Directive 89/107. The framework directive covered three separate directives on color, sweeteners, and all other additives.

Risk management responsibility for food additives within the EU lies with the Directorate General for Health and Consumer Protection (DG Sanco). Specific risk assessment for the safety of food ingredients is reviewed under the European Food Safety Authority (EFSA). EFSA has two panels. It is the panel on Food Additives and Nutrient Sources Added to Food (ANS) that deals with questions of safety in the use of food additives, nutrient sources, and other substances deliberately added to food, excluding flavorings and enzymes. The ANS panel applies the "Low Risk philosophy," with an added "precautionary principle" philosophy. The precautionary principle is frequently called a "better safe than sorry" outlook. It has also been the subject of disagreement and controversy between the EU and the US in trade issues.

The United States and Japan employ the "No Risk philosophy." JECFA and the European Commission (EC) of the European Union (EU) employ the "Low Risk philosophy." EU food additive standards, in general, are built on the notion of an acceptable daily intake (ADI) for additives, including color compounds. This ADI-based approach is a mechanism used to spread out the risk of cancer over time. This is the basis of the "Low Risk" philosophy. It is also the "Low Risk philosophy" that is the hallmark of new and revised food safety regulations.

The EU for all practical purposes, diverging from the US approach, treat color additives according to its general food additives legislation (see Fig. 14.2). That is, there are no special provisions within EU legislation to give color additives special standing. This is not to say that they are not scrutinized or subject to a rigorous risk assessment procedure. In the EU, food colors are regulated as food additives under a comprehensive set of regulations for food improvement agents. Regulation (EC) No. 1331/2008 (EC, 2008a) sets out a common authorization procedure, and Regulation (EC) No. 1333/2008 (EC, 2008c) on food additives and its amendment, Regulation (EC) No. 1129/2011 (EC, 2011), includes the rules for food colors. The annexes of the Regulation (EC) No. 1333/2008 contain food categories and a positive list of colors permitted in the EU including maximum quantities and instructions for use. Regulation (EU) No. 231/2012 (EC, 2012) lays down the specifications for food additives listed in Annexes II and III to Regulation (EC) No. 1333/2008 (Lehto et al., 2017). This represents a major departure from the approach taken by the US. In the US, for instance, there are no positive lists, *per se,* or established food categories where certain color additives are allowed and with maximum established limits codified.

As a food additive, a food color is a substance not normally consumed as a food or a characteristic ingredient of food that can be used to add or restore color. The colors are artificial dyes or natural constituents of foods and natural sources. Also extracts derived from natural sources containing pigments selectively enriched relative to the nutritive or aromatic constituents are defined as food colors (EC, 2013). Substances considered as food, e.g., fruit or vegetable concentrate and saffron used because of their coloring properties, are defined as coloring foods which do not fall within the scope of the food additives regulation. These food-derived coloring compounds should be used in accordance with the rules of the Regulation (EC) No. 178/2002, i.e., the general food law (EC, 2002) and other applicable rules.

In the EU, ingredients that are added to foods to change their color are classified one of three ways: as a color additive, as a flavor, or as a coloring food. Regulation 1333/2008 has several tables of additives that can be used, additives that can

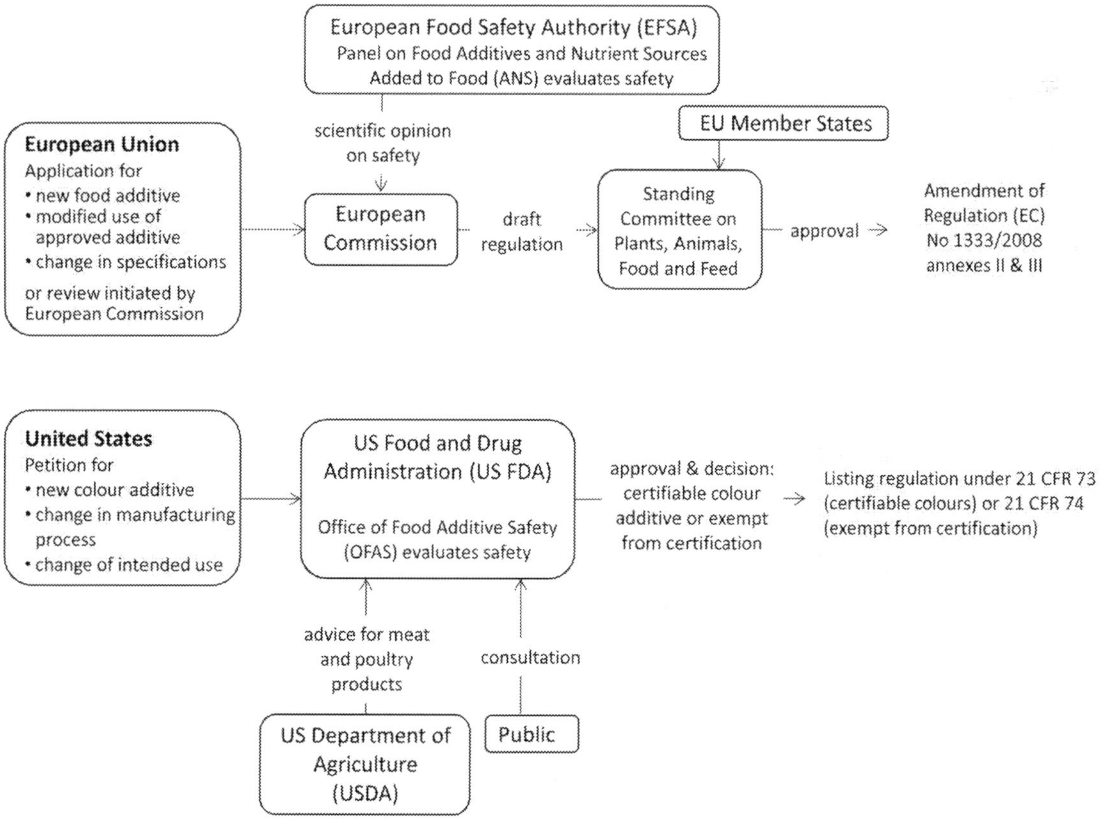

FIGURE 14.2 The approval processes of color additives in the EU and the US.

be used in additives and foods in which additives may be used, and the maximum usage level limits. Each additive has an "E" number which corresponds to the INS numbers listed in Codex (McAvoy, 2014). By definition, colors from synthetic sources, and selectively extracted colors from natural sources, as well as inorganic pigments, are regulated together. Some additives are allowed at *quantum satis* (GMP), while others have part per million levels based on the food category.

In the EU, color additives must be declared by the category name (color) and E number of the specific color, e.g., *Color (E 171)*. Coloring food shall be designated by its specific name, e.g., *beet juice concentrate/extract.* The term *coloring food is* not a legal category name nor a specific name for the relevant ingredient (McAvoy, 2014). Since 2008, these EU regulations have been amended over 30 times. Since it is new and often changing, it can feel disjointed and lead to some confusion (McAvoy, 2014).

14.1.3.3 China's color additive regulations

In China, color regulations are under GB2760-2011—National Food Safety Standard—Standard for uses of food additives. The latest version was implemented on June 20, 2011. Colors are registered and the toxicology data have to be submitted for new colors. Each color is petitioned for its use and usage level on an individual basis. Colors have their own purity specifications along the lines of Codex. However, there are some unusual importation regulations when you blend colors with other additives. GB 26687-2011, The National Standards for Food Safety, General Rules Regarding Compound Food Additives limits the combined level of lead to less than 2 ppm, and the combined level of arsenic to less than 2 ppm, despite the limit on the individual additive.

China's approach to regulating additive and color additive in particular is based in large measure on the legislation of the EU and of The Codex Alimentarius. Their approach is consistent with the "Low Risk philosophy" of the EU. Since 2011, China's food additives regulations have undergone major changes. On December 24, 2014, the National Health and Family Planning Committee of China issued the National Food Safety Standard for Food Additive Use (GB2760-2014), which was implemented on May 24, 2015. And much like the legislation of the EU, these regulations are replete with positive listings for color additives. The new regulations establish food category numbers; food category classification; and maximum use levels for each additive. Again consistent with the EU practice, use limits are per GMP or in ppm. The GB2760-2014 regulations do not create special provisions or subsections of the law to specifically treat color compounds as do the regulations of the United States of America.

14.1.3.4 Japan's color additive regulations

Japan has established their own regulations based on their own knowledge, much in the manner of the United States. As reported previously, the US and Japan are adherents of the "No Risk Philosophy" in regulating color compounds allowed in food. The Specifications and Standards for Foods, Food Additives, etc., under the Food Sanitation Act are published in English by the Japan External Trade Organization. In the act, they list the colors and other additives that are permitted. Colors can be under the following: Food additives with standards of use; Food additives with no standards of use; Existing food additives; and Substances which are generally provided as Food and which are used as food additives. In addition, Japan has the standard definitions of food categories, along with two special food categories: Food for Specified Health Uses and Food with Nutrient Function Claims. The English translation of the Specifications and Standards for Foods is located at http://www.jetro.go.jp/en/reports/regulations/pdf/foodext201112e.pdf.

14.1.4 Prohibited and banned substances

Globally, the finding of the prohibited and or banned compounds Melamine and Sudan Red in human food, dairy products, and infant formula has garnered spectacular international headlines and resulted in massive amounts of food products being recalled and destroyed.

The finding of melamine in dairy products and infant formula in China was reported in 2008. The adulterated products are reported to have caused the death of six infant deaths. It was also reported that this act of economic adulteration was responsible for causing illness in 300,000 babies in China (Huang, 2014). The compound was allegedly added to dairy product to mimic protein content for purpose of economic gain. According to the WHO, melamine is an organic base chemical most commonly found in the form of white crystals rich in nitrogen content. In China, where adulteration has occurred, water has been added to raw milk to increase its volume. As a result of this dilution, the milk has a lower protein concentration. Companies using the milk for further production (e.g., of powdered infant formula) normally check the protein level through a test measuring nitrogen content. The addition of melamine increases the nitrogen content of the milk and therefore its apparent protein content.

Addition of melamine into food is not approved by the FAO/WHO Codex Alimentarius (food standard commission), or by any national authorities (http://www.who.int/csr/media/faq/QAmelamine/en/).

The WHO reports further that there are no direct human studies on the effect of melamine; however, data from animal studies can be used to predict adverse health effects. Melamine alone causes bladder stones in animal tests. When combined with cyanuric acid, which may also be present in melamine powder, melamine can form crystals that can give rise to kidney stones. These small crystals can also block the small tubes in the kidney potentially stopping the production of urine, causing kidney failure and, in some cases, death. Melamine has also been shown to have carcinogenic effects in animals in certain circumstances, but there is insufficient evidence to make a judgment on carcinogenic risk in humans (http://www.who.int/csr/media/faq/QAmelamine/en/).

In 2003, Sudan Red dye was reported in chili peppers, paprika, and other spice. The finding caused among the largest global product recalls in recent history. The dye was added to an assortment of spices and peppers to enhance their market appeal and for increased sales and economic benefit.

Sudan dyes are synthetic chemical dyes of similar chemical structure. They are aromatic compounds containing azo group (−N=N−). Sudan I, II, III, and IV are kinds of the Sudan red dyes. They can be generally applied for coloring substances such as hydrocarbon solvents, oils, fats, waxes, and plastics. Sudan II and III can also be used in cosmetics and drugs applied externally, whereas Sudan IV (also known as scarlet red) can be used in veterinary and human medicine as an ointment or dressings for stimulating wound healing. In recent years, the adulterated use of Sudan dyes in food has attracted much concern. Following the discovery of Sudan I adulterated chili products in France in May 2003, surveillance of food products contaminated with Sudan dyes has been conducted by food authorities worldwide (Centre for Food Safety, 2006).

Sudan I was considered by the Joint FAO/WHO Expert Committee on Food Additives (JECFA) in 1973 to be unsafe for use in food, on the basis of toxicological evidence. Although Sudan dyes have been reported as contact allergens and sensitizers, the greatest concern has been on their possible carcinogenicity. The International Agency for Research on Cancer (IARC) conducted evaluation on Sudan dyes in 1975 and considered that Sudan I was carcinogenic in mice following its subcutaneous administration, producing tumors of the liver and that it also produced bladder tumors in mice following its implantation into the urinary bladder. In its subsequent evaluation in 1987, IARC considered that there were no adequate data for carcinogenicity in humans and limited evidence in experimental animals for Sudan I and II, and inadequate evidence in experimental animals for Sudan III and IV. IARC considered that Sudan I, II, III, and IV were unclassifiable as to their carcinogenicity to humans (Group 3). In recent years, concerns on the genotoxic potentials of Sudan dyes have been raised. Several national bodies and food authorities consider Sudan dyes as genotoxic carcinogens, while others regard these Sudan dyes as possible carcinogens (Centre for Food Safety, 2006).

Internationally, substances proposed for addition to human food and feed are banned for many reasons but primary among these is that the substance represents a hazard or threat to public health. The hazard presents to the consumer in the form of allergies and other toxicological reactions. In terms of safety measures, the amount of food additive to be consumed is specified scientifically. But if the food additive is considered hazardous even at the recommended quantity, then it is considered as a banned food additive.

In the US, the Food Drug and Cosmetic Act Sec. 189.1 delineates those substances (additives and other) that are prohibited from use in human food. The food ingredients listed in this section have been prohibited from use in human food by the Food and Drug Administration because of a determination that they present a potential risk to the public health or have not been shown by adequate scientific data to be safe for use in human food. Use of any of these substances in violation of this section causes the food involved to be adulterated in violation of the act. This section includes only a partial list of substances prohibited from use in human food, for easy reference purposes, and is not a complete list of substances that may not lawfully be used in human food. The regulations go on to affirm that no substance may be used in human food unless it meets all applicable requirements of the Food Drug and Cosmetic Act. This statement of comprehensive compliance with the FD&CA is compelling and raises a number of issues. First and perhaps foremost are the regulations at section 402 (a) (1) and 402(a) (2) of the FD&CA. These regulations provide the definition of adulterated. According to provisions of section 402 (a) (1), "a food shall be deemed adulterated if it bears or contains any poisonous or deleterious substance which may render the food injurious to health." However, if the substance(s) *is not added*, the food will not be considered adulterated "if the quantity of such substance(s) in such food does not ordinarily render it injurious to health." This latter clause is indeed very important as it covers the *naturally occurring toxicants* sometimes found even in common foods. Aflatoxin produced by species of *Aspergillus* is an excellent example of a naturally occurring toxicant occurring in food stuff worldwide. Maximum contamination levels have been established for Aflatoxin in a number of food and beverages consistent with the requirement that it is present at a level not ordinarily rendering the food injurious to the health.

Section 402 (a) (2) is concerned with *added* substances and states that a food shall be deemed adulterated "if it bears or contains any added poisonous or deleterious substance" which is unsafe within the meaning of the other applicable sections of the FD&CA. Specifically, the added substances referred to in this section are (1) materials used in production and processing which cannot be avoided by GMPs; (2) pesticides; (3) food additives; (4) color additives; and (5) new animal drugs. Substances in each of these groups are permitted in food if and only if they are not "unsafe."

Another important consideration, in terms of US laws and regulations, pertaining to banned substances in human food is the Delaney Clause of the FD&CA.

The Delaney Clause is legislation passed by the US Congress in 1958 that forbids the addition to food any additives shown to be carcinogenic in any species of animal or in humans. It has been criticized as being too restrictive by setting a *zero level of risk*. In fact, it applies only to approximately 400 of the 2700 substances intentionally added to foods in the United States, many of which are GRAS. If any GRAS substance is found to be carcinogenic, it would no longer be considered GRAS and would fall under the legal definition of a food additive, thereby becoming subject to the Delaney Clause. In 1958, the case for the Delaney Clause was simply stated. Its advocates maintained that the effects of carcinogens are insidious, irreversible, and cumulative. Further they argued that because there is no reliable method to demonstrate that any amount of a given carcinogen is safe, the limits of the threshold of harm are at present indeterminate (Blank, 1974). Since no level, however conservative, is known to be safe, carcinogens should be banned from all foods for human consumption.

A prime source of dissatisfaction with the Delaney clause is that it prohibits the FDA from following the long-standing practice it normally follows with respect to noxious substances other than carcinogens. For other kinds of substances, the FDA has discretion to establish upper limits of contamination or adulteration, based on such considerations as the relative toxicity of each particular substance and its utility in producing or processing the particular food in which it is present (Blank, 1974). Consider the example offered previously, for Aflatoxin, a noxious substance, for which the FDA has established critical upper limits for protecting public health.

The situation in the EU is not so straight forward, if in the first instance one considers the matter in the US straight forward, and there are no specific provisions of EU law that expressly prohibit food additives. Rather, using the system of food categories and approved additives if a substance (additive) is not on the positive list is banned by default. This is also the case with WHO/FAO and Codex.

A key element of the European Union's food safety management and public health protection policies, and one that clearly distinguishes the EU's approach from that of the US federal government, is what is called the precautionary principle.

This principle, in the words of the European Commission, "aims at ensuring a higher level of environmental and public health protections through preventative" decision-making. In other words, it says that when there is substantial, credible evidence of danger to human or environmental health, protective action should be taken despite continuing scientific uncertainty. In contrast, the US federal government's approach to food additive management sets a very high bar for the proof of harm that must be demonstrated before regulatory action is taken.

What does this mean in practice? In the case of Red Dye No. 40, Yellow Dye No. 5, and Yellow Dye No. 6, it means that after considering the same evidence, a 2007 double-blind study by UK researchers found that eating artificially colored food appeared to increase children's hyperactivity (McCann, 2007)—European and US authorities reached different conclusions. In the United Kingdom, the study persuaded authorities to bar use of these dyes as food additives. The EU chose to require warning labels on products that contained these dyes. In the US, the study's findings prompted the CSPI to petition the Food and Drug Administration for a ban on a number of food colorings. But in its review of these dyes, presented in 2011, the FDA found the study inconclusive because it looked at effects of a mixture of additives rather than individual colorings—and so these colors remain in use in the US

While FDA approval is required for food additives, the agency relies on studies performed by the companies seeking approval of chemicals they manufacture or want to use in making determinations about food additive safety. The standing US law that covers these substances is the 1958 Food Additives Amendment to the 1938 Federal Food, Drug, and Cosmetic Act. Reliance on voluntary measures, by industry, is a hallmark of the US approach to Food Safety regulation.

Currently, there are a number of food additives allowed in the US that other countries have deemed unsafe. Among these are "dough conditioners," additives to enhance flour's strength, or elasticity. The IARC considers one such chemical, potassium bromate, a possible carcinogen. This has led the EU, Canada, China, Brazil, and other countries to ban its use. Although the FDA limits the amount of these compounds that can be added to flour and has urged bakers to voluntarily discontinue their use, it has not banned them (Fig. 14.3) (Tables 14.3 and 14.4).

Banned Food Additives in US
Calamus extract
Calamus oil
Calcium cyclamate
Chlorofluorocarbons
Cinnamyl anthranilate
Cobaltous chloride
Cobalt sulfate
Coumarin
Cyclamate
Diethyl pyrocarbonatec
Dulcin
Fd&c green no. 1
Fd&c green no. 2
Fd&c red no. 3, aluminum lake
CFd&c red no. 3, calcium lake
Fd&c red no. 1
Fd&c red no. 2
Fd&c red no. 4
Fd&c violet no. 1
Magnesium cyclamate
Nordihydroguaiaretic acid
Potassium cyclamate
P-4000
Safrole
Sodium cyclamate
Thiourea

FIGURE 14.3 US FDA partial list of banned food additives.

TABLE 14.3 UK-approved color compounds banned by Australia and New Zealand.

UK food additives banned in Australia and New Zealand
Color
E131 Patent Blue V
E154 Brown FK
E161g Canthaxanthin
E180 Litholrubine BK

TABLE 14.4 UK-approved preservatives banned by Australia and New Zealand.

Preservatives
E214 Ethyl p-hydroxybenzoate
E215 Sodium ethyl p-hydroxybenzoate
E219 Sodium methyl p-hydroxybenzoate
E226 Calcium sulfite
E227 Calcium hydrogen sulfite
E230 Biphenyl; diphenyl
E231 Orthophenyl phenol
E232 Sodium orthophenyl phenol
E239 Hexamethylene tetramine
E284 Boric acid
E285 Sodium tetraborate; borax
E356 Sodium adipate antioxidant

14.1.5 Conclusion

Among the priority roles of most national governments are the advancement of commerce and trade, preservation of public health, and ensuring domestic tranquility. Achieving these priorities is fundamental to creating and preserving the wealth of nations. Countries like the Netherlands, Canada, Germany, Japan, and the United States, for example, have very stable governments, are leaders in trade and commerce, and enjoy high standards of public health. It is not an accident or coincidence that these nations are also among the world's wealthiest. Attainment of these national priorities, especially those related to promoting trade in food stuff and also in preserving public health (food safety), would benefit greatly from international efforts in harmonizing food safety regulations and legislation. "As our world transforms and becomes increasingly globalized, we must come together in new, unprecedented, even unexpected, ways to build a public health safety net for consumers around the world," FDA Commissioner Margaret A. Hamburg, M.D. (FDA, 2012; Food Product Development, 1980).

Currently, a great chasm of discord exists between and among national legislation globally related to this most fundamental aspect of food safety. Because absolute safety of any substance used in human food can never be proven, regulatory agencies and governments must determine if an additive is safe under the proposed conditions of use, based on the best scientific *knowledge available.* Yes, effective regulation of food additives and other substances allowed in human food is the foundation for public health protection and for assuring food safety. Moreover, the discordant international legislation is also a great source of conflict insofar as international trade is concerned. At present, there are many number of substances allowed for direct and indirect addition of food supply and many of these compounds are undeclared. Processing aids are an excellent example of a special class of substance permitted in human food that, in the regulations for many countries, are not required to be declared on consumer packaging. Processing aids are also very controversial and the definition of this class of food additives varies greatly by country. Let us be crystal clear in saying that most countries do under take risk assessments for food additives. It is evident that the safety of additives is thoroughly evaluated prior to approval in the EU, the US, Canada, Japan, and in many other countries. Despite the similarity of requirements, however, the evaluation of available safety data results in different decisions concerning their authorization and use. For instance, some of the synthetic color compounds are banned in the US on the basis of claims of carcinogenicity, applying the Delaney Clause, but are still permitted in the EU as later evaluations by the JECFA and the EFSA concluded their use is safe.

The approach to risk assessment is often different. The US approach is to confirm harm before promulgating regulations. By contrast, the EU and other countries use a more preemptive approach to food safety assurance. The precautionary principle is an approach where legislation is written when the scientific record is incomplete. In other words, the approach to risk assessment is much more conservative with the precautionary principle. Moreover, it is not uncommon that risk assessment data, on the safety of food additives, when reviewed by different countries result in a difference in the interpretation of risk. As a result, there are many number of food additives approved in one country but not in another. For example, there are more than 30 food additives approved in the US that are not allowed in the EU. Among these are "dough conditioners," additives to enhance flour's strength, or elasticity. The IARC considers one such chemical, potassium bromate, a possible carcinogen. This predicament causes confusion for consumers and is a source of great consternation for those companies engaged in international commerce.

The legislation of many countries does not require, or is mute on the subject, labeling for a raft of compounds that are approved for human food. Not only are these undeclared substances a bother for public health, they too are a great impediment to trade. Undeclared allergens for example are a leading cause of product recalls. Reconciling the discordant legislation or seeking to harmonize the standards and regulations pertaining to substance in human food would be a boon to food safety (public health) and likely promote the fair and safe trade in foods.

The US FDA estimates that US consumers purchased $2 trillion worth of imported products in 2007. These products came from 825,000 importers through more than 300 US ports of entry. And the volume of imports could double every 5 years, according to FDA staff. FDA Commissioner Margaret Hamburg has reported an estimated 24 million food import entries into the US during 2011 compared to about 12 million entries in 2007.

The FDA's office of regulatory affairs has reported that, in 2010, its inspectors physically examined 2.06% of all food-related imports. As reported previously, FDA anticipated 24 million agency-regulated products entering the US in 2011. While actual inspection rates are not available for these shipments, the FDA had previously projected that it would inspect only 1.59% of them. Further, this downward trend in inspection rates was expected to continue in 2012 as the FDA indicated that only 1.47% of all food imports will be examined (FDA, 2012; CAC, 2007). Even with exceptional inspection capability, the US FDA would not, in all likelihood, be able to confirm the presences of illegal or banned food additives in food shipments arriving at its ports. However, if the regulations of its supply chain partners and others were harmonized, then inspection activities could focus on other aspects of food safety assurance.

In recent decades, great changes in the world economy, together with expanded working relationships of regulatory agencies around the globe, have resulted in increased interest in international harmonization of regulatory requirements relative to all components of the food chain. Increased international commerce, opportunities to enhance public health through cooperative endeavors, and scarcity of government resources for regulation have resulted in efforts by the regulatory agencies of different nations to work together on standards and harmonize their regulatory requirements. The FDA has reported that "such harmonization, potentially, enhances public health protection and improves government efficiencies by reducing both unwarranted contradictory regulatory requirements and redundant applications of similar requirements by multiple regulatory bodies." FDA's publicly stated goals in participating in international harmonization are (2012) (Food Product Development, 1980):

- To safeguard global public health,
- To assure that consumer protection standards and requirements are met
- To facilitate the availability of safe and effective products
- To develop and utilize product standards and other requirements more effectively
- To minimize or eliminate inconsistent standards internationally.

Food additives are highly regulated by most countries or at the very least there is the intent to provide oversight and regulation of the substances added to food. Despite the different regulatory frameworks and underlying principles, the overall approach to ensuring food safety is similar, applying well-established risk assessment procedures and risk management measures. Nevertheless, some differences worthy of attention in the context of free movement of goods can be found in the details and implementation of discordant international regulations. Failure to comply gives rise to regulatory action for adulteration, misbranding or noncompliance, rejection at the border or removal, or recall from the market. A great deal of perfectly sound, safe food is also destroyed as a result of the discord in international regulatory interpretation of food safety.

Some of these pernicious effects could be overcome by aligning national regulations better with the internationally agreed JECFA specifications and safety assessments and their updates. While regional particularities, e.g., local dietary intake patterns, need to be considered, the safety margins may be able to accommodate a more harmonized approach to food safety assurance. Therefore, implementing the already existing policies on international harmonization by all concerned parties, collaboration in international fora and the uptake of international standards should be strongly encouraged.

References

Centre for Food Safety - Hong Kong, August, 2006.

Codex, 1991. The Codex General Standard for the Labeling of Prepackaged Foods.

Codex Alimentarius Commission, 2001. Guidelines and Principles on the Use of Processing Aids (Codex Discussion Paper CCFAC). New Zealand Delegation.

Codex Alimentarius Commission, 2007. Guidelines and Principles on the Use of Processing Aids (Codex discussion paper CX/FAC 02/0905).

https://www.insideeulifesciences.com/2021/08/06eu-plans-ban-on-titanium-dioxide-in-food/.

Food Additives Vegetable Gums Tracer Gas Thickeners Sweeteners Stabilizers Sequestrants Seasonings Propellants Preservatives Mineral Salts Flavors Flavor Enhancers Firming Agents Emulsifiers Coloring Agents Color Retention Agents Color Fixative Bulking Agents Antioxidants Antifoaming Agents Anti-caking Agents Acidity Regulator Humectants Glazing Agents Food Acids Gelling Agents Flour Treatment Agents Fig. 1. Showing Classification of Food Additives.

Food Product Development, December 1980, pp. 36–40.

Fredonia, 2004. www.pharmpro.com/ShowPR.aspx?PUBCODE=021&ACCT.

General Standard for Food Additives Codex STAN 192-1995, Rev. 3-2001.

Huang, Y., 2014. www.forbes.com/sites/yanzhonghuang/2014.

Lehto, S., Buchweitz, M., Klimm, A., Straßburger, R., Bechtold, C., Ulberth, F., 2017. Comparison of food colour regulations in the EU and the US: a review of current provisions. Food Addit. Contam. https://doi.org/10.1080/19440049.2016.1274431.

McCann, D., 2007. Food additives and hyperactive behavior in 3-year-old and 8/9-year-old children in the community: a randomized, double-blinded, placebo-controlled trial. Lancet. www.ncbi.nlm.nih.gov/pubmed/17825405.

Redlinger, P., Nelson, D., 1993.

Salminen, J., 2005. Chemical Health Hazard Assessment Division Health Products and Food Branch Health Canada. Personal Communications.

Schultz, H.W., 1981. Food Law Handbook, AVI Publishing Westport CT, USA13. US Code of Federal Regulations 21CFR101.100 (A)(3), US Food and Drug Administration 1986.

Chapter 15

Food contact materials legislation: sanitary aspects

Alejandro Ariosti[1,2]

[1]National Institute of Industrial Technology (INTI) – Plastics Center, Buenos Aires, Argentina; [2]Department of Food Science, Faculty of Pharmacy and Biochemistry, University of Buenos Aires (UBA), Buenos Aires, Argentina

15.1 Introduction

15.1.1 Scope

This chapter will cover regulations applying to the sanitary aspects of food contact materials (FCMs), a category of objects that comprises packaging, utensils, and other articles intended to come into contact with foodstuffs. Legislation on labeling (except cases directly related to sanitary aspects of FCMs) and metrology of packaged foodstuffs (both subject to food legislation), and legislation on FCMs and their waste (subject to environmental legislation), are beyond the scope of this chapter.

Before addressing the regulatory aspects of FCMs, this chapter includes a discussion on the interactions between them, foods, and the surrounding environment. This information is necessary for the understanding of international regulations on the subject.

15.1.2 Food–packaging–environment interactions

Interactions between a foodstuff, its package, and the surrounding environment to which it is exposed during storage have been extensively studied. An effort to understand the physicochemical principles involved, and thus being able to control these interactions, results in the assurance of a longer product shelf-life, improved food nutritional and sensory quality, and better consumer health protection. The main interactions for each type of FCMs are briefly described in the following sections.

Geueke et al. (2018) have reviewed publications that address safety issues related to common recycled FCMs (plastics, paper and board, metals (tinplate and aluminum), glass, and multimaterial multilayers) in the modern circular economy.

15.1.2.1 Plastic and elastomeric materials and coatings

Plastic and elastomeric materials and coatings intended to come into contact with foodstuffs are composed mainly of the "basic polymer" or "resin" and "nonpolymeric components." Resins comprise, for instance the following:

- conventional petroleum-based non(biodegradable/compostable) materials such as polyethylene (PE), polypropylene (PP), polystyrene (PS), polyvinyl chloride (PVC), polyethylene terephthalate (PET), etc., in the case of plastics;
- natural and synthetic rubber, nitrile rubber, thermoplastic elastomers (TPEs), etc., in the case of elastomeric materials; and
- epoxy, phenolic, and epoxy-phenolic resins, etc., in the case of coatings.

In the last 3 decades, the developments and use of food contact bioplastics have been increasing, though the applications at present are minimal compared with those of conventional food contact plastics. Bioplastics comprise bio- and petroleum-based biodegradable/compostable polymers, and bio-based non(biodegradable/compostable) polymers (Auras et al., 2004; Byun and Kim, 2014; Kale et al., 2007; Nakajima et al., 2017; Peelman et al., 2013; Robertson, 2012; Siracusa et al., 2008).

Ensuring Global Food Safety. https://doi.org/10.1016/B978-0-12-816011-4.00007-0

Main nonpolymeric components generally consist of the following:

(a) "polymerization residues" such as monomers, oligomers, catalysts, solvents, emulsifiers, etc.
(b) "additives," which are a type of intentionally added substances (IAS) to the basic polymer, either to facilitate the manufacture of commercial materials or articles (e.g., stabilizers, antioxidants, lubricants) or to impart to them certain technical and desired final properties (e.g., impact modifiers, plasticizers, pigments, colorants).

Though some polymers can be used as additives to the basic polymer in low percentages, their probability of migrating into the foodstuffs and of being physiologically absorbed in the gastrointestinal tract is generally considered as negligible, and thus as not toxicologically relevant. This consideration applies also in the case of substances with a molecular mass (i.e., molecular weight) greater than 1000 g mol^{-1} (Da) (EFSA, 2017; Franz and Störmer, 2008; ILSI, 2007; Ossberger, 2015). However, Groh et al. (2017) consider that further research is needed taking into account population subgroups that may exhibit increased intestinal permeability and thus a greater uptake of low and high molecular mass substances.

In this sense, polymers are rather inert as it is considered that the macromolecular chains of high molecular mass form a polymeric matrix in the FCMs and do not migrate to foodstuff. In the case of thermoplastic materials (e.g., PE, PP, PS, PVC, PET), the principal ways in which the polymeric macromolecules interact with each other to form the matrix are weak mechanical entanglements in the amorphous regions and crystals formation in the crystalline regions; both types of interactions being noncovalent in nature. In the case of thermoset materials (e.g., epoxy resins, polyurethanes, unsaturated polyesters), those interactions are strong covalent bonds (cross-linkings) between the main macromolecules.

However, in recent years, several scientific studies have reported the presence of microparticles of plastic materials with a size distribution ranging from 1 μm to 5 mm, known as microplastics, and of nanoplastics (size $<$ 100 nm) in freshwater, marine environments, soil, and air (Dris et al., 2016; Horton et al., 2017). Also the presence of microplastics, but not of nanoplastics, has been discussed in foods, mineral water, and other beverages (Ossman et al., 2018; Schymanski et al., 2018; Welle and Franz, 2018). Welle and Franz (2018) and Ossman et al. (2019) have critically reviewed the scarce scientific literature on microplastics contamination in foods and beverages. In their review, Welle and Franz (2018) concluded that quantitative determination of microplastics in foods and beverages is a challenging task, and that errors can arise due to sample preparation, laboratory air contamination, and lack of validated methods and of adequate reference materials and blank samples. The presence of microplastics in beverages can also be attributed to the plastic packaging itself (e.g., PET bottles), for instance, due to mechanical forces applied, mainly to refillable bottles, during the washing, filling, closing, and other processing steps, rather than to chemical diffusive migration. Furthermore, Welle and Franz (2018) analyzed in detail the available data for the risk assessment related to consumer exposure to microplastics from bottled mineral water, including toxicokinetics considerations (mainly absorption, distribution, and excretion of microplastics), and conclude that the reported amounts present in this beverage do not pose any safety concern for consumers.

Generally, nonpolymeric components are low molecular mass substances that, depending on contact time(s) and temperature(s), may migrate into foodstuffs. It is important to note that the US Food and Drug Administration (FDA) regulates both basic polymers and their additives as "indirect food additives."

The main food—package—environment interactions in these materials are as follows:

- "Permeability": transference of gases, water vapor, and aromas, from the environment to the foodstuff or vice versa, through the package or container wall.
- "Migration": transference of substances from the package or container wall to the contained foodstuff or to the environment. From the sanitary point of view, only migration into the foodstuff is important, and consequently, these are the substances to be determined quantitatively, by means of analytical validated methods.
- "Sorption": dissolution of major food components (e.g., water, oil, fat, blood) or minor food components (e.g., aromas, essential oils) into the package or container wall. The first case is known as "swelling" which normally alters the polymeric matrix, and the second case is known as "scalping" which does not usually alter it.
- "Desorption": transference of sorbed substances from the package or container to a product during refilling (in the case of refillable plastic packages) (Ariosti 2002a, 2018a; Devlieghere et al., 1997; Feron et al., 1994; Franz et al., 2004c; ILSI, 1993; Jetten et al., 1999; Nielsen et al., 1997; Padula et al., 2002; Safa and Bourelle, 1999; Widén et al., 2005); or in the packaging of foods in packages or containers manufactured with postconsumer recycled (PCR) plastics (see Section 15.1.4.3). The sorbed substances may be food components as mentioned previously; but also potentially harmful substances such as pesticides, herbicides, cleaning agents, etc., associated with misuse of the package or container by the consumer.

These are submicroscopic physicochemical phenomena, involving mass transferences that do not require a macroscopic discontinuity in the plastic or elastomeric material or coating (pore, micropore, fracture, or crack), due to a diffusion mechanism as described by Fick's laws, and that can be estimated by mathematical models (Brandsch, 2017; Hernández and Gavara, 1999; Hoekstra et al., 2015; Piringer, 2007; Piringer and Baner, 2008; Robertson, 2012).

Experimental determinations of known components (i.e., IAS) migration have been widely described in technical literature, such as books (e.g., Barnes et al., 2007; Bolzoni, 2015 (phthalates); Catalá and Gavara, 2002; Katan, 1996; Ossberger, 2015; Veraart, 2015), reports (e.g., FSANZ, 2010), and research articles and reviews (e.g., Ballesteros-Gómez et al., 2009 (bisphenol A (BPA)); Bhunia et al., 2013; Gallart-Ayala et al., 2013; Gehring and Welle, 2018; Genualdi et al., 2014 (styrene); Jakubowska et al., 2014; Martínez-Bueno et al., 2019; Paseiro-Cerrato et al., 2010 (polyfunctional amines); Sanchis et al., 2017; Sendón García et al., 2006).

Furthermore, in the last years, scientific interest has shifted toward the identification and determination of nonintentionally added substances (NIAS) present in or migrating from food contact plastics and other FCMs (see Section 15.1.4.4).

In summary, it is very important for packaging designers, FCMs manufacturers, food processors, public and private laboratories, universities, research and development institutes, and public health authorities to measure, assess and use data on

- permeability: in the design of packaging and prediction of shelf-life of packaged foodstuffs;
- migration: to establish compliance with FCMs regulations;
- sorption and desorption: to establish compliance with additional FCMs regulations in the case of refillable plastic packages and PCR plastics intended to come into contact with foodstuffs.

For an overview of the worldwide regulatory situation on food contact plastics, see LeNoir (2015); and on food contact rubbers, see Sidwell (2015).

15.1.2.2 Metallic materials (tinplate, tin-free steel, aluminum)

15.1.2.2.1 Corrosion

A galvanic cell develops when two metals acting as electrodes (e.g., tin (Sn) and iron (Fe), in the case of tinplate), and the canned foodstuff acting as an electrolyte, come into contact, due to, for example, discontinuities in the tin coating and the polymeric lacquer or varnish. A redox pair of reactions takes place, leading to the oxidation of the metal with a more electronegative potential (normally tin, dissolving as Sn^{2+}); and thus protecting the metal with a less electronegative potential (inert electrode, normally iron). The oxidation process releases electrons, and the electron flow on the metal surface corresponds to an ion flow in the foodstuff (Buculei et al., 2012; Kassouf et al., 2013; Montanari, 2015; NORDEN, 2015; Robertson, 2012).

Sometimes, the nature of canned foodstuff, or the degree of advancement of the corrosion process, produces the inversion or depolarization of the electrodes, changing the pattern of the oxidation. In the case of tinplate, under a depolarized corrosion process, tin does not protect the iron base, which dissolves into the product (as Fe^{2+}).

In aluminum and tin-free steel (TFS) (also known as electrolytically chromium-coated steel (ECCS)) cans, aluminum (Al^{3+}) and chromium (Cr^{3+}) ions, respectively, appear in the foodstuff due to corrosion.

The consequences of the presence of these ions are mostly sensory changes in the products: metallic taste, sulfide-black stains (due to the reaction of sulfur (S^{2-}) ions—generated, for instance, by heat treatment of food proteins with sulfur-containing amino acids—with Sn^{2+} and Fe^{2+}), discoloration of anthocyanins and betanin by Sn^{2+} in canned vegetables and fruits, etc. The use of a thicker metal (tin or chromium) coating and the use of polymeric sanitary lacquers or varnishes on the internal surface of the metallic can help to diminish, but not to completely eliminate, the corrosion process.

As another consequence of the corrosion process, some impurities present in the basic metals can contaminate the food. Heavy metals and other elements (arsenic (As), cadmium (Cd), mercury (Hg), lead (Pb), etc.) are of special toxicological interest in canned foodstuffs. Food regulations establish maximum levels for these elements and for tin in food products. Two recent cases of interest on metals interactions are discussed in the following paragraphs.

The Council of Europe (CoE) has issued nonmandatory Resolution CM/Res (2013) 9 of June 11, 2013, on metals and alloys used in the manufacture of FCMs. This Resolution was published along with the "Technical guide on metals and alloys used in FCMs and articles" (EDQM-CoE, 2013). Food contact metals have not yet been harmonized at the European Union (EU) level. Chapter 3 of the Technical Guide is a guideline on analytical methods for testing migration from FCMs made from metals and alloys (Ariosti, 2016a).

In 2008 the European Food Safety Authority (EFSA) adopted a Scientific Opinion of its Food Additives, Flavourings, Processing Aids and Food Contact Materials Panel (AFC Panel), establishing a tolerable weekly intake (TWI) of 1 mg (of Al) per kg body weight per week (EFSA, 2008a). In 2016, the European Commission (EC) sanctioned Regulation (EU) 2016/1416 (seventh. amendment), that amends Regulation (EU) 10/2011 on food contact plastics, and establishes a specific migration limit (SML) of 1 mg (of Al) kg^{-1} (of food or simulant), obtained by calculations from the TWI for Al. The 28th recital of Regulation (EU) 2016/1416 considers that the latter is an adequate SML for Al for FCMs.

15.1.2.3 Glass and ceramics

15.1.2.3.1 Leaching

The most common type of glass used in food packaging is soda-lime glass. Its basic ingredients are vitreous silicon oxide (SiO_2); alkaline oxides (mainly of sodium (Na) and potassium (K)) to lower the working temperature in the glass-melting furnace; and alkaline-earth oxides (mainly of calcium (Ca), magnesium (Mg), and elements of higher nuclear charges or "valences") to stabilize the glass structure and prevent excessive dissolution of glass into water or aqueous solutions. Recycled glass, known as cullet, is commonly used in high percentages (up to 70%−80%) in the manufacture of soda-lime glass. It is very important to control the source of cullet, in order to keep the heavy metals and metalloids content as low as possible (Mahinka et al., 2013).

In soda-lime glasses, the chemical bond to oxygen (O) in the case of Si−O is covalent in nature. Besides, alkalis are involved in ionic bonds with oxygen in the vitreous matrix. When glass comes into contact with water or acidic or basic solutions, as in the case of foodstuffs and beverages, Na^+ and K^+ ions take part in a mass transference process associated with an interchange of charged species. These ions migrate from the package or container glass wall to the foodstuff in contact; and protons (H^+), hydrated protons ($H^+.H_2O$), or hydroxyls (OH^-) enter the vitreous mass, altering its structure and resulting (to a certain degree) in its dissolution. The greater the amount of alkali in soda-lime glass, the less resistant it is to chemical attack (i.e., it has lesser hydrolytic resistance). In the case of divalent ions (e.g., Ca^{2+}, Mg^{2+}) or higher valence ions, the overall ionic bonds to oxygen are stronger than in the case of the monovalent alkaline ions, and their migration is lower. This process of differential migration of charged species through the interface glass−foodstuff is usually called "leaching" (Mari, 2002; Peltzer et al., 2015a; Tingle, 1996).

In the case of glazed substrates (glass, ceramics, or metals), vitreous enamels or glazes are used to coat the food contact surface. These glazes are also glasses, but with the difference that some substances are used to impart a special transparency or color, to modify the viscosity of the glaze during the coating process, or to lower its melting temperature. These substances are commonly Pb- and Cd-based compounds, and therefore the main regulations (e.g., EU, the Common Market of the South (MERCOSUR) in South America, Japan) establish SMLs for these ions in the case of glazed glass, ceramics, or metals (Peltzer et al., 2015b). Nonglazed porous ceramics are generally forbidden in the manufacture of food packages and containers (Mari, 2002).

Crystal glasses (containing a minimum of 10% of Pb and barium (Ba) oxides) and lead crystals (containing a minimum of 24% of Pb oxide) are used in the manufacture of high-quality tableware or hollowware intended for brief contact times with foodstuffs (CoE, 2004; Mari, 2002; Tingle, 1996). However, they are generally forbidden for the manufacture of food packages, due to the potential migration of Pb and Ba during longer contact times (MERCOSUR, 1992).

15.1.2.4 Cellulosic materials (paper and board)

15.1.2.4.1 Extraction

Cellulose-derived materials, mainly paper and board, are manufactured with pulps (e.g., mechanical, semichemical, chemical) from wood, sugar cane bagasse, and other natural resources.

The composition of cellulose-derived materials includes first-use cellulose fibers (primary, virgin, or "fresh" fibers), recycled cellulose fibers (secondary or reclaimed fibers), as well as synthetic fibers (plastic fibers), various types of additives (including pigments), and inorganic fillers. The tendency of fibers to bind together by chemical bonds is enhanced with the aid of additives (sizing agents), forming thus a cellulose matrix or network with pores. The diameters of these pores can be reduced by applying mechanical surface treatment operations (calendering or supercalendering), thus rendering the cellulose substrate less permeable to gases, water vapor, water, and fat. Sizing agents and oil repellents are also used to improve resistance to the penetration of water and grease.

Food contact cellulose-derived materials are basically intended to come into direct contact with dry solid foodstuffs without fat or oil on their surface. Nevertheless, if (due to deficient design or manufacture, or mechanical damage or misuse of the package), water, fat, or oil come into contact with the cellulose substrate, they can penetrate through the pores by capillarity, changing the cellulose network and resulting in the migration of additives, fillers, and fiber fragments into foods. This process is called "extraction" (Söderhjelm and Sipiläinen-Malm, 1996).

Apart from heavy metals, dioxins, pentachlorophenol, and polychlorinated biphenyls (Söderhjelm and Sipiläinen-Malm, 1996), several other contaminants—due mainly to additives (for instance, perfluorocarboxylic acids and perfluoroalkyl sulfonates (Xu et al., 2013)) or residues of printing inks in recycled fibers—may be present in paper and board, and have arisen concern in recent years (Biedermann et al., 2013; Grob, 2017; Koivikko et al., 2010; Lago et al., 2015; Lorenzini et al., 2013) (see more details in Section 15.1.4.4).

It is very important to control the source of recycled fibers, in order to keep the concentration of these contaminants as low as possible in order to minimize the risk of sanitary and sensory problems (BfR, 2017; CoE, 2009; MERCOSUR, 2015; US-FDA, 2018).

For an overview of the worldwide regulatory situation on food contact paper and board, see Baughan (2015).

15.1.3 Importance of assessing and controlling the interactions

In summary, every type of FCM interacts with the contained foodstuffs. The general phenomenon is known as "migration," though, as described, the mechanisms involved in the mass transference vary, depending on the type of material. A schematic summary of these interactions can be seen in Fig. 15.1 and Table 15.1.

Elements and substances migrated from FCMs to foodstuffs are ingested by the consumer. Therefore, to control the migration phenomenon, a realistic risk—exposure evaluation is of vital importance (Arvidson et al., 2007; Cheeseman, 2013; Muncke et al., 2017; Oldring et al., 2014), and legislation on FCMs is necessary, as well as its improvement and enforcement (Baughan and Attwood, 2010; Grob, 2017; Karamfilova et al., 2016; Ossberger, 2015; Schäfer, 2010; Simoneau et al., 2016).

15.1.4 Hygienic requirements of FCMs

FCMs regulations worldwide (e.g., EU, 2004, 2011; MERCOSUR, 1992; US-FDA 21 CFR 174.5) establish general requirements (i.e., "the three basic hygienic principles") that, while acknowledging that migration occurs, set limits to it. Migrants from FCMs must not:

(a) change the nutritional composition of food;
(b) pose a risk to human health, depending on the exposure of consumers to these substances;
(c) cause taints problems in food, with undesirable changes to their sensory characteristics.

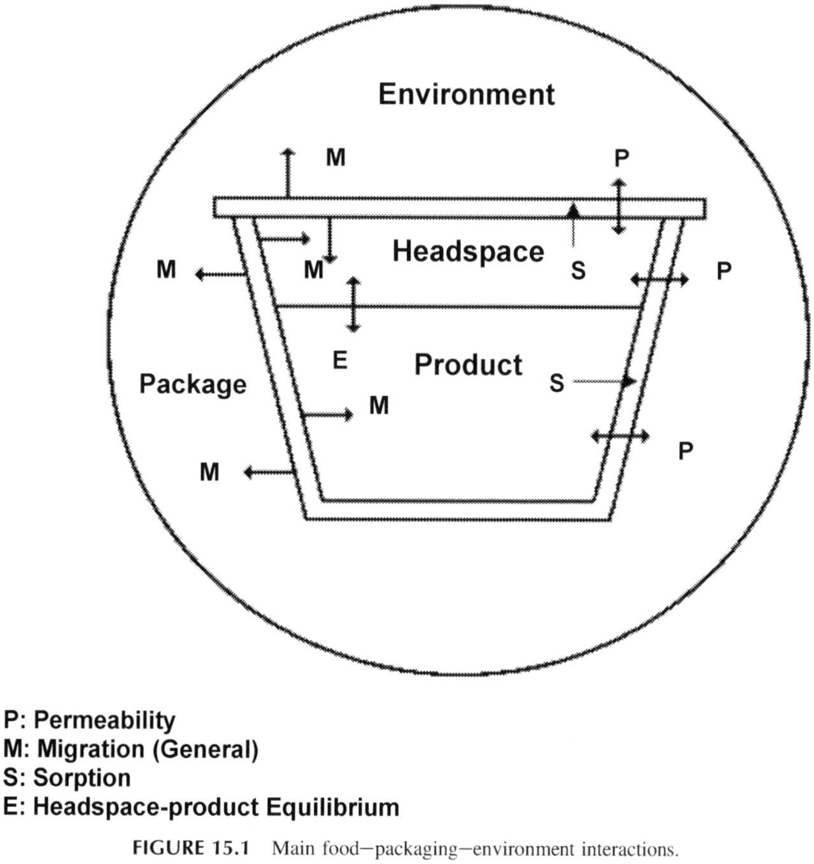

FIGURE 15.1 Main food—packaging—environment interactions.

TABLE 15.1 Main migrants from food contact materials to foods or simulants.

Migrants	Examples	FCMs
Gases and vapors	Ethylene monomer	Polyethylene
	Vinyl chloride monomer	PVC
	Acetaldehyde	PET
	Solvents	Plastic laminates Printed substrates Coated substrates
Ions	Na, K, Li, Ca, Mg	Soda-lime glass
	Pb, Cd	Glazed substrates (ceramic, glasses, and metals)
	Pb, Ba	Crystals Lead crystals
	Sn, Fe	Tinplate
	Al	Aluminum
	Cr	TFS (ECCS) Stainless steel
Low molecular mass compounds	Additives NIAS	Plastics Varnishes or lacquers Coatings Rubbers Paper and board Regenerated cellulose

ECCS, electrolytically chromium-coated steel; *FCM*, food contact material; *NIAS*, nonintentionally added substances; *PET*, polyethylene terephthalate; *PVC*, polyvinyl chloride; *TFS*, tin-free steel.

These three basic hygienic requirements should be developed in detail by specific regulations, depending on the application.

15.1.4.1 Basic hygienic requirements of FCMs

In particular, food contact plastics have been the subject of ample hygiene regulations in the past decades. Some of the concepts described in this section for food contact plastics may apply to other types of FCMs (e.g., positive lists, migration limits).

To have a general understanding of these hygienic requirements, three similar legislations on food contact plastics were reviewed: the Chinese, the EU, and the MERCOSUR regulations. The hygienic requirements can be summarized as follows:

- "Positive lists": the basic polymer or resin (China, MERCOSUR), or its monomer(s) and other starting substances (EU, MERCOSUR), and the additives (China, EU, MERCOSUR) used are regulated by positive lists of authorized substances. These lists are mandatory and contain substances that have been subjected to a process of risk assessment—in accordance with internationally recognized scientific principles—that comprises, among others, migration tests and toxicological evaluations (in vivo, in vitro, in silico). The use of substances included in the positive lists, frequently subject to specific restrictions, is considered safe for FCMs applications.
- "Overall migration limit(s)" (OMLs): maximum allowed quantities of nonpolymeric components released from a food contact plastic to the foodstuffs or food simulants. To verify the compliance of the food contact plastic with these limits, overall migration tests must be performed according to standardized and validated methods using foods, or preferably their simulants, under various conditions of contact time(s) and temperature(s).

A "simulant" is a substance or a simple mixture of substances that during the migration tests, under standardized conditions of contact time(s) and temperature(s), has an interaction with a food contact plastic, considered equivalent to that of a foodstuff or a group of foodstuffs. Simulants for aqueous nonacidic, aqueous acidic, alcoholic, fatty, dairy, and dry foodstuffs are described in regulations (e.g., China, EU, MERCOSUR, US-FDA), and standards (e.g., EN standards issued by the European Committee for Standardization (CEN)).

The OMLs are expressed in the following units:

- mg (of nonpolymeric components) kg^{-1} (of food or simulant);
- mg (of nonpolymeric components) dm^{-2} (of contact surface area).
- "Specific migration limit(s)" (SMLs): maximum allowed quantities of a particular nonpolymeric component (e.g., monomer, additive, heavy metal) released from a food contact plastic to the foodstuffs or food simulants. To verify the compliance of the food contact plastic with the SMLs, specific migration tests must be performed as described in the case of overall migration tests. To the same end, calculations of the potential migration of the specific migrant (assuming a 100% transference to the foodstuff or simulant) or predictions of its migration using a validated mathematical model are allowed.

The SMLs are expressed in the following units:

- mg (of nonpolymeric component) kg^{-1} (of food or simulant).
- "Concentration limit(s)" (QMs): maximum quantity of a particular nonpolymeric component allowed in the very food contact plastic. To verify the compliance of the later with these limits, no specific migration tests are performed, because the determination of the nonpolymeric component concentration is done on the food contact plastic directly.

The QMs (i.e., "quantity in material" according to the EU regulation) and the concentration limits per contact surface area (QMAs) are expressed in the following units:

- mg (of nonpolymeric component) kg^{-1} (of FCM) (QM);
- mg (of nonpolymeric component) 6 dm^{-2} (of contact surface area) (QMA).
- "Group limit(s)": this category of additional limits comprises the group migration limits (SML(T)) and the group concentration limits (QM(T) and QMA(T)). These are the maximum allowed quantities of related substances that contain a common chemical group (e.g., isocyanate, glycol).

The group limits are expressed in the following units:

- mg (of nonpolymeric components migrated with the same chemical group) kg^{-1} (of food or simulant) (SML(T));
- mg (of nonpolymeric components with the same chemical group) kg^{-1} (of FCM) (QM(T));
- mg (of nonpolymeric components with the same chemical group) 6 dm^{-2} (of contact surface area) (QMA(T)).
- "Sensory characteristics of the foodstuffs": the color, aroma, taste, flavor, and texture of the foodstuffs must not be adversely affected by migration of chemical species from FCMs.

Taints problems affecting packaged foods can be studied by instrumental analysis (e.g., GC, GC-MS), sensory analysis (e.g., ISO Standard 13302:2003, IRAM Standard 20021:2004), or by combined techniques such as GC-olfactometry-MS (GC-O-MS) (Callejón et al., 2016; Lord, 2003; Ridgway et al., 2010; Vera et al., 2014). These techniques are very useful for the prevention and correction of taints problems (see definition 3.5 on "taints" of BS ISO 13302:2003 (BS ISO, 2003)).

Other requirements may also be established:

- regulated use of a substance in certain types of plastics;
- regulated use of a substance in plastics in contact with certain types of foodstuffs;
- purity criteria for certain substances.

15.1.4.2 Pigments and colorants

Requisites on pigments and colorants for FCMs have been established, for instance, for food contact plastics, by regulations (e.g., MERCOSUR Resolution GMC 15/10, US-FDA (21CFR 178.3297)) and recommendations (e.g., German Federal Institute for Risk Assessment (BfR) Recommendation IX (2015), CoE Resolution AP (89)1).

15.1.4.3 Functional barriers, threshold of regulation, and postconsumer recycled plastics

Apart from IAS included in positive lists, "nonlisted substances" (NLS) (or "noninventoried materials" (NIM)) can be used as components of FCMs, except if they are carcinogenic, mutagenic, or toxic to reproduction (CMR) compounds. Providing that NLS comply with this first restriction, and that they are not in nanoform, they can be used—for instance, in a multilayer structure—in a layer (e.g., plastic, paper) separated from food by another layer (e.g., plastic, bio-based coating, aluminum foil) called a "functional barrier," that minimizes their migration from the layer where they are included to food up to a maximum limit of 0.01 mg (of a particular substance) kg^{-1} (of food or simulant) (i.e., 10 μg kg^{-1}) (Biedermann-Brem et al., 2017; EU, 2011; Genualdi et al., 2015; Guazzotti et al., 2014, 2015; Heckman, 2005). The functional barrier does not need to be into direct contact with food; for instance, aluminum foil, which is used as a layer separated from food, generally by a plastic layer.

The concept of functional barrier is now a well-established theoretical tool with technological applications, and in the US-FDA regulation, it is related to the concept of "no migration." It has been established, for instance, by Regulation (EU) 10/2011 on food contact plastics, by the Chinese national standard GB 4806.1-2016 on general safety requirements for FCMs, etc. In the MERCOSUR, it is included in Resolution GMC 25/99 on trilayer PET soft drinks bottles with a middle layer containing PCR-PET and an inner functional barrier of virgin PET, and in Resolution GMC 30/07 on PCR-PET multilayer and monolayer bottles (Ariosti, 2002b, 2018a; Bayer, 1997, 2002; Feigenbaum et al., 2005; Franz et al., 1996, 1997, 2003, 2004a, b; Piringer et al., 1998; Welle, 2011; Welle and Franz, 2007).

For instance, in a multilayer plastic package manufactured with PCR plastic material in the middle layer, the functional barrier is the layer that:

- is manufactured with virgin plastic;
- is generally (but not necessarily) in contact with food;
- separates food from an outer layer manufactured with PCR plastic that may contain potential residual contaminants (even after a process of washing, decontamination, and recycling);
- is efficient in reducing the migration of these potential contaminants to food, during its shelf-life, to quantities that neither pose any toxicological risk to human health nor alter the nutritional and sensory characteristics of the foodstuff.

In the last 2 decades, there has been an increasing industrial interest in the development of decontamination technologies to manufacture food grade PCR plastics, mainly PCR-PET which is the most widely recycled one for food contact applications. It is generally mixed with virgin PET for the manufacture of monolayer parisons and sheets, as precursor articles to obtain bottles and trays, respectively, by means of well-known conversion technologies. Postconsumer PET is collected from different sources, and used as input by plants to manufacture PCR-PET applying decontamination technologies. These technologies must be validated by means of "challenge tests" (using "surrogates" or model contaminants) or other appropriate scientific evidence (e.g., mathematical modelisation), and authorized/recognized by the sanitary authorities (e.g., EU, MERCOSUR, US-FDA) (Ariosti, 2018a; Barthélémy et al., 2014; Bayer, 1997, 2002; Dutra et al., 2014; EFSA, 2008b, 2011; EU, 2008; Franz et al., 2003, 2004a, 2004b; ILSI, 1998; MERCOSUR, 2007; US-FDA, 2021; Welle, 2011; Welle and Franz, 2007).

Taking into account certain assumptions described by the US-FDA Guidance (US-FDA, 2021), two limits can be derived from the "threshold of regulation" (TOR), established since 1995 by the US-FDA (21 CFR 170.39) as 0.5 μg (of the substance) kg^{-1} (of human diet):

- one limit comparable to a QM, a maximum quantity of surrogate per kg of plastic;
- one limit comparable to an SML, a maximum specific migration of surrogate per kg of food or simulant.

In the case of food grade PCR-PET, the values of these limits established by the US-FDA are, respectively, 220 μg (of surrogate) kg^{-1} (of PCR-PET) and 10 μg (of surrogate) kg^{-1} (of food or simulant).

If as a result of the challenge test (or mathematical modelisation), the PCR plastic complies with either of the TOR-derived limits, the decontamination technology is cleared and the US-FDA issues a "No-objection letter" (NOL) to its use for the manufacture of PCR plastic.

In general, if a manufacturer can demonstrate to the US-FDA that a noncarcinogenic substance present in an FCM ends up in the human diet in quantities below the TOR, it is considered a component of nontoxicological concern, and therefore, it is not a food additive, and it is not subject to further regulation (Cheeseman, 2005; Baughan and Atwood, 2010).

The EU approach for the validation of a decontamination technology for the manufacture of food grade PCR plastics differs from the US-FDA's one. For a detailed description of the assessment of the decontamination efficiency and the authorization procedure of a technology at the EU level, see Barthélémy et al. (2014), EFSA (2008b, 2011) and Welle (2013, 2016).

During the development of MERCOSUR Resolution GMC 30/07 on food grade PCR-PET, both approaches were recognized as valid, the two TOR-derived limits were adopted, and therefore technologies authorized/recognized by the EU and the US-FDA are accepted. Not only in MERCOSUR, but also in the majority of the rest of the South American countries (e.g., Bolivia, Colombia, Ecuador, Peru), in Mexico, etc., PCR-PET manufactured by means of validated technologies is widely used.

15.1.4.4 Nonintentionally added substances and the threshold of toxicological concern

NIAS, less frequently called ORPI (oligomers, reaction products, and impurities), comprise several types of substances that appear in FCMs, food, and food simulants, due to different sources, for instance:

- impurities of IAS and NLS (or their degradation/reaction products) used as raw materials for the manufacture of FCMs, oligomers in plastics;
- substances formed as reaction or breakdown products during the manufacture of FCMs (conversion processes);
- residual contaminants from recycled plastics, paper, and board;
- substances formed during the application of industrial preservation technologies to packaged foods (e.g., retortable or irradiated food packaging);
- substances formed during use by consumers (e.g., microwave and conventional oven heating at home, aging of reusable containers).

The sources and challenges in the identification and determination of NIAS have been described in detail elsewhere (Alin and Hakkarainen, 2011, 2013; Bach et al., 2013; Bayer, 2002; Biedermann et al., 2013; Bignardi et al., 2017; de Oliveira et al., 2012; de la Cruz García et al., 2014; Driffield et al., 2014; Dutra el al., 2014; Félix et al., 2012; Franz et al., 2004b; Gallart-Ayala et al., 2013; Geueke, 2018; Grob, 2017; Isella et al., 2013; Koster et al., 2014, 2015; Lorenzini et al., 2013; Martínez-Bueno et al., 2017; Nerín et al., 2013, 2016; Parisi et al., 2015; Peters et al., 2019; Petersen et al., 2008; Sanchis et al., 2015; Song et al., 2019; Vera et al., 2018).

For a long time, NIAS were not considered when assessing FCMs safety, but since 2003, several reports were published on migration into food of:

- semicarbazide from lids gaskets (Stadler et al., 2004);
- mineral oil hydrocarbons (MOHs) (comprising mineral oil aromatic hydrocarbons (MOAHs) and mineral oil saturated hydrocarbons (MOSHs)), UV-printing inks photo-initiators (PIs), and certain phthalates (plasticizers) from recycled paper and board (Biedermann et al., 2013; Grob, 2017; Koivikko et al., 2010; Lago et al., 2015; Lorenzini et al., 2013).
- UV-printing inks PIs, such as isopropylthioxanthone (ITX), benzophenone (BP), and 4-methylbenzophenone (4-MBP), from FCMs in general (Jung et al., 2010).

These cases of unexpected food contamination attracted attention from regulators, scientists, public, and the media, and thus regulatory actions were taken. For instance, at the EU level, Regulation (EC) 2023/2006 (EU, 2006) on good manufacturing practice (GMP) for the manufacture of FCMs was sanctioned. This Regulation's Annex establishes that printing inks must be applied to the nonfood contact side of FCMs, and that their formulations and the printing process must ensure that their components are not transferred to the food contact side of FCMs, in quantities no greater than the regulatory admissible levels. Later on, Regulation (EU) 10/2011 on food contact plastics (EU, 2011) established that NIAS must be risk assessed by manufacturers all along the supply chain. The findings of the NIAS risk assessment must be documented in the internal supporting documentation of the "Declaration of Compliance" (DoC), by manufacturers all along the supply chain.

At present, it is recognized that NIAS assessment is an analytically challenging and time-consuming effort—involving highly trained human resources and complex equipment— due to lack of worldwide harmonization of testing methods (Geueke, 2018; Grob, 2017; Gude, 2018). The Netherlands Organization for Applied Scientific Research (TNO) proposed a pragmatic analytical approach (including in vitro tests) based on applying the "threshold of toxicological concern" (TTC) concept to NIAS (except those that belong to defined categories of highly toxic substances), at low-exposure levels (Cheeseman, 2013; Gelbke et al., 2019; Gergely and Cheeseman, 2015; Koster et al., 2014; Nerín et al., 2013).

The EFSA and the World Health Organization (WHO) published a review of the TTC approach and development of a decision tree for the safety assessment of chemicals from different sources, including FCMs, when hazard data are incomplete and human exposure can be estimated (EFSA/WHO, 2016).

Other examples of initiatives for the risk assessment of NLS, NIAS, and mixtures of migrants, at the EU level are as follows:

- the Matrix European Exposure Project (2005–11) approach applied to plastic food packaging (Eisert, 2011);
- the exposure modeling tool developed by the Flavors, Additives and Food Contact Materials (FACET) Exposure Task—EU FP7 Project (2008–12) (Oldring et al., 2014);
- the MigraTox Project (launched in 2017) lead jointly by the Austrian Research Institute for Chemistry and Technology (OFI) and the FH Campus Wien—University of
  Applied Sciences, Vienna (Austria);

- the Guidance on best practices on the risk assessment of NIAS published by the International Life Sciences Institute (ILSI) Europe (Koster et al., 2015);
- Guidelines on the assessment of NIAS issued by European Industry Associations (e.g., PlasticsEurope in 2014; the Food Contact Additives Sector Group of the European Chemical Industry Council (FCA-CEFIC) in 2018; the Italian Packaging Institute (III) in 2018; the European Printing Ink Association (EuPIA) in 2019).

15.1.4.5 Active and intelligent materials

The interest in applications of active and intelligent materials (AIMs) to the manufacture of FCMs is increasing worldwide. To ensure their safe application, for instance, they were included in the list of specific FCMs that must be regulated in the EU (Annex I to Regulation (EC) 1935/2004) (EU, 2004).

Active FCMs and articles are intended to extend the shelf-life of packaged foods, maintaining or improving their quality. They are designed to deliberately incorporate components or devices that can release compounds (e.g., antioxidants, antimicrobials) into food, or absorb substances from the packaged food (e.g., ethylene) or the environment surrounding the food (e.g., oxygen permeating from outside the packaging). On the other hand, intelligent FCMs and articles are intended to monitor the conditions of packaged food (e.g., sensors) or the environment surrounding the food (e.g., temperature abuse indicators) (Ahmed et al., 2017; Dainelli, 2015; Fang et al., 2017; Ghaani et al., 2016; Gómez-Estaca et al., 2014; McMillin, 2017; Realini and Marcos, 2014; Restuccia et al., 2010; Schumann and Schmid, 2018; Vanderroost et al., 2014; Wu et al., 2018; Zhai et al., 2017).

Active materials may release substances to the food under certain circumstances. The released substances must be authorized in the Food Laws (such as allowed additives or flavorings) according to the approved usage and quantities. Consumers must not be misled through changes in food or information given about the quality condition of food. For instance, in the EU, AIMs must be declared in the label, and a special symbol must identify nonedible parts such as sachets with absorbers (e.g., oxygen or ethylene scavengers) placed inside the packages.

15.1.4.6 Food contact nanomaterials

Nanoscience and nanotechnology applied to FCMs are new disciplines in fast development that arise much interest, mainly in the academic and industrial research and development fields. Nanomaterials are those materials that contain natural, incidental, or intentionally added components, particles, structures, or devices that have at least one dimension that is approximately in the order of 1—100 nm in length (Veraart, 2017).

Some innovative developments of nanomaterials are commercially used in the manufacture of FCMs, due to specific enhanced properties: barrier to gases, aromas, light, and UV radiation; mechanical resistance; heat stability; etc. Nanomaterials may be cellulose based, or most commonly, plastic based. The plastic food contact nanomaterials (FCNMs) may be manufactured with conventional plastics or bioplastics, and may be used as films, containers, can coatings, active (e.g., carriers of antimicrobial and antioxidant components) and intelligent (e.g., sensors) materials, etc. On the other hand, it is widely recognized that nanoforms constituents (e.g., silver (Ag), copper (Cu), metal oxides, cellulose crystals, or titanium nitride (TiN) nanoparticles; carbon nanotubes) of FCNMs that may be present in food due to migration, and thus possibly gaining access to the human body, may have, at the nanoscale level, physical, chemical, and biological properties that differ from the ones that the same substances have in molecular, micro-, macro-, or bulk-form (Byun and Kim, 2014; Dilmaçünal, 2017; Dudefoi et al., 2018; Duncan, 2011; Gatos, 2016; Lagaron and Lopez-Rubio, 2011; Mihindukulasuriya and Lim, 2014; Peelman et al., 2013; Rhim et al., 2013; Robertson, 2012; Veraart, 2017; Youssef and El-Sayed, 2018).

This is the reason why the EFSA (EFSA Scientific Committee, 2011, 2018), the US-FDA, and other sanitary authorities worldwide require the safety assessment of FCNMs within the present regulatory framework in force in each jurisdiction, taking into account their nanoscale structure.

In this sense, several studies and reviews have been performed on the potential migration and risk assessment of nanoparticles from FCNMs into foods and their simulants (Bott et al., 2014a, b, 2018; Bouwmeester et al., 2014; Bumbudsanpharoke and Ko, 2015; Bumbudsanpharoke et al., 2019; Burello, 2017; Ćwiek-Ludwicka and Ludwicki, 2017; Drasler et al., 2017; Echegoyen and Nerín, 2013; Hannon et al., 2015, 2016, 2018; Huang et al., 2015; Iñiguez-Franco et al., 2017; Metak et al., 2015; Oomen et al., 2018; Polat et at., 2018; Su et al., 2017; Veraart, 2017; Wu et al., 2017).

Though at present some migration results are sometimes contradictory, it is recognized in general that nanoparticles have no or scarce potential to migrate by diffusion at sizes greater than approximately 5 nm, or if they are completely embedded in the polymeric matrix. However, other mechanisms different from diffusive migration may explain the presence of nanoparticles or their constituents in food or food simulants. The proposed mechanisms are the following: (a) desorption of the nanoparticles located at the FCNM surface in contact with food or food simulants, due to high temperature, vibration, physical abrasion, or release from cut edges of samples during migration tests ("cutting edge effect"); (b) dissolution of nanoparticles (e.g., Ag, zinc oxide (ZnO)) in contact with food or food simulants into ions, sometimes followed, for instance, by the reduction of Ag^+ in favorable conditions to metallic Ag and aggregation into new nanoparticles; and (c) release of nanoparticles by dissolution of the polymer matrix surrounding them, due to physical abrasion, high temperature, exposure to UV radiation, hydrolysis of biodegradable polymers (e.g., polylactic acid (PLA)), etc (Garcia et al., 2018; Gomes Lauriano Souza et al., 2016; Jokar et al., 2017; Noonan et al., 2014; Störmer et al., 2017).

15.2 FCMs legislation in the European Union

Since January 31, 2020, when the United Kingdom of Great Britain and Northern Ireland (UK) left the European Union, after a process called Brexit, the EU has 27 Member States. Two types of FCMs legislation coexist: the harmonized EU legislation and the national legislations of the Member States that apply only in the fields not harmonized at the EU level. In this chapter, only the EU legislation on FCMs is discussed (Heckman, 2005; Irvine, 2018, 2019; Schäfer, 2007, 2010; Semail, 2019; Simoneau et al., 2016).

In the EU, harmonized legislation was enacted to give a similar level of health protection to all the EU citizens, and to overcome possible technical barriers to trade between the Member States. It comprises mainly of two types of legal instruments: Directives and Regulations. The former come in force in the Member States only after a process of transposition (internalization) into their national legislations, and can have deadlines up to 18 months (Schäfer, 2010); the latter do not need transposition, so they can be implemented more quickly.

The authorization of FCMs substances undergoes a risk assessment process guided by the EFSA, located in Parma, Italy (EU, 2002, 2004, 2011), followed by a risk management decision from the EC or the Council of the EU and the European Parliament.

Within the EFSA, the Scientific Committee and the scientific-technical commissions or Panels—composed of independent scientific experts with a 3-year mandate—perform scientific assessment and develop assessment methodologies. Since July 1, 2018, there are 10 Panels, one being the Panel on Food Contact Materials, Enzymes, and Processing Aids (CEP Panel), formerly known as the Panel on Food Contact Materials, Enzymes, Flavourings and Processing Aids (CEF Panel). Since that date, the responsibility for evaluation of flavorings is in charge of the Panel on Food Additives and Flavourings (FAF Panel), previously known as the Panel on Food Additives and Nutrient Sources Added to Food (ANS Panel).

The EFSA's final resolutions are issued to comply with the mandates of EU regulations; as a response to requests from the EC, the European Parliament, or a Member State; or in some cases on its own initiative.

In summary, at the EU level, FCMs are regulated by the following harmonized legislation (that does not cover fixed public or private water supply equipment):

- Regulation (EC) 1935/2004, which establishes general requirements for all FCMs, and is known as the Framework Regulation;
- Regulation (EC) 2023/2006 on GMP in the manufacture of FCMs;
- Legislation on specific materials covering groups of FCMs listed in the Framework Regulation;
- Legislation on specific substances used in the manufacture of FCMs;
- Legislation on kitchenware made of melamine or polyamide originating or consigned from China or Hong Kong;
- EC Recommendation (EU) 2019/794 on a coordinated control plan of certain substances migrating from FCMs by Member States.

Due to Brexit, the EU primary and secondary law have ceased to apply to the United Kingdom, which is now considered a "third country." This date was later modified, due to an Agreement between the EC and the United Kingdom reached on April 11, 2019, and set at November 1, 2019 ("the withdrawal date").

Detailed information on FCMs research, analysis, and methodology harmonization is available from the website of the European Reference Laboratory for Food Contact Materials (EURL-FCM) of the EU Joint Research Center, located in Ispra, Italy. The EURL-FCM is supported by the EU Network of National Reference Laboratories.

15.2.1 EU Framework Regulation on FCMs

In October 2004, Directive 89/109/EEC was substituted by Regulation (EC) 1935/2004 that establishes general requirements (among them, the three basic hygienic principles described in Section 15.1.4), for all FCMs and is known as the Framework Regulation.

It also maintains the principle of "positive labeling" introduced by the previous Framework legislation, according to which all FCMs and articles (with exceptions) must indicate the words "for food" or an appropriate symbol, such as the one defined in Annex II to the Regulation (see Fig. 15.2).

Regulation (EC) 1935/2004 sets mechanisms to ensure traceability of FCMs at all stages in the supply chain, to facilitate the control and recall of products, and to identify responsibilities. Special consideration is given to GMP when producing FCMs that comply with specifications, and the important role of the food industry to cooperate with national authorities for product recall. FCMs must comply with the specific requirements, and must be accompanied by a DoC (Article 16), supported by documentation that must be available to the sanitary authorities.

It also establishes a list of materials for which the EC can issue specific EU legislation. The list mentions 17 different materials (including glass, metals and alloys, adhesives, printing inks, ion exchange resins, silicones, varnishes and coatings, wood, etc.). At present (October 2021), only a few of them are covered by harmonized EU legislation (plastics, ceramics, regenerated cellulose (cellophane), recycled plastics, and AIMs).

It is important to remember that, whereas no EU legislation exists, each Member State may apply its own national legislation, following the general rules described in Regulation (EC) 1935/2004. The FCMs thus cleared in the harmonized and nonharmonized fields can be commercialized in the whole EU under the principle of mutual recognition (Heckman, 2005; O'Keeffe, 2018, 2019; Schäfer, 2007, 2010; Semail, 2018; Simoneau et al., 2016).

In March 2019, the European Parliament and the Council sanctioned Regulation (EU) 2019/515 on the mutual recognition of goods lawfully marketed in Member States that repeals Regulation (EC) 764/2008 on the matter, and that shall apply from April 19, 2020.

15.2.2 EU regulation on GMP

Regulation (EC) 2023/2006 applies to GMP for all types of FCMs. It establishes that the FCMs manufacturers must ensure that general rules on GMP (such as requirements on the quality assurance system, the quality control system, and the supporting documentation, etc.), as well as specific rules related to printed FCMs (set in the Annex), are applied.

15.2.3 EU legislation on specific FCMs

15.2.3.1 Plastics

In January 2011, the EC sanctioned Regulation (EU) 10/2011 on food contact plastics (also known as the Plastics Implementation Measure (PIM)), which repeals the previous legislation on this topic. Up to October 2021, it has been amended by the following: (1) Regulation (EU) 321/2011; (2) Regulation (EU) 1282/2011; (3) Regulation (EU) 1183/2012 (corrected by Corrigendum, OJ L 349, December 19, 2012, p. 77); (4) Regulation (EU) 202/2014; (5) Regulation (EU) 865/2014 (that corrects a translation mistake in the Spanish version of Regulation (EU) 10/2011); (6) Regulation (EU) 2015/174; (7) Regulation (EU) 2016/1416; (8) Regulation (EU) 2017/752; (9) Regulation (EU) 2018/79; (10) Regulation (EU) 2018/213; (11) Regulation (EU) 2018/831; (12) Regulation (EU) 2019/37; (13) Regulation (EU) 2019/988; (14) Regulation (EU) 2019/1338; and (15) Regulation (EU) 2020/1245. (From now on, the denomination 'Regulation (EU) 10/2011' means the original Regulation and its amendments).

FIGURE 15.2 Mandatory symbol for labeling of food contact materials in the European Union (according to Regulation (EU) 10/2011).

Regulation 10/2011 applies to mono- and multilayer food contact plastics, plastics layers in multimaterial multilayer FCMs, and plastic layers or plastic coatings forming gaskets in caps and closures. It does not apply to ion exchange resins, rubber, silicones, printing inks, adhesives, and coatings.

In summary, for the manufacture of food contact plastics, there is now in Table 1 of Annex I of Regulation (EU) 10/2011, a new positive list (i.e., "the Union list") of monomers, other starting substances, macromolecules obtained from microbial fermentation, additives (except colorants), substances in nanoform, and certain "polymer production aids" (PPAs) (except solvents). This positive list does not include either "aids to polymerization" (AP) (such as catalytic systems) or NIAS, which may be present in food contact plastics and migrate to food. Tables 1, 2, 3, and 4 of Annex I include specific restrictions (mainly SMLs, SML(T)s, QMs, QMAs, restrictions of use, specifications, etc.) for certain listed substances.

Substances other than the listed (colorants, PPAs not included in the Union list, solvents, compounds included in the "Provisional list of additives used in plastics" (11th Update October 28, 2011) under evaluation by the EFSA) can be used in the manufacture of food contact plastics, provided that they comply with national legislations of the Member States. A plastic layer separated from food by a functional barrier may be manufactured with substances not listed in the Union list or the provisional list of additives (except in the case of vinyl chloride monomer, CMR compounds, or substances in nanoform), if it can be demonstrated that they migrate up to a maximum limit of 0.01 mg kg^{-1}.

Substances not included in the Union list and mentioned in the last two paragraphs may be used provided that, according to Article 19, they are evaluated in accordance with internationally recognized scientific principles on risk assessment.

Article 12 establishes the following OMLs:

- 10 mg (of nonpolymeric components) dm^{-2} (of contact surface area);
- 60 mg (of nonpolymeric components) kg^{-1} (of food or simulant), in the case of plastic materials in contact with food intended for infants and young children.

Written and signed DoC for food contact plastic articles, intermediate products, and raw materials (polymers, additives, adhesives, etc.) are required from all stakeholders throughout the supply chain (except at the retail level) (Article 15 and Annex IV). The aim of the DoC system is to ensure an adequate exchange of information on compliance of the different substances and materials between customers and suppliers. The supporting documentation must be available to the sanitary authorities (Article 16).

SMLs for heavy metals and other elements and for primary aromatic amines (PAAs) that may migrate from food contact plastics are set in Annex II.

Annex III defines the following food simulants for the overall and specific migration tests necessary to verify compliance with the established limits (simulant A: ethanol 10% (v/v) in aqueous solution; simulant B: acetic acid 3% (m/v) in aqueous solution; simulant C: ethanol 20% (v/v) in aqueous solution; simulant D1: ethanol 50% (v/v) in aqueous simulant D2: vegetable oil; simulant E: poly(2,6-diphenyl-p-phenylene oxide), also known as modified polyphenylene oxide (MPPO)).

Annex III also sets rules and includes Tables 2 and 3 for the assignment of food simulants for different foods, for instance, simulant A for aqueous nonacidic foods (pH $\geq$ 4.5); simulant B for aqueous acidic foods (pH $\leq$ 4.5); simulant C for alcoholic foods with an ethanol content up to 20% (v/v) and aqueous foods of a lipophilic character; simulant D1 for alcoholic foods with an ethanol content greater than 20% (v/v) and for dairy products; simulant D2 for fatty foods; simulant E only for specific migration tests in the case of dry foods, and not for overall migration test.

Real foods may be simulated by different combinations of these simulants, according to the rules set in Annex III.

Table 2 of Annex III assigns one or more simulants for different foods or groups of foods, indicating with the symbol "X" if the migration test with the simulant is required. For certain fatty foods, where "X/n" appears ("n" being a figure), it means that the value of the migration test with simulant D2 must be divided by "n," which is the "simulant D2 reduction factor" (DRF) for fatty foods.

Annex V establishes the general rules for the overall and specific migration tests, such as sample preparation; selection of time and temperature of contact with simulants according to real use with food; type of use (single or repeated); use of substitute simulants (ethanol 95% (v/v) in aqueous solution, or isooctane) instead of simulant D2 (if its use is not technically feasible) in the overall migration test; etc. This Annex also explains the use of the "fat reduction factor" (FRF)—assigned for certain lipophilic substances by Table 1 of Annex I, to correct the specific migration value of these substances into foods containing more than 20% (m/m) of fat—alone or combined with the DRF, if applicable, to obtain a total reduction factor.

Annex V also specifies, in certain cases, three alternatives to specific migration tests, in order to verify compliance with SMLs:

- to use the overall migration test result, in the case of a nonvolatile substance, if it is no greater than the SML;
- to calculate the potential migration of a substance assuming 100% migration, if its content in the plastic material is known (e.g., an additive), or can be determined experimentally (e.g., a monomer, a heavy metal);
- to apply generally recognized mathematical models based on scientific evidence that they never underestimate real levels of specific migration.

15.2.3.2 Recycled plastics

Regulation (EC) 282/2008 establishes requisites on PCR plastic materials and articles intended to come into contact with food, on the processes of manufacture of PCR plastics (i.e., the validated decontamination technologies used), and an authorization procedure for their use. In support of this Regulation, the EFSA has issued scientific opinions on the criteria of safety evaluation of decontamination technologies for PCR food contact plastics (EFSA 2008b, 2011), as previously mentioned in Section 15.1.4.3.

15.2.3.3 Active and intelligent materials

The main Articles of Regulation (EC) 450/2009 on food contact AIMs are related to the following:

- definitions (Article 3);
- requisites for placing AIMs on the market (Article 4);
- EU list of substances that may be used as active and intelligent components (Articles 5−8);
- specific labeling (Article 11); the nonedible parts or components must be labeled with the words "DO NOT EAT," and always where technically possible, must be accompanied by the symbol reproduced in Fig. 15.3;
- DoC in accordance with Article 16 of Regulation (EC) 1935/2004 (Article 12);
- requisites on supporting documentation (Article 13).

For an overview of the worldwide regulatory situation on food contact AIMs, see Dainelli (2015).

15.2.3.4 Ceramics

Ceramics are regulated by Council Directive 84/500/EEC as amended by Commission Directive (2005)/31/EC. The heavy metals oxides used in the production of ceramic articles intended to come into contact with foods may pose a risk for the consumer. Pb and Cd are of major concern, due to the possibility of their release from decoration and/or glazing enamels. SMLs for these elements, as well as the analytical methods for the determination of their specific migration (using as simulant acetic acid 4% (v/v) in aqueous solution), are established by the Council Directive. Lower SMLs for Pb and Cd, and new ones for additional elements, are currently under study (Semail, 2019). The Commission Directive establishes that it is mandatory for stakeholders to issue DoC as specified in Article 16 of Regulation (EC) 1935/2004, at all the market stages, in this case, including the retail stage.

15.2.3.5 Regenerated cellulose films

Regenerated cellulose (RC) films (i.e., cellophane) are regulated by Directive (2007)/42/EC. RC cellulose casings are excluded and must comply with the corresponding national regulations. This Directive applies to (a) uncoated RC films; (b) coated RC films with cellulose-based coatings; and (c) coated RC films with plastic-based coatings.

FIGURE 15.3 Mandatory symbol for labeling of nonedible parts or components of active and intelligent food contact materials in the European Union (according to Regulation (EC) 450/2009).

The two first types of RC films and the RC films prior to plastic coating must be manufactured with substances established in the positive list of this Directive. Coloring agents (dyes and pigments) and adhesives, which are not included in the positive list, can be used if there is no migration of these substances to food, detectable by a validated method. DoC according to Article 16 of Regulation (EC) 1935/2004 is required for RC films all along the supply chain (except at the retail level) (Ariosti, 2015).

For an overview of the worldwide regulatory situation on food contact RC, see Ariosti (2015).

15.2.4 EU legislation on specific substances

Directive 93/11/EEC establishes maximum levels for the release of N-nitrosamines and N-nitrosatable substances from rubber teats and soothers:

- 0.01 mg (in total of N-nitrosamines released) kg^{-1} (of the parts of teat or soother made of elastomer or rubber);
- 0.1 mg (in total of N-nitrosatable substances) kg^{-1} (of the parts of teat or soother made of elastomer or rubber).

Regulation 1895/2005/EC sets restrictions for epoxy derivatives of toxicological concern, such as BADGE (bisphenol A diglycidyl ether), BFDGE (bisphenol F diglycidyl ether), and NOGE (novolac glycidyl ethers), in food contact plastics, adhesives, and coatings of epoxy base. BADGE and its hydrolysis products, BADGE.H_2O and BADGE.$2H_2O$, have an SML(T) of 9 mg kg^{-1} (of food or food simulant); BADGE.HCl, BADGE.2HCl, and BADGE.H_2O.HCl have an SML(T) of 1 mg kg^{-1} (of food or food simulant), whereas the use and presence of BFDGE and NOGE in the manufacture of FCMs and articles are banned.

Regulation (EU) 2018/213, already mentioned as the 10th amendment of Regulation (EU) 10/2011, sets for BPA (2,2-bis(4-hydroxyphenyl)propane, CAS 80-05-7), released from food contact plastics, varnishes, or coatings, an SML of 0.05 mg (of BPA) kg^{-1} (of simulant or food). BPA must not to be used for the manufacture of plastic feeding bottles for infants and other drinking devices for infants and young children. It also required "no-migration" of BPA (i.e., a non-detectable migration level, determined with a detection limit of 0.01 mg (of BPA) kg^{-1} (of simulant or food)) in the case of certain packaged foods for infants and young children.

On January 17, 2017, the EC issued Recommendation (EU) 2017/84 launching a plan of MOH monitoring in certain foods and the FCMs used for these products, by the Member States with the active involvement of business operators and other interested parties, during 2017–18. The last monitoring data should be delivered to the EFSA by February 28, 2019.

15.2.5 Legislation on kitchenware made of melamine or polyamide originating or consigned from China or Hong Kong

Since July 1, 2011, kitchenware made of melamine or polyamide originating or consigned from China or Hong Kong must comply with the import requirements established by Regulation (EU) 284/2011. Consignments must be notified to the competent authorities at the EU entry points at least two working days before arrival, and must have a declaration and a laboratory report on the analysis of PAAs for polyamide articles, and of formaldehyde for melamine articles.

15.2.6 EC recommendation on the coordinated control plan of migrating substances from FCMs

In May 2019, the EC issued Recommendation (EU) 2019/794 that establishes a coordinated control plan of specific migration tests of substances (PAAs, formaldehyde, melamine, phenol, bisphenols (including BPA and bisphenol S (BPS)), phthalates and other plasticizers, fluorinated compounds, and metal elements) in different FCMs and of overall migration tests of nonconventional kitchenware and tableware, such as reusable coffee cups containing bamboo-derived components. Controls made by Member States should take place during the second semester of 2019, and the results should be reported by February 29, 2020.

15.3 The Council of Europe technical recommendations on FCMs

The CoE was founded by the Treaty of London in May 1949, being thus the oldest European Organization. Its headquarters is located in Strasbourg (France), and at present (October 2021), it has 47 Member States, including all the EU Member States, as well as other countries in Europe and Asia. From the early 1960s up to 2008, the technical work on FCMs was in charge of the "Partial Agreement in the Social and Public Health Field" (adopted in November 1959) (www.coe.int/en/web/disability/partial-agreement), through its "Committee of Experts on materials coming into contact with food." This Committee prepared Resolutions and Technical Documents, with the advice of "Ad-hoc Working Groups." Resolutions, after an approval of the CoE "Public Health Committee," were adopted by the "Committee of Ministers." The latter is the CoE decision-making body, composed by the Member States Foreign Ministers or their Strasbourg-based deputies (ambassadors or permanent representatives). Technical Documents were adopted by the "Committee of Experts" (Bolle, 2016; Rossi, 2010).

In 2008, the Committee of Ministers dissolved this Partial Agreement (Accord Partiel (AP), in French), and transferred its activities in the fields of cosmetics and food and drug packaging to the European Directorate for the Quality of Medicines and HealthCare (EDQM) which is dealing with these subjects since 2009 (Bolle, 2016).

Since 2009, at the EDQM the "Committee for the Protection of Consumers' Health" (CD-P-SC) was in charge of the management of the work program on FCMs. Within it, the specific group of experts dealing with FCMs was the "Committee of Experts on Food and Pharmaceutical Contact Materials" (P-SC-EMB). Its tasks were to examine health topics and to issue reports and recommendations. "Ad-hoc groups" formed to study specific subjects not harmonized at the EU level (e.g., metal and alloys, glass, printing inks). Rapporteur Countries participating in the EDQM are reviewing some recommendations on specific FCMs: paper and board (Austria), ion exchange resins (Belgium), coatings (Belgium and the Netherlands), cork (Portugal), and inks (Greece) (Bolle, 2016). In 2018 the European Committee for Food Contact Materials and Articles (CD-P-MCA) was created to be in charge of these activities (https://www.edqm.eu/en/food-contact-materials-and-articles).

In summary, the results of the harmonization process on FCMs are the following types of recommendations: Policy Statements, Resolutions, Guidelines, and Technical Documents.

The CoE recommendations are not binding for the Member States, unless they are transposed into the National Laws. Several European countries have instructed the use of the CoE recommendations as reference documents to enforce the clause on safety of Article 3 of the EU Framework Regulation (EC) 1935/2004 on FCMs (Rossi, 2010).

The CoE has traditionally taken initiatives in technical issues not yet harmonized by the EU, and the CoE recommendations are considered as appropriate references in the absence of EU or National Laws (Bolle, 2016; Rossi; 2010; Schäfer, 2007). Several recommendations have been issued over the years. They are summarized in Table 15.2, where only the last versions for each kind of FCMs appear. Previous versions are not referenced in Table 15.2, but can be found on the EDQM website.

15.4 FCMs legislation in the United States

The Federal Food, Drug, and Cosmetic Act (FFDCA) was sanctioned in 1938, banning to put in the market unsafe foods and nonsanitary packages. The FDA was also created then. But it was not until 1958, with the passage of the Food Additives Amendment Act, that the formal food packaging regulation in the United States was introduced (Baughan and Attwood, 2010; Heckman, 2005; Twaroski et al., 2007).

The Code of Federal Regulations (CFR) publishes the legal requirements of the United States, divided into Titles or areas of control. Title 21 corresponds to the FDA and contains all the regulations issued by this agency to assure the compliance of the FFDCA and the Fair Packaging and Labeling Act. FCMs are regulated mainly in two Acts: the Food Additives Amendment of the FFDCA of 1958 and the National Environmental Policy Act (NEPA) of 1969.

In 1958, the US Congress gave authority to the FDA to regulate food additives, defined as any substance whose intended use could result or reasonably is expected to result, direct or indirectly, as part of the food or affect its characteristics (FFDCA section 201(s)). This definition excludes explicitly the following categories: substances generally recognized as safe (GRAS); pesticide chemical residues on raw agricultural commodities or in processed foods; pesticide chemicals; color additives; prior sanctioned substances (the use of which was accepted before September 6, 1958); new animal drugs; and dietary supplements. Besides, section 409(h)(6) of the FFDCA defines a food contact substance (FCS) as "any substance intended for use as a component of materials used in manufacturing, packing, packaging, transporting, or holding food if such use is not intended to have a technical effect in such food." According to these definitions, the FDA considers that FCSs (i.e., FCMs) are indirect food additives, if any of their components can migrate into food.

As detailed in the FFDCA, an indirect food additive is considered unsafe unless (Baughan and Attwood, 2010; Greenberg, 2018; Heckman, 2005; Misko, 2017; Twaroski et al., 2007):

- it is included in a regulation list (positive list of the CFR), according to the standard Food Additive Petition (FAP) process;
- it has an effective Food Contact Notification (FCN);
- it is subject to any exemption, for instance: GRAS; prior sanctioned; migration below the TOR; "no migration" (i.e., "not a component of food," if it can be demonstrated that the substance does not migrate or migrates in negligible quantities, in association with the functional barrier concept).

TABLE 15.2 Council of Europe Technical Recommendations on food contact materials.

Subject	Recommendation	Last version dated	Content
Coatings	Policy statement concerning coatings intended to come into contact with foodstuffs.	Version 3 February 12, 2009	- Framework Resolution ResAP (2004) 1 on coatings intended to come into contact with foodstuffs. - Technical document N° 1. List of substances to be used in the manufacture of coatings intended to come into contact with foodstuffs (version 3).
Colorants for plastic materials	Resolution AP (89) 1 on the use of colorants in plastic materials coming into contact with food.	September 13, 1989	Resolution AP (89) 1.
Cork	Policy statement concerning cork stoppers and other cork materials and articles intended to come into contact with foodstuffs.	Version 2 September 5, 2007	- Resolution ResAP (2004) 2 on cork stoppers and other cork materials and articles intended to come into contact with foodstuffs. - Technical document N° 1. List of substances to be used in the manufacture of cork stoppers and other cork materials and articles intended to come into contact with foodstuffs (July 12, 2007). - Technical document N° 2. Test conditions and methods of analysis for cork stoppers and other cork materials and articles intended to come into contact with foodstuffs.
Glass	Policy statement concerning lead leaching from glass tableware into foodstuffs.	Version 1 dated September 22, 2004	- Guidelines for lead leaching from glass tableware into foodstuffs. - Appendix 1. Parameters influencing lead leaching. - Appendix 2. Excerpt for lead from Resolution AP (96) 4 on maximum and guideline levels and on source-directed measures aimed at reducing the contamination of food by lead, cadmium, and mercury.
Inks	Policy statement concerning packaging inks applied to the nonfood contact surface of food packaging.	Version 2 dated October 10, 2007	- Resolution ResAP (2005) 2 on packaging inks applied to the nonfood contact surface of food packaging materials and articles intended to come into contact with foodstuffs. - Technical document N° 1. Requirements for the selection of packaging ink raw materials applied to the nonfood contact surface of food packaging materials and articles intended to come into contact with foodstuffs (version 1, December 21, 2006). - Technical document N° 2. Part 1. GMP for the production of packaging inks formulated for use on the nonfood contact surface of food packaging materials and articles intended to come into contact with foodstuffs (prepared by CEPE). - Technical document N° 2. Part 2. Code of GMP for flexible and fiber-based packaging for food (prepared by FPE in cooperation with CITPA). - Technical document N° 3. Guidelines on test conditions for packaging inks applied to the nonfood contact surface of food packaging materials and articles intended to come into contact with foodstuffs.

Continued

TABLE 15.2 Council of Europe Technical Recommendations on food contact materials.—cont'd

Subject	Recommendation	Last version dated	Content
Metals and alloys	Resolution CM/Res (2013) 9 on metals and alloys used in food contact materials.	Adopted June 11, 2013	- Council of Ministers Recommendation to Member states. - Technical guide on metals and alloys used in food contact materials and articles (prepared by the Committee of experts on packaging materials for food and pharmaceutical products (P-SC-EMB)).
Paper and board	Policy statement on paper and board used in food contact materials and articles intended to come into contact with foodstuffs.	Version 2021	- Council of Europe Resolution CM/Res (2020) 9 on the safety and quality of materials and articles for contact with food, 7 October 2020. - Technical Guide on paper and board used in food contact materials and articles (Version 2021).
Plastics polymerization aids	Resolution AP (92) 2 on control of aids to polymerization (technological coadjuvants) for plastic materials and articles intended to come into contact with foodstuffs.	October 19, 1992	Resolution AP (92) 2
Resins for ion exchange and adsorption	Policy statement concerning ion exchange and adsorbent resins in the processing of foodstuffs.	Version 3 dated August 21, 2009	- Resolution ResAP (2004) 3 on ion exchange and adsorbent resins used in the processing of foodstuffs. - Technical document N° 1. List of substances to be used in the manufacture of ion exchange and adsorbent resins used in the processing of foodstuffs (version 3).
Rubbers	Policy statement concerning rubber products intended to come into contact with foodstuffs.	Version 1 dated June 10, 2004	- Resolution ResAP (2004) 4 on rubber products intended to come into contact with foodstuffs. - Technical document N° 1. List of substances to be used in the manufacture of rubber products intended to come into contact with foodstuffs (to be prepared). - Technical document N° 2. Practical guide for users of Resolution ResAP (2004) 4 on rubber products intended to come into contact with foodstuffs. - Appendix 1. Inventory list of substances to be used in the manufacture of rubber products intended to come into contact with foodstuffs.
Silicones	Policy statement concerning silicones used for food contact applications.	Version 1 dated June 10, 2004	- Resolution ResAP (2004) 5 on silicones to be used for food contact applications. - Technical document N° 1. List of substances to be used in the manufacture of silicones used for food contact applications.
Tissue paper	Policy statement concerning tissue paper kitchen towels and napkins.	Version 1 dated September 22, 2004	- Guidelines (covering: Specifications, raw materials, test conditions and methods of analysis, use of recycled fibers, GMP). - 4 technical appendixes.

AP, Partial Agreement (Accord Partiel, in French); *CEPE*, European Council of Paints, Printing Inks and Artist's Colors Industries (Conseil Européen des Producteurs de Peintures, d'Encres d'Imprimerie et de Couleurs pour Artistes, in French); *CITPA*, International Confederation of Paper and Board Converters in Europe (Confédération Internationale des Transformateurs de Papier et Carton en Europe, in French); *CM*, Committee of Ministers; *FPE*, Flexible Packaging Europe; *GMP*, good manufacturing practice.

Indirect food additives in the United States must get a premarket evaluation and clearance by the US-FDA before entering into interstate commerce. The FDA established the Division of Food Contact Notification (DFCN) within the Office of Food Additive Safety (OFAS), at the Center for Food Safety and Applied Nutrition (CFSAN), to ensure that components of FCMs, including food packaging and processing equipment, are safe for their intended use.

Under the standard FAP system, the process begins with a manufacturer's petition of the premarket approval of an indirect food additive and concludes in the publication of the corresponding regulation in the CFR. The regulation under the FAP system is of general application, and is not specific for the petitioner. The petition process is described in Title 21 Part 171 of the CFR. The general and specific regulations for all food ingredients and packaging materials can be found in Title 21 CFR—Parts 170 through 190 (see Table 15.3).

Of special interest are, for instance, Part 175 (adhesives and coatings), Part 176 (paper and paperboard), Part 177 (Polymers), and Part 178 (Adjuvants, production aids, and sanitizers), since all of them are substances intended to come into contact with food as part of FCMs, and the components of which can migrate to food (i.e., they are indirect food additives).

In 1997, the Food and Drug Administration Modernization Act (FDAMA) amended the FFDCA to renew the way in which the FDA conducted the premarket approvals. One of the new procedures established to accomplish this goal was the FCN system, implemented since 2000. This is a simplified procedure, faster than the FAP system, and is intended to replace the latter as the primary means for authorizing new uses of FCSs that are indirect food additives (21 CFR Sections 170.100 through 170.106). However, discretion is given to the FDA for deciding when the FAP process is more appropriate for evaluating data submitted by petitioners to provide an adequate safety assurance.

An FAP must be submitted to the FDA for an indirect food additive when its "estimated daily intake" (EDI), or, alternatively, its "cumulative estimated daily intake" (CEDI) is greater than 3 mg per person per day (equivalent to a dietary concentration—supposing a consumption of 3 kg of food per person per day—of 1 mg (of substance) kg^{-1} (of food)); or than 0.6 mg per person per day for biocidal compounds; or when the FDA has not reviewed bioassays which are not clearly negative for carcinogenic effects. When the EDI or the CEDI are equal to or less than those values, the FCN system is recommended (Baughan and Attwood, 2010; Honigfort, 2018; Twaroski et al., 2007).

TABLE 15.3 Title 21 (FDA) of the US code of federal regulations—parts 170–190.

Part	Description
170	Food additives
171	Food additive petitions
172	Food additives permitted for direct addition to food for human consumption
173	Secondary direct food additives permitted in food for human consumption
174	Indirect food additives: General
175	Indirect food additives: Adhesives and components of coatings
176	Indirect food additives: Paper and paperboard components
177	Indirect food additives: Polymers
178	Indirect food additives: Adjuvants, production aids, and sanitizers
179	Irradiation in the production, processing and handling of food
180	Food additives permitted in food or in contact with food on an interim basis pending additional study
181	Prior-sanctioned food ingredients
182	Substances generally recognized as safe
184	Direct food substances affirmed as generally recognized as safe
186	Indirect food substances affirmed as generally recognized as safe
189	Substances prohibited from use in human food
190	Dietary supplements

Available from: www.ecfr.gov/cgi-bin/text-idx?SID=4ec8f91e9d7761feaa2871762c48972c&mc=true&tpl=/ecfrbrowse/Title21/21cfrv3_02.tpl#0.

One of the possible alternatives to an FAP or an FCN is the TOR exemption process, established by the FDA in 1995 (21 CFR 170.39). It exempts certain substances used in FCMs from the requirement of an authorizing regulation before its use. To obtain a TOR exemption, the EDI (or CEDI) of the substance must be no greater than 1.5 μg per person per day (equivalent to a dietary concentration of 0.5 μg (of substance) kg^{-1} (of food)). The substance must be noncarcinogenic and must not contain certain carcinogenic impurities (Baughan and Attwood, 2010; Honigfort, 2018; Misko, 2017; Twaroski et al., 2007).

US-FDA Guidances for industry on FAPs, FCNs, TOR and other exemptions are available from: www.fda.gov/Food/GuidanceRegulation/GuidanceDocumentsRegulatoryInformation/IngredientsAdditivesGRASPackaging/default.htm.

In contrast with a TOR exemption and an FAP regulation, an FCN is a proprietary authorization for only the notifier and the manufacturer or supplier listed. Besides, all FAPs, FCNs, and TOR exemptions are subjected to the NEPA regulations on environmental assessment (Baughan and Atwood, 2010; Twaroski et al., 2007). The FDA maintains a list of effective FCNs (www.accessdata.fda.gov/scripts/fcn/fcnNavigation.cfm?rpt=fcsListing) and TOR exemptions (www.accessdata.fda.gov/scripts/fdcc/?set=TOR) on its website.

The limitations of use of an FCS may include the concentration limits in the FCM itself, as well as the type of foods, the time(s) and temperature(s) conditions under which these FCMs must be used, the FCMs thicknesses, etc. For instance, Table 1 of section 21 CFR 176.170 "Components of paper and board in contact with aqueous and fatty foods" describes nine types of foods (fresh or processed), classified as aqueous nonacid, aqueous acid, alcoholic, dairy, fatty, and dry solid foodstuffs. Conditions of use are also defined regarding thermal treatment and storage. For migration tests ("end tests" to verify compliance with CFR regulations of FCS already cleared), Table 2 of section 21 CFR 176.170 establishes the use of different food simulants, such as distilled water, ethanol 8% (v/v) in aqueous solution, ethanol 50% (v/v) in aqueous solution, and n-heptane, for different types of foods and conditions of real filling, processing, and storage. It also sets the different conditions of time and temperature of contact for the end tests. Another example can be found in section 21 CFR 175.300 "Resinous and polymeric coatings" that describes eight types of foods and establishes distilled water, ethanol 8% (v/v) in aqueous solution, and n-heptane, as food simulants, and different conditions of time and temperature of contact for the end tests. The end tests described in the CFR sections must not be confounded with migration tests performed to assess exposure of consumers to a new FCS in the FAP or FCN submissions that are detailed in the US-FDA Guidance for industry "Preparation of premarket submissions for food contact substances (Chemistry Recommendations)—December 2007" (www.fda.gov/regulatory-information/search-fda-guidance-documents/guidance-industry-preparation-premarket-submissions-food-contact-substances-chemistry) and that differ in simulants, conditions of time and temperature of contact, etc.

In January 2011, the FDA Food Safety Modernization Act (FSMA) (www.fda.gov/Food/GuidanceRegulation/FSMA/ucm247548.htm#SEC301) was enacted to amend the FFDCA with respect to the safety of the food supply chain. It gives the FDA greater authority to prevent food safety problems (Title I), detect and respond to food safety problems (Title II), and to improve safety of imported food (Title III). Most of the FSMA requirements apply to registered food facilities, and do not apply to FCS manufacturers. There are two key rules: the Hazard Analysis Risk−Based Preventive Controls (HARPC) Rule (Title I) and the Foreign Supplier Verification Program (FSVP) Rule (Title III). The HARPC Rule applies to food facilities that must register under the Bioterrorism Act, but not to FCS manufacturers, as they are not required to register.

The FSVP applies to food and food additives, and thus to FCS. So it impacts on importers of FCMs that are indirect food additives. However, the FDA, in its January 2018 announcement (www.fda.gov/Food/NewsEvents/ConstituentUpdates/ucm590667.htm), stated that it shall exercise enforcement discretion to not require importers of FCS to meet the FSVP provisions (though this policy might be revised) (Koh-Fallet, 2018).

15.5 FCMs legislation in the MERCOSUR

The structure and updates of the legislation on FCMs in the MERCOSUR have been addressed in detail elsewhere (Ariosti, 2016b, 2018a, b, 2019a, b, 2021a, b; Ariosti and Olivera Carrión, 2014; Ariosti and Padula, 2017; Padula, 2010, 2019). It seems that among the different blocks into which the Latin American and Caribbean countries have organized themselves the MERCOSUR has made the greater effort to harmonize and enforce regulations on FCMs (LeNoir, 2015).

The MERCOSUR (MERCOSUL, in Portuguese) was created by the Treaty of Asunción (Paraguay), signed on March 26, 1991. The original Member States were Argentina, Brazil, Paraguay, and Uruguay. The Protocol of Ouro Preto (Brazil) (December 17, 1994) fixed the present institutional structure of the MERCOSUR. The maximum decision-making political body is the Common Market Council (CMC) and the executive body is the Common Market Group (GMC). The Administrative Secretariat has its headquarters in Montevideo, Uruguay.

Venezuela entered the MERCOSUR in July 2006, after signing the Protocol of Adhesion in December 2005. At present (October 2021), Venezuela is suspended as a Member State by application of Article 5, paragraph 2, of the Protocol of Ushuaia (Argentina). Bolivia signed its Protocol of Adhesion to the MERCOSUR in July 2015, and it is at present following this integration process. The Associated States to the MERCOSUR are Chile, Colombia, Ecuador, Guyana, Peru, and Suriname.

The activities on FCMs legislation are organized in the following hierarchical way:

- 'Packaging Group': which discusses and prepares draft regulations on technical issues;
- 'Food Commission': which coordinates the work of the Packaging Group and other Groups within the Commission (that deal with food contaminants, food additives, nutritional labeling, etc.);
- 'Working Sub-Group 3 (SGT 3)': which coordinates the technical work of the Food Commission and other Commissions within the SGT 3 (that deal with pharmaceutical products, toys, etc.);
- 'Common Market Group (GMC)': which sanctions the Resolutions, after a process of national and international consultation.

The harmonized GMC Resolutions must be transposed (incorporated) into the national legislation of the Member States. For instance, in Argentina, GMC Resolutions on FCMs must be incorporated into Chapter IV of the Argentine Food Code (Código Alimentario Argentino, in Spanish) (www.argentina.gob.ar/anmat/codigoalimentario); and in Brazil, into its federal legislation, as Decrees (Portarias, in Portuguese), and, more recently, as Resolutions of the Collegiate Directorate of the National Agency of Sanitary Surveillance (ANVISA RDC) (www.anvisa.gov.br).

The harmonization work began in 1991 on the basis of the two main FCMs legislations available in the region, the Argentine and Brazilian ones. Since 1991, the author has been participating within the Packaging Group—Food Commission, in technical discussions with his colleagues from other Member States, at the periodical meetings held in the Member States. The main technical references are the EU and US-FDA FCMs regulations. These are used to periodically update the MERCOSUR FCMs legislation. In special cases, the EU Member States legislation (for instance, the Spanish Royal Decree 847/2011 on plastics and coatings, and the Dutch Commodities Act Regulation on Packaging and Consumer Articles (Warenwetregeling) of December 20, 2016) and recommendations issued by internationally recognized institutions are taken as references (e.g., CoE (for pigments and colorants for plastics), BfR (for paper and board)).

The MERCOSUR FCMs legislation is, in general, similar to the EU FCMs legislation. This is due to the fact that the Argentine and Brazilian FCMs legislations previous to MERCOSUR were both based on the FCMs Italian legislation (Italian Ministerial Decree of March 21, 1973, and amendments). But it is rather eclectic, because it also incorporates some concepts, such as the TOR, and certain positive lists of substances and their restrictions, such as those corresponding to rubber and certain polymeric coatings from the US-FDA FCMs legislation.

The positive lists are adapted from the international references previously mentioned. For a new substance to be incorporated into the MERCOSUR FCMs positive lists, it must be previously included in the FCMs positive list of one of the international references recognized by the MERCOSUR. Substances covered by FCNs issued by the US-FDA, and not included in the 21CFR positive lists (after the FAP process), can also be included in the MERCOSUR positive lists.

The MERCOSUR Framework Resolution GMC 3/92 on FCMs establishes a premarket approval of food packages and articles before its commercialization. It means that the FCMs manufacturers must demonstrate the compliance of their products with the legislation, to the sanitary authorities, which then issue an approval/authorization certificate, case by case. Food manufacturers must use only the FCMs already approved by the sanitary authorities. Since the 1980s, the system has been in full application in Argentina, even before the enforcement of the MERCOSUR legislation (from 1995 on). Presently in Brazil, however, only FCMs and articles manufactured with recycled materials must be subjected to this system of premarket approval, according to ANVISA Resolutions 22/2000 and 23/2000, and ANVISA RDC 27/2010. Paraguay and Uruguay have also implemented the premarket approval system. In all the cases, the MERCOSUR legislation applies to national products and imports. The premarket approval requisites also apply to national products and imports, except in Brazil (if no recycled materials are used).

The main similarities with the EU FCMs legislation are the existence of positive lists, OMLs (60 mg (of non-polymeric components) kg^{-1} (of food or simulant) and 10 mg (of non-polymeric components) dm^{-2} (of contact surface area)), QMs, SMLs and SML(T)s. The MERCOSUR OMLs are a little bit different from the US-FDA OMLs ($50\ mg\ kg^{-1}$ and $0.5\ mg\ in^{-2}$ ($=7.75\ mg\ dm^{-2}$)), and a little bit different from the EU OMLs ($60\ mg\ kg^{-1}$ and $10\ mg\ dm^{-2}$).

A summary of the GMC Resolutions in force can be found in Table 15.4.

The GMC Resolutions are issued in Spanish and in Portuguese, and can be found on the MERCOSUR websites; but some unofficial translations into English are available for purchase from specialized companies.

TABLE 15.4 MERCOSUR GMC Resolutions on food contact materials.

Material	Subject	GMC Resolution N°
General	Framework Resolution: General requisites for FCMs	3/92
	Reference analytical methodology for the control of FCMs	32/99
Plastic FCMs	General requisites	56/92 and 20/21
	Positive list of resins and polymers	02/12 and 19/21
	Positive list of additives	39/19, 62/19 and 11/20
	Migration tests	32/10
	Colorants and pigments	15/10
	Fluorinated polyethylene	56/98
	Polymeric and resinous coatings for foods	55/99
	Refillable PET packages for carbonated nonalcoholic beverages	16/93
	Multilayer PET packages, with central layer containing recycled material, for carbonated nonalcoholic beverages	25/99
	PCR-PET for food packages (multilayer and monolayer packages)	30/07
Metallic FCMs	General requisites	46/06 and 16/20
Glass and ceramic FCMs	General requisites	55/92
Cellulosic FCMs (paper and board)	General requisites	40/15
	Papers for hot filtration and cooking	41/15
	Papers for cooking and heating in oven	42/15
Regenerated cellulose FCMs	Films	55/97
	Casings	68/00
Rubber FCMs (elastomers)	General requisites	54/97
	Positive list of components	28/99
Adhesives	General requisites	27/99
Paraffins for food contact	General requisites	67/00

Nomenclature: GMC Resolutions are referred to by the designation xx/yy, where xx is the GMC Resolution number, and yy are the two last digits of the year of sanction. Acronyms: *FCM*, food contact material; *GMC*, Common Market Group (Grupo Mercado Común, in Spanish; Grupo Mercado Comum, in Portuguese); *PCR*, postconsumer recycled; *PET*, polyethylene terephthalate.

In December 2007, the technical document on PCR-PET for food grade packages was sanctioned as GMC Resolution 30/07. It covers multilayer packages with functional barrier and monolayer packages (MERCOSUR, 2007). GMC Resolution 15/10 regarding colorants and pigments for food contact plastics is based on the CoE "Resolution AP (89)1 on the use of colorants in plastic materials coming into contact with food (dated September 13, 1989)." GMC Resolution 32/10 on migration tests establishes that the CEN standards on this subject must be followed for the overall and specific migration tests. The last updates of the positive list of monomers, other starting substances and polymers for food contact plastics and coatings, were sanctioned in 2012 and 2021 as GMC Resolution 02/12 and GMC Resolution 19/21, respectively. The last update of several previous GMC Resolutions on cellulosic FCMs (paper and board) resulted in the sanction of GMC Resolutions 40/15, 41/15, and 42/15, that are based, respectively, on the German BfR "Recommendation XXXVI. Paper and board for food contact," "Recommendation XXXVI/1. Cooking papers, hot filter papers, and filter layers," and "Recommendation XXXVI/2. Paper and paperboard for baking purposes." GMC Resolutions 39/19, 62/19 and 11/20 establish requirements for additives for food contact plastics and coatings. GMC Resolution 46/06 (in force) for food contact metals and alloys has been amended by GMC Resolution 16/20.

15.6 FCMs legislation in Japan

The basic characteristics of the Japanese FCMs regulations and updates have been described elsewhere (Hong, 2016; JETRO, 2009, 2011; Kawamura, 2015; Mori, 2010; Mutsuga, 2012; Otsuka, 2015; Shigekura, 2018). JETRO is the Japan External Trade Organization, a government-related agency. One of its main objectives is to promote and facilitate mutual trade and investment between Japan and the rest of the world.

The following regulations on food safety deal with FCMs (JETRO, 2011):

(a) two National Laws:
- **(1)** Chapter III "Apparatus and Containers and Packaging" of the Food Sanitation Act (Act 233 of December 24, 1947; last reported revision by Act 49 of June 5, 2009 (JETRO, 2011); last amend in June 2018 (MHLW, 2018a, b, 2019a), which is under the jurisdiction of the Ministry of Health, Labour and Welfare (MHLW) and
- **(2)** Article 8 of the Food Safety Basic Act (Act 48 of May 23, 2003; last reported revision by Act 8, 2007 (JETRO, 2011), which is under the jurisdiction of the Food Safety Commission of Japan (FSCJ).

(b) two Ministerial Regulations:
- **(1)** Annex 4 "The Standards of equipment or containers/packages of milk, etc., or their raw materials and the Standards of manufacturing" of the Ministerial Ordinance on "Milk and milk products concerning compositional standards, etc." (MHLW Ordinance 52, 1951; last reported revision by MHLW Ordinance 132 of October 30, 2007 (JETRO, 2011); last amend in 2018 (Overbeek, 2019) and
- **(2)** Section 3 "Apparatus and containers/packages" of the "Specifications and Standards for food and food additives, etc." (MHLW Notification 370, 1959; last reported revision by MHLW Notification 336 of September 6, 2010 (JETRO, 2011)).

Of particular importance are two documents published by JETRO in English version (unofficial translations):

- Document 1: "Specifications and standards for foods, food additives, etc., under the Food Sanitation Act (Abstract) 2010" (JETRO, 2011), is an outline of specifications and standards for foods; milk and milk products; food additives; apparatus and containers/packages (Chapter IV, which summarizes Section 3 of the "Specifications and Standards for food and food additives, etc." (MHLW Notification 370); toys; and cleaning agents (detergents); and covers regulations (a) (1), (b) (1), and (b) (2) (last reported revisions (JETRO, 2011);
- Document 2: "Specifications, standards, and testing methods for foodstuffs, implements, containers, and packaging, toys, detergents 2008" (JETRO, 2009) contains the standards and testing methods for foods; implements, containers and packaging (Chapter II, which summarizes Section 3 of the "Specifications and Standards for food and food additives, etc." (MHLW Notification 370); toys; and detergents; and covers regulations (a) (1) (last reported revision by Act 53 of June 7, 2006 (JETRO, 2009)) and (b) (2) (last reported revision by MHLW Notice 529 of November 27, 2008 (JETRO, 2009).

The Food Sanitation Act specifies general hygienic requirements on apparatus and containers/packages (as defined by Article 4) intended to come into contact with foodstuffs. The category "apparatus" includes tableware, kitchen utensils and machine implements, and other articles used for collecting, producing, processing, cooking, storing, transporting, displaying, delivering or consuming food or additives, except machines, implements, and other articles used for harvesting food in agriculture and fisheries.

The objective of the Food Sanitation Act, stated in Article 1, is the protection of human health from hazards arising from food consumption. The introductory Articles are outstanding in promoting science, education, training, and industrial voluntary effort:

- Article 2 summons the State to take measures through educational activities and public relations, in order to promote research relating to food sanitation, to improve control and inspection facilities concerning food sanitation, to train human resources and enhance their capabilities, to conduct imports inspection, etc.
- Article 3 states that food and FCMs manufacturers shall ensure the safety of their products, and shall make a voluntary effort to get knowledge and techniques relating to the safety of their products and to the practice of self-imposed control.

Chapter III (comprising Articles 15–18) of the Food Sanitation Act is specifically dedicated to FCMs. Article 18 states that the MHLW may establish standards for FCMs (e.g., (b) (1), (b) (2)).

The Food Sanitation Act requires that food contact final articles comply with any labeling standards established by the MHLW (e.g., recycling symbols, statements of promotion of effective utilization of resources) (Article 19) (Hong, 2016). The Act also establishes a case by case obligatory control system of food and FCMs subject to regulations and mandatory standards and specifications, and the obligatory use of a label stating that they passed the technical inspections and tests performed by the national or local governments or Registered Laboratories (Article 25). It also defines the procedures for official inspections of food and FCMs (Articles 26 and 28), and a notification system of imported goods (Article 27).

With respect to the Food Safety Basic Act, Article 8 summons operators of business dealing with food and FCMs to take measures to ensure the safety of their products along the supply chain, in collaboration with national and local governments, and to provide accurate and appropriate information on their products. Articles 9 and 19 state that consumers must have an active role in ensuring food safety, through training in knowledge of the principles of food safety, and by expressing their opinions on policies to ensure food safety. Article 16 promotes the establishment of tests and research systems to ensure food safety, research and development programs, dissemination of results, and training of researchers, in order to ensure scientific knowledge on food safety. Article 22 establishes the FSCJ in the Cabinet Office.

The official mandatory specifications and standards for FCMs are based in a system of negative lists (i.e., lists that indicate only substances that are banned or have restrictions) and establish requisites for all FCMs types (safety principles, testing procedures and migration limits, instructions on the preparation of reagents and solutions). Requisites are also established for specific materials: glass; ceramics; enameled ware; synthetic resins (i.e., plastics) in general; specific plastics (PVC, PE, PP, PS, polyvinylidene chloride (PVDC), PET, polymethyl methacrylate (PMMA), polyamide (PA), poly(4-methyl-1-pentene) (PMP), polycarbonate (PC), polyvinyl alcohol (PVA), PLA); phenol-, melamine-, and urea-formaldehyde resins; rubber (including nursing utensils); and metal cans. SMLs for elements and substances in different materials are established (e.g., Cd and Pb in glass and ceramics; As, Cd, and Pb in metal cans; antimony (Sb) in PET; vinyl chloride monomer in PVC; caprolactam in PA; BPA in PC; epichlorohydrin in coated metal cans). For synthetic resins in general (except phenol-, melamine-, and urea–formaldehyde resins), a limit for the consumption of potassium permanganate ($KMnO_4$), the objective of which is to quantify organic matter migrated from FCMs, is established (10 $\mu g\ mL^{-1}$) (JETRO, 2011).

The general OML in the case of resins, plastics, and coated metal cans is 30 $\mu g\ mL^{-1}$. If n-heptane (the extractive capacity of which is recognized as higher than other fatty food simulants) is used as fatty foods simulant, OMLs are greater, for instance (JETRO, 2011):

- for PS: 240 $\mu g\ mL^{-1}$;
- for PVC: 150 $\mu g\ mL^{-1}$;
- for coated metal cans: 90 $\mu g\ mL^{-1}$.

In May 2012, the MHLW issued two guidelines for the use of recycled plastics and recycled papers in food contact articles (Mutsuga, 2012).

The FCMs legislation establishes also standards by applications, specifically on pressure and heat-sterilized packaged foods (except canned and bottled foods); soft drinks (except fruit juice as ingredient) packages manufactured with glass, metal, plastics, and composite materials; vending machines; implements used in the production of frozen confection; and stock solutions of soft drinks. It is important to note that for some categories, mechanical properties specifications and performance requisites are established.

Finally, the FCMs legislation specifies requisites on manufacturing standards and dairy products packages; in the last case, for milk, fermented milk, prepared milk powder, cream, etc.

In the regulated field, the MHLW is the risk management authority (Food Sanitation Act, Article 18), and the FSCJ is the risk assessment authority (Food Safety Basic Act, Articles 23 and 24) (Otsuka, 2015; Overbeek, 2019; Shigekura, 2018).

In the voluntary field, FCMs manufacturers also apply voluntary standards issued by Industrial Hygienic Associations that include positive lists (such as the Japan Hygienic Olefin and Styrene Plastics Association (JHOSPA), the Japan Hygienic PVC Association (JHPA), the Japan Hygienic Association of Vinylidene Chloride (JHAVC), the Japan Rubber Manufacturers Association (JRMA), the Japan Petrolatum Wax Industry Association (JPWIA), etc.) or negative lists (such as the Japan Adhesive Industry Association (JAIA), the Japan Printing Ink Makers Association (JPIMA), the Japan Paper Association (JPA), etc.). These associations manage certification systems and issue certificates of compliance of FCMs for their associates (Kawamura, 2015; Otsuka, 2015; Overbeek, 2019; Shigekura, 2018; Thompson, 2019a).

In summary, today in Japan there are mandatory regulations on FCMs, and the government manages control systems of food contact final articles; and in the voluntary field, there are standards and certification systems developed by Industrial Hygienic Associations that imply controls from raw materials to food contact final articles all along the supply chain (Kawamura, 2015; Otsuka, 2015; Shigekura, 2018).

Several important regulatory developments have taken place since 2015 (Hong, 2016; Kawamura, 2015; Overbeek, 2019; Thompson, 2019a, 2021a). That year the MHLW announced a possible transition to an official system of positive lists, such as the ones in force in the United States and the EU. On July 10, 2017, the MHLW issued the "Guideline for safety assurance in the manufacture, etc., of food utensils, containers, and packaging" (MHLW, 2017). On June 13, 2018, the MHLW issued the "Amendment of the Food Sanitation Act" (MHLW, 2018a, b, 2019a). The main changes have been (a) the introduction of the principle of the official positive list system in Article 18, intended to replace in the future the voluntary positive lists issued by the Industrial Hygienic Associations; (b) the requirement of general sanitary management and GMP in facilities all along the supply chain (Article 52 newly added); (c) information sharing requirements between business operators (including raw materials manufacturers and importers) in the supply chain (e.g., DoC) (Article 53 newly added); (d) implementation of a notification system of domestic business operators (Article 57 newly added); etc (MHLW, 2019a). On April 24, 2019, Japan issued an overview (in English) of the amendments to the Food Sanitation Act of 2018 (MHLW, 2019a); and the draft of the official positive list of synthetic resins (i.e., plastics) that comprises several parts: "Table 1.1—Base polymers (plastics)," "Table 1.2—Coating resins," "Table 1.3—Minor monomers that can be used in the manufacture of base polymers," and "Table 2—Additives, etc." (in Japanese with the essential explanations in English) (MHLW, 2019b). The positive list draft is based on the voluntary lists of the Industrial Hygienic Associations that deal with plastics, and comments on it were received by the MHLW (MHLW, 2019a; Overbeek, 2019; Thompson, 2019a, 2021a). On April 28, 2020, the MHLW issued Notification 196, that established June 1, 2020, as the positive list effective date (with a period of grace of five years for certain substances), and that formal binding lists are expected as of May 31, 2025 (Thompson, 2021a).

Meanwhile, some of the issues on which further research and discussions between government and industry going on are the following: the expansion of the official positive list system to other FCMs (paper, rubber, metal, and glass); the review of official standards, specifications, and test methods (including migration simulants and tests conditions); the risk assessment procedures, including the calculation of exposure of consumers to FCS using consumption factors, calculations of specific migrations assuming a 100% transference of migrants to the foodstuff or simulant (but not predictions using mathematical models by the moment), and toxicological tests, similar, but not equal, to those adopted by the US FDA; synthetic resins used on paper; adhesives; printing inks; AIMs; etc (MHLW, 2019a; Thompson, 2019a, 2021a).

15.7 FCMs legislation in China

The initial steps of the People's Republic of China (PRC) legislation on FCMs have been summarized by Li and Bian (2010). Due to the fast market developments in China, it has been changing its legislation regularly in the last years. This process and the most recent updates have been described elsewhere (Gu and Zhang, 2019; Keithline, 2017; Thompson, 2017, 2019b, 2019c, 2021b; Thompson and Gu, 2019; Zhang, 2018; Zhang, 2019; Zheng, 2017; Zhu, 2019). Official information can be found on the suggested websites, but it is available only in Chinese.

The Framework regulation is the Food Safety Law, which was issued on June 1, 2009, replaced the Food Hygiene Law of 1995, and was lately amended in December 2018 and April 2021. It requires regulation of all "food-related products" (i.e., food packaging materials and containers, detergents, disinfectants, utensils, and equipment used in food manufacture), and is the basis for China's National Food Safety Standards. These standards are known as "GB standards." GB is the acronym of "Guobiao" (in Chinese, meaning "National Standards"). GB standards are mandatory and GB/T standards are voluntary or recommended. The Food Safety Law establishes that food manufacturers must ensure compliance for all raw materials and food packaging and requires certification documents and inspection by sanitary authorities.

FCMs are regulated by two basic mechanisms: the premarket approval of raw materials (e.g., additives, resins), and the compliance of FCMs and final articles with testing standards.

The State Council of the PRC has empowered mainly four agencies to manage FCMs:

- the National Health Commission (NHC)
- the Center for Food Safety Risk Assessment (CFSA)
- the State Administration for Market Regulation (SAMR)
- the General Administration of Customs (GAC)

Both the NHC and the CFSA perform risk assessment and risk management; the SAMR is in charge of enforcement of regulations, market supervision, inspection, samples testing, and product licensing; and the GAC is responsible for supervision and testing of imported goods.

The Ministry of Health sanctioned the "Management Rules for the Administrative Approval of New Varieties of Food-Related Products" (in short, "Management Rules"), which became effective on June 1, 2011. In case that manufacturers are interested in getting cleared a new FCS or FCM not covered by any existing GB standard or NHC Announcements, the "Management Rules" established a Food Contact Petition process, consisting of several stages. The applicant (local manufacturer or importer) submits a petition with supporting documentation, which is reviewed by an NHC Reception Desk. If it is accepted, then it is evaluated by an Expert Panel of the CFSA (in charge of this task since 2016). The Expert Panel is composed by delegates from government and research institutes, and is managed by the CFSA. If the petition passes the technical evaluation, the CFSA publishes a draft approval on its website for a comment period of 1 month. Finally, the NHC issues a final approval, and makes an "Announcement of New Approval," which has the same effect than that of a GB Standard. Announcements are compiled until a new GB Standard covering them is issued. Afterward the Announcements that entered the new GB standard are repealed. A point to take into account is that the reviewers generally may require that the petitioned FCS or FCM be cleared in at least two international jurisdictions (e.g., Canada, EU, Japan, United States) (Gu and Zhang, 2019; Keithline, 2017; Thompson, 2017, 2021b; Thompson and Gu, 2019; Zhang, 2018).

An overview of the GB standards on FCMs is presented in Table 15.5. GB Standards are usually classified into four main types (General application (Horizontal), Commodity, Manufacturing Practice, and Compliance Testing Methods).

The following is a summary of the main features of some of the GB standards in force at present (October 2021).

15.7.1 GB standards of general application (horizontal)

15.7.1.1 GB 4806.1-2016 "general safety requirements for FCMs and articles"

It applies to all FCMs (including inks, adhesives, lubricant oil, etc.) that may come directly or indirectly in contact with food, except detergents, disinfectants, and public water facilities. It establishes the three basic hygienic principles for FCMs; the safety assessment of NIAS; and the use of a functional barrier that ensures migration of substances form a layer beyond it to food up to 0.01 mg kg^{-1}, subjected to the same restrictions as in the EU.

Manufacturers are requested to operate according to GMP (GB 31603-2015), to establish a product traceability system involving suppliers of raw materials and customers, to keep relevant information available, and to issue DoC similar to the required by the EU. It also establishes the mandatory use of the "for food contact use" label or the "spoon and chopsticks" logo (with exceptions) (see Fig. 15.4). FCMs must comply with use restrictions, OMLs, SMLs, SML(T)s, and QMs, established in GB standards. Substances used in FCMs manufacture must be included in the positive lists of GB 9685-2016 and GB standards specific of each material, or must be provisionally cleared by "Announcements of New Approvals" by the NHC. Migration tests must follow GB 31604.1-2015 and GB 5009.156-2016 requirements.

15.7.1.2 GB 9685-2016 "additives used in FCMs and articles"

It establishes the positive lists of additives that are allowed to be used for different materials. NLS can also be used (mixtures of approved additives; sodium, potassium, and calcium salts of listed acids, phenols, or alcohols; etc.). Requirements on pigments and colorants are established, similar to those of CoE Resolution AP (89)1. Appendix A to this GB contains seven positive lists of additives for different materials (with SMLs, SML(T)s, and QMs):

Table A.1 Plastics; Table A.2 Coatings; Table A.3 Rubber; Table A.4 Printing inks; Table A.5 Adhesives; Table A.6 Paper and paperboard; and Table A.7 Silicone rubber and other materials. Appendix C to this GB establishes SMLs for certain elements (Ba, cobalt (Co), Cu, Fe, lithium (Li), manganese (Mn), and zinc (Zn)) (Thompson, 2019b). In June 2017 the CFSA isued the GB 9685-2016 Implementation Guidance, which describes the CFSA interpretations of this standard's scope and requirements (Thompson, 2021b).

TABLE 15.5 Main GB standards for food contact materials in force in China.

Standard type	GB number year	Subject	Main features
General application (horizontal)	GB 4806.1-2016	General safety requirements for FCMs and articles	Applies to all FCMs (with exceptions). Basic requirements to be established by GBs (positive lists with SMLs, SML(T)s, QMs, etc., OMLs). NIAS assessment. Functional barrier. Traceability, DoC, documentation. GMP. Mandatory label or logo (see Fig. 15.4). Migration tests to be established by GBs.
	GB 9685-2016	Additives used in FCMs and articles	Positive lists of additives. SMLs, SML(T)s, QMs, other restrictions. Pigments and colorants (similar to CoE Res.AP (89)1). Annex A: Table A.1 Plastics Table A.2 Coatings Table A.3 Rubber Table A.4 Printing inks Table A.5 Adhesives Table A.6 Paper and paperboard Table A.7 Silicone rubber and other materials Annex C: Elements SMLs.
Commodity	GB 4806.2-2015	Rubber nipple	Applies to natural rubber and silicone rubber nipples. Does not apply to pacifiers.
	GB 4806.3-2016	Enamelware	Applies to enamelware (flat ware, hollow ware, hold-up vessel).
	GB 4806.4-2016	Ceramic ware	
	GB 4806.5-2016	Glass materials and articles	
	GB 4806.6-2016	Plastic resins	Applies to plastics, their blends, and nonvulcanized TPEs. Positive list of monomers (SMLs, SML(T)s, QMs). Additives must comply with GB 9685-2016 Table A.1. Sensory requirements.
	GB 4806.7-2016	Plastic materials and articles	Applies to articles manufactured with plastics, their blends and non-vulcanized TPEs. Monomers and resins must comply with GB 4806.6-2016. Additives must comply with GB 9685-2016 Table A.1. Sensory requirements. OMLs, $KMnO_4$ consumption limit, heavy metals limit, negative decoloration test. Mandatory label or logo GB 4806.1-2016.
	GB 4806.8-2016	Paper and paperboard materials and articles	Applies to paper, paperboard, and paper articles including pulp molded articles. Additives must comply with GB 9685-2016 Table A.6. Sensory requirements. OML; limits for As, Pb, formaldehyde, and fluorescent substances. $KMnO_4$ consumption limit, heavy metals extraction limit. Microbial specifications. Mandatory label or logo GB 4806.1-2016.

Continued

TABLE 15.5 Main GB standards for food contact materials in force in China.—cont'd

Standard type	GB number year	Subject	Main features
	GB 4806.9-2016	Metal materials and articles	Applies to metals, alloys and metal plating. Restrictions of use of certain materials in direct contact with acidic foods. Specific simulants and migration conditions (time and temperature). SMLs of elements for stainless steel, and for other metals and alloys. Mandatory label or logo GB 4806.1-2016; and specific rules for labeling of standardized types of stainless steel and aluminum alloys, and for plating and organic coatings.
	GB 4806.10-2016	Coatings	Applies to coatings on surfaces in direct contact with foods. Does not apply to paper coatings. Additives must comply with GB 9685-2016 Table A.2. Sensory requirements. OMLs, $KMnO_4$ consumption limit, heavy metals limit, Mandatory label or logo GB 4806.1-2016.
	GB 4806.11-2016	Rubber materials and articles	Applies to natural rubber, synthetic rubber (including vulcanized TPEs), and silicone rubber. Additives must comply with GB 9685-2016 Tables A.3. and A.7. Sensory requirements. OML, $KMnO_4$ consumption limit, heavy metals extraction limit. Mandatory label or logo GB 4806.1-2016.
	GB/T 33320-2016	Adhesives	Applies to production, specifications, management and testing of adhesives, etc.
	GB 14934-2016	Disinfected dinner and drinking sets	Applies to catering service providers; cleaning and disinfection service agencies for dinner and drinking sets; other disinfected food containers, food production tools and equipment; etc.
Manufacturing practice	GB 31603-2015	General health code for production of FCMs and articles	GMP for FCMs.
Compliance testing methods	GB 31604.1-2015	General rules for migration test of FCMs and articles	Horizontal standard, applies to all FMCs, except special provisions in materials standards. Food simulants. Contact conditions (time and temperature). Use of FRF.
	GB 5009.156-2016	General principle of migration test pretreatment method of FCMs and articles	Horizontal standard, applies to all FMCs. Methods, sampling, equipment, calculations, results report.
	GB 31604.2-2016 to GB 31604.9-2016	Physicochemical tests.	
	GB 31604.10-2016 to GB 31604.49-2016	Residue and migration (specific and overall) tests.	

CoE, Council of Europe; *DoC*, Declaration of Compliance; *FCM*, food contact material; *FRF*, fat reduction factor; *GB*, mandatory National Standard; *GB/T*, voluntary or recommended National Standard; *GMP*, good manufacturing practice; *$KMnO_4$*, potassium permanganate; *NIAS*, nonintentionally added substances; *OML*, overall migration limit; *QM*, quantity in material; *SML(T)*, group migration limit; *SML*, specific migration limit; *TPE*, thermoplastic elastomer.

FIGURE 15.4 Mandatory symbol for labeling of food contact materials in China (according to GB 4806.1-2016). *Source: www.ChineseStandard.net, with permission.*

15.7.2 Commodity GB standards

At present (October 2021), commodity GB standards drafts are being developed for several materials, for instance: adhesives, printing inks, composite materials (i.e., multilayer structures), regenerated cellulose (cellophane), bamboo and wood, starch-based materials and articles, etc., (Thompson, 2019c, 2021b; Zhu, 2019). Several commodity GB standards are under revision (plastics, paper and paperboard, metals, coatings and rubbers); and it has been announced that GB standards are going to be developed for certain FCMs (lubricating oils, wax, textiles and AIMs) (Thompson, 2021b).

The following is a summary of the main features of some commodity GB standards in force:

15.7.2.1 GB 4806.6-2016 "plastics resins"

It applies to plastics, their blends, and nonvulcanized TPEs. It establishes a positive list of resins from previous GB standards and clearances, with SMLs, SML(T)s, and QMs for monomers. Additives must comply with GB 9685-2016 Table A.1 and relevant announcements. Sensory requirements are established (Thompson, 2017; O'Keeffe and Gu, 2019).

15.7.2.2 GB 4806.7-2016 "plastic materials and articles"

It applies to FCMs and articles manufactured with plastics, their blends, and nonvulcanized TPEs. Monomers and resins must comply with GB 4806.6-2016, and additives with GB 9685-2016 Table A.1, and relevant announcements. It establishes sensory requirements; OMLs (10 mg dm^{-2}, and 60 mg kg^{-1} in the case of FCMs for infant foods) (test method GB 31604.8); a limit of 10 mg kg^{-1} for the consumption of $KMnO_4$ in an extract in water (60°C, 2 h) (test method GB 31604.2); a limit of 1 mg kg^{-1} for heavy metals (expressed as Pb) extracted by an aqueous solution of acetic acid 4% (v/v) (60°C, 2 h) (test method GB 31604.9); and a negative result for the decoloration test (test method GB 31604.7). The label or logo established by GB 4806.1-2016 must be used (Thompson, 2017; Zheng, 2017).

15.7.2.3 GB 4806.8-2016 "paper and paperboard materials and articles"

It covers paper, paperboard, and paper articles including pulp molded products. Additives must comply with GB 9685-2016 Table A.6, and relevant announcements. It establishes limits for As (1 mg kg^{-1}), Pb (3 mg kg^{-1}), formaldehyde (1 mg kg^{-1}), and fluorescent substances (negative residue); an OML (10 mg dm^{-2}); a limit of 40 mg kg^{-1} for the consumption of $KMnO_4$ in an extract in water (60°C, 2 h); a limit of 1 mg kg^{-1} for heavy metals (expressed as Pb) extracted by an aqueous solution of acetic acid 4% (v/v) (60°C, 2 h) (only for papers in contact with aqueous food or food with free water on its surface); sensory requirements; and microbial specifications (coliforms and *Salmonella*: counts of colony-forming units (CFU) 50 cm^{-2}, not detectable; molds: no more than 50 CFU g^{-1}). The label or logo established by GB 4806.1-2016 must be used (O'Keeffe and Gu, 2019; Thompson, 2017).

15.7.2.4 GB 4806.9-2016 "metals and alloys materials and articles"

It applies to metals and alloys materials and articles and metal platings (i.e., metal layers formed on the surface of various materials and articles using the plating technology). Articles of aluminum and aluminum alloys, copper and copper alloys without organic coatings on the food contact surface, and metal plating are not allowed in direct contact with acidic food. Migration tests are generally in line with GB 31604.1-2015 and GB 5009.156-2016, but there are special rules for the selection of simulants and test conditions (time and temperature). SMLs of elements for stainless steel (As, Cd, Pb, Cr, and nickel (Ni)) and for other metals and alloys (As, Cd, and Pb) are established. The label or logo established by GB 4806.1-2016 must be used; the base materials types (e.g., "stainless steel," "aluminum alloy") must be identified with their international standardized denomination; and if used in the food contact surface, plating layers and polymeric coatings must be identified in the labeling (O'Keeffe and Gu, 2019).

15.7.2.5 GB 4806.10-2016 "coatings"

It applies to coatings in direct contact with foods, but not to paper coatings. It establishes a positive list of coatings (e.g., epoxy coatings, phenolic coatings) from previous GB standards and clearances. Additives must comply with GB 9685-2016 Table A.2, and relevant announcements. It establishes sensory requirements; an OML (10 mg dm^{-2}); a limit of 10 mg kg^{-1} for the consumption of $KMnO_4$ in an extract in water (60°C, 30 min); and a limit of 1 mg kg^{-1} for heavy metals (expressed as Pb) extracted by an aqueous solution of acetic acid 4% (v/v) (60°C, 30 min) The label or logo established by GB 4806.1-2016 must be used (O'Keeffe and Gu, 2019).

15.7.2.6 GB 4806.11-2016 "rubber materials and articles"

It covers natural rubber, synthetic rubber (including vulcanized thermoplastic elastomers), and silicone rubber and establishes a positive list of polymers. Additives must comply with GB 9685-2016 Table A.3 and Table A.7, and relevant announcements. It establishes sensory requirements; an OML (10 mg dm^{-2}); a limit of 10 mg kg^{-1} for the consumption of $KMnO_4$ in an extract in water (60°C, 30 min); and a limit of 1 mg kg^{-1} for heavy metals (expressed as Pb) extracted by an aqueous solution of acetic acid 4% (v/v) (60°C, 30 min). The label or logo established by GB 4806.1-2016 must be used, and the label must identify natural latex if it is contained in the product (O'Keeffe and Gu, 2019; Thompson, 2017).

15.7.3 GB 31603-2015 "general health code for production of FCMs and products"

This standard on GMP establishes the hygiene requirements and management rules on facilities and personnel in the manufacturing of FCMs and articles, including raw materials purchase, processing, packaging, storage, and transportation. The requisites are similar to those established by Regulation (EC) 2023/2006 and the US-FDA (21CFR 174.5), but they are developed in more detail in this GB (Zheng, 2017).

15.7.4 Compliance testing methods

15.7.4.1 GB 31604.1-2015 "general rules for migration test of FCMs and products"

This horizontal standard applies to all FCMs, except in special provisions established in the materials GB standards. It establishes food simulants and contact conditions for the overall and specific migration tests (time and temperature). In general, the simulants and the test conditions are similar to those established by Regulation (EU) 10/2011, except that the acidic simulant (for foods with $pH \leq 5$) is acetic acid 4% (v/v) in aqueous solution, and that there is no simulant for dry foods. This GB introduces the use of the FRF as a correction factor of the specific migration results. Calculations assuming a 100% transference of migrants to the foodstuff or simulant to verify compliance with their SMLs are allowed, though predictions of their specific migrations using validated mathematical models (as in the case of Regulation (EU) 10/2011) are not mentioned in this GB (Keithline, 2017; Semail and Coneski, 2019; Zheng, 2017).

15.7.4.2 GB 5009.156-2016 "general principle of migration test pretreatment method of FCMs and their products"

This is a horizontal standard that applies to all FCMs. It describes requirements related to migration tests, such as sampling, samples handling, preparation of food simulants, equipment, area of contact, methods, calculations, results report, etc (Keithline, 2017; Zheng, 2017).

15.8 Comparison of FCMs legislations

Due to the diversity of regulatory approaches in different jurisdictions (of which only a few examples were addressed in this chapter), a general scenario will be described in this section.

The Codex Alimentarius, an internationally recognized set of standards, codes of practice, guidelines, etc., on food-related issues, does not establish requisites on FCMs, except in the case of general recommendations (in the voluntary field) for Sn in canned foodstuffs and vinyl chloride monomer in packaging materials.

The CoE has issued voluntary recommendations mainly on subjects not covered yet by the EU harmonized legislation, considered of great technical value, that follow the EU approach to safety assessment, and that have been taken as references by several EU Member States (Bolle, 2016; Rossi; 2010; Schäfer, 2007, 2010), and by the MERCOSUR when developing their regulations (Ariosti, 2016b, 2018a, b, 2019a, b, 2021a, b).

The MERCOSUR FCMs harmonized legislation is based mainly both in the EU, EU Member States (e.g., the Netherlands, Spain) and the US-FDA legislation, and in the CoE and BfR recommendations. It is based on the system of positive lists, and takes topics from the EU regulations (list of monomers and additives, OMLs, SMLs, SML(T)s, QMs, functional barrier (in multilayer PET bottles), etc.) and the US-FDA regulations (lists of polymers and additives (from the CFR and US-FDA FCNs), functional barrier (in multilayer PET bottles), TOR, etc.) (Ariosti, 2016b, 2018a, b, 2019a, b, 2021a, b).

The Japanese regulatory outstanding point is that, in order to assure public health, FCMs (that include by definition packages, containers, utensils, and tableware) are covered not only by legislation in the mandatory field, but also by voluntary standards issued by several private Industrial Hygienic Associations. The Japanese legislation traditionally established horizontal requisites on FCMs, by the way of negative lists, valid for different kinds of foodstuffs, and some vertical requisites on FCMs for special types of foodstuffs, such as milk and milk products, and certain types of packages and equipment. On the other hand, Industrial Hygienic Associations voluntary standards are based on positive and negative lists. There is a government program to shift the focus of legislation toward the system of positive lists, with the help of the ones already issued by the Industrial Hygienic Associations. So Japan has adopted officially in 2020 the positive list system for plastics as the ones in force in other jurisdictions (e.g., China, EU, MERCOSUR, United States). The Japanese legislation is recognized as a reference by the Member States of the Association of Southeast Asian Nations (ASEAN) and South Korea (Kenny, 2018).

The FCMs regulatory panorama in China is changing very fast, trying to take pace with the quick development of its markets. Due to the wide scope and complexity of China's FCMs regulations and GB standards, along with their enforcement (inspection, supervision, certification, etc.), it is considered one of the most detailed legislations of the world that requires a very thorough compliance testing (Zheng, 2017). This legislation is based on the principle of the positive lists, and in several fields is similar to the EU harmonized regulations (OMLs, SMLs, SML(T)s, QMs, NIAS, functional barrier, etc.). It also takes as reference the US-FDA legislation (Kenny, 2018), and incorporates topics from the Japanese regulation (e.g., the $KMnO_4$ consumption test). For the petition of approvals of new substances, clearances in other jurisdictions are generally required (e.g., Canada, EU, Japan, United States) (Keithline, 2017; Thompson, 2017, 2021b; Thompson and Gu, 2019).

The focus of the comparisons in the available publications is centered, no doubt due to their international standing, in the differences between the EU and the US-FDA FCMs legislations (de la Cruz García et al., 2014; Eisert, 2008; Heckman, 2005; Rossi, 2014; Schäfer, 2007, 2010; Twaroski et al., 2007). A brief summary related to both legislations follows.

According to the US-FDA, polymers and additives as well as other FCSs are considered as indirect food additives when present in the human diet due to migration, in quantities greater than the TOR. Excluded from the indirect food additives definition are several types of FCS (e.g., colorants, GRAS substances, prior sanctioned substances). In addition, houseware and utensils, as well as drinking water supply equipment, are exempt from the FCMs legislation. In contrast, the EU legislation does not define FCMs as food additives, and it applies to all the FCMs (including houseware and utensils), except those used for the transport and storage of drinking water. Different types of FCMs have been regulated by both legislations.

As discussed before, there are two main ways established by the US-FDA by means of which FCSs can be regulated: the FAP system (positive list of substances of general use) and the FCN one (positive list of substances of proprietary use). In the EU legislation, the substances enumerated in the positive lists are of general use: that means that manufacturers may buy the cleared FCSs or FCMs from different producers, as long as they comply with the established restrictions. However, in the EU, food contact PCR plastics and AIMs are subjected to a case by case (proprietary) clearance.

The conceptual basis of the FCSs clearance for their inclusion in the positive lists differs in both legislations. The US-FDA approach is an exposure-based risk assessment of the FCSs, which takes into account two statistical parameters:

- the "consumption factors" (CFs), which are the fractions of the daily diet (assumed as 3 kg per day per person) of a 60 kg body weight consumer, expected to contact each specific FCM; and
- the "food-type distribution factors" (f_Ts), defined as the fractions, for each FCM, of different types of foodstuff (aqueous, acidic, alcoholic, and fatty) packaged in the FCM.

These factors (CFs and f_Ts) and the migration tests data for FCSs in different simulants are used to calculate their "dietary concentrations" (DCs) and theirs EDIs. If different uses of the FCS are already regulated for similar or different technological purposes, a CEDI must be calculated. The EDI or CEDI values, and a comparison of them with the FCS acceptable daily intake (ADI), determine the kind of toxicological data needed to be submitted to the US-FDA. The intended use for each FCM, either for single or repeated use, the type(s) of foodstuff(s), and the contact time(s) and temperature(s) conditions, an estimated ratio of mass of foodstuff/contact surface area of FCM (M/S) and an estimated FCM thickness, must also be declared. The cleared FCSs are then included in the positive lists of the CFR or the FCNs list, along with the main physico-chemical parameters that allow their easy characterization by their purchaser (e.g., FCM manufacturer, food manufacturer). If the purchaser checks that the FCS complies with these specifications, the FCS is considered to pose no health risk if used in the conditions established in the positive lists, and that correspond to the assumptions made in the exposure evaluations. No other uses are allowed for the FCS other than the stated in the positive lists.

The EU approach, instead, is a migration-based risk assessment, that is, migration of an FCS determines the type of toxicological data that must be submitted to the EFSA. These studies involve, when feasible, the determination of the no-observed effect-level (NOEL) and the tolerable daily intake (TDI) of the FCS under evaluation (monomer, additive, etc.). A conservative assumption is made, that is, that a 60 kg body weight person consumes daily 1 kg of foodstuff packaged in an FCM containing this FCS. For certain substances, the FCS TDI and the mentioned assumption are used to calculate an SML. No exposure considerations involving the statistical parameters CFs and f_Ts and EDIs or CEDIs are made. Generally, the use of the cleared FCS is allowed in different types of FCMs and in different situations of use (e.g., type of plastic, type of foodstuff, contact time(s), and temperature(s) conditions), providing that the specific migration of the FCS complies with its SML, or other restrictions, in the case they exist. This last point can pose a considerable analytical effort for the FCMs manufacturer. Some kinds of exposure considerations have been introduced in Regulation (EU) 10/2011 relating to the specific migration determination of certain lipophilic substances, by means of the use of correction factors: the FRF, the DRF, and, when it corresponds, the product of both of them.

These two divergent approaches, naturally, may pose different types of problems and uncertainties to the FCMs and food manufacturers. In the United States, the procedure adopted minimizes the initial analytical effort for the FCS manufacturer for a foreseen use, but may imply a reduction of flexibility in the case of possible novel uses (e.g., type of food, contact time(s) and temperature(s) conditions, FCM thickness). In the EU, the procedure adopted maximizes the consumer protection, assuming a greater exposure to the FCSs, but may imply a greater analytical effort for the manufacturer to verify SMLs. On the other hand, through the conceptual dissociation of the FCS and the FCM in which it can be used (neither CFs nor f_Ts are applied), there are generally more options of final uses, including the novel ones. In the United States, the petitioner first makes an exposure evaluation (EDIs or CEDIs), and according to its result proceeds, if necessary, to perform toxicological studies. In the EU, the petitioner is expected to provide toxicological data according to the level of migration of the FCS, which are the basis of SMLs calculations, without an exposure evaluation (Baughan and Attwood, 2010; Eisert, 2008; Heckman, 2005; Schäfer, 2007, 2010).

Finally, it must be noted that the US-FDA has a double role, because it performs risk assessments by evaluating, for instance, FAPs or FCNs, and risk managements by issuing, respectively, regulations (included in the CFR) or an accepted FCN list. In the EU, the risk assessment is the duty of the EFSA, and the risk management is the task of the EC, the EU Council of Ministers, and the EU Parliament.

A comparative summary of the regulatory situation in the jurisdictions discussed in this chapter is presented in Table 15.6.

15.9 Conclusions—harmonization, mutual recognition, and new legislations

As has been seen, for instance, in the EU, a harmonization process in the block precedes the sanction of the Union FCMs legislation, and in the nonharmonized fields, there is also a system of mutual recognition of the products cleared in the origin Member States (Eisert, 2008; Heckman, 2005; O'Keeffe, 2018; Schäfer, 2007).

TABLE 15.6 Comparison of food contact materials legislations.

Subject	European Union	US-FDA	MERCOSUR	Japan	China
Level	Supranational (27 Member States).	National.	Supranational (4 full Member States; 1 in integration process; 1 suspended; 6 Associated States).	National.	National.
Legal status of FCMs	Regulated. Drinking water supply equipment is excluded from the regulation. Houseware and utensils are not excluded.	Regulated. FCMs are considered as indirect food additives. Drinking water supply equipment is excluded from the regulation. Houseware and utensils are excluded.	Regulated. Drinking water supply equipment is excluded from the regulation. Houseware and utensils are not excluded.	Regulated. Covered also by voluntary standards issued by Hygienic Industrial Associations. Drinking water supply equipment is regulated by MHLW Ordinance 123 (2012), and Japan industrial Standard JIS S 3200-7:2004. Houseware and utensils are not excluded.	Regulated. Drinking water supply equipment is excluded from the regulation. Houseware and utensils are not excluded.
Type of legislation related to FCMs hygienic requirements	Directives (must be transposed into the national legislations). Regulations (direct application without transposition).	Federal law	Resolutions (must be transposed into the national legislations).	National laws. Ministerial Ordinances. Official specifications and standards.	National law. Mandatory GB standards.
General FCMs (framework) legislation/regulation	Regulation (EC) 1935/2004	FFDCA 21 CFR 174.5	Resolution GMC 3/92	Food Sanitation Act. Food Safety Basic Act.	Food Safety Law. GB 4806.1-2016.
Specific regulation on GMP	Regulation (EC) 2023/2006	21 CFR 174.5	General requirement in Resolution GMC 3/92 and resolutions on specific FCMs.	Food Sanitation Act (art. 52).	GB 31603-2015

Continued

TABLE 15.6 Comparison of food contact materials legislations.—cont'd

Subject	European Union	US-FDA	MERCOSUR	Japan	China
Specific regulated FCMs	Plastics Ceramics Regenerated cellulose (films) PCR plastics Active and intelligent materials	Plastics Coatings Paper and board Rubbers Adhesives PCR plastics Active and intelligent materials	Plastics Coatings Paper and board Rubbers Metals Glass Ceramics Regenerated cellulose (films and casings) Adhesives PCR PET	Plastics Rubbers Metal cans Glass Ceramics Guideline for use of recycled plastics in food contact articles (2012). (Voluntary standards issued by industrial hygienic associations on plastics, paper, rubber, adhesives, printing inks, etc.).	Plastics Coatings Paper and board Rubbers and rubber nipples Metals Glass Ceramics Adhesives Enamelware
Cleared FCM logo or label	- Mandatory and standardized logo or obligatory label (with exceptions). - Voluntary label on FCMs manufactured with PCR plastics. - Mandatory label and standardized logo (this last where possible) for active and intelligent materials. DO NOT EAT	–	– Mandatory label on refillable and recycled PET packages (i.e., manufactured with PCR-PET), indicating these conditions.	Mandatory and standardized label.	Mandatory and standardized logo or obligatory label (with exceptions).

Legal obligation for the FCMs manufacturer	Compliance with legislation. Case by case mandatory approval system for PCR-plastics and AIMs.	Compliance with legislation.	Compliance with legislation. Case by case mandatory approval system for final FCMs that comply with legislation (partially applied in Brazil).	Compliance with legislation. Compliance with voluntary standards issued by industrial hygienic associations. Case by case mandatory control system for FCMs that comply with legislation.	Compliance with legislation and mandatory GB standards. Case by case mandatory approval/licensing system for FCMs that comply with legislation.
DoC	Required (Regulation (EC) 1935/2004)	Not required	Not required	Information sharing required in the supply chain (Food Sanitation Act—Art. 53).	Required (GB 4806.1-2016)
Positive lists	General use positive lists, nonproprietary.	General use positive lists (CFR), nonproprietary (under the FAP system). Case by case positive list (FCNs list), proprietary (under the FCN system).	General use positive lists, nonproprietary.	Official negative list, replaced in 2020 by positive list system in the case of plastics. (Positive and negative lists in voluntary standards issued by industrial hygienic associations).	General use positive lists, nonproprietary.
Plastics positive list	Monomers and other starting substances Additives SML (based on toxicological data (NOEL, TDI)) SML(T) QM Purity criteria and specifications of use.	Polymers Additives Purity criteria and specifications of use (based on exposure assessment, CF, f_T, EDI, CEDI, ADI).	Monomers and other starting substances. Polymers. Additives SML (=LME) SML(T) (=LME(T)) QM (=LC) Purity criteria and specifications of use transposed from EU and US-FDA FCMs legislations.	Official negative lists: Restrictions on certain monomers (SMLs), heavy metals, etc.; replaced in 2020 by positive list system in the case of plastics. Purity criteria and specifications of use. (Positive lists of polymers and additives in voluntary standards issued by industrial hygienic associations).	Polymers with restrictions for certain monomers. Additives SML SML(T) QM Purity criteria and specifications of use.
Plastics OMLs	10 mg dm^{-2} 60 mg kg^{-1} (plastics for food for infants and young children)	7.75 mg dm^{-2} (= 0.5 mg in $^{-2}$) 50 mg kg^{-1}	10 mg dm^{-2} 60 mg kg^{-1} (plastics for food for infants and young children)	30 mg kg^{-1} (general); different values for different plastics when using n-heptane as fatty food simulant.	10 mg dm^{-2} 60 mg kg^{-1} (plastics for food for infants and young children)
TOR	Not established	0.5 μg kg^{-1} (dietary base)	0.5 μg kg^{-1} (dietary base), only in the case of PCR-PET.	Not established	Not established

Continued

TABLE 15.6 Comparison of food contact materials legislations.—cont'd

Subject	European Union	US-FDA	MERCOSUR	Japan	China
Functional barrier concept	Adopted	Adopted	Adopted (only for PCR-PET multilayer bottles, and some substances listed in the positive list of additives for plastics).	Adopted	Adopted
Nontoxicological concern migration limit (s)	10 $\mu g\ kg^{-1}$	Derived from the TOR, different values for different plastics.	Derived from the TOR, and accepted only for PCR-PET (= 10 $\mu g\ kg^{-1}$).	10 $\mu g\ kg^{-1}$	10 $\mu g\ kg^{-1}$
Specific migration of substances determination by: - Laboratory tests - 100% migration calculation - Mathematical modeling	Yes Yes Yes	Yes Yes Yes	Yes Yes Yes	Yes Yes No	Yes Yes No
Risk assessment authority	EFSA	US-FDA	Food Commission-SGT 3.	Government (FSC). Industrial Hygienic Associations.	NHC CFSA
Risk management authority	EC EU Council of Ministers EU Parliament	US-FDA	GMC	Government (MHLW). Industrial Hygienic Associations.	NHC CFSA
Reference legislations and recommendations	EU level: EU MS level: other national MS legislations; recommendations (e.g., CoE, BfR), for developing national legislation on nonharmonized FCMs.	–	EU, EU MS, US-FDA legislations; CoE and BfR recommendations.	EU, US-FDA.	EU, Japan, US-FDA.

ADI, acceptable daily intake; *AIMs*, active and intelligent materials; *BfR*, German Federal Institute for Risk Assessment (Bundesinstitut für Risikobewertung, in German); *CEDI*, cumulative estimated daily intake; *CF*, consumption factor; *CFR*, Code of Federal Regulations (United States); *CFSA*, National Center for Food Safety Risk Assessment (China); *CoE*, Council of Europe; *DoC*, declaration of compliance; *EC*, European Commission (EU); *EDI*, estimated daily intake; *EFSA*, European Food Safety Authority (EU); *EU*, European Union; *FAP*, Food Additive Petition (US-FDA); *FCM*, food contact material; *FCN*, Food Contact Notification (US-FDA); *FDA*, Food and Drug Administration (United States); *FFDCA*, Federal Food, Drug, and Cosmetic Act (United States); *FSC*, Food Safety Commission (Japan); f_T, food-type distribution factor; *GB*, National Standard (Guobiao, in Chinese); *GMC*, Common Market Group (Grupo Mercado Común, in Spanish; *GMP*, good manufacturing practice; Grupo Mercado Comum, in Portuguese); *MHLW*, Ministry of Health, Labour and Welfare (Japan); *MS*, Member State; *NHC*, National Health Commission (China); *NOEL*, no-observed effect-level; *OML*, overall migration limit; *PCR*, postconsumer recycled; *PET*, polyethylene terephthalate; *QM*, quantity in material (LC, acronym in Spanish and Portuguese); *SGT 3*, Working Sub-Group 3 (acronym in Spanish and Portuguese); *SML(T)*, group migration limit (LME(T), acronym in Spanish and Portuguese); *SML*, specific migration limit (LME, acronym in Spanish and Portuguese); *TDI*, tolerable daily intake; *TOR*, threshold of regulation.

The differences between the US-FDA and EU legislations on FCMs are so significant that they pose real hurdles to the harmonization process of these two major regulatory bodies, and to global trade. A harmonization process seems rather difficult to achieve quickly, due to the differences commented in the previous section. An alternative step should be a mutual recognition system. But also in this subject, possibilities of noncompliance of products with legislations with a different basis may pose severe concerns to the sanitary authorities of the EU and the United States (Eisert, 2008; Heckman, 2005).

Therefore, no clear scenario can be envisaged for the near future, as has been discussed extensively at international conferences and symposia on FCMs held during the last 10 years mainly in the EU and the United States. In the meantime, a mutual recognition system has raised an increasing interest between the actors involved all over the world. Also a better sharing of information between them has been improving. For the emerging FCMs legislations of countries in Africa, Asia, and Latin America and the Caribbean, it is important to recognize the global benefits of harmonization, to analyze the two main international legislative bodies with their strengths and drawbacks, in order to be able to decide if they can be adopted as references, with adaptations if necessary, to the local realities. In order to facilitate a future harmonization process or a mutual recognition system, the less divergent legislations there are, the easier the implementation of these strategies will be.

This is being applied in Western Europe, in countries that do not belong to the EU, such as Iceland, Liechtenstein, Norway, and Switzerland that form the European Free Trade Association (EFTA) since 1960. In 2004, the EU and the EFTA countries, except Switzerland, created the European Economic Area (EEA), and thus these three EFTA Member States can participate in the preparation of EU regulations, and transpose them into their own legislations. Switzerland has its own FCMs regulations, increasingly aligned with the EU legislation since the 2017 amendment, and can sign agreements with the EU (Ariosti, 2015; O'Keeffe, 2017; Rossi, 2007).

In summary, in the last 2 decades, for the development or update of their FCMs regulations, different countries and blocks (e.g., ASEAN, Australia–New Zealand, Canada, China, Colombia, the Gulf Cooperation Council (GCC) in the Middle East, India, Japan, MERCOSUR, South Korea) have been taking as references, or manifesting interest in, the regulations in force in the EU and the United States, or both. These are based on two different principles of performing risk assessment. Harmonization then seems very difficult at this moment, and being a complex process, could be only a long-term approach. Mutual recognition at a global level, based on the acceptance of equivalent—rather than similar—food safety assurance and consumer protection systems in different countries and blocks, does not seem to be feasible in the short term, due mainly to the fact of colliding requirements of both major reference legislations. The work on FCMs harmonization at the Codex Alimentarius Commission level is another possibility that could be explored also in the long term.

List of acronyms

4-MBP 4-methylbenzophenone
ADI acceptable daily intake
AFC Panel Panel on Food Additives, Flavourings, Processing Aids and Food Contact Materials (EFSA)
AIMs active and intelligent materials
ANS Panel Panel on Food Additives and Nutrient Sources Added to Food (EFSA)
ANVISA National Agency of Sanitary Surveillance (Agência Nacional de Vigilância Sanitária, in Portuguese) (Brasilia, Brazil)
AP aids to polymerization
AP Partial Agreement (Accord Partiel, in French) (CoE)
ASEAN Association of Southeast Asian Nations
BADGE bisphenol A diglycidyl ether
BFDGE bisphenol F diglycidyl ether
BfR German Federal Institute for Risk Assessment (Bundesinstitut für Risikobewertung, in German) (Berlin)
BP benzophenone
BPA bisphenol A
BPS bisphenol S
CD-P-MCA European Committee for Food Contact Materials and Articles (CoE-EDQM)
CD-P-SC Committee for the Protection of Consumers' Health (CoE-EDQM)
CEDI cumulative estimated daily intake (US-FDA)
CEF Panel Panel on Food Contact Materials, Enzymes, Flavourings and Processing Aids (EFSA).
CEN European Committee for Standardization (Comité Européen de Normalisation, in French) (Brussels, Belgium)
CEP Panel Panel on Food Contact Materials, Enzymes and Processing Aids (EFSA)

CEPE European Council of Paints, Printing Inks and Artist's Colors Industries (Conseil Européen des Producteurs de Peintures, d'Encres d'Imprimerie et de Couleurs pour Artistes, in French)
CF consumption factor (US-FDA)
CFR Code of Federal Regulations (United States)
CFSA National Center for Food Safety Risk Assessment (China)
CFSAN Center for Food Safety and Applied Nutrition (US-FDA)
CFU colony-forming units
CITPA International Confederation of Paper and Board Converters in Europe (Confédération Internationale des Transformateurs de Papier et Carton en Europe, in French)
CM Committee of Ministers (CoE)
CMC Common Market Council (MERCOSUR)
CMR carcinogenic, mutagenic, or toxic to reproduction
CoE Council of Europe
DC dietary concentration (US-FDA)
DFCN Division of Food Contact Notification (US-FDA)
DoC Declaration of Compliance
DRF simulant D2 reduction factor
EC European Commission (EU)
ECCS electrolytically chromium-coated steel
EDI estimated daily intake (US-FDA)
EDQM European Directorate for the Quality of Medicines and HealthCare (Strasbourg, France)
EEA European Economic Area
EFSA European Food Safety Authority (EU) (Parma, Italy)
EFTA European Free Trade Association
EU European Union
EU FP7 EU 7th. Framework Program for Research and Technological Development
EuPIA European Printing Ink Association (Brussels, Belgium)
EURL-FCM European Reference Laboratory for Food Contact Materials (EU-JRC) (Ispra, Italy)
FACET Flavors, Additives and Food Contact Materials Exposure Task (EU)
FAF Panel Panel on Food Additives and Flavourings (EFSA)
FAP Food Additive Petition (US-FDA)
FCA-CEFIC Food Contact Additives Sector Group—The European Chemical Industry Council) (Brussels, Belgium)
FCM food contact material
FCN Food Contact Notification (US-FDA)
FCNM food contact nanomaterial
FCP Food Contact Petition (China)
FCS food contact substance
FDA Food and Drug Administration (Title 21 of CFR) (United States)
FDAMA Food and Drug Administration Modernization Act (United States)
FFDCA Federal Food, Drug, and Cosmetic Act (United States)
FPE Flexible Packaging Europe
FRF fat reduction factor
FSANZ Food Standards Australia New Zealand
FSCJ Food Safety Commission of Japan
FSMA Food Safety Modernization Act (United States)
FSVP Foreign Supplier Verification Program (FSMA—United States)
$\mathbf{f_T}$ food-type distribution factor
GAC General Administration of Customs (China)
GB mandatory National Standard (Guobiao, in Chinese) (China)
GB/T voluntary or recommended National Standard (China)
GC-O-MS gas chromatography—olfactometry—mass spectrometry
GCC Gulf Cooperation Council (Middle East)
GMC Common Market Group (Grupo Mercado Común, in Spanish; Grupo Mercado Comum, in Portuguese) (MERCOSUR)
GMP good manufacturing practice
GRAS generally recognized as safe
HARPC hazard analysis risk-based preventive controls (FSMA—United States)
IAS intentionally added substances
III Italian Packaging Institute (Istituto Italiano Imballaggio, in Italian) (Milan)

ILSI-Europe International Life Sciences Institute (Brussels, Belgium)
IRAM Argentine Standardization and Certification Institute (Instituto Argentino de Normalización y Certificación, in Spanish) (Buenos Aires)
ISO International Organization for Standardization
ITX isopropylthioxanthone
JAIA Japan Adhesive Industry Association
JETRO Japan External Trade Organization
JHAVC Japan Hygienic Association of Vinylidene Chloride
JHOSPA Japan Hygienic Olefin and Styrene Plastics Association
JHPA Japan Hygienic PVC Association
JPA Japan Paper Association
JPIMA Japan Printing Ink Makers Association
JPWIA Japan Petrolatum Wax Industry Association
JRC Joint Research Center (EU)
JRMA Japan Rubber Manufacturers Association
M/S ratio of mass of foodstuff/contact surface area of FCM
MERCOSUL The Common Market of the South (Mercado Comum do Sul, in Portuguese)
MERCOSUR The Common Market of the South (Mercado Común del Sur, in Spanish)
MHLW Ministry of Health, Labour and Welfare (Japan)
MOAH mineral oil aromatic hydrocarbons
MOH mineral oil hydrocarbons
MOSH mineral oil saturated hydrocarbons
MPPO modified polyphenylene oxide (i.e., poly(2,6-diphenyl-p-phenylene oxide))
NEPA National Environmental Policy Act (United States)
NHC National Health Commission (China)
NIAS nonintentionally added substances
NIM noninventoried materials
NLS nonlisted substances
NOEL no-observed effect-level
NOGE novolac glycidyl ethers
NOL No-objection letter (US-FDA)
NORDEN Nordic Council of Ministers (Copenhagen, Denmark)
OFAS Office of Food Additive Safety (US-FDA)
OFI Austrian Research Institute for Chemistry and Technology (Vienna)
OML overall migration limit
ORPI oligomers, reaction products and impurities
P-SC-EMB Committee of Experts on Food and Pharmaceutical Contact Materials (CoE-EDQM)
PA polyamide
PAA primary aromatic amine
PC polycarbonate
PCB polychlorinated biphenyl
PCR postconsumer recycled
PE polyethylene
PET polyethylene terephthalate
PFAS perfluoroalkyl sulfonate
PFCA perfluorocarboxylic acid
PI photo-initiator
PIM Plastics Implementation Measure (Regulation (EU) 10/2011)
PLA polylactic acid
PMMA polymethyl methacrylate
PMP poly(4-methyl-1-pentene)
PP polypropylene
PPA polymer production aid
PRC The People's Republic of China
PS polystyrene
PVA polyvinyl alcohol
PVC polyvinyl chloride
PVDC polyvinylidene chloride
QM quantity in material

QM(T) group concentration limit
QMA quantity in material per contact surface area
QMA(T) group concentration limit per contact surface area
RC regenerated cellulose
RDC Resolution of the Collegiate Directorate (Resolução da Diretoria Colegiada, in Portuguese) (ANVISA, Brazil)
SAMR State Administration for Market Regulation (China)
SGT 3 Working Sub-Group 3 (MERCOSUR)
SML specific migration limit
SML(T) group migration limit
TDI tolerable daily intake
TFS tin free steel
TNO The Netherlands Organization for Applied Scientific Research
TOR threshold of regulation (US-FDA)
TPE thermoplastic elastomer
TTC threshold of toxicological concern
TWI tolerable weekly intake
WHO World Health Organization

Acknowledgment

This chapter is an adapted and updated review of "Kopper, G., Ariosti, A. 2010. Food Packaging Legislation: Sanitary Aspects (Chapter 14). In: Boisrobert, C.E., Stjepanovic, A., Oh, S., Lelieveld H.L.M. (Eds.), Ensuring Global Food Safety—Exploring Global Harmonization, Academic Press/Elsevier, United States, 2010." I wish to thank Gisela Kopper, M.Sc., at present Director of the Program of Accreditations of LSQA before the US-FDA, for her contribution to the original chapter as coauthor.

References

Ahmed, I., Lin, H., Zou, L., et al., 2017. A comprehensive review on the application of active packaging technologies to muscle foods. Food Control 82, 163–178.

Alin, J., Hakkarainen, M., 2011. Microwave heating causes rapid degradation of antioxidants in polypropylene packaging, leading to greatly increased specific migration to food simulants as shown by ESI-MS and GC-MS. J. Agric. Food Chem. 59 (10), 5418–5427.

Alin, J., Hakkarainen, M., 2013. Combined chromatographic and mass spectrometric toolbox for fingerprinting migration from PET tray during microwave heating. J. Agric. Food Chem. 61 (6), 1405–1415.

Ariosti, A., 2002a. Aptitud sanitaria de botellas de PET retornables para bebidas gaseosas. In: Catalá, R., Gavara, R. (Eds.), Migración de componentes y residuos de envases en contacto con alimentos. Instituto de Agroquímica y Tecnología de Alimentos (IATA), Valencia, pp. 233–247.

Ariosti, A., 2002b. Uso de materiales plásticos reciclados en contacto con alimentos. Barreras funcionales. In: Catalá, R., Gavara, R. (Eds.), Migración de componentes y residuos de envases en contacto con alimentos. IATA, Valencia, pp. 261–279.

Ariosti, A., 2015. Global legislation for regenerated cellulose materials in contact with food. In: Baughan, J.S. (Ed.), Global legislation for Food Contact Materials. Woodhead Publishing Ltd., Cambridge, pp. 109–139.

Ariosti, A., 2016a. Managing contamination risks from packaging materials. In: Lelieveld, H., Gabrić, D., Holah, J. (Eds.), Handbook of Hygiene Control in the Food Industry, second ed. Woodhead Publishing, Duxford, pp. 147–177.

Ariosti, A., 2016b. From Mexico to MERCOSUR: an updated panorama of FCMs regulations in Latin America. In: Proceedings of the Food Packaging Law Seminar US 2016, Arlington, VA, United States, October 19–20, 2016. Keller and Heckman LLP, Washington DC, United States, p. 77.

Ariosti, A., 2018a. MERCOSUR legislation on food contact materials. In: Reference Module in Food Science. Elsevier, p. 19. https://doi.org/10.1016/B978-0-08-100596-5.21879-9.

Ariosti, A., 2018b. Food contact materials regulations – practical aspects in MERCOSUR and other Latin American countries. In: Proceedings of the Global Food Contact Conference 2018, Bethesda, MD, United States, May 9–11, 2018. Smithers Pira, Akron, OH, United States, p. 32.

Ariosti, A., 2019a. Food contact materials regulatory overview: MERCOSUR and other Latin American countries. In: Proceedings of the Global Food Contact Conference 2019, Lisbon, Portugal, May 14–16, 2019. Smithers Pira, Leatherhead, Surrey, United Kingdom, p. 37.

Ariosti, A., 2019b. Overview of regulations for food packaging across MERCOSUR and Colombia – an update. In: Proceedings of the 14th. Biennial International Symposium on Worldwide Regulation of Food Packaging, Baltimore, MD, United States, June 11–14, 2019. The Plastics Industry Association (PLASTICS), Washington DC, United States, p. 24.

Ariosti, A., 2021a. Food contact materials regulations updates in MERCOSUR and Latin America. In: Proceedings of the Global Food Contact Conference 2021 on-line, June 14–16, 2021. Smithers Pira, Leatherhead, Surrey, United Kingdom, p. 19.

Ariosti, A., 2021b. Sustainable and recyclable food contact materials in the MERCOSUR countries. In: Proceedings of the Food Contact Compliance Conference on-line, September 29–30, 2021. The Italian Institute of Packaging, Milan, Italy, p. 17.

Ariosti, A., Olivera Carrión, M., 2014. Argentina. In: Kirchsteiger-Meier, E., Baumgartner, T. (Eds.), Global Food Legislation: An Overview. Wiley-VCH Verlag GmbH & Co. KGaA., Weinheim, pp. 1–32.

Ariosti, A., Padula, M., 2017. Use of nanomaterials for food packaging in Latin American and Caribbean countries. In: Veraart, R. (Ed.), The Use of Nanomaterials in Food Contact Materials — Design, Application, Safety. DEStech Publications Inc., Lancaster, PA, pp. 165—209.

Arvidson, K.B., Cheeseman, M.A., McDougal, A.J., 2007. Toxicology and risk assessment of chemical migrants from food contact materials. In: Barnes, K.A., Sinclair, C.R., Watson, D.H. (Eds.), Chemical Migration and Food Contact Materials. Woodhead Publishing Ltd., Cambridge, pp. 158—179.

Auras, R., Harte, B., Selke, S., 2004. An overview of polylactides as packaging materials. Macromol. Biosci. 4, 835—864.

Bach, C., Dauchy, X., Severin, I., et al., 2013. Effect of temperature on the release of intentionally and non-intentionally added substances from polyethylene terephthalate (PET) bottles into water: chemical analysis and potential toxicity. Food Chem. 139, 672—680.

Ballesteros-Gómez, A., Rubio, S., Pérez-Bendito, D., 2009. Review — analytical methods for the determination of bisphenol A in food. J. Chromatogr. A 1216, 449—469.

Barnes, K.A., Sinclair, C.R., Watson, D.H. (Eds.), 2007. Chemical Migration and Food Contact Materials. Woodhead Publishing Ltd., Cambridge.

Barthélémy, E., Spyropoulos, D., Milana, M.R., et al., 2014. Safety evaluation of mechanical recycling processes used to produce polyethylene terephthalate (PET) intended for food contact applications. In: Ariosti, A. (Ed.), Proceedings of the 5th. ILSI International Symposium on Food Packaging 'Scientific Developments Supporting Safety and Innovation', Berlin, Germany, November 14—16, 2012, vol. 31. Food Additives and Contaminants, pp. 490—497. Part A, 3.

Baughan, J.S., 2015. Global legislation for paper and board materials in contact with food. In: Baughan, J.S. (Ed.), Global Legislation for Food Contact Materials. Woodhead Publishing Ltd., Cambridge, pp. 201—210.

Baughan, J.S., Attwood, D., 2010. Food packaging law in the United States. In: Rijk, R., Veraart, R. (Eds.), Global Legislation for Food Packaging Materials. Wiley-VCH Verlag GmbH & Co. KGaA., Weinheim, pp. 223—242.

Bayer, F.L., 1997. The threshold of regulation and its application to indirect food additive contaminants in recycled plastics. Food Addit. Contam. 14 (6—7), 661—670.

Bayer, F.L., 2002. Polyethylene terephthalate recycling for food-contact applications: testing, safety and technologies: a global perspective. In: Gilbert, J., López de Sá, A. (Eds.), Proceedings of the 2nd. ILSI International Symposium 'Food Packaging: Ensuring the Safety and Quality of Foods', Vienna, Austria, November 8—10, 2000, vol. 19. Food Additives and Contaminants, pp. 111—134. Suppl.

BfR, 2017. Recommendation XXXVI. Paper and Board for Food Contact. Bundesinstitut für Risikobewertung, Berlin. Available from: https://bfr.ble.de/kse/faces/resources/pdf/360-english.pdf. (Accessed 18 May 2018).

Bhunia, K., Sablani, S.S., Tang, J., et al., 2013. Migration of chemical compounds from packaging polymers during microwave, conventional heat treatment, and storage. Ins. Food Technol. — Compr. Rev. Food Sci. Food Saf. 12, 523—545. https://doi.org/10.1111/1541-4337.12028.

Biedermann, M., Ingenhoff, J.-E., Zurfluh, M., et al., 2013. Migration of mineral oil, photoinitiators and plasticizers from recycled paperboard into dry foods: a study under controlled conditions. Food Addit. Contam. 30 (5), 885—898.

Biedermann-Brem, S., Biedermann, M., Grob, K., 2017. Taped barrier test for internal bags used in boxes of recycled paperboard: the role of the paperboard and its consequence for the test. Packag. Technol. Sci. 30, 75—89.

Bignardi, C., Cavazza, A., Laganà, C., et al., 2017. Release of non-intentionally added substances (NIAS) from food contact polycarbonate: effect of ageing. Food Control 71, 329—335.

Bolle, F., 2016. Council of Europe harmonization of non-plastics food contact. In: Proceedings of the Seminar Main Laws on FCM in the World, Milan, Italy, May 24—27, 2016. Istituto Italiano Imballaggio, Milan, Italy, p. 117.

Bolzoni, L., 2015. Plasticisers used in PVC for foods: assessment of specific migration. In: Barone, C., Bolzoni, L., Caruso, G., et al. (Eds.), Food Packaging Hygiene. Springer International Publishing, AG Switzerland, pp. 43—61. https://doi.org/10.1007/978-3-319-14827-4.

Bott, J., Störmer, A., Franz, R., 2014a. Migration of nanoparticles from plastic packaging materials containing carbon black into foodstuffs. Food Addit. Contam. 31 (10), 1769—1782.

Bott, J., Störmer, A., Franz, R., 2014b. A model study into the migration potential of nanoparticles from plastics nanocomposites for food contact. Food Packag. Shelf Life 2, 73—80.

Bott, J., Störmer, A., Albers, P., 2018. Investigation into the release of nanomaterials from can coatings into food. Food Packag. Shelf Life 16, 112—121.

Bouwmeester, H., Brandhoff, P., Marvin, H.J.P., et al., 2014. State of the safety assessment and current use of nanomaterials in food and food production. Trends Food Sci. Technol. 40, 200—210.

Brandsch, R., 2017. Probabilistic migration modelling focused on functional barrier efficiency and low migration concepts in support of risk assessment. In: Ariosti, A., Guest (Eds.), Proceedings of the 6th. ILSI International Symposium on Food Packaging 'Scientific Developments Supporting Safety and Innovation', Barcelona, Spain, November 16—18, 2016, vol. 34. Food Additives and Contaminants: Part A, pp. 1743—1766, 10.

BS ISO, 2003. British Standard ISO 13302:2003 Sensory Analysis — Methods for Assessing Modifications to the Flavor of Foodstuffs Due to Packaging. British Standards Institution, London, p. 36.

Buculei, A., Gutt, G., Sonia, A., et al., 2012. Study regarding the tin and iron migration from metallic cans into foodstuff during storage. J. Agroaliment. Processes Technol. 18 (4), 299—303.

Bumbudsanpharoke, N., Ko, S., 2015. Nano-food packaging: an overview of market, migration research, and safety regulations. Institute of Food Technologists — concise Reviews in Food Science. J. Food Sci. 80 (5), 910—923. https://doi.org/10.1111/1750-3841.12861.

Bumbudsanpharoke, N., Choi, J., Park, H.J., et al., 2019. Zinc migration and its effect on the functionality of a low density polyethylene-ZnO nanocomposite film. Food Packag. Shelf Life 20, 100301.

Burello, E., 2017. Review of (Q)SAR models for regulatory assessment of nanomaterials risks. NanoImpact 8, 48—58.

Byun, Y., Kim, Y.T., 2014. Bioplastics for food packaging: Chemistry and physics. In: Han, J.H. (Ed.), Innovations in Food Packaging, second ed. Academic Press/Elsevier, United Kingdom/United States, pp. 353–368.

Callejón, R.M., Ubeda, C., Ríos-Reina, R., et al., 2016. Recent developments in the analysis of musty odour compounds in water and wine: a review. J. Chromatogr. A 1428, 72–85.

Catalá, R., Gavara, R. (Eds.), 2002. Migración de componentes y residuos de envases en contacto con alimentos. IATA, Valencia.

Cheeseman, M., 2005. Thresholds as a unifying theme in regulatory toxicology. Food Addit. Contam. 22 (10), 900–906.

Cheeseman, M., 2013. Toxicological threshold of concern: a tool for global food contact material compliance. In: Proceedings of the Global Food Contact Conference 2013, Barcelona, Spain, May 14–16, 2013. Smithers Pira, Leatherhead, Surrey, United Kingdom, p. 29.

CoE, 2004. Policy Statement Concerning Lead Leaching from Glass Tableware into Foodstuffs (Version 1). Strasbourg. Available from: www.edqm.eu/en/resolutions-policy-statements. (Accessed 24 May 2018).

CoE, 2009. Technical document No. 3 Guidelines on paper and board materials and articles, made from recycled fibres, intended to come into contact with foodstuffs (Version 2). In: Policy Statement Concerning Paper and Board Materials and Articles Intended to Come into Contact with Foodstuffs (Version 4). Strasbourg. Available from: www.edqm.eu/en/resolutions-policy-statements. (Accessed 24 May 2018).

Ćwiek-Ludwicka, K., Ludwicki, J.K., 2017. Nanomaterials in food contact materials; considerations for risk assessment. Rocz. Panstw. Zakl. Hig. 68 (4), 321–329.

Dainelli, D., 2015. Global legislation for active and intelligent materials. In: Baughan, J.S. (Ed.), Global Legislation for Food Contact Materials. Woodhead Publishing Ltd., Cambridge, pp. 183–199.

de la Cruz García, C., Sánchez Moragas, G., Nordqvist, D., 2014. Food contact materials. In: Motarjemi, Y., Lelieveld, H. (Eds.), Food Safety Management – A Practical Guide for the Food Industry. Academic Press/Elsevier, United Kingdom/United States, pp. 397–419.

de Oliveira, C., Rodríguez, A., Ferreira Soares, N., et al., 2012. Multiple headspace-solid-phase microextraction as a powerful tool for the quantitative determination of volatile radiolysis products in a multilayer food packaging material sterilized with gamma-radiation. J. Chomatogr. A 1244, 61–68.

Devlieghere, F., De Meulenaer, B., Sekitoleko, P., et al., 1997. Evaluation, modelling and optimization of the cleaning process of contaminated plastic food refillables. Food Addit. Contam. 14 (6–7), 671–683.

Dilmaçünal, T., 2017. Intelligent systems in the food packaging industry: contaminant sensors and security/anticounterfeiting devices. In: Oprea, A.E., Grumezescu, A.M. (Eds.), Nanotechnology Applications in Food – Flavor, Stability, Nutrition and Safety. Academic Press/Elsevier, United Kingdom/United States, pp. 287–306.

Drasler, B., Sayre, P., Steinhäuser, K.G., et al., 2017. In vitro approaches to assess the hazard of nanomaterials. NanoImpact 8, 99–116.

Driffield, M., Bradley, E.L., Leon, I., et al., 2014. Analytical screening studies on irradiated food packaging. In: Ariosti, A., Guest (Eds.), Proceedings of the 5th. ILSI International Symposium on Food Packaging 'Scientific Developments Supporting Safety and Innovation', Berlin, Germany, November 14–16, 2012. Food Additives and Contaminants: Part A, vol. 31, pp. 556–565, 3.

Dris, R., Gasperi, J., Saad, M., et al., 2016. Synthetic fibers in atmospheric fallout: a source of microplastics in the environment? Mar. Pollut. Bull. 104, 290–293.

Dudefoi, W., Villares, A., Peyron, S., et al., 2018. Nanoscience and nanotechnologies for biobased materials, packaging and food applications: new opportunities and concerns. Innovat. Food Sci. Emerg. Technol. 46, 107–121.

Duncan, T.V., 2011. Applications of nanotechnology in food packaging and food safety: barrier materials, antimicrobials and sensors. J. Colloid Interface Sci. 363, 1–24.

Dutra, C., Freire, M.T. de A., Nerín, C., et al., 2014. Migration of residual nonvolatile and inorganic compounds from recycled post-consumer PET and HDPE. J. Braz. Chem. Soc. 25 (4), 686–696.

Echegoyen, Y., Nerín, C., 2013. Nanoparticle release from nano-silver antimicrobial food containers. Food Chem. Toxicol. 62, 16–22.

EDQM-CoE, 2013. Metals and Alloys Used in Food Contact Materials and Articles. A Practical Guide for Manufacturers and Regulators. Committee of Experts on Packaging Materials for Food and Pharmaceutical Products (P-SC-EMB) – European Directorate for the Quality of Medicines and HealthCare. EDQM) – Council of Europe, Strasbourg, p. 218.

EFSA, 2008a. Scientific opinion of the Panel on food additives, flavourings, processing aids and food contact materials (AFC) on the safety of aluminium from dietary intake. EFSA J. 754, 1–34.

EFSA, 2008b. Scientific Opinion of the Panel on Food Additives, Flavourings, Processing Aids and Food Contact Materials (AFC) on Guidelines on submission of a dossier for safety evaluation by the EFSA of a recycling process to produce recycled plastics intended to be used for manufacture of materials and articles in contact with food. EFSA J. 717, 1–12.

EFSA, 2011. Scientific Opinion on the criteria to be used for safety evaluation of a mechanical recycling process to produce recycled PET intended to be used for manufacture of materials and articles in contact with food. EFSA Panel on food contact materials, enzymes, flavourings and processing aids (CEF). EFSA J. 9 (7), 1–25, 2184.

EFSA Scientific Committee, 2011. Scientific Opinion on Guidance on the risk assessment of the application of nanoscience and nanotechnologies in the food and feed chain. EFSA J. 9 (5), 2140.

EFSA, 2017. Note for Guidance for the preparation of an application for the safety assessment of a substance to be used in plastic food contact materials, updated on 23 March 2017. EFSA Panel on food contact materials, enzymes, flavourings and processing Aids (CEF). EFSA J. 6 (7), 21r.

EFSA Scientific Committee, Hardy, A., Benford, D., et al., 2018. Guidance on risk assessment of the application of nanoscience and nanotechnologies in the food and feed chain: Part 1, human and animal health. EFSA J. 16 (7), e05327.

EFSA/WHO, 2016. Review of the Threshold of Toxicological Concern (TTC) Approach and Development of New TTC Decision Tree. EFSA Supporting Publication 2016: EN-1006, p. 50.

Eisert, R., 2008. Comparing food contact legislation. In: Proceedings of the Intertech-Pira Global Legislation for Food Contact Packaging Conference, Alexandria, VA, United States, April 2–4, 2008, p. 24. Leatherhead, Surrey, United Kingdom.

Eisert, R., 2011. EU exposure matrix Project – results. In: Proceedings of the Global Food Contact Conference 2011, Frankfurt Am Main, Germany, June 14–16, 2011. Pira International, Leatherhead, Surrey, United Kingdom, p. 22.

EU, 2002. Regulation (EC) No 178/2002 of the European Parliament and of the Council laying down the general principles and requirements of food law, establishing the European Food Safety Authority and laying down procedures in matters of food safety. OJ. 31, 1–24, 1.2.2002.

EU, 2004. Regulation (EC) No. 1935/2004 of the European Parliament and of the Council of 27 October 2004 on materials and articles intended to come into contact with food and repealing Directives 80/590/EEC and 89/109/EEC. OJ. 338, 4–17, 13.11.2004.

EU, 2006. Commission Regulation (EC) No. 2023/2006 of 22 December 2006 on good manufacturing practice for materials and articles intended to come into contact with food. OJ. 384, 75–78, 29.12.2006.

EU, 2008. Commission Regulation (EC) No. 282/2008 of 27 March 2008 on recycled plastic materials and articles intended to come into contact with foods and amending Regulation (EC) No 2023/2006. OJ. 86, 9–18, 28.3.2008.

EU, 2011. Commission Regulation (EU) No. 10/2011 of 14 January 2011 on plastic materials and articles intended to come into contact with food. OJ. 12, 1–89, 15.1.2011.

Fang, Z., Zhao, Y., Warner, R.D., et al., 2017. Active and intelligent packaging in meat industry. Trends Food Sci. Technol. 61, 60–71.

Feigenbaum, A., Dole, P., Aucejo, S., et al., 2005. Functional barriers: properties and evaluation. Food Addit. Contam. 22 (10), 956–967.

Félix, J.S., Isella, F., Bosetti, O., et al., 2012. Analytical tools for identification of non-intentionally added substances (NIAS) coming from polyurethane adhesives in multilayer packaging materials and their migration into food simulants. Anal. Bioanal. Chem. 403, 2869–2882.

Feron, V.J., Jetten, J., de Kruijf, N., et al., 1994. Polyethylene terephthalate bottles (PRBs): a health and safety assessment. Food Addit. Contam. 11 (5), 571–594.

Franz, R., Huber, M., Piringer, O., et al., 1996. Study of functional barrier properties of multilayer recycled poly(ethylene terephthalate) bottles for soft drinks. J. Agric. Food Chem. 44 (3), 892–897.

Franz, R., Huber, M., Piringer, O., 1997. Presentation and experimental verification of a physico-mathematical model describing the migration across functional barrier layers into foodstuffs. Food Addit. Contam. 14 (6–7), 627–640.

Franz, R., Bayer, F.L., Welle, F., 2003. Guidance and criteria for safe recycling of post-consumer polyethylene terephthalate (PET) into new food packaging applications. In: Program on the Recyclability of Food Packaging Materials with Respect to Food Safety Considerations – Polyethylene Terephthalate (PET), Paper and Board and Plastics Covered by Functional Barriers. Fraunhofer Institute for Process Engineering and Packaging (IVV), Freising, p. 32.

Franz, R., Bayer, F.L., Welle, F., 2004a. Guidance and Criteria for Safe Recycling of Post-consumer Polyethylene Terephthalate (PET) into New Food Packaging Applications. Report EU-Project FAIR-CT98-4318 Recyclability. European Commission, Brussels, p. 37.

Franz, R., Mauer, A., Welle, F., 2004b. European survey on post-consumer poly(ethylene terephthalate) materials to determine contamination levels and maximum consumer exposure from food packages made from recycled PET. Food Addit. Contam. 21 (3), 265–286.

Franz, R., Palzer, G., Gawlik, B.M., et al., 2004c. Certification of a Refillable PET Bottle Material with Respect to Chemical Inertness Behaviour According to a Pr-CEN Standard Method – BCR-712. European Commission – Directorate-General Joint Research Center – Institute for Reference Materials and Measurements, Brussels, p. 68.

Franz, R., Störmer, A., 2008. Migration of plastic constituents. In: Piringer, O.G., Baner, A.L. (Eds.), Plastic Packaging – Interactions with Food and Pharmaceuticals, second ed. Wiley-VCH Verlag GmbH & Co. KGaA, Weinheim, pp. 349–415.

FSANZ, 2010. Survey of Chemical Migration from Food Contact Packaging Materials in Australian Food. Food Standards Australia New Zealand, p. 35. Available from: www.foodstandards.gov.au/science/surveillance/pages/surveyofchemicalmigr5148.aspx. (Accessed 14 May 2018).

Gallart-Ayala, H., Núñez, O., Lucci, P., 2013. Recent advances in LC-MS analysis of food-packaging contaminants. Trends Anal. Chem. 42, 99–124.

Garcia, C.V., Shin, G.H., Kim, J.T., 2018. Metal oxide-based nanocomposites in food packaging: applications, migration, and regulations. Trends Food Sci. Technol. 82, 21–31.

Gatos, K.G., 2016. Potential of nanomaterials in food. In: Grumezescu, A.M. (Ed.), Novel Approaches of Nanotechnology in Food – Nanotechnology in the Agri-Food Industry Volume 1. Academic Press/Elsevier, United Kingdom/United States, pp. 587–621.

Gehring, C., Welle, F., 2018. Migration testing of polyethylene terephthalate: comparison of regulated test conditions with migration into real food at the end of shelf life. Packag. Technol. Sci. 31, 771–780.

Gelbke, H.-P., Banton, M., Block, C., et al., 2019. Risk assessment for migration of styrene oligomers into food from polystyrene food containers. Food Chem. Toxicol. 124, 151–167.

Genualdi, S., Nyman, P., Begley, T., 2014. Updated evaluation of the migration of styrene monomer and oligomers from polystyrene food contact materials to foods and food simulants. Food Addit. Contam. 31 (4), 723–733. https://doi.org/10.1080/19440049.2013.878040.

Genualdi, S., Addo Ntim, S., Begley, T., 2015. Suitability of polystyrene as a functional barrier layer in coloured food contact materials. Food Addit. Contam. 32 (3), 395–402.

Gergely, A., Cheeseman, M., 2015. TTC: practical and legally sound approach to evaluate safety and regulatory compliance. In: Proceedings of the Global Food Contact Conference 2015, Rome, Italy, May 19–21, 2015. Smithers Pira, Leatherhead, Surrey, United Kingdom, p. 22.

Geueke, B., 2018. Dossier – Non-intentionally Added Substances (NIAS), second ed. Food Packaging Forum, Zurich. https://doi.org/10.5281/zenodo.1265331.

Geueke, B., Groh, K., Muncke, J., 2018. Food packaging in the circular economy: overview of chemical safety aspects for commonly used materials. J. Clean. Prod. 193, 491–505.

Ghaani, M., Cozzolino, C.A., Castelli, G., et al., 2016. An overview of the intelligent packaging technologies in the food sector. Trends Food Sci. Technol. 51, 1–11.

Gomes Lauriano Souza, V., Fernando, A.L., 2016. Nanoparticles in food packaging: biodegradability and potential migration to food – a review. Food Packag. Shelf Life 8, 63–70.

Gómez-Estaca, J., López-de-Dicastillo, C., Hernández-Muñoz, P., et al., 2014. Advances in antioxidant active food packaging. Trends Food Sci. Technol. 35, 42–51.

Greenberg, E.F., 2018. GRAS from a different angle. In: Pre-Global Food Contact Conference 2018 Workshop, May 9, 2018, Bethesda, MD. United States. Smithers Pira, Akron, OH, United States, p. 90.

Grob, K., 2017. The European system for the control of the safety of food-contact materials needs restructuring: a review and outlook for discussion. Food Addit. Contam. 34 (9), 1643–1659.

Groh, K.J., Geueke, B., Muncke, J., 2017. Food contact materials and gut health: implications for toxicity. Food Chem. Toxicol. 109, 1–18.

Gu, Z., Zhang, H., 2019. Administrative licensing of FCM in China. In: Proceedings of the 14th. Biennial International Symposium on Worldwide Regulation of Food Packaging, Baltimore, MD, United States, June 11–14, 2019. The Plastics Industry Association (PLASTICS), Washington DC, United States, p. 23.

Guazzotti, V., Marti, A., Piergiovanni, L., et al., 2014. Bio-based coatings as potential barriers to chemical contaminants from recycled paper and board for food packaging. In: Ariosti, A. (Ed.), Proceedings of the 5th. LSI International Symposium on Food Packaging 'Scientific Developments Supporting Safety and Innovation', Berlin, Germany, November 14–16, 2012. Food Additives and Contaminants: Part A, vol. 31, pp. 402–413, 3.

Guazzotti, V., Limbo, S., Piergiovanni, L., et al., 2015. A study into the potential barrier properties against mineral oils of starch-based coatings on paperboard for food packaging. Food Packag. Shelf Life 3, 9–18.

Gude, T., 2018. Harmonization of (NIAS) testing – is that possible?. In: Proceedings of the Global Food Contact Conference 2018, Bethesda, MD, United Sates, May 9–11, 2018. Smithers Pira, Akron, OH, United States, p. 24.

Hannon, J.C., Kerry, J., Cruz-Romero, M., et al., 2015. Advances and challenges for the use of engineered nanoparticles in food contact materials. Trends Food Sci. Technol. 43, 43–62.

Hannon, J.C., Kerry, J.P., Cruz-Romero, M., et al., 2016. Human exposure assessment of silver and copper migrating from an antimicrobial nanocoated packaging material into an acidic food simulant. Food Chem. Toxicol. 95, 128–136.

Hannon, J.C., Kerry, J.P., Cruz-Romero, M., et al., 2018. Migration assessment of silver from nanosilver spray coated low density polyethylene or polyester films into milk. Food Packag. Shelf Life 15, 144–150.

Heckman, J.H., 2005. Food packaging regulation in the United States and the European Union. Regul. Toxicol. Pharmacol. 42, 96–122.

Hernández, R.J., Gavara, R., 1999. Plastics Packaging. Methods for Studying Mass Transfer Interactions. PIRA International, Leatherhead, Surrey.

Hoekstra, E.J., Brandsch, R., Dequatre, C., et al., 2015. Practical Guidelines on the Application of Migration Modeling for the Estimation of Specific Migration – in Support of Regulation (EU) 10/2011 on Plastic Food Contact Materials. EU Joint Research Center (JRC) Technical Reports, p. 40. https://doi.org/10.2788/04517. EUR 27529 EN.

Hong, S., 2016. Food contact regulations and requirements in Japan. In: Proceedings of the Global Food Contact Conference 2016, Bethesda, MD, United States, May 11–13, 2016. Smithers Pira, Akron, OH, United States, p. 31.

Honigfort, P., 2018. FDA Update on regulatory issues related to food contact materials. In: Proceedings of the Global Food Contact Conference 2018, Bethesda, MD, United States, May 9–11, 2018. Smithers Pira, Akron, OH, United States, p. 30.

Horton, A.A., Walton, A., Spurgeon, D.J., et al., 2017. Microplastics in freshwater and terrestrial environments: evaluating the current understanding to identify the knowledge gaps and future research priorities. Sci. Total Environ. 586, 127–141.

Huang, J.-Y., Li, X., Zhou, W., 2015. Safety assessment of nanocomposite for food packaging application. Trends Food Sci. Technol. 45, 187–199.

ILSI, 1993. White Paper on Refillable Plastic Packaging Made from PET (Polyethylene Terephthalate). Task Force on Refillable PET Packaging, Washington DC, p. 140.

ILSI, 1998. Recycling of Plastics for Food Contact Use. International Life Sciences Institute (ILSI) Europe Packaging Materials Task Force, Brussels, p. 24. Available from: http://ilsi.org/publication/recycling-of-plastics-for-food-contact-use/. (Accessed 25 May 2018).

ILSI, 2007. Guidance for Exposure Assessment of Substances Migrating from Food Packaging Materials. Report of an ILSI-Europe Expert Group, Brussels, p. 79. Available from: http://ilsi.eu/wp-content/uploads/sites/3/2016/06/O2007Gui_Exp.pdf. (Accessed 26 July 2019).

Iñiguez-Franco, F., Auras, R., Rubino, M., et al., 2017. Effect of nanoparticles on the hydrolytic degradation of PLA nanocomposites by water-ethanol solutions. Polym. Degrad. Stabil. 146, 287–297.

Irvine, A., 2018. European landscape, in the food contact materials industry. In: Proceedings of the Global Food Contact Conference 2018, Bethesda, MD, United States, May 9–11, 2018. Smithers Pira, Akron, OH, United States, p. 29.

Irvine, A., 2019. EU: core developments and changes relating to European food contact legislation. In: Proceedings of the Global Food Contact Conference 2019, Lisbon, Portugal, May 14–16, 2019. Smithers Pira, Leatherhead, Surrey, United Kingdom, p. 30.

Isella, F., Canellas, E., Bosetti, O., et al., 2013. Migration of non intentionally added substances from adhesives by UPLC–Q-TOF/MS and the role of EVOH to avoid migration in multilayer packaging materials. J. Mass Spectrom. 48, 430–437.

Jakubowska, N., Beldì, G., Peychès Bach, A., et al., 2014. Optimisation of an analytical method and results from the inter-laboratory comparison of the migration of regulated substances from food packaging into the new mandatory European Union simulant for dry foodstuffs. In: Ariosti, A. (Ed.), Proceedings of the 5th. LSI International Symposium on Food Packaging 'Scientific Developments Supporting Safety and Innovation', Berlin, Germany, November 14–16, 2012. Food Additives and Contaminants: Part A, vol. 31, pp. 546–555, 3.

Jetten, J., de Kruijf, N., Castle, L., 1999. Quality and safety aspects of reusable plastic food packaging materials: a European study to underpin future legislation. Food Addit. Contam. 16 (1), 25–36.

JETRO, 2009. Specifications, Standards and Testing Methods for Foodstuffs, Implements, Containers and Packaging, Toys, Detergents 2008. Japan External Trade Organization, Tokyo, p. 149. Available from: www.jetro.go.jp/en/reports/regulations/. (Accessed 9 July 2018).

JETRO, 2011. Specifications and Standards for Foods, Food Additives, etc., under the Food Sanitation Act (Abstract) 2010. Japan External Trade Organization, Tokyo, p. 190. Available from: www.jetro.go.jp/en/reports/regulations/. (Accessed 9 July 2018).

Jokar, M., Pedersen, G.A., Loeschner, K., 2017. Six open questions about the migration of engineered nano-objects from polymer-based food-contact materials: a review. Food Addit. Contam. 34 (3), 434–450.

Jung, T., Simat, T.J., Altkofer, W., 2010. Mass transfer ways of ultraviolet printing ink ingredients into foodstuffs. Food Addit. Contam. 27 (7), 1040–1049.

Kale, G., Kijchavengkul, T., Auras, R., et al., 2007. Compostability of bioplastic packaging materials: an Overview. Macromol. Biosci. 7, 255–277.

Karamfilova, E., Sacher, M., Sabbati, G., 2016. Food Contact Materials Regulation (EC) 1935/2004 – European Implementation Assessment. Secretariat of the European Parliament, p. 142. https://doi.org/10.2861/162638.

Kassouf, A., Chebib, H., Lebbos, N., et al., 2013. Migration of iron, lead, cadmium and tin from tinplate-coated cans into chickpeas. Food Addit. Contam. 30 (11), 1987–1992.

Katan, L.L. (Ed.), 1996. Migration from Food Contact Materials. Blackie Academic and Professional, London.

Kawamura, Y., 2015. Current status of food contact regulations in Japan. In: Proceedings of the Global Food Contact Conference 2015, Rome, Italy, May 19–21, 2015. Smithers Pira, Leatherhead, Surrey, United Kingdom, p. 35.

Keithline, J., 2017. Update on food-contact regulations in China. In: Proceedings of the Plastics and Papers in Contact with Foodstuffs Conference 2017, Berlin, Germany, December 4–7, 2017. Smithers Pira, Leatherhead, Surrey, United Kingdom, p. 50.

Kenny, K.C., 2018. Asia panorama. In: Proceedings of the Global Food Contact Conference 2018, Bethesda, MD, United States, May 9–11, 2018. Smithers Pira, Akron, OH, United States, p. 67.

Koh-Fallet, S., 2018. FDA's Foreign Supplier Verification Program for food contact manufacturers. In: Proceedings of the Global Food Contact Conference 2018, Bethesda, MD, United States, May 9–11, 2018. Smithers Pira, Akron, OH, United States, p. 20.

Koivikko, R., Pastorelli, S., Rodríguez-Bernaldo de Quirós, A., et al., 2010. Rapid multi-analyte quantification of benzophenone, 4-methylbenzophenone and related derivatives from paperboard food packaging. Food Addit. Contam. 27 (10), 1478–1486.

Koster, S., Rennen, M., Leeman, W., et al., 2014. A novel safety assessment strategy for non-intentionally added substances (NIAS) in carton food contact materials. In: Ariosti, A., Guest (Eds.), Proceedings of the 5th. ILSI International Symposium on Food Packaging 'Scientific Developments Supporting Safety and Innovation', Berlin, Germany, November 14–16, 2012. Food Additives and Contaminants: Part A, vol. 31, pp. 422–443, 3.

Koster, S., Bani-Estivals, M.-H., Bonuomo, M., et al., 2015. Guidance on Best Practices on the Risk Assessment of Non Intentionally Added Substances (NIAS) in Food Contact Materials and Articles. ILSI Europe Packaging Materials Task Force, Brussels, p. 72. Available from: http://ilsi.org/publication/guidance-on-best-practices-on-the-risk-assessment-of-non-intentionally-added-substances-nias-in-food-contact-materials-and-articles/. (Accessed 25 May 2018).

Lagaron, J.M., Lopez-Rubio, A., 2011. Nanotechnology for bioplastics: opportunities, challenges and strategies. Trends Food Sci. Technol. 22, 611–617.

Lago, M.A., Rodríguez-Bernaldo de Quirós, A., Sendón, R., et al., 2015. Photoinitiators: a food safety review. Food Addit. Contam. 32 (5), 779–798.

LeNoir, R.T., 2015. Global legislation for plastic materials in contact with food. In: Baughan, J.S. (Ed.), Global Legislation for Food Contact Materials. Woodhead Publishing Ltd., Cambridge, pp. 77–108.

Li, C., Bian, S., 2010. In: Rijk, R., Veraart, R. (Eds.), Global Legislation for Food Packaging Materials. Wiley-VCH Verlag GmbH & Co. KGaA., Weinheim, pp. 319–335.

Lord, T., 2003. Packaging materials as a source of taints. In: Baigrie, B. (Ed.), Taints and Off-Flavours in Food. Woodhead Publishing Ltd., Cambridge, United Kingdom, pp. 64–111. CRC Press LLC, Boca Raton, FL.

Lorenzini, R., Biedermann, M., Grob, K., et al., 2013. Migration kinetics of mineral oil hydrocarbons from recycled paperboard to dry food: monitoring of two real cases. Food Addit. Contam. 30 (4), 760–770.

Mahinka, S.P., Reisin Miller, A., Vaughn, J.L., 2013. Compliance of Glass Packaging with Human and Environmental Health and Safety Toxics – in – Packaging Requirements. Glass Packaging Institute, Arlington, VA, p. 50. Available from: www.gpi.org/sites/default/files/Compliance%20of%20Glass%20Packaging%20with%20Human%20and%20Environmental%20Health%20and%20Safety%20Toxics.pdf. (Accessed 14 May 2018).

Mari, E.A., 2002. Migración en envases de vidrio y de cerámica esmaltada. In: Catalá, R., Gavara, R. (Eds.), Migración de componentes y residuos de envases en contacto con alimentos. IATA, Valencia, pp. 329–346.

Martínez-Bueno, M.J., Hernando, M.D., Uclés, S., et al., 2017. Identification of non-intentionally added substances in food packaging nano films by gas and liquid chromatography coupled to orbitrap mass spectrometry. Talanta 172, 68–77.

Martínez-Bueno, M.J., Gómez Ramos, M.J., Bauer, A., et al., 2019. An overview of non-targeted screening strategies based on high resolution accurate mass spectrometry for the identification of migrants coming from plastic food packaging materials. Trends Anal. Chem. 110, 191–203.

McMillin, K.W., 2017. Advancements in meat packaging – Review. Meat Sci. 132, 153–162.

MERCOSUR, December 15, 1992. Resolución GMC 55/92 – envases y equipamientos de vidrio y cerámica destinados a entrar en contacto con alimentos.

MERCOSUR, December 11, 2007. Reglamento Técnico MERCOSUR sobre envases de polietilentereftalato (PET) postconsumo reciclado grado alimentario (PET-PCR grado alimentario) destinados a estar en contacto con alimentos.

MERCOSUR, September 23, 2015. Resolución GMC 40/15 – Reglamento Técnico MERCOSUR sobre materiales, envases y equipamientos celulósicos destinados a estar en contacto con alimentos.

Metak, A.M., Nabhani, F., Connolly, S.N., 2015. Migration of engineered nanoparticles from packaging into food products. LWT Food Sci. Technol. 64, 781–787.

MHLW, 2017. Guideline for Safety Assurance in the Manufacture Etc. Of Food Utensils, Containers and Packaging. Ministry of Health, Labour and Welfare of Japan, Tokyo. Available from: www.mhlw.go.jp/stf/seisakunitsuite/bunya/kenkou_iryou/shokuhin/kigu/index_00003.html. (Accessed 21 May 2019).

MHLW, 2018a. Outline of the Act on the Partial Amendment of the Food Sanitation Act. Ministry of Health, Labour and Welfare of Japan, Tokyo. Available from: www.mhlw.go.jp/stf/seisakunitsuite/bunya/kenkou_iryou/shokuhin/yunyu_kanshi/index_00016.html. (Accessed 23 August 2019).

MHLW, 2018b. The Summary of the Amendment to the Food Sanitation Act. Ministry of Health, Labour and Welfare of Japan, Tokyo. Available from: https://members.wto.org/crnattachments/2018/SPS/JPN/18_0425_00_e.pdf. (Accessed 23 August 2019).

MHLW, 2019a. Overview of Amendments to the Food Sanitation Act. Ministry of Health, Labour and Welfare of Japan, Tokyo. Available from: www.mhlw.go.jp/content/11130500/000537823.pdf. (Accessed 23 August 2019).

MHLW, 2019b. The Draft of the Positive List. Ministry of Health, Labour and Welfare of Japan, Tokyo. Available from: www.mhlw.go.jp/stf/newpage_06143.html. (Accessed 23 August 2019).

Mihindukulasuriya, S.D.F., Lim, L.-T., 2014. Nanotechnology development in food packaging: a review. Trends Food Sci. Technol. 40, 149–167.

Misko, G.G., 2017. The basics of FDA food-contact regulation – an Overview. In: Proceedings of the Food Packaging Law Seminar US 2017, Arlington, VA, United States. October 10–11, 2017. Keller and Heckman LLP, Washington DC, United States, p. 20.

Montanari, A., 2015. Basic principles of corrosion of food metal packaging. In: Barone, C., Bolzoni, L., Caruso, G., et al. (Eds.), Food Packaging Hygiene. Springer International Publishing AG Switzerland, pp. 105–132. https://doi.org/10.1007/978-3-319-14827-4.

Mori, Y., 2010. Rules on food contact materials and articles in Japan. In: Rijk, R., Veraart, R. (Eds.), Global Legislation for Food Packaging Materials. Wiley-VCH Verlag GmbH & Co. KGaA., Weinheim, pp. 291–317.

Muncke, J., Backhaus, T., Geueke, B., et al., 2017. Scientific challenges in the risk assessment of food contact materials. Environ. Health Perspect. 125 (9), 1–9.

Mutsuga, M., 2012. An update on efforts for regulation of food packaging in Japan. In: Proceedings of the Global Food Contact Conference 2012, Baltimore, MD, United States, May 15–17, 2012. Smithers Pira, Leatherhead, Surrey, United Kingdom, p. 28.

Nakajima, H., Dijkstra, P., Loos, K., 2017. Review – the recent developments in biobased polymers toward general and engineering applications: polymers that are upgraded from biodegradable polymers, analogous to petroleum-derived polymers, and newly developed. Polymers 9 (523), 26. https://doi.org/10.3390/polym9100523.

Nerín, C., Alfaro, P., Aznar, M., et al., 2013. The challenge of identifying non-intentionally added substances from food packaging materials: a review. Anal. Chim. Acta 775, 14–24.

Nerín, C., Aznar, M., Carrizo, D., 2016. Food contamination during food process. Trends Food Sci. Technol. 48, 63–68.

Nielsen, T., Damant, A.P., Castle, L., 1997. Validation studies of a quick test for predicting the sorption and washing properties of refillable plastic bottles. Food Addit. Contam. 14 (6–7), 685–693.

Noonan, G.O., Whelton, A.J., Carlander, D., et al., 2014. Measurement methods to evaluate engineered nanomaterial release from food contact materials. Compr. Rev. Food Sci. Food Saf. 13, 679–692.

NORDEN, 2015. Food Contact Materials – Metal and Alloys – Nordic Guidance for Authorities, Industry and Trade. TemaNord 2015:522. Nordic Council of Ministers, Copenhagen, p. 66. Available from: http://norden.diva-portal.org/smash/get/diva2:816816/FULLTEXT02.pdf. (Accessed 14 May 2018).

O'Keeffe, H., 2017. Update on National legislative developments. In: Proceedings of the Food Packaging Law Seminar 2017, Brussels, Belgium, March 8–9, 2017. Keller and Heckman LLP, Washington DC, United States, p. 32.

O'Keeffe, H., 2018. Overview and updates of EU National laws in non-harmonized food-contact sectors. In: Proceedings of the Food Packaging Law Seminar 2018, Brussels, Belgium, March 6–7, 2018. Keller and Heckman LLP, Washington DC, United States, p. 31.

O'Keeffe, H., 2019. Overview and updates of EU National laws in non-harmonized food-contact sectors. In: Proceedings of the Food Packaging Law Seminar 2019, Brussels, Belgium, March 13–14, 2019. Keller and Heckman LLP, Washington DC, United States, p. 27.

O'Keeffe, H., Gu, E., 2019. Material standards. In: Proceedings of the Workshop Food Contact Regulations in China, Brussels, Belgium, March 12, 2019. Keller and Heckman LLP, Washington DC, United States, p. 33.

Oldring, P.K.T., OMahony, C., Dixon, J., et al., 2014. Development of a new modeling tool (FACET) to assess exposure to chemical migrants from food packaging. In: Ariosti, A., Guest (Eds.), Proceedings of the 5th. LSI International Symposium on Food Packaging 'Scientific Developments Supporting Safety and Innovation', Berlin, Germany, November 14–16, 2012. Food Additives and Contaminants: Part A, vol. 31, pp. 444–465, 3.

Oomen, A.G., Steinhäuser, K.G., Bleeker, E.A.J., et al., 2018. Risk assessment frameworks for nanomaterials: scope, link to regulations, applicability, and outline for future directions in view of needed increase in efficiency. NanoImpact 9, 1–13.

Ossberger, M., 2015. Food migration testing for food contact materials. In: Baughan, J.S. (Ed.), Global Legislation for Food Contact Materials. Woodhead Publishing Ltd., Cambridge, pp. 3–41.

Ossman, B.E., Sarau, G., Holtmannspötter, H., et al., 2018. Small-sized microplastics and pigmented particles in bottled mineral water. Water Res. 141, 307–316.

Ossman, B., Schymanski, D., Ivleva, N., et al., 2019. Comment on exposure to microplastics (<10 mm) associated to plastic bottles mineral water consumption: the first quantitative study. Water Res. 162, 516–517.

Otsuka, M., 2015. Overview of Japanese regulations of food contact utensils and containers/packages. In: Proceedings of the 12th. Biennial International Symposium on Worldwide Regulation of Food Packaging, Baltimore, MD, United States, June 16–18, 2015. The US Society of Plastics Industry (SPI), Washington DC, United States, p. 28.

Overbeek, R.A., 2019. Food contact regulations in Japan. In: Proceedings of the Global Food Contact Conference 2019, Lisbon, Portugal, May 14–16, 2019. Smithers Pira, Leatherhead, Surrey, United Kingdom, p. 30.

Padula, M., 2010. Food packaging legislation in South America and Central America. In: Rijk, R., Veraart, R. (Eds.), Global Legislation for Food Packaging Materials. Wiley-VCH Verlag GmbH & Co. KGaA., Weinheim, pp. 255–282.

Padula, M., 2019. Overview of food packaging regulations in Brazil and other South American countries. In: Proceedings of the 14th. Biennial International Symposium on Worldwide Regulation of Food Packaging, Baltimore, MD, United States, June 11–14, 2019. The Plastics Industry Association (PLASTICS), Washington DC, United States, p. 24.

Padula, M., García, E.E.C., Segantini Saron, E., 2002. Migración en envases retornables de agua mineral. In: Catalá, R., Gavara, R. (Eds.), Migración de componentes y residuos de envases en contacto con alimentos. IATA, Valencia, pp. 249–260.

Parisi, S., Barone, C., Caruso, G., 2015. The influence of chemical composition of food packaging materials on the technological suitability: a matter of food safety and hygiene. In: Barone, C., Bolzoni, L., Caruso, G., et al. (Eds.), Food Packaging Hygiene. Springer International Publishing AG Switzerland, pp. 1–16. https://doi.org/10.1007/978-3-319-14827-4.

Paseiro-Cerrato, R., Rodríguez-Bernaldo de Quirós, A., Sendón, R., et al., 2010. Chromatographic methods for the determination of polyfunctional amines and related compounds used as monomers and additives in food packaging materials: a state-of-the-art review. Compr. Rev. Food Sci. Food Saf. 9, 676–694. https://doi.org/10.1111/j.1541-4337.2010.00133.x.

Peelman, N., Ragaert, P., De Meulenaer, B., et al., 2013. Application of bioplastics for food packaging. Trends Food Sci. Technol. 32, 128–141.

Peltzer, M.A., Beldì, G., Jakubowska, N., et al., 2015a. Scoping investigations on the release of metals from crystalware – in support of the revision of the ceramics directive 84/500/EEC. EU Joint Research Center (JRC) Technical Reports. EUR 27180 EN 26. https://doi.org/10.2788/885263.

Peltzer, M.A., Beldì, G., Jakubowska, N., et al., 2015b. Scoping investigations on the release of metals from the rim area of decorated articles – in support of the revision of the Ceramics Directive 84/500/EEC. EU Joint Research Center (JRC) Technical Reports. EUR 27178 EN 34. https://doi.org/10.2788/484454.

Peters, R.J.B., Groeneveld, I., Lopez Sanchez, P., et al., 2019. Review of analytical approaches for the identification of non-intentionally added substances in paper and board food contact materials. Trends Food Sci. Technol. 85, 44–54.

Petersen, H., Biereichel, A., Burseg, K., et al., 2008. Bisphenol A diglycidyl ether (BADGE) migrating from packaging material 'disappears' in food: reaction with food components. Food Addit. Contam. 25 (7), 911–920.

Piringer, O., 2007. Mathematical modelling of chemical migration from food contact materials. In: Barnes, K.A., Sinclair, C.R., Watson, D.H. (Eds.), Chemical Migration and Food Contact Materials. Woodhead Publishing Ltd., Cambridge, pp. 180–202.

Piringer, O.G., Baner, A.L. (Eds.), 2008. Plastic Packaging – Interactions with Food and Pharmaceuticals, second ed. Wiley-VCH Verlag GmbH & Co. KGaA, Weinheim.

Piringer, O., Franz, R., Huber, M., et al., 1998. Migration from food packaging containing a functional barrier: mathematical and experimental evaluation. J. Agric. Food Chem. 46, 1532–1538.

Polat, S., Fenercioğlu, H., Glüçü, M., 2018. Effects of metal nanoparticles on the physical and migration properties of low density polyethylene films. J. Food Eng. 229, 32–42.

Realini, C.E., Marcos, B., 2014. Active and intelligent packaging systems for a modern society. Meat Sci. 98, 404–419.

Restuccia, D., Spizzirri, U.G., Parisi, O.I., et al., 2010. New EU regulation aspects and global market of active and intelligent packaging for food industry applications. Food Control 21, 1425–1435.

Rhim, J.-W., Park, H.-M., Ha, C.-S., 2013. Bio-nanocomposites for food packaging applications. Prog. Polym. Sci. 38, 1629–1652.

Ridgway, K., Lalljie, S.P.D., Smith, R.M., 2010. Analysis of food taints and off-flavours: a review. Food Addit. Contam. 27 (2), 146–168.

Robertson, G.L., 2012. Food Packaging – Principles and Practice, third ed. CRC Press, Boca Raton, FL.

Rossi, L., 2007. Key regulatory differences in non-Member European States. In: Proceedings of the Global Food Contact Conference 2007, Barcelona, Spain, July 10–12, 2007. Pira International, Leatherhead, Surrey, United Kingdom, p. 49.

Rossi, L., 2010. Council of Europe resolutions. In: Rijk, R., Veraart, R. (Eds.), Global Legislation for Food Packaging Materials. Wiley-VCH Verlag GmbH & Co. KGaA., Weinheim, pp. 49–65.

Rossi, L., 2014. Plastics regulations in EU, USA, China and Japan: the main different approaches. In: Proceedings of the Plastics and Paper in Contact with Foodstuffs Conference, Munich, Germany, December 2–5, 2014. Smithers Pira, Leatherhead, Surrey, United Kingdom, p. 48.

Safa, H.L., Bourelle, F., 1999. Studies on polyester packaging. Effects of basic washing on multi-use PET and PEN bottles. Packag. Technol. Sci. 12 (2), 67–74.

Sanchis, Y., Coscollà, C., Roca, M., et al., 2015. Target analysis of primary aromatic amines combined with a comprehensive screening of migrating substances in kitchen utensils by liquid chromatography-high resolution mass spectrometry. Talanta 138, 290–297.

Sanchis, Y., Yusà, V., Coscollà, C., 2017. Analytical strategies for organic food packaging contaminants. J. Chromatogr. A 1490, 22–46.

Schäfer, A., 2007. Regulation of food contact materials in the EU. In: Barnes, K.A., Sinclair, C.R., Watson, D.H. (Eds.), Chemical Migration and Food Contact Materials. Woodhead Publishing Ltd., Cambridge, pp. 43–63.

Schäfer, A., 2010. EU legislation. In: Rijk, R., Veraart, R. (Eds.), Global Legislation for Food Packaging Materials. Wiley-VCH Verlag GmbH & Co. KGaA., Weinheim, pp. 1–25.

Schumann, B., Schmid, M., 2018. Packaging concepts for fresh and processed meat – recent progresses. Innovat. Food Sci. Emerg. Technol. 47, 88–100.

Schymanski, D., Goldbeck, C., Humpf, H.-U., et al., 2018. Analysis of microplastics in water by micro-Raman spectroscopy: release of plastic particles from different packaging into mineral water. Water Res. 129, 154–162.

Semail, R., 2018. The mutual recognition principle to market food contact materials and articles EU-wide. In: Proceedings of the Plastics and Papers in Contact with Foodstuffs Conference 2018, Vienna, Austria, December 3–6, 2018. Smithers Pira, Leatherhead, Surrey, United Kingdom, p. 29.

Semail, R., 2019. Basics of EU food-contact legislation. In: Proceedings of the Brussels Food Packaging Law Seminar 2019, Brussels, Belgium, March 13–14, 2019. Keller and Heckman LLP, Washington DC, United States, p. 36.

Semail, R., Coneski, P.N., 2019. Comparison of China and EU food-contact legislations. In: Proceedings of the Workshop Food Contact Regulations in China, Brussels, Belgium, March 12, 2019. Keller and Heckman LLP, Washington DC, United States, p. 21.

Sendón García, R., Sanches Silva, A., Cooper, I., et al., 2006. Revision of analytical strategies to evaluate different migrants from food packaging materials. Trends Food Sci. Technol. 17, 354–366.

Shigekura, M., 2018. Food packaging regulations in Japan. In: Proceedings of the Food Packaging Law Seminar US 2018, Arlington, VA, United States, October 10–11, 2018. Keller and Heckman LLP, Washington DC, United States, p. 42.

Sidwell, J., 2015. Global legislation for rubber materials in contact with food. In: Baughan, J.S. (Ed.), Global Legislation for Food Contact Materials. Woodhead Publishing Ltd., Cambridge, pp. 141–160.

Simoneau, C., Raffael, B., Garbin, S., et al., 2016. Non-harmonised Food Contact Materials in the EU: Regulatory and Market Situation – Baseline Study – Final Report. Joint Research Center – European Commission, p. 332. https://doi.org/10.2788/234276. EUR 28357 EN.

Siracusa, V., Rocculi, P., Romani, S., et al., 2008. Biodegradable polymers for food packaging: a review. Trends Food Sci. Technol. 19, 634–643.

Söderhjelm, L., Sipiläinen-Malm, T., 1996. Paper and board. In: Katan, L.L. (Ed.), Migration from Food Contact Materials. Blackie Academic and Professional, London, pp. 159–180.

Song, X.-C., Wrona, M., Nerín, C., et al., 2019. Volatile non-intentionally added substances (NIAS) identified in recycled expanded polystyrene containers and their migration into food simulants. Food Packag. Shelf Life 20, 11, 100318.

Stadler, R.H., Mottier, P., Guy, P., et al., 2004. Semicarbazide is a minor thermal decomposition product of azodicarbonamide used in the gaskets of certain food jars. Analyst 129, 276–281.

Störmer, A., Bott, J., Kemmer, D., et al., 2017. Critical review of the migration potential of nanoparticles in food contact plastics. Trends Food Sci. Technol. 63, 39–50.

Su, Q.-Z., Lin, Q.-B., Chen, C.-F., et al., 2017. Effect of organic additives on silver release from nanosilver-polyethylene composite films to acidic food simulant. Food Chem. 228, 560–566.

Thompson, M., 2017. All wrapped up: food-contact regulations in China. In: Proceedings of the Food Packaging Law Seminar US 2017, Arlington, VA, United States, October 10–11, 2017. Keller and Heckman LLP, Washington DC, United States, p. 30.

Thompson, M., 2019a. All eyes on the Rising Sun: industry perspective on Japan's new positive list system. In: Proceedings of the 14th. Biennial International Symposium on Worldwide Regulation of Food Packaging, Baltimore, MD, United States, June 11–14, 2019. The Plastics Industry Association (PLASTICS), Washington DC, United States, p. 37.

Thompson, M., 2019b. Basics of China food-contact legislation. In: Proceedings of the Workshop Food Contact Regulations in China, Brussels, Belgium, March 12, 2019. Keller and Heckman LLP, Washington DC, United States, p. 39.

Thompson, M., 2019c. New developments and draft standards. In: Proceedings of the Workshop Food Contact Regulations in China, Brussels, Belgium, March 12, 2019. Keller and Heckman LLP, Washington DC, United States, p. 25.

Thompson, M., 2021a. Regulation of food packaging in Japan. In: Proceedings of the Food Packaging Law Seminar US 2021 on line, October 19–20, 2021. Keller and Heckman LLP, Washington DC, United States, p. 17.

Thompson, M., 2021b. Regulation of food packaging in China. In: Proceedings of the Food Packaging Law Seminar US 2021, Arlington, VA, United States, October 19-20, 2021. Keller and Heckman LLP, Washington DC, United States, p. 19.

Thompson, M., Gu, E., 2019. Petitioning process in China. In: Proceedings of the Workshop Food Contact Regulations in China, Brussels, Belgium, March 12, 2019. Keller and Heckman LLP, Washington DC, United Sates, p. 26.

Tingle, V., 1996. Glass. In: Katan, L.L. (Ed.), Migration from Food Contact Materials. Blackie Academic and Professional, London, pp. 145–158.

Twaroski, M.L., Batarseh, L.I., Bailey, A.B., 2007. Regulation of food contact materials in the USA. In: Barnes, K.A., Sinclair, C.R., Watson, D.H. (Eds.), Chemical Migration and Food Contact Materials. Woodhead Publishing Ltd., Cambridge, pp. 17–42.

US-FDA, 2021. Guidance for Industry: Use of Recycled Plastics in Food Packaging (Chemistry Considerations). Washington DC. Available from: www.fda.gov/RegulatoryInformation/Guidances/ucm120762.htm. (Accessed 28 October 2021).

US-FDA, 2018. Code of Federal Regulations. Title 21, Part 176 Indirect Food Additives: Paper and Paperboard Components. Section 176.260 Pulp from Reclaimed Fiber. Washington DC. Available from: www.ecfr.gov/cgi-bin/text-idx?SID=e956d645a8b4e6b3e34e4e5d1b690209&mc=true&node=pt.21.3.176&rgn=div5. (Accessed 30 June 2018).

Vanderroost, M., Ragaert, P., Devlieghere, F., et al., 2014. Intelligent food packaging: the next generation. Trends Food Sci. Technol. 39, 47–62.

Vera, P., Canellas, E., Nerín, C., 2014. Migration of odorous compounds from adhesives used in market samples of food packaging materials by chromatography olfactometry and mass spectrometry (GC–O–MS). Food Chem. 145, 237–244.

Vera, P., Canellas, E., Nerín, C., 2018. Identification of non volatile migrant compounds and NIAS in polypropylene films used as food packaging characterized by UPLC-MS/QTOF. Talanta 188, 750–762.

Veraart, R., 2015. Compliance testing for food contact materials. In: Baughan, J.S. (Ed.), Global Legislation for Food Contact Materials. Woodhead Publishing Ltd., Cambridge, pp. 43–64.

Veraart, R. (Ed.), 2017. The Use of Nanomaterials in Food Contact Materials – Design, Application, Safety. DEStech Publications Inc., Lancaster, PA.

Welle, F., 2011. Twenty years of PET bottle to bottle recycling – an overview. Resour. Conserv. Recycl. 55, 865–875.

Welle, F., 2013. Is PET bottle-to-bottle recycling safe? Evaluation of post-consumer recycling processes according to the EFSA guidelines. Resour. Conserv. Recycl. 73, 41–45.

Welle, F., 2016. Investigation into cross-contamination during cleaning efficiency testing in PET recycling. Resour. Conserv. Recycl. 112, 65–72.

Welle, F., Franz, R., 2007. Recycled plastics and chemical migration into food. In: Barnes, K.A., Sinclair, C.R., Watson, D.H. (Eds.), Chemical Migration and Food Contact Materials. Woodhead Publishing Ltd., Cambridge, pp. 205–227.

Welle, F., Franz, R., 2018. Microplastic in bottled natural mineral water – literature review and considerations on exposure and risk assessment. Food Addit. Contam. 35 (12), 2482–2492.

Widén, H., Leufvén, A., Nielsen, T., 2005. Identification of chemicals, possibly originating from misuse of refillable PET bottles, responsible for consumer complaints about off-odours in water and soft drinks. Food Addit. Contam. 22 (7), 681–692.

Wu, L.-B., Su, Q.-Z., Lin, Q.-B., et al., 2017. Impact of migration test method on the release of silver from nanosilver-polyethylene composite films into an acidic food simulant. Food Packag. Shelf Life 14, 83–87.

Wu, S., Wang, W., Yan, K., et al., 2018. Electrochemical writing on edible polysaccharide films for intelligent food packaging. Carbohydr. Polym. 186, 236–242.

Xu, Y., Noonan, G.O., Begley, T.H., 2013. Migration of perfluoroalkyl acids from food packaging to food simulants. Food Addit. Contam. 30 (5), 899–908.

Youssef, A.M., El-Sayed, S.M., 2018. Bionanocomposites materials for food packaging applications: concepts and future outlook. Carbohydr. Polym. 193, 19–27.

Zhai, X., Shi, J., Zou, X., et al., 2017. Novel colorimetric films based on starch/polyvinyl alcohol incorporated with roselle anthocyanins for fish freshness monitoring. Food Hydrocoll. 69, 308–317.

Zhang, H., 2018. China system and food contact regulations. In: Proceedings of the Food Contact Regulations USA 2018 Conference, Arlington, VA, United States, March 8–9, 2018. Chemical Watch Research, Ltd., Shrewsbury, United Kingdom, p. 23.

Zhang, R., 2019. Food contact materials industrial compliance strategy of China. In: Proceedings of the 14th. Biennial International Symposium on Worldwide Regulation of Food Packaging, Baltimore, MD, United States, June 11–14, 2019. The Plastics Industry Association (PLASTICS), Washington DC, United States, p. 50.

Zheng, N., 2017. China's food contact regulations related to plastics. In: Proceedings of the Global Food Contact Conference 2017, Rome, Italy, May 15–17, 2017. Smithers Pira, Leatherhead, Surrey, United Kingdom, p. 27.

Zhu, L., 2019. Updates and trends of food contact standards in China. In: Proceedings of the 14th. Biennial International Symposium on Worldwide Regulation of Food Packaging, Baltimore, MD, United States, June 11–14, 2019. The Plastics Industry Association (PLASTICS), Washington DC, United States, p. 36.

Websites of interest

ANVISA (Brazil). www.anvisa.gov.br.

Argentine Food Code. www.argentina.gob.ar/anmat/codigoalimentario.

BfR. http://www.bfr.bund.de/en/bfr_recommendations_on_food_contact_materials-1711.html.

CEN—EN Standards. www.cen.eu.

China

NHC. www.nhc.gov.cn/; http://en.nhc.gov.cn/.

CFSA. www.cfsa.net.cn/; http://en.cfsa.net.cn/.

Council of Europe – General

www.coe.int/en/web/portal.

Council of Europe—Resolutions on FCMs

www.edqm.eu/en/resolutions-policy-statements.

www.edqm.eu/en/food-contact-materials.

EFSA

www.efsa.europa.eu.

European Union—FCMs regulations

https://ec.europa.eu/food/safety/chemical_safety/food_contact_materials_en.

EURL-FCM (JRC)

https://ec.europa.eu/jrc/en/eurl/food-contact-materials.

FSANZ (Australia—New Zealand)

Food Standards. www.foodstandards.gov.au/consumer/chemicals/foodpackaging/Pages/default.aspx.

ILSI. http://ilsi.org/.

Japan—Ministry of Health, Labour and Welfare (MHLW)

Ministry of Health, Labour and Welfare. www.mhlw.go.jp/english/policy/health-medical/food/index.html.

Ministry of Health, Labour and Welfare. www.mhlw.go.jp/english/topics/foodsafety/containers/index.html.

Japan—Food Sanitation Act (English and Japanese versions). www.japaneselawtranslation.go.jp/law/detail/?ft=2&re=01&dn=1&yo=%E9%A3%9F%E5%93%81%E8%A1%9B%E7%94%9F%E6%B3%95&ia=03&x=25&y=17&ky=&page=1.

Japan—Food Safety Basic Act (English and Japanese versions)

Japanese Translation. http://www.japaneselawtranslation.go.jp/law/detail/?id=1839&vm=04&re=02.

Japan—Japan External Trade Organization (JETRO)

JETRO. www.jetro.go.jp/en/reports/regulations/.

MERCOSUR—General. www.mercosur.int.

MERCOSUR—Resolutions

MERCOSUR. www.mercosur.int/innovaportal/v/527/2/innova.front/resoluciones.

Punto Focal. www.puntofocal.gob.ar.

MERCOSUR—Situation of Venezuela and Protocol of Ushuaia. www.mercosur.int/innovaportal/v/7823/11/innova.front/paises-del-mercosur.

US-FDA. www.fda.gov.

US-FDA—eCFR Title 21—Parts 170–190. www.ecfr.gov/cgi-bin/text-idx?SID=4ec8f91e9d7761feaa2871762c48972c&mc=true&tpl=/ecfrbrowse/Title21/21cfrv3_02.tpl#0.

US-FDA—Guidances for industry (general). www.fda.gov/Food/GuidanceRegulation/GuidanceDocumentsRegulatoryInformation/IngredientsAdditivesGRASPackaging/default.htm.

US-FDA—Guidance for industry (Chemistry recommendations for FAPs and FCNs submissions). www.fda.gov/Food/GuidanceRegulation/GuidanceDocumentsRegulatoryInformation/ucm081818.htm.

Chapter 16

Nanotechnology and food safety

Syed S.H. Rizvi[1], Carmen I. Moraru[1], Hans Bouwmeester[2], Frans W.H. Kampers[3] and Yifan Cheng[1]
[1]Department of Food Science, Cornell University, Ithaca, NY, United States; [2]Division of Toxicology, Wageningen University and Research, Wageningen, the Netherlands; [3]Wageningen UR, Wageningen, the Netherlands

16.1 Introduction

Nanoscale objects are not new; they have been known to exist for decades. Yet, it was the ability of scientists to see and engineer nanostructures via self- or directed assembly in the 1980s that catalyzed their rapid development. Nanotechnology has now evolved into a convergent discipline involving a variety of sciences (physical, chemical, biological, engineering, and electronic) designed to understand and manipulate structures and devices at nanoscale. The use of nano-based consumer products is growing rapidly and many such products are available in the market. To date, more than 1800 consumer products that are self-identified by the manufacturers as containing nanotechnology are included in the public database (Project on Emerging Nanotechnologies, 2018). Nano-based goods are projected by various sources to be an estimated $2.6 trillion global industry by 2014 (ScienceDaily, 2007) and a nano-dominated future is not too distant.

A nanomaterial is generally defined as "a discrete entity that has one or more dimensions of the order of 100 nm or less" (SCENIHR, 2007b). Unique properties of these materials arise at the nanoscale where they do not behave like their macroscale counterparts. The physico-chemical properties of nanostructures are not governed by the same laws as larger structures, but by quantum mechanics. Color, solubility, diffusivity, material strength, toxicity, and other properties will be very different at the nanoscale as compared to the macroscale. Other important properties include increased reactivity because of quantum mechanical effects in combination with the high (relative) surface area of nanostructures, which allows the creation of genuinely new properties and materials with desired functionalities. The potential benefits for application of nanotechnologies in food have been widely discussed and cover many aspects such as efficient nutrient delivery, formulations with improved bioavailability, novel antimicrobials, new tools for molecular, and cellular detection of contaminants and food packaging materials (Chaudhry et al., 2008; Moraru et al., 2003, 2009; Feng et al., 2014, 2015; Chen et al., 2006; Das et al., 2008; Weiss et al., 2006; Luksiene, 2017).

As new applications emerge, there is also growing concern, both from the public and the scientific community, about the potential risks and toxicity of nanotechnology products, or the environmental and personal safety aspects of their use. The fact that size matters as it influences efficacy and safety, the question arises whether a nanoscale particle that has unique material properties should be deemed new or nonnatural for purposes of safety evaluation. A lack of knowledge of how nanoscale structures may interact with biological systems and potentially create safety issues is often cited as the last hurdle to overcome in their acceptability in food and pharmaceutical applications.

The purpose of this chapter is to explore the global safety and regulatory issues associated with the application of nanotechnology to food systems. Emphasis is placed on the toxicological knowledge needed to perform a hazard assessment and the lack of measurement technologies for exposure assessment. Additionally, the current state of the regulatory food safety framework is described, and the consequences of the scientific knowledge gaps that exist in the application and enforcement of the current regulations are highlighted.

16.2 Nanotechnology and food systems

Food science and technology has made many micro- and often nano-size food particles by utilizing either the top-down (e.g., grinding, microparticulation, micronizing) or bottom-up (e.g., molecular aggregation) approaches. The advent of nanotechnology has ushered in new scientific and technological opportunities for the food industry, but the uncertainty

Ensuring Global Food Safety. https://doi.org/10.1016/B978-0-12-816011-4.00016-1

remains whether small sized material should be treated as new entities when compared to their larger forms (Chau et al., 2007). Efforts are underway worldwide to most responsibly realize the benefits of nanomaterials without exposing the public and environment to harm. Controlled cross-linking, inhibition of droplets coalescence, and generation of multilayered structures are just a few generic examples of the role of nanotechnology. To date, the key areas of food nanotechnology research, development, and application include the following:

16.2.1 Structure and function characterization and modification

Macro components of food, protein, carbohydrates, and fat constitute a set of nanostructures that are ideally suited for targeted advances via nanotechnology. A vast majority of food carbohydrates and lipids are one-dimensional nanostructures of less than 1 nm in thickness, while globular proteins are nanoparticles of 10–100 nm in size. An understanding of the behavior and functionality of these nanostructures can be profitably used to set up processing strategies to improve food structure. Availability of new physical tools like the atomic force microscope (AFM) to study nanostructures has proven invaluable in understanding food structure–function relationships. AFM has been successfully used to quantify properties like stiffness, hardness, friction, elasticity, or adhesion at the molecular level. The ability to manipulate individual biomolecules using AFM has allowed the study of structural and phase transitions, nanorheological, and nanotribological properties of polymers (Strick et al., 2000; Terada et al., 2000; Morris et al., 2001; Nakajima et al., 2001; Boskovic et al., 2002). The gelation ability, the mechanisms of gelation, and the microstructure of the resulting gels were studied for biopolymers such as gums and proteins (Gajraj and Ofoli 2000; Ikeda et al., 2001; Morris et al., 2001). Recent development in AFM imaging modes such as multiparametric, molecular recognition, multifrequency, correlative, and high-speed imaging further expands the capabilities of this powerful tool in understanding structure–function relationships in complex biological and food systems (Dufrêne et al., 2017).

The development of new generation scanning probe microscopy (SPM) instrumentation, including the near-field scanning optical microscope, scanning thermal microscope, scanning capacitance microscope, magnetic force and resonance microscopes, and scanning electrochemical microscope, further enhanced the capability of investigations at the nanoscale. By combining SPM and AFM with other imaging, mechanical, and spectroscopic methods, it is now possible to quantitatively characterize the structures of polymers at the micrometer and nanometer level, as well as the intra- and intermolecular forces that stabilize such structures. Nonintrusive determination of local phase behavior and structure in complex biopolymer matrices could ultimately lead to improved control and design of the quality and stability of foods.

Fabrication at the nanoscale opened windows of opportunity for the creation of new, high-performance materials with applications in food processing, packaging, and storage. For example, nanotechnology approaches have been used to develop nanoparticles enhanced polymeric membranes for applications such as purification of ethanol and methanol (Jelinski, 1999; Kingsley, 2002) and packaging materials with low gas permeability (Duncan, 2011). Another novel solution for enhancing membrane functionality is based on the use of nanotubes.

Nanotubes are long and thin tubes that can be assembled in extremely stable, strong, and flexible honeycomb structures. Nanotubes are the strongest fibers known—one nanotube is estimated to be 10–100 times stronger than steel per unit weight. By functionalizing nanotubes in a desired manner, membranes could be tailored to efficiently separate molecules both on the basis of their molecular size and shape and on their chemical affinity. High selectivity nanotube membranes can be used both for analytical purposes, as part of sensors for molecular recognition of enzymes, antibodies, proteins and DNA, or for the membrane separation of biomolecules (Huang et al., 2002; Lee and Martin 2002; Rouhi, 2002; Siwy et al., 2005).

Another application of nanotubes is the fabrication of nanotube-reinforced composites with high fracture and thermal resistance. Such materials could replace conventional materials in the manufacture of a wide range of machinery, including food-processing equipment (Gorman, 2003; Zhan et al., 2003; Raviathul Basariya et al., 2014; Chen et al., 2017). While such technologies are still too expensive for commercial scale food applications, it can be foreseen that they would become feasible for food-related applications in the not so distant future.

16.2.2 Nutrient delivery systems

Nanostructures in foods can be designed for the targeted delivery of nutrients in the body for the most beneficial effects. By facilitating a precise control of properties and functionality at the molecular level, nanotechnology enabled the development of highly effective encapsulation and delivery systems. Examples include nanometer-sized association colloids such as surfactant micelles, vesicles, bilayers, reverse micelles, or liquid crystals. Such systems could be used in food applications as carrier or delivery systems for vitamins, antimicrobials, antioxidants, flavorings, colorants, or preservatives (Weiss et al., 2006; Jafari and McClements, 2017).

Nanospheres have been proven to have superior encapsulation and release efficiency as compared to traditional encapsulation systems (Riley et al., 1999; Weiss et al., 2006; Jafari and McClements, 2017). Nanoscale encapsulation systems can be produced using food biopolymers such as proteins or polysaccharides, which then can be used to encapsulate functional ingredients and release them in response to specific environmental triggers. Dendrimer-coated particles and cochleates can also be used as efficient encapsulation and delivery systems (Santangelo et al., 2000; Khopade and Caruso, 2002; Gould-Fogerite et al., 2003). Cochleates can be used for the encapsulation and delivery of many bioactive materials, including compounds with poor water solubility, protein and peptide drugs, and large hydrophilic molecules (Gould-Fogerite et al., 2003). Another solution for encapsulation of functional components is via nanoemulsions. The advantage of nanoemulsions is that they can enable the slowdown of chemical degradation by engineering the properties of the interfacial layer surrounding them (McClements and Decker, 2000). Such systems could potentially be used for the encapsulation and targeted delivery and controlled release of functional food molecules.

16.2.3 Sensing and safety

Nanotechnology has benefited the area of food safety mostly through the development of highly sensitive biosensors for pathogen detection and the development of novel antimicrobial solutions. Fellman (2001) reported the development of a method to produce nanoparticles with a triangular prismatic shape for detecting biological threats such as anthrax, smallpox and tuberculosis, and a wide range of genetic and pathogenic diseases. Latour et al. (2003) investigated the ability of two types of nanoparticles to irreversibly bind to certain bacteria, inhibiting them from binding to and infecting their host. One type was based on inorganic nanoparticles functionalized with polysaccharides and polypeptides that promote the adhesion of the targeted bacterial cells. This research has the potential to reduce the infective capability of human food-borne enteropathogens such as *Campylobacter*, *Salmonella*, and *Escherichia coli* in poultry products (Latour et al., 2003). Kuo et al. (2008) successfully developed a bioconjugation procedure that allows the attachment of water-soluble cadmium tellurium semiconductor quantum dots to anti-*E.coli* antibody. Such quantum dots are promising probe materials in the development of antibody-based immunosensors with high stability, sensitivity, and reproducibility, that could allow the detection of a single pathogenic cell (Kuo et al., 2008). Jin et al. (2009) showed that quantum dots made out of zinc oxide could be effective at inhibiting pathogens such as *Listeria monocytogenes, Salmonella enteritidis*, and *Escherichia coli* O157:H7, which further demonstrated the promise of nanoparticles for food safety applications.

16.2.4 Antimicrobials

Advances in nanoparticle synthesis have given rise to a variety of nano-enabled or nano-enhanced antimicrobials that hold great promise in improving food safety. Compared to their bulk counterparts, nanoparticles-based antimicrobials have increased activity owing both to their small size, which facilitates penetration of cellular membranes, and to a high surface-area-to-volume ratio, which increases the number of active sites and/or boosts the release rate of active components (Lemire et al., 2013; Bastarrachea et al., 2015).

Metal-based nanoparticles have garnered increasing attention due to their excellent antimicrobial effects. For instance, silver and copper in their bulk solid form have been long known for their capability of disinfecting water and preserving food (Castellano et al., 2007). Recent development of metal oxide nanoparticles, including silver, gold, copper, aluminum, zinc oxide, and titanium oxide, started to reveal their exceptional potential as alternatives to conventional chemical antimicrobials in food systems (Luksiene, 2017). The currently accepted antimicrobial mechanisms of metallic nanoparticles include generation of reactive oxygen species, release of metal ions, and nonoxidative mechanisms (Wang et al., 2017). One important advantage of metallic nanoparticles over conventional chemical-based antimicrobials is that metallic nanoparticles can effectively lower the risk of developing antimicrobial resistance by attacking a broad range of targets of the microbial cells, including DNA and some intracellular proteins, which limits the chance of developing mutations necessary for the microorganisms to survive (Lemire et al., 2013; Luksiene, 2017). As the threat of multidrug-resistant bacteria is increasing globally (Norrby et al., 2005; Roca et al., 2015), metallic nanoparticles may become a very attractive alternative to some of the existing antimicrobials that are known to induce microbial resistance.

Nanoparticles can also enhance the antibacterial activity of chemical antimicrobials, due to the dramatic increase in the number of active sites, enabled by the high specific surface area of nanoparticles. Nanoparticle-associated antimicrobials have been found to exhibit superior stability over their free-form counter-parts, which is important for their long-term biocidal activity (Jain et al., 2014). One such example is the combination of nanoparticles with quaternary ammonium (QA) compounds, which are among the most promising antimicrobial chemicals for food and agriculture systems, thanks to their high biocidal activity, long-term effectiveness, and environmental friendly nature. Song et al. (2011) showed that silica nanoparticles functionalized with QA exhibited enhanced antimicrobial efficacy against growth of *E. coli* and *S. aureus* compared to silica nanoparticles alone, and the biocidal performance of QA-modified silica nanoparticles improved with decreasing size of the nanoparticles. Nanoscale structures such as dendrimers can further enhance the surface density of antimicrobial compounds. Wen et al. (2012) synthesized a core shell type dendrimer nanoparticles featuring a high-density QA shell that improved water solubility and antimicrobial ability. Future research on the synergistic effects between antimicrobial compounds and various nanostructures may provide novel, highly effective bactericidal solutions.

In spite of their great promise, one caveat in using nanoscale metallic antimicrobials in the food industry is that their effectiveness can be impaired by the interactions with organic compounds that exist in abundance in food systems (Noyce et al., 2006). The other limitation lies in public health and environmental concerns about migration of metallic nanoparticles (Lemire et al., 2013; Llorens et al., 2012; Marambio-Jones and Hoek 2010; Rai et al., 2009). For instance, although some silver-based coatings are listed by the US Food and Drug Administration (FDA) in their Inventory of Effective Food Contact Substance Notifications (FDA, 2018), the use of silver and silver zeolite—based antimicrobial materials in food applications is currently limited.

16.2.5 Food packaging and tracking

The use of nanostructured materials, particularly nanocomposites, could considerably enhance the functional properties of packaging materials, and thus improve the shelf life of packaged foods. Nanocomposites are made out of nanoscale structures with unique morphology, increased modulus, and strength, as well as good barrier properties. For example, a packaging material made out of potato starch and calcium carbonate that has good thermal insulation properties, lightweight, and biodegradability was proposed as a replacement for the polystyrene "clam-shell" used for fast food (Stucky, 1997). Nanocomposites are also regarded as a potential solution for plastic beer bottles (Moore, 1999). Natural smectite clays, particularly montmorillonite, a volcanic material that consists of nanometer-thick platelets, can be used as an additive that makes plastics lighter, stronger, more heat resistant, with improved oxygen, carbon dioxide, moisture, and volatile barrier properties (Quarmley and Rossi, 2001). Nanocomposites based on starch and reinforced with tunicin whiskers (Mathew and Dufresne, 2002) or clay nanocomposites (Park et al., 2003) have also been developed in recent years. A nanocomposite material based on chitosan and reinforced with exfoliated hydroxyapatite layers was developed by Weiss et al. (2006).

Coatings or films for food packaging materials could also be made using nanolaminates and nanofibers. Nanolaminates consist of two or more layers of material with nanometer dimensions physically or chemically bonded to each other. Nanolaminates can be made using the layer-by-layer deposition technique, which allows precise control of the thickness and properties of the nanolaminates (Weiss et al., 2006). Nanofibers are polymeric strands of submicrometer diameters produced by interfacial polymerization and electrospinning. Electrospun polymer fibers have unique mechanical, electrical, and thermal properties, and have applications in filtration, manufacturing of protective clothing, and biomedical applications. Production of nanofibers from food biopolymers in the future might increase their use in the food industry for a range of applications, including packaging materials (Weiss et al., 2006). In addition to pursuing improved functional properties of packaging materials, nanocomposite research has also expanded into developing smart packaging (e.g., enzyme immobilization, indicators of temperature abuse of product during its manufacturing and shelf life), green synthesis of nanomaterials, antimicrobial packaging, or materials with improved recyclability and biodegradability (Rhim et al., 2006, 2013; Yoksan and Chirachanchai, 2010).

Food traceability is another important area where nanotechnology is playing an increasingly important role. Within the context of agri-food system, according to Costa et al. (2013), traceability can be defined as "the ability to locate an animal, commodity, food product, or ingredient and follow its history in the supply chain forward (from source to consumer) or backward (from consumer to source)." For example, some efforts in nanotechnology-based detection and tracking targeted the meat industry, with nanobarcodes that were incorporated into animals through feed or direction injection to create unique identification that would last through the lifecycle of the animals and the resulting products (Kuzma, 2010). Such nanobarcodes could allow tracking of feed contamination, tracing animal products from farm to fork for supply chain management, and ensuring food safety. Despite the promising applications, nanobarcodes incorporated into food have raised concerns in some consumer groups, as such a tracking system may infringe upon consumer and company privacy; additionally, the environmental and health ramifications of the nanoparticles used in this technology need to be carefully studied before implementation on a large scale (Kuzma, 2010).

Product tracking can also be accomplished via packaging, which circumvents most of the abovementioned concerns associated with nanosized tags inside food products. Among the available tracking technologies compatible with food packaging applications, Radio Frequency Identification (RFID) technology has shown the most promise (Costa et al., 2013), thanks to its long detection range, fast reaction speed, minimal user effort, unique identification of every individual item/product, and simultaneous reading of information from more than one tag. Despite all the advantages of RFID technology over traditional barcodes, its market penetration is largely hampered by its high cost, which a few years ago started at 7 US cents per RFID tag, much higher than the "less than 1 US cent" target for it to become economically feasible for food packaging (BRIDGE, 2007). Yet, as it is the case with any new technology, costs are coming down as the technology develops. For instance, Jung et al. (2010) successfully printed the electronics of an RFID tag with single-walled carbon nanotube (SWCNT)—based ink and demonstrated a practical way to scale up the production with a roll-to-roll process on plastic foils. The exceptional electric properties of SWCNT and their high dispensability in solvent enabled the scalable and deterministic printing of RFID tags (Jung et al., 2010; Cao et al., 2008; Ahn et al., 2006). This represents a clear example of how nanotechnological breakthroughs in fields not directly related to food can help solve one of the most challenging problems in the agri-food system including food traceability and food safety.

16.3 Current status of regulation of nanomaterials in food

The need for information and scientific advice on the safety implications that may arise from the use of nanotechnology in food and agriculture was recognized by the World Health Organization (WHO) and the Food and Agriculture Organization (FAO) of the United Nations. These organizations decided to work together on identifying knowledge gaps in areas related to food safety, risk assessment procedures, as well as on developing global guidance on adequate and accurate methodologies to assess potential food safety risks that may arise from nanoparticles. This collaborative effort will focus on both the application of nanotechnology in the primary production of foods, in food processing, packaging, and distribution, as well as the use of nano-diagnostic tools for detection and monitoring in the food and agriculture production. According to information posted on the FAO website (http://www.fao.org), the issues that will be addressed jointly by the two agencies include: (a) on-going research and development on nanotechnologies for use in the food and agriculture sectors that are expected to reach market within the next 10 years; (b) investigations of nanoparticle migration from food contact materials into foods; (c) purity, particle size distribution, and properties of nanoparticulate substances for use in foods and food contact surfaces; (d) mechanistic understanding of the behavior of nanoparticles in the body; (e) nanoforms of vitamins and nutrients in relation to their bioavailability, interference with the absorption of other nutrients, and consideration of safelimits; (f) interactions of nanoparticles with biomolecules, nutrients and contaminants, and their relevance to human health; (g) techniques for detecting, characterizing, and measuring nanoparticles in foods and food contact materials; (h) risk assessments of nanomaterials for use in foods and food contact surfaces; (i) information on nanodiagnostic tools in the food and agriculture sectors; (j) public perceptions of the applications of nanotechnologies to the food and agriculture sectors; and (k) identification of needs and priority areas for scientific advice needed in safety management and regulation by national authorities.

The last issue listed above is of particular importance, as regulation of nanotechnology is still in its infancy, and there is a great deal of variability in how this topic is addressed from country to country. A brief overview of the current status of regulating nanotechnology products that are relevant for the food and agriculture sectors is provided below.

16.3.1 North America

In the United States, the government agency responsible for regulating food, dietary supplements, and drugs is the FDA. However, it must be noted that dietary supplements fall under a different set of regulations than those covering "conventional" foods and drug products. The fact that some of these products, especially dietary supplements, are currently manufactured using nanotechnology creates an additional layer of complexity, as size has not been addressed so far by existing regulations. Under current US legislation, an ingredient or substance that will be added to foods is subject to premarket approval by FDA, unless its use is generally recognized as safe (GRAS). Yet, at this point, there is no information or guidance on how existing listings for food additives and GRAS substances apply to nanoscale materials. Clarification in this area has been identified by many policy experts as an urgent need, because otherwise products that contain nanoscale ingredients could be placed on the market without FDA clearance.

In a statement posted on its official website (http://www.fda.gov), the FDA states that it regulates products, not technologies, and that "nanotechnology products will be regulated as 'Combination Products' for which the regulatory pathway has been established by statute." At the same time, recognizing the challenges associated with the development of nanoproducts, the FDA issued in July 2007 a "Nanotechnology Task Force Report." The report is public and available at http://www.fda.gov. This report acknowledges that nanoscale materials could be potentially used in most product types regulated by the FDA and that such materials present challenges because properties relevant to product safety and effectiveness may change at the nanoscale. This report recommends that FDA provides guidance regarding when the use of nanoscale materials changes the regulatory status of foods, food additives, food contact substances, or dietary supplements.

Efficient and strict regulations about nanotechnology cannot be passed, however, without a proper understanding of the potential risks associated with nanoproducts. The House Science and Technology Committee is currently looking into the need to strengthen federal efforts to learn more about the potential environmental, health, and safety risks posed by engineered nanomaterials. This was done following Environmental Protection Agency (EPA) recommendations for improving federal risk research and oversight of engineered nanomaterials by the EPA, FDA, and the Consumer Product Safety Commission. The report, "Nanotechnology Oversight: An Agenda for the Next Administration," published by the Project on Emerging Nanotechnologies (PEN) (Davies, 2008), offers a range of proposals on how Congress, federal agencies, and the White House can improve oversight of engineered nanomaterials.

The report "Review of the Federal Strategy for Nanotechnology-Related Environmental, Health and Safety Research" (National Research Council, 2008) identified serious weaknesses in the National Nanotechnology Initiative (NNI) plan for research on the potential health and environmental risks posed by nanomaterials. Among other observations, the report states that the NNI plan fails to identify important areas that should be investigated, such as a more comprehensive evaluation of how nanomaterials are absorbed and metabolized by the body and how toxic they are at realistic exposure levels. Significant criticism stemmed from the fact that the NNI plan does not address the current lack of studies on how to manage consumer and environmental risks, or mitigate exposure through consumer products. The report called for a revamped and comprehensive national strategic plan to minimize the potential risks of nanotechnology, which will allow the society to fully benefit from the discoveries of this technology in areas like medicine, energy, transportation, and communications.

In a significant development, Canada is planning to become the first country in the world to require companies to detail their use of engineered nanomaterials. This information is meant to help evaluate the risks of engineered nanomaterials and will be used toward the development of a regulatory framework (Heintz, 2009). This action came shortly after the Office of Pollution Prevention and Toxics of the US EPA released in January 2009 an interim report on the Nanoscale Materials Stewardship Program (NMSP) (EPA, 2009). NMSP was developed to help provide a firmer scientific foundation for regulatory decisions by encouraging submission and development of information about nanoscale materials. Under the NMSP Basic Program, EPA invited participants to voluntarily report information on the engineered nanoscale materials they manufacture, import, process, or use. Under the NMSP In-Depth Program, EPA invited participants to work with the Agency and others on a plan for the development of data on representative nanoscale materials over a longer time frame.

16.3.2 Europe

It is clear from a number of regulatory reports that there is currently no nano-specific regulation in the European Union (EU) (Chaudhry et al., 2008), or other countries (Hodge et al., 2007). However, the EU's approach to nanotechnology is that "nanotechnology must be developed in a safe and responsible manner" (EC recommendation, 2008). To that end, the EU has commissioned the Scientific Committee on Emerging and Newly Identified Health Risks (SCENIHR) to make an inventory to check whether nanotechnologies are already covered by other community legislation, thus defining the legislative framework, considering both implementation and enforcement tools for this specific framework. It was concluded that the EU regulatory framework in principle also covered nanotechnologies (SCENIHR, 2007a). In line with this, EU member states will aim to modify existing laws and rules as and when developments within the fields of nanoscience and nanotechnology render such measures necessary (Franco et al., 2007; Health Council Netherlands, 2006). Recently, the European Food Safety Authority EFSA was asked for a scientific opinion on the need for specific risk assessment approaches for technologies, processes, and applications of nanoscience and nanotechnologies in the food and feed area. While EFSA recognized the limited knowledge on possible food applications and the limited knowledge on the nanotoxicology, it considered the currently used risk assessment paradigm applicable for nanoparticles in food (EFSA, 2021).

16.3.2.1 Nano-size and regulations

The European General Food Law (GFL) is the umbrella of EU's food safety regulations (EC/178/2002, 2002). According to the GFL, all foods placed on the Community market must be safe ("Food shall not be placed on the market if it is unsafe"; where unsafe is defined as "injurious to health" or "unfit for human consumption" article 14 sub 1). In making an assessment of food safety, producers, among others, are required to take into account the probable immediate, short-term and/or long-term effects on the consumer and subsequent generations. The GFL stipulates that it is the responsibility of "food business operators" to ensure that their foods satisfy the requirements of food law. This regulation clearly stipulates that in decision-making, scientific risk assessment should be central. The GFL stipulates that if after assessing the available information a possibility of harmful effects on health is identified but scientific uncertainty persists, risk management measures to ensure a high level of health protection may be adopted. Pending gathering and developing further scientific information for a more comprehensive risk assessment, the GFL allows the application of the precautionary principle.

Chemicals or substances intentionally added to food need to be authorized, meaning that in general a safety assessment of the material has been made before its market entry. In order to conduct a safety assessment, sufficient toxicological hazard information should be made available by the producers of the substance. This will also be the case for nano-substances subjected to authorization. However, in the existing European food safety legislation, no reference is made to nanotechnology or the nanosize of chemicals.

Authorization procedures, legislation, guidelines, and guidance documents describe how and which toxicity tests should be performed. Adjustments of legislation in particular, guidelines and guidance documents concerning the testing of nanoparticles are considered to be necessary (SCENIHR, 2007a). In particular, information concerning the physico-chemical parameters—e.g., particle size, particle form, surface properties, and other properties that may impact the toxicity of the substance—should be included. In addition, the validity of currently used toxicological assays—such as an OECD (Organisation for Economic Co-operation and Development) safety test protocols—for the detection of "novel" nanoparticle related effects needs to be determined. The currently used assays are validated for the toxicity testing of bulk chemicals. Furthermore, appropriate dose metrics to use in the hazard characterization and consumer exposure assessments of nanostructured materials should be developed. Thresholds or limits already set may not be appropriate for nanosized variants of the particular substances.

If a substance in its conventional form has been evaluated, reevaluation of the nano-sized form may be necessary. One should be aware that each new nano-sized form of a certain chemical probably has to be considered as a separate new compound, as long as size–effect relationships are not established for that compound. This underscores the need for taking into account the effect of particle size (including distribution of the size) in toxicological studies.

16.3.2.2 Monitoring the products containing nanotechnology on the market

A requirement in the GFL is that member states should monitor to verify if the requirements of food law are fulfilled by food business operators. Also the EFSA should establish monitoring procedures to identify emerging risks. The monitoring of nanoparticles will require the development of new analytical detection and confirmation techniques.

The Novel Food Regulation (EC/258/97) can be very relevant for nanotechnology in food. This regulation addresses "production processes not currently used," making it likely that this regulation also covers nanotechnology because of its novelty. It is not clear, however, whether the use of nano variants of chemicals in foods already on the market makes these foods "novel" and thus requiring authorization. The Novel Food Regulation is under revision at this moment, which creates an opportunity to sort out nanotechnology related issues.

In conclusion, the current food safety legislative system should be adapted but not rewritten to cover nanotechnologies, while it continues to protect the European consumer. The discussions on definitions of nanotechnologies and nanoparticles continue within, for example, the scientific committees of the European Commission but also globally within ISO (IRGC, 2008). The outcome will have a direct effect on the regulatory framework within Europe. However, there are serious concerns on the sensitivity of current toxicity assays; these concerns are addressed in the next section.

16.4 Hurdles in evaluation and regulation of the use of nanotechnology in foods

Food safety regulations require scientific safety assessments of foods and their ingredients, and this applies to nano-sized substances as well. While EFSA considered the currently used risk assessment paradigm also applicable to nanoparticles in food (EFSA, 2009), it is also clear that this is severely hampered by the limited scientific knowledge of the biological interactions of nanoparticles and on consumer exposure (Bouwmeester et al., 2009; EFSA 2009). The following section will focus on the main scientific knowledge gaps that currently hinder the safety assessment of nanoparticles in food and related products.

16.4.1 Lack of a good definition

The lack of a good definition of nanotechnology is problematic from a governance point of view. There are currently many definitions of nanotechnology available. Unfortunately, these either are too rigid to be applicable for food applications, or they are too flexible to be useful in legislation since they do not specify clear boundaries. The rigid definitions, e.g., the ones that specifically mention that nanotechnology refers to dimensions of less than 100 nm, open the possibility that applications that use structures of slightly more than 100 nm need not conform to the regulation. If the definition is too flexible—e.g., if it refers to sizes of "about 100 nm," although it is scientifically more accurate—it cannot be used in legal texts. Something as simple as labeling cannot be enforced at the moment because of the lack of a clear definition of nanomaterials, which allows industry to maintain a lack of transparency. For instance, the food industry is actively exploring the applicability of nanotechnology in food products, but is reluctant to admit to that.

16.4.2 Detection of manmade nanomaterials in complex matrices, including foods

Governance of applications of nanotechnology requires regulation of its use. Regulation in turn requires legislation and enforcement. Without means to enforce the regulation, the governance is useless and only constitutes an administrative process that does not really provide the protection against unwanted effects that the governance is seeking. Unfortunately, one key issue that hinders the enforcement of regulations related to the use of nanotechnology products is the capacity to detect nanostructures. Enforcement implies that manmade nanomaterials can be detected even if manufacturers of the products deny the use of nanotechnology.

Whereas the characterization of bulk chemical compounds in foods is usually relatively straightforward, characterization of nanoparticles is much more complex, due to several reasons. First, from an analytical point of view, there is not a single (or a handy) analytical toolkit that allows the full characterization of nanoparticles in food. At present, there is a vast array of analytical techniques available to characterize nanoparticles, both single-particle techniques and techniques for characterizing the assembly of engineered nano materials (ENMs) is simple solvents (Powers et al., 2006; Hassellov et al., 2008; Luykx et al., 2008; Tiede et al., 2008; Minelli et al., 2019). Food and other biological samples and agricultural samples are heterogeneous mixtures. Characterizing the ENMs from these matrices requires separation or pretreatment to isolate the ENM from the interfering matrix components (Bouwmeester et al., 2014). Due to their high reactivity, nanoparticles can change in composition and size as a response to changes in their environment. Ideally, sample preparation is kept minimal (Szakal et al., 2014). An example of a powerful approach that has been developed in recent years is single particle ICP-MS (sp-ICPMS) in which nanoparticles can be detected with limited sample preparation. In sp-ICPMS individual nanoparticles are atomized and ionized using ICP plasma and the resulting plume of ions is detected by the MS. Detection limits are generally in the ng/L range. A great advantage of sp-ICPMS is that it determines a number-based size distribution (Peters et al., 2014, 2018). Generally, these methods are only able to determine one single characteristic; and currently, it is practically impossible to fully characterize nanoparticles. Therefore, research should primarily focus on method development for the detection and characterization of nanoparticles. Ideally, such methods should be relatively easily performed, and use equipment that is currently present at laboratories equipped for detection of chemicals in food. Interestingly, some of the definitions of nanoparticles introduce the specific functionalities of nanoparticles compared to the larger scale equivalents. While it might be difficult to define specific functionalities in general terms, this opens an alternative avenue for the characterization of nanoparticles: effect characterization. For this approach, in vitro assays searching for biomarkers for exposure might be a very elegant alternative.

Secondly, at the moment it is virtually impossible—apart from some very specific cases—to distinguish between manmade and natural nanoscale structures. Multielement analysis of single nanoscale structures might be a direction to further investigate (Naasz et al., 2018). Lastly, from a toxicological point of view, there is a lack of knowledge on how to describe biological dose–response relations, i.e., which metrics need to be used to express these relations. Up to now, it has not been possible to establish a single dose–describing parameter that best describes the possible toxicity. It is likely that mass alone is not the good metric (SCENIHR, 2007a), but other characteristics such as size specific surface area, surface charge (Zeta potential), as well as the number of particles per particle size might be very useful for describing the dose (Hagens et al., 2007; McNeil, 2009; Oberdorster et al., 2007a). Given the complexity of the matter, it is reasonable to say that the scientific requirements for analytical tools for nanostructures cannot be fully formulated yet.

16.4.3 Assessment of exposure to nanoparticles

Exposure assessment is defined as the qualitative and/or quantitative evaluation of the likely intake of biological, chemical, or physical agents via food, as well as exposure from other sources if relevant (FAO/WHO, 1997). The reliability of the exposure assessment is critically depending on the availability of analytical tools to determine the presence or absence of nanoparticles in food. Basically, the principle of assessing exposure to nanoparticles via food will be comparable to the exposure assessment of conventional chemicals. Usually one of the following three approaches is applied for integration of data: (1) point estimated; (2) simple distributions; and (3) probabilistic analyses (Kroes et al., 2002). Issues like food sampling and variability within composite samples, variation in concentrations between samples, and consumption data on specific food products are not different from those encountered when assessing the exposure to conventional chemicals. Alternatively, food processors should provide reliable data on the use of nanoparticles in their products. The quality and reliability of the exposure assessment will be greatly improved if the concentration data are collected in an occurrence database. The last step in performing exposure assessment is the integration of occurrence and food consumption data. The procedures used here are not different from the ones used for conventional chemicals.

16.4.4 Toxicity of nanoparticles

Knowledge on the potential toxicity of nanoparticles is rapidly growing. Original work suggested that nanoparticles may have a deviating toxicity profile when compared to their bulk equivalents (Oberdorster et al., 2005a, 2007b; Donaldson et al., 2001; Nel et al., 2006).

After nanomaterials enter the body via the oral route, they are subjected to conditions that are very different from those encountered via other exposure routes. The extremely low pH of the stomach and the high ionic strength in the stomach and intestine will critically affect ENM properties, potentially yielding products with differing toxicity profiles. Further, pH changes in the small intestine, mucus, and resident microbiota in the GIT lumen add to the complexity (Bouwmeester et al., 2018). Various models are available that closely simulate human physiology and have been used for decades to assess the solubility, digestion, and epithelial permeability of conventional food and drug components in the stomach, small intestine, and colon. These models are now being explored and used for the assessment of food-relevant nano materials (NMs) (Lefebre et al., 2015; Braakhuis et al., 2015). In the end, a combination or battery of models may be required depending on the data gaps addressed. An example approach for assessing NM uptake in the gastrointestinal tract was proposed earlier in this chapter. This included a tiered testing strategy with a thorough physical and chemical characterization of the NM in the matrix as dosed to the test system. In such testing, it is important to assess NM stability, taking into account that NM changes over time are heavily influence by the local environment of the NM (i.e., the matrix). Following this physical and chemical characterization, toxicological characterization of NM should be initiated by using alternative testing strategies (in vitro or in silico). Ultimately, this would prioritize materials for further testing in vivo (EFSA, 2018).

The suitability of studies to underpin a risk assessment, however, is still disputable, severely limiting the use of this information for risk assessment purposes (EFSA 2009). Usually, the quality of nanomaterial characterization is the limiting component. For example, in most studies only a single sized, poorly characterized nanoparticle is used or nanoparticles are administered at unrealistically high doses, or a narrow range of effects are generally studied (Oberdorster et al., 2007b). In addition, when evaluating the plethora of in vitro studies with nanoparticles, caution has to be exercised when extrapolating their results or mechanisms for the hazard characterization to subsequent human risk assessment (Oberdorster et al., 2007b). The in vitro studies might be suitable for searching mechanistic explanations of toxic effects, or as screening methods in combination with profiling studies in a tiered hazard assessment approach (Balbus et al., 2007; Lewinski et al., 2008).

One of the most important questions for the safety assessment is the sensitivity and validity of currently used test assays. While the knowledge on potential toxicity of nanoparticles is growing, so far only studies following acute (single dose) oral exposure are available. There is a great demand for studies using chronic oral exposure to nanoparticles combined with a broad screen for potential effects. Information from toxicity studies with other routes of exposure indicates that several systemic effects on different organ systems may occur after long-term exposure to nanoparticles, including the immune, inflammatory, and cardiovascular system. Effects on the immune and inflammatory systems may include oxidative stress and/or activation of proinflammatory cytokines in the lungs, liver, heart, and brain. Effects on the cardiovascular system may include prothrombotic effects and adverse effects on the cardiac function (acute myocardial infarction and adverse effects on the heart rate). Furthermore, genotoxicity, and possible carcinogenesis and teratogenicity may occur, but no data on the latter are available as yet (Bouwmeester et al., 2009).

16.4.5 Characteristics and behavior of nanoparticles in food

According to the Woodrow Wilson Center's Project on Emerging Nanotechnologies, 118 consumer products from the food and beverage sector are currently available on the market at the time of writing this chapter (Project on Emerging Nanotechnologies, 2018). The list contains a range of items that come in direct contact with food, such as from aluminum foil or antibacterial kitchenware, but also dietary supplements (i.e., Nanoceuticals Artichoke Nanoclusters) or canola oil fortified with free phytosterols. Such products allow, directly or indirectly, nanoparticles to enter the human body via ingestion.

Engineered nanoparticles in food may encompass many forms. Here the focus will be on persistent nanoparticles, i.e., nonsoluble or biodegradable particles, since potential risks are predominantly associated with these types of particles. It is likely that nanoparticles are used in foods in an agglomerated form, but it cannot be excluded that these agglomerates will break down and that the consumer will finally be exposed to free nanoparticles. Due to their specific chemico-physical properties, it is to be expected that nanoparticles could interact with proteins, lipids, carbohydrates, nucleic acids, ions, minerals, and water in food, feed, and biological tissues. Experimental data available so far indicate that the characteristics of nanoparticles are likely to influence their absorption, metabolism, distribution, and excretion (ADME) (Ballou et al., 2004; des Rieux et al., 2006; Florence, 2005; Jani et al., 1990; Roszek et al., 2005; Singh et al., 2006). For nanoparticles present in food, their interactions with proteins are important (Linse et al., 2007; Lynch and Dawson, 2008). Protein adsorption to engineered nanomaterials may enhance membrane crossing and cellular penetration (John et al., 2001, 2003; Pante and Kann, 2002). Furthermore, interaction with engineered nanomaterials may affect the tertiary structure of a protein, resulting in malfunctioning (Lynch et al., 2006). Therefore, it is important that the effects and interactions of nanoparticles are characterized in the relevant food matrix (Oberdörster et al., 2005b; Powers et al., 2006; The Royal Society and the Royal Academy of Engineering, 2004).

Translocation of particles through the gastrointestinal wall is a multistep process, involving diffusion through the mucus lining the gut wall, contact with enterocytes or M-Cells, cellular or paracellular transport, and posttranslocation events (des Rieux et al., 2006; Hoet et al., 2004). After passage of the intestinal epithelium, nanoparticles can enter the capillaries and enter the portal circulation to the liver, a major site for metabolism, or they can enter the lymphatic system which empties directly into the systemic blood circulation. The interactions with blood components might itself affect the fate of the nanoparticles. Unfortunately, there is little information regarding the distribution of nanoparticles following oral exposure (Hagens et al., 2007). But following other exposure routes, a widespread distribution of nanoparticles has been identified, where as a general pattern it appears that the smallest nanoparticles have the most widespread distribution (De Jong et al., 2008; Hillery et al., 1994; Hillyer and Albrecht, 2001; Hoet et al., 2004; Jani et al., 1990). Information on the potential of the nanoparticles to cross natural barriers like cellular, blood—brain, placenta, and blood—milk barriers is important for the safety assessors.

Very little is known regarding biotransformation of nanoparticles after oral administration. The metabolism of nanoparticles should depend, among other properties, on their surface chemical composition. Polymeric nanoparticles can be designed to be biodegradable. For metal and metal oxide nanoparticles, the slow dissolution will be of importance. Even less is known about the excretion of nanoparticles. As indicated, the potency of nanoparticles to interact with normal food constituents has raised speculation whether some nanoparticles may act as carriers (a "Trojan horse" effect) of contaminants or foreign substances present in food (Shipley et al., 2008). This could result in aberrant exposure to these compounds, with severe implications on consumer health.

Generally, the focus is placed on nonsoluble free nanoparticles. But another category of nanotechnology applications in food is represented by nanoencapsulates. These are specially designed to deliver their content with increased bioavailability. This type of application also needs to be considered by safety assessors.

To perform a robust safety assessment, more information needs to be gathered on the mechanism of ADME of nanoparticles and other nanostructures. Only when this information is available it will be possible to initiate extrapolation and modeling approaches that will allow a more generalized safety assessment of nanostructured particles in food. Due to the potential impact of toxicological effects, special attention needs to be paid to the possibility that certain nanoparticles can cross the barriers (e.g., gastrointestinal barrier, cellular barrier, blood—brain barrier, placenta barrier, blood—milk barrier).

16.5 Future developments and challenges

At the first International Food Nanotechnology Conference organized by the Institute of Food Technologists (IFT) in 2006, participants agreed that nanotechnology is still in its infancy, with food applications being rather in a preinfancy state, but also recognized a great amount of enthusiasm and anticipation surrounding this technology (Bugusu et al., 2006). Consumer acceptance and the regulatory issues will dominate and dictate its growth. In the absence of mandatory product labeling anywhere in the world, it is not easy to pinpoint exactly how many commercial products now contain nano ingredients. It is clear that applications of nanotechnology such as sensors or process innovations have very different risk profiles than those where nanostructures are added to food products and are ingested by the consumer. Likewise, applications where nanotechnology is used to improve certain properties of the packaging material of food products should be considered differently. The nanomaterial first has to migrate from the packaging material into the foodstuff and in the absences of migration consumers will not be affected. Of course also these applications need to be assessed for possible unexpected effects, but the impact will be different than in the case of nanomaterials directly added to food products. The type of governance of applications of nanotechnologies in food and food industry should be dependent on the type of application of nanotechnologies.

Although potential beneficial effects of nanotechnologies are generally well described, the potential (eco)toxicological effects and impacts of nanoparticles have so far received little attention. The high speed of introduction of nanoparticle-based consumer products observed nowadays urges the need to generate a better understanding of the potential negative impacts that nanoparticles may have on biological systems. The main concerns stem from the lack of knowledge about the potential effects and impacts of nano-sized materials on human health and the environment (Bouwmeester et al., 2009). In addition to the scientific risk assessment—related concerns, the consumers concerns regarding nanotechnology application in food products are mainly related to safety issues. It is recognized that the public concerns about the safety of products derived from new technologies may differ from those using established technologies (Siegrist et al., 2008).

Nanotechnologies used to improve certain properties of food products can range from the use of so called soft nanomaterials like micelles and vesicles to encapsulate nutrients and deliver them to specific locations in the gastrointestinal tract, to the use of nano formulated substances to improve the flow behavior of powdered foodstuffs. It is generally agreed among toxicologists that the supramolecular structures that are designed to break down within the gastrointestinal tract constitute relatively low risks, assuming that the molecules used to make these structures are safe. Also, nanoparticles that easily dissolve in water or are biodegradable will most likely not be very hazardous. Most of the concerns of applications of nanotechnologies in food are focused on nonsoluble free and persistent nanoparticles that potentially can pass certain barriers and enter the body, and subsequently enter certain tissues or even individual cells. Because of their persistent nature, they can stay there for prolonged periods and induce harmful effects. A special cause of concern is represented by nanoformulations designed to increase the bioavailability of the bulk equivalent. This might impact on the toxic profile of these compounds and needs to be assessed. To ensure that consumers are not subjected to unacceptable risks and that foods that incorporate nanotechnology products are as safe as those foods that do not contain nanomaterials, governance should focus on nonsoluble free manmade nanomaterials that have functional characteristics different from their bulk equivalents in food products.

The general public strongly associates nanotechnology with nanoparticles and therefore assumes that the risks of all applications of nanotechnologies are comparable with the risks of nonsoluble free nanoparticles. Since nanotechnology is an enabling technology, the actual form in which the consumer is exposed to the products of nanotechnology can be wide ranging. It is therefore important to educate the public and help it distinguish between the various forms and uses of nanotechnology, as well as the differences in risks between these applications. Unfortunately, the application of other state-of-the-art technologies in the past has shown that it takes time for this type of information to become widely accepted in society.

Proper regulation and monitoring of nanotechnology can help this process, since it would help build trust among users. Regulation implies that at least one impartial and objective body has reviewed and analyzed the specific application of the technology and has concluded that it is safe.

Globally, the scientific and industry communities need to come together to resolve the key issues of safety and public perception of nanotechnology. To fully exploit the benefits of nanomaterials without exposing the public to harm requires a judicious risk analysis and management. For the benefit of the humanity at large, the most expedient and efficient way of doing it is through global harmonization of the regulations of nanomaterials.

References

Ahn, J.-H., Kim, H.-S., Lee, K.J., Jeon, S., Kang, S.J., Sun, Y., Rogers, J.A., 2006. Heterogeneous three-dimensional electronics by use of printed semiconductor nanomaterials. Science 314 (5806), 1754—1757. https://doi.org/10.1126/science.1132394.

Balbus, J.M., Maynard, A.D., Colvin, V.L., Castranova, V., Daston, G.P., Denison, R.A., Dreher, K.L., Goering, P.L., Goldberg, A.M., Kulinowski, K.M., et al., 2007. Meeting report: hazard assessment for nanoparticles—report from an interdisciplinary workshop. Environ. Health Perspect. 115 (11), 1654—1659.

Ballou, B., Lagerholm, B.C., Ernst, L.A., Bruchez, M.P., Waggoner, A.S., 2004. Noninvasive imaging of quantum dots in mice. Bioconjugate Chem. 15 (1), 79—86.

Bastarrachea, L.J., Denis-Rohr, A., Goddard, J.M., 2015. Antimicrobial food equipment coatings: applications and challenges. Annu. Rev. Food Sci. Technol. 6 (1), 97—118. https://doi.org/10.1146/annurev-food-022814-015453.

Boskovic, S., Chon, J.W.M., Mulvaney, P., Sader, J.E., 2002. Rheological measurements using cantilevers. J. Rheol. 46 (4), 891—899.

Bouwmeester, H., Dekkers, S., Noordam, M.Y., Hagens, W.I., Bulder, A.S., de Heer, C., ten Voorde, S.E.C.G., Wijnhoven, S.W.P., Marvin, H.J.P., Sips, A.J.A.M., 2009. Review of health safety aspects of nanotechnologies in food production. Regul. Toxicol. Pharmacol. 53 (1), 52—62.

Bouwmeester, H., van der Zande, M., Jepson, M.A., 2018. Effects of food-borne nanomaterials on gastrointestinal tissues and microbiota. Wiley Interdiscip. Rev. Nanomed. Nanobiotechnol. 10 (1). https://doi.org/10.1002/wnan.1481.

Bouwmeester, H., Puck, B., Marvin Hans, J.P., Stefan, W., Peters Ruud, J.B., 2014. State of the safety assessment and current use of nanomaterials in food and food production. Trends Food Sci. Technol. 40, 200—210.

Braakhuis, H.M., Kloet, S.K., Kezic, S., Kuper, F., Park, M.V., Bellmann, S., van der Zande, M., Le Gac, S., Krystek, P., Peters, R.J., Rietjens, I.M., Bouwmeester, H., 2015. Progress and future of in vitro models to study translocation of nanoparticles. Arch. Toxicol. 89 (9), 1469–1495. https://doi.org/10.1007/s00204-015-1518-5.

BRIDGE, 2007. European Passive RFID Market Sizing 2007-2022.

Bugusu, B., Bryant, C., Cartwright, T.T., Chen, H., Schavemaker, J.M., Davis, S., Hunter, K., Irudayaraj, J., Mohanty, A., Moraru, C.I., Weiss, J., 2006. Report on the First IFT International Food Nanotechnology Conference. June 28-29, 2006, Orlando, FL. Available online at: http://members.ift.org/IFT/Research/ConferencePapers/firstfoodnano.htm (Accessed March 2008).

Cao, Q., Kim, H., Pimparkar, N., Kulkarni, J.P., Wang, C., Shim, M., Rogers, J.A., 2008. Medium-scale carbon nanotube thin-film integrated circuits on flexible plastic substrates. Nature 454 (7203), 495–500. https://doi.org/10.1038/nature07110.

Castellano, J.J., Shafii, S.M., Ko, F., Donate, G., Wright, T.E., Mannari, R.J., Robson, M.C., 2007. Comparative evaluation of silver-containing antimicrobial dressings and drugs. Int. Wound J. 4 (2), 114–122. https://doi.org/10.1111/j.1742-481X.2007.00316.x.

Chau, C.F., Wu, S.H., Yen, G.C., 2007. The development of regulations for food nanotechnology. Trends Food Sci. Technol. 18 (3), 269–280.

Chaudhry, Q., Scotter, M., Blackburn, J., Ross, B., Boxall, A., Castle, L., Aitken, R., Watkins, R., 2008. Applications and implications of nanotechnologies for the food sector. Food Addit. Contam. 25 (3), 241–258.

Chen, B., Shen, J., Ye, X., Jia, L., Li, S., Umeda, J., Kondoh, K., 2017. Length effect of carbon nanotubes on the strengthening mechanisms in metal matrix composites. Acta Mater. 140, 317–325. https://doi.org/10.1016/J.ACTAMAT.2017.08.048.

Chen, H., Weiss, J., Shahidi, F., 2006. Nanotechnology in nutraceuticals and functional foods. Food Technol. 60 (3), 30–36.

Costa, C., Antonucci, F., Pallottino, F., Aguzzi, J., Sarriá, D., Menesatti, P., 2013. A review on agri-food supply chain traceability by means of RFID technology. Food Bioprocess Technol. 6 (2), 353–366. https://doi.org/10.1007/s11947-012-0958-7.

Das, M., Saxena, N., Dwivedi, P.D., 2008. Emerging trends of nanoparticles application in food technology: safety paradigms. Nanotoxicology 99999 (1), 1–9.

Davies, C.J., 2008. Nanotechnology Oversight: An Agenda for the Next Administration. Woodrow Wilson International Center for Scholars. Project on Emerging Nanotechnologies (PEN). Available online at: http://www.nanotechproject.org (Accessed February 2009).

De Jong, W.H., Hagens, W.I., Krystek, P., Burger, M.C., Sips, A.J., Geertsma, R.E., 2008. Particle size-dependent organ distribution of gold nanoparticles after intravenous administration. Biomaterials 29 (12), 1912–1919.

des Rieux, A., Fievez, V., Garinot, M., Schneider, Y.J., Preat, V., 2006. Nanoparticles as potential oral delivery systems of proteins and vaccines: a mechanistic approach. J. Contr. Release 116 (1), 1–27.

Donaldson, K., Stone, V., Clouter, A., Renwick, L., MacNee, W., 2001. Ultrafine particles. Occup. Environ. Med. 58 (3), 211–216, 199.

Dufrêne, Y.F., Ando, T., Garcia, R., Alsteens, D., Martinez-Martin, D., Engel, A., Müller, D.J., 2017. Imaging modes of atomic force microscopy for application in molecular and cell biology. Nat. Nanotechnol. 12 (4), 295–307.

Duncan, T.V., 2011. Applications of nanotechnology in food packaging and food safety: barrier materials, antimicrobials and sensors. J. Colloid Interface Sci. 363 (1), 1–24. Available at: https://doi.org/10.1016/j.jcis.2011.07.017.

EC/178/2002, 2002. Regulation (EC) No 178/2002 of the European Parliament and of the Council of 28 January 2002 laying down the general principles and the requirements of food law, establishing the European Food Safety Authority and laying down procedures in matters of food safety. OJEU 31.

European Commission, 2008. Commission Recommendation of 07/02/2008 on a Code of Conduct for Responsible Nanoscience and Nanotechnologies Research. Available online at: http://ec.europa.eu/nanotechnology/index_en.html.

EFSA, 2018. Guidance on risk assessment of the application of nanoscience and nanotechnologies in the food and feed chain: Part 1, human and animal health. EFSA J. 16 (7), 5327. https://doi.org/10.2903/j.efsa.2018.5327.

EFSA, 2021. Guidance on technical requirements for regulated food and feed product applications to establish the presence of small particles including nanoparticles. EFSA J. 19 (8), 6769. https://doi.org/10.2903/j.efsa.2021.6769.

Environmental Protection Agency, 2009. Nanoscale Materials Stewardship Program (NMSP). Interim Report. U.S. Environmental Protection Agency. The Office of Pollution Prevention and Toxics. Available online at: www.epa.gov (Accessed February 2009).

FAO/WHO, 1997. Food consumption and exposure assessment of chemicals: report of a FAO/WHO consultation. World Health Organization, Geneva, Switzerland, 10–14 February 1997. https://www.fao.org/3/w4982e/w4982e00.htm.

FDA (U.S. Food Drug Admin.), 2018. Inventory of Effective Food Contact Substance (FCS) Notifications. Silver Spring. US Food Drug Administration, MD. http://www.accessdata.fda.gov/scripts/fdcc/?set=FCN.

Fellman, M., 2001. Nanoparticle Prism Could Serve as Bioterror Detector. Avaliable online at. http://unisci.com/stories/20014/1204011.htm (Accessed 28 May 2002).

Feng, G., Cheng, Y., Wang, S., Borca-Tasciuc, D.A., Worobo, R.W., Moraru, C.I., 2015. Bacterial attachment and biofilm formation on surfaces are reduced by small-diameter nanoscale pores: how small is small enough? NPJ Biofilms Microbiomes 1 (December).

Feng, G., Cheng, Y., Wang, S., Hsu, L.C., Feliz, Y., Borca-Tasciuc, D., Moraru, C.I., 2014. Alumina surfaces with nanoscale topography reduce attachment and biofilm formation by *Escherichia coli* and Listeria spp. Biofouling 30 (November), 1253–1268.

Florence, A.T., 2005. Nanoparticle uptake by the oral route: fulfilling its potential? Drug Discov. Today Technol. 2 (1), 75–81.

Franco, A., Hansen, S.F., Olsen, S.I., Butti, L., 2007. Limits and prospects of the "incremental approach" and the European legislation on the management of risks related to nanomaterials. Regul. Toxicol. Pharmacol. 48 (2), 171–183.

Gajraj, A., Ofoli, R., 2000. Quantitative technique for investigating macromolecular adsorption and interactions at the liquid-liquid interface. Langmuir 16, 4279–4285.

Gorman, J., 2003. Fracture protection: nanotubes toughen up ceramics. Sci. News 163, 1.

Gould-Fogerite, S., Mannino, R.J., Margolis, D., 2003. Cochleate delivery vehicles: applications to gene therapy. Drug Deliv. Technol. 3 (2), 40–47.

Hagens, W.I., Oomen, A.G., de Jong, W.H., Cassee, F.R., Sips, A.J., 2007. What do we (need to) know about the kinetic properties of nanoparticles in the body? Regul. Toxicol. Pharmacol. 49 (3), 217–229.

Hassellov, M., Readman, J.W., Ranville, J.F., Tiede, K., 2008. Nanoparticle analysis and characterization methodologies in environmental risk assessment of engineered nanoparticles. Ecotoxicology 17 (5), 344–361.

Health Council Netherlands, 2006. Health Significance of Nanotechnologies. Health Council of the Netherlands, The Hague. Publication no. 2006/06.

Heintz, M.E., 2009. National Nanotechnology Regulation in Canada? Available online at: www.nanolawreport.com (Accessed March 2009).

Hillery, A.M., Jani, P.U., Florence, A.T., 1994. Comparative, quantitative study of lymphoid and non-lymphoid uptake of 60 nm polystyrene particles. J. Drug Target. 2 (2), 151–156.

Hillyer, J.F., Albrecht, R.M., 2001. Gastrointestinal persorption and tissue distribution of differently sized colloidal gold nanoparticles. J. Pharmaceut. Sci. 90 (12), 1927–1936.

Hodge, G., Bowman, D., Ludlow, K. (Eds.), 2007. New Global Frontiers in Regulation: The Age of Nanotechnology. Edward Elgar Publishing Ltd, Cheltenham UK.

Hoet, P., Bruske-Hohlfeld, I., Salata, O., 2004. Nanoparticles - known and unknown health risks. J. Nanobiotechnol. 2 (1), 12.

Huang, W., Taylor, S., Fu, K., Lin, Y., Zhang, D., Hanks, T.W., Rao, A.M., Sun, Y.P., 2002. Attaching proteins to carbon nanotubes via diimide-activated amidation. Nano Lett. 2 (4), 311–314.

Ikeda, S., Morris, V., Nishinari, K., 2001. Microstructure of aggregated and non aggregated k-carrageenan helices visualized by atomic force microscopy. Biomacromolecules 2, 1331–1337.

International Risk Governance Council (IRGC), 2008. A Report for IRGC Risk Governance of Nanotechnology Applications in Food and Cosmetics. International Risk Governance Council, Geneva. Available online at: http://www.irgc.org/IMG/pdf/IRGC_PBnanofood_WEB.pdf.

Jafari, S.M., McClements, D.J., 2017. Nanotechnology Approaches for Increasing Nutrient Bioavailability, pp. 1–30. https://doi.org/10.1016/bs.afnr.2016.12.008.

Jain, A., Duvvuri, L.S., Farah, S., Beyth, N., Domb, A.J., Khan, W., 2014. Antimicrobial polymers. Adv. Healthc. Mater. 3 (12), 1969–1985. https://doi.org/10.1002/adhm.201400418.

Jani, P., Halbert, G.W., Langridge, J., Florence, A.T., 1990. Nanoparticle uptake by the rat gastrointestinal mucosa: quantitation and particle size dependency. J. Pharm. Pharmacol. 42 (12), 821–826.

Jelinski, L., 1999. Biologically related aspects of nanoparticles, nanostructured materials and nanodevices. In: Siegel, R.W., Hu, E., Roco, M.C. (Eds.), Nanostructure Science and Technology. A Worldwide Study. Prepared under the Guidance of the National Science and Technology Council and the Interagency Working Group on NanoScience, Engineering and Technology. Available online at: http://www.wtec.org (Accessed May 2002).

Jin, Z.T., Zhang, H.Q., Sun, D., Su, J.Y., Sue, H., 2009. Antimicrobial efficacy of zinc oxide quantum dots against Listeria monocytogenes, Salmonella enteritidis and *Escherichia coli* O157:H7. J. Food Sci. 74 (1), M46–M52.

John, T.A., Vogel, S.M., Minshall, R.D., Ridge, K., Tiruppathi, C., Malik, A.B., 2001. Evidence for the role of alveolar epithelial gp60 in active transalveolar albumin transport in the rat lung. J. Physiol. 533 (Pt 2), 547–559.

John, T.A., Vogel, S.M., Tiruppathi, C., Malik, A.B., Minshall, R.D., 2003. Quantitative analysis of albumin uptake and transport in the rat microvessel endothelial monolayer. Am. J. Physiol. Lung Cell Mol. Physiol. 284 (1), L187–L196.

Jung, M., Kim, J., Noh, J., Lim, N., Lim, C., Lee, G., Cho, G., 2010. All-Printed and roll-to-roll-printable 13.56-MHz-operated 1-bit RF tag on plastic foils. IEEE Trans. Electron. Dev. 57 (3), 571–580. https://doi.org/10.1109/TED.2009.2039541.

Khopade, A.J., Caruso, F., 2002. Electrostatically assembled polyelectrolyte/dendrimer multilayer films as ultrathin nanoreservoirs. Nano Lett. 2 (4), 415–418.

Kingsley, D., 2002. Membranes Show Pure Promise. ABC Science Online. Available online at: www.abc.net.au (Accessed 22 August 2003).

Kroes, R., Muller, D., Lambe, J., Lowik, M.R.H., van Klaveren, J., Kleiner, J., Massey, R., Mayer, S., Urieta, I., Verger, P., et al., 2002. Assessment of intake from the diet. Food Chem. Toxicol. 40 (2–3), 327–385.

Kuo, Y.C., Wang, Q., Ruengruglikit, C., Huang, Q.R., 2008. Antibody-conjugated CdTe quantum dots for *E. coli* detection. J. Phys. Chem. C. 112 (13), 4818–4824 (in press).

Kuzma, J., 2010. Nanotechnology in animal production-Upstream assessment of applications. Livest. Sci. 130, 14–24. https://doi.org/10.1016/j.livsci.2010.02.006.

Latour, R.A., Stutzenberger, F.J., Sun, Y.P., Rodgers, J., Tzeng, T.R., 2003. Adhesion-specific Nanoparticles for Removal of Campylobacter Jejuni from Poultry. CSREES Grant (2000-2003). Clemson University (SC) (Accessed June 2003). http://www.clemson.edu.

Lee, S.B., Martin, C.R., 2002. Electromodulated molecular transport in gold-nanotube membranes. J. Am. Chem. Soc. 124 (40), 11850–11851.

Lefebvre David, E., Koen, V., Lourdes, G., Valerio Luis Jr., G., Jayadev, R., Bondy Genevieve, S., Hans, B., Paul, S.R., Clippinger Amy, J., Eva-Maria, C., Mehta, R., Stone, V., 2015. Utility of models of the gastrointestinal tract for assessment of the digestion and absorption of engineered nanomaterials released from food matrices. Nanotoxicology 9 (4), 523–542.

Lemire, J.A., Harrison, J.J., Turner, R.J., 2013. Antimicrobial activity of metals: mechanisms, molecular targets and applications. Nat. Rev. Microbiol. 11 (6), 371–384. https://doi.org/10.1038/nrmicro3028.

Lewinski, N., Colvin, V., Drezek, R., 2008. Cytotoxicity of nanoparticles. Small 4 (1), 26–49.

Linse, S., Cabaleiro-Lago, C., Xue, W.F., Lynch, I., Lindman, S., Thulin, E., Radford, S.E., Dawson, K.A., 2007. Nucleation of protein fibrillation by nanoparticles. Proc. Natl. Acad. Sci. USA 104 (21), 8691–8696.

Llorens, A., Lloret, E., Picouet, P.A., Trbojevich, R., Fernandez, A., 2012. Metallic-based micro and nanocomposites in food contact materials and active food packaging. Trends Food Sci. Technol. 24 (1), 19–29. https://doi.org/10.1016/J.TIFS.2011.10.001.

Luksiene, Z., 2017. Nanoparticles and Their Potential Application as Antimicrobials in the Food Industry. Elsevier Inc.

Luykx, D.M.A.M., Peters, R.J.B., van Ruth, S.M., Bouwmeester, H., 2008. A review of analytical methods for the identification and characterization of nano delivery systems in food. J. Agric. Food Chem. 56 (18), 8231–8247.

Lynch, I., Dawson, K.A., Linse, S., 2006. Detecting cryptic epitopes created by nanoparticles. Sci. STKE 2006 (327), pe14.

Lynch, I., Dawson, K.A., 2008. Protein-nanoparticle interactions. Nano Today 3 (1–2), 40–47.

Marambio-Jones, C., Hoek, E.M.V., 2010. A review of the antibacterial effects of silver nanomaterials and potential implications for human health and the environment. J. Nanoparticle Res. 12 (5), 1531–1551. https://doi.org/10.1007/s11051-010-9900-y.

Mathew, A.P., Dufresne, A., 2002. Morphological investigation of nanocomposites from sorbitol plasticized starch and tunicin whiskers. Biomacromolecules 3 (3), 609–617.

McClements, D.J., Decker, E.A., 2000. Lipid oxidation in oil-in-water emulsions: impact of molecular environment on chemical reactions in heterogeneous food systems. J. Food Sci. 65 (8), 1270–1282.

McNeil, S.E., 2009. Nanoparticle therapeutics: a personal perspective. Wiley Interdiscip. Rev. Nanomed. Nanobiotechnol. https://doi.org/10.1002/wnan.006.

Minelli, C., Dorota, B., Peters, R., Jenny, R., Anna, U., Aneta, S., Eva, S., Heidi, G.-I., Shard Alexander, G., 2019. Sticky measurement problem: number concentration of agglomerated nanoparticles. Langmuir 35 (14), 4927–4935.

Moore, S., 1999. Nanocomposite achieves exceptional barrier in films. Mod. Plast. 76 (2), 31–32.

Moraru, C.I., Huang, Q., Takhistov, P., Dogan, H., Kokini, J.L., 2009. In: Barbosa-Canova, G.V., Mortimer, A., Lineback, D., Walter, S., Buckle, K., Colonna, P. (Eds.), Food Nanotechnology: Current Developments and Future Prospects. Global Issues in Food Science and Technology. Associated Press, ISBN 978-0-12-374124-0, pp. 369–399.

Moraru, C.I., Panchapakesan, C.P., Huang, Q., Takhistov, P., Liu, S., Kokini, J.L., 2003. Nanotechnology: a new frontier in food science. Food Technol. 57 (12), 24–29.

Morris, V., Mackie, A., Wilde, P., Kirby, A., Mills, C., Gunning, P., 2001. Atomic force microscopy as a tool for interpreting the rheology of food biopolymers at the molecular level. Food Sci. Technol. 34, 3–10.

Naasz, S., Stefan, W., Olga, B., Andrius, S., Claudia, C., Undas Anna, K., Simeone Felice, C., Marvin Hans, J.P., Peters Ruud, J.B., 2018. Multi-element analysis of single nanoparticles by ICP-MS using quadrupole and time-of-flight technologies. J. Anal. Atomic Spectrom. 33 (5), 835–845.

Nakajima, K., Mitsui, K., Ikai, A., Hara, M., 2001. Nanorheology of single protein molecules. Riken Rev. 37, 58–62.

National Research Council, 2008. Review of the Federal Strategy for Nanotechnology-Related Environmental, Health and Safety Research. National Academies Press, ISBN 0309116996.

Nel, A., Xia, T., Madler, L., Li, N., 2006. Toxic potential of materials at the nanolevel. Science 311 (5761), 622–627.

Norrby, S.R., Nord, C.E., Finch, R., 2005. Lack of development of new antimicrobial drugs: a potential serious threat to public health. Lancet Infect. Dis. 5 (2), 115–119. https://doi.org/10.1016/S1473-3099(05)01283-1.

Noyce, J.O., Michels, H., Keevil, C.W., 2006. Use of copper cast alloys to control *Escherichia coli* O157 cross-contamination during food processing. Appl. Environ. Microbiol. 72 (6), 4239–4244. https://doi.org/10.1128/AEM.02532-05.

Oberdorster, G., Maynard, A., Donaldson, K., Castranova, V., Fitzpatrick, J., Ausman, K., Carter, J., Karn, B., Kreyling, W., Lai, D., others, 2005a. Principles for characterizing the potential human health effects from exposure to nanomaterials: elements of a screening strategy. Part. Fibre Toxicol. 2, 8.

Oberdorster, G., Oberdorster, E., Oberdorster, J., 2007a. Concepts of nanoparticle dose metric and response metric. Environ. Health Perspect. 115 (6), A290.

Oberdorster, G., Stone, V., Donaldson, K., 2007b. Toxicology of nanoparticles: a historical perspective. Nanotoxicology 1 (1), 2–25.

Oberdörster, G., Oberdörster, E., Oberdörster, J., 2005b. Nanotoxicology: an emerging discipline evolving from studies of ultrafine particles. Environ. Health Perspect. 113 (7), 823–839.

Pante, N., Kann, M., 2002. Nuclear pore complex is able to transport macromolecules with diameters of about 39 nm. Mol. Biol. Cell 13 (2), 425–434.

Park, H.M., Lee, W.K., Park, C.Y., Cho, W.J., Ha, C.S., 2003. Environmentally friendly polymer hybrids. Part I: mechanical, thermal, and barrier properties of the thermoplastic starch/clay nanocomposites. J. Mater. Sci. 38, 909–915.

Peters, R.J.B., Rivera, Z.H., van Bemmel, G., Marvin, H.J.P., Weigel, S., Bouwmeester, H., 2014. Development and validation of single particle ICP-MS for sizing and quantitative determination of nano-silver in chicken meat. Anal. Bioanal. Chem. 1–11.

Peters Ruud, J.B., Undas Anna, K., Joost, M., van Bemmel Greet, Sandra, M., Hans, B., Peter, N., Wobbe, S., van der Lee Martijn, K., 2018. Development and validation of a method for the detection of titanium dioxide particles in human tissue with single particle ICP-MS. Current Trends Anal. Bioanal. Chem. 2 (1), 74–84.

Powers, K.W., Brown, S.C., Krishna, V.B., Wasdo, S.C., Moudgil, B.M., Roberts, S.M., 2006. Research strategies for safety evaluation of nanomaterials. Part VI. Characterization of nanoscale particles for toxicological evaluation. Toxicol. Sci. 90 (2), 296–303.

Project on Emerging Nanotechnologies, 2018. Consumer Products. An Inventory of Nanotechnology-Based Consumer Products Currently on the Market. Available online at: www.nanotechproject.org (Accessed September 2018).

Quarmley, J., Rossi, A., 2001. Nanoclays. Opportunities in polymer compounds. Ind. Miner. 400 (47–49), 52–53.

Rai, M., Yadav, A., Gade, A., 2009. Silver nanoparticles as a new generation of antimicrobials. Biotechnol. Adv. 27 (1), 76–83. https://doi.org/10.1016/J.BIOTECHADV.2008.09.002.

Raviathul Basariya, M., Srivastava, V.C., Mukhopadhyay, N.K., 2014. Microstructural characteristics and mechanical properties of carbon nanotube reinforced aluminum alloy composites produced by ball milling. Mater. Des. 64, 542–549. https://doi.org/10.1016/j.matdes.2014.08.019.

Rhim, J.-W., Hong, S.-I., Park, H.-M., Ng, P.K.W., 2006. Preparation and characterization of chitosan-based nanocomposite films with antimicrobial activity. J. Agric. Food Chem. 54, 5814–5822. https://doi.org/10.1021/JF060658H.

Rhim, J.-W., Park, H.-M., Ha, C.-S., 2013. Bio-nanocomposites for food packaging applications. Prog. Polym. Sci. 38 (10–11), 1629–1652. https://doi.org/10.1016/J.PROGPOLYMSCI.2013.05.008.

Riley, T., Govender, T., Stolnik, S., Xiong, C.D., Garnett, M.C., Illum, L., Davis, S.S., 1999. Colloidal stability and drug incorporation aspects of micellar-like PLA-PEG nanoparticles. Colloids Surf. B Biointerfaces 16, 147–159.

Roca, I., Akova, M., Baquero, F., Carlet, J., Cavaleri, M., Coenen, S., Vila, J., 2015. The global threat of antimicrobial resistance: science for intervention. New Microbes New Infect. 6, 22–29. https://doi.org/10.1016/J.NMNI.2015.02.007.

Roszek, B., de Jong, W., Geertsma, R., 2005. Nanotechnology in Medical Applications: State-of-the-Art in Materials and Devices. RIVM Report 265001001/2005.

Rouhi, M., 2002. Novel chiral separation tool. Chem. Eng. News 80 (25), 13.

Santangelo, R., Paderu, P., Delmas, G., Chen, Z.W., Mannino, R., Zarif, L., Perlin, D.S., 2000. Efficacy of oral cochleate-amphotericin B in a mouse model of systemic candidiasis. Antimicrob. Agents Chemother. 44 (9), 2356–2360.

Scientific Committee on Emerging and Newly Identified Health Risk SCENIHR SCENIHR, 2007a. Opinion on: The Appropriateness of the Risk Assessment Methodology in Accordance with the Technical Guidance Documents for New and Existing Substances for Assessing the Risks of Nanomaterials European Commission Health & Consumer Protection Directorate-General. Directorate C - Public Health and Risk Assessment C7-Risk Assessment.

Scientific Committee on Emerging and Newly Identified Health Risk SCENIHR SCENIHR, 2007b. Opinion on: The Scientific Aspects of the Existing and Proposed Definitions Relating to Products of Nanoscience and Nanotechnologies European Commission Health & Consumer Protection Directorate-General. Directorate C - Public Health and Risk Assesment C7- Risk Assessment.

ScienceDaily, May 23, 2007. Project on Emerging Nanotechnology. Nanotechnology Now Used in Nearly 500 Everyday Products. http://www.sciencedaily.com/releases/2007/05/070523075416.htm.

Shipley, H.J., Yean, S., Kan, A.T., Tomson, M.B., 2008. Adsorption of arsenic to magnetite nanoparticles: effect of particle concentration, pH, ionic strenght and, temperature. Environ. Toxicol. Chem. 1.

Siegrist, M., Stampfli, N., Kastenholz, H., Keller, C., 2008. Perceived risks and perceived benefits of different nanotechnology foods and nanotechnology food packaging. Appetite 51 (2), 283–290.

Singh, R., Pantarotto, D., Lacerda, L., Pastorin, G., Klumpp, C., Prato, M., Bianco, A., Kostarelos, K., 2006. Tissue biodistribution and blood clearance rates of intravenously administered carbon nanotube radiotracers. Proc. Natl. Acad. Sci. USA. 103 (9), 3357–3362.

Siwy, Z., Trofin, L., Kohli, P., Baker, L.A., Trautmann, C., Martin, C.R., 2005. Protein biosensors based on biofunctionalized conical gold nanotubes. J. Am. Chem. Soc. 127 (14), 5001.

Song, J., Kong, H., Jang, J., 2011. Bacterial adhesion inhibition of the quaternary ammonium functionalized silica nanoparticles. Colloids Surf. B Biointerfaces 82 (2), 651–656. https://doi.org/10.1016/J.COLSURFB.2010.10.027.

Strick, T., Allemand, J., Croquette, V., Bensimon, D., 2000. Stress-induced structural transitions in DNA and proteins. Annu. Rev. Biophys. Biomol. Struct. 29, 523–543.

Stucky, G.D., 1997. Oral Presentation at the WTEC Workshop on R&D Status and Trends in Nanoparticles, Nanostructured Materials, and Nanodevices in the United States, May 8-9, Rosslyn, VA.

Szakal, C., Roberts, S.M., Westerhoff, P., Bartholomaeus, A., Buck, N., Illuminato, I., et al., 2014. Measurement of nanomaterials in foods: integrative consideration of challenges and future prospects. ACS Nano 8, 3128–3135.

Terada, Y., Harada, M., Ikehara, T., 2000. Nanotribology of polymer blends. J. Appl. Phys. 87 (6), 2803–2807.

The Royal Society and the Royal Academy of Engineering, 2004. Nanoscience and Nanotechnologies: Opportunities and Uncertainties. The Royal Society and the Royal Academy of Engineering, London, UK.

Tiede, K., Boxall, A.B., Tear, S.P., Lewis, J., David, H., Hassellov, M., 2008. Detection and characterization of engineered nanoparticles in food and the environment. Food Addit. Contam. 25 (7), 795–821.

Wang, L., Hu, C., Shao, L., 2017. The antimicrobial activity of nanoparticles: present situation and prospects for the future. Int. J. Nanomed. 12, 1227. https://doi.org/10.2147/IJN.S121956.

Weiss, J., Takhistov, P., McClements, J., 2006. Functional materials in food nanotechnology. J. Food Sci. 71 (9), R107–R116.

Wen, Y., Tan, Z., Sun, F., Sheng, L., Zhang, X., Yao, F., 2012. Synthesis and characterization of quaternized carboxymethyl chitosan/poly(amidoamine) dendrimer core–shell nanoparticles. Mater. Sci. Eng. C 32 (7), 2026–2036. https://doi.org/10.1016/J.MSEC.2012.05.019.

Yoksan, R., Chirachanchai, S., 2010. Silver nanoparticle-loaded chitosan–starch based films: fabrication and evaluation of tensile, barrier and antimicrobial properties. Mater. Sci. Eng. C 30 (6), 891–897. https://doi.org/10.1016/J.MSEC.2010.04.004.

Zhan, G.D., Kuntz, J., Wan, J., Mukherjee, A.K., 2003. Single-wall carbon nanotubes as attractive toughening agents in alumina-based nanocomposites. Nat. Mater. 2, 38–42.

Further reading

Barlow, P.G., Clouter-Baker, A.C., Donaldson, K., MacCallum, J., Stone, V., 2005. Carbon black nanoparticles induce type II epithelial cells to release chemotaxins for alveolar macrophages. Part. Fibre Toxicol. 2, 11–24.

Centers for Disease Control and Prevention/National Institute for Occupational Safety and Health (CDC/NIOSH), 2006. Approaches to Safe Nanotechnology: An Information Exchange with NIOSH. Available at: www.cdc.gov/niosh (Accessed February 2009).

Donaldson, K., Aitken, R., Tran, L., Stone, V., Duffin, R., Forrest, G., Alexander, A., 2006. Carbon nanotubes: a review of their properties in relation to pulmonary toxicology and workplace safety. Toxicol. Sci. 92 (1), 5–22.

Duffin, R., Tran, C.L., Clouter, A., Brown, D.M., MacNee, W., Stone, V., Donaldson, K., 2002. The importance of surface area and specific reactivity in the acute pulmonary inflammatory response to particles. Ann. Occup. Hyg. 46, 242–245.

Kuzma, J., VerHage, P., 2006. New Report on Nanotechnology in Agriculture and Food Looks at Potential Applications, Benefits and Risks. Available online at: http://www.nanotechproject.org/news/archive/new_report_on_nanotechnology_in/ (Accessed April 2008).

Lison, D., Lardot, C., Huaux, F., Zanetti, G., Fubini, B., 1997. Influence of particle surface area on the toxicity of insoluble manganese dioxide dusts. Arch. Toxicol. 71 (12), 725–729.

Maynard, A.D., Baron, P.A., Foley, M., Shvedova, A.A., Kisin, E.R., Castranova, V., 2004. Exposure to carbon nanotube material: aerosol release during the handling of unrefined single walled carbon nanotube material. J. Toxicol. Environ. Health 67 (1), 87–107.

Monteiro-Riviere, N.A., Nemanich, R.J., Inman, A.O., Wang, Y.Y., Riviere, J.E., 2005. Multi-walled carbon nanotube interactions with human epidermal keratinocytes. Toxicol. Lett. 155 (3), 377–384.

Nemmar, A., Hoet, P.H.M., Vanquickenborne, B., Dinsdale, D., Thomeer, M., Hoylaerts, M.F., Vanbilloen, H., Mortelmans, L., Nemery, B., 2002. Passage of inhaled particles into the blood circulation in humans. Circulation 105, 411–414.

Oberdörster, G., Ferin, J., Lehnert, B.E., 1994. Correlation between particle-size, in-vivo particle persistence, and lung injury. Environ. Health Perspect. 102 (S5), 173–179.

Oberdörster, G., Ferin, J., Gelein, R., Soderholm, S.C., Finkelstein, J., 1992. Role of the alveolar macrophage in lung injury—studies with ultrafine particles. Environ. Health Perspect. 97, 193–199.

Oberdörster, G., Sharp, Z., Atudorei, V., Elder, A., Gelein, R., Lunts, A., Kreyling, W., Cox, C., 2002. Extrapulmonary translocation of ultrafine carbon particles following whole-body inhalation exposure of rats. J. Toxicol. Environ. Health 65 Part A (20), 1531–1543.

Pritchard, D.K., 2004. Literature Review - Explosion Hazards Associated with Nanopowders. Health and Safety Laboratory, HSL/2004/12, United Kingdom. Available online at: http://www.hse.gov.uk (Accessed February 2009).

Ryman-Rasmussen, J.P., Riviere, J.E., Monteiro-Riviere, N.A., 2006. Penetration of intact skin by quantum dots with diverse physiochemical properties. Toxicol. Sci. 91 (1), 159–165.

Shvedova, A.A., Kisin, E.R., Mercer, R., Murray, A.R., Johnson, V.J., Potapovich, A.I., Tyurina, Y.Y., Gorelik, O., Arepalli, S., Schwegler-Berry, D., 2005. Unusual inflammatory and fibrogenic pulmonary responses to single walled carbon nanotubes in mice. Am. J. Physiol. Lung Cell Mol. Physiol. 289, L698–L708.

Takenaka, S., Karg, D., Roth, C., Schulz, H., Ziesenis, A., Heinzmann, U., Chramel, P., Heyder, J., 2001. Pulmonary and systemic distribution of inhaled ultrafine silver particles in rats. Environ. Health Perspect. 109 (Suppl. 4), 547–551.

Tran, C.L., Buchanan, D., Cullen, R.T., Searl, A., Jones, A.D., Donaldson, K., 2000. Inhalation of poorly soluble particles. II. Influence of particle surface area on inflammation and clearance. Inhal. Toxicol. 12 (12), 1113–1126.

Chapter 17

Monosodium glutamate in foods and its biological importance

Helen Nonye Henry-Unaeze
Department of Food, Nutrition and Home Science, Faculty of Agriculture, University of Port Harcourt, East-West Road Choba, Rivers, Nigeria

17.1 Introduction

Monosodium glutamate (MSG) is the most common food flavor-enhancer used extensively in almost all ethnic cuisines (Chaudhari et al., 2000; Smriga et al., 2010; Odunfa, 1985; Aidoo, 1986; Onoifok et al., 1996; Achi, 2005; Kurihara, 2015; Stanska and Krzeski, 2016). Chemically, it is the sodium salt of L-glutamic acid, the most abundant nonessential amino acids in protein foods, such as cheese, meat, fish, chicken, pork, milk, sea foods, shrimps, crab, clams, tomatoes, mushroom, onions, walnuts, garlic, peas, carrot, spinach, and potatoes (Kochem and Breslin, 2017; Shigemura et al., 2009a,b; Yamaguchi and Ninomiya, 2000; Ninomiya, 1998; Munro, 1979; Raiten et al., 1995; Walker and Lupien, 2000; Fernstrom and Smriga, 2017). Glutamate is also produced in the human body where it has evolved into the main substrate for energy production (in the gut cells), an intermediate substance for protein metabolism, a precursor for minor but important metabolites (e.g., glutathione or *N*-acetylglutamate), as well as a neurotransmitter in the central nervous system (CNS) (Kurihara, 2009; Meldrum, 2000; Newsholme, 2001).

Originally, glutamate was described in 1866 by Karl Heinrich Ritthausen, after isolating it from the wheat proteins (Sano, 2009). In 1908, a Japanese Professor Kikunae Ikeda found substantial amounts of glutamate in broth made from seaweed (kombu) (Ikeda, 1908; Yamaguchi, 1998) and described its taste properties. Glutamate is more abundant in plant proteins than in animal proteins, and more in human milk (0.18 g/serving) than in animal milk (0.016 g/serving of cow's milk) (United States Department of Agriculture (USDA, 2014). It exists in two forms: free (not part of protein) or bound form (making up protein) (Kuninaka et al., 1964a,b). Free glutamate [Table 17.1 (United States Department of Agriculture (USDA, 2014), (Yoshida, 1988; Skurray and Pucar, 1988; Ninomiya and Katsula, 2014)] has effect on glutamate receptors and is used as a taste stimulus but glutamic acid bound to proteins does not stimulate glutamate receptors. However, it can also become free (nonbound) during heat processing to bring out their flavor characteristic (Chaudhari et al., 2000; Loliger, 2000; Yamaguchi, 1991). The sodium form of free glutamate creates a rich meaty, broth-like or savory taste in foods (Kurihara, 2015; Li et al., 2002; Nelson et al., 2002), and its presence in traditional soup stocks, meat dishes, air dried ham, shellfish, aged cheeses, mushroom, and ripe tomatoes is unique in improvement of taste, mouth-feel, and smoothness.

MSG interacts with other flavors in food to produce an influential taste experience called umami; by implication, its sensory function is based on its ability to enhance the presence of other taste-active compounds. An appropriate concentration of MSG in foods improves not only the palatability of the food but also the overall preference for food due to its ability to improve specific food flavor characteristics like continuity, mouth fullness, mildness, and thickness. MSG consists specifically of sodium, glutamate, and water, readily soluble in water, resistant to humidity, stable in various storage conditions, and very much available and affordable to all consumers (Henry-Unaeze, 2010). It is odorless and does not have a distinct taste of its own, but brings to the diet a unique taste impression (umami) that is savory, complex, and wholly distinct. MSG flavor enhancing level complements more with sweet and salty taste and will not improve the taste of poor quality foods (International Food Information Council Foundation (IFICF, 2009)). In pure form (containing 99.6% of monosodium L-glutamate) as found in food seasonings, it can be produced in commercial quantities from natural sources like starch, corn sugar, or molasses from sugar cane or sugar beets through the natural fermentation process just like in wine or yogurt making (Food and Drug Administration (FDA, 2012)). MSG can also be found in many processed foods

Ensuring Global Food Safety. https://doi.org/10.1016/B978-0-12-816011-4.00022-7

TABLE 17.1 Free glutamate contents in various foods.

Food category	Food item	Free glutamic acids (mg/100g)
Plants – seaweeds	Kelp	1608
	Dried lever	1378
	Nori	1380
	Wakame (*Undaria pinnatifida*)	9
	Tamarillo	470–1200
Vegetables	Tomato, macambo	220–246
	Garlic	110–128
	Corn	106
	Green peas	106
	Chinese cabbage	40–90
	Carrot	40–80
	Shiitake mushroom (fresh)	71
	Onion	20–51
	Cabbage	50
	Green asparagus	49
	Spinach	48
	Mushroom	42
	Potato	10–100
Nuts	Walnut	757
Fruits	Almond	45
	Avocado	18
	Grape	5
	Kiwi	5
	Apple	4–13
Animal - seafood	Scallop	140
	Crab (Alaska, blue, snow etc.)	19–120
	Shrimp (Kuruma, white)	20–120
	Sea urchin	100
	Clam	90
Meat and poultry	Egg yolk	50
	Chicken	22
	Beef	10
	Pork	9
Milk and products	Cheese (emmenthaler, parmegiano reggiano, cheddar)	300–1680
	Human breast milk	19–21
	Goat	4
	Cow	1

TABLE 17.1 Free glutamate contents in various foods.—cont'd

Food category	Food item	Free glutamic acids (mg/100g)
Traditional foods	Fermented beans	136–1700
	Anchovies	630–1440
	Fish sauce	620–1383
	Soy sauce	410–1264
	Green tea	220–670
	Aged cured ham	340

Reproduced from United States Department of Agriculture USDA, 2014. USDA Agricultural Research Services. National Nutrient Database for Standard Reference, http://www.ars.usda.gov/ba/bhnrc/ndl. Accessed 9 September 2019, Ninomiya, M., Katsula, Y., 2014. Basic information and ways to learn more. In: Umami: The Fifth Taste. Japan Publications Trading. pp. 136–151, Yoshida, y., 1988. Umami taste and traditional seasonings. Food Rev. Int., 14, 213–246, Skurray, G.R., Pucar, N., 1988. L-glutamic acid content of fresh and processed Foods. Food Chem. 27, 177–180.

(Conn, 1992; Scopp, 1991), bouillon cubes, yeast extract, soy sauce, autolyzed yeast extract, hydrolyzed vegetable protein, sodium and calcium caseinates, textured proteins, etc.

17.2 Umami taste

The umami taste is a specific taste associated with natural foods (like seaweed, soy sauce, mushroom, tomatoes, meat, etc.) used to describe the taste of L-glutamate salt (Schiffman, 2000). This unique taste was identified as a basic taste independent of the four basic tastes (sour, sweet, salty, and bitter) by Prof. Ikeda (Ikeda, 1908) and referred to as umami—a Japanese word meaning savory, broth-like, meaty, or delicious. Umami, the fifth basic taste (Chaudhari et al., 2000; Li et al., 2002; Nelson et al., 2002), is not produced by any combination of other basic tastes in food (Yamaguchi, 1987), but has a distinct taste quality. Just as sweet, bitter, salty, and sour are, respectively, linked to sugar, coffee, anchovy, and lemon, so is umami linked to tomatoes or cheese. The basic nature of umami was further proven by psychophysical and electrophysiological studies; availability of umami specific receptors in humans; and the universality of glutamate presence in foods (Kunihara, 2009). Umami harmonizes with other basic flavors in food to achieve a greater intensity of natural food flavor, palatability, and acceptability, a characteristic attributed to the concentration of umami compounds (glutamate and 5′-ribonucleotides) and the food matrix (Yamaguchi and Ninomiya, 2000; Masic and Yeomans, 2013). Yamaguchi and Ninomiya (Yamaguchi and Ninomiya, 2000) described umami taste as a unique quality of taste-enhancing synergism between L-glutamic acid and 5′-ribonucleotides. These two compounds constitute the three main umami substances—MSG, inosine-5′-monophosphate (IMP), and guanosine-5′-monophosphate (GMP) (Stanska and Krzeski, 2016; Kochem and Breslin, 2017; Kitamura et al., 2010a,b; Wifall et al., 2007), suggesting that in addition to free glutamates widely distributed in many natural protein foods, and as additives in some food sources like soups, sauces, canned vegetables, processes foods, bouillon cubes, and salty snacks (Conn, 1992; Scopp, 1991), more umami substances can also be obtained in higher concentration by proteolysis during fermentation. Inosinate, for instance, was isolated from Skipjack and is produced from decomposition of Adenosine-tri-phosphate (ATP), while guanylate from dried mushroom (Kodama, 1913; Kuninaka, 1960) can be produced by decomposition of ribonucleic acid—RNA (Kurihara, 2015). Umami substances have weak umami taste individually, but their combination similar to the combination of different foods items like meat, fish, vegetables, seaweeds, cheese, tomatoes, and sea foods in certain dishes can greatly enhance the umami potency (Beauchamp, 2009; Yagamuchi and Kimizuka, 1979; Kuninaka et al., 1964a,b). The umami taste intensity of MSG can be markedly enhanced by IMP and GMP (Yamaguchi and Ninomiya, 2000; Wifall et al., 2007; Festring and Hofmann, 2011; Nakamura et al., 2011; Narukawa et al., 2011; Tsurugizawa et al., 2011; Shigemura et al., 2009a,b), as they intensify transport inhibitor responses 1 (TIR1) and 3 (TIR3) stimulation by glutamate (Kochem and Breslin, 2017; Blonde and Spector, 2017; Kinnamon and Vandenbeuch, 2009). The sodium in MSG has also been reported to stimulate glutamate to produce umami effect (Hengebart, 1992); this may be similar to the improvement in odor and flavor characteristics of food when NaCl is used with MSG thus intensifying umami effect on palatability as reported (Ventanas et al., 2010). GMP can additionally (2.3 times) intensify MSG activation than IMP due to its ability to yield 6 times more umami molecules (Festring and Hofmann, 2011). Umami substances are selective in enhancing food palatability, and can improve the taste of canned vegetables, meat and fish dishes, but not those of cereals, milk products, and sweet-flavored recipes (Girardot and Peryam, 1954). Also, the umami substances were found to reduce sour and bitter taste (Shim et al., 2015).

The content of natural umami substances in food varies depending on foodstuff, state of preservation, aging, and measurement method. For instance, the level of free glutamates increases significantly after ripening or seasoning of certain foods. A higher concentration of umami substances in food does not bring a strong taste but rather a taste harmony with other tastes in foods that brings out mildness and deliciousness (Kurihara, 2015). The umami detection threshold is 0.7–3.0/mmol (Roper, 2016), only a small quantity of MSG is required to augment the umami potency, and additional quantities will not enhance the food taste (Geha et al., 2000a; Beyreuther et al., 2007) but will diminish the palatability of food (Yamaguchi and Takahashi, 1984). Umami taste preference is reported to be affected by physiological and nutritional condition. Evidence suggests that poorly nourished subjects prefer foods with high MSG concentration more than healthy subjects and umami taste sensitivity determines ones preference for protein food (Bellisle et al., 1996). Umami taste substances have also been found to increase the nutritional status of individuals as it can enhance taste quality and decrease the desire for saltiness (Yamaguchi, 1987). Its addition in low fat diet can also maintain the palatability of the foods (Yamaguchi and Ninomiya, 2000).

Umami taste has two main independent taste receptor systems: taste receptor type 1 member 1 and member 3—T1R1 and T1R3 (Yamaguchi, 1987). These taste receptors have special affinity for free glutamate and are present in the oral tissues and the gut (stomach and intestinal epithelium) where they function in nutrient sensing and digestion (Sasano et al., 2010; Tsurugizawa et al., 2009; San Gabriel et al., 2007; Bezencon et al., 2007). Glutamate receptors are also distributed in the CNS where they regulate metabolic and autonomic function. Moreover, they can also be found in the pancreas, liver, adipose tissue, and skeletal muscles (Kochem and Breslin, 2017; Kitamura et al., 2010a,b; Tsurugizawa et al., 2011; Blonde and Spector, 2017; Maruyama & Yamaguchi, 1994, 1996; Kawamura, 1993).

Umami taste can also be obtained from varied animal- and plant-based products depending on the region. In African countries, umami taste can be obtained from fermented soybean, locust bean, or castor oil seeds; in Asian countries, Hajeb and Jinap (Hajeb and Jinap, 2015) listed fermented and dried sea foods, beans, grains, dried and fresh mushrooms, tea, sauces made from fish, seafood, soybean, and kombu or dried bonito in Japan. Kombu/Dashi, a Japanese stock (prepared by dipping dried Kelp into boiling water) used to enhance food palatability, is similar to African use of fermented locust bean/castor oil bean—iru/dadawa (Achi, 2005; Isichei and Achinewhu, 1988), fish sauce in ancient Greece, Rome, and south-East Asia (Yamaguchi and Ninomiya, 2000; Curtis, 1991), and glutamate-rich soy sauce in China, Korea, and Philippines (Yoshida, 1988).

17.3 Glutamate in human metabolism

Glutamate is very important in the human body where it is distributed in various organs and tissues. About 10 g of free glutamate is distributed across the brain (2.3g), muscle (6.0g), liver (0.7g), kidney (0.7g), and plasma (0.04g) (Rassini et al., 1978). The human breast milk contains 21.6 mg/100g free glutamate—more than 50% of free amino acids in total milk content (Rassini et al., 1978). The human system also produces about 50g glutamate on a daily basis and glutamate constitutes 10% of the total amino acid content that enters into the system daily as a natural constituent of protein foods (Fernstrom, 2007). Dietary consumption studies show a range of 10–20g of bound glutamate and 1g of free glutamate from dietary sources daily (Giacometti, 1979; Maga and Yamaguchi, 1983; Brosnan et al., 2014; Fernstrom, 1994), with an average daily intake of 0.3–3.0 g/day of added MSG depending on culinary origin (Henry-Unaeze, 2017). A daily estimate of 36 mg/kg of free glutamic acids in breast-fed infants was reported by Appaiah (Appaiah, 2010). Humans have the ability to metabolize higher doses of glutamate (Stegink et al., 1986) whose metabolism involves series of anabolic or catabolic processes where it either serves as a precursor or a metabolite depending on the tissues involved. The human body cannot differentiate or discriminate between naturally occurring glutamate and those added as seasoning (FASEB, 1995), and its metabolism is compartmentalized within the cells with no interorgan exchange through the placenta or blood brain barrier (Brosnan et al., 2014). During food intake, ingested glutamate is oxidized to carbon dioxide within the enterocytes in the small intestine (Blachier et al., 2009); absorbed in the lumen and used immediately to generate ATP for gut cells (Neame and Wiseman, 1957; Windmueller and Spaeth, 1975; Reeds et al., 1996, 2000; Uneyama et al., 2006; Burrin and Stoll, 2009), it is further converted to other amino acids like alanine (Burrin and Stoll, 2009; Stegink et al., 1979) or used as a precursor for some bioactive compounds (Wu, 2009; Wakabayashi et al., 1991). Amino acids like glutamine, proline, arginine, and histidine are converted to glutamic acid and then to α-ketoglutaric acids by related enzymes during energy metabolism (Hayamizu, 2017). Although amino acids are used as an energy source mainly in situations of insufficient carbohydrates or lipids, glutamate is the main energy source for the gut cells during digestion (Reeds et al., 1996). It is also a good source of nitrogen that could conserve essential amino acids from depletion (Featherstone, 2015). Cynober (Cynober, 2018) explained that glutamate is at the crossroad of nitrogen and energy metabolism. Further conversion of glutamate to alanine, glucose and lactate (Stegink et al., 1979), aspartate, and other minor metabolites (α-ketoglutarate, glutamine, γ-aminobutyrate, urea, and gluthatione) occurs in the liver. There is a report of an interorgan exchange of glutamate fluctuating from the liver to the muscle with extensive catabolism in the splanchnic area due to the important role of glutamate in nitrogen homeostasis and the body's ability to limit the bioavailability of glutamate (Cynober, 2018). Nearly all the enterally delivered glutamate is completely absorbed with $<5\%$ glutamate appearance in the systemic circulation (Brosnan et al., 2014; Reeds et al., 2000; Hays et al., 2007; Williams and Woessner, 2009). It was explained that glutamate metabolism in the gut is a saturable process that results in a dose-dependent appearance of glutamate in the systemic circulation on high doses (Cynober, 2018). This increase in glutamate blood levels after absorption was attributed to ingestion of high doses (>12 g) on empty stomach or through nonoral routes that overwhelm normal glutamate metabolism (Walker and Lupien, 2000). All in all, considering that glutamate is hardly consumed in isolation, its self-limiting quality results in the unlikely consumption of high doses as well as the efficient barrier posed by the gastrointestinal tract to the penetration of glutamate into the rest of the body (Maga and Yamaguchi, 1983); it was not surprising that Battezzati, Brillon, and Matthews (Battezzati et al., 1995) attributed this increase in plasma glutamate after absorption to catabolism of glutamine by intestinal glutaminase rather than from the absorption of dietary glutamate.

During infancy, breast-fed infants are reported to ingest more free glutamate per kilogram of body weight than during any other period of life, because the human breast milk contains high concentration of glutamate (Steiner, 1987, 1993). There is evidence that infants can metabolize glutamate efficiently Stegink et al. (1986); Tung and Tung (1980); Joint FAO/WHO Expert Committee on Food Additives (JECFA, 1988; Remesy et al., 1997), which is attributed to the important role of the placenta and the fetal liver in amino acid transport and metabolism. Infant liver has the ability to convert glutamine to glutamate that is subsequently exported into the fetal circulation; while the placenta uses glutamate derived from both maternal and fetal blood as energy and also has the ability to prevent glutamate from transiting into the fetal circulation (Battaglia, 2000). MSG was found to have no effect on lactation, poses no risk to breast-fed infants (American Academy of Pediatrics' Committee on Drugs, 1994), and can be metabolized efficiently by both children and adults (Stegink et al., 1979; Ball et al., 2002).

The difference in the concentration of glutamate in the human brain and plasma allows for natural control of glutamate movement only from the brain to periphery, because the brain has far greater concentration of glutamate (12.0 mml/g) than those found in the plasma (Meldrum, 2000; Hawkins and Vina, 2016) and the blood brain barrier is impermeable to glutamate even at high concentrations (Burrin and Stoll, 2009).

17.4 Nutritional studies

MSG ability to enhance food palatability, improve salivary secretion, and interfere with carbohydrate metabolism could restore food intake in individuals with loss of appetite. Dietary consumption of MSG at low doses stimulates appetite, increases palatability, and improves sensory-specific postsymptomatic satiety (Brosnan et al., 2014; Di Sebastiano et al., 2013; Masic and Yeomans, 2014b). Adding umami substances to high protein foods enhances the satiety signal of proteins (Masic and Yeomans, 2014b). The increase in satiety is specific to protein meals and not carbohydrate meals (Kochem and Breslin, 2017). Improved glucose metabolism with MSG supplementation in healthy males (Di Sebastiano et al., 2013), and increased hunger and food intake, was observed in human subjects (Yeomans et al., 2008; Bellisle et al., 1991; Rogers and Blundell, 1990) without affecting the overall energy intake due to reduced consumption of non-MSG-enriched foods (Carter et al., 2011; Luscombe-Marsh et al., 2008). An opposite increase in energy intake was reported on MSG consumption (Imada et al., 2014; Luscombe-Marsh et al., 2009). The variance between the last report and the former ones was attributed to the different macronutrient content of the test diets used in the different studies that determines MSG influence on appetite (Zanfirsecu et al., 2019). Individuals with reduced protein status were found to prefer higher concentrations of MSG indicating that the taste detection threshold for MSG correlated with a preference for protein foods and habitual protein intake (Luscombe-Marsh et al., 2008; Masic and Yeomams, 2017). It was explained that the satiety observed on MSG intake with protein foods was due to the presence of leucine that has the ability to suppress food intake (Casperson et al., 2012; Laeger et al., 2014). Also consumption of high-fat foods and MSG sweet snacks reduced intake of energy and added sugar (Boutry et al., 2010).

MSG role in memory is underscored in its involvement in cognitive executive processes regulating eating behaviors and food choices, as soups with added MSG were found to stimulate the brain region associated with self-control in dietary choices among healthy women (Magerowski et al., 2018). An association between MSG intake and increased body mass index (bib74He et al., 2008, 2011), as well as prevalence of metabolic syndrome, dose dependently, irrespective of total energy intake and physical activity level, was reported (Insawang et al., 2012). Other studies found no influence on weight gain with MSG intake (Shi et al., 2010; Thu Hien et al., 2012). The inconsistency of the above results was due to the different protocols, and incorrect measurement methods employed.

MSG and its substitutes can significantly intensify the sensory qualities of meals (Wang et al., 2019). Increasing the umami taste in dishes can reduce both salt and fat intake while retaining meal acceptance and good health. MSG can substitute NaCl in low-sodium meals, as it contains one-third of sodium as table salt and will considerably reduce sodium content (20%–40%) of diets without losing pleasantness (International Food Information Council Foundation (IFICF, 2009; Jinap et al. (2016); Mauly et al. (2017); Institute of Medicine (IOM, 2010; Roininen et al., 1996; Yamaguchi and Takahashi, 1984; Chi and Chen, 1992). Its interaction with other flavors can produce a transformative gustatory experience (Mouritsen and Stybaek, 2014). Umami substances can stimulate salivary secretion, enhance appetite, intensify food palatability, participate in metabolism, and suppress obesity by increasing satiety and reducing postingestive recovery of hunger (Stanska and Krzeski, 2016).

Finally, the taste enhancing properties of umami can improve the acceptability and palatability of foods for older persons (Bellisle et al., 1991; Yamamoto et al., 2009; Sasano et al., 2014). Its unique distinct flavor perception in humans, impact on food palatability, satiety, and postmeal recovery of hunger depending on meal composition has been affirmed (Zanfirsecu et al., 2019).

17.5 Toxicological studies

MSG has been well studied in terms of safety (Zanfirsecu et al., 2019; Husarova and Ostatnikova, 2013; Geha et al., 2001). A comprehensive review of the alleged MSG effects on biochemical parameters (total plasma proteins, HDL, LDL, glucose, triglycerides, oxidative stress biomarkers, insulin leptin), histology, and morphology of several organs (heart, brain ovaries, liver) as well as the CNS at doses between 0.04 and 8 g/kg administered orally, intragastrically, subcutaneously, and intraperitoneally was recently reported (Zanfirsecu et al., 2019). Chronic MSG exposure was implicated in biochemical and morphological alterations of the heart tissue and cardiac rhythm (Kumar and Bhandari, 2013; Liu et al., 2013; Baky et al., 2009; Singh and Pushpa, 2005). Metabolic syndrome in human population (Insawang et al., 2012) as well as hypertension and oxidative stress of tissues in experimental animals was also associated with MSG intake (Contini et al., 2017; Onyema et al., 2006; Farombi and Onyema, 2006; Diniz et al., 2004). Oxidative stress was attributed to formation of free radicles and altered antioxidant reactions due to MSG-induced hyperglycemia (Koya et al., 2003) or increased influx of substance into the kidney (Diniz et al., 2004). The MSG doses (0.5 and 1.5 g/kg body weight) and route of administration (subcutaneous, intraperitoneal, or intravenous) employed in these studies were not consistent with human normal dietary intake and thus cannot be extrapolated to humans.

Some studies observed changes in liver morphology and antioxidant defense following MSG administration through different routes. Hepatotoxic effects like increased hepatic lipid peroxidation, reduced antioxidant and enzymes activity, diabetes, obesity, steatosis, inflammation, infiltration of liver cells, nodular lesions, bile duct deterioration as well as liver scarring and tumor formation were reported on chronic MSG administration (Onyema et al., 2006; Paul et al., 2012; El-Meghawry El-Kenawy et al., 2012; Eweka et al., 2011; Nakanishi et al., 2008). Oxidative liver damage in rats treated with MSG 0.6 mg/g body weight for 10 consecutive days and 0.6 and 1.6 mg/kg body weight MSG for 2 weeks was reported by Onyema et al. (Onyema et al., 2006) and Tawfik and Al-Badr (Tawfik and Al-Badr, 2012), respectively. Nonalcoholic fatty liver and steatohepatitis were also observed in mice treated MSG (Nakanishi et al., 2008). Increased adiposity, obesity, hyperinsulinemia, elevated triglycerides, and low density lipoproteins cholesterol were observed in MSG treated rats with unrestricted dietary regimen (Fujimoto et al., 2014). The relevance of these studies was undermined by the dosage and route of administration employed that were not similar to human normal metabolic pathway as well as the inability of the study protocol to efficiently quantify the actual MSG ingested.

Inconsistencies were observed in studies investigating MSG consumption and body weight. Metabolic changes and obesity were reported on chronic MSG intake (Diniz et al., 2005; Hirata et al., 1997), increase in food intake correlated positively with obesity in MSG users independent of physical activity and total energy intake (Insawang et al., 2012; He et al., 2008a,b), MSG interferes with leptin processing and causes obesity (Shi et al., 2010; Nicholas, 2010), and Hirata et al. (Hirata et al., 1997) associated MSG intake with higher energy intake and weight gain. When physical activity, type of food item, and total energy intake were adjusted, Shi et al. (Shi et al., 2010) found diminished MSG effect on obesity. Kondoh and Torii (Kondoh and Torii, 2008) reported negative correlation of MSG intake and body weight. The inconsistency in these results has been attributed to differences in macronutrient composition of the test diets. Increased abdominal fat depots and liver weight as well as liver steatosis (metabolic risk factors) were observed in obese adult rats fed MSG containing chow (Olugin et al., 2018). Nagata et al. (Nagata et al., 2006) observed glycosuria, hyperglycemia, hyperinsulinemia, increased lipid levels, and hypertrophic pancreas leading to diabetes at 29 weeks in new born mice injected with an MSG dose of 2 mg/g body weight. Iwase et al. (Iwase et al., 2000) induced obesity and high triglyceride levels in hypertensive newborn rats intraperitoneally injected with 4 mg/kg of MSG for 5 days.

Chronic MSG administration was also implicated in changes in behavior and physiology of experimental animals, modified behavioral, and physiological alterations (increased aggressively, decreased locomotor activity, and weakness (Campos-Sepulveda et al., 2009)). Significant changes in the neuronal redox homeostasis (increased lipid peroxidation levels, nitrite concentration, decreased levels of antioxidants) and in the neuronal histology of the hippocampus with an increase of brain and serum cholinesterase (ChE) levels (Onaolapo et al., 2016; Sadek et al., 2016) were observed in MSG administration. MSG intake was associated with neurotoxicity (Izumi et al., 2009; Rivera-Cervantes et al., 2014; Swamy et al., 2014; Weil et al., 2008). Increased glutamate concentration in the CNS causes brain damage (Meldrum, 2000). Severe damage of the hypothalamic nuclei with resultant increase in body weight, fat deposition, reduced motor activity, and secretion of growth hormone was observed in neonatal rats exposed to MSG (Nakagawa et al., 2000). Excess glutamate in the CNS has been implicated in severe neuronal damage, ischemia, brain injury, lateral and multiple sclerosis as well as Parkinson's disease (Lau and Tymianski, 2010). Subcutaneous administration of MSG in the brain of neonatal rats causes alterations in the biochemical environment with behavioral changes (Lopez-Perez et al., 2010). Altered hippocampal integrity was observed in MSG treated neonatal rats (Beas-Zarate et al., 2002). Degenerated cortex was also observed in rats treated with 3 g/kg/day (Hashem et al., 2002). Overactivation of glutamate pathways due to reduced γ-aminobutyric levels following MSG administration was associated with increased aggressiveness (Campos-Sepulveda et al., 2009; Rivera-Cervantes et al., 2014). Free radical—induced dopaminergic neurodegeneration due to MSG administration causes decreased locomotor activity (Rivera-Cervantes et al., 2014). MSG effect on the brain cells resulted in increased food intake, body weight, and lipid metabolism associated with obesity and metabolic syndrome (Brosnan et al., 2014). As a neurotransmitter, excess MSG has been linked to neurological symptoms like migraines, seizures, autism, attention deficit disorder, hyperactivity, Alzheimer's disease, Lou Gehrig's disease, multiple sclerosis and Parkinson's disease (Nicholas & et al., 2001; Rivera & et al., 2004), and cell death (Pelaez, 2001). Histopathological and morphological alterations in the brain associated with MSG administration include changes in antioxidant defense homeostasis (Swamy et al., 2014), changes in hippocampus (Dief et al., 2014), upregulation of proapoptotic Bax protein (Iwase et al., 2000), and neuronal damage in the cerebrum and cerebellum (Onaolapo et al., 2016; Swamy et al., 2014). The non-oral route, high doses, experimental models that have poorly developed blood brain barrier (Walker and Lupien, 2000), and insufficient enzymatic equipment for glutamate metabolism (Blood et al., 1969) employed in these studies, as well as the efficient metabolism of MSG in humans (Henry-Unaeze, 2017) demean the relevance of these studies in human dietary intake.

Hypertension, oxidative stress, and pathologic changes in the kidney and kidney functions in tissues of experimental animals (Contini et al., 2017; Onyema et al., 2006; Farombi and Onyema, 2006; Diniz et al., 2004) were reported on MSG administration. The excessive doses and nonoral route of MSG administration used in these studies eliminate the validity of these effects in extrapolation to human MSG intake.

MSG-induced obesity has been associated with tumor development in experimental animals. MSG-induced obese rats were observed to have chronic inflammatory state (Hernandez-Bautista et al., 2014; Tehkonia et al., 2010), insulin resistance, hyperglycemia and hypercholesterolemia (Hata et al., 2012), impaired glycemic control, and elevated serum levels of resistin and leptin (Roman-Ramos et al., 2011). Metabolic alterations with differential gender vulnerability (Hernandez-Bautista et al., 2014), hepatic steatosis, and inflammation related to liver carcinogenesis (Park et al., 2010; Siegel and Zhu, 2009), preneoplastic lesions (Nakanishi et al., 2008), increased oxidative stress and hepatic lipogenesis (Siegel and Zhu, 2009; Ratziu et al., 2002), decreased antioxidant defense, and diminished chemo-protection (Dluzen and Lazarus, 2015; Matouskova et al., 2015) were also observed on MSG administration. The relevance of these preclinical studies associating MSG with tumor progression is doubtful as majority employed poorly designed protocols not similar to MSG dietary consumption in humans.

Fertility and fetal abnormalities associated with chronic MSG intake include morphological changes in testes and sperm abnormalities (Nosseir et al., 2012; Das and Ghosh, 2011a, 2011b), pathological changes in oocytes (Eweka and Om'i-niabohs, 2011), and fallopian tubes (Eweka et al., 2010), reduced convulsion threshold, and negative effects in offspring (Yu et al., 1997, 2006). These studies have already been criticized for inconsistency in the doses employed, species difference, poor statistics as well as nonspecification of pathological changes per group (Zanfirsecu et al., 2019).

MSG could dose dependently exert genotoxic effect on human peripheral blood lymphocytes (Ataseven et al., 2016), induce and enhance oxidative stress and apoptosis rate in immune cells (Farombi and Onyema, 2006; Malik et al., 2016; Jovic et al., 2009; Pavlovic et al., 2006, 2007), alter metabolic and neuro-endocrine functions in newborn rats (Castrogiovanni et al., 2008), cause inflammation and damage to the lining of the small intestine (Kohan et al., 2016), impair the defense mechanism against pathogenic microorganism (Nakadate et al., 2016), and reduce blood lymphocytes levels (Hriscu et al., 1997), dysfunction of T-cells (Kato et al., 1986), depression of natural killer cell, and reduced lymphocytes number (Belluardo et al., 1990). These studies employed nonoral route of MSG administration, high MSG dose, as well as inappropriate design and, consequently, cannot be extrapolated to human dietary MSG intake.

17.6 Sensitivity

MSG symptom complex, Chinese restaurant syndrome (CRS), glutamate-induced asthma, hot dog headache, or MSG syndrome are terms used to describe a cluster of symptoms that could be mild or severe. The symptoms included headache, nausea, weakness, sweating, chest pain, burning sensation, facial pressure/numbness, flushing, dizziness, difficulty in breathing, muscle tightness, abdominal discomfort, skin rashes, abnormal heart rhythms, asthma, neuropathy, atopic dermatitis, and syncope (Geha et al., 2000a; Niaz et al., 2018; Freeman, 2006; Walker, 1999; Yang et al., 1997; Schaumburg et al., 1969). They were observed following consumption of MSG by certain sensitive individuals (Schaumburg et al., 1969; Simon, 2006) estimated to be below 1% of the general public (Schaumburg et al., 1969).

Urticaria, angioedema, rhinitis, sneezing, rhinorrhea, and nasal itching have been linked to MSG intake (Freeman, 2006). No effect of high dose MSG was found on chronic urticaria patients (Tarasoff and Kelly, 1993). Asthmatic symptoms, for instance, observed to present 1–2 h after ingestion had no substantial evidence associating with MSG dietary consumption (Walker, 1999). Geha et al. (Geha et al., 2001) found negligible relation between MSG intake and CRS. Although many people perceived themselves as sensitive to MSG, these symptoms are anecdotal because well-conducted double blind placebo clinical trials have not been able to consistently trigger same reactions in such individuals (Geha et al., 2000a; Williams and Woessner, 2009; Niaz et al., 2018; Freeman, 2006; Walker, 1999; Yang et al., 1997; Bawaskar et al., 2017). MSG at levels customarily consumed with food had no evidence linking it to long-term reactions or chronic diseases, brain lesions, or nerve damage in humans (FASEB, 1995). Although some short-term, transient, and generally mild symptoms, such as headache, numbness, flushing, tingling, palpitations, and drowsiness could present in some sensitive individuals who consume 3 g or more of MSG without food (Geha et al., 2001; Freeman, 2006; Yang et al., 1997; Bawaskar et al., 2017), a typical serving of a food with added MSG contains less than 0.5 g of MSG. Consuming more than 3 g of MSG without food at one time is unlikely. Recent suggestions to consider the cumulative effect of other additives (Kostoff et al., 2018) and poorly designed studies (Zanfirsecu et al., 2019) have been made.

17.7 Health effects

Decreased taste acuity, poor appetite, reduced dietary intake, and weight loss are common manifestations in aging. Woschnagg et al. (2002) reported loss of gustatory function accompanied by poor appetite, reduced dietary intake, and weight loss in older persons. Specific loss of umami taste perceptions resulted in loss of appetite, weight loss, and poor overall health in older persons (Satoh-Kuwariwada et al., 2010). The taste-enhancing properties of umami can improve the acceptability and palatability of food for older persons (Bellisle et al., 1991; Yamamoto et al., 2009; Yamamoto et al., 2009, 2009) and Umami-rich foods can improve the severe condition of dry mouth in older persons because it promotes salivation (Schiffman, 2000; Satoh-Kuriwadk et al., 2009), more than sour taste (Hodson and Linden, 2006; Hayakawa et al., 2008), increases salivary flow rate (because of the gustatory–salivary reflex), and promotes reflexive salivation, taste function, appetite, weight, and overall health in older persons (Sasano et al., 2015; Pangborn and Chung, 1981; Horio and Kawamura, 1989). The increase of several chronic noncommunicable diseases due to consumption of diets high in sodium could be reduced if NaCl is substituted with MSG as it contains one-third of the sodium content of table salt (Mauly et al. (2017); Institute of Medicine (IOM, 2010)). Although MSG stimulates appetite, food use of MSG at customary levels poses no health risk for the general public (Husarova and Ostatnkov, 2007; Selvakumar et al., 2006).

17.8 Other effects

MSG is associated with increased food consumption, metabolic dysfunction with elevated blood glucose, triacylglycerols, insulin, leptin, and homeostasis model assessment index (Diniz et al., 2005), and altered brain activation pattern affecting emotional learning and memory (Meyer-Gerspach et al., 2016). MSG intake enhances insulin secretion and glucose tolerance (Bertrand et al., 1997), interferes with gene expression causing obesity, steatosis, impairment of insulin secretion, and alterations in the gene expression associated with lipid metabolism (Tomankova et al., 2015). It also alters musculoskeletal pain sensitivity (Cairns et al., 2007), and causes significant reduction in thermal nociceptive threshold (Zanfirescu et al., 2017). Although these studies demonstrated that MSG could influence metabolism, weight, and pain sensitivity, the mega doses employed are not consistent with human MSG regular dietary intake. Some studies have implicated MSG intake in atopic dermatitis (Worm et al., 2000; Van Bever et al., 1989). Other studies stress the need to view the effect holistically as a lot of allergens, food additives environmental factors, and limited double-blinded placebo-controlled studies may be considered (Zanfirsecu et al., 2019; Overgaard et al., 2017; Kim, 2015).

MSG has been associated with headache, migraines, elevated systolic blood pressure, spontaneous pain, and fibromyalgia (Yang et al., 1997; Tarasoff and Kelly, 1993; Shimada et al., 2016; Vellisca and Latorre, 2013; Kitamura et al., 2010a,b; Baad-Hansen et al., 2009; Geha et al., 2000b). These results are doubtful as the small sample size, undefined taste threshold dose, and the potential effect of other food molecules could undermine their validity (Zanfirsecu et al., 2019; Taheri, 2017; Finocchi and Sivori, 2012). Also Holton et al. (Holton et al., 2012) could not reduce fibromyalgia symptoms on removal of MSG. Zanfirescu et al. (Zanfirsecu et al., 2019) noted that most reports of negative effects are controversial as they are flawed with poor design and wrong protocols thus have no relevance in human dietary intake.

17.9 Safety evaluations

Several food safety regulatory agencies consider food intake of MSG safe. The United States Food and Drug Administration [FDA (Food and Drug Administration (FDA, 2012)] listed MSG as Generally Recognized as Safe (GRAS) substance similar to common ingredients like salt and baking powder. This decision was echoed by the Joint FAO/WHO Expert Committee on Food additives [JECFA (Remesy et al., 1997)] and the European Food Safety Authority (EFSA) based on its wide use in food before 1958 as well as its safety at levels customarily consumed in food as confirmed by scientific toxicological reports. MSG as a flavor enhancer is safe for human consumption. European Community's Scientific Committee for Food designates MSG as safe, and concluded that is safe for all. The American Medical Association (American Medical Association Council on Scientific Affairs (AMASCA, 1992)) endorses MSG safety at customary levels in the diet, and Federation of American Societies for Experimental Biology (FASEB, 1995) confirms MSG safety. Select Committee on GRAS Substances (Select Committee on GRAS Substances (SCOGS, 1980)) found no evidence linking MSG with public health hazard at levels customarily consumed. Sequel to the update of GRAS substances and advances in toxicological testing, ESFA (EFSA Panel on Food Additives and Nutrient Sources Added to Food, 2017) reevaluated MSG safety and established an acceptable daily intake of 30 mg/kg by weight (a level far above the NOAEL—no adverse effect level) based on a neuro-developmental toxicity study that observed no adverse effect at 3.2mgMSG/kg body weight. Some population group may exceed the proposed ADI and levels associated with adverse effects in human.

17.10 Labeling issues

Various government agencies like the FDA, the European Commission (EC), and Food Standard Australia New Zealand FSANZ (Food Standard Australia New Zealand (FSANZ, 2003)) require that MSG added foods and those containing naturally occurring MSG (hydrolyzed vegetable proteins, protein isolate, hydrolyzed yeast, yeast extract, autolyzed yeast, soy extract, tomatoes, cheese, etc.) should be specified on the ingredient label of the product. In addition, "No MSG" or "No added MSG" should not be claimed for foods containing natural MSG and MSG cannot be listed as spices and flavoring.

17.11 Future perspective

In view of the on-going advances in scientific investigations, MSG researchers with intensions of contributing relevance to human MSG dietary intake should plan to overcome the study defects adequately outlined by Zanfirsecu et al. (2019) and employ good standard protocols (double-blind placebo-controlled clinical trials) to investigate holistically the impact of the umami taste on the nutritional status of all life-cycle stages.

References

Achi, O.K., 2005. Traditional fermented protein condiments in Nigeria. Afr. J. Biotechnol. 4, 1612–1621.

Aidoo, K.E., 1986. Lesser-known fermented plant foods. Trop. Sci. 26, 249–258.

American Academy of Pediatrics' Committee on Drugs, 1994. The transfer of drugs and other chemicals into human milk. Pediatrics 93, 137–150.

American Medical Association Council on Scientific Affairs, (AMASCA, 1992. Report D of the Council on Scientific Affairs on Food and Drug Administration Regulations Regarding the Inclusion of Added L- Glutamic Acid Content on Food Labels. Association House of Delegates American Medical Report.

Appaiah, K.M., 2010. Monosodium glutamate in foods and its biological effects. In: Boisrobert, C.E., Oh, S., Stjepanovic, A., Lelieveld, H.L.M. (Eds.), Ensuring Global Food Safety, Exploring Global Harmonization.

Ataseven, N., Yuzbsioglu, D., Keskin, A.C., Unal, I., 2016. Genotoxicity of monosodium glutamate. Food Chem. Toxicol. 91, 8–18.

Baad-Hansen, L., Cairns, B., Ernberg, M., Svensson, P., 2009. Effect of systemic monosodium glutamate (MSG) on headache and pericranial muscle sensitivity. Cephalalgia 30 (1), 68–76.

Baky, N.A., Mohamed, A.M., Faddah, L.M., 2009. Protective effect of N-acetyl cysteine and/or pro vitamin A against monosodium glutamate-induced cardiopathy in rats. J. Pharmacol. Toxicol. 4 (5), 178–193.

Ball, P., Woodward, D., Beard, T., Shoobridge, Ferrier, M., 2002. Calcium di-glutamate improves taste characteristics of lower salt soup European. J. Clin. Nutr. 56, 519–523.

Battaglia, F.C., 2000. Glutamine and glutamate exchange between the fetal liver and the placenta. J. Nutr. 130, 975S–977S.

Battezzati, A., Brillon, D.J., Matthews, D.E., 1995. Oxidation of glutamic acid by the splanchnic bed in humans. Am. J. Physiol. 269 (2 Pt 1), E269–E276.

Bawaskar, H.S., Bawaskar, P.H., Bawaskar, P.H., 2017. Chinese restaurant syndrome. Indian J. Crit. Care Med. 21 (1), 49–50.

Beas-Zarate, c., Perez-Vega, M., Gonzalez-Burgos, I., 2002. Neonatal exposure to monosodium L-glutamate in hippocampal CA1 pyramidal neurons of adult rats. Brain Res. 952, 275–281.

Beauchamp, G.K., 2009. Sensory and receptor responses to umami: an overview of pioneering work. Am. J. Clin. Nutr. 90, 723–727.

Bellisle, F., Monneuse, M.O., Chabert, M., Laure-Achagiotis, C., Lantaume, M.T., Louis-Sylvestre, J., 1991. Monosodium glutamate as a palatability enhancer in the European diet. Physiol. Behav. 49, 869–874.

Bellisle, F., Dalix, A.M., Chapppuis, A.S., Rossi, F., Fiquet, P., Gaudin, V., Slama, G., et al., 1996. Monosodium glutamate affects mealtime food selection in diabetic patients. Appetite 26, 267–276.

Belluardo, N., Mudo, G., Bindoni, M., 1990. Effects of early destruction of the mouse arcuate nucleus by monosodium glutamate on age-dependent natural killer activity. Brain Res. 534 (1–2), 225–233.

Bertrand, G., Ravier, M., Puech, R., Loubatieres Mariani, M.M., Bockaert, J., 1997. Effects of glutamate on glucose tolerance and insulin secretion in a rat model of type II diabetes. Diabetologia 40, 514–514.

Beyreuther, K., Biesalski, H.K., Fernstrom, J.D., Grimm, P., Hammes, W.P., Heinemann, U., Walker, R., 2007. Consensus meeting: monosodium glutamate – an update. Eur. J. Clin. Nutr. 61, 304–313.

Bezencon, C., le Coutre, J., Damak, S., 2007. Taste signaling proteins are coexpressed in solitary intestinal epithelial cells. Chem. Senses 32 (1), 41–49.

Blachier, F., Boutry, C., Bos, C., Tome, D., 2009. Metabolism andfunctions of l-glutamate in the epithelial cells of the small and large intestines. Am. J. Clin. Nutr. 90 (3), 814S–821S.

Blonde, G.D., Spector, A.C., 2017. An examination of the role of l-glutamate and inosine 5-monophosphate in hedonic taste-guided behavior by mice lacking the T1R1 + T1R3 receptor. Chem. Senses 42 (5), 393–404.

Blood, F.R., Oser, B.L., White, P.L., Olney, J.W., 1969. Monosodium glutamate. Science 165 (3897), 1028–1029.

Boutry, C., Matsumoto, H., Airinei, G., Benamouzig, R., Tome, D., Blachier, F., Bos, C., 2010. Monosodium glutamate raises antral distension and plasma amino acid after a standard meal in humans. Am. J. Physiol. Gastrointest. Liver Physiol. 300 (1), G137–G145.

Brosnan, J.T., Drewnowski, A., Friedman, M.I., 2014. Is there a relationship between dietary MSG and Obesity in animals or humans? Amino Acids 46, 2075–2089.

Burrin, G.D., Stoll, B., 2009. Metabolic fate and function of dietary glutamate in the gut. Am. J. Clin. Nutr. 90 (3), 850–856.

Cairns, B.E., Dong, X., Mann, M.K., Svensson, P., Sessle, B.J., Arendt-Nielsen, L., McErlane, K.M., 2007. Systemic administration of monosodium glutamate elevates intramuscular glutamate levels and sensitizes rat masseter muscle afferent fibers. Pain 132 (1–2), 33–41.

Campos-Sepulveda, A.E., Martinez Enriquez, M.E., Rodriguez Arellanes, R., Pelaez, L.E., Rodriguez Amezquita, A.L., Cadena Razo, A., 2009. Neonatal monosodium glutamate administration increases aminooxyacetic acid (AOA) susceptibility effects in adult mice. Proc. West. Pharmacol. Soc. 52, 72–74.

Carter, B.E., Monsivais, P., Perrigue, M.M., Drewnowski, A., 2011. Supplementing chicken broth with monosodium glutamate reduces hunger and desire to snack but does not affect energy intake in women. Br. J. Nutr. 106 (9), 1441–1448.

Casperson, S.L., Sheffield-Moore, M., Hewlings, S.J., Paddon-Jones, D., 2012. Leucine supplementation chronically improves muscle protein synthesis in older adults consuming the RDA for protein. Clin. Nutr. 31 (4), 512–519.

Castrogiovanni, D., Gaillard, R.C., Giovambattista, A., Spinedi, E., 2008. Neuroendocrine, metabolic, and immune functions during the acute phase response of inflammatory stress in monosodium l-glutamate-damaged, hyperadipose male rat. Neuroendocrinology 88 (3), 227–234.

Chaudhari, N., Landin, A.M., Roper, S.D., 2000. A metabotropic glutamate receptor variant functions as a taste receptor. Nat. Neurosci. 3, 113–119.

Chi, S.P., Chen, T.C., 1992. Predicting optimum monosodium glutamate and sodium chloride concentrations in chicken broth as affected by spice addition. J. Food Process. Preserv. 16, 313–326.

Conn, H., 1992. "Umami": the fifth basic taste. Food Sci. 92 (2), 21–23.

Contini, M., Fabro, A., Millen, N., Benmelej, A., Mahieu, S., 2017. Adverse effects in kidney function, antioxidant systems and histopathology in rats receiving monosodium glutamate diet. Exp. Toxicol. Pathol. 69 (7), 547–556, 5.

Curtis, R.I., 1991. Garum and Salsementa. Production and Commerce in Material Medical, Studies in Ancient Medicine, E.J. British Academic Publisher, Leiden, the Netherlands.

Cynober, L., 2018. Metabolism of dietary glutamate in adults. Ann. Nutr. Metabol. 73 (5), 5–14.

Das, R.S., Ghosh, S.K., 2011a. Long-term effects in ovaries of the adult mice following exposure to monosodium glutamate during neonatal life - a histological study. Nepal Med. Coll. J. 13 (2), 77–83.

Das, R.S., Ghosh, S.K., 2011b. Long-term effects of monosodium glutamate on spermatogenesis following neonatal exposure in albino mice—a histological study. Nepal Med. Coll. J. 12 (3), 149–153.

Di Sebastiano, M.K., Bell, K.E., Barnes, T., Weeraratne, A., Premji, T., Mourtzakis, M., 2013. Glutamate supplementation is associated with improved glucose metabolism following carbohydrate ingestion in healthy males. Br. J. Nutr. 110 (12), 2165–2172.

Dief, A.E., Kamha, E.S., Baraka, A.M., Elshorbagy, A.K., 2014. Monosodium glutamate neurotoxicity increases beta amyloid in the rat hippocampus: a potential role for cyclic AMP protein kinase. Neurotoxicology 42, 76–82.

Diniz, Y.S., Fernandes, A.A., Campos, K.E., Mani, F., Ribas, B.O., Novelli, E.L., 2004. Toxicity of Hyper caloric diet and monosodium glutamate: oxidative stress and metabolic shifting in hepatic tissue. Food Chem. Toxicol. 42, 319–325.

Diniz, Y.S., Faine, L.A., Galhardi, C.M., Rodrigues, H.G., Ebaid, G.X., Burneiko, R.C., Novelli, E.L., 2005. Monosodium glutamate in standard and high-fiber diets: metabolic syndrome and oxidative stress in rats. Nutrition 21 (6), 749–755.

Dluzen, D.F., Lazarus, P., 2015. MicroRNA regulation of the major drug-metabolizing enzymes and related transcription factors. Drug Metabol. Rev. 47 (3), 320–334.

EFSA Panel on Food Additives and Nutrient Sources Added to Food, 2017. Re-evaluation of glutamic acid (E 620), sodium glutamate; (E 621), potassium glutamate (E 622), calcium glutamate; (E 623), ammonium glutamate (E 624) and magnesium; glutamate (E 625) as food additives. EFSA J. 15 (7), 4910.

El-Meghawry El-Kenawy, A., Osman, H.E., Daghestani, M.H., 2012. The effect of vitamin C administration on monosodium glutamate induced liver injury. An experimental study. Exp. Toxicol. Pathol. 65 (5), 513–521.

Eweka, A., Om'iniabohs, F., 2011. Histological studies of the effects of monosodium glutamate on the ovaries of adult wistar rats. Ann. Med. Health Sci. Res. 1 (1), 37–43.

Eweka, A.O., Eweka, A., Om'iniabohs, F.A., 2010. Histological studies of the effects of monosodium glutamate of the fallopian tubes of adult female Wistar rats. N. Am. J. Med. Sci. 2 (3), 146–149.

Eweka, A., Igbigbi, P., Ucheya, R., 2011. Histochemical studies of the effects of monosodium glutamate on the liver of adult wistar rats. Ann. Med. Health Sci. Res. 1 (1), 21–29.

Farombi, E.O., Onyema, O.O., 2006. Monosodium glutamate-induced oxidative damage and genotoxicity in the rat: modulatory role of vitamin C, vitamin E and quercetin. Hum. Exp. Toxicol. 25 (5), 251–259.

FASEB, 1995. Analysis of Adverse Reactions to Monosodium Glutamate (MSG) Report, Life Science Research Office. Federation of American Societies for Experimental Biology Washington DC July.

Featherstone, S., 2015. Ingrdients used in preparation of canned foods. In: A Complete Course in Canning and Related Processes, fourteenth ed.

Fernstrom, J.D., Smriga, M., 2017. Letter-to-the-. In: Shannon, M., et al. (Eds.), Toxicol. Lett, vol. 272, pp. 101–102 (2017).

Fernstrom, J.D., 1994. Dietary amino acids and brain function. J. Am. Diet Assoc. 94, 71–77.

Fernstrom, J.D., 2007. Health issues relating to monosodium glutamate use in the diet. In: Kilcast, D., Angus, F. (Eds.), Reducing Salt in Foods. CRC Press, Boca Raton, pp. 55–76.

Festring, D., Hofmann, T., 2011. Systematic studies on the chemical structure and umami enhancing activity of Maillard-modified guanosine 5-monophosphates. J. Agric. Food Chem. 59 (2), 665–676.

Finocchi, C., Sivori, G., 2012. Food as trigger and aggravating factor of migraine. Neurol. Sci. 33 (Suppl. 1), S77–S80.

Food and Drug Administration (FDA, 2012. Question and Answers on Monosodium Glutamate (MSG). United State Food and Drug Administration. https://www.fda.gov/food/food-additives-petitions/questions-and-answers-monosodium-glutamate-msg. accessed 09/09/2019.

Food Standard Australia New Zealand, (FSANZ, 2003. Monosodium Glutamate. A Safety Assessment, Technical Report Series No. 20. Food Standards Australia New Zealand, Canberra.

Freeman, M., 2006. Reconsidering the effects of monosodium glutamate: a literature review. J. Am. Acad. Nurse Pract. 18 (10), 482–486.

Fujimoto, M., Tsuneyama, K., Nakanishi, Y., Salunga, T.L., Nomoto, K., Sasaki, Y., Selmi, C., 2014. A dietary restriction influences the progression but not the initiation of MSG-Induced nonalcoholic steatohepatitis. J. Med. Food 17 (3), 374–383.

Geha, R.S., Beiser, A., Ren, C., Patterson, R., Greenberger, P.A., Grammer, L.C., Saxon, A., 2000a. Multicenter, double-blind, placebo-controlled, multiple-challenge evaluation of reported reactions to monosodium glutamate. J. Allergy Clin. Immunol. 106 (5), 973–980.

Geha, R.S., Beiser, A., Ren, C., Patterson, R., Greenberger, P.A., Grammer, L.C., Saxon, A., 2000b. Review of alleged reaction to monosodium glutamate and outcome of a multicenter double-blind placebo-controlled study. J. Nutr. 130 (4S Suppl. 1), 1058S–1062S.

Geha, R.S., Beiser, A., Ren, C., Patterson, R., Grammar, L.C., Ditto, A.M., Harris, K.E., 2001. Review of allergic reaction to monosodium glutamate and outcome of a multicenter double blind placebo-controlled study. J. Nutr. 130, 1032–1038.

Giacometti, T., 1979. Free and bound glutamate in natural products. In: Filer Jr., L.J., Garattini, S., Kare, M.R., Reynolds, W.A., Wurtman, R.J. (Eds.), Glutamic Acid: Advances in Biochemistry and Physiology. Raven Press, New York, pp. 25–34.

Girardot, N.F., Peryam, D.R., 1954. MSG's power to park up foods. Food Eng. 26 (71–72), 182–185.

Hajeb, P., Jinap, S., 2015. Umami taste components and their sources in asian foods. Crit. Rev. Food Sci. Nutr. 778–791.

Hashem, H.E., Safwat, M.D.E., Algaidi, S., 2002. The effect of monosodium glutamate on the cerebellar cortex of male albino rats and the protective role of vitamin C (Histological and immunohistochemical study). J. Mol. Histol. 43 (2), 179–186.

Hata, K., Kubota, M., Shimizu, M., Moriwaki, H., Kuno, T., Tanaka, T., Hirose, Y., 2012. Monosodium glutamate-induced diabetic mice are susceptible to azoxymethane-induced colon tumorigenesis. Carcinogenesis 33 (3), 702–707.

Hawkins, R.A., Vina, J.R., 2016. How glutamate is managed by the blood brain barrier. Biol. 5, 37.

Hayakawa, Y., et al., 2008. The effect of umami taste on saliva secretion. Jpn. J. Taste Smell Res. 14, 443–446.

Hayamizu, K., 2017. Amino acids and energy metabolism: amino acids that produce α-Ketoglutaric acid via Glutamic Acid. In: Sustained Energy for Enhanced Human Functions and Activity.

Hays, S.P., Ordonez, J.M., Burrin, D.G., Sunehag, A.L., 2007. Dietary glutamate is almost entirely removed in its first pass through the splanchnic bed in premature infants. Pediatr. Res. 62, 353–356.
He, K., Zhao, L., Daviglus, M.L., Dyer, A.R., Van Horn, L., Garside, D., Stamler, J., 2008a. Association of monosodium glutamate intake with overweight in Chinese adults: the INTERMAP Study. Obesity 16 (8), 1875–1880.
He, K., Zhao, L., Daviglus, M.L., Dyer, A.R., Van Horn, L., Garside, D., Stamler, J., 2008b. Association of monosodium glutamate intake with overweight in Chinese adults: the INTERMAP Study. Obesity 16 (8), 1875–1880.
He, K., Du, S., Xun, P., Sharma, S., Wang, H., Zhai, F., Popkin, B., 2011. Consumption of monosodium glutamate in relation to incidence of overweight in Chinese adults: China Health and Nutrition Survey (CHNS). Am. J. Clin. Nutr. 93 (6), 1328–1336.
Hengebart, 1992
Henry-Unaeze, H.N., 2010. Consumer knowledge attitude and practice towards the use of monosodium glutamate and food grade bouillon cubes as dietary constituents. Pakistan J. Nutr. 9 (1), 76–80.
Henry-Unaeze, H.N., 2017. Update on food safety of monosodium l-glutamate (MSG). Pathophysiology 24 (4), 243–249.
Hernandez-Bautista, R.J., Alarcon-Aguilar, F.J., Del, C.E.-V.M., Almanza-Perez, J.C., Merino-Aguilar, H., Fainstein, M.K., Lopez-Diazguerrero, N.E., 2014. Biochemical alterations during the obese-aging process in female and male monosodium glutamate (MSG)-treated mice. Int. J. Mol. Sci. 15 (7), 11473–11494.
Hirata, A.E., Andrade, I.S., Vaskevicius, P., Dolnikoff, M.S., 1997. Monosodium glutamate (MSG)-obese rats develop glucose intolerance and insulin resistance to peripheral glucose uptake. Braz. J. Med. Biol. Res. 30 (5), 671–674.
Hodson, N., Linden, R., 2006. The Effect of monosodium glutamate on parotid salivary flow in comparison to the response nto the representatives of the other four basic taste. Physiol. Behav. 89, 711–717.
Holton, K.F., Taren, D.L., Thomson, C.A., Bennett, R.M., Jones, K.D., 2012. The effect of dietary glutamate on fibromyalgia and irritable bowel symptoms. Clin. Exp. Rheumatol. 30 (6 Suppl. 74), 10–17.
Horio, T., Kawamura, Y., 1989. Salivary secretion induced by umami taste. Jpn. J. Oral Biol. 31, 107–111.
Hriscu, M., Saulea, G., Vidrascu, N., Baciu, I., 1997. Effects of monosodium glutamate on blood neutrophils phagocytic activity and phagocytic response in mice. Rom. J. Physiol. 34 (1–4), 95–101.
Husarova, V., Ostatnikova, D., 2013. Monosodium glutamate toxic effects and their implications for human intake: a review. J. Med. Res. 1–12.
Husarova, V., Ostatnkov, D., 2007. Monosodium glutamate toxic effects and their implications for human intake: a review. Eur. J. Clin. Nutr. 34, 758–765.
Ikeda, K., 1908. On a new seasoning. J. Tokyo Chem. Soc. 30, 820–836.
Imada, T., Hao, S.S., Torii, K., Kimura, E., 2014. Supplementing chicken broth with monosodium glutamate reduces energy intake from high fat and sweet snacks in middle-aged healthy women. Appetite 79, 158–165.
Insawang, T., Selmi, C., Cha'on, U., Pethlert, S., Yongvanit, P., Areejitranusorn, P., Hammock, B.D., 2012. Monosodium glutamate (MSG) intake is associated with the prevalence of metabolic syndrome in a rural Thai population. Nutr. Metabol. 9 (1), 50.
Institute of Medicine (IOM, 2010. Strategies to Reduce Sodium Intake in the United States. National Academy Press, Washington, D.C.
International Food Information Council Foundation (IFICF, 2009. Everything You Need to Know about Glutamate and Monosodium Glutamate. Food insight. https://foodinsight.org/everything-you-need-to-know-about-glutamate-and-monosodium-glutamate/. accessed 09/09/2019.
Isichei, M., Achinewhu, S.C., 1988. The nutritive value of African oilbean seed (*Pentaclethra macrophylla*). Food Chem. 30, 83–92.
Iwase, M., Ichikkawa, K., Tashiro, K., Iino, K., Shinohara, N., Ibayashi, S., Fujishima, M., 2000. Effects of monosodium glutamate-induced obesity in spontaneously hypertensive rats vs Wistar Kyoto Rats: serum leptin and blood flow to brown adipose tissue. Hypertens. Res. 23 (5), 503–510.
Izumi, Y., Yamamoto, N., Matsuo, T., Wakita, S., Takeuchi, H., Kume, T., Akaike, A., 2009. Vulnerability to glutamate toxicity of dopaminergic neurons is dependent on endogenous dopamine and MAPK activation. J. Neurochem. 110 (2), 745–755.
Jinap, S., Hajeb, P., Karim, R., Norliana, S., Yibadatihan, S., Abdul-Kadir, R., 2016. Reduction of sodium content in spicy soups using monosodium glutamate. Food Nutr. Res. 60, 30463.
Joint FAO/WHO Expert Committee on Food Additives (JECFA, 1988. L-glutamic acid and its ammonium, calcium, monosodium and potassium salts. In: Toxicological Evaluation of Certain Food Additives and Contaminants, vol. 22. Cambridge University Press, New York, pp. 97–161.
Jovic, Z., Veselinovic, M., Vasic, K., Stankovic-Djordjevic, D., Cekic, S., Milovanovic, M., Sarac, M., 2009. Monosodium glutamate induces apoptosis in naive and memory human B cells. Bratislava Med. J. 110 (10), 636–640.
Kato, K., Hamada, N., Mizukoshi, N., Yamamoto, K., Kimura, T., Ishihara, C., Matsuura, N., 1986. Depression of delayed-type hypersensitivity in mice with hypothalamic lesion induced by monosodium glutamate: involvement of neuroendocrine system in immunomodulation. Immunology 58 (3), 389–395.
Kawamura, Y., 1993. Significance and history of research on umami. In: Kawamura, Y., Omura, Y., Kimura, S., Konosu, S. (Eds.), Umami. Kyoristu, Shuppan Tokyo, Japan, pp. 1–16. Japanese.
Kim, K., 2015. Influences of environmental chemicals on atopic dermatitis. Toxicol. Res. 31 (2), 89–96.
Kinnamon, S.C., Vandenbeuch, A., 2009. Receptors and transduction of umami taste stimuli. Ann. N. Y. Acad. Sci. 1170, 55–59.
Kitamura, A., Sato, W., Uneyama, H., Torii, K., Niijima, A., 2010a. Effects of intragastric infusion of inosine monophosphate and L-glutamate on vagal gastric afferent activity and subsequent autonomic reflexes. J. Physiol. Sci. 61 (1), 65–71.
Kitamura, A., Sato, W., Uneyama, H., Torii, K., Niijima, A., 2010b. Effects of intragastric infusion of inosine monophosphate and L-glutamate on vagal gastric afferent activity and subsequent autonomic reflexes. J. Physiol. Sci. 61 (1), 65–71.
Kochem, M., Breslin, P.A., 2017. Clofibrate inhibits the umami-savory taste of glutamate. PLoS One 12 (3), e0172534.

Kodama, S., 1913. Separating methods of inosinic acid. J. Tokyo Chem. Soc. 34, 751–757.

Kohan, A.B., Yang, Q., Xu, M., Lee, D., Tso, P., 2016. Monosodium glutamate inhibits the lymphatic transport of lipids in the rat. Am. J. Physiol. Gastrointest. Liver Physiol. 311 (4), G648–G654.

Kondoh, T., Torii, K., 2008. MSG intake suppresses weight gain, fat deposition, and plasma leptin levels in male Sprague–Dawley rats. Physiol. Behav. 95 (1–2), 135–144.

Kostoff, R.N., Goumenou, M., Tsatsakis, A., 2018. The role of toxic stimuli combinations in determining safe exposure limits. Toxicol. Rep. 5, 1169–1172.

Koya, D., Jirousek, M.R., Lin, Y.W., Ishii, H., Kuboki, K., King, G.L., 2003. Effects of antioxidants in diabetes-induced oxidative stress in the glomeruli of diabetic rats. J. Am. Soc. Nephrol. 8 (3), S250–S253. Koya D1, Hayashi K, Kitada M, Kashiwagi A, Kikkawa R, Haneda M.

Kumar, P., Bhandari, U., 2013. Protective effect of Trigonella foenum-graecum Linn. on monosodium glutamate-induced dyslipidemia and oxidative stress in rats. Indian J. Pharmacol. 45 (2), 136–140.

Kunihara, K., 2009. Glutamate from discovery as a food flavor to role as a basic taste (umami). Am. J. Clin. Nutr. 90, 3.

Kuninaka, A., Kibi, M., Sakaguchi, K., 1964a. History and development of flavor nucleotides. Food Technol. 18, 287–293.

Kuninaka, A., Kibi, M., Sakaguchi, K., 1964b. History and development of of flavor neoclitides. Food Technol. 3 (1949), 287–293.

Kuninaka, A., 1960. Studies on taste of Ribonucleic acid derivatives. J. Agric. Chem. Soc. Jpn. 34, 487-472.

Kurihara, k., 2009. Glutamate: from discovery as a food flavor to role as a basic taste (umami). Am. J. Clin. Nutr. 90, 719S–722S.

Kurihara, K., 2015. Umami the fifth basic taste: history of studies on receptor mechanisms and role as a food flavor. BioMed Res. Int. 189402, 2015.

Laeger, T., Reed, S.D., Henagan, T.M., Fernandez, D.H., Taghavi, M., Addington, A., Morrison, C.D., 2014. Leucine acts in the brain to suppress food intake but does not function as a physiological signal of low dietary protein. Am. J. Physiol. Regul. Integr. Comp. Physiol. 307 (3), R310–R320.

Lau, A., Tymianski, M., 2010. Glutamate receptors, neurotoxicity and neurodegeneration. Eur. J. Physiol. 460 (2), 525–542.

Li, X., Staszewski, L., Xu, H., Durick, K., Zoller, M., Adler, E., 2002. Human receptors for sweet and umami taste. Proc. Nat. Acad. Sci. US 99, 4692–4696.

Liu, Y., Zhou, L., Xu, H.F., Yan, L., Ding, F., Hao, W., Gao, X., 2013. A preliminary experimental study on the cardiac toxicity of glutamate and the role of alpha-amino-3-hydroxy-5-methyl-4-isoxazolepropionic acid receptor in rats. Chinese Med. J. 126 (7), 1323–1332.

Loliger, J., 2000. Functions and importance of glutamate for savory foods. J. Nutr. 130, 915–920.

Lopez-Perez, S.J., Urena-Guerrero, M.E., Morales-Villagran, A., 2010. Monosodium glutamate neonatal treatment as a seizure and excitotoxicity model. Brain Res. 1317, 246–281.

Luscombe-Marsh, N.D., Smeets, A.J., Westerterp-Plantenga, M.S., 2008. Taste sensitivity for monosodium glutamate and an increased liking of dietary protein. Br. J. Nutr. 99 (4), 904–908.

Luscombe-Marsh, N.D., Smeets, A.J., Westerterp-Plantenga, M.S., 2009. The addition of monosodium glutamate and inosine monophosphate-5 to high-protein meals: effects on satiety, and energy and macronutrient intakes. Br. J. Nutr. 102 (6), 929–937.

Maga, J.A., Yamaguchi, S., 1983. Flavor potentiators. CRC Crit. Rev. Food Sci. Nutr. 18, 231–312.

Magerowski, G., Giacona, G., Patriarca, L., Papadopoulos, K., Garza-Naveda, P., Radziejowska, J., Alonso-Alonso, M., 2018. Neurocognitive effects of umami: association with eating behavior and food choice. Neuropsychopharmacology 43 (10), 2009–2016.

Malik, S.S., Nawaz, G., Masood, N., 2016. Genotype of GSTM1 and GSTT1: useful determinants for clinical outcome of bladder cancer in Pakistani population. Egypt. J. Med. Human Genet. https://doi.org/10.1016/j.ejmhg.2016.03.001.

Maruyama, I., Yamaguchi, S., 1994. In: Proceedings of the 24th Symposium on Sensory Evaluation. Japanese Union of Scientista snd Engineers, Tokyo, Japan, p. 181.

Maruyama, I., Yamaguchi, S., 1996. Proceedings of the 30th Japanese symposium on taste and smell. Jpn. J. Taste Smell Res. 3, 632–635.

Masic, U., Yeomams, M.R., 2017. Does acute or habitual protein deprivation influence liking for monosodium glutamate? Physiol. Behav. 171, 79–86.

Masic, U., Yeomans, M.R., 2013. Does monosodium glutamate interact with macronutrient composition to influence subsequent appetite? Physiol. Behav. 116–117, 23–29.

Masic, U., Yeomans, M.R., 2014b. Umami flavor enhances appetite but also increases satiety. Am. J. Clin. Nutr. 100 (2), 532–538.

Matouskova, P., Bartikova, H., Bousova, I., Levorova, L., Szotakova, B., Skalova, L., 2015. Drug-metabolizing and antioxidant enzymes in monosodium L-glutamate obese mice. Drug Metab. Dispos. 43 (2), 258–265.

Mauly, H.D., Arisseto-Bragotto, A.O., Reyes, F.G., 2017. Monosodium glutamate as a tool to reduce sodium in foodstuffs: technological and safety aspects. Food Sci. Nutr. 5, 1039–1048.

Meldrum, B.S., 2000. Glutamate as a neurotransmitter in the brain: review of physiology and pathology. J. Nutr. 130 (4S Suppl. l), 1007S–1015S.

Meyer-Gerspach, A.C., Suenderhauf, C., Bereiter, L., Zanchi, D., Beglinger, C., Borgwardt, S., Wolnerhanssen, B.K., 2016. Gut taste stimulants alter brain activity in areas related to working memory: a pilot study. Neurosignals 24 (1), 59–70.

Mouritsen, O.G., Stybaek, K., 2014. Umami: Unlocking the Secrets of the Fifth Taste. Columbia University Press.

Munro, H.N., 1979. Factors in the regulation of glutamate metabolism. In: Filer Jr., L.J., Garattini, S., Kare, M.R., Reynolds, W.A., Wurtman, R.J. (Eds.), Glutamic Acid: Advances in Biochemistry and Physiology. Raven Press, New York, pp. 55–68.

Nagata, M., Suzuki, W., Lizuka, S., Tabuchi, M., Maruyama, H., Takeda, S., Miyamoto, K., 2006. Type 2 diabetes mellitus in obese mouse model induced by monosodium glutamate. Exp. Anim. 55, 109–115.

Nakadate, K., Motojima, K., Hirakawa, T.T., Tanaka-Nakadate, S., 2016. Progressive depletion of rough endoplasmic reticulum in epithelial cells of the small intestine in monosodium glutamate mice model of obesity. BioMed Res. Int. 5251738, 2016.

Nakagawa, T., Ukai, K., Ohyama, T., Gomita, Y., Okamura, H., 2000. Effect of chronic administration of sibutramine on body weight, food intake and Motor activity in neonatal monosodium glutamate-treated obese female rats: relationship of antiobesity effect with monoamines. Exp. Anim. 49 (4), 239–249.

Nakamura, Y., Goto, T.K., Tokumori, K., Yoshiura, T., Kobayashi, K., Nakamura, Y., Yoshiura, K., 2011. Localization of brain activation by umami taste in humans. Brain Res. 1406, 18–29.

Nakanishi, Y., Tsuneyama, K., Fujimoto, M., Salunga, T.L., Nomoto, K., An, J.L., Gershwin, M.E., 2008. Monosodium glutamate (MSG): a villain and promoter of liver inflammation and dysplasia. J. Autoimmun. 30 (1–2), 42–50.

Narukawa, M., Morita, K., Uemura, M., Kitada, R., Oh, S.H., Hayashi, Y., 2011. Nerve and behavioral responses of mice to various umami substances. Biosc. Biotech. Biochem. 75 (11), 2125–2131.

Neame, K.D., Wiseman, G., 1957. The transamination of glutamate and aspartate during absorption by the small intestine of the dogs in vivo. J. Physiol. 135, 442–450.

Nelson, G., Chandrashakar, J., Hoon, M.A., Ferng, L., Zhao, G., 2002. An amino acid taste receptor. Nature 416 (2002), 199–202.

Newsholme, P., 2001. Glutamine metabolism: nutritional and clinical significance. J. Nutr. 131 (9 Suppl. l), 2515–2522.

Niaz, K., Zaplatic, E., Spoor, J., 2018. Extensive use of monosodium glutamate: a threat to public health? Exp. Clin. Sci. 17, 273–278.

Nicholas, et al., 2001. Glutamate transporters in neurologic disease. J. Nutr. Health 58, 365–370.

Nicholas, B., 2010. Consumption of Monosodium Glutamate, the Widely Used Food Additive, May Increase the Likehood of Being Overweight, a New Study Says. New York Times accessed 1/10/2019.

Ninomiya, M., Katsula, Y., 2014. Basic information and ways to learn more. In: Umami: The Fifth Taste. Japan Publications Trading, pp. 136–151.

Ninomiya, K., 1998. Natural occurrence. Food Rev. Int. 14, 177–212.

Nosseir, N.S., Ali, M.H.M., Ebaid, H.M., 2012. A histological and morphometric study of monosodium glutamate toxic effect on testicular structure and potentiality of recovery in adult albino rats. Res. J. Biol. Sci. 2 (2), 66–78.

Odunfa, S.A., 1985. African fermented foods, Microbiology of fermented foods. BJB Wood Elsevier Appl. Sci. 2, 155–191.

Olugin, M.C., Posadas, M.D., Revalant, G.C., Marinzzi, D., Labourdette, V., Venezia, M.R., 2018. Monosodium glutamate affects metabolic syndrome risk factors on obese adult rats: a preliminary study. J. Obes. Weight-loss Med. 4, 023.

Onaolapo, O.J., Onaolapo, A.Y., Akanmu, M.A., Gbola, O., 2016. Evidence of alterations in brain structure and antioxidant status following 'low-dose' monosodium glutamate ingestion. Pathophysiology 23 (3), 147–156.

Onoifok, N., Nnanyelugo, D.O., Ukwondi, B.E., 1996. Usage pattern and contributions of fermented foods to the nutrient intake of low-income households in Emene, Nigeria. Plant Foods Hum. Nutr. 49, 199–211.

Onyema, O.O., Farombi, E.O., Emerole, G.O., Ukoha, A.I., Onyeze, G.O., 2006. Effect of vitamin E on monosodium glutamate induced hepatotoxicity and oxidative stress in rats. Indian J. Biochem. Biophys. 43 (1), 20–24.

Overgaard, L.E.K., Main, K.M., Frederiksen, H., Stender, S., Szecsi, P.B., Williams, H.C., Thyssen, J.P., 2017. Children with atopic dermatitis and frequent emollient use have increased urinary levels of low-molecular-weight phthalate metabolites and parabens. Allergy 72 (11), 1768–1777.

Pangborn, R., Chung, C., 1981. Parotid salivation in response to sodium chloride and monosodium glutamate in water and in broths. Appetite 2, 380–385.

Park, E.J., Lee, J.H., Yu, G.Y., He, G., Ali, S.R., Holzer, R.G., Karin, M., 2010. Dietary and genetic obesity promote liver inflammation and tumorigenesis by enhancing IL-6 and TNF expression. Cell 140 (2), 197–208.

Paul, M.V., Abhilash, M., Varghese, M.V., Alex, M., Nair, R.H., 2012. Protective effects of alpha-tocopherol against oxidative stress related to nephrotoxicity by monosodium glutamate in rats. Toxicol. Mech. Methods 22 (8), 625–630.

Pavlovic, V., Cekic, S., Sokolovic, D., Djindjic, B., 2006. Modulatory effect of monosodium glutamate on rat thymocyte proliferation and apoptosis. Bratislava Med. J. 107 (5), 185–191.

Pavlovic, V., Pavlovic, D., Kocic, G., Sokolovic, D., Jevtovic-Stoimenov, T., Cekic, S., Velickovic, D., 2007. Effect of monosodium glutamate on oxidative stress and apoptosis in rat thymus. Mol. Cell. Biochem. 303 (1–2), 161–166.

Pelaez, B., 2001. Lectin histochemistry and ultra-structure of microbial response to monosodium glutamate-mediated neirotoxicity in the arcuate nucleus. Arch. Neurol. 14, 165–174.

Raiten, D.J., Talbot, J.M., Fisher, K.D., 1995. Excutive summary from the report: analysis of adverse reactions to monosodium glutamate (MSG). J. Nutr. 125, 2892S–2906S.

Rassini, D.K., Sturman, J.A., Gauli, G.E., 1978. Taurine and other amino acids in milk and other mammals. Early Hum. Dev. 2, 1–15.

Ratziu, V., Bonyhay, L., Di Martino, V., Charlotte, F., Cavallaro, L., Sayegh-Tainturier, M.H., Poynard, T., 2002. Survival, liver failure, and hepatocellular carcinoma in obesity-related cryptogenic cirrhosis. Hepatology 35 (6), 1485–1493.

Reeds, P.J., Burrin, D.G., Jahoor, F., Wykes, L., Henry, J., Frazer, E.M., 1996. Enteral glutamate is almost completely metabolized in first pass by the gastrointestinal tract of infant pigs. Am. J. Physiol. 270, E413–E418.

Reeds, P.J., Burrin, D.G., Stoll, B., Jahoor, B.F., 2000. Intestinal glutamate metabolism. J. Nutr. 130, 978S–982S.

Remesy, A., Moundras, C., M orand, C., Demigne, C., 1997. Glutamine or glutamate release by the liver constitutes a major mechanism for nitrogen salvage. Am. J. Physiol. 272, G257–G264.

Rivera, M.C., et al., 2004. NMDA and AMPA receptor expression and cortical neuronal death are associated with p38 in glutamate-induced excitotoxicity in vivo. J. Neurosci. Res. 76, 678–685.

Rivera-Cervantes, M.C., Castaneda-Arellano, R., Castro-Torres, R.D., Gudino-Cabrera, G., Feria y Velasco, A.I., Camins, A., Beas-Zarate, C., 2014. P38 MAPK inhibition protects against glutamate neurotoxicity and modifies NMDA and AMPA receptor subunit expression. J. Mol. Neurosci. 55 (3), 596–608.

Rogers, P.J., Blundell, J.E., 1990. Umami and appetite: effects of monosodium glutamate on hunger and food intake in human subjects. Physiol. Behav. 48 (6), 801–804.

Roininen, K., Lahteenmaki, L., Tuorila, H., 1996. Effect of umami taste on pleasantness of low-salt soups during repeated testing. Physiol. Behav. 60 (3), 953–958.

Roman-Ramos, R., Almanza-Perez, J.C., Garcia-Macedo, R., Blancas-Flores, G., Fortis-Barrera, A., Jasso, E.I., Alarcon-Aguilar, F.J., 2011. Monosodium glutamate neonatal intoxication associated with obesity in adult stage is characterized by chronic inflammation and increased mRNA expression of peroxisome proliferator-activated receptors in mice. Basic Clin. Pharmacol. Toxicol. 108 (6), 406–413.

Roper, 2016. Taste: Mammalian taste bud Physiology. In the Curated Reference Collection in Neuroscience and Biobehavioral Psychology, pp 887–893.

Sadek, K., Abouzed, T., Nasr, S., 2016. Lycopene modulates cholinergic dysfunction, Bcl-2/Bax balance, and antioxidant enzymes gene transcripts in monosodium glutamate (E621) induced neurotoxicity in a rat model. Can. J. Physiol. Pharmacol. 94 (4), 394–401.

San Gabriel, A.M., Maekawa, T., Uneyama, H., Yoshie, S., Torii, K., 2007. mGluR1 in the fundic glands of rat stomach. Fed. Eur. Biochem. Soc. Lett. 581 (6), 1119–1123.

Sano, C., 2009. History of glutamate production. Am. J. Clin. Nutr. 90, 728–732.

Sasano, T., Satoh-Kuriwada, S., Shoji, N., Seikine-Hayakawa, Y., Kawai, M., Uneyama, H., 2010. Application of umami taste stimulation to remedy for hypogeusia based on reflex salivation. Biol. Pharm. Bull. 33, 1791–1795.

Sasano, T., Satoh-Kuriwada, S., Shoji, N., Iikubo, M., Kawai, M., Uneyama, H., Sakamoto, M., 2014. Important role of umami taste sensitivity in oral and overall health. Curr. Pharmaceut. Des. 20, 2750–2754.

Sasano, T., Satoh-Kuriwada, S., Shoji, N., 2015. The important role of umami taste in oral and overall health. Flavour 4 (1), 10.

Satoh-Kuriwadk, S., Shoji, N., Kawai, M., Uneyama, H., Kaneta, N., Sasano, T., 2009. Hyposalivation strongly influences hypogeusia in the elderly. J. Health Sci. 55, 689–698.

Satoh-Kuwariwada, S., Kawai, M., Shoji, N., Sekine, Y., Uneyama, H., Sasano, T., 2010. Loss of appetite and weight among elderly patients with umami-taste disorder. Jpn. J. Oral Diagn. Oral Med. 23, 195–200.

Schaumburg, H.H., Byck, R., Gerstl, R., Mashman, J.H., 1969. Monosodium L-glutamate: its pharmacology and role in the Chinese restaurant syndrome. Science 163 (3869), 8266–8828.

Schiffman, S.S., 2000. Intensification of sensory properties of foods for the elderly. J. Nutr. 130, 927S–930S.

Scopp, A.L., 1991. MSG and hydrolyzed vegetable protein induced headache: review and case studies. Headache 31 (2), 107–110.

Select Committee on GRAS Substances (SCOGS, 1980. Evaluation of the Health Aspects of Certain Glutamates as a Food Ingredient. Paper Presented to U.S. Food and Drug Administration. SCOGS- 37a- Suppl.

Selvakumar, E., Prahalathan, C., Sudharsan, P.T., Varalakshmi, P., 2006. Chemoprotective effect of lipoic acid against cyclophosphamide-induced changes in the rat sperm. Toxicology 217 (1), 71–78.

Shi, Z., Luscombe-Marsh, N.D., Wittert, G.A., Yuan, B., Dai, Y., Pan, X., Taylor, A.W., 2010. Monosodium glutamate is not associated with obesity or a greater prevalence of weight gain over 5 years: findings from the Jiangsu Nutrition Study of Chinese adults. Br. J. Nutr. 104 (3), 457–463.

Shigemura, N., Shirosaki, S., Sanematsu, K., Yoshida, R., Ninomiya, Y., 2009a. Genetic and molecular basis of individual differences in human umami taste perception. PLoS One 4 (8), e6717.

Shigemura, N., Shirosaki, S., Sanematsu, K., Yoshida, R., Ninomiya, Y., 2009b. Genetic and molecular basis of individual differences in human umami taste perception. PLoS One 4 (8), e6717.

Shim, J., Son, H.J., Kim, Y., Kim, K.H., Kim, J.T., Moon, H., Rhyu, M.R., 2015. Modulation of sweet taste by umami compounds via sweet taste receptor subunit hT1R2. PLoS One 10 (4), e0124030.

Shimada, A., Castrillon, E.E., Baad-Hansen, L., Ghafouri, B., Gerdle, B., Wahlen, K., Svensson, P., 2016. Increased pain and muscle glutamate concentration after single ingestion of monosodium glutamate by myofascial temporomandibular disorders patients. Eur. J. Pain 20 (9), 1502–1512.

Siegel, A.B., Zhu, A.X., 2009. Metabolic syndrome and hepatocellular carcinoma: two growing epidemics with a potential link. Cancer 115 (24), 5651–5661.

Simon, R.A., 2006. Additive-induced urticarial: experience with monosodium glutamate (MSG). J. Nutr. 130 (4S), 1063S–1066S.

Singh, K., Pushpa, A., 2005. Alteration in some antioxidant enzymes in cardiac tissue upon monosodium glutamate [MSG] administration to adult male mice. Indian J. Biochem. Biophys. 20 (1), 43–46.

Skurray, G.R., Pucar, N., 1988. L-glutamic acid content of fresh and processed Foods. Food Chem. 27, 177–180.

Smriga, M., Mizukoshi, T., Iwahata, D., Eto, S., Miyano, H., Kimura, T., Curtis, R.I., 2010. Amino Acids and Minerals in Ancient Remnants of Fish Sauce (Garum) Sampled in the "Garum Shop" of Pompeii, Italy.

Stanska, K., Krzeski, A., 2016. The umami taste: from discovery to clinical use. Otolaryngol. Pol. 70 (4), 10–15.

Stegink, L.D., Filer Jr., L.J., Baker, L., Mueller, S.M., Wu-Rideout, M.Y.-C., 1979. Factors affecting plasma glutamate levels in normal adult subjects. In: Filer, L.J., Garattini, S., Kare, M.R., Reynolds, W.A., Wurtman, R.J. (Eds.), Glutamic Acid: Advances in Biochemistry. Raven Press, New York, NY, pp. 333–351.

Stegink, L.D., Filer, L.J., Baker, G.L., Bell, E.F., 1986. Plasma glutamate concentration in 1-year-old infants and adults ingesting monosodium L-glutamate in consommé. Pediatr. Res. 20, 53–58.

Steiner, J.E., 1987. What the neonate can tell us about umami. In: Kawamura, Y., Kare, M.R. (Eds.), Umami: A Basic Taste. Marcel Dekker, New York NY, pp. 97–123.

Steiner, J.E., 1993. Behavioral responses to taste and odors in man and animals. In: Proceedings of the Umami International Symposium. Society for Research on Umami Taste, Tokyo Japan, July, pp. 30–43.

Swamy, A.H., Patel, N.L., Gadad, P.C., Koti, B.C., Patel, U.M., Thippeswamy, A.H., Manjula, D.V., 2014. Neuroprotective activity of Pongamia pinnata in monosodium glutamate-induced neurotoxicity in rats. Indian J. Pharmaceut. Sci. 75 (6), 657–663.

Taheri, S., 2017. Effect of exclusion of frequently consumed dietary triggers in a cohort of children with chronic primary headache. Nutr. Health (Bicester) 23 (1), 47–50.

Tarasoff, L., Kelly, M.F., 1993. Monosodium L-glutamate: a double-blind study and review. Food Chem. Toxicol. 31 (12), 1019–1035.

Tawfik, M.S., Al-Badr, N., 2012. Adverse effects of monosodium glutamate on liver and kidney functions in adult rats and potential protective effect of vitamins C and E. Food Nutr. Sci. 3, 651–659.

Tehkonia, T., Morbeck, D.E., Von Zglinick, T., Van Daeursen, J., Lustgarten, J., Scable, H., Kirkland, J.L., 2010. Fat tissue, aging and cellular senescence. Aging Cell 9 (5), 667–684.

Thu Hien, V.T., Thi Lam, N., Cong Khan, N., Wakita, A., Yamamoto, S., 2012. Monosodium glutamate is not associated with overweight in Vietnamese adults. Publ. Health Nutr. 16 (5), 922–927.

Tomankova, V., Liskova, B., Skalova, L., Bartikova, H., Bousova, I., Jourova, L., Anzenbacherova, E., 2015. Altered cytochrome P450 activities and expression levels in the liver and intestines of the monosodium glutamate-induced mouse model of human obesity. Life Sci. 133, 15–20.

Tsurugizawa, T., Uematsu, A., Nakamura, E., Hasumura, M., Hirota, M., Kondoh, T., Torii, K., 2009. Mechanisms of neural response to gastrointestinal nutritive stimuli: the gut-brain axis. Gastroenterology 137, 262–273.

Tsurugizawa, T., Uematsu, A., Uneyama, H., Torii, K., 2011. Different BOLD responses to intragastric load of l-glutamate and inosine monophosphate in conscious rats. Chem. Senses 36 (2), 169–176.

Tung, T.C., Tung, K.S., 1980. Serum free amino acid levels after oral glutamate intake in infant and adult Humans. Nutr. Rep. Int. 22 (43), 1–442.

Uneyama, H., Niijima, A., SanGabriel, A., Torii, K., 2006. Luminal amino acid sensing in the rat gastric mucosa. Am. J. Physiol. Gastrointest. Liver Physiol. 291, G1163–G1170.

United States Department of Agriculture (USDA, 2014. USDA Agricultural Research Services. National Nutrient Database for Standard Reference accessed 9/9/19. http://www.ars.usda.gov/ba/bhnrc/ndl.

Van Bever, H.P., Docx, M., Stevens, W.J., 1989. Food and food additives in severe atopic dermatitis. Allergy 44 (8), 588–594.

Vellisca, M.Y., Latorre, J.I., 2013. Monosodium glutamate and aspartame in perceived pain in fibromyalgia. Rheumatol. Int. 34 (7), 1011–1013.

Ventanas, S., Mustonen, S., Puolanne, E., Tuorila, H., 2010. Odour and flavor perception in flavored model systems: influence of sodium chloride, umami compounds and serving temperature. Food Qual. Prefer. 21 (5), 453–462.

Wakabayashi, Y., Iwashima, A., Yamada, E., Yamada, R., 1991. Enzymological evidence for the indispensability of small intestine in the synthesis of arginine from glutamate. II. N-acetylglutamate synthase. Arch. Biochem. Biophys. 291 (1), 9–14.

Walker, R., Lupien, J.R., 2000. The safety evaluation of monosodium glutamate. J. Nutr. 130, 1048S–1052S.

Walker, R., 1999. The significance of excursions above the ADI. Case study: monosodium glutamate. Regul. Toxicol. Pharmacol. 30 (2 Pt 2), S119–S121.

Wang, S., Zhang, S., Adhikari, K., 2019. Influence of monosodium glutamate and its substitutes on sensory characteristics and consumer perceptions of chicken soup. Foods 8 (2), 0071.

Weil, Z.M., Norman, G.J., DeVries, A.C., Nelson, R.J., 2008. The injured nervous system: a Darwinian perspective. Prog. Neurobiol. 86 (1), 48–59.

Wifall, T.C., Faes, T.M., Taylor-Burds, C.C., Mitzelfelt, J.D., Delay, E.R., 2007. An analysis of 5-inosine and 5-guanosine monophosphate taste in rats. Chem. Senses 32 (2), 161–172.

Williams, A.N., Woessner, K.M., 2009. Monosodium glutamate 'allergy': menace or myth? Clin. Exp. Allergy 39 (5), 640–646.

Windmueller, H.G., Spaeth, A.E., 1975. Intestinal metabolism of glutamine and glutamate from the lumen as compared to glutamine from blood. Arch. Biochem. Biophys. 171, 662–672.

Worm, M., Ehlers, I., Sterry, W., Zuberbier, T., 2000. Clinical relevance of food additives in adult patients with atopic dermatitis. Clin. Exp. Allergy 30 (3), 407–414.

Woschnagg, H., Stollberger, C., Finsterer, J., 2002. Loss of taste is loss of weight. Lancet 359, 891.

Wu, G., 2009. Amino acids: metabolism, functions, and nutrition. Amino Acids 37 (1), 1–17.

Yagamuchi, S., Kimizuka, A., 1979. Psychometric studies on the taste of monosodium glutamate. In: Filer, L.J., Garattini, S., Kare, M.R., Reynolds, W.A., Wurtman, R.J. (Eds.), Glutamic Acid: Advances in Biochemistry. Raven Press, New York, NY, pp. 35–54.

Yamaguchi, S., Ninomiya, K., 2000. Umami and food palatability. J. Nutr. 130 (4S Suppl. l), 921S–926S.

Yamaguchi, S., Takahashi, C., 1984a. Interactions of monosodium glutamate and sodium chloride on saltiness and palatability of a clear soup. J. Food Sci. 65 (1), 82–85.

Yamaguchi, S., Takahashi, C., 1984b. Interactions of monosodium glutamate and sodium chloride on saltiness and palatability of a clear soup. J. Food Sci. 65 (1), 82–85.

Yamaguchi, S., 1987. Fundamental properties of umami in human taste sensation. In: Kawamura, Y., Kare, M.R. (Eds.), Umami: A Basic Taste. Marcel Dekker, New York NY, pp. 41–73.

Yamaguchi, S., 1991. Basic properties of Umami and effects on humans. Physiol. Behav. 49, 833–841.

Yamaguchi, S., 1998. Technical committee, umami manufacturers association of Japan. What is Umami? Food Rev. Int. 14, 123–138.

Yamamoto, S., Tomoe, M., Toyama, K., Kawai, M., Uneyama, H., 2009. Can dietary supplementation of monosodium glutamate improve the health of elderly? Am. J. Clin. Nutr. 90, 844–849.

Yang, W.H., Drouin, M.A., Herbert, M., Mao, Y., Karsh, J., 1997. The monosodium glutamate symptom complex: assessment in a double-blind, placebo-controlled, randomized study. J. Allergy Clin. Immunol. 99 (6 Pt 1), 757–762.

Yeomans, M.R., Gould, N.J., Mobini, S., Prescott, J., 2008. Acquired Flavor acceptance and intake facilitated by monosodium glutamate as a palatability enhancer in the European diet. Physiol. Behav. 93 (4), 958–966.

Yoshida, y., 1988. Umami taste and traditional seasonings. Food Rev. Int. 14, 213–246.

Yu, T., Zhao, Y., Shi, W., Ma, R., Yu, L., 1997. Effect of maternal oral administration of monosodium glutamate at a late stage of pregnancy on developing mouse fetal brain. Brain Res. 747 (2), 195–206.

Yu, L., Zhang, Y., Ma, R., Bao, L., Fang, J., Yu, T., 2006. Potent protection of ferulic acid against excitotoxic effects of maternal intragastric administration of monosodium glutamate at a late stage of pregnancy on developing mouse fetal brain. Europ. Neuropsycholpharmacol. 16 (3), 170–177.

Zanfirescu, A., Cristea, A.N., Nitulescu, G.M., Velescu, B.S., Gradinaru, D., 2017. Chronic monosodium glutamate administration-induced hyperalgesia in mice. Nutrients 10 (1).

Zanfirsecu, A., Ungurianu, A., Tsatsakis, A.M., Nitulescu, G.M., Kouretas, D., Veskoukis, A., Margina, A., 2019. Areview of the alleged health hazards of monosodium glutamate. Compr. Rev. Food Sci. Food Saf. 18, 1111–1134.

Chapter 18

Responding to incidents of low-level chemical contamination and deliberate contamination in food

Elizabeth A. Szabo[1], Elisabeth J. Arundell[3], Hazel Farrell[2], Alison Imlay[1], Thea King[1], Craig Shadbolt[1] and Matthew D. Taylor[4]

[1]*The New South Wales Department of Primary Industries, Silverwater, NSW, Australia;* [2]*The New South Wales Department of Primary Industries, Taree, NSW, Australia;* [3]*The New South Wales Department of Primary Industries, Orange, NSW, Australia;* [4]*The New South Wales Department of Primary Industries, Taylors Beach, NSW, Australia*

18.1 Introduction

There is a community expectation in Australia and New Zealand that food will be safe, and, in general, for most of the people most of the time, this expectation is met. The safety of food, however, is dependent on many factors, not all of which can be controlled through government legislation and regulations. Much of the shared responsibility for food safety lies with the agricultural sector and the processed food industry to ensure that reliable, preventative procedures are in place to produce consistently safe primary produce and processed foods. Part of this shared responsibility also lies with food outlets and consumers to ensure food is handled and prepared in ways that do not introduce new risks.

Food hazards refer to any agents with the potential to cause adverse health consequences to the consumers. These agents can be mainly classified into three categories: physical, chemical, and biological hazards. Chemical hazards can contaminate the food supply chain at any point, may be persistent and can bioaccumulate in animals and humans, as well as biomagnify to increasing concentrations in the tissues of organisms at successively higher levels in the food chain. Chemicals may be present in foods through the intentional and legitimate use of various chemicals, such as pesticides, veterinary medicines and other agricultural chemicals, and via chemicals used in food production, such as food additives and processing aids. Chemical contamination may also occur through environmental pollution by heavy metals, because of chemicals found naturally in foods such as some plant toxins, via the migration of chemicals from packaging materials, and as a result of food processing. Seemingly unrelated activities to food production can also have an impact on the food supply. For example, per- and polyfluoroalkyl substances (PFASs) were historically used in firefighting foams and their use was phased out within Australia around a decade ago. However, PFAS from historical land use activities, such as at some firefighting training grounds, may be present in high levels in soil and leach to surface water and groundwater for decades.

The potential human health effects, though infrequently realized, of chemical contaminants found in food are as diverse as the variety of contaminants. This ranges from relatively modest changes such as effects on enzyme activity and bodyweight, to more significant changes such as developmental effects on the fetus or cancer formation (WHO, 2021). The toxicity of a chemical contaminant will depend on a number of factors, including the absorbed dose and the duration of exposure, i.e., acute or chronic. However, acute impact on health is rare and it is well accepted that the largest potential impact on human health is through low-level repeated exposure. While the amount of a chemical contaminant consumed within a single meal may be below the threshold of toxicological concern, repeated dietary exposure to the chemical may lead to bioaccumulation of the contaminant and the threshold being exceeded in time. The time at which the toxicity threshold concentration is reached will depend on the rate of uptake relative to the rate of removal from the pool via elimination and storage. Therefore, the onset of health effects and association with a chemical foodborne contaminant might occur months to years after initial exposure. This makes the link between exposure and ill-health very difficult to

Ensuring Global Food Safety. https://doi.org/10.1016/B978-0-12-816011-4.00010-0

establish, but there is increasing concern that chemical exposure may play a major role in the etiology of many disorders (Bergman et al., 2013). The ability to accurately determine the concentration of a particular contaminant in a food matrix is critical for the evaluation of potential risk to the consumer. Advances in chemical analytical tools with enhanced specificity and sensitivity have resulted in the ability to detect chemicals in increasingly minute amounts and to the discovery of new chemicals. Advancements in scientific fields, including toxicology, systems biology, and chemical genomics, are also providing unprecedented opportunities for a multidisciplinary approach to address a number of challenges in chemical safety assessment.

Risk-based approaches are used worldwide for the assessment of avoidable and unavoidable chemical contaminants in food, whether intentionally added or nonintentionally present, and whether natural or man-made. The intention of this chapter is to explore strategies for responding to low-level chemical contamination of food. We use case studies of incidents of environmental (per- and poly-fluoroalkyl substances; PFASs) and naturally occurring (ciguatoxins in seafood) chemical contamination. In the absence of a recent deliberate case of chemical contamination of food within Australia, an example of a deliberate case of physical contamination (needles in strawberries) is used to illustrate response approaches. There are important differences among hazards of different classes (microbiological, chemical, or physical), which require somewhat different approaches to risk analysis (for more detail, see FAO, 2006). In regard to chemicals in food, the risks to health mostly depend on the duration, frequency, and level of exposure. Low-level exposures are often of no or negligible risk, with the likelihood of risk increasing as exposure increases and, consequently, thresholds for triggering of toxicological effects are exceeded. In contrast, physical hazards in food have the potential to cause physical danger due to their size, shape or consistency (Edwards, 2014). Hard objects such as glass fragments, metal, and bone pose the biggest food safety concern because they can cause injuries such as cuts, broken teeth, choking, and intestinal perforation (Edwards, 2014). Inhalation of foreign matter into the lungs may also result in partial lung collapse, secondary infection, or destruction of lung tissue from retained foreign material (Edwards, 2014). In the following section, each issue will be discussed in accordance with the widely accepted method of risk analysis, which identifies, assesses, and manages food-related health risks within a structured framework.

18.2 Risk analysis

Risk analysis is used to develop an estimate of the risks to human health and safety, to identify and implement appropriate measures to control the risks, and to communicate with stakeholders about the risks and measures applied. It can be used to support and improve the development of standards, as well as to address food safety issues that result from emerging hazards or breakdowns in food control systems. Risk analysis can be used across a broad range of circumstances and can lead to effective management strategies even when the available data are limited. The framework used in Australia and New Zealand is based on the general framework endorsed by the Codex Alimentarius Commission (Codex, 2004). The risk analysis framework is comprised of three distinct but interrelated components namely risk assessment, risk management, and risk communication. The components of risk analysis are summarized in Fig. 18.1 and are discussed briefly below and in more detail elsewhere in this book.

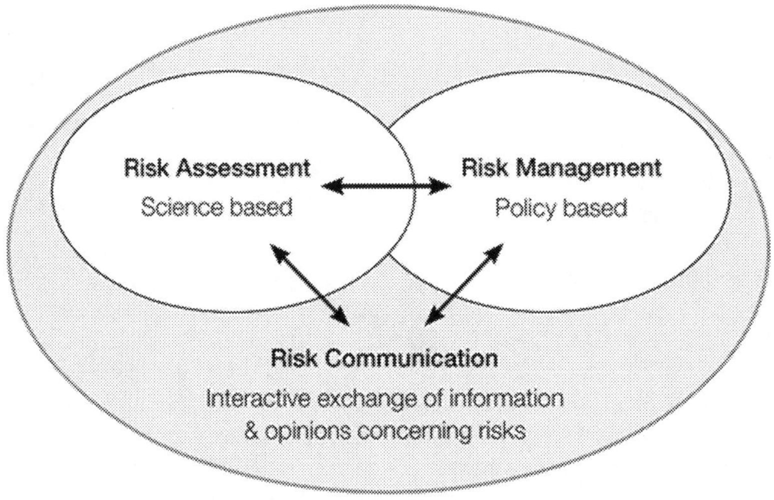

FIGURE 18.1 The Codex risk analysis framework.

Risk assessment involves a science-based approach that utilizes experimental and other available data to characterize the risk and arrive at a conclusion regarding the potential risk associated with a food or food ingredient. Risk assessment consists of four steps (Codex, 1999):

- hazard identification — the identification of biological, chemical, and physical agents capable of causing adverse health effects and which may be present in a particular food or group of foods.
- exposure assessment — the qualitative and/or quantitative evaluation of the likely intake of biological, chemical, and physical agents via food as well as exposures from other sources if relevant.
- hazard characterization — the qualitative and/or quantitative evaluation of the nature of the adverse health effects associated with the hazards.
- risk characterization — the process of determining the qualitative and/or quantitative estimation, including attendant uncertainties, of the probability of occurrence and severity of known or potential adverse health effects in a given population based in hazard identification, hazard characterization, and exposure assessment.

Risk management assists in defining the risk assessment scope and questions to be addressed, considers options for managing identified food risks in the broader context, taking into account the potential benefits of the food as well as relevant policy, consumer behaviors, and economic issues associated with use of the food. The risk assessment and risk management components of the Codex risk analysis framework operate together as an iterative process, with active communication between risk assessors and risk managers.

Risk communication is the interactive exchange of information and opinions regarding risks, risk-related factors, and risk perceptions among all concerned parties, or stakeholders, throughout the entire risk analysis process. It is an ongoing process that engages stakeholders and the public in decision-making to the maximum extent possible. Risk communication is also important to assist in bridging the gap which sometimes exists between the scientific assessment of the health risk and consumers' perception of the health risk.

Any risk analysis framework needs to be supported by guiding principles and these might include:

- *Use the best available data and methodologies*
  Scientific, economic, and other data and information come from both published and unpublished sources, but in both cases, data should be as credible and as objective as possible. Critical evaluation of the available data is an essential element in establishing the basis for the safety of food and subsequent risk management decisions. Where possible, collaboration with other experts or organizations, both national and international, should be sought.
- *Recognize uncertainty in risk analysis*
  It is inevitable that decisions in relation to the safety of food will be made in the presence of scientific uncertainty. In deciding on the risk management options, it is appropriate to recognize, document, and address scientific uncertainty. Depending on the level and nature of uncertainty, a cautious approach to risk management options, such as proposed changes to current food regulations, may be taken to ensure that the overall risk remains low. However, uncertainty in the scientific data should not be used as a reason for inaction when there is reasonable evidence to indicate a potential health risk. Where high levels of uncertainty exist, further information or data may need to be collected, and a revised risk assessment conducted, before risk management options can be considered. As new evidence emerges over time, risk management options need to be revised and updated.
- *Tailor the risk management approach to the risk*
  In managing food-related health risks, there are generally a number of options available, depending on the nature of the risk. Quantifying and comparing different risks is difficult, but qualitative comparisons are generally possible using criteria such as the severity of the outcome and the likelihood of the risk. In deciding on the risk management approach, consideration needs to be given to the level of potential risk, which in the case of food will also depend on the importance of the food in the context of the total diet or for a particular population subgroup. Another factor influencing the level of protection in a particular case will be the level of risk that is acceptable to the community.
- *Involve interested and affected groups*
  The involvement of groups that have an interest in the outcome of a risk analysis can enhance the process through the provision of scientific data, by identifying relevant social, ethical, and economic factors, and by suggesting alternative management approaches. While the process and rules for such involvement need to be clear, involving interested and affected groups can provide opportunities for building trust as well as helping to lend credibility to the ultimate risk management decisions leading to their successful implementation.
- *Communicate in an open and transparent manner*
  Documents outlining risk management options prepared in relation to food-related health risks should generally be publicly available and public submissions on these documents taken into account in the regulatory decisions. Confidential

commercial information can be protected. Dialogue with industry, consumers, and health professionals on food regulatory matters is integral to the risk analysis process and is facilitated, including encouraging stakeholders to comment on documents outlining risk management options.

- *Review the regulatory response*
  In some cases, it is not easy to predict with certainty the outcome of a regulatory decision regarding food and it is necessary to examine the impact of the regulation after a certain period, to ensure that the predicted outcome was achieved and/or that the assumptions used in the assessment were correct. Surveys of the food supply and key stakeholder groups affected by regulatory changes, such as the food industry, health professionals, enforcement officers, or consumers, can generally provide information to evaluate the outcome and determine whether further regulatory action is required.

The general principles supporting the risk analysis framework are applicable to responding to rapidly emerging incidents; however, time constraints might impact on the sequence of steps undertaken within this framework, which is determined on a case by case basis. The FAO/WHO have produced a guide for application of risk analysis principles and procedures during food safety emergencies (FAO/WHO, 2011).

18.3 General control measures for chemicals

Many countries have established regulatory levels for the control of chemical contaminants in food. This can be either a maximum residue limit (MRL) in the case of an agricultural or veterinary chemical or a maximum level (ML) for a contaminant or natural toxicant in the food supply. For example, the Australia New Zealand Food Standards Code (the Code) outlines the limits that relate to agricultural and veterinary residues or contaminants in food. Food containing chemicals at levels greater than the MRL or ML are viewed as being in breach of the Code and subject to regulatory enforcement action at the border and within the nation.

18.3.1 Maximum residue limits for agricultural and veterinary residues in food

The MRL is the highest concentration of a chemical residue that is legally permitted or accepted in a food. The MRL does not indicate the amount of chemical that is always present in a food but it does indicate the highest residue that could possibly result from the registered conditions of use. Testing for agricultural and veterinary residue levels in food assists in indicating whether an agricultural or veterinary chemical product has been used according to its registered use and if the MRL is exceeded then this indicates a likely misuse of the chemical product. In addition, MRLs, while not direct public health limits, act to protect public health and safety by minimizing residues in food while enabling the effective control of pests and diseases.

Standard 1.4.2 of the Code lists the limits for agricultural and veterinary chemical residues that may occur in foods. Standard 1.4.2 specifies that if an MRL is not listed for a particular agricultural or veterinary chemical—commodity combination, there must be no detectable residues of that agricultural or veterinary chemical in that food. This general prohibition means that in the absence of the relevant MRL in Standard 1.4.2, legitimately treated produce may not be sold where there are detectable residues. In the current Australian regulatory system, there may be a time gap between the approval of MRLs for agricultural and veterinary chemical use on crops or animals and the adoption of these MRLs into the Code that applies to food offered for sale.

In assessing the public health and safety implications of agricultural and veterinary residues, the dietary exposure to chemical residues from all potentially treated foods in the diet is compared to the relevant reference health standards such as the acceptable daily intake (ADI) and/or the acute reference dose (ARfD). The ADI of a chemical is the estimate of the amount of a substance in food or drinking water, expressed on a body weight basis, that can be ingested daily over a lifetime without appreciable health risk to the consumer on the basis of all the known facts at the time of the evaluation (WHO, 1997). The ADI is expressed in milligrams of the chemical per kilogram of body weight (WHO, 1997). The ARfD is derived in a manner analogous to the ADI, but with attention paid to short-term effects and the estimate based on the amount of a substance ingested in a period of 24 h or less (JMPR, 2002). The object of the ARfD is to provide a measure of exposure that relates to the hazards occurring during short-term exposure. The ARfD is particularly appropriate for pesticide residues, because the nonuniform distribution of residues within food crops may result in some individuals being exposed to above average levels for short periods of time. The steps undertaken in conducting a dietary exposure assessment are:

- Determination of the residues of a chemical in a treated food and
- Calculation of the dietary exposure to that food using food consumption data from national nutrition surveys and comparing this to the acceptable reference health standard.

18.3.2 Maximum levels for contaminants in foods

MLs for contaminants and natural toxicants are listed in Standard 1.4.1 of the Code, where it has been established that an ML can serve an effective risk management function and only for those foods that provide a significant contribution to the total dietary exposure. Foods not listed in Standard 1.4.1 may contain low levels of contaminants or natural toxicants. As a general principle, regardless of whether or not an ML exists, the levels of contaminants and natural toxicants in all foods should be kept **A**s **L**ow **A**s **R**easonably **A**chievable (the ALARA principle) (Abbott et al., 2003). A difficulty and point of reservation for some groups on the ALARA principle is how to define "reasonably" in each case to ensure that reasonably achievable levels are indeed "reasonably achievable," based on good practices that are possible in commercial practice (IFT Expert Report, 2009). In this context, stakeholder involvement has gradually emerged as a key issue in the decision-making processes to ensure that social, technical, economic, practical, and public policy considerations are taken into account. In assessing food risks, there is a need to ensure that the benefits of a nutritious and well-balanced diet are recognized. For some, if not most, foods, low levels of undesirable chemicals may exist without causing any appreciable health risk.

In practice, MLs are set at a level that is slightly higher than the normal range of variation in levels in foods in order to avoid undue disruptions of food production and trade (ANZFA, 1998). However, MLs are not direct safety limits and cannot fulfill the purpose of being direct safety limits. This is because it is the exposure to a contaminant from the entire food supply that is most relevant to estimating public health implications not exposure from a specific food.

Therefore, for many contaminant/food combinations, an ML has not been established or cannot be justified when a risk assessment indicates a low health risk. This does not mean that chemical contamination events for which there is no regulatory level are ignored and left unmanaged. In circumstances of chemical contamination where there are no established MLs, it is necessary to undertake an assessment of the risk for that particular situation, usually within a short timeframe. The first step in assessing the risk in relation to a chemical contaminant is understanding the nature of the potential adverse health effects associated with exposure to the contaminant and, if possible, to establish a safe level of dietary exposure. In cases where the toxic effect of the chemical contaminant results from a mechanism that has a threshold, hazard characterization usually results in the establishment of a safe level of intake for contaminants. For many contaminants, there are insufficient reliable data on which to identify and characterize the potential hazards, and thus to establish a safe level of human exposure.

Where data are available, it may be possible to establish a reference health standard or so-called "tolerable intake," which can be calculated on a daily, weekly, or longer basis. These reference values are usually established internationally by the Joint (FAO/WHO) Expert Committee on Food Additives (JECFA). The tolerable intake is also generally referred to as "provisional" due to the lack of data on the consequences of human exposure at low levels, and new data may result in a change to the tolerable level.

For contaminants that accumulate in the body over time such as lead, cadmium, dioxin, and mercury, the provisional tolerable weekly intake (PTWI) or monthly intake (PTMI) are used as reference values in order to minimize the significance of daily variations in dietary exposures. For contaminants that do not accumulate in the body such as arsenic, provisional tolerable daily intake (PTDI) is used. The PTDI, PTWI, and PTMI apply to total exposure, i.e., food and nonfood exposures, and all sources of exposure need to be considered when comparing exposure to the reference health standard (WHO, 1987).

Many contaminants do not have an established reference health standard and data limitations might make it impossible to establish a provisional tolerable intake (PTI). In these cases, it might be possible to use a margin-of-exposure approach to determine the potential level of risk, by comparing the lowest level at which adverse effects occur (or the benchmark dose level) with the estimated level of exposure.

Estimation of dietary exposure to contaminants relies on concentration data for the contaminant in various foods together with data on the consumption level for each of these foods among various population groups. The more robust and extensive the input data, the more accurate the estimate of exposure. This information is then compared to the reference health standard, if available, or to a level known to cause adverse effects in animal or human studies resulting in a quantitative estimation of the level of risk from the diet. This becomes one input into risk management decision-making on the safety and suitability of food and is equally applicable to the control of chemicals and during incident management.

18.4 Case study 1

18.4.1 Naturally occurring contamination: ciguatoxins

18.4.1.1 The issue

Naturally occurring marine toxins called ciguatoxins cause ciguatera fish poisoning. Ciguatoxins are produced by some species of dinoflagellates (microalgae), which are typically found in tropical and subtropical coastal reef systems. Ciguatoxins enter the food chain when predatory fish feed on smaller grazing species that have fed on the toxic microalgae. Eating fish contaminated with ciguatoxins can have serious debilitating effects on human health.

Since 2014, the NSW Food Authority (the food regulatory body in the Australian state of New South Wales) investigated 13 incidents of ciguatera fish poisoning, which affected 49 individuals (Table 18.1). The fish associated with these incidents were either tropical reef fish caught in locations outside NSW (tropical Australian or international waters) or Spanish mackerel caught in NSW coastal waters. The relatively high number of outbreaks within a short time period indicated that ciguatera fish poisoning was an emerging issue for NSW, given that previous reports of the illness in the state were uncommon (Farrell et al., 2017). Reports of ciguatera fish poisoning cases linked to imported fish were few and, although the coastal climate of northern New South Wales is considered subtropical, prior to 2014 there was only one documented outbreak of ciguatera fish poisoning from fish caught in the region (Farrell et al., 2016a).

18.4.1.2 The hazard

The marine benthic dinoflagellate genus *Gambierdiscus* is commonly found in tropical and subtropical coastal reef environments. Yasumoto et al. (1977) were the first to propose a dinoflagellate species as the suspected agent of ciguatera fish poisoning (Yasumoto et al., 1977). This dinoflagellate was later described as *Gambierdiscus toxicus* (Adachi and Fukuyo, 1979). While initially the genus *Gambierdiscus* was considered monotypic, there are now 18 described species of *Gambierdiscus*. Species of *Gambierdiscus* can produce an array of complex polyether compounds (as reviewed in Murray et al., 2018). More than 90 fish species and other marine fauna, the majority from tropical regions, have tested positive for ciguatoxins (Kohli et al., 2015). Ciguatoxins can bioaccumulate and biotransform in the marine food chain, typically when

TABLE 18.1 Summary of ciguatera fish poisoning incidents in NSW since 2014 (Farrell et al., 2017 and NSW Food Authority previously unpublished data).

Incident	Date	Cases	Fish species/origin
1	February 2014	4	Spanish Mackerel/Evans Head, NSW
2	March 2014	9	Spanish Mackerel/Scotts Head, NSW
3	April 2015	4	Spanish Mackerel/South West rocks, NSW
4	September 2015	3	Redthroat Emperor/Capel Bank Seamount, high seas
5	September 2015	1	Purple Rockcod/Capel Bank Seamount, high seas
6	February 2016	5	Green Jobfish/Capel Bank Seamount, high seas
7	March 2016	3	Spanish Mackerel/Crowdy Head, NSW
8	April 2016	4	Spanish Mackerel/Crescent Head, NSW
9	January 2017	4	Grouper/Between Cooktown and Lizard is., QLD
10	February 2018	4	Spanish Mackerel/Coffs Harbor, NSW
11	April 2018	3	Spanish Mackerel/Wooli, NSW
12	July 2019	3	Redthroat Emperor/Capel Bank Seamount, high seas
13	July 2019	2	Redthroat Emperor/Capel Bank Seamount, high seas

the toxic dinoflagellates are consumed by herbivorous fish that are then preyed upon by larger carnivorous fish. In this process, the toxic compounds produced by the microalgae are metabolized into more toxic forms (Holmes et al., 1991; Ikehara et al., 2017; Lehane and Lewis, 2000).

Ciguatoxins are lipophilic, heat-stable, cyclic polyether ladder-shaped compounds. There are more than 20 ciguatoxin analogues described to date. Ciguatoxins (CTXs) are grouped based on their chemical structures and origin as P-CTXs (Pacific region) (Yogi et al., 2011), C-CTXs (Caribbean region) (Vernoux and Lewis, 1997), and I-CTXs (Indian Ocean) (Diogene et al., 2017; Hamilton et al., 2002). The molecular structure of I-CTXs is the least well characterized and have not been fully elucidated. The main ciguatoxin found in carnivorous fish in the Pacific region (P-CTX-1B, Fig. 18.2) was originally described by Murata et al. (1990). This analogue is also known as P-CTX-1 as described by Lewis et al. (1991) and is the most toxic ciguatoxin compound identified to date (potency 0.25 μg/kg based on a median lethal dose (LD_{50}) following intraperitoneal injection in mice) (Lewis et al., 1991). The availability of standardized reference material for ciguatoxins is limited to a small number of research groups globally and P-CTX-1B was the only reference material available for testing samples from the NSW illness investigations.

Ciguatoxins are potent neurotoxins that disrupt mammalian cell function by activating voltage-gated sodium channels. Normal cell function is disturbed when the sodium channel is persistently activated from resting membrane potential. The resulting influx of sodium (Na+) ions can cause an imbalance in the routine processes of affected cells. The effects of ciguatoxins can include increased neuronal excitability and neurotransmitter release, impaired synaptic vesicle recycling, and swelling of cells. There is also evidence that ciguatoxins can inhibit neuronal potassium channels. The combined effects of ciguatoxins on sodium channel activation and neuronal potassium inhibition are considered the cause of the varied and complex symptoms of ciguatera fish poisoning (see reviews by Lewis and Vetter, 2014; Nicholson and Lewis, 2006).

While ciguatera fish poisoning is the most common nonbacterial seafood-related illness globally, it is acknowledged that cases are substantially underreported. In part, this is due to the wide range of gastrointestinal, neurological, and cardiovascular symptoms that ciguatera fish poisoning presents, which makes diagnosis difficult. While symptoms of ciguatera fish poisoning can include diarrhea, vomiting, extremity paresthesias, and bradycardia, and symptoms can vary between individuals (Table 18.2), a key indicator of the disease is a sensation of hot/cold temperature reversal, which can be described as a painful sharpness when in contact with or drinking cold water (Farrell et al., 2019). In areas where ciguatera fish poisoning is not commonly encountered, cases can be misdiagnosed or confused with other illness types (e.g., other seafood-related illnesses or chronic illnesses such as multiple sclerosis or chronic fatigue syndrome). Ciguatera fish poisoning usually occurs within 24 h of consuming fish contaminated with ciguatoxins. The duration of symptoms can vary, generally lasting 1–4 days. While the onset of symptoms will vary depending on portion size, fish type, ciguatoxin level, and general wellbeing of the case, a review of cases by Edwards et al. (2019) identified that symptoms progressed from an initial feeling of being unwell (1–2 h), gastrointestinal symptoms (within 6 h), and neurological symptoms (within 3–6 days) (Edwards et al., 2019). In extreme cases, symptoms of ciguatera fish poisoning may persist for weeks or months (see Farrell et al., 2019, 2017, 2016a). Fatalities as a result of contracting ciguatera fish poisoning are rare, but severe chronic cases can impact individuals for many years postexposure.

In general, the treatment of ciguatera fish poisoning is symptomatic and supportive. The provision of mannitol via intravenous drip as a treatment for those affected by ciguatera fish poisoning has had varied success and can depend on

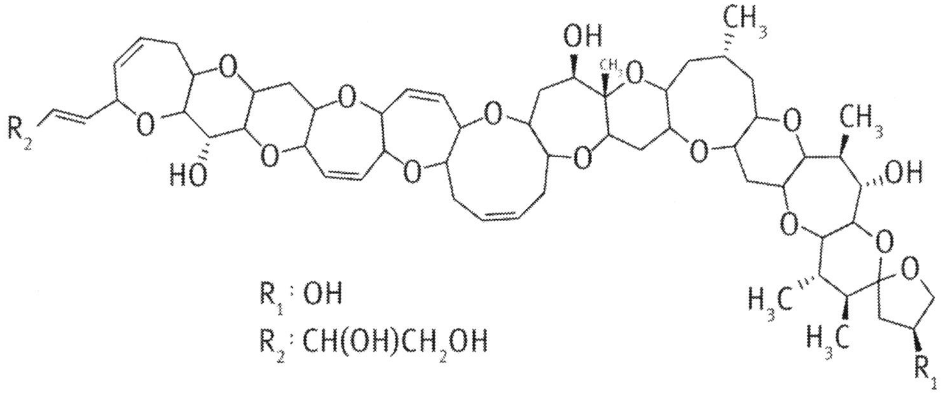

FIGURE 18.2 Chemical structure of Pacific ciguatoxin 1 (P-CTX-1). Alternative names in the literature for this compound are P-CTX-1B and CTX. The structure was originally characterized from Moray Eel (*Gymnothorax javanicus*) samples (Murata et al., 1990).

how soon (within 72 h) the treatment is administered (Friedman et al., 2017). Previous exposure to ciguatoxins is a key factor in individual susceptibility, as this can increase sensitivity to ciguatoxins and result in more severe symptoms. It is particularly important for those who have previously suffered from ciguatera fish poisoning to avoid further exposure to potentially contaminated fish. There is no test to determine previous exposure levels in an individual or the amount of ciguatoxins that would trigger illness or impact the severity of symptoms for a case that was subsequently exposed to ciguatoxins.

The United States Food and Drug administration have set guidance limits for Pacific (0.01 μg/kg P-CTX-1 equivalent) and Caribbean (0.1 μg/kg C-CTX-1 equivalent) ciguatoxins (USFDA, 2019b). To date, regulatory limits for ciguatoxins have not been established. The establishment of such limits has been restricted by the lack of available standardized reference material. As described above, this is due to the complexity of the chemical compounds and processes involved, and this has constrained the development of reliable routine testing for ciguatoxins. Consequently, risk management strategies for ciguatoxins in seafood have focused on fishing bans in locations known to be ciguatera fish poisoning "hot spots" and prohibiting the catch and sale of certain species of fish and size ranges of fish species known to be high risk. An example of this is the *"Schedule of Ciguatera High-Risk Areas & Species Size Limits"* provided in the Seafood Handling Guidelines (Sydney Fish Market Pty Ltd, 2015). Awareness of ciguatera fish poisoning in seafood consumers can also help to prevent illness outbreaks; however, knowledge may be limited in areas where ciguatera is not common or in previously unaffected locations.

18.4.1.3 The risks

While the true incidence of ciguatera fish poisoning is unknown, global estimates of cases range from 50,000 to 500,000 cases per year (Friedman et al., 2017). Australian national data reports between 2 and 10 incidents per year between 2001 and 2015 ($n = 102$), affecting 442 individuals (Department of Health, 2018). Since 2014, there has been an apparent increase in ciguatera fish poisoning illnesses reported in NSW. These incidents were associated with fish imported to NSW from other regions (Grouper, Redthroat Emperor, Green Jobfish, Purple Rockcod) and Spanish mackerel (*Scomberomorus commerson*) caught in NSW coastal waters. The incidents related to Spanish mackerel coincided with the peak annual fishing season (February−April) for this species. Table 18.3 summarizes the recent (since 2014) cases of ciguatera fish poisoning in NSW, relative to reports of the illness since the 1980s. Early reports of ciguatera fish poisoning were linked to imported fish and were generally only documented if many individuals were impacted. The first record of ciguatera fish poisoning linked to Spanish mackerel caught in northern NSW waters was in 2002. There were no further reports following this until 2014, and since then 7 ciguatera fish poisoning incidents (31 individuals) have been associated with Spanish mackerel caught in coastal waters off the mid-north and northern NSW. Concurrently (2015−19), 6 incidents (18 individuals) have been associated with fish imported from locations outside NSW.

Until recently, ciguatera fish poisoning has been considered as a disease that occurs only in tropical areas. The apparent increase in ciguatera fish poisoning cases in NSW and in other regions can be attributed to several variables. It is considered that climate change-related factors may be influencing the distribution of the causative microalgae and/or changing the migratory patterns of the contaminated fish. For example, in NSW, our investigations demonstrated a perceptible southern extension of the catch locations of the implicated Spanish mackerel along the NSW coast, which may be associated with the intensification of the Eastern Australian Current (Farrell et al., 2016b). Additionally, increased market demand for imported seafood products may see a notable increase in ciguatera fish poisoning outbreaks, if catch locations coincide with new or expanding ciguatoxin "hot-spots."

TABLE 18.2 Summary of ciguatera fish poisoning symptoms reported from NSW cases (Farrell et al., 2019, 2017, 2016a).

Gastrointestinal symptoms
Nausea and vomiting, diarrhea, loose stool, and abdominal cramps
Neurological symptoms
Reverse temperature sensation, tingles, and a burning sensation or sharp pain when touching something cold or drinking cold water, paraesthesia of the hands and lips/tingling or numbness in hands and around the mouth, metallic taste (up to 2 months postexposure), swollen lips, aching teeth, myalgia, headaches, itchiness (arms and legs), lethargy/disorientation.
Cardiovascular symptoms
Bradycardia, hypotension, chest tightness/pain or described a "heavy chest," racing heart/anxiety

TABLE 18.3 Summary of ciguatera fish poisoning incidents in NSW since the 1980s (Farrell et al., 2017) and NSW Food Authority previously unpublished data).

Implicated fish	1980s[a]	1990s[a]	2000s	2010s
Imported tropical reef fish and Spanish mackerel	1984: 40 cases 1987: 63 cases (Spanish mackerel)	1994: >30 cases (Spanish mackerel) 1997 and 1998: 26 cases from 3 clusters (tropical reef fish)	2005 and 2009: 2 incidents (tropical reef fish)	2015–19: 18 cases from 6 incidents (tropical reef fish: Grouper, Redthroat Emperor, Green Jobfish, Purple Rockcod)
Spanish mackerel caught in NSW coastal waters	No documented reports	No documented reports	2002: First documented illness from Spanish mackerel caught in NSW	2014–18: 31 cases from 7 incidents

[a]*Published accounts primarily focused on incidents with large case numbers.*

Without specific chemical testing, ciguatoxins cannot be detected in contaminated fish. The toxins are odorless and tasteless. Cooking, freezing, or other processing methods cannot remove ciguatoxins from contaminated fish. The cumulative impact of ciguatoxins, and potentially the other toxins produced by *Gambierdiscus*, on human health is not clear. Further, depending on the region that the toxin is from (Pacific, Caribbean, or Indian), there may be a different pattern or range of symptoms observed in affected individuals (Friedman et al., 2008). In NSW, as in other outbreaks of ciguatera fish poisoning elsewhere, the severity of an individual's symptoms depended on the age and overall wellbeing of the person, size of fish portion consumed, part of fish portion consumed, and previous exposure to ciguatoxins (Farrell et al., 2019, 2017, 2016a). As ciguatoxins can accumulate at higher concentrations in the internal organs of fish, consumption of fish meal that includes the viscera of the fish (or if the fish is cooked whole) can increase exposure to ciguatoxins. In addition, eating large portions of contaminated fish will increase exposure. During the NSW case investigations, it was apparent that affected individuals experienced a greater severity of symptoms if a larger portion was consumed (Farrell et al., 2019, 2017).

The guidance level for P-CTX-1 provided by the USFDA of 0.01 μg/kg is based on exposure estimates in Lehane (2000). In Pacific regions, mild symptoms of ciguatera fish poisoning could be expected after consuming fish containing 0.1 μg/kg P-CTX-1. By applying a safety factor of 10 and allowing for inherent variabilities (meal size, individual response, testing variability at low levels etc.), a "safe fish" is considered to contain no more than 0.01 μg/kg P-CTX-1 equivalent.

Recently, the first baseline study of ciguatoxins in Spanish mackerel in NSW (Kohli et al., 2017) found that 1 in 71 (prevalence 1.4%, 95% confidence interval 1%–4%) samples of fish tissue and 5 in 71 (prevalence 7%, 95% confidence interval 1%–12%) samples of fish liver were positive for P-CTX-1B (<0.1 and < 0.4 μg/kg P-CTX-1B in tissue and liver, respectively). Although the baseline study mentioned above (Kohli et al., 2017) demonstrated that there was not a clear relationship between the length or weight of Spanish mackerel and P-CTX-1B content, we found that reports of ciguatera fish poisoning from Spanish Mackerel in NSW were associated with large fish >10 kg (Farrell et al., 2016a,b). Further work is ongoing by the same research group to understand the trend observed that the toxic specimens from the study were lighter for their length.

Where samples were available from the imported fish cases investigated by the NSW Food Authority since 2015, P-CTX-1B concentrations ranged from 0.006 to 0.069 μg/kg (Farrell et al., 2019, 2017). Spanish mackerel tissue samples collected following illness investigations in NSW since 2014 were found to contain between 0.1 and 1 μg/kg P-CTX-1B (Farrell et al., 2017), although it is considered that the upper range value is a substantial underestimate given significant matrix suppression from the fish tissue during the analysis (Farrell et al., 2016a). For these analyses, it was not possible to determine the number of ciguatoxin analogues present and their contribution to the total toxicity of the fish. While other ciguatoxin analogues may have been present in the fish implicated in NSW illness cases since 2014 and the baseline study by Kohli et al. (2017), the only available reference material at the time of testing was P-CTX-1B.

18.4.1.4 The response

The response to each of the recent ciguatera food poisoning incidents involved, where possible, interviews with affected individuals, supplier traceback, and determination of catch location. As more frequent cases emerged, the findings were reviewed to refine the risk management process and to identify knowledge gaps (see Farrell et al., 2017). The main limitation to risk management has been the limited capability for cost-effective, accurate, and efficient routine testing.

In lieu of this testing capability, risk management strategies for ciguatoxins in NSW have focused on the promotion of educational material for seafood consumers and recreational and commercial fishers (e.g., Farrell et al., 2018; NSW Food Authority, 2017). In response to the illnesses associated with Spanish mackerel in 2014, the regional Fishermen's Co-op at Ballina New South Wales and Sydney Fish Markets Pty Ltd. updated their risk management strategies to implement a ban on the sale of all fish over 10 kg in the mackerel species (ABC, 2014; Farrell et al., 2016a) and this is also reflected in the NSW Food Authority factsheet on ciguatera fish poisoning (NSW Food Authority, 2017). When the ciguatera fish poisoning incident (affecting four, Table 18.1) occurred in NSW during February 2014, an educational outreach campaign, specifically targeted at 117,000 recreational fishers, was facilitated through the NSW Department of Primary Industries (DPI) and highlighted the location of the outbreak, fish involved, and the range of symptoms associated with ciguatera fish poisoning (NSW Food Authority, 2014). Advice to minimize exposure to ciguatoxins recommends that seafood consumers avoid warm water fish from known or high risk ("hot spot") ciguatera areas, to vary the type of warm water fish eaten and to avoid eating multiple portions or servings of the same fish meal. Medical advice and treatment should be sought following exposure to ciguatoxins. Two manuscripts describing recent ciguatera food poisoning illnesses in NSW (Farrell et al., 2019, 2016a) were published in the national Australian journal "*Communicable Disease Intelligence*" to highlight the circumstances of the cases and the importance of sample collection to medical professionals, as ciguatera fish poisoning is not a notifiable disease in NSW.

Since 2015, liquid chromatography mass spectrometry (LC-MS) capability to test for ciguatoxins was established at the Sydney Institute of Marine Science and the fish samples (Spanish mackerel and tropical fish) from subsequent investigations were tested here. The NSW Food Authority actively supported the establishment of this facility and the baseline study of Spanish mackerel conducted by the same research group (Kohli et al., 2016, 2017).

SafeFish (an Australian body concerned with issues of seafood safety and market access for seafood: https://www.safefish.com.au/) recently established the SafeFish Ciguatera Working Group, of which NSW is a member. The group collated and submitted available Australian data on ciguatera fish poisoning to the FAO and WHO following a JECFA call for data and information as part of the 2018 FAO and WHO Expert Meeting on Ciguatera Poisoning (FAO/WHO, 2020). Through SafeFish, the working group facilitated an Australian workshop (March 2019) to coordinate a national response for ciguatera risk management, research priorities, and capability development going forward, which was presented as the Australian National Ciguatera Fish Poisoning Research Strategy (Beatty et al., 2019). The meeting participants included members of the national working group, seafood industry representatives, public health agencies, food safety regulators, research scientists, and invited international experts. It is acknowledged in Australia, and elsewhere, that risk management of ciguatoxins is highly complex. While there have been advances in research, standardized and efficient routine testing is not yet available. A bottleneck to our understanding of and subsequent advances in detection and risk management of ciguatoxins is the lack of availability of standardized reference material.

Further research and collaboration are essential to develop appropriate testing capability for ciguatoxins. The NSW DPI is a partner in an ongoing project led by the University of Technology Sydney supported by an Australian Research Council Linkage Grant ("*Seafood safety: high throughput diagnostics for ciguatoxin risk assessment*" LP180100001 2019—21). The project aims to provide new knowledge of ciguatoxin risk in known "hot spot" locations in Australia and extensively test field—based methods to detect ciguatoxin producing microalgae and ciguatoxins in situ. The resulting data will inform policy to safeguard the seafood industry and consumers.

18.5 Case study 2

18.5.1 Deliberate tampering of strawberries with needles

18.5.1.1 The issue

In September 2018, multiple detections of needles found in strawberries produced by growers in Queensland (an Australian state) were received within a number of days of each other. The detections were confirmed to be an act of deliberate tampering that resulted in a joint investigation and response by police and food safety authorities, which rapidly escalated into a major national incident. The resulting media attention and consumer concern had a devastating impact on the entire strawberry industry, including export markets.

18.5.1.2 The hazard

Foreign matter complaints (such as needles, but also other hazards including glass, plastic, mold, insects) are reported frequently to various authorities and food businesses across Australia. As an example, over 2014—18, an average of 339 foreign matter complaints were reported each year directly to the NSW Food Authority alone (including 361 in 2018).

All complaints are assessed and investigated according to the level of risk, business compliance history, and whether there are any trends or patterns in complaint reporting. Most complaints tend to be isolated in nature, with no clear evidence of system failure or breakdown in processing where foods were manufactured or handled, or any threat of tampering at retail level.

Where an investigation does identify a trend or significant public health issue, action will be taken to reduce the risk—usually through a food recall and media, with businesses required to document any corrective action taken.

In some instances, there might be a suspicion of deliberate tampering and/or threat made against a business, such as the 2005 extortion of popular chocolate bars. For issues such as these, police will typically be the lead agency for investigation with assistance provided by food safety agencies.

18.5.1.3 The risk

Mass tampering of the food supply and subsequent injury from needles in strawberries or other fruit is unlikely. The time and effort required to insert needles (or other foreign matter) into large numbers of fruit makes this a prohibitive exercise itself, and in the majority of circumstances, businesses will have systems in place (such as metal detectors or visual checks) to detect this. Consumers would also be likely to see (either visually or when cutting fruit) any significant level of deliberate tampering.

The most significant level of risk or harm from deliberate tampering relates to the insidious nature or threat of injury. While there may be one or two instances of consumer harm, such as damaged gums or a swallowed needle, which will be very distressing for the victim (including the parent or carer in the case of a child), the psychological nature and perception of risk is likely to be far more damaging. As in the case of the 2005 Mars Bar extortion issue, the threat of deliberate tampering can result in significant media attention, and removal of a product from the market for a considerable length of time, causing brand damage and potential loss of exports (SMH, 2005).

18.5.1.4 The response

The first report of a needle embedded in a strawberry was reported to Queensland health authorities on September 9, 2018 (FSANZ, 2018b). Shortly after this, deliberate tampering was confirmed in two brands of strawberries, and Queensland health and police authorities held a joint media conference on 12 September, alerting consumers and advising them to discard any of the affected products. This was expanded to include a third Queensland brand on 14th September.

As the initial allegations of tampering implicated farms in Queensland, an emergency coordination center was established in that State to coordinate police and health department resources. The Bi-National (Australian and New Zealand) Food Safety Network (BFSN) was engaged to facilitate the sharing of information and media, communication with other jurisdictions, including New Zealand. Sharing of information through the BFSN was essential for coordination of the investigation and incident response. The Australian Government Department of Agriculture also instigated export conditions (such as use of metal detectors for all exporters) to assist with market access and confidence of overseas trading partners.

An unforeseen consequence of media alerts to consumers was the substantial numbers of "copycat," or false reports of tampering made to retailers, police, and food safety authorities across Australia. By the end of September 2018 alone, well over 200 allegations of strawberry and fresh produce tampering with needles were reported implicating brands from all Australian states and territories. This included brands that had no links with those that were credibly associated with the initial acts of deliberate tampering. In November 2018, one farm worker was charged with seven counts of deliberate tampering of strawberries (ABC, 2018).

The copycat reports resulted in a significant escalation of police and food safety regulatory agency responses to follow up and investigate each report. A particular feature of the incident was the reporting of needles in strawberries by affected consumers via social media, rather than going to police or food safety agencies in the first instance. This resulted in ongoing media attention and damage to the strawberry industry through reduced sales and a decision by retailers to instigate a precautionary withdrawal of all products on 15th September. Strawberry farmers across Australia dumped tonnes of fruit, regardless of any association with the tampering incident.

In many instances, it was found that copycat reports were instigated by juveniles, who withdrew allegations when questioned further by police. At least one adult was charged by police for making a false allegation of tampering (The Advertiser, 2018). As penalties for making false allegations of tampering were increased and charges were laid, reports of needles in fruit declined.

18.5.1.5 Key lessons learnt

18.5.1.5.1 Greater coordination between food safety regulators and police

The strawberry tampering incident was unique, being the first national incident requiring extensive coordination and collaboration between various state and territory food safety regulators and police. The response to the incident was hampered, in part, by there being no agreed lead agency or coordination of messaging and information. This resulted in a media release by one police agency where three brands of strawberries were named as potentially affected by deliberate tampering. This advice was later withdrawn after 6 days, but during this time resulted in confusion and loss of confidence in the market. The need for greater linkages between police and food safety agencies was highlighted during the incident debrief and substantial changes have been made in many jurisdictions as a result (FSANZ, 2019b).

18.5.1.5.2 Traceability of farms, produce, workers

The response and investigation of the incident were complicated by a lack of traceability. There is currently no regulation for strawberry farms to be registered or inspected. This made identifying the location of farms across all jurisdictions problematic and relied on ad hoc networks of industry contacts which slowed down identification of the movement of strawberries. It is common for fresh produce, including strawberries, to be comingled at common points in the supply chain (such as packhouses, wholesale markets). This further complicates identification of potential source farms where product may have been tampered. A consequence of this is that fruit from multiple farms may need to be discarded in the event of an incident, rather than limiting product loss to a single point in the supply chain. Investigation of workers on farm was also problematic due to the transient nature of labor across farms. Workers may be moved from farm to farm across different days, depending on need. Accurate records of farm workers were not always maintained across strawberry farms. The lack of traceability will be considered in future regulatory initiatives by Food Standards Australia New Zealand (FSANZ) (FSANZ, 2019a).

18.5.1.5.3 Incident response capacity in the horticulture industry

The horticulture industry in Australia comprises a wide variety of fruit, nut, and vegetable growers. The level of organization and resourcing of each individual commodity sector is highly variable. In this incident, the capability and coordination of strawberry growers to respond to a crisis was low. This made communication with government difficult, consistent messaging and media coordination impossible, and resulted in additional confusion and reduced consumer confidence. There was also a low awareness of supply chain vulnerabilities for deliberate tampering of strawberries by growers. This incident resulted in major retailers requiring all growers to implement metal detection systems to minimize future risks of this nature. The strawberry industry is working to improve coordination in the event of another incident of this nature, including greater awareness of "food defense" systems and avenues for product tampering.

18.6 Case study 3

18.6.1 Environmental contamination—per- and poly-fluoro alkyl substances

18.6.1.1 The issue

PFASs are a group of manufactured chemicals found widely in natural and urban environments. Due to their widespread use, these chemicals have been detected at trace levels in soil, water, and biota in Australia and across the world. The issue has been of considerable public interest in Australia with particular concern from people living near sites where PFASs have been detected. Limited knowledge of their potential impacts, an evolving response from governments to this new contaminant, and reported loss of property values at some sites has increased community concern and led to two Commonwealth senate enquiries.

By design, PFASs are generally stable molecules, with the carbon—fluorine bond being both extremely strong and stable. It is these two key properties that have made these chemicals so useful across so many applications, particularly the combination of a high degree of chemical and thermal stability with other useful properties such as hydro- and lipophobicity (Buck et al., 2011). However, it is these same properties that make PFASs extremely persistent, with limited breakdown over extremely long periods in the environment. The relative solubility of PFASs in water (Kucharzyk et al., 2017) also contributes to their mobility from contaminant point sources.

PFASs can thus move considerable distances where groundwater or surface water is mobile, although concentrations tend to rapidly decrease with increasing distances from point sources (AECOM, 2016).

As an emerging contaminant, health risks from exposure to PFAS are unclear. The Australian PFAS Expert Health Panel (Expert Health Panel for PFAS, 2018) concluded that while there is no current evidence that suggests an increase in overall health risk related to PFAS exposure, the effects cannot be ruled out.

Consistent with national guidance (enHealth, 2019), the NSW Government is taking a precautionary approach in investigations and providing dietary advice to people living in affected areas to minimize exposure, as the health effects are not clear and the chemical takes a long time to break down.

Internationally, everyone generally has low levels of PFAS in their blood, due to the chemical's persistence and widespread presence in the environment (Expert Health Panel for PFAS, 2018). The most recent data show that in Australia, the general population's exposure to PFAS is low and declining. The Food Regulation Standing Committee (FRSC) found no evidence that low-level food exposure has been harmful to human health (FRSC, 2017).

In Australia, investigations into PFAS contamination have focused on sites where large quantities of PFAS have been used in the past; this legacy contamination largely related to PFAS containing fire-fighting foams used in training. Three PFAS chemicals, Perfluorooctane Sulfonate (PFOS), Perfluorooctanoic Acid (PFOA) and Perfluorohexane Sulfonate (PFHxS), were the predominant ingredients of fire-fighting foam and have been the focus of investigations. Measures are underway in Australia to phase-out use of these chemicals in fire-fighting foam.

Investigations in NSW are being led by NSW Environment Protection Authority (NSW EPA), with actions informed and guided by the NSW Expert panel and NSW Technical Advisory Group. NSW Department of Primary Industries, which includes the NSW Food Authority, supports the government's response in providing advice and assistance to government and community, including impacted fishers and farmers. Over 45 sites are currently under investigation in NSW, including Australian Defence Department bases, airports, local fire-fighting training sites, industrial sites, and power stations. PFAS has been found in soil, sediment, groundwater, surface water, and biota. At some sites, remediation measures are being developed to reduce the amount of PFAS at the source and to reduce further migration of the chemicals into the environment.

18.6.1.2 The hazard

Humans are commonly exposed to PFASs in the environment through ingestion of contaminated drinking water (e.g., Hölzer et al., 2008) or consumption of plants or animals that have accumulated the contaminant (e.g., Chen et al., 2018; Fromme et al., 2009; Vestergren and Cousins, 2009). In terrestrial plants and animals, exposure to PFASs and bioaccumulation also occur through ingestion of contaminated water, or to a lesser extent ingestion of contaminated soil. In aquatic animals, exposure and bioaccumulation can occur through contaminated sediments, and for animals that respire using gills, through respiration in contaminated water. While trophic level magnification of PFASs can occur in air breathing animals, the magnification of PFAS through trophic pathways in aquatic animals is presently unclear, with inconsistent data in the literature. The rate at which PFASs are accumulated also depends on species-specific toxicokinetics.

The uptake of PFAS into plant and animal food products has been studied and reported in many countries, including Australia. Levels of PFAS in the general food supply and levels in foods grown and raised on contaminated land are considered separately because the background levels in the general supply are assumed to be low and declining as PFAS use is better managed, and levels at contaminated sites result from the particular conditions at each site, so are considered on a case by case basis.

The food standards setting agency Food Standards Australia New Zealand (FSANZ) conducted the 24th Australian Total Diet Study (ATDS), Phase 2, in 2011 (FSANZ, 2016). The study analyzed perfluorinated compounds in a range of foods in the Australian diet as part of a general survey of the occurrence of packaging chemicals in foods. There were no detections for PFOA and two detections for PFOS from 50 foods (306 samples in total) tested. The concentrations of PFOS were at very low levels (1 part per billion) and similar to those reported internationally for the same foods. Foods were sampled from a range of different retail outlets representing the buying habits of most of the community, including supermarkets, corner stores, delicatessens, markets, and takeaway shops. FSANZ concluded that the Theoretical Maximum Daily Exposure to PFOS was low in comparison to the Tolerable Daily Intake (TDI) (33% at the maximum concentration assuming PFOS was present in all foods). FSANZ also concluded that this indicated a negligible public health and safety risk.

To further understand and characterize the hazard of PFAS to Australian consumers, another survey is underway and due for completion by 2021: the 27th ATDS will monitor PFASs in the general food supply.

Approaches to PFAS Risk Analysis vary from country to country and continue to evolve. This includes the development of Health-Based Guidance Values (HBGVs). The HBGVs in Australia were developed by FSANZ for the

Commonwealth Department of Health and published in 2017 in the form of Tolerable Daily Intakes for PFOS (20 ng/kg bw/day) and PFOA (160 ng/kg bw/day). Insufficient data existed to recommend a TDI for PFHxS. No regulatory limits (Maximum Limits or MLs) have been set for these chemicals, but FSANZ recommended trigger points for investigation of PFOS + PFHxS combined and for PFOA. If food testing found levels at or above the trigger points, further investigation or risk management action may be required, evaluated on a case by case basis. In Australia, FSANZ also monitors international developments, to advise on the approach to guidance for individual residents affected near contaminated sites, to monitor, and evaluate international and local developments and communicate accordingly.

The USFDA (USFDA, 2019a) has taken several approaches to assessing foods for PFAS, including:

- in foods from specific areas affected by environmental contamination;
- in certain foods that may have an increased likelihood of PFAS contamination, but not associated with a specific site; and
- in foods more generally

and in June 2019, the USFDA published its findings from a testing program of milk and a range of produce foods sampled in the United States of America, from specific areas with known environmental contamination (USFDA, 2019a). The USFDA decided to remove the milk from one farm from the general supply. The farm was located near a United States Defense base, and levels of PFAS in the milk were determined to be a potential human health concern.

In the United States, the Agency for Toxic Substances and Disease Registry (ATSDR) has also published draft minimal risk levels as tools for risk assessors to look at particular sites where exposures to a chemical may have occurred (FSANZ, 2018a).

The European Food Safety Authority (EFSA) published HBGVs in 2008 and revised them down in 2018, publishing provisional new Tolerable Weekly Intake (TWI) values of 13 ng/kg bw for PFOS and 6 ng/kg bw for PFOA (EFSA Panel on Contaminants in the Food Chain et al., 2018).

In NSW testing of foods sampled from properties near contaminated sites has been conducted by the Australian Defence Department contractors as part of the environmental assessment process. In some cases, individual dietary advice has been provided for home consumers on these properties. Home-grown vegetables and eggs are the most common foods for which advice has been given. The advice given has included to moderate or avoid consumption of eggs and in the case of vegetables, or to seek alternate ways of growing them, for example, in raised garden beds with clean soil, and to irrigate with town water.

Many species of freshwater and marine aquatic biota have been shown to bioaccumulate PFASs, especially perfluoroalkyl acids (PFAAs) with six or more fluorinated carbons, such as PFOS, PFHxS, and perfluorodecane sulfonate (PFDS) (e.g., Baduel et al., 2014; Ding and Peijnenburg, 2013; Fang et al., 2016; Martin et al., 2003; Naile et al., 2010; O'Connor et al., 2018; Taylor, 2019b; Taylor and Johnson, 2016).

Many of these aquatic species are also exploited through commercial and recreational fisheries or aquaculture, either being consumed directly (by the fishers themselves) or on-sold into the food supply. This generally includes a broad range of species spanning molluscs, prawns (= shrimps), crabs, and fishes (hereafter called "seafood") (Taylor, 2019a,b, 2018). However, the concentrations in edible tissues can be highly variable across seafood species (Taylor, 2018, 2019b), with concentrations influenced by a diverse suite of physiological, ecological, and environmental factors (Taylor, 2019a). Furthermore, several seafood species possess the ability to depurate contaminants from their edible tissues over very short periods once PFAS exposure ceases (O'Connor et al., 2018; Taylor et al., 2017), which means there is limited scope to predict concentrations in seafood species given knowledge of concentrations in environmental sources. Consequently, comprehensive sampling and testing programs have been required to properly quantify the scale and extent of potential exposure risk through consumption of seafood (Taylor et al., 2018).

Within New South Wales, a comprehensive framework for testing seafood species for PFAS was implemented in 2015. This framework had some origins in the model applied for testing of seafood species for dioxin in Sydney Harbour (see Manning et al., 2017), and had dual objectives to both inform modeling of the exposure risk associated with seafood consumption, and to inform the spatial scale of any dietary advice that may need to be issued to manage exposure risk. Early work showed that concentrations in the edible tissues of fish, prawns, and crabs were highly variable among individuals (Taylor and Johnson, 2016), meaning that a large number of individuals needed to be collected from each location within the survey design. Due to the high cost of chemical analyses for PFAS in biota samples, testing most commonly involved composites, with a target of 40 animals per location (or site) whose tissues were pooled into 4 composites of 10 individuals. Modeling (see Taylor, 2019b) indicated that the collection and testing of muscle tissue from 40 individuals would give a 90%–95% probability that testing provided an estimate within 20% of the actual population mean or median at those sites. The ±20% reflects the threshold criteria for analytical precision for PFAAs using LC/MS/MS (Shoemaker et al., 2008).

Notwithstanding the general principles outlined above, the design of sampling programs was necessarily specific to the estuary or waterbody under investigation. Existing databases of commercial and/or recreational fisheries catch were used as a quantitative basis for species selection. Catches from the waterbody were reviewed for recent years and used to generate a target species list which nominally encompassed >80% of the total biomass (for commercial fisheries) or number (for recreational fisheries) of seafood harvested from that particular waterbody. Spatially, the size and scale of the water body was considered alongside the number of potential contaminant sources into the estuary and the migratory habits of the target species. Usually, 4–5 locations or sites were sampled, and collection of 40 individuals from each species (as outlined above) was targeted from each of these locations. Capture methods specific for the target species were selected, which usually included mesh or haul nets for fish species (electrofishing was employed for freshwater fishes), traps for crustaceans, trawls for prawns, and jigs for cephalopods.

Cultured oysters support a large open water aquaculture industry in New South Wales, and sampling of the individuals did not follow the approach above. In the estuaries targeted in the initial investigation, oysters were sampled at increasing distances from the point source and tested. It was found that oysters generally accumulated only low concentrations of PFAS (well below the relevant FSANZ threshold), even when in reasonably close proximity to the contaminant point source (O'Connor et al., 2018). Oysters were also shown to rapidly depurate PFASs from their systems. Similarly, low concentrations were confirmed in other contaminated estuaries, which supported a broad conclusion that there was negligible exposure risk associated with consumption of oysters cultured in estuaries affected by PFAS.

18.6.1.3 The risk

Management of PFAS contamination in NSW has been informed by two scientific advisory groups, the NSW PFAS Expert Panel, formed in September 2015 (formerly the Williamtown Expert Panel), and the NSW Technical Advisory Group.

Members include technical experts in contamination and public health, animal science, food safety, and hydrology. The specialist groups provide advice on the health risks posed by the levels of PFAS in produce grown or harvested from impacted sites, including seafood and aquatic species caught in local waterways. Exposure of the general population to PFAS is considered to be low and declining (FRSC, 2017) since PFOS was withdrawn from sale in 2000. However, people living near contaminated sites have been identified as at greater risk of exposure to PFAS and dietary exposure assessments have been conducted to assist affected communities, including frequent fishers, to manage their PFAS exposure. Tailored dietary advice, based on individual circumstances, is provided as a precaution.

Residents near contaminated sites who are exposed to contaminated ground or surface water or who use these to grow or produce their own food will have greater exposure to PFAS than the rest of the population. It is prudent for these people to reduce their exposures where possible, usually achieved by a combination of avoiding contaminated water sources, using town supplied water for drinking, watering home produce and stock watering, by reducing or avoiding consumption of home-grown produce if likely to be contaminated from water or soil, and by choosing and eating food from a variety of sources.

Dietary exposure assessments have been conducted from test results of food samples collected from contaminated sites. The FSANZ approach to dietary modeling uses chemical concentration data and National Nutrition Survey data to estimate food consumption for identified consumer groups.

For PFAS dietary exposure assessments at contaminated sites, the modeling approach includes estimates for high consumers. High consumers are those people who consume:

- a lot of one food that contains a chemical of interest
- smaller amounts of a number of different foods that all contain the same chemical or
- small amounts of a food which contains a high concentration

FSANZ estimates chronic dietary exposure using the mean and 90th percentile "consumers only" food consumption estimates (FSANZ, 2014).

18.6.1.4 The response

When Aqueous Film Forming Foam derived PFAS contamination was first detected in seafood species in New South Wales in the Port Stephens and Hunter River estuary (URS Australia, 2015), initial short-term closures were put in place in the areas adjacent to point sources while additional data were collected and the risk further considered. At the same time, fishers holding prawn trawl endorsements in the Hunter River elected to implement a voluntary agreement to cease harvesting of School Prawn throughout all fished areas of the Hunter River estuary, while further data were collected and exposure risks were considered. All closures were lifted around 9 months later, after more comprehensive data became available to inform risk assessment.

Ongoing management of PFAS exposure risk through seafood consumption, in Williamtown and across all other affected sites, is achieved through the development and communication of precautionary dietary advice. This advice is informed by the data collection and modeling outlined earlier in this chapter. Advice is intended to allow regular consumers of seafood potentially affected by PFAS, who are at the highest risk of exposure, to manage their exposure through time. This approach includes both recreational and commercial fishers, and complements the existing general advice issued to Australian consumers to eat 2–3 serves of seafood per week. At some sites, additional, low-level sampling work has been commissioned to keep a watching brief on PFAS concentrations in seafood through time, to support adaptive management of advice as required.

Clear and consistent communication to the public, and those directly affected in particular, is important. Unified messaging is developed and agreed through a communication and engagement plan and distributed to relevant government departments. This approach is important to ensure messaging is factual and consistent, and that there is no conflicting information provided by different government departments. Once messaging and dietary advice have been agreed, the information is communicated through a number of avenues, depending on the stakeholder group or targeted audience. When new advice becomes available, a press release may be issued by relevant government departments to generate publicity through the regular media. Relevant information is also posted on government web pages, and flyers may be produced for distribution. For fishers, direct engagement occurs at the same time as information is released through other avenues. Commercial fishers who hold endorsements for the affected areas are alerted via an SMS directing them to the presence and location of the dietary advice. Recreational fishers are alerted through a post on the Department's Facebook page with the relevant information. Also, a short article is usually written alerting fishers to the new dietary advice for the Department's e-newsletter "Newscast," which is sent by email to recreational fishing license holders. For some water bodies, signs have also been erected at access points. There is also a section published in the NSW Recreational Fishing Guide alerting fishers to the risks posed by contamination and directing them where to find the relevant information.

18.7 Conclusion

The need to respond to low-level chemical contaminants is often prompted by an emerging food incident that requires a response at local, national, or international levels within a short timeframe. The case studies presented illustrate how risk analysis offers a structured and flexible framework in considering the risks associated with food. Incorporating the key components of risk assessment, risk management, and risk communication, risk analysis provides a systematic and disciplined approach to establishing and implementing risk management options applicable to the severity of risk and local circumstances.

Illustrated within each case study was the need for risk managers to recognize that, in all cases, risk assessment results contained a degree of uncertainty depending on the quality of the available data. Advances in scientific knowledge will have an impact on these risk management decisions. Subsequently, a review of risk management decisions is essential in determining whether advance knowledge has made an impact on scientific advice previously used to make decisions resulting in policy, regulation, or action.

Acknowledgments

The authors praise the professionalism, dedication, and commitment of national and international regulatory colleagues and scientific experts involved in responding to the issues presented as case studies. Some of the material featured in this chapter appears in the public domain on government websites: the New South Wales Food Authority (www.foodauthority.nsw.gov.au) and the Food Standards Australia New Zealand (www.foodstandards.gov.au). Full links to the material are provided in the reference section.

References

Abbott, P., Baines, J., Fox, P., Graf, L., Kelly, L., Stanley, G., Tomaska, L., 2003. Review of the regulations for contaminants and natural toxicants. Food Control 14 (6), 383–389. https://doi.org/10.1016/S0956-7135(03)00040-9.

ABC, 2014. Nine Victims of Ciguatera Poisoning from Fish Caught off Scotts Head. https://www.abc.net.au/news/2014-03-04/nine-victims-of-ciguatera-poisoning-from-fish-caught-off-scotts/5296808.

ABC, 2018. Strawberry Needle Contamination: Accused Woman Motivated by Spite, Court Hears. https://www.abc.net.au/news/2018-11-12/strawberry-needle-contamination-woman-to-face-court/10486770.

Adachi, R., Fukuyo, Y., 1979. The thecal structure of a marine toxic Dinoflagellate *Gambierdiscus toxicus* gen. et sp. nov. Collected in a Ciguatera-endemic Area. Bull. Jpn. Soc. Sci. Fish. 45 (1), 67–71.

AECOM, 2016. Off-Site Human Health Risk Assessment – July 2016 – RAAF Base Williamtown Stage 2B Environmental Investigation. Retrieved from: https://www.defence.gov.au/Environment/PFAS/docs/Williamtown/Reports/HHRAReports/2016HHRAFullReport.pdf.

ANZFA, 1998. The Regulation of Contaminants and Other Restricted Substances in Food: Policy Paper. Canberra.

Baduel, C., Lai, F.Y., Townsend, K., Mueller, J.F., 2014. Size and age-concentration relationships for perfluoroalkyl substances in stingray livers from eastern Australia. Sci. Total Environ. 496, 523–530. https://doi.org/10.1016/j.scitotenv.2014.07.010.

Beatty, P., Boulter, M., Carter, S., Chinain, M., Doblin, M., Farrell, H., Gatti, C., Hallegraeff, G., Harwood, T., Sandberg, S., Lewis, R., Llewellyn, L., Murray, S., Murray, S., Poole, S., Poole, E., Robertson, A., Sparrow, L., Stafford, R., Stamp, T., Stanley, G., Zammit, A., Seger, A., Dowsett, N., Turnbull, A., 2019. In: Seger, A., Dowsett, N., Turnbull, A. (Eds.), National ciguatera research strategy: reducing the incidence of ciguatera in Australia through improved risk management. South Australian Research and Development Institute, Adelaide, pp. 1–32.

Bergman, Å., Heindel, J.J., Jobling, S., Kidd, K.A., Zoeller, R., 2013. State of the Science of Endocrine Disrupting Chemicals - 2012: An Assessment of the State of the Science of Endocrine Disruptors Prepared by a Group of Experts for the United Nations Environment Programme and World Health Organization. WHO, Geneva, Switzerland.

Buck, R.C., Franklin, J., Berger, U., Conder, J.M., Cousins, I.T., de Voogt, P., et al., 2011. Perfluoroalkyl and polyfluoroalkyl substances in the environment: terminology, classification, and origins. Integr. Environ. Assess. Manag. 7 (4), 513–541. https://doi.org/10.1002/ieam.258.

Chen, W.-L., Bai, F.-Y., Chang, Y.-C., Chen, P.-C., Chen, C.-Y., 2018. Concentrations of perfluoroalkyl substances in foods and the dietary exposure among Taiwan general population and pregnant women. J. Food Drug Anal. 26 (3), 994–1004. https://doi.org/10.1016/j.jfda.2017.12.011.

Codex, 1999. Principles and Guidelines for the Conduct of Microbiological Risk Assessment.

Codex, 2004. Working principles for risk analysis for application in the framework of the Codex Alimentarius. In: Codex Alimentarius Commission Procedural Manual, fourteenth ed. Joint FAO/WHO Food Standards Programme, Rome.

Department of Health, 2018. OzFoodNet Reports. Retrieved from: https://www1.health.gov.au/internet/main/publishing.nsf/Content/cdna-ozfoodnet-reports.htm.

Ding, G., Peijnenburg, W.J.G.M., 2013. Physicochemical properties and aquatic toxicity of poly- and perfluorinated compounds. Crit. Rev. Environ. Sci. Technol. 43 (6), 598–678. https://doi.org/10.1080/10643389.2011.627016.

Diogene, J., Reverte, L., Rambla-Alegre, M., Del Rio, V., de la Iglesia, P., Campas, M., et al., 2017. Identification of ciguatoxins in a shark involved in a fatal food poisoning in the Indian Ocean. Sci. Rep. 7 (1), 8240. https://doi.org/10.1038/s41598-017-08682-8.

Edwards, M., 2014. Other significant hazards: physical hazards in foods. In: Encyclopedia of Food Safety, vol. 3. Campden BRI, Chipping Campden, UK, pp. 117–123.

Edwards, A., Zammit, A., Farrell, H., 2019. Four recent ciguatera fish poisoning incidents in New South Wales, Australia linked to imported fish. Commun. Dis. Intell. 43. https://doi.org/10.33321/cdi.2019.43.4.

EFSA Panel on Contaminants in the Food Chain, Knutsen, H.K., Alexander, J., Barregård, L., Bignami, M., Brüschweiler, B., et al., 2018. Risk to human health related to the presence of perfluorooctane sulfonic acid and perfluorooctanoic acid in food. EFSA J. 16 (12), e05194. https://doi.org/10.2903/j.efsa.2018.5194.

enHealth, June 2019. FACT SHEET: Revised enHealth Guidance Statements on Per-and Poly-Fluoroalkyl Substances (PFAS). Retrieved from: https://www1.health.gov.au/internet/main/publishing.nsf/Content/44CB8059934695D6CA25802800245F06/$File/FactSheet-PFAS-june19-enHealth.pdf.

Expert Health Panel for PFAS, 2018. Expert Health Panel for PFAS: Summary. Retrieved from: https://www1.health.gov.au/internet/main/publishing.nsf/Content/C9734ED6BE238EC0CA2581BD00052C03/$File/summary-panels-findings.pdf.

Fang, S., Zhang, Y., Zhao, S., Qiang, L., Chen, M., Zhu, L., 2016. Bioaccumulation of perfluoroalkyl acids including the isomers of perfluorooctane sulfonate in carp (*Cyprinus carpio*) in a sediment/water microcosm. Environ. Toxicol. Chem. 35 (12), 3005–3013. https://doi.org/10.1002/etc.3483.

FAO, 2006. Food safety risk analysis. A guide for national food safety authorities. FAO Food Nutr. Pap. 87 (ix-xii), 1–102.

FAO/WHO, 2011. FAO/WHO Guide for Application of Risk Analysis Principles and Procedures during Food Safety Emergencies.

FAO/WHO, 2020. Report of the expert meeting on ciguatera poisoning: Rome, 19–23 November 2018. World Health Organization. License: CC BY-NC-SA 3.0 IGO. https://apps.who.int/iris/handle/10665/332640.

Farrell, H., Edwards, A., Bowman, R., Bonello, G., Murray, S., January/March 2018. Fishy business. Food Aust. 70, 44.

Farrell, H., Edwards, A., Zammit, A., 2019. Four recent ciguatera fish poisoning incidents in New South Wales, Australia linked to imported fish. Comm. Dis. Intell. 43. https://doi.org/10.33321/cdi.2019.43.4.

Farrell, H., Murray, S.A., Zammit, A., Edwards, A.W., 2017. Management of ciguatoxin risk in Eastern Australia. Toxins 9 (11). https://doi.org/10.3390/toxins9110367.

Farrell, H., Zammit, A., Harwood, D.T., Murray, S.M., 2016a. Is Ciguatera moving south in Australia? Harmful Algae News 54, 5–6.

Farrell, H., Zammit, A., Manning, J., Shadbolt, C., Szabo, L., Harwood, D.T., et al., 2016b. Clinical diagnosis and chemical confirmation of ciguatera fish poisoning in New South Wales, Australia. Comm. Dis. Intell. 40 (1), E1–E6.

Friedman, M.A., Fernandez, M., Backer, L.C., Dickey, R.W., Bernstein, J., Schrank, K., et al., 2017. An updated review of ciguatera fish poisoning: clinical, epidemiological, environmental, and public health management. Mar. Drugs 15 (3). https://doi.org/10.3390/md15030072.

Friedman, M.A., Fleming, L.E., Fernandez, M., Bienfang, P., Schrank, K., Dickey, R., et al., 2008. Ciguatera fish poisoning: treatment, prevention and management. Mar. Drugs 6 (3), 456–479. https://doi.org/10.3390/md20080022.

Fromme, H., Tittlemier, S.A., Volkel, W., Wilhelm, M., Twardella, D., 2009. Perfluorinated compounds–exposure assessment for the general population in Western countries. Int. J. Hyg Environ. Health 212 (3), 239–270. https://doi.org/10.1016/j.ijheh.2008.04.007.

FRSC, 2017. Food Regulation Standing Committee Statement: Per- and Poly-Fluoroalkyl Substances (PFAS) and the General Food Supply.

FSANZ, 2014. Protecting 'high Consumers'. https://www.foodstandards.gov.au/science/exposure/Pages/protectinghighconsum4441.aspx.

FSANZ, 2016. 24th ATDS Phase 2. https://www.foodstandards.gov.au/publications/Pages/24th-ATDS-Phase-2.aspx.

FSANZ, 2018a. Perfluorinated Compounds. Retrieved from: http://www.foodstandards.gov.au/consumer/chemicals/Pages/Perfluorinated-compounds.aspx.

FSANZ, 2018b. Strawberry Tampering Incident: Report to Government. Retrieved from: https://www.foodstandards.gov.au/publications/SiteAssets/Pages/Strawberry-tampering-incident/FSANZ%20Strawberry%20Report%20doc.pdf.

FSANZ, 2019a. Proposal P1052 – Primary Production and Processing Requirements for High-Risk Horticulture. Retrieved from: https://www.foodstandards.gov.au/code/proposals/Pages/P1052.aspx.

FSANZ, 2019b. Strawberry Tampering Incident Debrief May 1 2019, Follow-Up Report to Government. Retrieved from: https://www.foodstandards.gov.au/publications/SiteAssets/Pages/Strawberry-tampering-incident/FSANZ%20Strawberry%20Debrief%20Report%20doc.pdf.

Hamilton, B., Hurbungs, M., Jones, A., Lewis, R.J., 2002. Multiple ciguatoxins present in Indian Ocean reef fish. Toxicon 40 (9), 1347–1353. https://doi.org/10.1016/s0041-0101(02)00146-0.

Holmes, M.J., Lewis, R.J., Poli, M.A., Gillespie, N.C., 1991. Strain dependent production of ciguatoxin precursors (gambiertoxins) by *Gambierdiscus toxicus* (Dinophyceae) in culture. Toxicon 29 (6), 761–775. https://doi.org/10.1016/0041-0101(91)90068-3.

Hölzer, J., Midasch, O., Rauchfuss, K., Kraft, M., Reupert, R., Angerer, J., et al., 2008. Biomonitoring of perfluorinated compounds in children and adults exposed to perfluorooctanoate-contaminated drinking water. Environ. Health Perspect. 116 (5), 651–657. https://doi.org/10.1289/ehp.11064.

IFT Expert Report, 2009. Making decisions about the risks of chemicals in foods with limited scientific information. Compr. Rev. Food Sci. Food Saf. 8 (3), 269–303. https://doi.org/10.1111/j.1541-4337.2009.00081.x.

Ikehara, T., Kuniyoshi, K., Oshiro, N., Yasumoto, T., 2017. Biooxidation of ciguatoxins leads to species-specific toxin profiles. Toxins 9 (7). https://doi.org/10.3390/toxins9070205.

JMPR, 2002. Further Guidance on Derivation of the ARfD. Pesticide Residues in Food—2002. Report of the JMPR 2002. FAO Plant Production and Protection Paper, p. 172.

Kohli, G.S., Farrell, H., Murray, S.A., 2015. *Gambierdiscus*, the cause of ciguatera fish poisoning: an increased human health threat influenced by climate change. In: Botana, L.M., Louzao, C., Vilariño, N. (Eds.), Climate Change and Marine and Freshwater Toxins. De Gruyter, Berlin, Germany, pp. 273–312.

Kohli, G.S., Harwood, D.T., Laczka, O., Boulter, M., Murray, S.A., 2016. Safeguarding Seafood Consumers in New South Wales from Ciguatera Fish Poisoning (FRDC Project No 2014–035). Retrieved from: https://www.frdc.com.au/Archived-Reports/FRDC%20Projects/2014-035%20-%20DLD.pdf.

Kohli, G.S., Haslauer, K., Sarowar, C., Kretzschmar, A.L., Boulter, M., Harwood, D.T., et al., 2017. Qualitative and quantitative assessment of the presence of ciguatoxin, P-CTX-1B, in Spanish Mackerel (*Scomberomorus commerson*) from waters in New South Wales (Australia). Toxicol. Rep. 4, 328–334. https://doi.org/10.1016/j.toxrep.2017.06.006.

Kucharzyk, K.H., Darlington, R., Benotti, M., Deeb, R., Hawley, E., 2017. Novel treatment technologies for PFAS compounds: a critical review. J. Environ. Manag. 204, 757–764. https://doi.org/10.1016/j.jenvman.2017.08.016.

Lehane, L., 2000. Ciguatera update. Med. J. Aust. 172 (4), 176–179.

Lehane, L., Lewis, R.J., 2000. Ciguatera: recent advances but the risk remains. Int. J. Food Microbiol. 61 (2–3), 91–125. https://doi.org/10.1016/s0168-1605(00)00382-2.

Lewis, R.J., Sellin, M., Poli, M.A., Norton, R.S., MacLeod, J.K., Sheil, M.M., 1991. Purification and characterization of ciguatoxins from moray eel (*Lycodontis javanicus*, Muraenidae). Toxicon 29 (9), 1115–1127. https://doi.org/10.1016/0041-0101(91)90209-a.

Lewis, R.J., Vetter, I., 2014. Ciguatoxin and ciguatera. In: Gopalakrishnakone, P., Haddad Jr., V., Kem, W.R., Tubaro, A., Kim, E. (Eds.), Marine and Freshwater Toxins: Marine and Freshwater Toxins. Springer Netherlands, Dordrecht, pp. 1–19.

Manning, T.M., Roach, A.C., Edge, K.J., Ferrell, D.J., 2017. Levels of PCDD/Fs and dioxin-like PCBs in seafood from Sydney Harbour, Australia. Environ. Pollut. 224, 590–596. https://doi.org/10.1016/j.envpol.2017.02.042.

Martin, J.W., Mabury, S.A., Solomon, K.R., Muir, D.C., 2003. Dietary accumulation of perfluorinated acids in juvenile rainbow trout (*Oncorhynchus mykiss*). Environ. Toxicol. Chem. 22 (1), 189–195.

Murata, M., Legrand, A.M., Ishibashi, Y., Fukui, M., Yasumoto, T., 1990. Structures and configurations of ciguatoxin from the moray eel *Gymnothorax javanicus* and its likely precursor from the dinoflagellate *Gambierdiscus toxicus*. J. Am. Chem. Soc. 112 (11), 4380–4386. https://doi.org/10.1021/ja00167a040.

Murray, J.S., Boundy, M.J., Selwood, A.I., Harwood, D.T., 2018. Development of an LC-MS/MS method to simultaneously monitor maitotoxins and selected ciguatoxins in algal cultures and P-CTX-1B in fish. Harmful Algae 80, 80–87. https://doi.org/10.1016/j.hal.2018.09.001.

Naile, J.E., Khim, J.S., Wang, T., Chen, C., Luo, W., Kwon, B.-O., et al., 2010. Perfluorinated compounds in water, sediment, soil and biota from estuarine and coastal areas of Korea. Environ. Pollut. 158 (5), 1237–1244. https://doi.org/10.1016/j.envpol.2010.01.023.

Nicholson, G.M., Lewis, R.J., 2006. Ciguatoxins: cyclic polyether modulators of voltage-gated iion channel function. Mar. Drugs 4 (3), 82–118.

NSW Food Authority, 2014. Advice to Fishers on North NSW Coast [Press release]. Retrieved from: http://www.foodauthority.nsw.gov.au/news/newsandmedia/departmental/2014-03-04-advice-to-fishers-nsw-coast.

NSW Food Authority, 2017. Ciguatera Fish Poisoning. Retrieved from: http://www.foodauthority.nsw.gov.au/rp/fish-ciguatera-poisoning.

O'Connor, W.A., Zammit, A., Dove, M.C., Stevenson, G., Taylor, M.D., 2018. First observations of perfluorooctane sulfonate occurrence and depuration from Sydney Rock Oysters, *Saccostrea glomerata*, in Port Stephens NSW Australia. Mar. Pollut. Bull. 127, 207–210. https://doi.org/10.1016/j.marpolbul.2017.11.058.

Shoemaker, J.A., Grimmett, P.E., Boutin, B.K., 2008. Determination of Selected Perfluorinated Alkyl Acids in Drinking Water by Solid Phase Extraction and Liquid Chromatography/Tandem Mass Spectrometry (LC/MS/MS). Retrieved from Washington, DC: U.S: https://cfpub.epa.gov/si/si_public_file_download.cfm?p_download_id=525468.

SMH, 2005. Police Reveal MasterFoods Extortion Target. https://www.smh.com.au/national/police-reveal-masterfoods-extortion-target-20050720-gdlpop.html.

Sydney Fish Market Pty Ltd, 2015. Seafood Handling Guidelines.

Taylor, M.D., 2018. First reports of per- and poly-fluoroalkyl substances (PFASs) in Australian native and introduced freshwater fish and crustaceans. Mar. Freshw. Res. 69 (4), 628–634. https://doi.org/10.1071/MF17242.

Taylor, M.D., 2019a. Factors affecting spatial and temporal patterns in perfluoroalkyl acid (PFAA) concentrations in migratory aquatic species: a case study of an exploited crustacean. Environ. Sci. Processes Impacts. https://doi.org/10.1039/c9em00202b.

Taylor, M.D., 2019b. Survey design for quantifying perfluoroalkyl acid concentrations in fish, prawns and crabs to assess human health risks. Sci. Total Environ. 652, 59–65. https://doi.org/10.1016/j.scitotenv.2018.10.117.

Taylor, M.D., Beyer-Robson, J., Johnson, D.D., Knott, N.A., Bowles, K.C., 2018. Bioaccumulation of perfluoroalkyl substances in exploited fish and crustaceans: spatial trends across two estuarine systems. Mar. Pollut. Bull. 131 (Pt A), 303–313. https://doi.org/10.1016/j.marpolbul.2018.04.029.

Taylor, M.D., Bowles, K.C., Johnson, D.D., Moltschaniwskyj, N.A., 2017. Depuration of perfluoroalkyl substances from the edible tissues of wild-caught invertebrate species. Sci. Total Environ. 581–582, 258–267. https://doi.org/10.1016/j.scitotenv.2016.12.116.

Taylor, M.D., Johnson, D.D., 2016. Preliminary investigation of perfluoroalkyl substances in exploited fishes of two contaminated estuaries. Mar. Pollut. Bull. 111 (1–2), 509–513. https://doi.org/10.1016/j.marpolbul.2016.06.023.

The Advertiser, 2018. Adelaide Man Charged over Strawberry Contamination Claim. https://www.adelaidenow.com.au/news/law-order/adelaide-man-charged-over-strawberry-contamination-claim/news-story/07df1a7705958e2850df4ea663212ccc.

URS Australia, 2015. AFFF PFAS, RAAF Base Williamtown, Williamtown NSW: Stage 2 Environmental Investigation. URS Australia Pty Ltd, Sydney: Prepared for Defence Environmental Remediation Programs, p. 48.

USFDA, 2019a. FDA Statement: Statement on FDA's Scientific Work to Understand Per- and Polyfluoroalkyl Substances (PFAS) in Food, and Findings from Recent FDA Surveys.

USFDA, 2019b. Fish and Fishery Products Hazards and Controls Guidance, fourth ed. Retrieved from: https://www.fda.gov/food/seafood-guidance-documents-regulatory-information/fish-and-fishery-products-hazards-and-controls-guidance.

Vernoux, J.P., Lewis, R.J., 1997. Isolation and characterisation of Caribbean ciguatoxins from the horse-eye jack (*Caranx latus*). Toxicon 35 (6), 889–900. https://doi.org/10.1016/s0041-0101(96)00191-2.

Vestergren, R., Cousins, I.T., 2009. Tracking the pathways of human exposure to perfluorocarboxylates. Environ. Sci. Technol. 43 (15), 5565–5575. https://doi.org/10.1021/es900228k.

WHO, 1987. Principles for the Safety Assessment of Food Additives and Contaminants in Food. International Programme on Chemical Safety.

WHO, 1997. Guidelines for Predicting Dietary Intake of Pesticide Residues/Prepared by the Global Environment Monitoring System - Food Contamination Monitoring and Assessment Programme (GEMS/Food).

WHO, 2021. Chemical Risks in Food. Retrieved from: http://www.who.int/foodsafety/chem/en/.

Yasumoto, T., Nakajima, I., Bagnis, R., Adachi, R., 1977. Finding of a dinoflagellate as a likely culprit of ciguatera. Bull. Jpn. Soc. Sci. Fish. 43, 1021–1026.

Yogi, K., Oshiro, N., Inafuku, Y., Hirama, M., Yasumoto, T., 2011. Detailed LC-MS/MS analysis of ciguatoxins revealing distinct regional and species characteristics in fish and causative alga from the Pacific. Anal. Chem. 83 (23), 8886–8891. https://doi.org/10.1021/ac200799j.

Chapter 19

Nutraceuticals: possible future ingredients and food safety aspects

M.A.J.S. van Boekel
Food Quality & Design Group, Wageningen University & Research, Wageningen, the Netherlands

19.1 Introduction

Basic nutrients are needed for growth, maintenance, and wellness of the body and are normally supplied by food in the form of fats, carbohydrates, proteins, vitamins, minerals (partly used as energy source as well). Next to that there are also nonnutrients such as fibers, antioxidants, inducers of beneficial enzyme activities, prebiotics, and probiotics. The human body is well capable of utilizing all these molecules from the food. It is, however, a very complex process that is far from being understood completely. The "classic" nutrients fat, carbohydrates, proteins, and minerals have been well studied with their roles more or less established, the main remaining question being the role of the food matrix in which they are present. This so-called matrix effect implies that nutrients may behave differently biochemically, and/or have a different physiological effect, when isolated in comparison to being present in a food. A recent example of that are saturated fats, for which the consensus was that they are undesired, but recent research shows that they have a neutral or even a positive effect on health depending on the food they are in (Astrup et al., 2019). Next to "normal" foods, nutraceuticals have come up in the past 2 decades or so. This chapter aims to give a brief overview of what nutraceuticals are, to discuss whether or not they may be useful and whether or not there is concern about possible safety issues.

19.2 What are nutraceuticals?

The word "nutraceutical" is a merger from the words "nutrition" and "pharmaceuticals." It is therefore no surprise that knowledge on nutraceuticals stems partly from the pharmaceutical literature, perhaps even more than from food science literature. Even though the name nutraceutical does not exist for a long time yet (30 years or so), compounds that would now perhaps be classified as nutraceuticals are already used for ages, especially in traditional medicines, for instance, in China, Africa, and South America. Local ethnic knowledge is of potential interest, usually passed on from generation to generation over hundreds of years. However, science-based claims are usually missing.

When bioactive compounds are isolated from a food matrix, they tend to be called "nutraceuticals," a term coined by DeFelice in 1992 and defined as "food or parts of food that provide medical or health benefits, including the prevention and treatment of disease." Next to nutraceuticals there are also functional foods and supplements and there is a gray area here: what is what? With functional foods, a specific effect is targeted, such as calcium supplementation, vitamin supplementation, reduction of cholesterol levels, prebiotics, probiotics, but all within a food matrix. A clear distinction between supplements and nutraceuticals on the one hand and functional foods on the other is that functional foods are still foods (containing bioactive compounds), whereas nutraceuticals and supplements are not foods but isolated compounds or extracts. Supplements are usually connected to deficiencies (Aronson, 2017). Another term one can find in the literature is "fortified foods" or "designer foods," which apparently are foods to which health promoting compounds are added (for instance, folic acid to bread, or iodine to salt). The diets of humans consist of many foods, and health effects should indeed be ascribed to diets and not to individual foods; it is the combination of all the components that end up in the digestive tract

Ensuring Global Food Safety. https://doi.org/10.1016/B978-0-12-816011-4.00015-X

that will have an effect. Nutraceuticals go one step beyond functional foods by taking out the bioactive compounds from a food matrix. The danger then becomes that an "overdose" of bioactive molecules is ingested. With functional foods, this is less likely because of the very fact that the intake is limited to the food and there will be a limit of what people eat. Here we focus on nutraceuticals but we acknowledge the fact that there is a gray area of unclear terminology.

Food components that may be considered as potential sources of nutraceuticals are (e.g., Das et al., 2012):

1. Dietary fibers
2. Probiotics: live bacteria that are supposed to positively influence microbiota in the gut
3. Prebiotics: substances that help the development of a beneficial microbiota system
4. Poly-unsaturated fatty acids: omega-3 and omega-6
5. Antioxidants: vitamins, carotenoids
6. Polyphenols
7. Glucosinolates

This list is not exhaustive. A relatively new source could, for instance, be marine organisms (Ansar Rasul Suleria et al., 2015). These authors mention secondary metabolites derived from marine invertebrates such as tunicates, sponges, molluscs, bryozoans, and sea slugs. These bioactive molecules are supposed to act as antibiotics, antiparasitic, antiviral, antiinflammatory, antifibrotic, and anticancer, or act as inhibitors or activators of critical enzymes and transcription factors, among others.

In this chapter, nutraceuticals are considered as bioactive compounds that are extracted from their original food matrix. They are meant to promote health and to prevent the occurrence of diseases. They are usually not considered as medicines because they are not meant to cure diseases. However, one of the problems of nutraceuticals is that they are not really well defined and this is a potential source of confusion.

19.3 Supposed health effects

Nutraceuticals are commonly mentioned in relation to prevention of cardiovascular disease, cancer, diabetes, obesity, high blood pressure, arthritis, allergy, eye disorders, inflammation, Parkinson's disease, and some more; this is not to say that all nutraceuticals act like this; their activity depends of course on the type of molecules that are present in a specific formulation. This long list may raise the question whether or not this is all clinically proven. Here we discuss recent scientific results obtained on nutraceuticals. There are several forces that shape health effects and nutritional food formulations and their role in prevention of diseases and promoting well-being, so it is a complex matter that is very difficult to investigate. Bioactive molecules reach the target organ through various ways. Biochemical effects in the digestive tract (including the mouth) and in organs, indirect effects of prebiotics and probiotics, may modify such molecules. As mentioned, whether or not molecules are present in a food matrix may have a considerable effect on the eventual biological effect.

It is remarkable that DeFelice, the person who gave a definition of nutraceuticals, has made the remark that nutraceuticals may not be as effective as originally thought. As quoted in Aronson (2017), DeFelice has apparently made this statement in 2014: *"Within the past decade, the past ten years, many studies now have been published on dietary supplements and diets … and most of them have proven that these things do not work. Not proven. The results of clinical studies have shown that they do not work. They may work. But the studies may not have been designed properly. … Now, is it due to poorly designed clinical trials? Perhaps. Is it due to the fact that they don't work? I have problems with that. But I will say 'perhaps'. I have to be intellectually honest. You know, I can't be an advocate of something I believe in when the proof's not there."* A recent review (Jain et al., 2018) concludes that nutraceuticals hold promises but also that "*large-scale randomized, placebo-controlled, double-blind clinical trials are needed to confirm the health benefit claims about nutraceuticals and herbal products to establish their long-term safety and to resolve the controversy about the role of clinical nutrition in curing lifestyle diseases.*"

Yet another study about effects of nutraceuticals on elderly (D'Cunha et al., 2018) concluded: "*As only five included studies revealed notable benefits, presently based on the specific compounds explored here, there is not compelling evidence to support the use of nutraceuticals to improve cognition in the elderly. Future long-term trials of nutraceuticals should investigate interactions with lifestyle, blood biomarkers, and genetic risk factors.*"

This more or less summarizes the scientific literature of the past decade: there is not so much scientific evidence in the form of clinical or epidemiological studies showing health effects of nutraceuticals. On the other hand, there is also not much evidence of harmful effects, and the conclusion at this stage must be that a lot is still unknown.

19.4 Challenges

The question that needs to be discussed is whether nutraceuticals can indeed significantly contribute to health. According to the present author, this is not at all clear from the literature. There are two main challenges:

1. Biochemical and clinical studies are badly needed to investigate health claims
2. Legislation is needed to create order in the chaos about what is meant with the term nutraceuticals. Safety cannot be guaranteed as long as there is ambiguity about what they are, and what their mode of action is.

What is definitely needed is a thorough risk–benefit analysis. It may seem at first sight that isolated bioactive compounds are almost by definition "healthy," but this is not always true. The prime example that corroborates this is the Finland study in which carotenoids were supplied to a group of people (smokers and nonsmokers) based on the presumed antioxidant activity of these compounds in foods. However, the experiment had to be stopped prematurely because of the damage done by carotenoids to the smokers. The carotenoids were supposed to act as antioxidants but they did not function as such (Omenn et al., 1996; Satia et al., 2009; The Alpha-Tocopherol, Beta-Carotene Cancer Prevention Study Group, 1994). This shows that bioactive compounds may act completely differently when they are taken out of their natural environment. While we cannot generalize—based upon this one example—it definitely indicates that synergistic effects that occur in the food matrix but are absent when compounds are in isolated form, can be of great importance.

Research needs to be done at several levels. A clear understanding of the chemistry of a molecule or a group of molecules is essential. The next level is that of molecule–gene interaction. Nutrigenomics is now well established as a scientific field to provide evidence-based effects of nutraceuticals. The way molecules interact at a cellular level among the individual genotype or phenotype, either with a single gene polymorphism or with a multigene polymorphic interaction, is important information to design new foods and to study the impact of nutraceuticals. Several parameters need to be considered here, including the metabolic profile, structure of the molecule, function, pathways, metabolic pool, and phyto-genetic relations. Another level is with clinical studies to investigate the effects of nutraceuticals on health, either with animals or with humans. And yet another level would be epidemiology. All this information combined should ultimately lead to a holistic approach to food, medicine, health, wellness, exercise, and age. This stage has clearly not been reached yet.

Another challenge is the production and analysis of nutraceuticals (Durazzo and Lucarini, 2019). To what extent does processing affect the biochemical potential of nutraceuticals? What is the natural variation of compounds in the raw material? The level in which they are present in the foods must also reflect the level in which they are being substituted and/or fortified, otherwise a toxic level may be reached upon ingestion of a single component rather than microdoses of multiple components that may otherwise have synergy. The synergistic effect of one component with another should not be overlooked. In that sense, the food-based approach is always an easier way than a component-based approach.

Before offering these products to consumers, the positive effects must be clear and there must be no ill effects. The Codex specifications, as well as the local regulatory systems, need to be followed when validating to the customer and consumer in the ultimate market. All this requires very good chain coordination and governance, which is not always easy.

19.5 Regulations and safety issues

Unclear terminology and lack of scientific evidence about possible health effects may contribute to the consumer's confusion. Consumers probably assume that nutraceuticals are safe when they are marketed, but there is, as yet, little proof of safety of these substances and this remains therefore a very important issue. The question pops up whether or not nutraceuticals should be treated in the same way as pharmacological products, which are subject to very strict regulations and premarket research. Some authors reason that nutraceuticals are not the same as drugs (no government sanction and usually not patented) and therefore should not be treated in the same way (e.g., Nasri et al., 2014). Others claim that more strict regulations are needed (e.g., Santini and Novellino, 2017, Santini et al., 2018). These last authors argue that: "*The specificity of nutraceuticals would need a premarket approval system substantiated by scientific data (*in vivo *clinical trials) showing their complete safety and efficacy profile. The regulatory approach could be based on the one adopted for drugs, which is stricter and more complex.*" Another argument for stricter regulation is that nutraceuticals can be used by anyone and people tend not to tell that when they are medically treated, with a big danger for adverse effects (Ronis et al., 2018).

19.6 Conclusion

Judged by marketing, nutraceuticals promise a lot for positive health effects, but there is very little scientific evidence to substantiate this. These market possibilities are huge and there is a potential danger here that commercial interests are considered more important than safety concerns. This is not to say that nutraceuticals are not safe, it is just to say that much is still unclear in terms of health effects. It is in the interest of the consumer to have stricter rules and regulations on the one hand to reduce confusion about nomenclature, and well-designed clinical studies on the other hand to substantiate health effects on the one hand, and to study adverse effects on the other hand.

Whether or not nutraceuticals have potential in improving the nutritional status of populations remains to be seen, and whether or not there is additional benefit to ingest nutraceuticals on top of a normal diet. It is not impossible that they can have additional effects for specific groups, such as elderly, but also that is unclear. Science should support claims with a mandate of quality. Such an approach requires a thorough risk—benefit analysis that may be different from that used for foods today; since we talk about very bioactive compounds, the presently available methodologies may not be sufficient. It should ensure adequate scientific evidence for the benefits and claims that are going to be made. A global harmonization approach needs to be developed to make a sustainable solution possible for the beneficial use of nutraceuticals in society.

References

Ansar Rasul Suleria, H., Osborne, S., Masci, P., Gobe, G., 2015. Marine-based nutraceuticals: an innovative trend in the food and supplement industries. Mar. Drugs 13 (10), 6336—6351. https://doi.org/10.3390/md13106336.

Aronson, J.K., 2017. Defining 'nutraceuticals': neither nutritious nor pharmaceutical. Br. J. Pharmacol. 83, 8—19. https://doi.org/10.1111/bcp.12935.

Astrup, A., et al., 2019. WHO draft guidelines on dietary saturated and trans fatty acids: time for a new approach? BMJ 366, l4137. https://doi.org/10.1136/bmj.l4137.

Das, L., Bhaumik, E., Raychaudhuri, U., Chakraborty, R., 2012. Role of nutraceuticals in human health. J. Food Sci. Technol. 49, 173—183. https://doi.org/10.1007/s13197-011-0269-4.

DeFelice, S.L., 1992. Nutraceuticals: opportunities in an emerging market. Scrip Mag. 9.

Durazzo, A., Lucarini, M., 2019. Extractable and non-extractable antioxidants. Molecules 24, 1933.

D'Cunha, N.M., Georgousopoulou, E.N., Dadigamuwage, L., Kellett, J., 2018. Effect of long-term nutraceutical and dietary supplement use on cognition in the elderly: a 10-year systematic review of randomised controlled trials. Br. J. Nutr. 119, 280—298.

Jain, S., Buttar, H.S., Chintameneni, M., Kaur, G., 2018. Prevention of cardiovascular diseases with anti-inflammatory and anti- oxidant nutraceuticals and herbal products: an overview of pre-clinical and clinical studies. Recent Pat. Inflamm. Allergy Drug Discov. 12 (2), 145—157. https://doi.org/10.2174/1872213X12666180815144803.

Nasri, H., Baradaran, A., Shirzad, H., Rafieian-Kopaei, M., 2014. New concepts in nutraceuticals as alternative for pharmaceuticals. Int. J. Prev. Med. 5, 1487—1499.

Omenn, G.S., Goodman, G.E., Thornquist, M.D., et al., 1996. Risk factors for lung cancer and for intervention effects in CARET, the Beta-Carotene and Retinol Efficacy Trial. J. Natl. Cancer Inst. 88, 1550—1559.

Ronis, M.J.J., Pedersen, K.B., Watt, J., 2018. Adverse effects of nutraceuticals and dietary supplements. Annu. Rev. Pharmacol. Toxicol. 58, 583—601.

Santini, A., Novellino, E., 2017. To nutraceuticals and back: rethinking a concept. Foods 6 (9), 74. https://doi.org/10.3390/foods6090074.

Santini, A., Cammarata, S.M., Capone, G., Ianaro, A., Tenore, G.C., Pani, L., Novellino, E., 2018. Nutraceuticals: opening the debate for a regulatory framework. Br. J. Clinical Pharmacol. 84, 659—672.

Satia, J.A., Littman, A., Slatore, C.G., et al., 2009. Long- term use of β-carotene, retinol, lycopene, and lutein supplements and lung cancer risk: results from the VITamins and Lifestyle (VITAL) study. Am. J. Epidemiol. 169 (7), 815—828.

The Alpha-Tocopherol, Beta-Carotene Cancer Prevention Study Group, 1994. The effect of vitamin E and beta carotene on the incidence of lung cancer and other cancers in male smokers. N. Engl. J. Med. 330, 1029—1035.

Chapter 20

Nutrition and bioavailability: sense and nonsense of nutrition labeling

Adelia C. Bovell-Benjamin
Food and Nutritional Sciences, Tuskegee University, Tuskegee, AL, United States

20.1 Introduction

Chronic diseases are a leading cause of preventable deaths in the United States (U.S.) and the rest of the world, and the burden is increasing rapidly (DHHS, 2001). Older data indicated that chronic diseases accounted for roughly 60% and 46% of the 56.5 million total reported deaths in the world and the global burden of disease, respectively (The World Health Report, 2002). Furthermore, this year, 2020, the proportion of the burden of chronic diseases is expected to increase to 57%. Approximately 50% of the total chronic disease deaths are ascribable to cardiovascular diseases (CVDs); obesity; and diabetes (The World Health Report, 2002).

It is recognized that chronic diseases are diet related, which can be managed or prevented through appropriate diet and lifestyle practices. Changes in dietary patterns require that sufficient information be provided at the point of food purchase. This information can be provided in a number of ways; however, the food label is one of the most immediate and direct sources of information, which can be supported with education and advertising. The section of information on the food label which declares nutrient content is termed nutrition labeling, nutrition panel, or nutrition facts. The current Nutrition Facts box that appears on food labels was conceived as an important public health tool to reduce diet-related diseases. The Nutrition Facts panel that currently appears on food labels includes the percent daily value (% DV) and is a critical tool for consumer use to make informed food choices.

Since 1941, nutrition labeling in the U.S. has reflected the current scientific knowledge on the relationship between diet and health. For example, the changes reflected in nutrition labeling regulations promulgated by the Food and Drug Administration (FDA) in 1973 required that both positive and negative aspects of the nutrient content of food appear on the label to emphasize the relationship between diet and health (Hutt, 1981). The Nutrition Facts panel and the related nutrition information on the label continued this effort to encourage healthier food choices. To achieve this health goal, the 1993 U.S. version of nutrition labeling included a new tool, the % DV, which enables consumers to rapidly and efficiently understand how a particular food fits in the context of a healthy diet (FDA, 1993a).

Nutrition labeling is a population-based approach, which provides information to consumers about the nutrient content of a food; its intent is to make the food selection environment more conducive to healthy choices (Cowburn and Stockley, 2004). The U.S., Canada, Australia, and New Zealand have implemented mandatory nutrition labeling, although the format and content of the labels differ in these countries (Sibbald, 2003; Curran, 2002). Post December 2016, in the European Union, Regulation (EU) No 1169/2011 requires the vast majority of prepacked foods to bear a nutrition declaration, which provides the energy value, the amounts of fat, saturates, carbohydrate, sugars, protein, and salt of the food (European Commission, 2020a) According to The Food and Drugs (Composition and Labeling) (Amendment) Regulation enacted by the Hong Kong government, it is mandatory that all prepackaged foods sold in the country should carry a Nutrition Information Panel (NIP). The NIP should indicate the values of energy, protein, fat, saturated fat, trans fat, carbohydrates, sugars, and sodium. If there is any nutrition claim, the nutrition content declared must be listed in the panel (China Dragon, 2020). China requires all food and drink products, including imported food products, to be labeled to comply with Chinese food labeling regulations. The Nutritional Food Product Labeling regulation is GB 28,050−2011, which is being revised 2019 to 2020 (China-Britain Business Council, 2020).

Ensuring Global Food Safety. https://doi.org/10.1016/B978-0-12-816011-4.00011-2

Nutrition labeling can be used to help prevent the progress of poor nutrition, obesity, and other chronic diseases. However, it is crucial that nutrition labeling is combined with education on healthy lifestyles, including clear advice about the contribution that all foods make to a healthy diet, and the importance of physical activity. Table 20.1 shows that, in general, countries can be characterized as having one of four types of nutrition labeling regulatory environment: (i) mandatory nutrition labeling on all prepackaged food products; (ii) voluntary nutrition labeling, which becomes

TABLE 20.1 Nutritional labeling regulations in selected countries and areas, by category (from Hawkes, 2004; Grunert, 2013; Mitchell, 2020; Taillie et al., 2020).

Mandatory	Date implemented	Voluntary except a nutrition claim is made	Voluntary, except certain foods with special dietary uses (b)	No documented regulations the author could find
Argentina	2006	Brunei Darussalam	Bahrain	Bahamas
Australia	2002	Switzerland[a]	Costa Rica	Bangladesh
Brazil	2006		India	Barbados
Canada	2003		Kuwait (GCC)	Belize
Chile	2016		Mauritius (Codex)	Bermuda
China	2011, revision		Morocco	Bosnia and Herzegovina
	2019		Nigeria	Botswana
European Union	2018		Oman (GCC)	Dominican Republic
Hong Kong	2010		Philippines	Egypt
Israel	1993		Qatar (GCC)	El Salvador
Japan	2015		Saudi Arabia (GCC)	Guatemala
Malaysia	2003 approved		United Arab Emirates (GCC)	Honduras
Mexico	2020		Venezuela	Jordan
New Zealand	2002			Kenya
Paraguay	2006			Nepal
Peru	2019			Netherlands Antilles
	2010, revisions			Pakistan
	2019			Turkmenistan
Republic of Korea	2003, revised			
	2019			
Singapore	2016			
South Africa	2017			
Thailand	1994			
United States	2006, FoP			
Uruguay	2020			
Vietnam	2014			

FoP, Front-of-pack label.
GCC, regulations based on the Gulf Cooperation Council Standard (GS) September 1995 on nutrition labeling.
Codex, regulations developed taking guidance from the Codex Guidelines on Nutrition Labeling.
[a]*Conforms with EU, not mandatory except for food and drinks.*

mandatory on foods where a nutrition claim is made; (iii) voluntary nutrition labeling, which becomes mandatory on foods with special dietary uses; and (iv) no regulations on nutrition labeling (Hawkes, 2004; Grunert, 2013; Mitchell, 2020; Taillie et al., 2020).

One of the intents of nutrition labeling is to make the food selection environment more conducive to healthy choices. An ultimate outcome of healthier food choices in populations is savings through increased productivity; lower health care costs associated with cancer, diabetes, CVD, and other chronic diseases; and overall improved health over time. The increase in prepackaged and processed foods means that the nature of food is not always clear from visual inspection and information is required about contents, storage, and preparation. The nutrition label provides information for consumer protection. For example, the increased use of additives in the food system over the past few decades requires governmental regulatory activities. Consumers who are allergic to certain product ingredients could be protected if the food label is read. Many food label regulations prohibit the use of misleading advertising or health claims, putting the burden of proof of nutritive or any other health benefit on the manufacturer while protecting the consumer.

Epidemiological evidence on the relationships between changing patterns of diet and disease has underscored the importance between nutrition and health. Many countries have suggested reductions in the mean intake of nutrients such as sugar, salt, total, and saturated fat. On the other hand, increased consumption of whole-grain cereals, vegetables, and fruits is being encouraged. Nutrition labels are meant to educate consumers regarding the extent of salt, sugar, fat, cholesterol, minerals, some vitamins, and protein in processed foods.

Currently, absorption and bioavailability rates do not have to be tested under any of the nutrition labeling requirements worldwide. A main drawback with nutrition labeling is that the nutrient content is not supported by information regarding its bioavailability. Knowledge of the quantity of nutrients in a food product is of very limited use when assessing its nutritional adequacy. For example, two food products could have nutrients in the same quantities, but very different levels of bioavailability as is clearly demonstrated with calcium. Calcium in milk is more easily absorbed by the intestine than the calcium from vegetables and cereals. Phytates and oxalates present in cereals, bean, and pulses and in leafy vegetables, respectively, and long-chain saturated fatty acids and dietary fiber can reduce the bioavailability of calcium by forming insoluble calcium complexes (Fairweather-Tait et al., 1989).

Furthermore, there are a large number of dietary and host-related factors, which affect the absorption of nutrients, especially minerals (Fairweather-Tait, 1992). Added information regarding the bioavailability of nutrients in the Nutrition Facts will be much more useful to consumers in their quest for healthier food choices. Information about bioavailability of the nutrients which they are consuming will be important to consumers because of the direct relationship between dietary consumption and chronic diseases. It is against this background that the present objective for this chapter was formulated. Therefore, the objective of the chapter was to briefly review and update the current state of knowledge on nutrition and bioavailability with respect to nutrition labeling.

20.2 Scope

The chapter is limited to a discussion of the Nutrition Facts Panel (a portion of the food label) and not the entire food label. It focuses briefly on the information on the basic Nutrition Facts Panel and the inclusion of nutrient bioavailability on the panel and updates the information on Front-of-Pack labeling (FoPL), which is intended to replace the Nutrition Facts Panel on the back or side of the package. Discussion regarding dietary supplements, health, and nutritional claims are beyond the scope of this chapter. Details about methods to assess bioavailability and the mechanism of absorption are also beyond the scope of this chapter.

There is no consensus in the literature regarding a definition of bioavailability. However, researchers have coined several definitions for the term. Reeves and Chaney (2008) have defined bioavailability as the extent to which a nutrient, toxin, or other substances become available for body use or deposition after intake whether oral or otherwise. They further stated that in the instances of oral exposure to the nutrient, bioavailability usually involves absorption, body utilization, and/or deposition. Similarly, Welch and House (1984) defined bioavailability as the proportion of the total amount of a mineral element present in a nutrient medium, which is potentially absorbable in a metabolically active form. Southgate (1989) argued that bioavailability represents the response of the human, animal, or cells in culture to the diet or food, and is not an inherent property of the food as such. For example, in rats, zinc bioavailability from bread was variable and dependent on the zinc status of the body (Hallmans et al., 1987). House (1999) has indicated that the bioavailability of a trace mineral in a specific food is not static.

20.3 Methodology

A review of the existing literature was conducted. The literature surveyed covered scientific journals, trade journals, magazines, market reports, conference proceedings, books, and other published materials. Web page content was used, as necessary. Literature was collected using numerous search engines and databases (for example, EbscoHost, Ingenta, Science Direct, Google, Google Scholar, PubMed) in addition to library databases and internet libraries of international organizations.

20.4 Structure of the review

In this review, Sections 20.1−20.4 justify the topic, present the objective, and outline the scope and structure of the review. Section 20.5 presents an overview of the history of nutrition labeling focusing on the U.S., Canada, Australia and New Zealand, and developing countries. Section 20.6 is a more in-depth discussion regarding similarities and differences of nutrition labeling in different countries and FoPL. Section 20.7 reviews consumers' understanding and use of nutrition labels, while Section 20.8 covers bioavailability and nutrition labeling. Sections 20.9 to 20.10 consist of conclusion, future scope, acknowledgment, and references.

20.5 Overview of nutrition labeling

20.5.1 United States

In the following sections, the updated summarized information comes primarily from the respective countries food regulatory agencies. In the U.S., the regulatory agency for food nutrition labels on certain foods is the U.S. FDA. In 1972, nutrition labeling, which would allow some nutrients to be listed as percent of a daily dietary intake standard per serving, was proposed by the U.S. FDA (Pennington et al., 1997). The U.S. has experienced three major phases of nutrition labeling revisions during which different reference values were used on the label. From 1941 to 1972, Minimum Daily Requirements were used and from 1973 to 1993, U.S. Recommended Daily Allowances (US RDAs) were used. A further modification from 1993 required % DVs to be used. The objective was to allow consumers to understand the relative significance of food in the context of total daily intake of nutrients and use the label information to plan a healthy overall diet (Pennington et al., 1997). The FDA proposed a set of values called Recommended Daily Allowances for proteins, vitamins, and minerals using the 1968 Recommended Dietary Allowances (RDAs) as the basis. The RDAs suggest levels of intake adequate to meet the nutrient needs of all healthy persons. In 1973, the FDA established nutrition labeling of a single set of daily intake standards called US RDAs for protein, 12 vitamins and 7 minerals (Table 20.2).

The FDA proposed regulations to update and expand the daily intake standards for nutrition labeling of foods in 1990. The Nutrition Labeling and Education Act (NLEA), which required foods and dietary supplements to bear nutritional labeling, became law in 1990 (Pray, 2003). The FDA issued proposed rules to implement NLEA, stating that vitamins would be held to the same standards as other medications, and that any claims would be required to withstand scientific scrutiny because companies were making fraudulent claims (Pray, 2003). NLEA allows a format for preapproved health claims that make a nutrient or diet disease link. Only a limited number of such health claims are approved in the U.S. after the FDA conducts a thorough review of the scientific studies documenting the nutrient or diet-disease link. Outside the U.S., the inclusion of a disease in a health claim classifies the product as intended for medical uses. In such instances, classification and approval as a drug and not as a food product is required.

FDA summarized and reviewed the 1990 NLEA requiring that food labels provide nutrient information for 10 food components in terms of % DV per serving portion. These were published and implemented in 1993 and 1994, respectively. Mandatory food components included total fat, saturated fat, cholesterol, sodium, total carbohydrate, dietary fiber, calcium, iron, and vitamins A and C (Pennington et al., 1997). A sample U.S. Nutrition Facts Label is shown in Fig. 20.1. The reference daily intakes (RDIs) for vitamin K, selenium, manganese chromium, molybdenum, and chloride were established in 1995 (Pennington et al., 1997). In the 1995 final rule, the units of measure for biotin and folate were changed from mg to μg, and calcium and phosphorus from g to mg (Pennington et al., 1997). Label to include trans fat content became effective in 2006.

On May 20, 2016, the U.S. FDA announced a new Nutrition Facts label for packaged foods, which is intended to come into effect 2020 and 2021. Every FDA-regulated food package, which currently uses a nutrition panel, will be required to have changes made to the information in the panel, in addition to the format. New scientific information, reflecting the link between diet and chronic diseases such as obesity and heart disease, will have to be reflected on the label. It was envisaged that the new label will make it easier for consumers to make better informed food choices (Fig. 20.2).

TABLE 20.2 US recommended daily allowances (US RDAs) established by the food and drug administration in 1973 for nutrition labeling.

Nutrient	US RDA	Basis for US RDA
Mandatory		
Protein[a]	65g[b]	1968 Recommended Dietary Allowance (RDA) for men
Vitamin A[a]	5000 IU	1968 RDA for men
Vitamin C[a,c]	60 mg	1968 RDA for men
Thiamin(e)[a,c]	1.5 mg	1968 RDA for teenage boys; RDA for men was 1.4 mg
Riboflavin[a,c]	1.7 mg	1968 RDA for men
Niacin[a]	20 mg	1968 RDA for teenage boys; RDA for men 18 mg
Calcium[a]	1.0 mg	1968 RDA; higher than RDA for adults (800 mg) and lower than the RDA for teenagers (1.2 g)
Iron[a]	18 mg	1968 RDA for women
Optional		
Vitamin D	400 IU	1968 RDA for children and adults
Vitamin E	30 IU	1968 RDA for men
Vitamin B-6	2.0 mg	1968 RDA for men and women
Folic acid[c]	0.4 mg	1968 RDA for men and women
Vitamin B-12	6.0 µg	1968 RDA for men and women
Biotin	0.3 mg	Text in the 1968 RDA book
Pantothenic acid	10 mg	Text in the 1968 RDA book
Phosphorus	1.0 g	1968 RDA; higher than the RDA for adults (800 mg) and lower than the RDA for teenagers (1.2 g)
Iodine	150 µg	1968 RDA for men aged 14–18 years; 1968 RDA was 140 µg for men aged 18–35 years, 125 µg for men aged 35–55 years, and 110 µg for men older than 55 years
Magnesium	400 mg	1968 RDA for teenage boys; RDA for men was 350 mg
Zinc	15 mg	Text in the 1968 RDA book
Copper	2 mg	Text in the 1968 RDA book

[a] *Required component for the nutrition label in 1973; others were voluntary unless claims were made about them or they were added to foods.*
[b] *The protein US RDA was 45 g if the protein efficiency ratio of the total protein in the product was equal to or greater than that of casein, the protein US RDA was 65 g if the protein efficiency ratio was less than that of casein.*
[c] *Permitted synonyms for nutrition labeling were ascorbic acid for vitamin C, vitamin B-1 for thiamin, vitamin B-2 for riboflavin, and folacin for folic acid.*
Taken from Pennington et al. (1997).

The benchmarks for the new Nutrition Fact label stipulated that manufacturers with $10 million or more in annual sales must switch to the new label by January 1, 2020. Manufacturers with less than $10 million in annual food sales have until January 1, 2021 to comply. Manufacturers of most single-ingredient sugars such as honey, maple syrup, and certain cranberry products have until July 1, 2021 to make the changes. Manufacturers of certain flavored dried cranberries have until July 1, 2020 to bring the changes into effect. The FDA indicated its plans to work cooperatively with manufacturers to meet the new Nutrition Facts label requirements.

In addition, the FDA has finalized a new Nutrition Facts label for packaged foods, which will make it easier for consumers to make informed food choices to enable the consumption of healthier diets. The updated label has a fresh new design and reflects current scientific information, including the link between diet and chronic diseases. Table 20.3 reflects some of the changes.

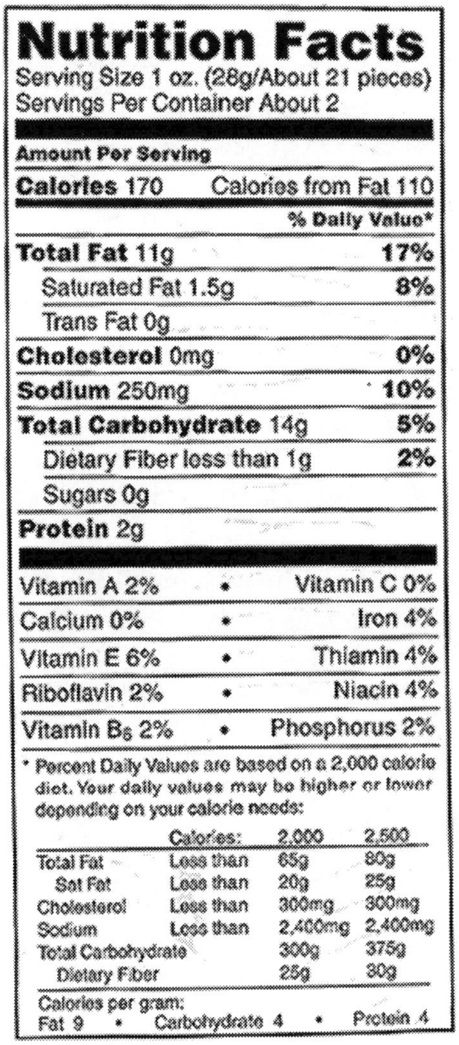

FIGURE 20.1 Sample United States nutrition facts label. *2003. Guidance on How to Understand and Use the Nutrition Facts Panel on Food Labels, Food and Drug Administration Center for Food Safety and Applied Nutrition, Washington D.C. http://www.cfsan.fda.gov/~dms/foodlab.html.*

20.5.2 Canada

The Food and Drugs Act is the primary federal decree governing the labeling of all foods sold in Canada (Canada, 2003). Regulations made under the Act include ingredient listing, nutrition labeling, and all types of claims. Health Canada and the Canadian Food Inspection Agency (CFIA) oversee the regulatory process of food labeling in Canada. Health Canada is responsible for setting health and safety standards and for developing food labeling policies related to health and nutrition under the Food and Drugs Act. CFIA is responsible for administering other food labeling policies and enforcing all food labeling regulations.

Nutrition labeling guidelines were introduced in Canada in 1988, along with amendments to the Food and Drug Regulations, which was voluntary, with a few exceptions (Canada, 2003). The guidelines on nutrition labeling regulated format, nutrient content information, and a declaration of serving size (Canada, 1989). The NIP was required to list amounts of vitamins and minerals as a percentage of a single set of nutrient reference values (NRVs), Recommended Daily Intakes, per serving of stated size (Canada, 1986). In 2003, Canada published and implemented new food labeling regulations (Canada, 2003). The new regulations established nutrition labeling on most prepackaged food, updated and consolidated permitted nutrient content claims, and introduced a new regulatory framework and process for diet-related health claims (Canada, 2003). A sample of the Canadian Nutrition Facts is shown in Fig. 20.3.

FIGURE 20.2 The new and improved nutrition facts label—key changes. *Source: https://www.fda.gov/files/food/published/The-New-and-Improved-Nutrition-Facts-Label-%E2%80%93-Key-Changes.pdf.*

In 2007, nutrition labeling for all prepackaged foods became mandatory in Canada. These labeling regulations updated from 2003 revised the requirements for nutrient-content claims and allowed health claims on diet–health relationships to be used on food labels. Such claims include: sodium and potassium and their association with blood pressure; calcium and vitamin D and their association with osteoporosis; saturated fat and *trans* fat and their association with heart disease; and vegetables and fruits and their association with some types of cancer (Canada, 2003). The regulations stipulate the prescribed wording for the permitted claims.

The Canadian regulations require *trans* fat to be incorporated with saturated fat with the % DV for the sum of saturated and *trans* fats being 20 g based on 10% of energy with a 2000-calorie dietary energy reference value. Expression of a % DV was considered important to assist consumers in understanding the relative significance of the amount of these nutrients in a food. The % DV for cholesterol is optional. There is no % DV for protein because protein intakes in Canada were not considered to be a public health concern. Explanatory footnotes related to the DV are similar to those used in the United States and may be included in the Nutrition Facts table. The graphic elements of the Nutrition Facts table are tightly regulated to ensure the use of a consistent and legible format. The Canadian regulations, unlike those of the U.S., do not include specific regulations to define the serving size except in the case of single-serving containers.

Canada has made improvements to the Nutrition Facts Table and List of Ingredients on food labels based on feedback from Canadians and stakeholders. In December 2016, the food industry was given a transition period of 5 years to make these changes. During the transition period, both preexisting and new food labels will be seen. The changes to the Nutrition Facts Table are shown in Table 20.4 and Fig. 20.4.

TABLE 20.3 Selected changes in the nutrition facts labels for packaged foods.

Category	Change made
Servings	The number of *"servings per container"* and the *"serving size"* declarations have been increased and are now in larger and/or bolder type. The serving sizes have been updated to reflect what people actually eat and drink. For example, the previous serving size for ice cream was ½ cup; it has been updated to 2/3 cup. There are also new requirements for certain size packages, which are between one and two servings, or are larger than a single serving but could be eaten in one or multiple sittings.
Calories	The word *"Calories"* is now written larger and bolder
Fats	*"Calories from Fat"* has been removed because research shows the type of fat consumed is more important than the amount.
Added Sugars	*"Added Sugars"* in grams and as a percent Daily Value (% DV) is now required on the label. Added sugars include sugars that are either added during the processing of foods, or are packaged as such (for example, a bag of table sugar), and also includes: (i) sugars from syrups and honey and (ii) sugars from concentrated fruit or vegetable juices. Scientific data show that it is difficult to meet nutrient needs while staying within calorie limits if you consume more than 10% of your total daily calories from added sugar.
Nutrients Footnote	The lists of nutrients that are required or permitted on the label have been updated. Vitamin D and potassium are now required on the label because Americans do not always get the recommended amounts. Vitamins A and C are no longer required since deficiencies of these vitamins are rare among Americans. For vitamin D, calcium, iron, and potassium, the actual amount (in milligrams or micrograms) in addition to the % DV must be listed. The daily values for nutrients have also been updated based on newer scientific evidence. The daily values are reference amounts of nutrients to consume or not to exceed and are used to calculate the % DV. The footnote at the bottom of the label has been changed to better explain the meaning of % DV. The % DV helps the consumer to understand the nutrition information in the context of a total daily diet.

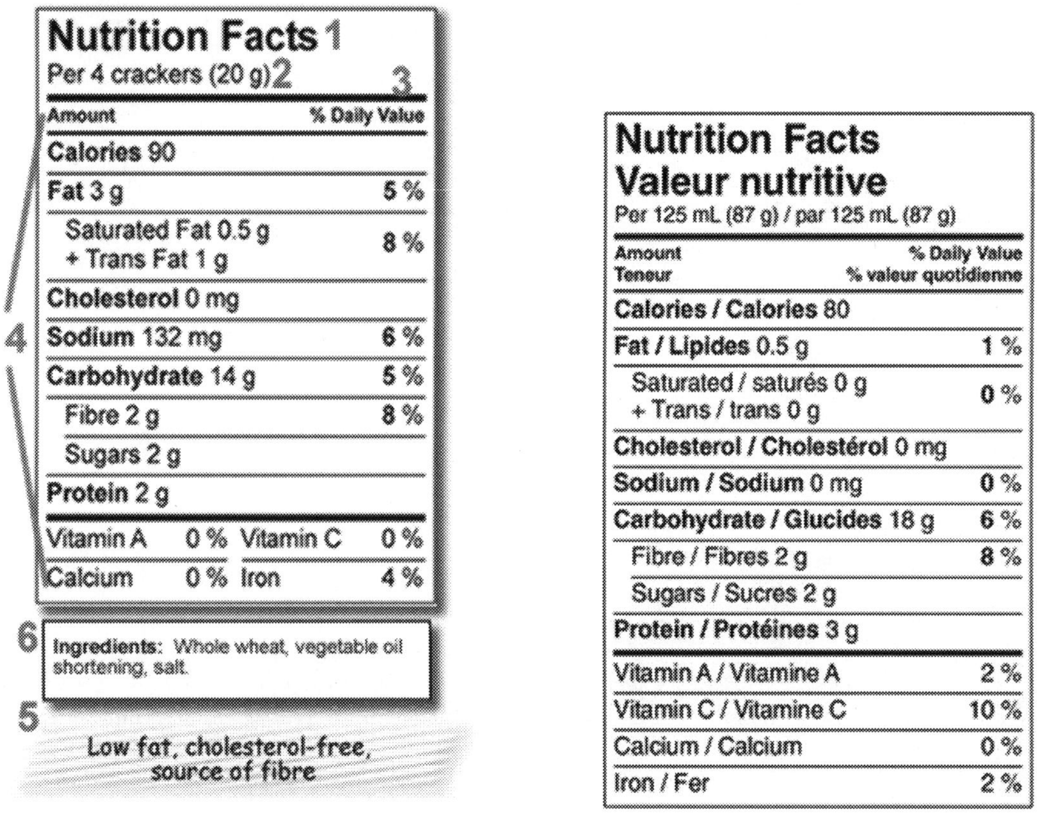

1 Nutrition Facts Table
2 Specific amount of Food
3 % Daily Value
4 Core Nutrients
5 Nutrition Claims
6 List of Ingredients

FIGURE 20.3 Sample Canada nutrition labels. *Source: http://www.hc-sc.gc.ca/fn-an/alt_formats/hpfb-dgpsa/pdf/label-etiquet/inl-eni-eng.pdf. Canada, 2003. SOR/2003-11. Regulations amending the Food and Drug Regulations (Nutrition labeling, nutrient content claims and health claims). Canada Gazette, Part II 137:154–405. http://www.inspection.gc.ca/english/fssa/labeti/guide/ch5e.shtml#a5_4.*

TABLE 20.4 Changes to the nutrition facts table, Canada.

- Making the serving size more:
  - consistent so that it's easier to compare similar foods
  - realistic so that it reflects the amount that Canadians typically eat in one sitting
- Making the information on serving size and calories easier to find and read by:
  - increasing the font size of serving size and calories
  - adding a bold line under the calories
- Revising the % daily values based on updated science
- Adding a new % daily value for total sugars
- Updating the list of nutrients to:
  - add potassium because:
    - -it is important for maintaining healthy blood pressure
    - -most Canadians are not getting enough of this nutrient
  - remove vitamin A and vitamin C because:
    - -most Canadians get enough of these nutrients in their diets
- Adding the amounts in milligrams (mg) for potassium, calcium, and iron
- Adding a footnote at the bottom of the table about % daily value
  - this will help consumers understand how much sugar and other nutrients (like sodium) are in their food and will explain that:
    - -5% or less is a little
    - -15% or more is a lot

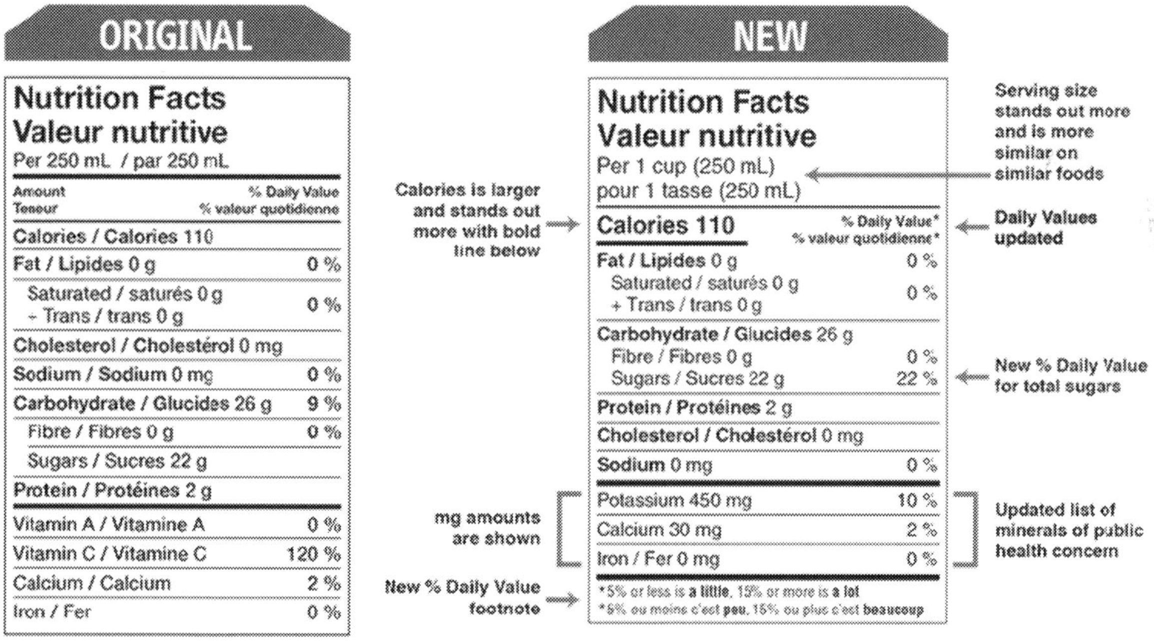

FIGURE 20.4 Sample of the nutrition facts table changes, Canada. *Source: https://www.canada.ca/en/health-canada/services/food-labelling-changes.html#a1.*

20.5.2.1 List of ingredients changes

Changes were made to the list of ingredients to enable consumers to find it easier and allow them to read and understand it better (Fig. 20.5). The changes to the list of ingredients include:

- Grouping sugars-based ingredients in brackets after the name "sugars"
  - this will help consumers identify all of the sources of sugars added to a food

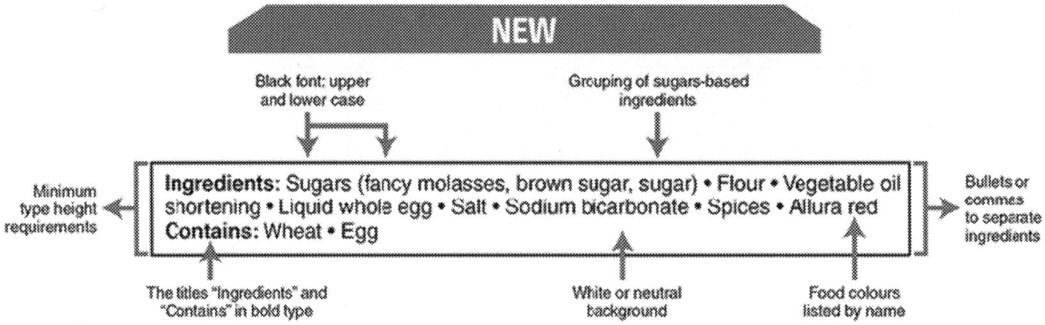

FIGURE 20.5 Sample of the list of ingredients in the updated label. *Source: https://www.canada.ca/en/health-canada/services/food-labelling-changes.html#a1.*

- Listing food colors by their individual common names
- Making the text in black font on white or neutral background
- Creating minimum type height requirements for ingredients
- Using bullets or commas to separate ingredients
- Using both upper- and lowercase letters for the ingredients in the list
  - the same format rules will apply to any "contains" statement indicating the presence or potential presence of:
    - Priority food allergens
    - Gluten sources
    - Added sulfites

20.5.2.2 Serving size

The changes in serving size are intended to better reflect the amount that Canadians eat in one sitting. Serving sizes will be based on regulated reference amounts. The changes are different for single serve and multiserving packages. Serving sizes will also be more consistent, making it easier to compare similar foods, and know how many calories and nutrients are being consumed.

20.5.2.2.1 Foods in single serving containers

On single serving packages containing up to 200% of the reference amount for that food, the serving size will be the amount in the whole container. For example, the reference amount for milk is 250 mL. For containers with up to 500 mL (200% of 250 mL), the serving size shown will be the amount of milk in the entire container. Fig. 20.6 demonstrates that on a 473 mL carton of milk, the serving size will be shown as *"Per 1 carton* (473 mL)."

20.5.2.2.2 Foods in multiserve packages

On multiserve packages, serving sizes will be in an amount as close as possible to the food's reference amount. For multiserve packages, serving sizes are based on the type of food, such as (i) foods that can be measured; (ii) foods that come in pieces or are divided; and (iii) quantities of foods that are typically eaten. The intent is that the food industry will make serving sizes more uniformed, and easier to compare similar foods. For foods that can be measured, such as yogurt,

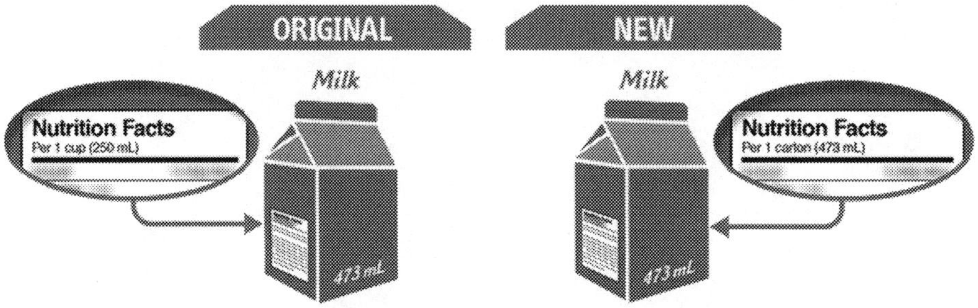

FIGURE 20.6 Foods in single serving containers. *Source: https://www.canada.ca/en/health-canada/services/food-labelling-changes.html#a1.*

the serving size will be shown as a common household measurement (for example, cup, teaspoon, and tablespoon). This will be paired with its metric equivalent in mL or grams (g). Similar products will have the same mL or gram amount which will make them easier to compare. To demonstrate, yogurt has a reference amount of 175 g, which might be normally consumed in one sitting. Therefore, the serving size on all cartons of yogurt will be based on 175 g (Figs. 20.7 and 20.8).

20.5.2.2.3 Foods that come in pieces or are divided

For foods like crackers, which come in pieces, or are divided into pieces before eating (for example, lasagna), the serving size will be shown as either (i) the number of pieces or (ii) as a fraction of the food in addition to its weight in grams. Similar products will have the same or very similar gram amounts. For example, 20 g is the reference amount for crackers; therefore, the serving size on cracker boxes will have to be as close as possible to 20 g. Although the number of crackers may vary from product to product, weights will remain very similar.

20.5.2.2.4 Amounts of foods that are typically eaten

For certain foods like sliced bread, the serving size will reflect the way they are typically consumed, followed by its weight in grams. For example, the serving size on a bag of bread will show two slices of bread and its weight in grams (Fig. 20.9). Most consumers usually eat two slices of bread at one time.

20.5.2.3 Sugars information

The changes to sugars include those in the Nutrition Facts Table and the list of ingredients. In the Nutrition Facts Table, a % daily value has been included for total sugars to help the consumer compare the sugars content of different foods, and identify sugary foods that should be limited, such as those with a sugars daily value of 15% or more. Table 20.5 provides examples of the sugars % daily value for some commonly eaten food items. Fig. 20.10 shows a sample of the changes to sugars label.

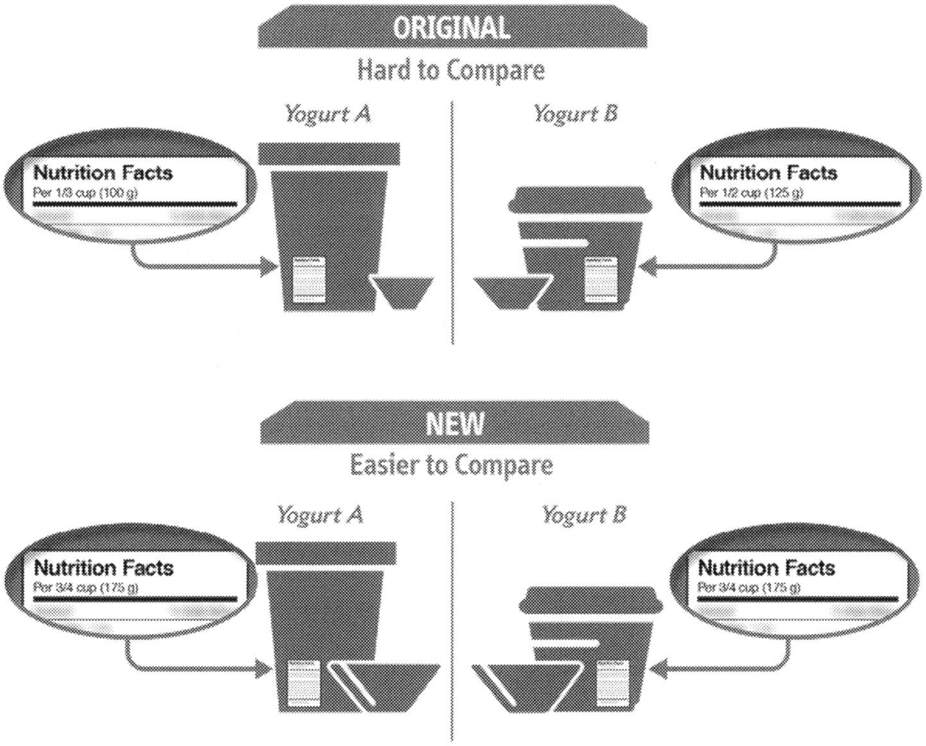

FIGURE 20.7 Sample label for foods that can be measured. *Source: https://www.canada.ca/en/health-canada/services/food-labelling-changes.html#a1.*

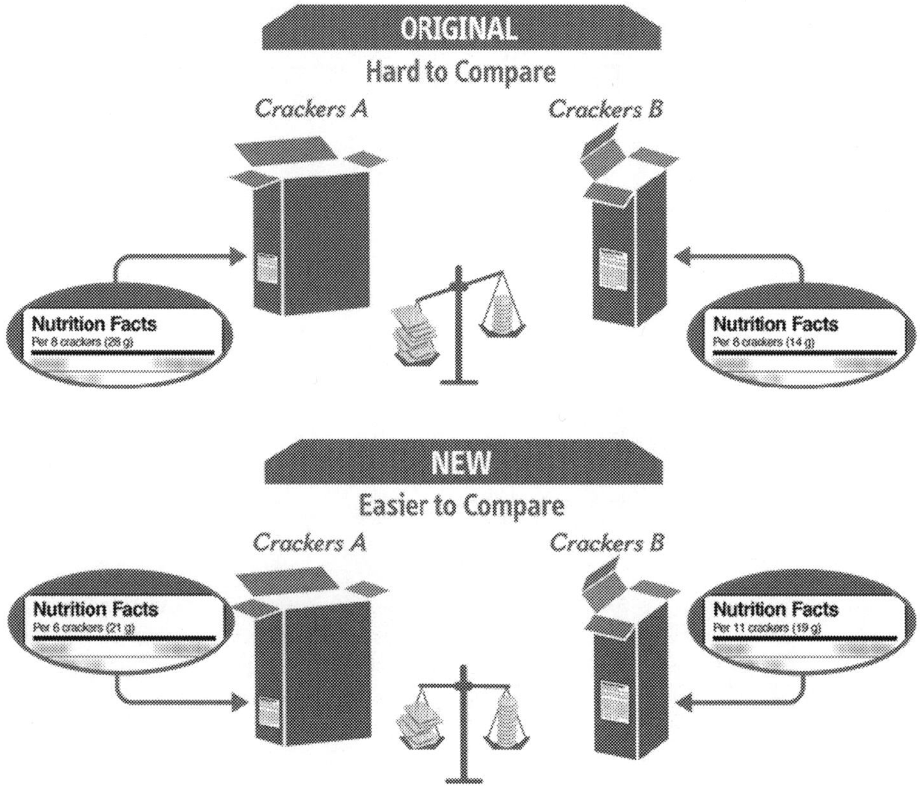

FIGURE 20.8 Sample label for foods that come in pieces or are divided. *Source: https://www.canada.ca/en/health-canada/services/food-labelling-changes.html#a1.*

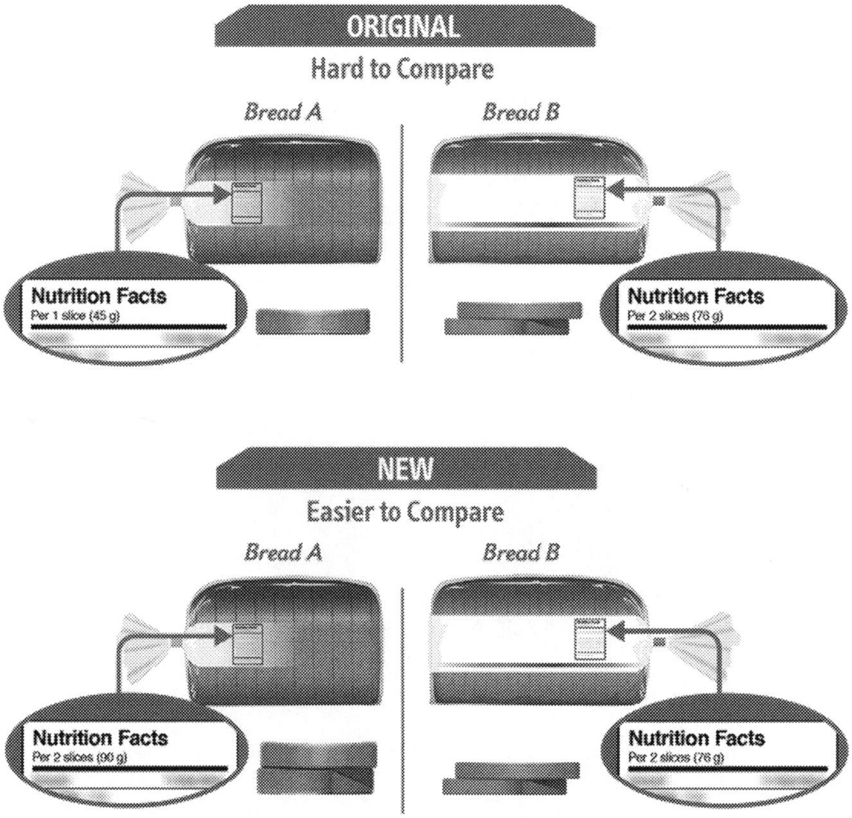

FIGURE 20.9 Sample label for foods that are typically consumed. *Source: https://www.canada.ca/en/health-canada/services/food-labelling-changes.html#a1.*

TABLE 20.5 Examples of the sugars % daily value for some commonly eaten foods.

Less than 15% daily value of sugars		More than 15% daily value of sugars	
Food	% DV Sugars	Food	% DV Sugars
Milk	13	Chocolate milk	26
Plain yogurt	12	Flavored yogurt	31
Canned fruit in water	10	Canned fruit in light syrup	21
Unsweetened frozen fruit	6	Fruit juice	25
Unsweetened oat cereal	1	Frosted oat cereal	18

FIGURE 20.10 Sample of the new sugars information label. *Source: https://www.canada.ca/en/health-canada/services/food-labelling-changes.html#a1.*

20.5.2.3.1 List of Ingredients

Sugars-based ingredients have been grouped in brackets in descending order by weight after the name *"sugars"* to help consumers see that sugars have been added to the food, quickly find the sources of sugars added to the food, and understand how much sugars are added to the food compared to other ingredients. Sugars can include (i) white, beet, raw, or brown sugar; (ii) agave syrup, honey, maple syrup, barley malt extract, or fancy molasses; (iii) fructose, glucose, glucose-fructose (also known as high fructose corn syrup), maltose, sucrose, or dextrose; (iv) fruit juice concentrates and purée concentrates that are added to replace sugars in foods; (v) fancy molasses by weight than brown sugar or sugar; and (vi) sugars in the food by weight than any other ingredient (Fig. 20.11).

20.5.3 Australia and New Zealand

Australia and New Zealand share joint food labeling standards developed by Food Standards Australia New Zealand (FSANZ). Two organizations, New Zealand Food Safety Authority and FSANZ, share the primary responsibility for protecting consumers. FSANZ was established by the Food Standards Australia New Zealand Act 1991. Fig. 20.12 shows

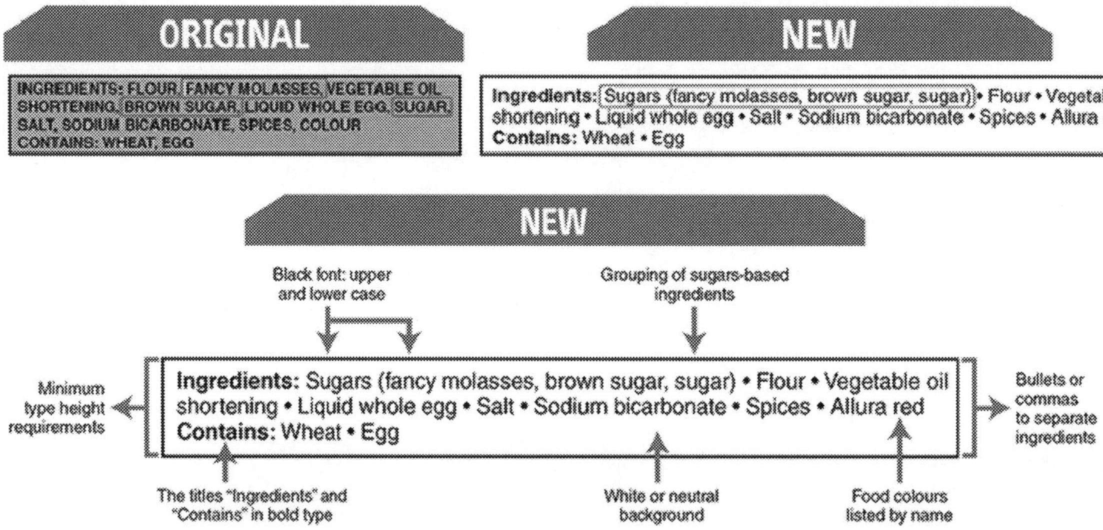

FIGURE 20.11 Sample label new sugar-based list of ingredients. *Source: https://www.canada.ca/en/health-canada/services/food-labelling-changes.html#a1.*

the Organization Chart of FSANZ. FSANZ, which was formerly the Australia New Zealand Food Standards Authority (ANZFA), develops food standards and joint codes of practice with industry, covering the content and labeling of food sold in Australia and New Zealand. The joint arrangement led to a joint Australia New Zealand Food Standards Code (the Code), which replaced the New Zealand Food Regulations made under the New Zealand Food Act 1981, and the Australian Food Standards Code. The Code is law that applies in Australia and New Zealand.

The main responsibility of FSANZ is to develop and administer the Code. The Code lists requirements for foods such as additives, food safety, labeling, and genetically modified (GM) foods. FSANZ's ultimate goal is to provide a safe food supply and well-informed consumers. The harmonization of food standards between the two countries is obvious because of the considerable amount of food trade between the two countries. According to FSANZ (2009), the NIP must be presented in a standard format which shows the amount per serving or 100 g or 100 mL of the food. Table 20.6 shows an example of an NIP and the nutrients which should be listed.

Some exceptions to products requiring an NIP include (i) very small packages which are about the size of a larger chewing gum packet; (ii) foods with no significant nutritional value, such as a single herb or spice; (iii) tea and coffee; (iv) foods sold unpackaged (unless a nutrition claim is made); and (v) foods made and packaged at the point of sale, for example, bread made in a local bakery.

20.5.3.1 Nutrition information panels

Australia and New Zealand food labels were updated in December 2015. Briefly, the NIPs provide information on the average amount of energy (in kilojoules or both in kilojoules and kilocalories), protein, fat, saturated fat, carbohydrate, sugars, and sodium in the food, in addition to any other claim, which requires nutrition information. If a food label claimed a "good source of fiber," then the amount of fiber in the food must be shown in the NIP. The NIP must be presented in a standard format which shows the average amount per serving and per 100 g (or 100 mL if liquid) of the food.

Foods lacking significant nutritional value such as herb or spice, mineral water, tea and coffee, foods sold unpackaged, foods made and packaged at the point of sale, such as bread made and sold in a local bakery do not require an NIP. However, if a claim is made about any of these foods, an NIP must be provided. Foods in small packages with a surface area of less than 100 mm squared are exempt from an NIP.

20.5.3.1.1 Serving size

The serving size listed in the NIP is determined by the food business. This explains why it sometimes varies from one product to the next. The *"per serving"* information is useful in estimating how much of a nutrient the consumer is eating. For example, if the consumer is limiting fat intake, the *"per serving"* amount can be used to calculate the daily total fat intake from packaged foods.

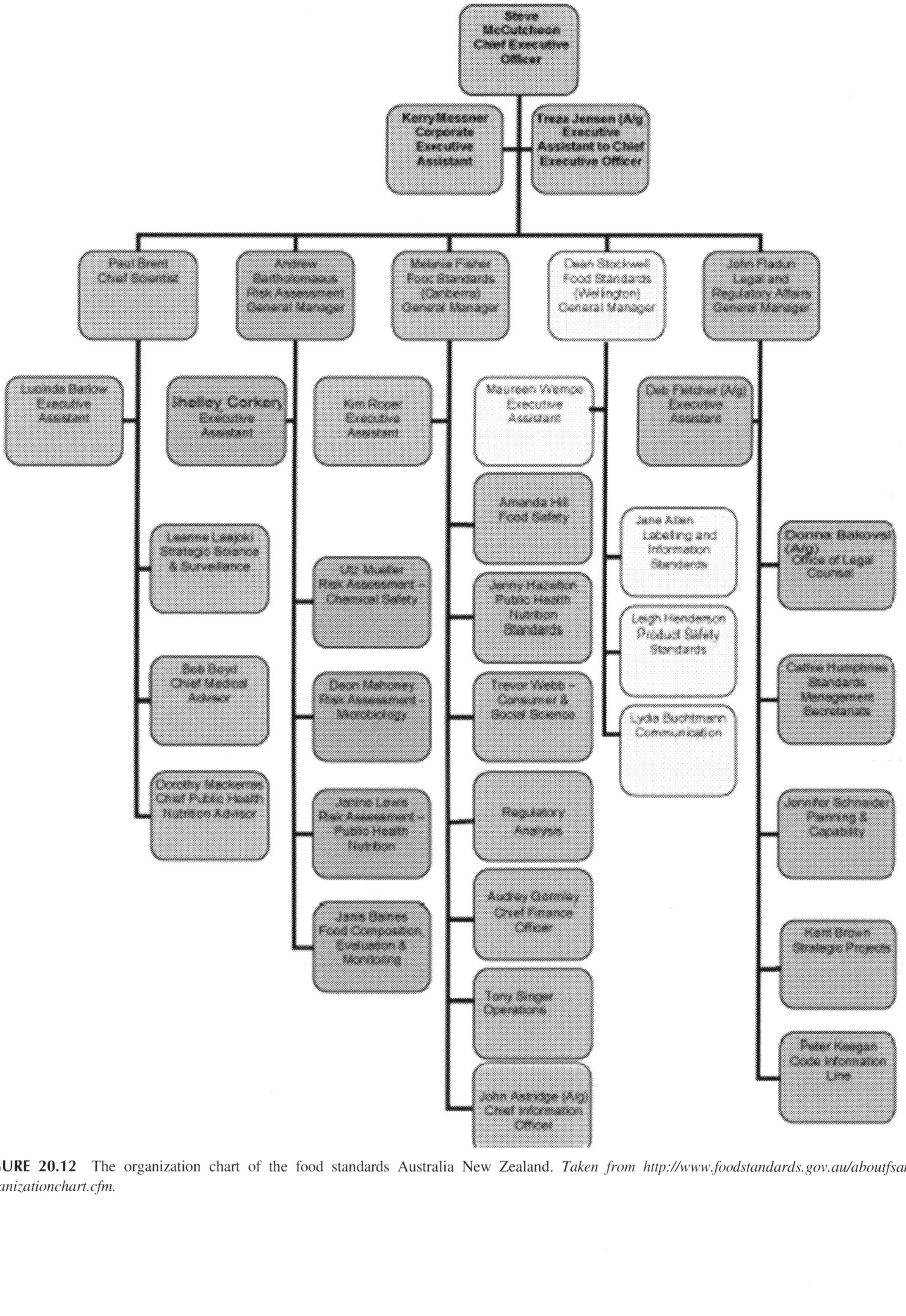

FIGURE 20.12 The organization chart of the food standards Australia New Zealand. *Taken from http://www.foodstandards.gov.au/aboutfsanz/organizationchart.cfm.*

TABLE 20.6 Example of a Food Standard Australia New Zealand (FSANZ) nutrition information panel and the nutrients which should be listed (FSANZ, 1990).
Nutrition Information (Average) *(http://www.kelloggs.com.au/Home/tabid/36/Default.aspx).*

Average serving size - 30g (2/3 metric cup[a])				
	Quantity per serving	% Daily intake[b] per serving	per serve with 1/2 cup skim milk	Quantity per 100 g
Energy	450 kJ	5%	650 kJ	1500 kJ
Protein	3.0 g	6%	7.6 g	9.9 g
Fat, total	0.5 g	0.6%	0.6 g	1.5 g
-Saturated	<0.1 g	0.4%	0.2 g	0.3 g
Carbohydrate	20.7 g	7%	27.2 g	68.9 g
-Sugars	4.9 g	5%	11.4 g	16.3 g
Dietary fiber	4.1 g	14%	4.1 g	13.5 g
-Soluble	0.7 g	–	0.7 g	2.3 g
-Insoluble	3.6 g	–	3.6 g	11.9 g
Sodium[d]	99 mg	4%	156 mg	330 mg
Potassium	137 mg	–	342 mg	455 mg
		%RDI[c]		
Thiamin (VIT B1)	0.28 mg	25%	0.33 mg	0.92 mg
Riboflavin (VIT B2)	0.4 mg	25%	0.7 mg	1.4 mg
Niacin	2.5 mg	25%	2.6 mg	8.3 mg
Vitamin B6	0.4 mg	25%	0.4 mg	1.3 mg
Folate	100 µg	50%	106 µg	333 µg
Iron	3 mg	25%	3.1 mg	10 mg
Magnesium	38 mg	12%	54 mg	128 mg
Zinc	1.8 mg	15%	2.3 mg	6 mg

Ingredients: Whole wheat (67%), sugar, wheat bran (12%), barley malt extract, salt, minerals (iron, zinc oxide), vitamins (niacin, riboflavin, thiamin, folate). Contains gluten containing cereals. May contain traces of peanuts and/or tree nuts.
Blé entier (67%), sucre, son de blé (12%), extrait de malt d'orge, sel, minéraux (fer, oxyde de zinc), vitamines (niacine, riboflavine, thiamine, folate). Contient gluten contenir les céréales. Peut contenir des traces de cacahuètes et/ou noix d'arbres. Contains 20% wheatbran* *from wheatbran (12%) and whole wheat (8%).
[a]*Cup measurement is approximate and is only to be used as a guide. If you have any specific dietary requirements, please weigh your serving.*
[b]*% Daily Intakes are based on an average adult diet of 8700 kJ. Your daily intakes may be higher or lower depending on your energy needs.*
[c]*% Recommended Dietary Intake (Aust/NZ) per serving.*
[d]*12% wheat bran and 8% wheat bran from whole wheat.*

20.5.3.1.2 Quantity per 100 g

The *"quantity per 100 g"* (or 100 mL if liquid) information is convenient for comparison of similar products. The figures in the *"quantity per 100 g"* column are the same as percentages. For example, if 20 g of fat is listed in the *"per 100 g"* column, it means that the product contains 20% fat.

20.5.3.1.3 Energy/kilojoules

The energy value is the total amount of kilojoules from protein, fat, carbohydrate, dietary fiber, and alcohol that is released when food is used by the body.

20.5.3.1.4 Fat

Fat is listed in the NIP as total fat (which is the total of the saturated fats, trans fat, polyunsaturated fats, and monounsaturated fats in the food). A separate entry must also be provided for saturated fat in the food. If a nutrition claim is made about cholesterol, saturated fats, trans fatty acids, polyunsaturated fats, monounsaturated fats or omega-3, omega-6, or omega-9 fatty acids, the NIP must also include the amount of trans fat, polyunsaturated fats and monounsaturated fats, and also omega fatty acids if claimed.

20.5.3.1.5 Carbohydrates

Carbohydrate in the NIP includes starches and sugars.

20.5.3.1.6 Sugars

Sugars are included as part of the carbohydrates, and separately in the NIP. The amount of sugars in the NIP include naturally occurring sugars and added sugar.

20.5.3.1.7 Dietary fiber

The NIP excludes fiber except a nutrition claim has been made on the label about fiber, sugar, or carbohydrate.

20.5.3.1.8 Sodium/salt

Sodium is included in the NIP because it is the component of salt, which has been linked to chronic diseases such as hypertension and stroke (Food Standards Australia New Zealand, 2020a).

20.5.3.2 Review of sugar labeling

Since 2017, the Australian Ministerial Forum on Food Regulation has been reviewing sugar labeling (Food Standards Australia New Zealand, 2020a). The intent is to determine how best to give Australian and New Zealand consumers useful information about added sugar in food, which will assist them in making more informed choices that support the dietary guidelines. In August 2019, FSANZ was requested to review labeling for added sugars because it was determined that information about added sugars on food labels is limited and/or unclear. This limitation prevented consumers from making food choices consistent with dietary guidance. It was agreed that a pictorial approach applied to sugary beverages/sugar-sweetened beverages should be looked at, along with other options, pending the response to the Health Star Rating 5-year review (Health Star Rating System Five Year Review Report, 2019).

In August 2019, requirements in the Food Standards Code for food labels were updated to include the total amount of sugars in the NIP. Total sugars include naturally occurring sugars and sugars that have been added as an ingredient to foods. The Code also contains requirements for foods that make claims about sugar, for example "low sugar." Foods that claim to be "low sugar" must contain $\leq$2.5 g of sugar per 100 mL of liquid food or 5 g per 100 g of solid food. There are also criteria in the Code for "reduced sugar," "x% sugar free," "no added sugar," and "unsweetened" claims. On July 17, 2020, this was endorsed, and a Review Implementation Plan and an implementation start date of November 15, 2020 were set.

20.5.3.3 Country of origin

The Australian Government has introduced a new country-of-origin food labeling system, which commenced on July 1, 2016. Under the new system, the country-of-origin labeling requirements come under Australian Consumer Law. Businesses were given a 2-year period to transition to the new food label law, which became mandatory on July 1, 2018. The Food Standards Code was amended in July 2018 to remove its country-of-origin labeling requirements.

Country of origin labeling is voluntary in New Zealand, and suppliers may choose not to display this information. There is an exception for grape wine, which must be labeled with its country or countries of origin. When suppliers do provide country of origin information, it must be accurate. All food must be labeled with the contact details of the food supplier in New Zealand or Australia, so consumers can contact the supplier and ask for details about the food (Food Standards Australia New Zealand, 2020a).

20.5.3.4 Food additives

Food additives in most packaged food must be listed in the statement of ingredients on the label. Most food additives must be listed by their class name followed by the name of the food additive or the food additive number. For example, Color (Caramel I) or Color (150a). Enzymes and most flavorings (or flavor) do not need to be named or identified by a food additive number and can be labeled by their class name only. The class name indicates the purpose of the food additive (Food Standards Australia New Zealand, 2020a).

20.5.3.5 Exemptions

Food additives in foods that are not required to be labeled with statements of ingredients are not required to be labeled. In instances where the food contains compound ingredients (an ingredient made up of two or more ingredients such as tomato paste), the compound ingredient does not have to be listed if it is less than 5% of the final food (Food Standards Australia New Zealand, 2020a).

20.5.3.6 Food additives and allergies

Certain food allergens must be declared at all times when present in food as an ingredient, including food additives (Food Standards Australia New Zealand, 2020a).

20.5.3.7 Genetically modified food labeling

FSANZ is responsible for approving GM foods and ingredients for use in the food supply in Australia and New Zealand as shown in Schedule 26 of the Food Standards Code (Food Standards Australia New Zealand, 2020a). FSANZ does not maintain a list of food products in the marketplace which contain GM foods or ingredients. Food businesses may be able to provide this information. GM foods and ingredients (including food additives and processing aids) that contain novel DNA or novel protein must be labeled with the words *"genetically modified."* Novel DNA and novel protein are defined in Standard 1.5.2 of the Food Standards Code:

Novel DNA and novel protein mean DNA or protein which, as a result of the use of gene technology, is different in chemical sequence or structure from DNA or protein present in counterpart food that has not been produced using gene technology, other than protein that:

- is used as a processing aid or used as food additive and
- has an amino acid sequence that is found in nature (Food Standards Australia New Zealand, 2020a).

20.5.3.8 Altered characteristics

Labeling is also required for GM foods that have an altered characteristic (e.g., altered nutritional profile) when compared to a counterpart non-GM food (e.g., soybeans with increased oleic acid content). These GM foods are listed in subsection S26—3(2) of Schedule 26 of the Food Standards Code and must be labeled with the words "genetically modified," as well as any additional labeling required by the Schedule, regardless of the presence of novel DNA or novel protein. More information about labeling of altered characteristics are presented below (Food Standards Australia New Zealand, 2020a). All GM foods and ingredients must undergo a safety assessment and be approved before they can be sold in Australia and New Zealand. GM labeling is not about safety. It is about helping consumers make an informed choice about the food they buy (Food Standards Australia New Zealand, 2020a). The decision on how GM foods are labeled was made by the ministers responsible for food regulation in 2001. In response to an independent review of food labeling in 2011 ministers agreed with the recommendation that the existing labeling provisions should remain (Blewett et al., 2011).

The following guidelines are used to determine if a GM food has an altered characteristic which would require it to be labeled as "genetically modified": (i) if the genetic modification has significantly altered the composition or nutritional qualities compared to the existing counterpart non-GM food; (ii) if the intended use of the GM food is different to the existing counterpart non-GM food; and (iii) FSANZ also considers if additional labeling about the nature of any altered characteristic is required. For instance, high lysine corn, which has been genetically modified, will require additional label (Food Standards Australia New Zealand, 2020a).

20.5.3.9 Position of the GM information on the label

The statement *"genetically modified"* will be found on the label either next to the name of the food (e.g., "genetically modified soy beans") or in association with the specific GM ingredient in the ingredient list (e.g., "soy flour" [genetically modified]). If the food is unpackaged, then the information must accompany or be displayed with the food (Food Standards Australia New Zealand, 2020a).

20.5.3.10 Exemptions from GM labeling

GM foods that do not contain any novel DNA or novel protein, and do not have an altered characteristic, do not require GM labeling. The decision not to label these foods was made because the composition and characteristics of these foods are exactly the same as the non-GM food. These foods are typically highly refined foods, such as sugars and oils, where processing has removed the DNA and protein from the food, including novel DNA and novel protein. GM flavorings that are present in food in a concentration of no more than 0.1% are also exempt from labeling (Food Standards Australia New Zealand, 2020a).

Labeling is also not required when there is no more than 1% (per ingredient) of an approved GM food unintentionally present in a non-GM food. To clarify, labeling is not required when a manufacturer genuinely orders non-GM ingredients but finds that up to 1% of an approved GM ingredient is accidently mixed with the non-GM ingredient (Food Standards Australia New Zealand, 2020a).

20.5.3.11 GMO labeling for restaurant foods

Food intended for immediate consumption that is prepared and sold from food premises and vending vehicles (e.g., restaurants, takeaway food outlets, caterers) is also exempt from GM food labeling requirements. In these cases, the consumer can seek information about the food from the food business. Information supplied by the food business must not be misleading or untruthful (Food Standards Australia New Zealand, 2020a).

20.5.3.12 Food from animals that have eaten GM feed

Animals that are fed with feed that has been produced using gene technology are not themselves genetically modified. The food products (e.g., meat, milk, eggs) derived from an animal which has been fed GM feed are not regarded as GM foods and are not required to be labeled (Food Standards Australia New Zealand, 2005).

20.5.3.13 "GM free" and "non-GM" claims

"GM free" and "non-GM" claims are made voluntarily by food manufacturers and are subject to relevant fair trading laws in Australia and New Zealand which prohibit representations about food that are, or likely to be, false, misleading or deceptive (Food Standards Australia New Zealand, 2020a).

20.5.3.14 Nutrition content claims and health

Nutrition content claims and health claims are voluntary statements made by food businesses on labels and in advertising about a food. Standard 1.2.7 sets out the rules for food businesses choosing to make nutrition content claims and health claims (Food Standards Australia New Zealand, 2020b). Nutrition content claims are claims about the content of certain nutrients or substances in a food, such as low in fat or good source of calcium. These claims will need to meet certain criteria set out in the Standard (Food Standards Australia New Zealand, 2020b). For example, with a "good source of calcium" claim, the food needs to contain at least the amount of calcium specified in the Standard.

Health claims refer to a relationship between a food and health. There are two types of health claims: (i) general level health claims refer to a nutrient or substance in a food, or the food itself, and its effect on health. For example: calcium for healthy bones and teeth. They must not refer to a serious disease or to a biomarker of a serious disease. Food businesses making general level health claims are able to base their claims on any one of the more than 200 preapproved food–health relationships in the Standard or self-substantiate a food–health relationship in accordance with detailed requirements set out in the Standard, including notifying FSANZ (Food Standards Australia New Zealand, 2017); (ii) high level health claims refer to a nutrient or substance in a food and its relationship to a serious disease or to a biomarker of a serious disease. For example, diets high in calcium may reduce the risk of osteoporosis in people 65 years and over. An example of a biomarker health claim is Phytosterols that may reduce blood cholesterol (FSANZ, 2017). High level health claims must be based on a food–health relationship preapproved by FSANZ (Food Standards Australia New Zealand, 2017). There are currently 13 preapproved food–health relationships for high level health claims listed in the Standard (Food Standards Australia New Zealand, 2017). All health claims are required to be supported by scientific evidence to the same degree of certainty, whether they are preapproved by FSANZ or self-substantiated by food businesses. Health claims are only permitted on foods that meet the nutrient profiling scoring criterion (Food Standards Australia New Zealand, 2017). For example, health claims will not be allowed on foods high in saturated fat, sugar, or salt. Endorsements that are nutrition content claims or health claims will be permitted, provided the endorsing body meets requirements set out in the Standard (Food Standards Australia New Zealand, 2017).

20.5.4 Developing countries—Codex Alimentarius

Several developing countries have adopted the *Codex Alimentarius* or Food Code guidelines for nutrition labeling. Currently, the Codex Alimentarius Commission has 189 Codex Members made up of 188 Member Countries and one Member Organization (The European Union) (FAO/WHO, 2020). The Commission resides within the framework of the Joint Food and Agriculture Organization (FAO)/World Health Organization (WHO) Food Standards Program. It was established by the FAO/WHO of the United Nations (UN) in 1963 (Codex Alimentarius Commission, 2003). Since 1963, Codex has worked to create harmonized international food standards to protect the health of consumers and ensure fair practices in the food trade.

The Commission compiles internationally adopted food standards, guidelines, codes of practice, and other recommendations called *Codex Alimentarius* (FAO/WHO, 2007). The Codex Guidelines on Nutrition Labeling were adopted by the Codex Alimentarius Commission at its 16th Session, 1985. The "Nutrient Reference Values for Food Labeling Purposes"; "Listing of Nutrients"; "Presentation of Nutrient Contents"; and "Definitions" were amended by the 20th, 26th, and 29th sessions of the Commission in 1993, 2003, and 2006, respectively (FAO/WHO, 2007). The guidelines on nutrition labeling (CAC/GL 2-1985) in Food Labeling (fifth edition) include text adopted by *Codex Alimentarius* up to 2007 (FAO/WHO, 2007). In 2015/2016, Nutrition Labeling Guideline, CAC/GL 2-1985 was revised to update the NRVs for selected vitamins and minerals. An underlying principle for the revision of the NRVs was the evidence for a relationship between chronic disease risk and sodium and potassium intake for adults and children.

The purpose of the *Codex Alimentarius's* guidelines is to ensure that nutrition labeling is effective in (i) providing the consumer with information about a food so that wise food choices can be made; (ii) providing a means for conveying information of the nutrient content of foods on the label; (iii) encouraging the use of sound nutrition principles in the formulation of foods which would benefit public health; (iv) providing the opportunity to include supplementary nutrition information on the label; and (v) ensuring that nutrition labeling does not present product information which is false, misleading, or deceptive and ensure that nutritional claims are appropriately labeled (FAO/WHO, 2007).

The Codex Alimentarius Commission has also defined several terms for the purposes of the guidelines including nutrition labeling, nutrition declaration, nutrition claim, nutrient and sugars, dietary fiber, polyunsaturated fatty acids, and trans fatty acids. Extensive details regarding calculation of energy and protein, presentation of nutrient content, and format of presentation and periodic review are also included in the guidelines. For example, numerical information on vitamins and minerals should be expressed in metric units and/or as a percentage of the NRVs per 100 g, 100 mL or per package if the package has a single portion (FAO/WHO, 2007). NRVs values are a set of numerical values based on scientific data for purposes of nutrition labeling and relevant claims. There are two types of NRVs: (i) NRVs Requirement which is based on levels of nutrients associated with nutrient requirement and (ii) NRVs NCD (noncommunicable diseases), which refer to NRVs based on levels of nutrients associated with the reduction in the risk of diet-related NCD, excluding nutrient deficiency diseases or disorders (Nishida, 2019). A sample NRV for nutrition labeling, which should be used for labeling purposes in the interests of international standardization and harmonization, is shown in Fig. 20.13.

Codex Alimentarius has also outlined principles for nutrition labeling, which consists of nutrient declaration and supplementary nutrition information. According to FAO/WHO (2007), the nutrient declaration should give consumers an accurate summation of the nutritionally important nutrients in the product. The information should be truthful and convey only the quantity of nutrients contained in the product. Nutrient declaration should be mandatory for foods which have nutrition claims and voluntary for all other foods. Energy value amounts of protein, available carbohydrate (excluding dietary fiber), and fat should be mandatory in the nutrient declaration (FAO/WHO, 2007). An extensive description regarding listing of nutrients can be obtained elsewhere (FAO/WHO, 2007).

Supplementary nutrition information on food labels should be optional and given in addition to the nutrient declaration. Furthermore, supplementary nutrition information on labels should be accompanied by consumer education programs to increase understanding and use of the information. The content of supplementary nutrition information will vary from one country to another and within countries from one target group to another based on the country's educational policy, and the needs of the target groups. In situations where the target populations have high illiteracy rate and/or limited nutrition knowledge, food group symbols or other pictorial or color presentations may be used without the nutrient declaration (FAO/WHO, 2007). Labeling regulations have been adopted in some of the developing world including Asia, Africa, the Middle East, Latin America, and the Caribbean (i.e., the global South) (Fig. 20.14) (Mandle et al., 2015).

Nigeria and South Africa require nutrition labeling only on prepackaged foods, foods with special dietary uses, and on foods for which a nutrition claim is made, respectively (NLIP Watch, 2020; Hawkes, 2004). The Food Regulations of 1999 made under the Food Act 1998 allowed Mauritius to introduce nutrition labeling. The regulations set out the specific nutrients that must be labeled for a series of selected nutrition claims (Ministry of Health and Quality of Life, 1998a,b).

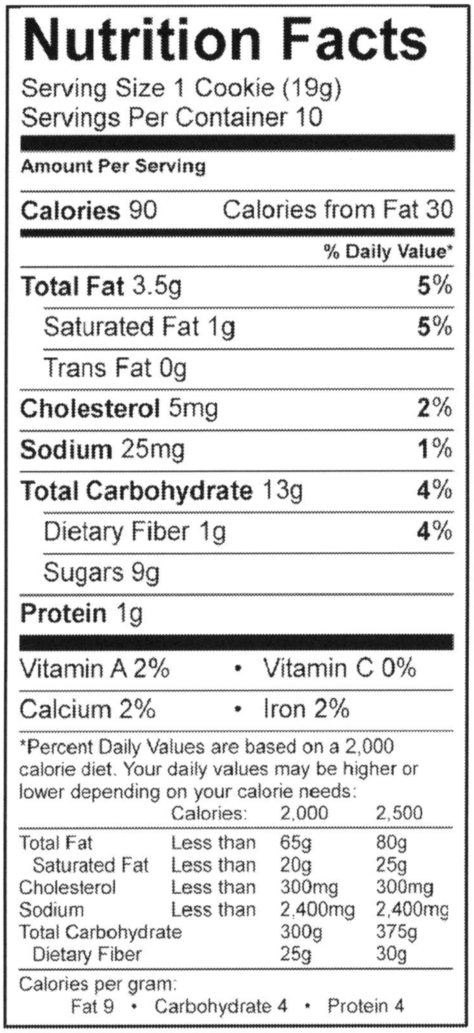

Nutrition Facts

Serving Size 1 Cookie (19g)
Servings Per Container 10

Amount Per Serving

Calories 90 Calories from Fat 30

% Daily Value*

Total Fat 3.5g	**5%**
Saturated Fat 1g	**5%**
Trans Fat 0g	
Cholesterol 5mg	**2%**
Sodium 25mg	**1%**
Total Carbohydrate 13g	**4%**
Dietary Fiber 1g	**4%**
Sugars 9g	
Protein 1g	

Vitamin A 2% • Vitamin C 0%

Calcium 2% • Iron 2%

*Percent Daily Values are based on a 2,000 calorie diet. Your daily values may be higher or lower depending on your calorie needs:

	Calories:	2,000	2,500
Total Fat	Less than	65g	80g
Saturated Fat	Less than	20g	25g
Cholesterol	Less than	300mg	300mg
Sodium	Less than	2,400mg	2,400mg
Total Carbohydrate		300g	375g
Dietary Fiber		25g	30g

Calories per gram:
Fat 9 • Carbohydrate 4 • Protein 4

FIGURE 20.13 Current nutrient reference value (NRV) for nutrition labeling (Codex Alimentarius). *Source: http://www.fao.org/fao-who-codex-alimentarius/roster/detail/zh/c/431721/.*

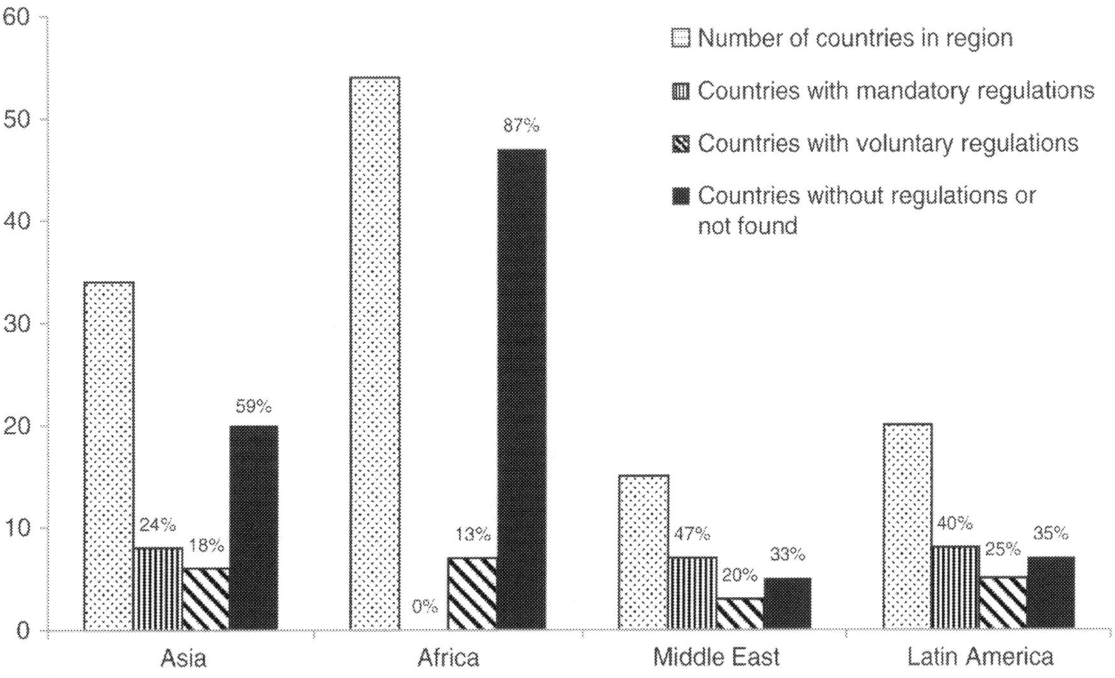

FIGURE 20.14 Food labeling regulations in the global South by region (Mandle et al., 2015).

It also mandates the labeling of protein, fat, carbohydrate, vitamin, and mineral content on infant foods, per 100g of the packaged food (Ministry of Health and Quality of Life, 1998a,b). Mauritius' 2009–10 National Nutrition Plan of Action proposed that the new labeling system on foodstuffs known as "Front of Pack Traffic Light Signpost Labeling" will be introduced. The label tells at a glance, the levels of sugars, fat, saturated fat, and salt using color codes where red means high, amber means medium, and green means low (Republic of Mauritius, 2009). Kenya and Botswana are in the process of developing nutrition labeling standards, drawing on the Codex Guidelines on Nutrition Labeling. In 2017, the Kenya Bureau of Standards presented the Draft Kenya Standard, which prescribed minimum labeling requirements of complementary foods and designated products. The standard applies to infant formula, follow-up formula, complementary foods, feeding bottles, pacifiers, spouts, breast milk fortifiers, and any other product that may be declared as a designated product in accordance to Breast Milk Substitutes (Regulations and Control) Act, 2012 of the laws of Kenya (KEBS, 2017). In Botswana, labeled food products in cans, jars, or boxes have to follow the labeling requirements laid down in the EU regulation 1169/2011 (Botswana Investment and Trade Center, 2020).

Nutrition labeling regulations in Latin American range from no regulations on nutrition labeling as in El Salvador and Guatemala) to mandatory labeling in Brazil (Hawkes, 2004, Table 20.1). Brazil passed legislation mandating labeling on all prepackaged foods in 2001 (Hawkes, 2004). In Venezuela, only foods for special dietary use are required to include nutrient information on the label. Nutritional information must be expressed as a percentage of the recommended daily requirement set by the National Nutrition Institute (INN). The label requires information on: energy value in calories, protein, digestible carbohydrate and fat contents (expressed in grams), and amounts of any nutrients claimed to have special nutritional value; for vitamins A and D, nutrient content must be expressed in international units, and for all other vitamins and minerals, nutrient content must be expressed in milligrams or micrograms. Baby foods, seasonings, colorings, flavorings, health claims, processed foods containing grain gluten, and products that may cause allergies must be labeled (Gilbert, 2015). India is preparing to finalize its new food labeling regulation. In 2019, the Food Safety and Standards Authority of India (FSSAI) indicated its intention to update to the new food labeling regulation (Verisk 3E, 2019). The draft proposed a mandatory declaration for products with high levels of fat, sugar, and salt, compelling manufacturers of prepackaged products to declare when the amounts of trans fat, sugar, or sodium exceed the specified threshold values on the FoP (Verisk 3E, 2019). FSSAI also stated that it expects to implement these measures over a 3-year phase-out period (Verisk 3E, 2019). Table 20.1 shows nutrition labeling status in several other countries.

20.6 Nutrition labeling in different countries

In the current global and economic framework, the need exists for a common framework of nutrition labeling so that products can be sold competitively worldwide. Harmonization of labeling regulations is desirable, although this should not supersede national standards. Such harmonization will make a positive contribution to global movements of foods and prevent the exploitation of different national labeling regulations (Marks, 1984). Even though there is variation among countries in the extent of labeling, most developed countries have evolved a similar minimum set of regulations.

For example, within the European Union (EU) and in various countries with regular trade links, have uniformity in the types of information which should appear on all prepackaged foods. As from December 2016, Regulation (EU) No 1169/2011 requires the vast majority of prepacked foods to bear a nutrition declaration. It must provide the energy value, the amounts of fat, saturates, carbohydrate, sugars, protein, and salt of the food. The declaration must be presented in a legible tabular format on the packaging (European Commission, 2020b).

The following demonstrates some of the similarities and minor differences between the nutrition labeling systems in Canada and the U.S. (Canada, 2003).

- Trans Fat: The *trans* fatty acid declaration is mandatory on Canadian and U.S. labels. Both countries require food manufacturers to list *trans* fatty acids, or *trans* fat, on the Nutrition Facts panel. However, the U.S. did not establish a reference standard for the sum of saturated and *trans* fats or for *trans* fats on their own, thus no % DV is declared in their table.
- Percent DV for Mandatory Vitamins and Minerals: In both countries, vitamins, and minerals must be declared as % DV. However, in the U.S. and Canada, the % DV is based on the 1968 U.S. RDIs and 1983 Recommended Daily Intakes for Canadians, respectively. There are differences in the DVs for 14 vitamins and minerals.
- Protein: The U.S. requires a DV for protein when a food is destined for children less than 4 years of age or when the protein is of low value, while this is not essential in Canada.
- Rounding to Zero: In Canada, total fat may be rounded to "0" only when the product contains $\leq$0.5 g fat and $\leq$0.2 g saturated and *trans* fats.

- Servings per container: This is mandatory and optional in U.S. and Canada, respectively. In Canada, a declaration of servings per container on the basis of "cups" or "tablespoons" is prohibited. The measuring systems in Canada (1 cup = 250 mL) and the U.S. (1 cup = 240 mL) are different.

20.7 Consumer understanding and use of nutrition labels

Nutrition Facts panel provides one of the most readily available sources of food choice information for identification of the relative healthfulness of packaged foods available to consumers (Blitstein and Evans, 2006; Hawthorne et al., 2006). Generally, the Nutrition Fact panel, which often includes amounts of protein, carbohydrate, and fat, is presented in tabular form on a food product's package (Vischers and Siegrist, 2009). The Nutrition Facts section defines a serving size and describes the weights of macronutrients (fat, carbohydrate, protein) in a serving. Fig. 20.15 demonstrates the details of the nutrition labeling information.

The intent is that people use and understand the information in the Nutrition Facts panel to assist them in making decisions about their dietary choices to reduce risk for chronic diseases such as some types of cancer, diabetes, obesity, and CVD (Vischers and Siegrist, 2009; U.S Food and Drug Administration, 2003). However, nutrition information is often misinterpreted (Pelletier et al., 2004; Hogbin and Hess, 1999). The European Heart Network (2003) and Higginson et al. (2002a,b) also noted that consumers pay little attention to nutrition labeling, although they frequently indicate doing so.

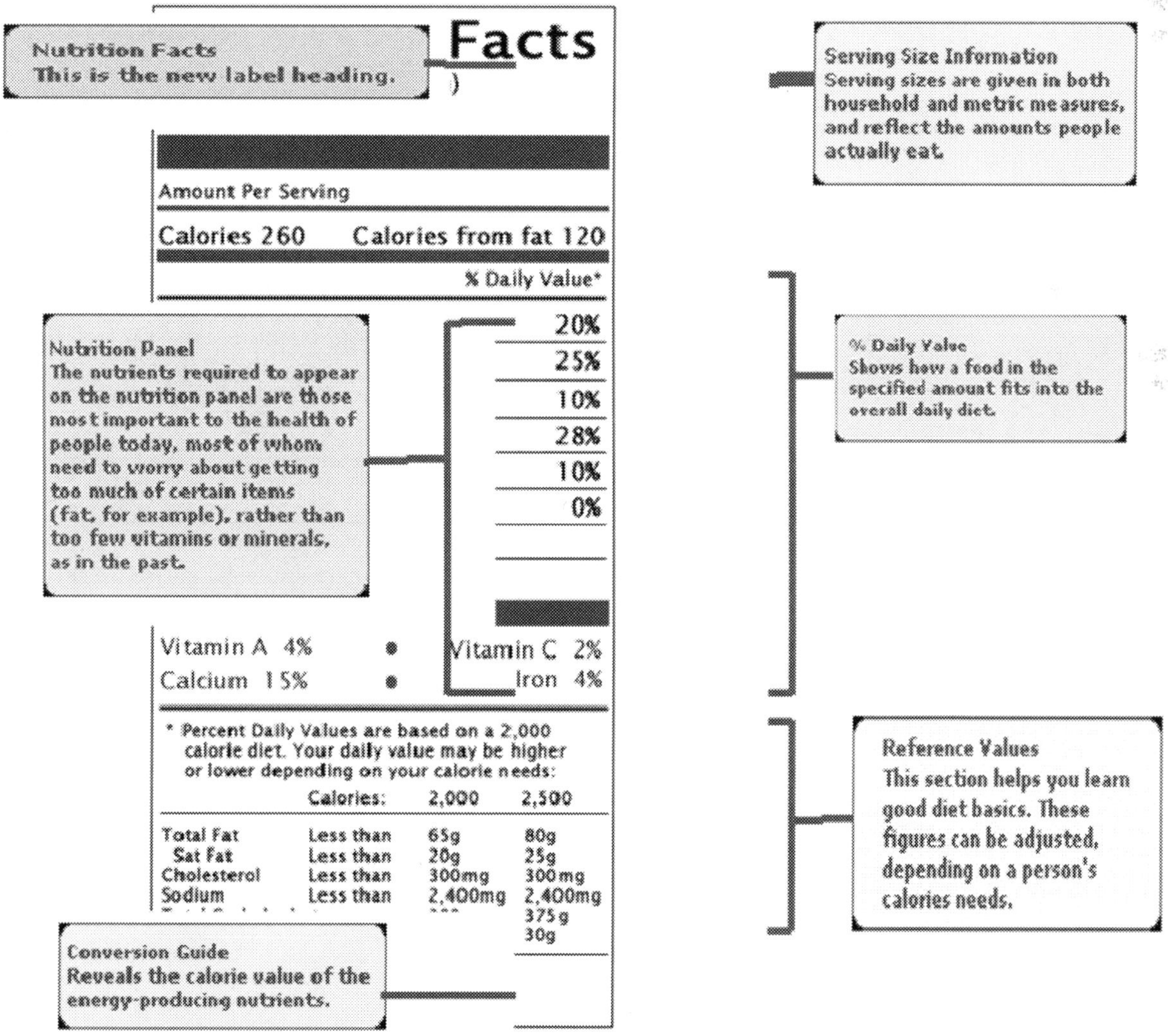

FIGURE 20.15 Details of the nutrition facts panel of a typical label. *2003. Guidance on How to Understand and Use the Nutrition Facts Panel on Food Labels, Food and Drug Administration Center for Food Safety and Applied Nutrition, Washington D.C. http://www.cfsan.fda.gov/~dms/foodlab.html.*

One reason consumers have given for not reading the nutrition label is poor comprehension. In one study, 70% of adults desired easier to understand food labels (Kristal et al., 1998; European Heart Network, 2003; Grunert and Wills, 2007).

In a systematic review of ≥100 studies, Cowburn and Stockley (2005) reported that nutrition labeling was confusing, with many consumers having difficulty interpreting serving size information. The European Heart Network (2003) reported that consumers are able to correctly use nutrition panel information for simple calculations and comparisons; however, calculating the total nutritional value of a product was more challenging for them. According to the Institute of Medicine (2003), consumers are expected to understand that calculation of sugar content must include the sugar listed on the snack food label as well as the fructose and corn syrup.

Furthermore, it was reported by Scheibehenne et al. (2007) that consumers' choice of food products were based on a single attribute rather than a multiattribute comparison of these products. Rothman et al. (2006) examined patient comprehension of food labels and the relationship of comprehension to their underlying literacy and numeracy skills. Respondents (89%) reported using food labels and 69% answered the food label questions correctly. Some reasons for inaccurate responses included misapplication of the serving size, confusion due to extraneous material on the food label, and incorrect calculations.

Huang et al. (2004) studied the relationship between reading nutrition labels and percent calorie intake from fat. In adolescent boys and girls, reading nutrition labels was associated with higher fat intake and no difference by frequency, respectively. Nutrition label reading does not necessarily translate into healthier diet in adolescents (Huang et al., 2004). Likewise, McCullum and Achterberg (1997) found that nutrition labels, compared with taste, habit, and price, ranked very low in how adolescents determine their food choices. This may reflect a lack of understanding of label information or inability to translate it into practical use.

Eight consumer focus groups in four U.S. cities were utilized to understand consumer interest in nutrition information on food labels and quick-service restaurant menu boards (Lando and Labiner-Wolfe, 2007). The results indicated that although participants were interested in having nutrition information available, they would not use it at every eating occasion. Respondents thought that food products typically consumed at one eating occasion should be labeled as a single serving; and they also indicated that more healthful options should be highlighted on labels and menu boards (Lando and Labiner-Wolfe, 2007).

In some instances, the nutrition label has been viewed as a useful educational tool, which helps consumers plan their diets. For example, Crane et al. (1999) stated that 48% and 30% of survey respondents reported changing their minds about buying or using a food product after reading the nutrition label. Jay et al. (2009) studied a multimedia intervention to improve food label comprehension in a sample of low-income patients in New York City. They concluded that a multimedia intervention is an effective way to improve short-term food label comprehension in patients with adequate health literacy. Further research is necessary to improve understanding of food labels in patients with limited health literacy.

Hawthorne et al. (2006) assessed the understanding and response of young adolescents to an educational program about Nutrition Facts panels. Initially, 55% of pretest questions were answered correctly, but scores increased to 70% in the posttest after the educational intervention. The researchers concluded that young adolescents could learn how to read and understand the Nutrition Facts labels through educational sessions. Sociodemographic variables and beliefs about the causes of obesity associated with reported use of Nutrition Facts panel information were examined by Blitstein and Evans (2006). 53% of the sample reported using NFP information on a consistent basis. Females, those with more education, and those currently married, were more likely to use NFP labels. The importance of knowledge in order to maintain healthy body weight was the only belief variable associated with use of nutrition labeling information.

Byrd-Bredbenner et al. (2000) evaluated and compared the abilities of 50 women residing in the United Kingdom (UK) to locate and manipulate information and to assess the accuracy of nutrient content claims on Nutrition Facts panels prepared in accordance with U.S. regulations, EU Directive, and UK food labeling Regulations 1996. The findings indicated that the women could locate and manipulate information on both labels equally well. However, they were significantly more able to assess nutrient content claims using the Nutrition Facts panel (U.S.). The researchers concluded that EU labeling changes, which may facilitate consumer use of labels in making dietary planning decisions, were warranted. Gorton et al. (2008) investigated understanding and preferences regarding nutrition labels among ethnically diverse shoppers at 25 supermarkets in New Zealand. The survey instrument assessed nutrition label use, understanding of the mandatory NIP, and preference for and understanding of four nutrition label formats. 66%−87% of the respondents reported reading food labels always, regularly, or sometimes. There were ethnic differences in ability to use the NIP to determine if a food was healthy.

Scott and Worsley (1997) determined the effectiveness of nutrition panel regulations using a nationwide postal survey in New Zealand to examine consumers' opinions and search behaviors relating to nutrition labeling. Of the 300 respondents, 60% claimed to have read food package labels for nutrition information in the last 10 days; most people sought information about nutrients which are generally regarded as being present in excess in Western diets. Respondents also had good knowledge of the recommendations for those nutrients but poor understanding of, and interest in, other nutrient terms. Most respondents were in favor of compulsory nutrition labeling on packaged foods. The researchers concluded that many consumers felt the information supplied in the nutrition panel was irrelevant and suggested further research to test how appealing and understandable the basic nutrition panel format is compared to other information on labels.

20.7.1 Front-of-pack nutrition labeling system

20.7.1.1 Describing front-of-pack labeling

FoPL was first introduced in the late 1980s by nonprofit organizations and government agencies (WHO, 2019). Front of Pack labels (FoPL) present abridged nutrition information on the front of the package. It has been proposed that FoPL are more visible than the conventional labeling on the backs of package (BOP), or the Nutrition Facts Panel (NFP) on the side panel. The WHO recommends FoPL labeling as a policy tool to tackle the global epidemic of obesity and diet-related noncommunicable diseases (Kanter et al., 2018; WHO, 2014). The WHO (2019) has outlined three major attributes of the FoPL: (i) presentation on the front of food packages; (ii) comprise an underlying nutrient profile model with consideration for the overall nutrient quality of the products, or the nutrients of concern for chronic diseases; and (iii) presentation of simple, usually graphic information on the nutrient content or nutritional quality of products to complement the more detailed nutrition declarations on the back of food packages. Thow et al. (2019) have broadly defined FoPL as the presentation of supplementary nutrition information on the main (front) panel of prepackaged foods which assist consumers in better understanding the nutritional quality of the product. FoPL is part of the comprehensive package of measures to improve diets and prevent obesity and noncommunicable diseases recommended by the WHO.

Several countries have implemented a variety of approaches to FoPL. Codex Alimentarius Commission (2017) Electronic Working Group defined FoPL as any system that would present simplified nutritional information, particularly on front of pack, with the intention of increasing the consumer's understanding of the overall nutritional value of the food, to assist in interpreting the nutrient declaration. Hawkes et al. (2013) and Van Trijp (2009) defined FoPL as the inclusion of simplified nutrition information on the front of food packages, further stating that it is a cost-effective strategy to help consumers make healthier choices and reduce the risk for chronic diseases at the population level. The Australian Medical Association describes FoPL as a simple and easy way to interpret and compare information about a food or drink product on the front of the product package (AMA, 2020). In addition, FoPL also communicates immediate information to consumers who would not habitually read food labels (AMA, 2020). Several other definitions are cited in the literature.

20.7.2 Global situation of FoPL

The WHO recommends FoPL as a policy tool to tackle the global epidemic of obesity and diet-related noncommunicable diseases (Kanter et al., 2018). Nevertheless, there is a lack of harmonization on mandatory FoPL and the specific format to be used when utilizing it (Kanter et al., 2018). In 2018, the Codex Alimentarius Commission agreed to develop guiding principles for FoPL (FAO, 2018). Guidance from Codex is likely to have a significant impact on global adoption of FoP nutrition labeling (Thow et al., 2019). At the 47th Session of Codex Alimentarius Commission Electronic Working Group, it was reported that in terms of FoPL, 20, eight, three and six countries have at least, one system implemented, one system proposed, systems implemented and proposed, and no systems defined, respectively (Codex Alimentarius, 2017). From the member countries responses, it was revealed that there were 16 different implemented FOPL systems in 23 countries. Of these, 6 and 10 were informative and interpretive, respectively. Interpretive system includes symbols, color codes, and graphic representations that facilitate interpretation by the consumer, while Informative (or noninterpretive) system only transfers part of the information considered relevant from the nutrient declaration, with no interpretation (Codex Alimentarius Commission, 2017).

Currently, more than 40 countries globally have some type of FoPL scheme in place. The majority of developing countries and other countries have implemented FoPL on a voluntary and mandatory basis, respectively (European Commission, 2020c). Countries such as the UK and France have implemented methods to communicate additional nutritional information to consumers in an easily understood manner. Warin (2016) and Emrich et al. (2017) have reported that the uniform color-coded interpretative labeling implemented in the UK and France has the potential to enhance consumers' understanding of nutritional information, and help them to make healthier food choices.

According to Goiana-da-Silva et al. (2019), the French Ministry of Health (MOH) developed and established the Nutri-Score model as the national reference model for food labeling. Other authors further stated that more than 90 food manufacturers and retailers have committed to use the NutriScore labels on their products (Goiana-da-Silva et al., 2019). The Belgian and Spanish MOH have also announced their endorsement of the NutriScore model (Goiana-da-Silva et al., 2019). Table 20.7 shows some of the different formats utilized for FoPL around the world. To reiterate, various countries in the developed or developing world have adopted regulations for FoPL, whether mandatory or voluntary. However, in the U.S., there are limited FoLP plans, and uncertainty about the future of a mandatory FoLP system (Becher et al., 2019). In the U.S., the FDA allows the voluntary FoPL Facts Up Front developed by the Grocery Manufacturers Association (GMA) and the Food Marketing Institute (FMI) (Becher et al., 2019). The Facts Up Front labeling scheme features essential information from the Nutrition Facts panel (Fig. 20.16).

20.7.3 Future directions of FoLP

According to WHO (2019), the principal aim of FoPL is to provide convenient, relevant, and readily understood nutrition information or guidance on food packs to assist consumers in making informed purchases and healthier eating choices. The WHO (2019) further stated that effective implementation of FoPL requires fundamental messaging on their use, vigorous consumer education, dietary advice, and country-specific nutrition tutelage. WHO's aim and underlying principles of the FoPL are inferring that simple-to-use FoPL is an excellent component of any nutrition labeling strategy to help improve dietary patterns and reduce the burden of diet-related chronic diseases.

Worldwide, organizations, whether government or nongovernmental, developing FoPLs utilize a gamut of systems and different strategies to promote consumer awareness, understanding, and use. The impact of the variety of systems and strategies on consumers' use of FoPLs remain unclear. Indeed, research on the ability of FoPLs to encourage more healthful food choices among consumers is still limited. Furthermore, Mørk et al. (2017) posited that with the limited available research on FoPLs, their effect would be minor without robust consumer education as stated by the WHO. Other studies have also provided evidence that FoPL nutrition labeling should be complemented with communication campaigns. Such campaigns can provide information about how to use the labeling scheme, guide interpretation, and encourage citizens to take nutrition labeling into account in their food choices (Ares et al., 2018). Research on the effects of the FoPLs on consumers and global harmonization of the formats, strategies, and support measures could therefore be a valuable contribution to help improve dietary patterns and reduce the burden of diet-related chronic diseases.

20.8 Bioavailability and nutrition label

Effective nutrition labels are part of a supportive environment, which encourages healthier food choices. However, globally, nutrition labels often focus on the total amount of nutrients in the particular product, giving no indication of the bioavailability of these nutrients once they are ingested. The bioavailability of the nutrients ingested in a product completes the whole picture about how nutritionally sound the product is. Nutrient content on a nutrition label does not guarantee its nutritional value. Nutrients in foods and food products can exist in different forms, which result in diverse levels of bioavailability; therefore, food manufacturers should be encouraged to provide information on the bioavailability of the nutrient. Additional information about bioavailability on nutrition labels would be more beneficial to consumers and enable them to make healthier food choices.

Bioavailability has no universally accepted definition; it has been defined in many ways. The FDA has defined bioavailability as the rate and extent to which the active substances or therapeutic moieties contained in a drug are absorbed and become available at the site of action (Shi and Le Maguer, 2000). Jackson (1997) described bioavailability as the "fraction of an ingested nutrient that is available for utilization in normal physiologic functions and for storage," although it should be recognized that there is no true storage of a water-soluble vitamin such as folate. Gregory et al. (2005) described bioavailability as the product of absorption efficiency (that is, fractional absorption, F1) and postabsorptive events involved in metabolic utilization (F2).

According to Fairweather-Tait (1992), bioavailability is the proportion of the total nutrient in a food, meal, or diet that is utilized for normal body functions, which involves various stages, each of which is affected by different dietary and physiological factors (Fig. 20.17). Briefly, factors which impact upon the bioavailability of nutrients include the quantity consumed including the chemical forms, dietary composition, and gastrointestinal secretions; the quantity available for absorption, which is affected by factors such as nutritional status and gut microflora; food preparation or processing; host factors; and the quantity absorbed. However, bioavailability still remains an important, but often vague notion linked to the effectiveness of absorption and metabolic utilization of consumed nutrients (Gregory et al., 2005; Solomons, 2001).

TABLE 20.7 Attributes of global front-of-pack labeling (FoLP) schemes.

FoPL format	Country	Status	Description
Traffic Light Labeling System	United Kingdom South Korea Ecuador India	Mandatory Voluntary Mandatory Considering mandatory	This system uses red, amber, and green signals to show consumers, at a glance, whether a product is high, medium, or low in fat, saturated fat, sugar, salt (and possibly overall energy), making it easy to identify healthier food choices by choosing products with green or amber lights, rather than red (AMA, 2020).
Healthy Star Rating	Australia New Zealand	Voluntary Voluntary	Rates the overall nutritional profile of packaged food and assigns it a rating from one-half star to five-star rating of nutritional quality via the front facings of food product packages. It provides a quick, easy, standard way to compare similar packaged foods. The more stars, the healthier the choice (Hamlin and McNeill, 2016).
Healthier Choices Logo	Malaysia Singapore Thailand Nigeria Zimbabwe Brunei	Voluntary Voluntary Voluntary Voluntary Voluntary Voluntary	Part of a bigger project, which motivates industries to reformulate food products and assist the consumers in making right food choices (Kanter et al., 2018). Food products may carry the Healthier Choices Logo if they meet the nutrient criteria specified by their respective Ministry of Health. Point system in six categories from worst = 0 to best = 5. Nutrients involved are sodium, total sugar, energy, total fat, saturated fat, protein, calcium, iron, trans fatty acid, and fiber (Kanter et al., 2020).
Warning System	Chile Mexico Uruguay Colombia Brazil Peru Canada Israel	Mandatory Mandatory (10/2020) Mandatory Adoption of the model with black octagons Mandatory Mandatory Mandatory Mandatory Mandatory	A system of black warning labels shaped like stop signs for food and drinks that exceed limits for three critical nutrients: sodium, saturated fats, total sugars, and total energy in kilocalories (Obesity Evidence Hub, 2020) Canada is proposing to set limits per serving rather than per 100 g and does not include energy (Kanter et al., 2018).
Nutri-Score	France Germany Spain Luxembourg	Voluntary Voluntary Voluntary Voluntary	A color coded and with letters from A to E, based on the United Kingdom profiling model with A the healthiest. The Nutri-score system indicates the overall nutritional quality of a given food item (Ereño, 2020; Kanter et al., 2018).
Keyhole System	Scandinavia Denmark Sweden Iceland	Voluntary Voluntary Voluntary Voluntary	A voluntary health symbol, the purpose of which is to highlight a "better choice" of food products within a product category. The Keyhole is awarded if the product meets the criteria established by the government in a nutrient profile model. It is integrated information about a range of nutrients; provides only a positive evaluative judgment to indicate which foods are "better for you" choices, often within a food category (Ereño, 2020; Kanter et al., 2018). Nutrients included saturated fat, total sugars, sodium, and fiber (Kanter et al., 2018; Mørk et al., 2017).
Positive Logo System	Finland Slovenia	Voluntary Voluntary	Ereño (2020)
NutrInform Battery	Italy	Voluntary	It is based on reference intake label indicating the amounts of energy and nutrients in a single serving as percentage of the daily intake (Ereño, 2020).
Choices Program	Netherlands Belgium Poland Czech Republic	Voluntary Voluntary Voluntary Voluntary	Nutrients included: saturated fat, trans fatty acids, added sugar, sodium, and energy (Kanter et al., 2018).

Continued

TABLE 20.7 Attributes of global front-of-pack labeling (FoLP) schemes.—cont'd

FoPL format	Country	Status	Description
Facts Up Front	United States of America	Voluntary	A simple and easy-to-use labeling system that displays key nutrition information on the front of food and beverage packages. It will add important nutrition information on calories, salt, sugar, and saturated fat to the front of packages of many food and beverage products. It can also include up to two other nutrients to encourage, such as potassium, fiber, vitamin A, vitamin C, vitamin D, calcium, iron, or protein when a product meets a "good source of" threshold. The information will be presented in a standardized, fact-based, simple, and easy-to-use format (Kennedy, 2018; Becher et al., 2019).

FIGURE 20.16 Example of the United States facts up front FoPL (front of pack nutrition label). *Taken from http://www.factsupfront.org/AboutTheIcons.html.*

Minerals, which are crucial for optimal health, are the most well-researched example when discussing bioavailability. Microminerals, which have been established to be either essential or beneficial for humans or animals, include iodine iron, selenium, and zinc (House, 1999). This section discusses calcium and iron bioavailability issues to highlight the importance of disclosing nutrient bioavailability information on nutrition labels. Heaney et al. (2005a) have demonstrated that calcium in products is absorbed with different efficiencies. The researchers fed women commercially marketed calcium-fortified orange juice and compared the bioavailability of calcium from calcium citrate malate and a combination of tricalcium phosphate and calcium lactate. The increase in calcium in the serum was measured up to 9 hours after ingestion of the test drink. Their results indicated that calcium citrate malate was better absorbed (more bioavailable to the body) than tricalcium phosphate/calcium lactate (Heaney et al., 2005a). Heaney et al. (2005b) have also reported that tricalcium phosphate in soy beverage was 25% less bioavailable than calcium in cow's milk.

Human diets contain two basic forms of iron: Fe(II) and Fe(III) and heme where iron is complexed to protoporphyrin IX (Bleackley et al., 2009; Benito and Miller, 1998). However, iron exists in oxidation states ranging from −II, as in the $Fe(CO)_4^{2-2}$ anion, to + VI, as in the ferrate ion FeO_4^{2-2}. Biochemically, Fe(II) and Fe(III) are the most relevant oxidation states (Bleackley et al., 2009). In an average diet, inorganic iron accounts for approximately 90% of total dietary iron content, whereas heme makes up the remaining 10% (Sharp and Srai, 2007; Anderson et al., 2005). Iron absorption varies significantly with diet composition, iron status of the individual, and, most importantly, bioavailability of the different iron forms. Bioavailability of inorganic iron is dependent on other dietary components. Enhancers, such as ascorbic acid, increase inorganic iron bioavailability by promoting the reduction of Fe(III) to soluble Fe (II). Inhibitors such as phytates in cereal and polyphenols in plants form insoluble complexes with inorganic iron and thereby reduce its absorption (Sharp and Srai, 2007). Additionally, inorganic iron absorption is influenced by the level of gastric acid secretion in the

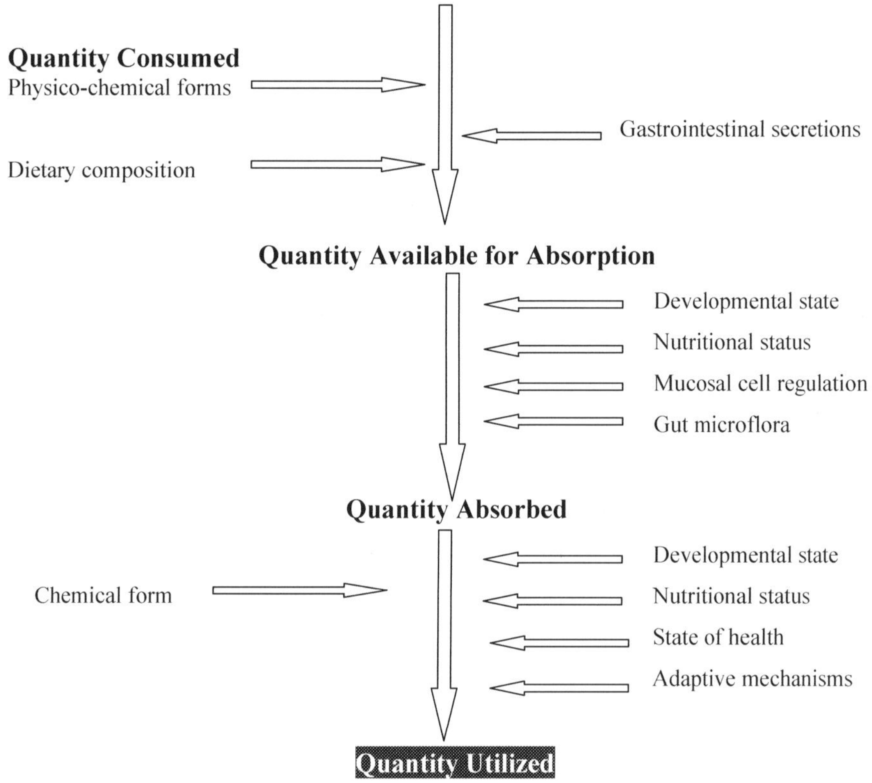

FIGURE 20.17 The different stages of bioavailability. *Taken from Fairweather-Tait, S.J., 1992. Bioavailability of trace elements. Food Chem. 43, 213–217.*

stomach with acidity increasing the solubility of inorganic iron. Dietary iron absorption occurs in the proximal small intestine by specialized epithelial cells called duodenal enterocytes.

Several researchers have shown that iron is absorbed with different efficiencies. Iron is only poorly absorbed from high-extraction flours because of the presence of phytate and other inhibitory factors (Hurrell et al., 2002, 2004). Findings from an efficacy trial in Thailand suggest that two forms of elemental iron, electrolytic iron and hydrogen-reduced iron, have bioavailability which is 50%–79% that of ferrous sulfate (Zimmerman et al., 2005). Two other forms of reduced iron, carbon monoxide–reduced and atomized iron, are poorly absorbed (Zimmerman and Hurrell, 2007). Sodium iron ethylenediaminetetraacetic acid (NaFeEDTA) has shown effectiveness as a fortificant in soy sauce in China (Huo et al., 2002), fish sauce in Vietnam (van Thuy et al., 2003), and maize flour in Kenya (Andag'o et al., 2007). NaFeEDTA is absorbed twice to three times more than ferrous sulfate from diets high in phytic acid (Bothwell and MacPhail, 2004).

20.9 Conclusion

Nutrition labels are intended to educate consumers regarding the extent of salt, sugar, fat, cholesterol, minerals, some vitamins, protein, and other nutrients in processed foods. The intent is that people use and understand the information in the Nutrition Facts panel and FoPL to assist them in making decisions about their dietary choices and to reduce risk for chronic diseases such as some types of cancer, diabetes, obesity, and CVD. It is crucial that nutrition labeling/FoPL is combined with education on healthy lifestyles, including clear advice about the contribution that all foods make to a healthy diet, and the importance of physical activity.

Food labels are one of the most immediate and direct sources of information, which can be supported with education and advertising. The Nutrition Facts box that appears on food labels was conceived as an important public health tool to reduce diet-related disease. Simple-to-use FoPL is an excellent component of any nutrition labeling strategy to help improve dietary patterns and reduce the burden of diet-related chronic diseases. Globally, there are four main types of nutrition labeling regulatory environment. The U.S., Canada, Australia, and New Zealand have implemented mandatory nutrition labeling, although the format and content of the labels differ in these countries. In the EU nutrition labeling is currently not mandatory unless a nutrition claim is made. Several developing countries have adopted the *Codex Alimentarius* guidelines for nutrition labeling. FoPL is mandatory in some, but not all countries in the world. The situation is the same for voluntary FoPL. More research on the effects of the FoPLs on consumers and global harmonization of the formats, strategies, and support measures could be a valuable contribution to help improve dietary patterns and reduce the burden of diet-related chronic diseases globally.

Nutrition labels including FoPL will be more useful as an important public health tool to reduce diet-related disease if they provide estimates of bioavailability in addition to nutrient. A main drawback with nutrition labeling is that the nutrient content is not supported by information regarding its bioavailability. Globally, nutrition labels often focus on the total amount of nutrients in the particular product, giving no indication of the bioavailability of these nutrients once they are ingested. The bioavailability of the nutrients ingested in a product completes the whole picture about how nutritionally sound it is. Nutrient content on a nutrition label does not guarantee its nutritional value. Nutrients in foods and food products can exist in different forms, which result in diverse levels of bioavailability; therefore, food manufacturers should be encouraged to provide information on the bioavailability of the nutrient listed on the label.

20.10 Future scope

When consumers read nutrition labels, it is assumed that the amount listed in the % DV is the amount that is utilized by the body; however, this is not always the case. The U.S. FDA and other global regulatory institutions should initiate and/or continue discussions regarding the inclusion of bioavailability information of nutrients, especially micronutrients on nutrition labels. Harmonization of labeling regulations is desirable especially in the current global and economic framework. A common framework of nutrition labeling complete with bioavailability information will contribute to products being sold more competitively worldwide. Harmonization will make a positive contribution to global movements of foods and help to minimize the exploitation of different national labeling regulations. Further research is needed to determine how appealing and understandable the basic nutrition panel format is compared to other information on labels. Additionally, more research on the effects of the FoPLs on consumers and global harmonization of the formats, strategies, and support measures could be a valuable contribution to help improve dietary patterns and reduce the burden of diet-related chronic diseases globally. Research is also needed to provide additional information about the inclusion of nutrient bioavailability information on nutrition labels and its benefit to consumers in terms of making healthier food choices.

Acknowledgments

The authors wish to thank Mr. Larry Keener for the pivotal role he played in making this chapter a reality and also the Food and Nutritional Sciences Advisory Board, Department of Food and Nutritional Sciences, Tuskegee University for their role in making the chapter possible.

References

AMA, 2020. Traffic Light Food Labelling – the Evidence. Australian Medical Association Limited. ABN 37 008 426 793. Available at: https://ama.com.au/system/tdf/documents/Traffic%20Light%20Food%20Labelling%20-%20The%20Evidence.pdf?file=1&type=node&id=36221 (Accessed 9 October 2020).

Andag'o, P.E., Osendarp, S.J., Ayah, R., et al., 2007. Efficacy of iron-fortified whole maize flour on iron status of schoolchildren in Kenya: a randomised controlled trial. Lancet 369, 1799–1806.

Anderson, G.J., Frazer, D.M., McKie, A.T., et al., 2005. Mechanisms of haem and non-haem iron absorption: lessons from inherited disorders of iron metabolism. Biometals 18, 339–348.

Ares, G., Aschemann-Witzel, J., Curutchet, M.R., Antúnez, L., Moratorio, X., Bove, I., 2018. A citizen perspective on nutritional warnings as front-of-pack labels: insights for the design of accompanying policy measures. Publ. Health Nutr. 21, 3450–3461.

Becher, S., Gao, H., Lai, J., Harrison, A., 2019. Improving Front-of-Package Food Health Labeling. The Regulatory Review. Available at: https://www.theregreview.org/2019/05/03/becher-gao-lai-harrison-front-package-food-labeling/ (Accessed 9 Novmber 2020).

Benito, P., Miller, D., 1998. Iron absorption and bioavailability: an updated review. Nutr. Res. 18, 581–603.

Bleackley, M.R., Wong, A.Y.K., Hudson, D.M., Wu, C.H.-Y., MacGillivray, R.T.A., 2009. Blood iron homeostasis: newly discovered proteins and iron imbalance. Transfus. Med. Rev. 23, 103–123.

Blewett, N., Goddard, N., Pettigrew, S., Reynold, C., Yeatman, H., 2011. Labelling Logic – the Final Report of the Review of Food Labelling Law and Policy. http://www.foodlabellingreview.gov.au/internet/foodlabelling/publishing.nsf/content/labelling-logic (Accessed 9 June 2020).

Blitstein, J.L., Evans, D.W., 2006. Use of Nutrition facts panels among adults who make household food purchasing decisions. J. Nutr. Educ. Behav. 38, 360–364.

Bothwell, T.H., MacPhail, A.P., 2004. The potential role of NaFeEDTA as an iron fortificant. Int. J. Vitam. Nutr. Res. 74, 421–434.

Botswana Investment and Trade Centre, 2020. Food Labelling. Available at: https://www.gobotswana.com/food-labelling (Accessed 9 September 2020).

Byrd-Bredbenner, C., Wong, A., Cottee, P., 2000. Consumer understanding of US and EU nutrition labels. Br. Food J. 102, 615–629.

Canada, 1986. Nutrition Labelling. Information Letter No. 713, July 24. Food Directorate, Health Protection Branch, Ottawa.

Canada, 1989. Guidelines on Nutrition Labelling. Food Directorate Guideline No. 2, November 30. Food Directorate, Health Protection Branch.

Canada, 2003. SOR/2003-11. Regulations amending the Food and Drug Regulations (Nutrition labeling, nutrient content claims and health claims). Canada Gazette, Part II 137, 154–405.

China Dragon, 2020. Hong Kong Food Labeling Consultation (For Foods Sold in Hong Kong. Available at: https://www.cdichk.com/index.php?option=com_content&view=article&id=24&Itemid=40&lang=en# (Accessed 9 May 2020).

China-Britain Business Council, 2020. China Labelling Regulations. Available at: http://www.cbbc.org/cbbc/media/cbbc_media/4.%20Files/CBBC-Labelling-in-China.pdf (Accessed 9 May 2020).

Codex Alimentarius Commission, 2003. Codex Alimentarius. Welcome. Online. http://www.codexalimentarius.net.

Codex Alimentarius Commission, 2017. Joint FAO/WHO food standards programme. Codex Committee on food labelling. In: 47th session, Asunción, Paraguay, 16–20 October 2017.

Cowburn, G., Stockley, L., 2004. Consumer understanding and use of nutrition labeling: a systematic review. Publ. Health Nutr. 8, 21–28.

Crane, N.T., Hubbard, V.S., Lewis, C.J., 1999. American diets and year 2000 goals. In: Frazao, E. (Ed.), America's Eating Habits: Changes and Consequences. US Department of Agriculture, Washington, DC, pp. 111–133. Agriculture Information Bulletin No. 750.

Curran, M.A., 2002. Nutrition labeling perspectives of a bi-national agency for Australia and New Zealand. Asia Pac. J. Clin. Nutr. 11, S72–S76.

Emrich, T.E., Qi, Y., Lou, W.Y., Abbe, M.R., 2017. Traffic-light labels could reduce population intakes of calories, total fat, saturated fat, and sodium. PLoS One 2017.

Ereño, D.P., 2020. Global Front-of-Pack Nutrition Labeling Schemes: Impact on Marketing Strategies. Regulatory Focus. Available at: https://www.raps.org/news-and-articles/news-articles/2020/6/global-front-of-pack-nutrition-labeling-schemes-im (Accessed 9 October 2020).

European Commission, 2020a. Which Nutrition Information Is Mandatory on Food Labels? Available at: https://ec.europa.eu/food/safety/labelling_nutrition/labelling_legislation/nutrition (Accessed 9 May 2020).

European Commission, 2020b. Nutrition Labelling. Available at: https://ec.europa.eu/food/safety/labelling_nutrition/labelling_legislation/nutrition-labelling_en (Accessed 9 September 2020).

European Commission, 2020c. Report from the Commission to the European Parliament and the Council Regarding the Use of Additional Forms of Expression and Presentation of the Nutrition Declaration. Available at: https://ec.europa.eu/food/sites/food/files/safety/docs/labelling-nutrition_fop-report-2020-207_en.pdf (Accessed 9 October 2020).

European Heart Network, 2003. A Systematic Review on the Research on Consumer Understanding of Nutrition Labeling. European Heart Network, Brussels.

Fairweather-Tait, S.J., 1992. Bioavailability of trace elements. Food Chem. 43, 213–217.

Fairweather-Tait, S.J., Johnson, A., Eagles, J., Ganatra, S., Kennedy, H., Gurr, M.I., 1989. Studies on calcium absorption from milk using a double-label stable isotope technique. Br. J. Nutr. 62, 379–388.

FAO and WHO, 2007. Guidelines on Nutrition Labeling CAC/GL 2-1985, fifth ed. In: Food Labeling. Codex Alimentarius, FAO and WHO of the United Nations, Rome, Italy ftp://ftp.fao.org/docrep/fao/010/a1390e/a1390e00.pdf.

FAO (Food and Agriculture Organisation), 2018. New Work on Front of Pack Nutrition Labelling and Date Marking Revision to Codex Standard. Available online. http://www.fao.org/news/story/en/item/1143286/icode/ (Accessed 9 October 2020).

FAO/WHO (Food and Agriculture Organisation/World Health Organisation), 2020. Codex Alimentarius International Standards. Members. Available at: http://www.fao.org/fao-who-codexalimentarius/about-codex/members/en/ (Accessed 9 September 2020).

Food Standards Australia New Zealand, 2005. GM Foods Safety Assessment of Genetically Modified Foods Safety Assessment of Genetically Modified Foods, ISBN 0 642 34599 6. Available at: https://www.foodstandards.gov.au/consumer/gmfood/safety/documents/GM%20Foods_text_pp_final.pdf (Accessed 9 June 2020).

Food Standards Australia New Zealand, 2017. Notifying a Self-Substantiated Food-Health Relationship. https://www.foodstandards.gov.au/industry/labelling/fhr/Pages/notifications.aspx.

Food Standards Australia New Zealand, 2020a. Food Additive Labelling. Available at: https://www.foodstandards.gov.au/consumer/labelling/Pages/Labelling-of-food-additives.aspx (Accessed 9 June 2020).

Food Standards Australia New Zealand, 2020b. Federal Register of Legislation. Available at: https://www.legislation.gov.au/Series/F2015L00394 (Accessed 9 July 2020).

Gilbert, A.J., 2015. Food and Agricultural Import Regulations and Standards – Narrative Venezuela. FAIRS Country Report. USDA-FAS/GAIN Report. USDA Foreign Agricultural Service. Available at: https://apps.fas.usda.gov/newgainapi/api/report/downloadreport.

Goiana-da-Silva, F., Cruz-e-Silva, D., Miraldo, M., Calhau, C., Bento, A., Cruz, D., Almeida, F., Darzi, A., Araújo, A., 2019. Front-of-pack labelling policies and the need for guidance. Lancet 4. Available at: www.thelancet.com/public-health (Accessed 9 October 2020).

Gorton, D., Ni Mhurchu, C., Chen, M.-H., Dixon, R., 2008. Nutrition labels: a survey of use, understanding and preferences among ethnically diverse shoppers in New Zealand. Publ. Health Nutr. https://doi.org/10.1017/S1368980008004059.

Gregory, J.F., Quinlivan, E.P., Davis, S.R., 2005. Integrated the issues of folate bioavailability, intake and metabolism in the era of fortification. Trends Food Sci. Technol. 20, 229–240.

Grunert, K.G., 2013. Nutrition labeling. In: Encyclopedia of Human Nutrition, third ed. Available at: https://www.sciencedirect.com/topics/agricultural-and-biological-sciences/nutrition-labeling# (Accessed 9 May 2020).

Grunert, K., Wills, J., 2007. A review of European research on consumer response to nutrition information on food labels. J. Public Health 15, 385–399.

Hallmans, G., Nilsson, U., SjoÈstroÈm, R., Wetter, L., Wing, K., 1987. The importance of the body's need for zinc in determining Zn availability in food: a principle demonstrated in the rat. Br. J. Nutr. 58, 59–64.

Hamlin, R., McNeill, L., 2016. Does the Australasian "health star rating" front of pack nutritional label system work? Nutrients 8, 327. Available at: https://www.ncbi.nlm.nih.gov/pmc/articles/PMC4924168/ (Accessed 9 October 2020).

Hawkes, C., 2004. Nutrition Labels and Health Claims: The Global Regulatory Environment. World Health Organization, Geneva, Switzerland.

Hawkes, C., Jewell, J., Allen, K., 2013. A food policy package for healthy diets and the prevention of obesity and diet-related non-communicable diseases: the NOURISHING framework. Obes. Rev. 14, 159–168.

Hawthorne, K., Moreland, K., Griffin, I.J., Abrams, S.A., 2006. An educational program enhances food label understanding of young adolescents. J. Am. Diet Assoc. 106, 913–916.

Health Star Rating System Five Year Review Report, 2019. Available at: http://www.healthstarrating.gov.au/internet/healthstarrating/publishing.nsf/Content/D1562AA78A574853CA2581BD00828751/$File/Health-Star-Rating-System-Five-Year-Review-Report.pdf (Accessed 12 March 2020).

Heaney, R.P., Rafferty, K., Dowell, M.S., Bierman, J., 2005a. Calcium fortification systems differ in bioavailability. J. Am. Diet Assoc. 105, 807–809.

Heaney, R.P., Dowell, M.S., Rafferty, K., Bierman, J., 2005b. Bioavailability of the calcium in fortified soy imitation milk, with some observations on method. Am. J. Clin. Nutr. 71, 1166–1169.

Higginson, C.S., Rayner, M.J., Draper, S., Kirk, T.R., 2002a. The nutrition label - which information is looked at? Nutr. Food Sci. 32, 92–99.

Higginson, C.S., Kirk, T.R., Rayner, M.J., Draper, S., 2002b. How do consumers use nutrition label information? Nutr. Food Sci. 32, 145–152.

Hogbin, M.B., Hess, M.A., 1999. Public confusion over food portions and servings. J. Am. Diet Assoc. 99, 21209–21211.

House, W.A., 1999. Trace element bioavailability as exemplified by iron and zinc. Field Crop. Res. 60, 115–141.

Huang, T., Kaur, H., Mccarter, K.S., Nazir, N., Choi, W.S., Ahluwalia, J.S., 2004. Reading nutrition labels and fat consumption in adolescents. J. Adolesc. Health 35, 399–401.

Huo, J., Sun, J., Miao, H., et al., 2002. Therapeutic effects of NaFeEDTA-fortified soy sauce in anaemic children in China. Asia Pac. J. Clin. Nutr. 11, 123–127.

Hurrell, R., Bothwell, T., Cook, J.D., et al., 2002. SUSTAIN Task Force. The usefulness of elemental iron for cereal flour fortification: a SUSTAIN task force report. Nutr. Rev. 60, 391–406.

Hurrell, R.F., Lynch, S., Bothwell, T., et al., 2004. Enhancing the absorption of fortification iron. Int. J. Vitam. Nutr. Res. 74, 387–401.

Jay, M., Adams, J., Herring, S.J., Gillespie, C., Ark, T., Feldman, H., Jones, V., Zabar, S., Stevens, D., Kalet, A., 2009. A randomized trial of a brief multimedia intervention to improve comprehension of food labels: a randomized trial of a brief multimedia intervention to improve comprehension of food labels. Prev. Med. 48, 25–31.

Kanter, R., Vanderlee, L., Vandevijvere, S., 2018. Front-of-package nutrition labelling policy: global progress and future directions. Publ. Health Nutr. 21, 1399–1408.

KEBS (Kenya Bureau of Standards), 2017. Labelling of Complementary Foods and Designated Products — Requirements. Available at: http://www.iso.org/sites/wtowp/kenya/kebs_standards_work_programme_bulletin.pdf (Accessed 9 September 2020).

Kennedy, P., 2018. Regulatory Round-Up: Will the U.S. Require Front-of-Pack Nutrition Labeling to Improve Public Health? Merieux NutriSciences. Available at: http://foodsafety.merieuxnutrisciences.com/2018/04/05/regulatory-round-up-will-united-states-require-front-pack-nutrition-labeling-improve-public-health/ (Accessed 9 November 2020).

Kristal, A.R., Levy, L., Patterson, R.E., Li, S.S., White, E., 1998. Trends in food label use associated with new nutrition labeling regulations. Am. J. Publ. Health 88, 1212–1215.

Lando, A.M., Labiner-Wolfe, J., 2007. Helping consumers make more healthful food choices: consumer views on modifying food labels and providing point-of-purchase nutrition information at quick-service restaurants. J. Nutr. Educ. Behav. 39, 157–163.

Mandel, J., Tugendhaft, A., Michalow, J., Hofman, K., 2015. Nutrition labelling: a review of research on consumer and industry response in the global South. Glob. Health Action 8.

Marks, L., 1984. What is in a label? Consumers, public policy, and food labels. Food Pol. 9, 252–258.

McCullum, C., Achterberg, C.L., 1997. Food shopping and label use behavior among high school-aged adolescents. Adolescence 32, 181–197.

Ministry of Health and Quality of Life, 1998a. Food Regulations Made under the Food Act 1998: Part I - Food Composition and Labeling. Republic of Mauritius. http://ncb.intnet.mu/moh/foodreg.htm.

Ministry of Health and Quality of Life, 1998b. Food Regulations Made under the Food Act 1998: Part XV - Special Purpose Food. Republic of Mauritius. http://ncb.intnet.mu/moh/foodreg.htm.

Mørk, T., Grunert, K.G., Fenger, M., Jørn Juhl, H., Tsalis, G., 2017. An analysis of the effects of a campaign supporting use of a health symbol on food sales and shopping behaviour of consumers. BMC Publ. Health 17, 239.

Nishida, C., 2019. Global Overview: Labelling into National Nutrition Policy. Available at:: file:///D:/tglobal.overview.labelling.nutrition.policy.pdf. (Accessed 9 September 2020).

NLIP Watch, 2020. Pre-packaged Food (Labelling) Regulations. Available at: https://nlipw.com/pre-packaged-food-labelling-regulations/ (Accessed 9 September 2020).

Obesity Evidence Hub, 2020. Front-of-pack Nutrition Labelling. Available at: https://www.obesityevidencehub.org.au/collections/prevention/front-of-pack-nutrition-labelling (Accessed 9 October 2020).

Pelletier, A.L., Chang, W.W., Delzell, J.E., McCall, J.W., 2004. Patients' understanding and use of snack food package nutrition labels. J. Am. Board Fam. Pract. 17, 319–323.

Pray, W.S., 2003. A History of Nonprescription Product Regulation. The Haworth Press Inc, Binghamton, NY, pp. 205–238.
Reeves, P.G., Chaney, R.L., 2008. Bioavailability as an issue in risk assessment and management of food cadmium: a review. Sci. Total Environ. 398, 13–19.
Republic of Mauritius, 2009. National Plan of Action for Nutrition 2009-2010 Final Report. Available at: http://www.africanchildforum.org/clr/policy%20per%20country/mauritius/mauritius_nutrition_2009-2010_en.pdf (Accessed 9 September 2020).
Rothman, R.L., Housam, R., Hilary Weiss, H., Davis, D., Gregory, R., Gebretsadik, T., Shintani, A., Elasy, T.A., 2006. Patient Understanding of food labels the role of literacy and numeracy. Am. J. Prev. Med. 31, 391–398.
Scheibehenne, B., Miesler, L., Todd, P.M., 2007. Fast and frugal food choices: uncovering individual decision heuristics. Appetite 49, 578–589.
Scott, V., Worsley, A., 1997. Consumer views on nutrition labels in New Zealand. Aust. J. Nutr. Diet 54, 6–13.
Sharp, P., Srai, S.K., 2007. Molecular mechanisms involved in intestinal iron absorption. World J. Gastroenterol. 13, 4716–4724.
Shi, J., Le Maguer, M., 2000. Lycopene in tomatoes: chemical and physical properties affected by food processing. Crit. Rev. Biotechnol. 20, 293–334.
Sibbald, B., 2003. Canada's nutrition labels: a new world standard? CMAJ (Can. Med. Assoc. J.) 168, 887.
Solomons, N., 2001. Bioavailability of nutrients and other bioactive components from dietary supplements. J. Nutr. 131, 1392S–1395S.
Southgate, D.A.T., 1989. Conceptual issues concerning the assessment of nutrient bioavailability. In: Southgate, D.A.T., Johnson, I.T., Fenwick, G.R. (Eds.), Nutrient Availability: Chemical and Biological Aspects. Special Publication No. 72. Royal Society of Chemistry, Cambridge, pp. 10–12.
Taillie, L.S., Reyes, M., Colchero, M.A., Popkin, B., Corvalań, C., 2020. An evaluation of Chile's law of food labeling and advertising on sugar sweetened beverage purchases from 2015 to 2017: a before-and-after study. PLoS Med. 17 (2), 1–22. https://doi.org/10.1371/journal. Available at:
The world health report, 2002. Reducing Risks, Promoting Healthy Life. World Health Organization, Geneva. Available at: https://www.who.int/nutrition/topics/2_background/en/index2.html. accessed 09/05/2020.
Thow, A., Jones, A., Huckel Schneider, C., Labonté, R., 2019. Global governance of front-of-pack nutrition labelling: a qualitative analysis. Nutrients 11, 268. https://doi.org/10.3390/nu11020268.
U.S. Food and Drug Administration, Center for Food Safety and Applied Nutrition, 2003. How to Understand and Use the Nutrition Facts Panel Label. U.S. Department of Health and Human Services, Washington, D.C.
van Thuy, P.V., Berger, J., Davidsson, L., et al., 2003. Regular consumption of NaFeEDTA-fortified fish sauce improves iron status and reduces the prevalence of anemia in anemic Vietnamese women. Am. J. Clin. Nutr. 78, 284–290.
Van Trijp, H.C., 2009. Consumer understanding and nutritional communication: key issues in the context of the new EU legislation. Eur. J. Nutr. 48, s41–48.
Verisk 3E, 2019. India to Finalize New Food Labeling Regulation. Available at: https://www.verisk3e.com/resource-center/blog/india-finalize-new-food-labeling-regulation (Accessed 9 September 2020).
Visschers, V.H.M., Siegrist, M., 2009. Applying the evaluability principle to nutrition table information. How reference information changes people's perception of food products. Appetite 52, 505–512.
Welch, R.M., House, W.A., 1984. Factors affecting the bioavailability of mineral nutrients in plant foods. In: Welch, R.M., Gabelman, W.H. (Eds.), Crops as Sources of Nutrients for Humans. American Society of Agronomy, Madison, WI, pp. 37–54. Special Publication No. 48.
WHO Regional Office for Europe, 2014. European food and nutrition action plan 2015–2020. In: Proceedings of the 64th Session Regional Committee for Europe, Copenhagen, Denmark, 15–18 September 2014.
WHO (World Health Organisation), 2019. Guiding Principles and Framework Manual for Front-of-Pack Labelling for Promoting Healthy Diet. Available at: www.guidingprinciples-labelling-promoting-healthydiet.pdf (Accessed 9 November 2020).
Zimmerman, M.B., Hurrell, R.F., 2007. Nutritional iron deficiency. Lancet 370, 511–520.
Zimmermann, M.B., Winichagoon, P., Gowachirapant, S., et al., 2005. Comparison of the efficacy of wheat-based snacks fortified with ferrous sulphate, electrolytic iron, or hydrogen-reduced elemental iron: randomized, double-blind, controlled trial in Thai women. Am. J. Clin. Nutr. 82, 1276–1282.

Further reading

Alexander, D., Hazel, J., 2008. Front-of-Pack Nutritional Labeling: Perspectives of the New Zealand Food Industry. New Zealand Food Authority. Available at: https://www.sciencedirect.com/science/article/pii/S0306919215001001 (Accessed 9 December 2020).
Cecchini, M., Warin, L., 2016. Impact of food labelling systems on food choices and eating behaviours: a systematic review and meta-analysis of randomized studies. Obes. Rev. 17, 201–210.
CFIA (Canadian Food Inspection Agency), 2001. Guide to Food Labelling and Advertising. http://www.inspection.gc.ca/english/bureau/labeti/guide/guidee.shtml.
Department of Health, 1988. Food Standards Committee: Terms of Reference. Wellington, New Zealand.
IOM (Institute of Medicine), 2003. Report - Health Literacy: A Prescription to End Confusion.
Michail, N., 2019. An 'outstanding Contribution' to Health: Peru Wins United Nations Award for Nutrition Label. Available at: https://www.foodnavigator-latam.com/Article/2019/09/26/United-Nations-awards-Peru-for-nutrition-label (Accessed 9 October 2020).
Michail, N., 2020. Warning Labels Set to Enter into Force in Uruguay: Is Your Product Compliant? Available at: https://www.foodnavigator-latam.com/Article/2020/02/24/Warning-labels-set-to-enter-into-force-in-Uruguay# (Accessed 9 May 2020).
New Zealand Department of Health, 1992. A Consolidation of the Food Regulations 1984 Incorporating Amendments 1 to 6. Government Printing Publication, Wellington.

Chapter 21

The first legislation for foods with health claims in Korea

Ji Yeon Kim[1], Sewon Jeong[2], Oran Kwon[3] and Sangsuk Oh[4]

[1]*Department of Food Science and Technology, Seoul National University of Science and Technology, Seoul, Korea;* [2]*BiofoodCRO, Seoul, Korea;* [3]*Department of Nutritional Science and Food Management, Graduate Program in System Health Science and Engineering, Ewha Womans University, Seoul, Korea;* [4]*Department of Food Science and Technology, Ewha Womans University, Seoul, Korea*

21.1 Background

The aging population, growing exponentially throughout the world, faces difficulties in meeting medical care costs because of the increase in chronic diseases, such as diabetes, cardiovascular disease, and cancer. These trends have compelled scientists to identify physiologically active components in foods, in turn stimulating the food industry to match consumers' desire for health benefits through food products, such as functional foods or food supplements. Therefore, information regarding the health benefits of foods is considered essential for both consumers and manufacturers; consumers can use this information to make purchasing decisions, and manufacturers can use it to promote sales. Therefore, labeling and advertising should be clear and correct to avoid any misunderstanding or exaggeration.

The requirement of protecting consumers and ensuring their right to accurate information regarding the physiological value of foods prompted governments to establish regulatory frameworks for the approval of foods with health claims in many nations. For example, in 1991, the Ministry of Health, Labor, and Welfare of Japan first established a regulatory system to review and approve label statements regarding the health benefits of foods under the Health Promotion Law (Shimizu, 2003; Ohama et al., 2006). In 1993, the Food and Drug Administration (FDA) of the United States promulgated new regulations implementing the provision authorizing disease prevention claims of foods under the Nutrition Labeling Education Act of 1990. However, the Dietary Supplement Health and Education Act of 1994 imposed substantial limitations on FDA authority over dietary supplements by exempting dietary supplement ingredients from regulation as "food additives"by the Federal Food, Drug, and Cosmetic Act (FD&C Act). Therefore, new and somewhat-more-tolerant safety requirements were established to regulate dietary supplements (Barton Hutt, 2000). In the meantime, the National Assembly of Korea enacted the Health/Functional Food Act (HFFA) in 2002, requiring the Korea FDA, which was renamed the "Ministry of Food and Drug Safety" (MFDS), to promulgate new regulations for the pre-market approval of food supplements and functional food (Ministry of Food and Drug Safety). These new regulations were promulgated in January 2004 (Ministry of Food and Drug Safety). This chapter reviews the regulatory framework for health/functional foods (HFFs) under the HFFA, focusing on the evaluation of new functional ingredients for HFFs.

21.2 Health/Functional Food Act

The HFFA's most significant accomplishment is the statutory definition of HFFs as food products and introducing new considerations for reviewing the safety and effectiveness of functional ingredients.

Article 3 of the Act defined HFFs as foods containing health benefit ingredients or components intended to enhance or preserve human health. This article also limited the form of HFFs only to tablets, capsules, powders, granules, pills, and liquids. Therefore, the term "HFF" was synonymous with food and dietary supplements or nutraceuticals. However, according to the 2008 revision of the HFFA, the scope of HFFs has been extended to include conventional foods (Amendment of Health/Functional Food Act, 2008).

Ensuring Global Food Safety. https://doi.org/10.1016/B978-0-12-816011-4.00013-6

Article 15 of the Act introduces the concept that all HFFs are subject to review safety and substantiation of claims based on their ingredients or physiologically active components before marketing. Authorization can be granted in two ways. Provision 1 shows how to list the authorized ingredients in the Code of HFFs by regulatory amendments, which is for broad use, although time consuming. Provision 2 shows the other way to issue a certificate for each ingredient: without regulatory amendments. Manufacturers or distributors are responsible for providing all evidence for backing up the claims of their products by developing substantiation or relying on existing information. The Act does not define what "health claims" are and what constitutes "substantiation" for a health claim made for an HFF. Instead, it gives the MFDS the exclusive authority to define health claims and review data.

Besides reviewing new functional ingredients, this Act is essential in the context of the validity of the HFF market in Korea. Article 16 states that any label and advertisement regarding the efficacy of HFFs shall be approved before marketing by the advisory committee. Moreover, for manufacturing and quality control of good HFFs, the MFDS may also designate good manufacturing practices (GMPs).

21.3 Health claims allowed for HFFs

The MFDS Regulation on the Labeling of HFFS describes permissible health claims. The only categories of claims allowed in labeling and advertising are nutrient function, enhanced function, and disease risk reduction. In addition, any claims that state or imply that the products possess the capacity to treat, prevent, or cure human diseases are prohibited. These definitions of health claims are compatible with those adopted by the Codex Alimentarius Commission in 2004 and summarized as follows(CCFL Codex Guidelines for Use of Nutritional and Health Claims, 2004):

- Nutrient function claims: These describe the physiological role of the nutrient in the growth, development, and normal functions of the body. Such claims should be applied to nutrients that have their own recommended dietary allowance and be based on current, university-level nutrition texts as a possible source of evidence.
- Enhanced function claims: These concern specific beneficial effects of HFFs on normal functions or biological activities of the body in the context of total diet. Such claims are related to a positive contribution to health, to the improvement of a function, or to modifying or preserving health.
- Reduction of disease risk claims: These describe the relationship between the consumption of HFFs (in the context of the total diet) and the reduction of the risk of developing a disease or health-related condition.

21.4 Scientific substantiation of health claims for HFFs

Two issues should take precedence over the substantiation of health claims: (1) Identification and stability of ingredients or components shall be evaluated by reviewing their origin, nature, composition, and processing methods and (2) Safety evaluation shall be done within a risk analysis framework.

21.4.1 Identification and stability of functional ingredients or components

Functional ingredients can be obtained through extraction, centrifugation, filtration, and/or fractionation. These are formulated as dietary supplements or added to conventional foods. In order to get a good quality product, standardization is necessary. Standardization is the process that optimizes the batch-to-batch consistency using index material; however, from the handling of raw material to the processing for final products, there are many factors affecting standardization. In order to ensure the quality of active ingredients of HFFs, the MFDS requires manufacturers to submit data on the contents and analytical methods applied to the index material. It is ideal for the index material to be an active compound. However, as there are many chemicals with similar activities in natural ingredients, it is difficult to determine the exact active compound. Instead of active compounds, marker compounds could be used as the index material. The marker compounds should have specificity, representativeness, stability, and generality characteristics, and an analytical method must be verified by method validation. The petitioner should establish selectivity, accuracy, ranges for limits of detection and quantification, and reproducibility. Determining the index material and validating the analytical method are key components in ensuring product quality and protecting consumers from misleading label information.

21.4.2 Safety evaluation of functional ingredients or components

All scientific information regarding the history of safe use, manufacturing process, exposure assessment, nutritional evaluation, bioavailability, and toxicological data is helpful for the safety evaluation of functional ingredients or components.

Safety evaluation should be done within a risk analysis framework. If safety is not assured, there is no need to review the effectiveness. No benefit–risk analysis is allowed under the Food Act. The basic principle of safety assessment is the confirmation of novelty. If the active ingredient in the food is traditionally consumed, it might be safe without further documentation. However, if a fraction of a plant is consumed as food, it should not be treated the same as a raw plant. The safety of the fraction should be substantiated using all available documentation.

To make a transparent and consistent evaluation, the MFDS provided guidelines on preparing safety documents using the decision tree approach, as presented in Fig. 21.1.

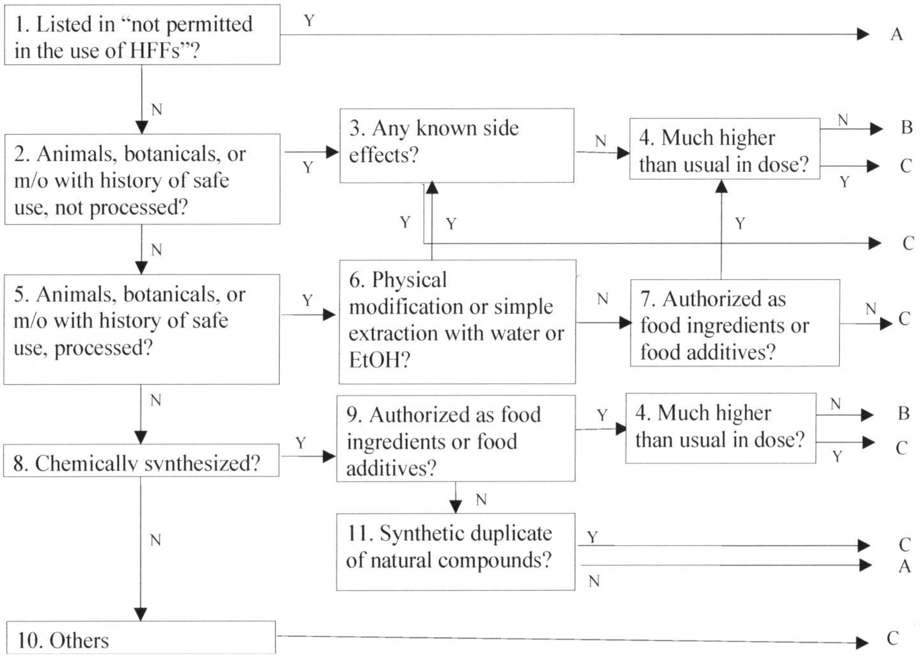

FIGURE 21.1 Decision tree for the preparation of safety data.

The three significant factors that were considered are the information regarding (1) raw materials (first column), (2) processing methods (second column), and (3) intake level (third column). According to the materials, manufacturing process, and intake level, there are four categories in the degree of scientific substantiation for safety. The first is the category that covers components that are not allowed as functional ingredients. These are the ingredients or components listed in the "regulations of prohibited ingredients as health/functional food." The second is the category that needs documentation for the history of use, safety information, and intake assessment. This is because these ingredients have traditionally been consumed as food, and there is no increase in the intake amount compared to the average intake level. The third is the category for which toxicological data, including all available documents such as the history of use, intake assessment, might be needed.

21.4.3 Review of scientific substantiation of health claims

The MFDS is applying a standard of "competent and reliable scientific evidence" to provide manufacturers or distributors with flexibility in the precise amount and type of evidence and help consumers preserve confidence in HFFs. Although no formula exists as to how many or what types of studies are needed to substantiate a claim, review articles, metaanalyses, and abstracts would be excluded because these do not contain sufficient information on the individual studies that were reviewed and, therefore, scientific conclusions cannot be drawn from this information.

First, the studies included will be reviewed independently in terms of design type and scientific quality. Competent and reliable scientific evidence that substantiates a claim would generally consist of information derived primarily from human studies. Especially, a randomized, double-blind, placebo-controlled intervention study is the "gold" standard. Other types of scientific evidence such as animal studies, in vitro studies, anecdotal evidence, metaanalysis, and review articles would generally be considered background information; alone, they may not be adequate to substantiate a claim. However, animal and in vitro studies that are relevant and sufficient enough to explain the biochemical and physiological mechanisms or show dose−response relationships would be strongly supportive data to health claims (US Food and Drug Administration; Health Canada, 2001).

Next, the scientific quality of each study will be reviewed based on several factors, including study design and conduct, study population, data collection, outcome measures, statistical analysis, and confounding variables(AHRQ, 2002). If the scientific study adequately addresses all or most of these factors, it would be considered of high quality. Finally, relevant biomarkers or clinical endpoints may be used to identify the proposed relationship between an ingredient and a component and health endpoint. Biomarkers can be classified according to whether they (1) are related to the exposure to the food component (can give some indication, but not absolute proof); (2) are related to the target function or biological response; or (3) are related to an appropriate intermediate endpoint (Asp and Contor, 2003; Richardson et al., 2003; Cummings et al., 2003; Howlett and Shortt, 2004; Aggett et al., 2005; FDA, 2005).

Lastly, the review of the totality of evidence will be followed to evaluate the strength of the evidence. Although the types and qualities of individual studies are essential, each piece should be considered in the context of all available information. The strength of the entire body of scientific evidence can be considered based on several criteria, including quantity, consistency, and relevance. The more data from independently conducted studies are developed or collected, the more persuasive the evidence is. If the evidence used to substantiate a claim agrees with the background information, it would be ideal. Conflicting or inconsistent results raise questions as to whether a particular claim is substantiated (Asp and Contor, 2003; Richardson et al., 2003; Cummings et al., 2003; Howlett and Shortt, 2004; Aggett et al., 2005; FDA, 2005).

21.4.4 Re-evaluation

Recently, the HFFA was revised to include a reevaluation article. According to Article 5, the MFDS can reevaluate functional ingredients, and depending on the result, may cancel or change acknowledgment. Reevaluation should be based on scientific evidence and an advisory committee should review the reevaluation results. The MFDS should list functional ingredients, which should be reevaluated until 1 year before the reevaluation. However, if new scientific evidence on safety or efficacy is identified, it is recognized that there is a need to re-evaluate urgently because of surging cases. Reevaluation is similar to the evaluation process of new functional ingredients. In 2018, based on reevaluation, manufacturing standards, specifications for heavy metals, and precautions were changed for eight functional ingredients.

21.4.5 Kinds of functional ingredients

There are about 301 ingredients, including 205 product-specific ingredients and 96 generic ingredients. There are two types of generic ingredients. One encompasses nutrients, including 14 vitamins, 11 minerals, proteins, essential fatty acids, and dietary fibers; the other includes functional ingredients with enhanced function claims. For dietary fibers and proteins, it could be both nutrients and functional ingredients. Nutrients can be any source of fiber or protein, whereas functional ingredients need to have the right ingredients listed. There are 14 functional fibers, such as guar gum/guar gum hydrolyzates, indigestible dextrin, polydextrose, soy fiber, oat fiber, and the like. Each fiber has its own claims and intake amounts. For example, guar gum or guar gum hydrolyzates have claims for maintaining healthy cholesterol levels, postprandial glucose levels, and gastrointestinal health, whereas psyllium husk could have been claimed only for maintaining good gastrointestinal health. As of July 2018, 205 ingredients have been approved as product-specific functional ingredients. There are various efficacies in HFFs, among which gastrointestinal health, body fat reduction, and bone/joint health occupy the most part (Fig. 21.2).

There have been two claims for reducing disease risk: one is the relationship between calcium and osteoporosis, and the other is between xylitol and dental caries. Based on the Korean Dietary Reference Intake, calcium is a nutrient Koreans need to consume more of. The daily intake of calcium from HFFs should be in the range of 210−800 mg. Labeling should include the following aspects: that the product contains adequate calcium and that intake of the amount of calcium contained in healthy meals with appropriate exercise may support the bone health of young women and reduce the risk of osteoporosis in the aged. Daily intake of xylitol from HFFs should be 10−25 g. The claim shall not imply that consuming noncariogenic carbohydrate sweetener-containing foods is the only recognized means of reducing dental caries risk. The other sweetener shall be included below 50% in the final product bearing claims about the effect of xylitol on dental caries. Moreover, when carbohydrates other than xylitol are present in HFF, the food shall not produce acid after consumption.

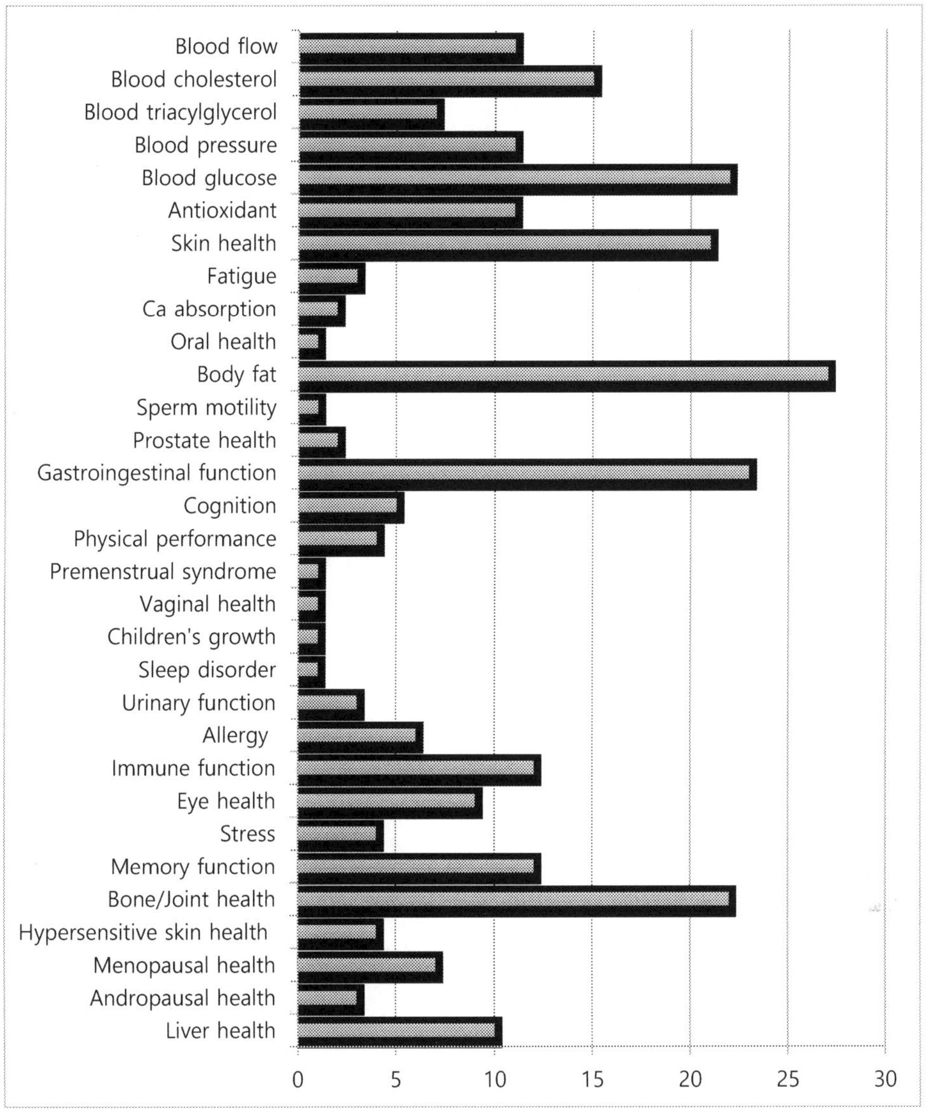

FIGURE 21.2 The kinds of efficacy for product-specific HFFs.

21.4.6 Connection of scientific evaluation to consumer understanding

The evidence-based ranking system linked the ranking of scientific evidence to the wording of relevant claims, where different levels of evidence would result in statements of varying levels of appropriateness. However, we surveyed consumer responses to the appropriate wording because there was concern regarding consumer confusion. We evaluated a range of product label formats and message language. About 2000 consumers were presented with several kinds of labels. According to the results, consumers cannot determine the level of strength of scientific evidence. To ensure consumer understanding, the translation from scientific to lay terms should be undertaken by the government and industry.

21.5 Future directions

The MFDS has collected all cases which are estimated as adverse events. Furthermore, they analyzed the relationships between the symptoms of adverse events and ingredients. If the results are significant, the MFDS can take actions such as urgent recall, prohibition of sale, and the addition of warnings on labels. In addition, to ensure the quality of HFFs, GMPs have been enforced as an obligation. Since 2017, for manufacturers with sales of more than 2 billion won, GMPs have been compulsory. Nonetheless, the most important thing is the consumer understanding of health claims. The ultimate goals of the HFFA are to reduce consumer confusion regarding health claims made by food manufacturers and to ensure that such claims are truthful and not misleading. To achieve these objectives, research on the consumer perceptions of health claims and development of the appropriate expression manner should focus in Korea.

References

Aggett, P.J., Antoine, J.M., et al., 2005. Process for the assessment of scientific support for claims on foods (PASSCLAIM): consensus on criteria. Eur. J. Nutr. 44 (Suppl), I/5—I/30.

AHRQ, 2002. Systems to Rate the Strength of Scientific Evidence.

Amendment of Health/Functional Food Act, March 21, 2008.

Asp, N.G., Contor, L., 2003. Process for the assessment of scientific support for claims on foods (PASSCLAIM): overall introduction. Eur. J. Nutr. 42 (Suppl), I/3—I/5.

Barton Hutt, P., 2000. US Government regulation of food with claims for special physiological value. In: Schmidl, M.K., Labuza, T.P. (Eds.), Essentials of Functional Foods. Aspen Publishers, Inc., Gaithersburg, MD, pp. 339—352.

CCFL, Codex Guidelines for Use of Nutritional and Health Claims, 2004. ALINORM 04/27/41. Codex Committee on Food Labeling.

Cummings, J.H., Pannemans, D., Persin, C., 2003. PASSCLAIM — report of the first plenary meeting including a set of interim criteria to scientifically substantiate claims on foods. Eur. J. Nutr. 42 (Suppl), I/112—I/119.

FDA, 2005. Substantiation for Dietary Supplement Claims Made under Section 403®(6) of the Federal Food, Drug, and Cosmetic Act.

Health Canada, 2001. Product-specific Authorization of Health Claims for Foods.

Howlett, J., Shortt, C., 2004. PASSCLAIM — report of the second plenary meeting: review of a wider set of interim criteria for the scientific substantiation of health claims. Eur. J. Nutr. 43 (Suppl), II/174—II/183.

Ministry of Food and Drug Safety. Health/Functional Food Act of 2002.

Ohama, H., Ikeda, H., Moriyama, H., 2006. Health foods and foods with health claims in Japan. Toxicology 221, 95—111.

Richardson, D.P., Affertsholt, T., et al., 2003. PASSCLAIM — synthesis and review of existing processes. Eur. J. Nutr. 42 (Suppl), I/96—I/111.

Shimizu, T., 2003. Health claims on functional foods: the Japanese regulations and an international comparison. Nutr. Res. Rev. 16, 241—252.

US Food and Drug Administration. Guidance for Industry: Evidence-based review system for the scientific evaluation of health claims. http://www.cfsan.fda.gov/~dms/hclmgui5.html. (Accessed 11 March 2008).

Chapter 22

Bioactivity, benefits, and safety of traditional and ethnic foods

Adelia C. Bovell-Benjamin
Food and Nutritional Sciences, Tuskegee University, Tuskegee, AL, United States

22.1 Introduction

Globally, there is growing research and consumer interest regarding diet-related (chronic) diseases and their influence on the health and well-being of communities (Day et al., 2008; Urquiaga and Leighton, 2000). Several factors have contributed to this interest. Firstly, industrialization, urbanization, and market globalization have impacted lifestyles, diets, and nutritional status of populations worldwide. For example, in developing countries, while urbanization has reduced undernutrition in metropolitan areas, it has increased physical inactivity and inadequate dietary patterns. Therefore, in many developing countries, undernutrition may coexist with a high prevalence of chronic diseases such as certain types of cancer, obesity, hypertension, cardiovascular diseases, and noninsulin dependent diabetes mellitus. Secondly, consumers in developing countries are experiencing escalating food prices, land scarcity, and population growth. As a result, interest in food composition has broadened beyond nutrients to include bioactive compounds in traditionally consumed foods, which may help to prevent malnutrition and chronic diseases.

Thirdly, in the developed world, growing environmental awareness and concerns about safer foods have resulted in increased demand for natural or organic foods, which are perceived to be healthier than conventionally grown or genetically modified foods (Saba and Messina, 2003). Another reason is the impact of global emerging ethnic food markets on dietary intake and ultimately on chronic diseases. The evidence indicates that total sales for ethnic foods in the United Kingdom (UK) have doubled between 2003 and 2006 (Leatherhead Food International, 2004, 2007). Leatherhead Food International (2007) has estimated that from 2007 to 2011, ethnic food markets will have annual growths of up to 10% in Ireland, and between 6% and 8% in Spain, the Netherlands, Denmark, Belgium, and Italy. Khokhar et al. (2009) have reported that increased consumption of traditional/ethnic foods will impact on nutrient intake.

These phenomena have led to increased consumer demands for nutritious foods with additional health-promoting activities, and also point to the need for foods to address the chronic diseases, which have become public health problems in various populations throughout the world. Foods contain nutrients which are essential for growth, maintenance, and repair of the body. More recently, scientists have reported that foods also contain nonessential bioactive compounds, which have proven or potential beneficial effects on human health. A growing number of studies have linked traditional/ethnic foods to bioactive compounds, which have the capacity to prevent various chronic diseases, and confer other putative health benefits in humans. Therefore, the notion of traditional/ethnic foods appears to be one approach to meet the worldwide challenge of chronic diseases. However, there are significant gaps in the data regarding the nutrient and nonnutrient composition of traditional/ethnic foods, especially those from developing countries. This shortage of information slows down effective health and disease prevention efforts. It is apparent that more information and better understanding of the composition of traditional/ethnic foods, which enable them to promote human health and prevent disease, are needed.

Ensuring Global Food Safety. https://doi.org/10.1016/B978-0-12-816011-4.00030-6

22.2 Objective

This chapter aims to review the available scientific literature related to the bioactivity, potential health benefits, and safety of traditional and ethnic foods from selected countries in different regions of the world, namely, Latin America, Africa, and Asia.

22.3 Scope

For the purposes of this chapter, bioactive food compounds are defined as naturally occurring nonessential constituents in or derived from plant, animal, or marine sources, which have the ability to modulate biochemical, physiological, and metabolic processes in the human body while exerting beneficial effects beyond basic nutritional functions (ADA, 2004; Denny and Buttriss, 2008; Gry et al., 2007; Health Canada, 2004; Tejasari, 2007). Bioactive compounds can be produced either in vivo or by industrial enzymatic digestion (food processing activities). In plant- and animal-derived foods, bioactive compounds are usually found in multiple forms such as glycosylated, esterified, thiolated, or hydroxylated. Bioactive compounds in plants are usually found in the leaves, stems, roots, tubers, buds, fruits, seeds, and flowers; they influence the color, flavor, structure, function, and defense system of plants (Cushnie and Lamb, 2005). The major classes of plant bioactives include flavonoids and other phenolic compounds, carotenoids, plant sterols, glucosinolates, and other sulfur compounds (Denny and Buttriss, 2008).

Traditional and ethnic foods (the two terms are used interchangeably) are defined as those that have been consumed as part of a usual diet locally or regionally for an extensive time period by specific group(s) of people. In general, "traditional/ethnic" foods should (i) be communicated from ancestors to descendants; (ii) have a long history of consumption; and (iii) be usually part of the history and culture of the population concerned. Traditional foods are valued for more than their sustenance role and are linked with cultural identity and civilization. Most traditional food processes have stemmed from the need to preserve food for the off season or to make it safe for consumption. Traditional food products are produced from indigenous crops and raw materials and are therefore typical to a certain region or area. Increased consumer awareness and demands about the role of diet and nutrition in chronic disease prevention have led to more research regarding the benefits and safety of traditional/ethnic foods.

Foods or food ingredients consumed as part of traditional/ethnic diets, eaten for health-promoting properties or not, are included, even when the existing scientific evidence for these benefits may not yet be substantial. In many instances, there is little documented information about the content and concentration of bioactive compounds of several traditional/ethnic foods. The scope is limited to mainly plant foods or drink, but animal source foods may be included. The review focuses on traditional/ethnic foods in the context of developing countries. The target developing regions are Latin America, Africa, and Asia. They were selected to represent different traditional/ethnic foods from varying parts of the world.

22.4 Methodology

A review of the existing literature was conducted. The literature surveyed covered scientific journals, trade journals, magazines, market reports, conference proceedings, books, and other published materials. Web page content was used, as necessary. Literature was collected using numerous search engines and databases (for example, EBSCOhost, Ingenta, ScienceDirect, Google, Google Scholar, PubMed) in addition to library databases and Internet libraries of international organizations.

22.5 Structure of the review

In this review, Sections 22.1–22.5 justify the topic, present the objective, and outline the scope and structure of the review. Sections 22.6 and 22.7 discuss the relationship between food and chronic diseases and summarize the biological mechanism of bioactive food compounds. Section 22.8 discusses the bioactive compounds and beneficial effects and safety of traditional/ethnic foods from Latin America in general and Mexico in particular; Africa (South Africa and Uganda); and Asia (India and Japan). Sections 22.9 and 22.10 consist of conclusion, future scope. Acknowledgments and a references section complete the chapter.

22.6 Food and chronic diseases

It has been well established that certain types of foods, including fruits and vegetables, are associated with a decrease in chronic diseases. For example, a metaanalysis of 26 studies found an association between the risk of breast cancer and intake of fruits and vegetables (Gandini et al., 2000). In another multiethnic (Japanese, African American, Chinese, and Caucasian) case control study consisting of 1619 and 1618 men as prostate cancer cases and controls, respectively, Kolonel et al. (2000) evaluated the protective effects of fruit and vegetable intake.

Cruciferous and yellow-orange vegetable intake was inversely related to prostate cancer.

Bazzano et al. (2002) reported that the incidences of stroke were decreased for individuals consuming more than three servings of fruits and vegetables daily. Similarly, participants who consumed less than one serving daily of fruits and vegetables had a higher incidence of stroke than those who consumed more than three servings. Kim and Kwon (2009) concluded that there is limited evidence to support a relationship between garlic consumption and reduced risk of colon, prostate, esophageal, larynx, oral, ovary, or renal cell cancers. Wen et al. (2009) prospectively evaluated the association of dietary carbohydrates, glycemic index, and glycemic load and dietary fiber with breast cancer risk to determine whether the effect of these dietary intakes is modified by age and selected insulin- or estrogen-related risk factors. It was concluded that a high carbohydrate diet with a high glycemic load may be associated with breast cancer risk in premenopausal women or women less than 50 years of age. In general, these studies suggest a relationship between certain components in food and chronic diseases. The following sections discuss some bioactive compounds in traditional/ethnic foods.

22.7 Biological mechanism of bioactive food compounds

Polyphenols or phenolic compounds are present in a variety of plants utilized as important components of human food systems (Crozier et al., 2000). Polyphenols, products of the secondary metabolism of plants, are the most abundant and widely distributed group of bioactive compounds (Crozier et al., 2000). Polyphenols constitute a range of substances with an aromatic ring and one or more hydroxyl substituents. In biological systems, polyphenols serve as free radical scavengers, metal chelators, and prevent inhibition of cell communication; all of which are precursors to chronic diseases (Fraga, 2007; Sigler and Ruch, 1993; Masibo and He, 2008). The presence of the phenolic groups confers their antioxidant capacity. These characteristics are due to the hydrogen of the phenoxyl groups, which is prone to be donated to a radical, and by the ensuing structure that is chemically stabilized by resonance (Fraga, 2007). The schematic in Fig. 22.1 relates polyphenols to potential health benefits.

As shown in Fig. 22.2, the basic structure of flavonoid compounds, the most common subgroup of plant phenolic compounds, is the 2-phenyl-benzo[α]pyrane or flavane nucleus, which consists of two benzene rings (A and B), each containing at least one hydroxyl, which are connected through a three-carbon "bridge" and become part of a six-member heterocyclic ring (C) (Beecher, 2003; Brown, 1980). Flavonoids can be classified according to biosynthetic origin. For example, flavanones and flavan-3-ols are intermediates in biosynthesis as well as end products, which have the ability to accumulate in plant tissues. Others such as anthocyanidins, flavones, and flavonols are end products of biosynthesis only (Crozier et al., 2006); and the isoflavones and isoflavonoids in which the 2-phenyl side chain of flavanone isomerizes to the 3-position. Fig. 22.3 shows the categories of flavonoids.

Most of the beneficial health effects of flavonoids are attributed to their antioxidant and chelating capacities (Heim et al., 2002). Antioxidants are substances, which inhibit oxidation and protect the body against the damaging effects of free radicals. Briefly, flavonoids protect the body against reactive oxygen species (ROS) by preventing injury caused by free radicals (Masibo and He, 2008; Pietta, 2000). Free radicals are species which contain unpaired electrons, making them

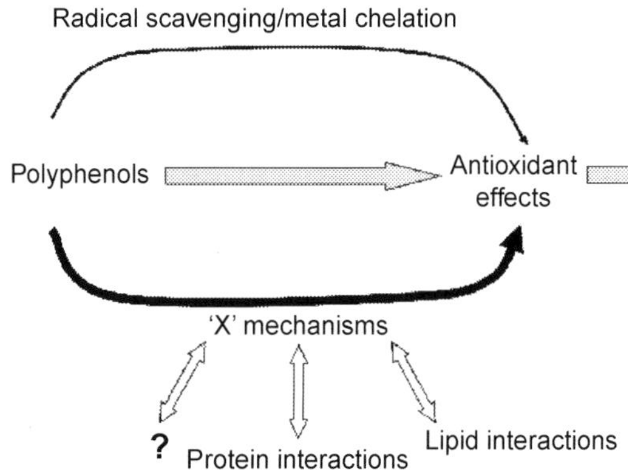

FIGURE 22.1 Scheme relating polyphenols with health benefits, through their observed antioxidant effects. The body of the *black arrows* indicates the "X" mechanism with free radical scavenging or metal chelating capacity of polyphenols (Fraga, 2007).

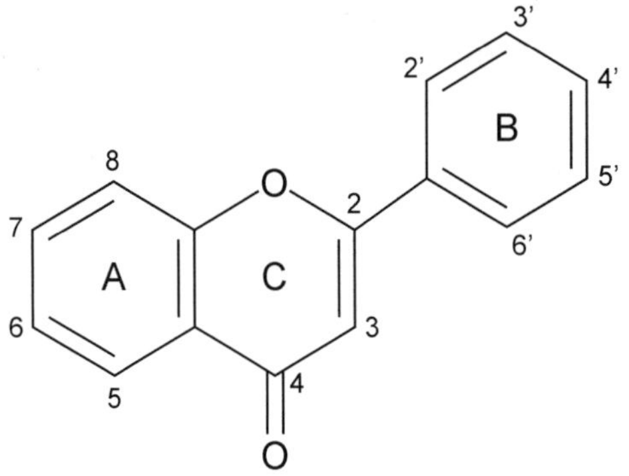

FIGURE 22.2 The skeleton structure of the flavones (a class of flavonoids) with rings named and positions numbered (Harborne and Baxter, 1999).

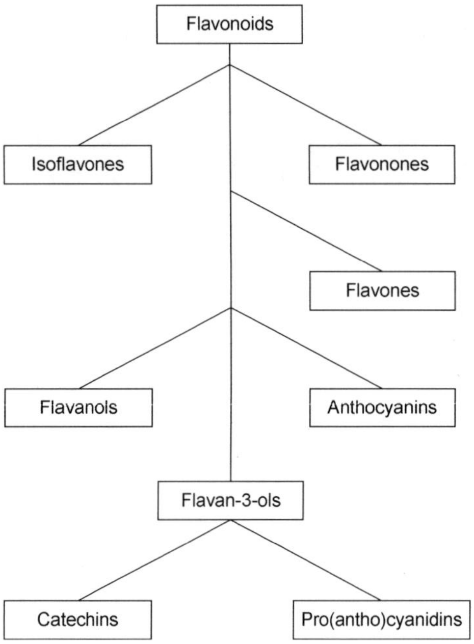

FIGURE 22.3 Categories of flavonoids. *Denny, A., Buttriss, J., 2008. Synthesis Report No. 4: Plant Foods and Health: Focus on Plant Bioactives. EuroFIR Project Management Office/British Nutrition Foundation.*

highly unstable and reactive. Free radicals cause cellular membrane damage, which is associated with chronic diseases such as cancer, diabetes, and coronary heart disease morbidity (Halliwell, 1994).

Flavonoids directly scavenge free radicals by stabilizing the ROS by reacting with the reactive compound of the radical as shown (Pietta, 2000):

$$FOH + R^{\cdot} \rightarrow FO^{\cdot} + RH$$

Where FOH is a flavonoid; $R^{\cdot}$ is a free radical; and $FO^{\cdot}$ is the less reactive flavonoid phenoxyl radical. Flavonoids may also suppress lipid oxidation by recycling other antioxidants by donating a hydrogen atom (McAnlis et al., 1999). The chemistry, metabolism, and structure—activity relationships of flavonoids have been extensively reviewed by Heim et al. (2002). Bioactive food compounds such as polyphenols and flavonoids have the capacity to prevent cellular membrane damage and suppress lipid oxidation, which all play a role in chronic disease prevention.

22.8 Bioactive food compounds in traditional/ethnic foods

22.8.1 Latin America

22.8.1.1 Beneficial effects of yerba mate

Latin American cultures have a tradition of using many native plants, for specific functional food purposes. Mate or yerba mate (*Ilex paraguariensis*), an indigenous plant, is used for the preparation of the most commonly consumed traditional tea-like beverage in South American countries (Filip et al., 2001). It is traditionally consumed by millions of South Americans as a tonic, a healthful alternative to coffee, a stimulant to reduce fatigue, and as an appetite suppressant (Cardozo et al., 2007; Di Gregorio et al., 2004). *I. paraguariensis* grows naturally in northeastern Argentina, southern Brazil, and eastern Paraguay where it is also cultivated (Gorzalczany et al., 2001). The habit of drinking yerba mate has remained unchanged, and it is widely consumed in Argentina, Paraguay, Uruguay, and Brazil, with a reportedly 30% of the population drinking more than one liter of mate infusions daily (Dellacassa and Bandoni, 2001; Filip et al., 2000; Filip et al., 2001). Yerba mate (Fig. 22.4) is usually prepared by pouring hot water unto a high concentration of the dried and minced leaves (50 g/L).

Yerba mate might be considered as one of the main antioxidant-rich beverages consumed in various South American countries. The leaves contain many bioactive compounds, such as chlorogenic acid and phenolic acids, which appear to

(a)

(b)

FIGURE 22.4 (A) *Agave americana* plant (B) Dried leaves of yerba mate. *Source: http://www.florahealth.com/flora/home/Canada/HealthInformation/Encyclopedias/Mate.htm.*

be responsible for the antioxidant activity of the tea, both in vivo and in vitro (Chandra and Gonzalez de Mejia, 2004; Bracesco et al., 2003; Schinella et al., 2000; Filip et al., 2000). Bastos et al. (2006) reported comparable antioxidant efficacy in green mate and mate tea infusions and BHT (a well-known phenolic antioxidant). da Silva et al. (2008) described yerba mate extract as a source of phenolic compounds, which contain in vitro antioxidant activities that may reduce cardiovascular disease risk. According to da Silva et al. (2008), yerba mate contains practically all of the vitamins and minerals necessary to sustain life. Additionally, Bixby et al. (2005) and Bracesco et al. (2003) have demonstrated that *I. paraguariensis* is a more potent antioxidant than red wine or green and black tea.

da Silva et al. (2008) examined the acute effects of the consumption of mate infusion on ex vivo plasma and low-density lipoprotein (LDL) oxidation, plasma antioxidant capacity, and platelet aggregation. Their results indicated that phenolic compounds from yerba mate infusion promoted plasma and LDL protection against ex vivo lipid peroxidation and also impacted significantly on the antioxidant capacity of plasma. Bravo et al. (2007) compared the total polyphenols and antioxidant activity of mate infusions with commonly consumed commercial beverages (orange juice, green and black teas, and red, rosé, and white wines). The total polyphenol content of mate was comparable to tea and orange juice. The antioxidant activity of mate was slightly higher than wines, orange juice, and black tea, but lower than green tea. Using rat models, Prediger et al. (2008) concluded that acute administration of hydroalcoholic extract of *I. paraguariensis* improved cognition by differentially modulating short- and long-term learning and memory. Overall, da Silva et al. (2008) indicated that, besides its stimulant and nutritional proprieties, yerba mate might be considered an important source of antioxidants to humans.

22.8.1.2 Safety of yerba mate

Literature regarding the safety of yerba mate is scarce, but consumption of yerba mate has been suggested as a risk factor for development of cancers of the oral cavity and upper aerodigestive tract, partly due to the very hot temperature at which this beverage is usually consumed (Goldberg and Brinckmann, 2000; Sewram et al., 2003). Di Gregorio et al. (2004) reported that yerba mate, yerba mate leaves, and tea manufactured in Apóstoles, Argentina, showed the presence of ^{137}Cs (radioactive contaminants from the Chernobyl nuclear fallout) contamination in concentrations of 710 Bq/kg. However, it was stated that these levels do not represent a public health risk. While it is recognized that this traditional food has excellent bioactive compounds, the presence of contaminants or its safety has not been extensively documented. However, it can be argued that the long history of use of traditional foods is in itself evidence of their safety; evidently, their use would have been discontinued if they were inexplicably unsafe. Furthermore, it can be argued that it is inappropriate to apply only scientific parameters to safety assessments of traditional foods without considering the traditional precautions, which are integral parts of their safe preparation and use.

22.8.1.3 Beneficial effects of pulque (Mexico)

Pulque, which is the most important traditional nondistilled alcoholic beverage of Mexico, was inherited from the Aztecs, and remains an essential part of the contemporary diet (Escalante et al., 2004). It is obtained by fermentation of the sap (*aguamiel*) from several species of the maguey plant such as *Agave atrovirens* and *Agave americana* (Fig. 22.4). The sap of the Agave plant consists of calcium oxalate crystals, an acrid volatile oil, Agave gum, and other compounds. Pulque processing is a complex succession of yeast and bacteria that produce ethanol, a diversity of chemical compounds, and some polymers that give a distinctive viscous consistency to the final product (Peña-Alvarez, 2004).

Pulque is viewed as healthy, and is usually consumed as a low-alcohol beverage and as a nutritional supplement. Backstrand et al. (2001) have postulated that alcoholic beverages, such as Pulque, could provide substantial amounts of vitamins and minerals to diets if they are not distilled or overly processed, and consumed in moderation. Pulque can provide important quantities of ascorbic acid, nonheme iron, riboflavin, folate, several other B vitamins, and some bioactive compounds (Backstrand et al., 2001). Overall, modest Pulque consumption has the potential to enhance iron status, increase intakes of key micronutrients, and improve nutritional status (Backstrand et al., 2001). Medicinal qualities have also been attributed to Pulque, and it is consumed for ailments such as renal infections, anorexia, and gastrointestinal disorders.

Saponins are natural glycosides of steroid or triterpene, which exhibit many different biological and pharmacological actions (Lacaille- Dubois, 2005; Sparg et al., 2004). Pulque contains significant quantities of steroidal saponins, many of which are bioactive. Several *Agave* species produce steroidal sapogenins and saponins, the raw materials for steroid hormone synthesis (Tinto et al., 2005). Saponins are being studied for their medicinal uses, including antispasmodic activity, and toxicity to cancer cells. The biological activities of saponins have been extensively reviewed elsewhere (Sparg et al., 2004; Lacaille-Dubois, 2005). Steroidal saponins have been described as the most important bioactive compounds in

TABLE 22.1 Biological activity of steroidal saponins.

Biological activity	References
Anticancer	Ravikumar et al. (1979), Sung et al. (1995)
Antithrombotic	Peng et al. (1996), Zhang et al. (1999)
Antiviral	Aquino et al. (1991)
Hemolytic	Zhang et al. (1999), Santos et al. (1997)
Hypercholesterolemic	Sauvaire et al. (1991), Malinow (1985)
Hypoglycemic	Kato et al. (1995)

yam (Yang et al., 2009) and several biological activities, such as anticancer, have been documented (Table 22.1). Saponins are said to make up the active major constituents of ginseng (*Panax ginseng*) (Sparg et al., 2004). The saponins present in *Agave* species have shown hypocholesterolemic, antiinflammatory, and antibiotic activity, but the functional effects of those in Pulque have not been detailed in the literature. Backstrand et al. (2002) concluded that consumption of Pulque predicted less risk of low ferritin and hemoglobin values for pregnant women in Solis Valley, Mexico.

22.8.1.4 Safety of pulque

Some γ-Proteobacteria have been identified in Pulque; although these are ubiquitous in fresh water, soil, vegetable surfaces, etc., some are also considered opportunistic human pathogens (Escalante et al., 2008; Waleron et al., 2002). Escalante et al. (2008) speculated that some of the γ -Proteobacteria detected in Pulque could have come from the natural bacterial diversity of the sap, whereas others could have been incorporated during its extraction, handling, and storage under nonaseptic conditions. Backstrand et al. (2001) reported that high consumption of Pulque was associated with poor infant outcomes among women in Solis Valley, Mexico. Consumption of Pulque by pregnant women was also linked to increased risk of the negative effects of fetal exposure to alcohol (Backstrand et al., 2001).

In summary, it has been shown that Pulque contains steroidal saponins and various other micronutrients. Although the biological activities of saponins have been extensively reviewed, their functional effect in Pulque has not been clearly elucidated; this warrants further research. The possibility of bacterial contamination during the processing of Pulque also needs to be addressed in future research studies.

22.8.2 Africa

22.8.2.1 Beneficial effects of rooibos (South Africa)

Traditionally, a number of plants such as *Aspalathus linearis* (Rooibos) and *Cyclopia intermedia* (Honeybush) have been used as teas in South Africa. This review discusses Rooibos. Rooibos tea originates from the leaves and fine stems of the indigenous South African plant *A. linearis* (Fig. 22.5). About three centuries ago, the indigenous Khoi-Khoi tribe of the

FIGURE 22.5 Mature Rooibos plants (*Aspalathus linearis*). Rooibos Ltd/SunnRooibos (2003) . *Source: www.rooibosltd. co.za.*

Western Cape Province, South Africa, used the leaves of the Rooibos plant as a tea, with an exceptional taste and aroma (Morton, 1983). It was consumed as a strong, hot brew with milk and sugar (Joubert et al., 2008). The leaves and stems of the plant were boiled in water and kept hot at low heat. Water, leaves, or stems were added to the pot after each serving. Surveys conducted by Oldewage-Theron et al. (2005) revealed that Rooibos tea was one of the 10 most frequently consumed foods in an informal settlement in the Vaal Triangle, South Africa. More recently, however, Rooibos is consumed as an herbal tea with or without milk and sugar, and is also used cold or hot.

Rooibos is naturally caffeine free with a low tannin content, which decreases the risk of poor iron bioavailability, a phenomenon frequently found in tea drinkers due to the formation of nonheme iron−tannin complexes (Erickson, 2003; Morton, 1983). The leaves and stems of *A. linearis* could be used as fermented or unfermented. The unfermented type is called green Rooibos; fermentation changes the Rooibos leaves from green to red due to oxidation of the constituent polyphenols; it is called red Rooibos or red tea (McKay and Blumberg, 2007). Today, the traditional fermented tea is processed in the same way it was done by the indigenous people centuries ago and also on a commercial basis.

Rooibos has been associated with antioxidant capacity, chemopreventive potential, modulated immune effect, and antiallergenic actions (Hesseling and Joubert, 1982; Kunishiro et al., 2001; Lamosova et al., 1997; Nakano et al., 1997a; Nakano, Nakashima and Itoh, 1997b; Schulz et al., 2003). Free radicals (unstable molecules that have lost an electron) can damage the DNA in cells and increase the risk for chronic and other diseases. Antioxidants, such as polyphenols, are capable of binding free radicals before they become harmful to the body. Polyphenols have subgroups such as flavonoids and phenolic acids, which are potent free radical scavengers (Erickson, 2003). Brewed green and red Rooibos teas are rich sources of different phenolic compounds (McKay and Blumberg, 2007; Dos et al., 2005). Red Rooibos is known to contain the following phenolic acids: caffeic, ferulic, p-coumaric, p-hydroxybenzoic, vanillic, and protocatechuic (Rabe et al., 1994). The major flavonoids found in Rooibos tea are aspalathin, iso-orientin, orientin, isovitexin, vitexin, isoquercitrin, hyperoside, quercetin, luteolin, chrysoeriol, and rutin (McKay and Blumberg, 2007; Shimamura et al., 2006). Aspalathin, a naturally occurring C-glycosyldihydrochalcone is found exclusively in *A. linearis* (Koeppen and Roux, 1965; Koeppen, 1970; Shimamura et al., 2006).

Generally, green Rooibos contains higher levels of total polyphenols. These differences are attributed to the enzymatic and chemical changes, which occur during the fermentation process (Joubert, 1996). For example, aspalathin is oxidized to dihydro-iso-orientin during fermentation, and its concentration was shown to reduce from 49.9 to 1.2 mg/g for green and red Rooibos, respectively (Bramati et al. 2003). Standley et al. (2001) also reported more total polyphenols in green than red Rooibos (41.0 vs. 35.0%). In 2% (w/v) solutions of Rooibos tea, green had more total polyphenols than red (41.2 vs. 29.7%), more flavonoids (28.1 vs. 18.8%), and 13.1 versus 10.9% nonflavonoids (Marnewick et al. 2000). Other flavonoids such as iso-orientin, orientin, isovitexin, and vitexin are also degraded during fermentation of *A. linearis*.

Several studies, mostly animal and in vitro, have investigated the potential health benefits of *A. linearis*. Shindo and Kato (1991) reported that weekly consumption of 1500 mL fermented red tea (0.2 g leaves/100 mL water, boiled for 20 min) was beneficial to patients with a wide range of dermatological diseases. It was also reported to decrease the incidence of *herpes simplex*, and the incurable human papilloma virus infection. In an animal model, Uličná et al. (2006) fed male Wistar rats a water extract (0.25 g leaves/100 mL water, boiled for 10 min, steeped for 20 min); an alkaline extract (10 g water-extracted leaves/100 mL 1% Na_2C0_3 extracted for 3 h at 45°C); ad libitum and a gavage of 5 mL/kg body weight. They reported that the extracts (red tea) significantly decreased plasma creatinine and lowered advanced glycation end products and lipid peroxidation in the plasma and lens. The water extract also decreased lipid peroxidation in the liver. Male Fischer rats were fed extracts of 2.0 g leaves/100 mL freshly boiled water, steeped for 30 min, freeze-dried (fermented and unfermented) by Sissing (2008). It was demonstrated that the extract reduced the number and size of methylbenzylnitrosamine-induced esophageal papillomas. In a recent study conducted by Juráni et al. (2008) using red tea, aged Japanese quails were fed 0.175 g leaves/100 mL water, boiled for 10 min, and steeped for 20 min, ad libitum or as supplemented feed with milled plant material (3.5 g/kg). They concluded that the extract prolonged productive life of the quails. Other potential health benefits of Rooibos are extensively discussed by Joubert et al. (2008).

22.8.2.2 Safety of rooibos

The general assumption is that Rooibos is safe because of its long history of traditional consumption by the South African population without documented adverse reports. It has been reported that chronic ingestion of green and red Rooibos extracts by rats for 10 weeks did not adversely affect the liver, body weight, and kidney parameters (Marnewick et al., 2003). Quality control of commercially produced Rooibos is limited to pesticide residues and microbial contamination (Joubert et al., 2008). Commercial producers are less reliant on traditional use as an indicator of safety and toxicity. Several manufacturers use varying standards to predict quality of the tea, for example, objective color measures, total polyphenol (TP), total antioxidant activity (TAA), and aspalathin content (Joubert, 1995). Much more scientific evidence is needed to conclusively establish the safety of Rooibos.

In summary, it has been shown that Rooibos contains bioactive compounds, which may help protect against free radical damage that can lead to cancer, heart attack, and stroke. However, it should be borne in mind that the evidence presented is primarily from animal models; human studies are very limited. Unfermented (green) Rooibos has a higher amount of polyphenols than traditionally fermented Rooibos, and generally demonstrates higher antioxidant and antimutagenic capabilities in vitro. The bioavailability, tissue distribution, and biological activities of the bioactive compounds in Rooibos need to be further researched. Further research is also needed to determine whether the bioactivity of Rooibos observed in vitro and in animals translate into health benefits for humans.

22.8.2.3 Beneficial effects of Java plum (Uganda)

In Central Uganda, the dried and powdered seeds of Java plum (*Syzygium cumini*) are traditionally consumed as herbal medicine to treat asthma and the fruit is also eaten, especially by young children (Stangeland et al., 2009). The Java plum was brought to Uganda in the early 1900s by the Indians who mainly ate the juicy pulp as ordinary fruit or used it in jam production (Fig. 22.6). In their examination of the bioactivity of several commonly eaten Ugandan fruits and vegetables, Stangeland et al. (2009) revealed that Java plum seeds and fruits have high antioxidant activity. In a preliminary study, Ndyomugyenyi (2008) demonstrated that the bioactive compounds in Java plum included sterols, triterpenes, coumarins, tannins, glycosides (cardiac and steroids), alkaloids, reducing compounds, anthocyanin pigments, and saponins.

Banerjee et al. (2005) concluded that the fruit skin of *S. cumini* has significant antioxidant activity, which may come in part from antioxidant vitamins, phenolics or tannins, and/or anthocyanins. In their study, Veigas et al. (2007) characterized and evaluated anthocyanin pigments from *S. cumini* fruit peels for their antioxidant efficacy and stability. Total anthocyanin content was 216 mg/100 mL of extract, equivalent to 230 mg/100 g fruit on a dry weight basis. Three anthocyanins were identified as glucoglucosides of delphinidin, and the researchers concluded that *S. cumini* extract was a more efficient free radical scavenger than BHA. Consumption of *S. cumini* fruit may supply substantial antioxidants, which may provide health promoting and disease preventing effects.

22.8.2.4 Safety of Java plum

Literature regarding the safety of Java plum is not forthcoming. In general, the traditionally eaten fruits are considered safe even though they may contain microorganisms, antinutrients, toxins, or allergens. Very often, special preparation or processing techniques have been associated with traditional foods to minimize any risks associated with them. The knowledge required to manage the risks associated with traditional foods has been acquired in the course of their long history of use. Much more research is needed to determine the safety of Java plum.

22.8.3 Asia

22.8.3.1 Beneficial effects of the mango (India)

Fruits from tropical and subtropical climates contain various bioactive compounds (Ribeiro et al., 2007). Mango (*Mangifera indica* L.) is a commonly eaten, traditional fruit in many tropical and subtropical countries throughout the world (Kim et al., 2009). *M. indica* has been cultivated in the Indian subcontinent for thousands of years; cultivation spread to East Asia and East Africa between the 5th–4th century BC and 10th century AD, respectively. Mangoes were

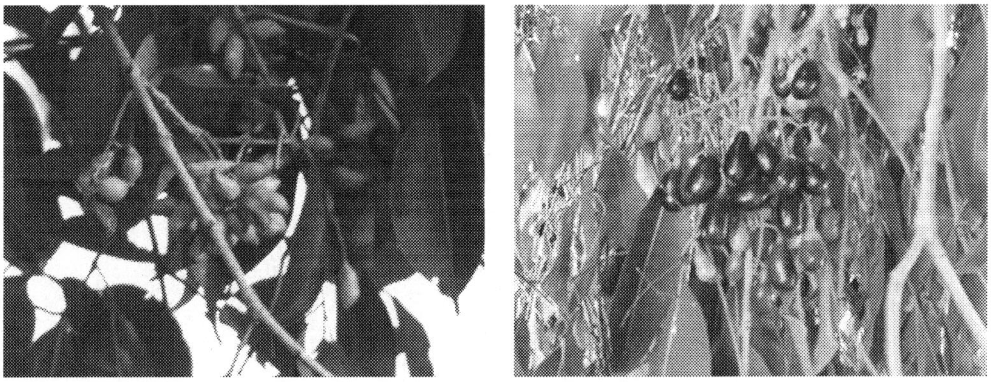

FIGURE 22.6 The tree and fruit of Java plum (*Syzygium cumini*). *Source: http://www.daleysfruit.com.au/forum/bugagali1/.*

FIGURE 22.7 Mature ripe and green mango (*Mangifera indica* L.). *Source: http://www.spicegrenada.com/.*

subsequently introduced to Brazil, West Indies, and Mexico (Kim et al., 2009). Brazil, Mexico, the Philippines, India, China, Nigeria, and Pakistan are among the world's major mango producing countries (Kim et al., 2009; Ribeiro et al., 2007; FAo, 2007). The ripe fruit is variable in size and color (Fig. 22.7). In the center of the fruit is a single flat oblong seed that can be fibrous or hairy on the surface, depending on the cultivar. The ripe fruits are usually eaten uncooked, while the unripe fruits or green mangoes can be made into pickles or chutneys. Table 22.2 shows some traditional dietary uses of mangoes in selected developing countries.

Mango is a rich source of bioactive compounds including phenols and carotenoids (Kim et al., 2007; Berardini et al., 2005; Godoy and Rodriguez-Amaya, 1989). Ribeiro et al. (2008) identified the following flavonoids and xanthone glycosides in the pulp of Brazilian Uba mango: mangiferin, mangiferin gallate, isomangiferin gallate, kaempferol, and quercetin. It was noted that the bioactive compound varied with variety of mango. Ribeiro et al. (2007) identified mango pulp as a rich source of antioxidants. Kim et al. (2009) reported free gallic acid and four gallotannins as the major polyphenols in mango. p—OH—benzoic acid, p-coumaric acid, ferulic acid, and catechin were also identified in mango at low concentrations by Kim et al. (2009). Other polyphenolic compounds which have been identified in mango include isoquercetin, ellagic acid, and 3-glucogallin (Schieber et al., 2000).

The main bioactive compound in mango is mangiferin or C-glucosyl xanthone (xanthones are some of the most potent antioxidants). Using cultured human peripheral blood lymphocytes, Jagetia and Baliga (2005) reported that mangiferin was protective against radiation-induced sickness and bone marrow mortality. Mangiferin has also been associated with antidiabetic, antiatherogenic, and antihyperlipidemic activities in rats (Yoshikawa et al., 2001; Muruganandan et al., 2002; Randle et al., 1963). It has also been reported that mangiferin has antibacterial and antifungal effects (Stoilova et al., 2005).

22.8.3.2 Safety of the mango

Human intake studies have indicated that consumption of 1 g total polyphenols daily has not resulted in any adverse effect or lethal toxicity (Scalbert and Williamson, 2000). Although this widely used traditional fruit has been reported to have excellent bioactive compounds, the question of its safety has not been extensively studied. In summary, mango and possibly its derived products, traditionally used in developing countries, are rich sources of bioactive compounds; however, the safety aspects and the effect of processing on the bioactive compounds have not been very well studied. Kim et al. (2009) studied the polyphenolic and antioxidant changes to mature green mangoes following different hot water treatment times and during storage. They indicated that gallic acid, gallotannins, and total soluble phenolics decreased as the length of hot water treatment increased. Further research is needed in this area to determine the safety and whether the bioactive compounds in the mango fruit are still intact in the derived products (Table 22.2).

TABLE 22.2 Traditional dietary uses of mango (*Mangifera indica* L.) in selected developing countries.

Food	Country	Description
Chutney	Indian subcontinent	Unripe mangoes with chili or limes
Aampadi	India	Ripe mango cut into thin layers, desiccated, folded then cut
Ayurvedic mango lassi	India	Sweet drink, which uses yogurt and milk as the base
Ras	India	Squeezed mango juice used on a variety of bread items
Unripe or green mangoes	India	Eaten with salt and chili
	Philippines	Eaten with bagoong, a salty paste made with fermented fish or shrimp
	Indonesia	
	Thailand	Sugar and salt and/or chili
Panha or Panna	India	Raw ripe mango drink
Mango	Mexico	Sliced mango with chili powder, juices, smoothies, fruit bars, etc.; whole in chili–salt mixture
Pickled mangoes	Southeast Asia	Pickled with fish sauce and wine vinegar
Amchur	Trinidad and Tobago, India, Southeast Asia	Dried unripe mango used as a condiment
Rujak or rojak	Malaysia, Singapore, Indonesia	Green mangoes in a sour salad

22.8.3.3 Beneficial effects of edible algae (Japan)

In Asian countries, Japanese are the main consumers of seaweed with an average annual consumption of 1.6 kg (dry weight) per capita (Fujiwara-Arasaki et al., 1984). A wide variety of sea algae such as wakame (*Undaria pinnatifida*) and hizikia (*Hizikia fusiforme*) are a part of staple diet in Japan from time immemorial (Dawczynski et al., 2007; Nagai and Yukimoto, 2003). Wakame is the dominant alga in Japan accounting for more than 50% consumption (Davidson, 1999). It is consumed in soups, salads, sweetened vinegar, cooked food, and ingredients. Wakame is rich in fucoxanthin, which is a xanthophyll characteristic of brown sea algae, and accounts for more than 10% of estimated total natural production of carotenoids (Miyashita andHosokawa, 2006, 2008; Shiratori et al., 2005). Fucoxanthin and its metabolites have been reported to possess antioxidative, anticancerous, antiobesity, and antiinflammatory properties (Miyashita and Hosokawa, 2008). Hizikia (*Hizikia fusiforme*), also called Hijiki, is eaten as fried food, sea lettuce, in sweetened vinegar, and green laver. It could be eaten fresh, but is mostly dried and rehydrated before use (Davidson, 1999). Data regarding the bioactive compounds of edible algae consumed in Japan are sparse.

Brown sea algae are known to contain more bioactive compounds than either green or red sea algae (Seafoodplus, 2008). Some of the bioactive compounds identified in brown sea algae include phylopheophylin, phlorotannins, fucoxanthin, and various other metabolites (Hosokawa et al., 2006). Nagai and Yukimoto (2003) investigated the preparation and functional properties of beverages made from brown sea algae. Four beverages were prepared from four sea algae (wakame, sea trumpet, hizikia, sea lettuce). The phenolic compounds in one of the beverages (48.3 mg/g dry matter) were similar to those in green tea (50.7 mg/g dry matter) of equivalent amounts. The antioxidant activity of the algae correlated with their polyphenol contents; the researchers concluded that the algae had high antioxidant properties. From their evaluation of three brown seaweeds from India, Chandini, Ganesan and Bhaskar (2008) concluded that sea algae can be utilized as a source of natural antioxidant compounds as their crude extracts and fractions exhibit antioxidant activity.

Another brown alga haba-nori (*Petalonia binghamiae* (J. Agaradh) Vinogradova) is consumed as a traditional food along the fisheries town areas in Japan (Kuda et al., 2006). It is usually dried then lightly roasted for consumption. There are limited reports regarding the bioactivity of dried algae plants (Kuda et al., 2005). Kuda et al. (2006) examined the beneficial properties of *P. binghamiae* for human food, and found that the water extract was a rich source of phenolic compounds with reducing power and radical scavenging activity. The radical scavenging activity was described as being promoted by heat treatment such as retorting and the antioxidant activities were dependent on the phenolic compounds (Kuda et al., 2006).

Kuda et al. (2005) also investigated another brown alga, *Scytosiphon lomentaria*. The dried product of *Scytosiphon lomentaria* (*Scytosiphonales*, *Phaeophyceae*) called kayamo-nori in Japan is consumed as a traditional food in the Noto, Ishikawa area of Japan (Kuda et al., 2005). It is eaten after drying and roasting lightly. Kuda et al. (2005) pointed out that *S. lomentaria* is not usually soaked before eating as is the case with other dried algae, which are usually eaten after swelling with 20–40 volumes of water. The researchers demonstrated that *S. lomentaria* contained total phenols and showed strong antioxidant properties. Jiménez-Escrig et al. (2001) reported that the radical scavenging activity of a brown alga *Fucus* was decreased by 98% after drying at 50°C for 48 h.

22.8.3.4 Safety of edible algae

It has been reported that the arsenic content of *H. fusiforme* extract is high (Nagai and Yukimoto, 2003). According to Araki (1983), drying and storage reduce the bioactive compounds in algal products. Prabhasankar et al. (2009) reported that fucoxanthin was not affected when pasta was prepared with wakame as well as an additional cooking step.

Bioactive compounds found in edible sea algae have great potential for increased use in diets and chronic disease prevention if further researched.

22.9 Conclusion

In most developing and developed countries, traditional foods are included in the daily diets; therefore, they could play an important role in the prevention of chronic diseases. This chapter supports the idea that some traditional/ethnic foods are good sources of bioactive compounds. Indeed, the foods discussed in this chapter have been reported to contain bioactive compounds in high levels. While it is recognized that yerba mate has excellent bioactive compounds, the presence of contaminants or its safety has not been extensively documented. Pulque contains steroidal saponins and various other micronutrients, while Rooibos contains bioactive compounds, which may help protect against free radical damage that can lead to cancer, heart attack, and stroke. Unfermented (green) Rooibos has higher amounts of polyphenols than traditionally fermented Rooibos, and generally demonstrates higher antioxidant and antimutagenic capabilities in vitro. The bioactive compounds reported for Java plum included sterols, triterpenes, coumarins, tannins, glycosides, alkaloids, reducing compounds, anthocyanin pigments, and saponins. Mango and possibly its derived products are rich sources of bioactive compounds; however, the safety aspects and effect of processing on these have not yet been very well studied. Bioactive compounds identified in brown sea algae, *Scytosiphon lomentaria* (*Scytosiphonales*, *Phaeophyceae*) and *Petalonia binghamiae* (J. Agaradh) Vinogradova include phylopheophylin, phlorotannins, fucoxanthin, and various other metabolites.

Even though the evidence is accumulating for the beneficial effects of bioactive compounds, the database remains incomplete and fragmented. Much of the evidence for the beneficial effects of bioactive compounds in traditional foods comes primarily from in vitro and animal models with short duration, which frequently use pharmacological doses; much more than is consumed in the diet. Furthermore, little is known about absorption, bioavailability, and safety of bioactive compounds in human cells. Bioavailability from different kinds of bioactive compounds will probably vary greatly, given the current variations in concentrations reported from food to food.

22.10 Future scope

The current chapter is a small effort in the direction of highlighting the benefits and safety of traditional/ethnic foods and their role in the prevention of chronic diseases. Although research has been done to evaluate the composition, beneficial effect, and safety of traditional/ethnic foods, much more work needs to be done to develop a comprehensive, globalized database. National and local governments, health agencies, scientists, international organizations, community-based organizations (CBOs), and nongovernmental organizations (NGOs) should come forward to ensure the research is conducted to meet this urgent need. Such database will have the potential to meet increasing consumer demand for foods to address and prevent chronic diseases.

In general, there is an urgent need for long-term systematic studies involving humans to evaluate the benefits, safety, and bioavailability of traditional/ethnic foods. Also, food preparation methods of traditional/ethnic foods need to be researched with emphasis on the best methods to preserve the bioactive compounds without impacting on sensory, nutritional, and consumer properties. Further research is needed to establish the safety of these foods and their potential as possible natural substitutes for the prevention of chronic disease.

Specifically, the functional effect of saponins in Pulque has not been clearly elucidated; this warrants further research. The possibility of bacterial contamination during the processing of Pulque also needs to be addressed in future research studies. The bioavailability, tissue distribution, and biological activities of the bioactive compounds in Rooibos need to be further researched. Further research is also needed to determine whether the bioactive compound of Rooibos observed in vitro and in animal translates into health benefits for humans. Much more research is needed to determine the safety of Java plum and mango, and to determine whether the bioactive compounds in the mango are still intact in the derived products. Bioactive compounds found in edible sea algae have great potential for increased use in diets and chronic disease prevention if further researched.

Much more research and information campaigns on bioactive compounds in these traditional/ethnic foods might have the potential to improve the health situation globally in reasonable and sustainable way. As the food system becomes more globalized and harmonized, traditional/ethnic foods, which have many health benefits, may be able to alleviate the development of chronic diseases. The notion of the bioactivity, benefits, and safety of traditional/ethnic foods constitutes a very interesting topic, which requires further research.

Acknowledgments

The author wishes to thank Mr. Larry Keener for the pivotal role he played in making this chapter a reality and also, the Food and Nutritional Sciences Advisory Board, Department of Food and Nutritional Sciences, Tuskegee University for their role in making the chapter possible.

References

ADA, 2004. Position paper of the American Dietetic association: functional foods. J. Am. Diet Assoc. 104, 814–826.

Araki, S., 1983. Processing of dried porphyra (nori). In: The Japanese Society of Scientific Fisheries. Biochemistry and Utilization of Marine Algae. Koseisha Koseikaku, Tokyo (In Japanese).

Aquino, R., Conti, C., deSimone, F., et al., 1991. Antiviral activity of constituents of *Tamus communis*. J. Chemother. 3, 305–309.

Backstrand, J.R., Allen, L.H., Black, A.K., et al., 2002. Diet and iron status of nonpregnant women in rural Central Mexico. Am. J. Clin. Nutr. 76, 156–164.

Backstrand, J.R., Allen, L.H., Martinez, E., Pelto, G.H., 2001. Maternal consumption of Pulque, a traditional central Mexican alcoholic beverage: relationships to infant growth and development. Publ. Health Nutr. 4, 883–891.

Banerjee, A., Dasgupta, N., De, B., 2005. In vitro study of antioxidant activity of *Syzygium cumini* fruit. Food Chem. 90, 727–733.

Bastos, D.H.M., Ishimoto, E.Y., Marques, M.O.M., et al., 2006. Essential oil and antioxidant activity of green mate and mate tea (*Ilex paraguariensis*) infusions. J. Food Compos. Anal. 19, 538–543.

Bazzano, L.A., He, J., Ogden, L.G., et al., 2002. Fruit and vegetable intake and risk of cardiovascular disease in US adults: the first National Health and Nutrition Examination Survey Epidemiologic Follow-up study. Am. J. Clin. Nutr. 76, 93–99.

Beecher, G.R., 2003. Overview of dietary flavonoids: nomenclature, occurrence and intake. J. Nutr. 133, 3248S–3254S.

Berardini, N., Fezer, R., Conrad, J., et al., 2005. Screening of mango (*Mangifera indica* L.) cultivars for their contents of flavonol O- and xanthone C-glycosides, anthocyanins, and pectin. J. Agric. Food Chem. 53, 1563–1570.

Bixby, M., Spieler, L., Menini, T., Gugliucci, A., 2005. *Ilex paraguariensis* extracts are potent inhibitors of nitrosative stress: a comparative study with green tea and wines using a protein nitration model and mammalian cell cytotoxicity. Life Sci. 77, 345–358.

Bracesco, N., Dell, M.R.A., Behtash, S., et al., 2003. Antioxidant activity of a botanical extract preparation of *Ilex paraguariensis:* prevention of DNA double-strand breaks in *Saccharomyces cerevisiae* and human low- density lipoprotein oxidation. J. Alternative Compl. Med. 9, 379–387.

Bramati, L., Aquilano, F., Pietta, P., 2003. Unfermented rooibos tea: quantitative characterization of flavonoids by HPLC-UV and determination of the total antioxidant activity. J. Agric. Food Chem. 50, 7472–7474.

Bravo, L., Luis, G., Lecumberri, E., 2007. LC/MS characterization of phenolic constituents of mate (*Ilex paraguariensis*, St. Hil.) and its antioxidant activity compared to commonly consumed beverages. Food Res. Int. 40, 393–405.

Brown, J.P., 1980. A review of the genetic effects of naturally occurring flavonoids, anthraquinones and related compounds. Mutat. Res. 75, 243–277.

Cardozo, E.L., Ferrarese-Filho, O., Filho, L.C., et al., 2007. Methylxanthines and phenolic compounds in mate (*Ilex paraguariensis* St. Hil.) progenies grown in Brazil. J. Food Compos. Anal. 20, 553–558.

Chandini, S.K., Ganesan, P., Bhaskar, N., 2008. In vitro antioxidant activities of three selected brown seaweeds of India. Food Chem. 107, 707–713.

Chandra, S., Gonzalez de Mejia, E., 2004. Polyphenolic compounds, antioxidant capacity, and quinone reductase activity of aqueous extract of *Ardisia compressa* in comparison to mate (*Ilex paraguariensis*) and green (*Camellia sinensis*) teas. J. Agric. Food Chem. 52, 3583–3589.

Crozier, A., Burns, J., Aziz, A., et al., 2000. Antioxidant flavonols from fruits, vegetables and beverages: measurements and bioavailability. Biol. Res. 33, 79–88.

Crozier, A., Jaganath, I.B., Clifford, M.N., 2006. Phenols, polyphenols and tannins: an overview. In: Crozier, A., Clifford, M.N., Ashihara, H. (Eds.), Plant Secondary Metabolites. Blackwell Publishing Ltd, Oxford, pp. 1–24.

Cushnie, T.P.T., Lamb, A.J., 2005. Antimicrobial activity of flavonoids. Int. J. Antimicrob. Agents 26, 343–356.

da Silva, E.L., Neiva, T.J.C., Shirai, M., et al., 2008. Acute ingestion of yerba mate infusion (*Ilex paraguariensis*) inhibits plasma and lipoprotein oxidation. Food Res. Int. 41, 973–979.

Davidson, A., 1999. The Oxford Companion to Food. Frome. Butler and Tanner Ltd, Somerset, Great Britain, pp. 831–832.

Dawczynski, C., Schubert, R., Jahreis, G., 2007. Amino acids, fatty acids, and dietary fibre in edible seaweed products. Food Chem. 103, 891–899.

Day, L., Seymour, R.B., Pitts, K.F., et al., 2008. Incorporation of functional ingredients into foods. Trends Food Sci. Technol.

Dellacassa, E., Bandoni, A.L., 2001. El mate. Revista de Fitoterapia 1, 269–278.

Denny, A., Buttriss, J., 2008. Synthesis Report No. 4: Plant Foods and Health: Focus on Plant Bioactives. EuroFIR Project Management Office/British Nutrition Foundation.

Di Gregorio, D.E., Huck, H., Aristegui, R., et al., 2004. ^{137}Cs contamination in tea and yerba mate in South America. J. Environ. Radioact. 76, 273–281.

Dos, A., Ayhan, Z., Sumnu, G., 2005. Effects of different factors on sensory attributes, overall acceptance and preference of rooibos (*Aspalathus linearis*) tea. J. Sensory Stud. 20, 228–242.

Erickson, L., 2003. Rooibos tea: research into antioxidant and antimutagenic properties. HerbalGram 59, 34–45.

Escalante, A., Giles-Gómez, M., Hernández, G., et al., 2008. Analysis of bacterial community during the fermentation of Pulque, a traditional Mexican alcoholic beverage, using a polyphasic approach. Int. J. Food Microbiol. 14, 126–134.

Escalante, A., Rodríguez, M.E., Martínez, A., et al., 2004. Characterization of bacterial diversity in Pulque, a traditional Mexican alcoholic beverage, as determined by 16S rDNA analysis. FEMS Microbiol. Lett. 235, 273–279.

FAO, 2007. FAO Statistical Database, Agriculture. http://apps.fao.org (Accessed 11 February 2009).

Filip, R., López, P., Giberti, P.G., et al., 2001. Phenolic compounds in seven South American *Ilex* species. Fitoterapia 72, 774–778.

Filip, R., Lotito, S.B., Ferraro, G., Fraga, C.G., 2000. Antioxidant activity of *Ilex paraguariensis* and related species. Nutr. Res. 20, 1437–1446.

Fraga, C.G., 2007. Plant polyphenols: how to translate their in vitro antioxidant actions to in vivo conditions. IUBMB Life 59, 308–315.

Fujiwara-Arasaki, T., Mino, N., Kuroda, M., 1984. The protein value in human nutrition of edible marine algae in Japan. Hydrobiologia 116/117, 513–516.

Gandini, S., Merzenich, H., Robertson, C., Boyle, P., 2000. Meta analysis of studies on breast cancer risk and diet: the role of fruit and vegetable consumption and the intake of associated micronutrients. Eur. J. Cancer 36, 636–646.

Godoy, H., Rodriguez-Amaya, D.B., 1989. Carotenoid composition of commercial mangoes from Brazil. Lebensm. Wiss. Technol. 22, 100–103.

Goldberg, B.M., Brinckmann, A., 2000. Herbal Medicine: Expanded Commission E Monographs. Newton, MA: Integrative Medicine Communications, pp. 249–252.

Gorzalczany, S., Filip, R., Alonso, M.-R., et al., 2001. Choleretic effect and intestinal propulsion of 'mate' (*Ilex paraguariensis*) and its substitutes or adulterants. J. Ethnopharmacol. 75, 291–294.

Gry, J., Eriksen, F.D., Pilegaard, K., et al., 2007. EuroFIR-BASIS: a combined composition and biological activity database for bioactive compounds in plant-based food. Trends Food Sci. Technol. 18, 434–444.

Halliwell, B., 1994. Free radicals, antioxidants and human disease: curiosity cause or consequence. Lancet 344, 721–724.

Harborne, J.B., Baxter, H., 1999. The Handbook of Natural Flavonoids, vols. 1–2. John Wiley and Sons, Chichester, UK.

Health Canada, 2004. Final Policy Paper on Nutraceuticals/functional Foods and Health Claims on Foods.

Heim, K.E., Tagliaferro, A.R., Bobilya, D.J., 2002. Flavonoid antioxidants: chemistry, metabolism and structure-activity relationships. J. Nutr. Biochem. 13, 572–584.

Hesseling, P.B., Joubert, J.R., 1982. The effect of rooibos tea on the type I allergic reaction. S. Afr. Med. J. 62, 1037–1038.

Hosokawa, M., Bhaskar, N., Sashima, T., Miyashita, K., 2006. Fucoxanthin as a bioactive and nutritionally beneficial marine carotenoid, A review. Carotenoid Sci. 10, 15–28.

Jagetia, G.C., Baliga, M.S., 2005. Radioprotection by mangiferin in Dbaxc57bl mice: a preliminary study. Phytomedicine 12, 209–215.

Jiménez-Escrig, A., Jiménez-Jiménez, I., Pulido, R., Saura- Calixto, F., 2001. Antioxidant activity of fresh and processed edible seaweeds. J. Sci. Food Agric. 81, 530–534.

Joubert, E., 1996. HPLC quantification of the dihydrochalcones, aspalathin and nothofagin in rooibos tea (*Aspalathus linearis*) as affected by processing. Food Chem. 55, 403–411.

Joubert, E., Gelderblom, W.C.A., Louw, A., de Beer, D., 2008. South African herbal teas: *Aspalathus linearis*, *Cyclopia* spp. and *Athrixia phylicoides*, A review. J. Ethnopharmacol. 119, 376–412.

Juráni, M., Lamosová, D., Mácajová, M., et al., 2008. Effect of rooibos tea (*Aspalathus linearis*) on Japanese quail growth, egg production and plasma metabolites. Br. Poultry Sci. 49, 55–64.

Kato, A., Miura, T., Fukunaga, T., 1995. Effects of steroidal glycosides on blood glucose in normal and diabetic mice. Biol. Pharm. Bull. 18, 167–168.

Khokhar, S., Gilbert, P.A., Moyle, C.W.A., et al., 2009. Harmonised procedures for producing new data on the nutritional composition of ethnic foods. Food Chem. 113, 816–824.

Kim, J.-Y., Kwon, O., 2009. Garlic intake and cancer risk: an analysis using the food and drug administration's evidence-based review system for the scientific evaluation of health claims. Am. J. Clin. Nutr. 89, 257–264.

Kim, Y., Brecht, J.K., Talcott, S.T., 2007. Antioxidant phytochemical and fruit quality changes in mango (*Mangifera indica* L.) following hot water immersion and controlled atmosphere storage. Food Chem. 105, 1327–1334.

Kim, Y., Lounds-Singleton, A.J., Talcott, S.T., 2009. Antioxidant phytochemical and quality changes associated with hot water immersion treatment of mangoes (*Mangifera indica* L.). Food Chem. 15 (3), 989–993.

Koeppen, B.H., April 1970. C-glycosyl compounds in rooibos tea. Food Ind. South Africa 49.

Koeppen, B.H., Roux, D.G., 1965. Aspalathin: a novel C-glycosylflavonoid from aspalathin linearis. Tetrahedron Lett. 39, 3497–3503.

Kolonel, L.N., Hankin, J.H., Whittemore, A.S., et al., 2000. Vegetables, fruits, legumes and prostate cancer: a multiethnic case-control study. Cancer Epidemiol. Biomark. Prev. 9, 795–804.

Kuda, T., Hishi, T., Maekawa, S., 2006. Antioxidant properties of dried product 'haba-nori', an edible brown alga, *Petalonia binghamiae* (J. Agaradh) Vinogradova. Food Chem. 98, 545–550.

Kuda, T., Tsunekawa, M., Hishi, T., Araki, Y., 2005. Antioxidant properties of driedkayamo-nori', a brown alga *Scytosiphon lomentaria* (Scytosiphonales, Phaeophyceae). Food Chem. 89, 617–622.

Kunishiro, K., Tai, A., Yamamoto, I., 2001. Effects of rooibos extract on antigen-specific antibody production and cytokine generation in vitro and in vivo. Biosci. Biotechnol. Biochem. 65, 2137–2145.

Lacaille-Dubois, M.-A., 2005. Bioactive saponins with cancer related and immunomodulatory activity: recent developments. Stud. Nat. Prod. Chem. 32, 209–246.

Lamosova, D., Jurani, M., Greksak, M., et al., 1997. Effect of Rooibos tea (*Aspalathus linearis*) on chick skeletal muscle cell growth in culture. Comp. Biochem. Physiol. C Pharmacol. Toxicol. Endocrinol. 116, 39–45.

Leatherhead Food International, 2004. The European Ethnic Foods Market Report, second ed. LFI, Leatherhead.

Leatherhead Food International, 2007. The European Ethnic Foods Market Report, third ed. LFI, Leatherhead.

Malinow, M.R., 1985. Effects of synthetic glycosides on cholesterol absorption. Ann. N. Y. Acad. Sci. 454, 23–27.

Marnewick, J.L., Gelderblom, W.C.A., Joubert, E., 2000. An investigation on the antimutagenic properties of South African herbal teas. Mutat. Res. 471, 157–166.

Marnewick, J.L., Joubert, E., Swart, P., et al., 2003. Modulation of hepatic drug metabolizing enzymes and oxidative status by green and black (*Camellia sinensis*), rooibos (*Aspalathus linearis*) and honeybush (*Cyclopia intermedia*) teas in rats. J. Agric. Food Chem. 51, 8113–8119.

Masibo, M., He, Q., 2008. Major mango polyphenols and their potential significance to human health. Compr. Rev. Food Sci. Food Saf. 7, 309–319.

McAnlis, G.T., McEneny, J., Pearce, J., Young, I.S., 1999. Absorption and antioxidant effects of quercetin from onions, in man. Eur. J. Clin. Nutr. 53, 92–96.

McKay, D.L., Blumberg, J.B., 2007. A review of the bioactivity of South African herbal teas: rooibos (*Aspalathus linearis*) and honeybush. Phytother Res. 21, 1–16.

Miyashita, K., Hosokawa, M., 2008. Beneficial health effects of seaweed carotenoid, fucoxanthin. In: Barrow, C., Shahidi, F. (Eds.), Marine Nutraceuticals and Functional Foods. CRC Press, Boca Raton, USA, pp. 297–320.

Morton, J.F., 1983. Rooibos tea, *Aspalathus linearis*, a caffeineless, low-tannin beverage. Econ. Bot. 37, 164–173.

Muruganandan, S., Gupta, S., Kataria, M., et al., 2005. Mangiferin protects the streptozotocin-induced oxidative damage to cardiac and renal tissues in rats. Toxicology 176, 165–173.

Nagai, T., Yukimoto, T., 2003. Preparation and functional properties of beverages made from sea algae. Food Chem. 81, 327–332.

Nakano, M., Itoh, Y., Mizuno, T., et al., 1997a. Polysaccharide from *Aspalathus linearis* with strong anti- HIV activity. Biosci. *Biotechnol. Biochem.* 61, 267–271.

Nakano, M., Nakashima, H., Itoh, Y., 1997b. Anti-human immunodeficiency virus activity of oligosaccharides from rooibos tea (*Aspalathus linearis*) extracts in vitro. Leukemia 11, 128–130.

Ndyomugyenyi, K.E., 2008. Nutritional Evaluation of Java Plum (Syzygium Cumini) Beans in Broiler Diets. MSc. Thesis. Available at: Makerere University Library, pp. 35–39.

Oldewage-Theron, W.H., Dicks, E.G., Napier, C.E., Rutengwe, R., 2005. Situation analysis of an informal settlement in the Vaal Triangle. Dev. South Afr. 22, 13–26.

Peña-Alvarez, A., Diaz, L., Medina, A., et al., 2004. Characterization of three *Agave* species by gas chromatography and solid-phase microextraction-gas chromatography-mass spectrometry. J. Chromatogr. A 1027, 131–136.

Peng, J.P., Chen, H., Qiao, Y.Q., et al., 1996. Two new steroidal saponins from *Allium sativum* and their inhibitory effects on blood coagulability. Acta Pharmacol. Sin. 31, 613–616.

Pietta, P.G., 2000. Flavonoids as antioxidants. J. Nat. Prod. 63, 1035–1042.

Prabhasankar, P., Ganesan, P., Bhaskar, N., et al., 2009. Edible Japanese seaweed, wakame (*Undaria pinnatifida*) as an ingredient in pasta: chemical, functional and structural evaluation. Food Chem. 15 (2), 501–508.

Prediger, R.D.S., Fernandes, M.S., Rial, D., et al., 2008. Effects of acute administration of the hydroalcoholic extract of mate tea leaves (*Ilex paraguariensis*) in animal models of learning and memory. J. Ethnopharmacol. 120, 465–473.

Rabe, C., Steenkamp, J.A., Joubert, E., et al., 1994. Phenolic metabolites from rooibos tea (*Aspalathus linearis*). Phytochemistry (Oxf.) 35, 1559–1565.

Randle, P.J., Garland, P.B., Hales, C.N., Newsholme, E.A., 1963. The glucose-fatty acid cycle, its role in insulin sensitivity and metabolic disturbances in diabetes mellitus. Lancet 1, 785–789.

Ravikumar, P.R., Hammesfahr, P., Sih, C.J., 1979. Cytotoxic saponins form the Chinese herbal drug Yunnan Bai Yao. J. Pharmaceut. Sci. 68, 900–903.

Ribeiro, S.M.R., Barbosa, L.C.A., Queiroz, J.H., et al., 2008. Phenolic compounds and antioxidant of Brazilian mango (*Mangifera indica* L.) varieties. Food Chem. 110, 620–626.

Ribeiro, S.M.R., de Queiroz, J.M., Lopes, M.A., et al., 2007. Antioxidant in mango (*Mangifera indica* L.) pulp. Plant Foods Hum. Nutr. 62, 13–17.

Saba, A., Messina, F., 2003. Attitudes towards organic foods and risk/benefit perception associated with pesticides. Food Qual. Prefer. 14, 637–645.

Santos, W.N., Bernardo, R.R., Pecanha, L.M.T., et al., 1997. Haemolytic activities of plant saponins and adjuvants. Effect of Periandra mediterranea saponin on the humoral response to the FML antigen of *Leishmania donovani*. Vaccine 15, 1024–1029.

Sauvaire, Y., Ribes, G., Baccou, J.C., Loubatierés-Mariani, M.M., 1991. Implication of steroid saponins and sapogenins in the hypocholesterolemic effect of fenugreek. Lipids 26, 191–197.

Scalbert, A., Williamson, G., 2000. Dietary intake and bioavailability of polyphenols. J. Nutr. 130, 2073–2085.

Schieber, A., Ullrich, W., Carle, R., 2000. Characterization of polyphenols in mango puree concentrate by HPLC with diode array and mass spectrometric detection. Innovat. Food Sci. Emerg. Technol. 1, 161–166.

Schinella, G.R., Troiani, G., Dävila, V., et al., 2000. Antioxidant effects of Na aqueous extract of Ilex paraguariensis. Biochem. Biophys. Res. Commun. 269, 357–360.

Schulz, H., Joubert, E., Schutze, W., 2003. Quantification of quality parameters for reliable evaluation of green rooibos. Eur. Food Res. Technol. 216, 539–543.

Seafoodplus, 2008. Seafoodplus Web Page. www.seafoodplus.org/fileadmin/files/news/2004-01-22SFRTD1launchBrussels. pdf (Accessed 11 February 2009).

Sewram, V., De Stefani, E., Brennan, P., Boffetta, P., 2003. Mate consumption and the risk of squamous cell esophageal cancer in Uruguay. Cancer Epidemiol. Biomark. Prev. 12, 508–513.

Shimamura, N., Miyase, T., Umehara, K., et al., 2006. Phytoestrogens from *Aspalathus linearis*. Biol. Pharm. Bull. 29, 1271–1274.

Shiratori, K., Ohgami, I., Ilieva, X., et al., 2005. Effects of fucoxanthin on lipopolysaccharide-induced inflammation in vitro and in vivo. Exp. Eye Res. 81, 422–428.

Sigler, K., Ruch, R.J., 1993. Enhancement of gap junctional intercellular communication in tumor promoter-treated cells by components of green tea. Cancer Lett. 69, 5–9.

Sissing, A., 2008. Investigations into the Cancer Modulating Properties of *Aspalathus Linearis* (Rooibos), *Cyclopia Intermedia* (Honeybush) and *Sutherlandia Frutescens* (Cancer Bush) in Oesophageal Carcinogenesis. MSc Thesis. Available at: The University of the Western Cape, Bellville, South Africa.

Sparg, S.G., Light, M.E., van Staden, J., 2004. Biological activities and distribution of plant saponins. J. Ethnopharmacol. 94, 219–243.

Stangeland, T., Remberg, S.F., Lye, K.A., 2009. Total anti- oxidant activity in 35 Ugandan fruits and vegetables. Food Chem. 113, 85–91.

Stoilova, I., Gargova, S., Stoyanova, A., Ho, L., 2005. Antimicrobial and antioxidant activity of the polyphenol mangiferin. Haematol. Pol. 51, 37–44.

Sung, M.K., Kendall, C.W.C., Rao, A.V., 1995. Effect of saponins and gypsophila saponin on morphology of colon carcinoma cells in culture. Food Chem. Toxicol. 33, 357–366.

Tejasari, D., 2007. Evaluation of ginger (*Zingiber officinale* Roscoe) bioactive compounds in increasing the ratio of T-cell surface molecules of CD3+ CD4:CD3+ CD8+ In-Vitro. Mal J. Nutrition 13, 161–170.

Tinto, W.F., Simmons-Boyce, J.L., McLean, S., Reynolds, W.F., 2005. Constituents of agave Americana and agave barbadensis. Fitoterapia 76, 594–597.

Ulicnä, O., Vancovä, O., Bozek, P., et al., 2006. Rooibos tea (*Aspalathus linearis*) partially prevents oxidative stress in streptozotocin-induced diabetic rats. Physiol. Res. 55, 157–164.

Urquiaga, I., Leighton, F., 2000. Plant polyphenol antioxidants and oxidative stress. Biol. Res. 33, 2–11.

Veigas, J.M., Narayan, M.S., Laxman, P.M., Neelwarne, B., 2007. Chemical nature, stability and bioefficacies of anthocyanins from fruit peel of *Syzygium cumini* Skeels. Food Chem. 105, 619–627.

Waleron, M., Waleron, K., Podhajska, A.J., Lojkowska, E., 2002. Genotyping of bacteria belonging to the former *Erwina* genus by PCR-RFLP analysis of a *recA* gene fragment. Microbiology 148, 583–595.

Wen, W., Shu, X.-O., Li, H., et al., 2009. Dietary carbohydrates, fiber, and breast cancer risk in Chinese women. Am. J. Clin. Nutr. 89, 283–289.

Yang, D.-J., Lu, T.-J., Hwang, L.S., 2009. Effect of endogenous glycosidase on stability of steroidal saponins in Taiwanese yam (*Dioscorea pseudo-japonica yamamoto*) during drying processes. Food Chem. 113, 155–159.

Yoshikawa, M., Nishida, N., Shimoda, H., et al., 2001. Polyphenol constituents from salacia species: quantitative analysis of mangiferin with glucosidase and aldose reductase inhibitory activities. Yakugaku Zasshi 121, 371–378.

Zhang, J., Meng, Z., Zhang, M., et al., 1999. Effect of six steroidal saponins isolated from *Anemarrhenae rhizome* on platelet aggregation and hemolysis in human blood. Clin. Chim. Acta 289, 79–88.

Further reading

Charmaine, S., 1998. Asia Food. Encyclopedia of Asian Food (Periplus Editions). New Holland Publishers Pty Ltd, Australia.

Henry, M., Harris, K.S., 2002. LMH Official Dictionary of Jamaican Herbs and Medicinal Plants and Their Uses. LMH Publishing Ltd.

Mitchell, S.A., Ahmad, M.H., 2006. A review of medicinal plant research at the University of the West Indies, Jamaica, 1948-2001. W. Indian Med. J. 55, 243–269.

Ricks, R.M., Vogel, P.S., Elston, D.M., Hivnor, C., 1999. Purpuric agave dermatitis. J. Am. Acad. Dermatol. 40, 356–358.

Stangeland, T., Remberg, S.F., Lye, K.A., 2007. Antioxidants in some Ugandan fruits. Afr. J. Ecol. 45, 29S–30S.

Vanisree, M., Alexander-Lindo, R.L., DeWitt, D.L., Nair, M.G., 2008. Functional food components of *Antigonon leptopus* tea. Food Chem. 106, 487–492.

Chapter 23

Water determination in food

Heinz-Dieter Isengard
University of Hohenheim, Institute of Food Science and Biotechnology, Stuttgart, Germany

23.1 Introduction

When goods are sold beyond international borders, they must meet the requirements of the receiving country. These may differ from the regulations existing in the delivering country. Therefore, when the laws or traditional consumer expectations are not the same in the countries involved, a compromise or an agreement between the trade partners is necessary. The best solution would, of course, be to have the same regulations in both countries and, by general extrapolation, in all countries.

These considerations are particularly relevant for trade of agricultural products and food. Certain components may be characteristic for a given product and may also be decisive for the price. The content or concentration of such compounds must then be analyzed. The method of determination may be not the same in the countries concerned. Depending on the situation, one of the partners would have an advantage if the regulation existing in one of the countries was taken as standard. Again, an agreement is necessary. Chapter 24 (Global Harmonization of Analytical Methods) gives a more general discussion on analytical methods used to establish the basic composition and quality of foods.

A further problem may be where a commonly accepted or internationally established method may exist, which may not be based on scientific research. Thus, although the results might be legally correct, they would not reflect the true value. If the existing method is replaced by another—scientifically correct—method, however, this procedure may meet the resistance of a partner should they want the existing method to be conserved. This chapter describes such a situation.

Indirect or secondary methods must be calibrated against a direct or primary method. When the primary method is not adequate, it may nevertheless be possible to establish a calibration, but the results derived from it may be erroneous or false, even though this might not be immediately obvious. Such an example is also given.

23.2 Water content

23.2.1 Importance of water content

Water is one of the most important substances in food (Isengard, 2001). It is present in every foodstuff, in a range that may vary from extremely low values in dried products to extremely high ones in beverages. Its content is of great significance in many respects, affecting properties like conductivity for heat and electrical current, density and rheological properties, or corrosiveness. All these factors must be taken into account for the design of technological processes. The amount of water in food is determinant for its nutritive value and taste. In some cases, it can also be considered as an impurity. In reference materials, the water content is important as far as specifications are given on the basis of either dry matter or initial mass (Rückold et al., 2001). Stability and shelf life of foodstuffs are highly dependent on water content (via water activity), since water is important for microbiological life and most enzymatic activities. Storage volume and mass depend also on water content, which can be reflected in transport costs. As water is relatively cheap, its presence in foodstuffs in general and in high-value products in particular is of commercial interest. For this and other reasons, rules and regulations concerning water content are imposed. As a consequence of all the above, the determination of water content is certainly the most frequent analysis performed on foodstuffs.

Ensuring Global Food Safety. https://doi.org/10.1016/B978-0-12-816011-4.00002-1

23.2.2 Methods to determine water content

The aim of correct water content analyses should be to detect water, all the water and nothing but the water in the sample. Direct or primary methods determine water as such. This can be done physically by separating the water contained in the sample and measuring its mass or its volume. Another possibility is to analyze water content by a selective chemical reaction. Indirect or secondary methods determine either a sample property that depends on water content, such as density, sound velocity, dielectric properties or electrical conductivity, or the response of the water molecules in the sample on a physical influence, which comprises spectrometric techniques like near-infrared (NIR) spectrometry, microwave resonance spectrometry, or time-domain nuclear magnetic resonance (TD-NMR). Indirect methods need a calibration against a direct method (Isengard, 1995).

23.2.3 Drying techniques

Water content (or what is believed to be water content)—and thus dry matter—is often determined by drying techniques, particularly by drying the product at a certain temperature for a certain time in a drying oven.

Drying techniques, be it the "classical" oven drying, vacuum drying, freeze drying, infrared, or microwave drying, do not distinguish between water and other volatile substances. The result of all of these methods is not water content but the mass loss the product undergoes under the conditions applied. These conditions (sample size, temperature, pressure, time, energy input, criteria to stop the analysis) can principally be freely chosen. The result depends very much on these conditions but may be very reproducible. This alone shows that this technique, leading to different results when the parameters are changed, cannot be the correct one, because water content is a sample property which has a certain, though unknown value. From the scientific point of view, the results of drying methods should therefore not be called "water content" but rather "mass loss on drying" with indication of the drying conditions. In the past years, the term "moisture content" has more often been used as a compromise. It means the relative mass loss by evaporation of water (though possibly not all of the water) and of other volatile compounds under the drying conditions.

The problem with all drying techniques is that they do not measure water specifically. All of the volatile compounds under the analytical conditions contribute to the mass loss, even compounds that are not originally contained in the sample but formed by chemical reactions during the analysis, particularly by decomposition reactions at higher temperatures. On the other hand, strongly bound water may escape detection.

These two opposite errors, inclusion of other volatiles on the one hand and water not detected on the other hand, may compensate each other when the drying parameters are chosen in an appropriate way (Isengard, 1995; Isengard and Walter, 1998; Isengard and Färber, 1999; Heinze and Isengard, 2001; Isengard and Präger, 2003). The appropriate choice of the parameters necessitates, of course, that the true water content has been analyzed before with a method selective for water as a primary method. The parameters of the secondary method must then be chosen in a way that the result corresponds to the water content determined with the primary method. Once the secondary method is calibrated in this way, it can be applied for this particular type of product. The calibration is product specific and the same parameters cannot be applied for other types of samples. This procedure is particularly interesting for rapid drying techniques like microwave or infrared drying.

23.2.4 Karl Fischer titration

The most important primary method to determine water content is the Karl Fischer titration (Scholz, 1984). It is based on a chemical reaction selective for water:

$$ROH + SO_2 + Z \rightarrow ZH^+ + ROSO_2^- \tag{23.1}$$

$$ZH^+ + ROSO_2^- + I_2 + H_2O + 2\,Z \rightarrow 3\,ZH^+ + ROSO_3^- + 2\,I^- \tag{23.2}$$

$$\text{Overall reaction: } 3\,Z + ROH + SO_2 + I_2 + H_2O \rightarrow 3\,ZH^+ + ROSO_3^- + 2\,I^- \tag{23.3}$$

Z is a base (very often imidazole), and ROH is an alcohol, usually methanol.

In the first step, the alcohol is esterified with sulfur dioxide to form alkyl sulfite. The base provides for a practically complete reaction, Eq. (23.1). In the second step, this alkyl sulfite is oxidized by iodine to form alkyl sulfate; this reaction requires water, Eq. (23.2). The overall reaction, Eq. (23.3), shows that the consumption of iodine is stoichiometrically equivalent to water present in the sample.

23.3 Water determination in dairy powders

Dairy powders are sold on the basis of dry matter. Water determination is therefore an important analysis in this field. Practically all dairy powders contain lactose. This component causes problems when drying techniques are used.

23.3.1 The lactose problem—scientific background

Lactose exists in different forms. The α-anomer is stable at temperatures below 93°C. It crystallizes with 1 mole water per mole lactose. This water content corresponds to 5.00% by mass. At higher temperatures, the anhydrous β-anomer is more stable. Lactose occurs also in amorphous form which may include small amounts of water. Depending on the production conditions, dried dairy powders contain mixtures of these modifications. In addition to the included water and water of crystallization, the product usually contains small quantities of surface water.

The usual drying temperature for moisture determination of dairy products in drying ovens is 102°C. At this temperature, the water of crystallization of α-lactose is not evaporated completely during the usual drying times. The separation of this water fraction from the matrix needs a high energy input (Rüegg and Mohr, 1987; Rückold et al., 2003). After the standard drying time of 2 h, only a part of this water is detected. The consequence is that drying techniques yield results that differ more or less from the true water content.

As lactose occurs in practically all dairy products and is also used in the pharmaceutical industry, this problem affects a wide range of products, particularly those with high lactose content like whey powders or lactose itself.

Usually the drying results are lower than water content. In special cases, however, they can be higher. This is possible if the lack in water detection is overcompensated by other volatile substances which are contained in the products or which are formed by chemical reactions of components and by decomposition processes during the drying process.

23.3.2 The lactose problem—economic aspects

Dairy powders are sold on the basis of dry matter, *DM*. This is (or should be) the mass of the product, m_0, minus the mass of the water, m_W, contained in it (Eq. 23.4). The mass of water can be calculated from water content, *WC* (Eq. 23.5). Eq. (23.6) gives the dependence of dry matter from water content.

$$DM = m_0 - m_W \tag{23.4}$$

$$WC = m_W/m_0 \quad \Rightarrow \quad m_W = WC \cdot m_0 \tag{23.5}$$

$$DM = m_0 - m_W = m_0 - WC \cdot m_0 = m_0 \cdot (1 - WC) \tag{23.6}$$

Dry matter decreases with increasing water content. If the analytical method yields a result lower than the true water content, dry matter is then overestimated. This would give the product supplier an unjustified advantage and the buyer would pay too much for the product. This is just the situation likely to happen in the trade of dairy powders.

23.3.3 Reference method for determining moisture in milk powders

The International Dairy Federation (IDF) has established a method for determining the moisture content in dried milk. For this purpose, a new drying device was specially designed (de Knegt and van den Brink, 1998). The method was also adopted by the International Organization for Standardization (ISO 5537. IDF 26, 2004).

The introduction of this method was strongly supported by the dairy industry. Scientific arguments brought forward and results of an international interlaboratory test (Grobecker et al., 1999; Rückold et al., 2000) were pushed aside. There were—apart from possible economic interests (see above)—particularly two arguments to introduce and establish this method. The first argument was that the results obtained by this new technique (description see below) were practically the same (but with smaller standard deviation for replicate samples) as those received by drying the samples according to the former method using an ordinary drying oven, independently from the geographic situation (altitude, air pressure, relative humidity of the environment). The second argument (against the objection that the result of this method is not the complete water content) was that the complete water content would not be the property of interest in milk powders. Rather, this would be the free water in the product (Isengard, 2006, 2008).

23.3.4 Mass loss, moisture content, and water content—comparison of results obtained by different methods for various dairy powders

Several dairy powders were analyzed for mass loss by drying and for water content: Lactose, skim milk powder, full cream milk powder, whey powder, and calcium caseinate.

Two drying techniques were used: The "classical" oven drying (OD) and the new "reference drying" method (RD). Water content was determined by Karl Fischer titration (KFT).

23.3.5 Oven drying

The experiments were carried out according to the former IDF standard method "Dried milk and dried cream, Determination of water content" (IDF 26A, 1993). It is remarkable that at that time the mass loss measured was defined as "water content," whereas the new method, which has officially replaced this one, determines "moisture content."

An amount of 1–3 g of the sample—for this investigation, approximately 2 g were used—are dried at 102 ± 2°C in a ventilated drying oven. The mass loss is measured by weighing the sample before and after 2 h of drying and cooling in a desiccator. According to the method, the sample is then to be dried for another hour and so forth until the difference between consecutive measurements is less than 0.5 mg. In this investigation, the samples were analyzed after different drying times to follow the drying process more closely (see below). The results after 2 h were then used to compare the results with each other and with those obtained through other methods. The analyses were carried out with the drying oven FD 115 from Binder, Tuttlingen, Germany.

23.3.6 Reference drying

The samples (5.0 ± 0.3 g) are placed in containers with a diameter of 20 mm and a height of 90 mm (plastics syringes without needle) between polyethylene filters and dried (up to eight in parallel per one analysis) in a heating block at 87 ± 1°C for 5 h. Dry compressed air is passed with a rate of 33 mL/min through the containers with the samples. The mass loss determined by weighing the sample and the container before and after the drying process (after cooling in a desiccator) is defined as moisture content. The method does not control if a constant mass has been reached. For this reference drying, according to the new standard method, the Referenztrockner RD eight from Funke-Dr. N. Gerber Labortechnik, Berlin, Germany, was used.

23.3.7 Karl Fischer titration

A KF Titrino 701 from Metrohm, Herisau, Switzerland, with a titration cell with thermostatic jacket was used. The two-component technique was applied with Hydranal-Titrant 2 as titrating solution and Hydranal-Solvent as working medium. All chemicals were from Sigma–Aldrich (now Honeywell), Seelze, Germany. The end point was detected using the voltametric technique with a polarizing current of 20 μA and a stop voltage of 100 mV, the stop criterion being the drift (5 μL/min above the drift measured before analysis). The minimal volume increment was set to 0.5 μL and the maximal titration rate to 5 mL/min. The sample size was between 0.1 and 0.5 g. In order to obtain a more rapid dissolution or dispersion of the samples in the working medium and, consequently, shorter titration times, the analyses were carried out at 50°C using a double-walled titration vessel (Isengard and Schmitt, 1995).

23.3.8 General procedure (Isengard et al., 2006a)

The different samples were analyzed on the same day by the three methods. Five Karl Fischer titrations were carried out for every sample. The "reference drying" was started with eight portions of each sample, two each were analyzed as duplicates after 3, 4, 5, and 6 h (the "official" drying time being 5 h). 12 portions of each sample were placed in the drying oven. Two each were analyzed in parallel after 60, 80, 100, 120, 150, and 180 min (the "official" drying time being 120 min).

The dependence of the "reference drying" results on the drying parameters was also examined (Isengard et al., 2006b).

23.3.9 Results and discussion

Table 23.1 (derived from Isengard et al., 2006a) gives a juxtaposition of the results obtained by Karl Fischer titration (KFT), by conventional oven drying (OD) according to the former standard method and by the new standard method ("reference drying," RD). The OD results are—for better comparison—those obtained after 2 h, and the RD results are the values after the "official" time of 5 h. Values for other drying times both for oven drying and for "reference drying" are given below. The shape of the Karl Fischer titration curves indicated a complete and correct determination of water for all the samples.

The results for the two milk powders are very close to each other. The KFT and the RD results are not significantly different. The OD results (obtained after 2 h drying time) come closer to the KFT results when the drying times are longer: (99.49 ± 0.26%) after 2.5 h for skim milk powder and (97.36 ± 0.38%) after 3 h for full cream milk powder (see Fig. 23.3 below). In the other cases, the results for water content and mass loss differ clearly. For lactose and whey powder, the differences are very high.

Concerning the standard deviations, it must be considered that the sample sizes for the Karl Fischer results are much smaller (0.1−0.5 g) than those for oven drying (approximately 2 g) and those for reference drying (approximately 5 g). Absolute weighing errors have therefore a higher relative influence in Karl Fischer analyses. In addition, the Karl Fischer titrations were carried out one after the other and are, thus, real replicates, whereas in the drying experiments the samples are analyzed at the same time in the same apparatus. They are therefore not real replicates but experiments in parallel.

Various drying times were tested for the two drying methods to determine the evolution of the results in the course of time. The results of these experiments are depicted in Figs. 23.1−23.5 (from Isengard, 2008).

The lactose sample (Fig. 23.1) is a technical product and obviously contains not only α-lactose but also anhydrous modifications. The water content found by Karl Fischer titration is slightly below 5 g/100 g. This value is by far not reached by the drying techniques because the water of crystallization is strongly bound.

A very important finding is that the drying techniques do not detect the "free" water only (which is usually in the range of 0.1 g/100 g) but also a part of the "bound" water. It cannot, therefore, be claimed that the new reference method will detect free water only. The measured entity is in fact not defined; it is neither free nor total water. This is a serious disadvantage for a reference method.

For the two milk powder samples (Figs. 23.2 and 23.3), the mass loss by drying corresponds approximately to the water content determined by Karl Fischer titration. The "lack" of water detection can be compensated by the determination of volatile substances formed by decomposition at higher temperatures (see above). This is obvious from the results by "reference drying" which rise to numbers above the Karl Fischer results when the drying process is longer than the "official" 5 h. The value for both milk powders after 5 h is, however, in very good consistence with the water content.

The whey powder (Fig. 23.4) contains approximately 85% lactose by mass. A part of it is crystallized. Consequently, the mass loss by drying does not reach the water content. Other components with high water binding capacity may contribute to this effect.

The calcium caseinate sample (Fig. 23.5) does not contain lactose. The reason for the too-low drying results, therefore, may be a slow diffusion of the water from the core of the particles to the surface. The airflow in the "reference dryer" is obviously advantageous for the drying process as it keeps the partial pressure of water above the sample extremely low. The Karl Fischer value would probably be reached if the drying time was longer.

TABLE 23.1 Results for water content by Karl Fischer titration (KFT) and for mass loss by oven drying (OD) after 2 h and by "reference drying" (RD); n = number of replicates. Values are referred to the Karl Fischer results (100%).

Sample	Water content by KFT (n = 5)	Mass loss by OD (n = 2)	Mass loss by RD (n = 2)
Lactose	100% ± 4.27%	55.06% ± 2.92%	23.37% ± 0.67%
Skim milk powder	100% ± 1.79%	98.29% ± 0.00%	100.51% ± 3.32%
Full cream milk powder	100% ± 1.89%	92.83% ± 0.75%	102.64% ± 5.28%
Whey powder	100% ± 1.12%	47.53% ± 0.22%	50.22% ± 1.57%
Calcium caseinate	100% ± 1.78%	90.79% ± 0.48%	92.57% ± 0.32%

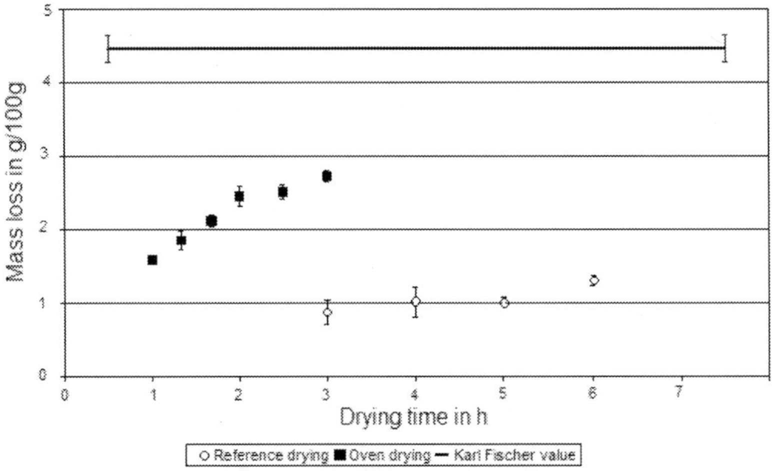

FIGURE 23.1 Mass loss by "reference drying" and oven drying after various drying times of crystallized lactose and—for comparison and reference—the water content by Karl Fischer titration (also in g/100 g). *From Isengard, H.-D., 2008. Water determination – scientific and economic dimensions. Food Chem. 106, 1393–1398.*

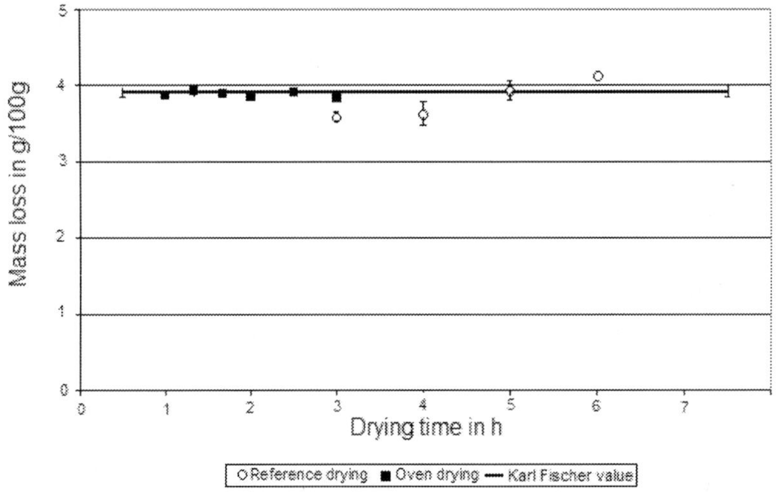

FIGURE 23.2 Mass loss by "reference drying" and oven drying after various drying times of skim milk powder and—for comparison and reference—the water content by Karl Fischer titration (also in g/100 g). *From Isengard, H.-D., 2008. Water determination – scientific and economic dimensions. Food Chem. 106, 1393–1398.*

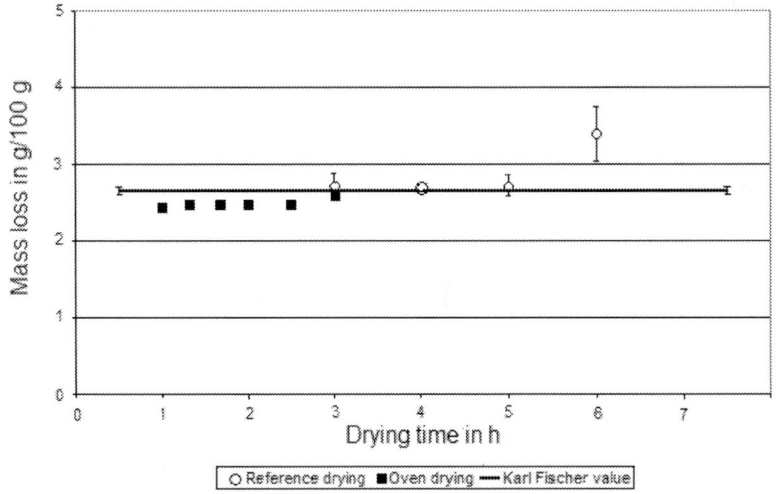

FIGURE 23.3 Mass loss by "reference drying" and oven drying after various drying times of full cream milk powder and—for comparison and reference—the water content by Karl Fischer titration (also in g/100 g). *From Isengard, H.-D., 2008. Water determination – scientific and economic dimensions. Food Chem. 106, 1393–1398.*

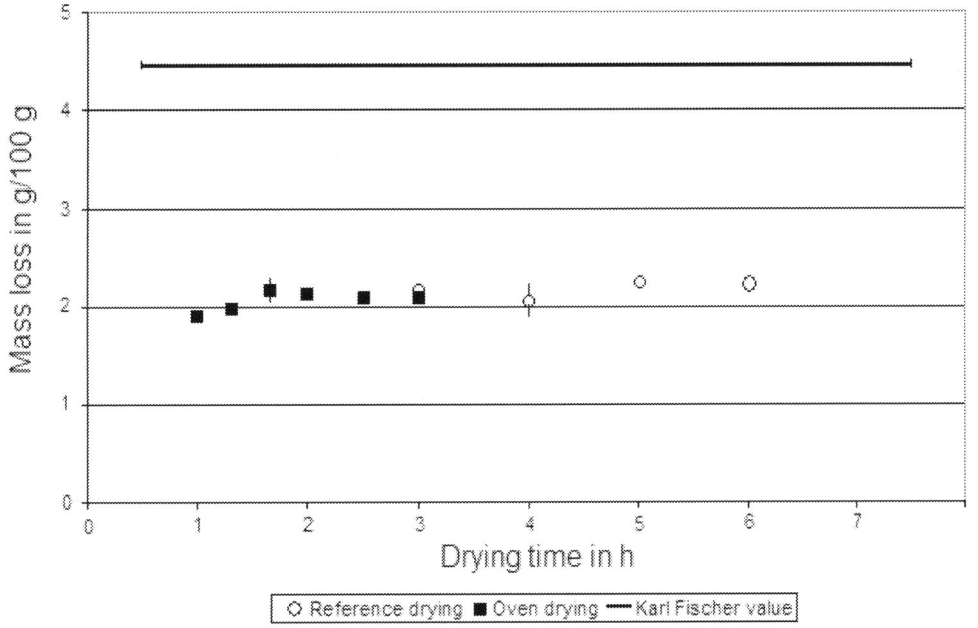

FIGURE 23.4 Mass loss by "reference drying" and oven drying after various drying times of whey powder and—for comparison and reference—the water content by Karl Fischer titration (also in g/100 g). *From Isengard, H.-D., 2008. Water determination – scientific and economic dimensions. Food Chem. 106, 1393–1398.*

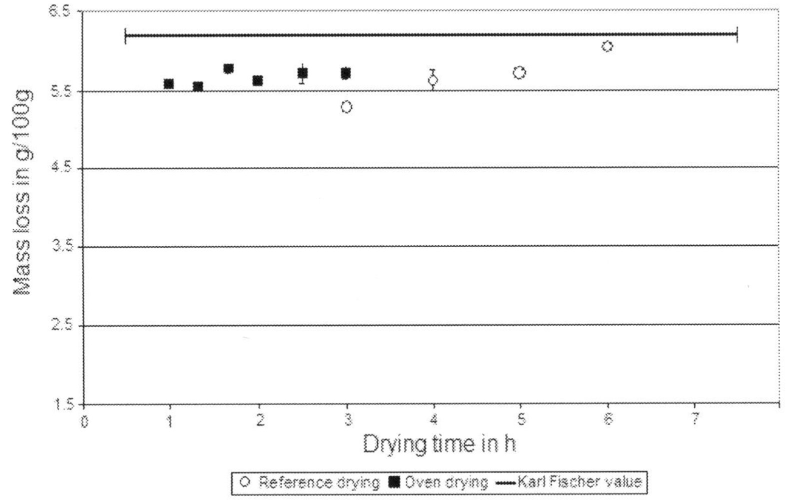

FIGURE 23.5 Mass loss by "reference drying" and oven drying after various drying times of calcium caseinate and—for comparison and reference—the water content by Karl Fischer titration (also in g/100 g). *From Isengard, H.-D., 2008. Water determination – scientific and economic dimensions. Food Chem. 106, 1393–1398.*

Figs. 23.1–23.5 reveal that the results obtained by "reference drying" depend very much on drying time. Other parameters like air flow, sample size, and temperature have an influence on the results as well. This is shown in Tables 23.2–23.4 (from Isengard et al., 2006b).

The mass loss detected increases with increasing flow rates. Mass loss decreases with sample size. The mass loss found at 102°C is higher than at 87°C and time dependent. The total water content (i.e., the Karl Fischer value) is not even found at 102°C after 7 h.

TABLE 23.2 Mass loss of dairy powder samples A to D (received from Nestlé, Vers-chez-les-Blanc, Switzerland, without further information or specification) determined by "reference drying" at 87°C after 5 h at different air flow rates and water content by Karl Fischer titration (KFT), n = number of replicates.

	Mass loss in g/100 g at air flow rate		
Sample	33 mL/min	>33 mL/min[a]	Water content by KFT (g/100 g)
A	2.44 ± 0.54 (n = 3)	2.83 ± 0.19 (n = 2)	2.58 ± 0.02 (n = 3)
B	3.38 ± 0.17 (n = 3)	4.35 ± 0.29 (n = 2)	3.39 ± 0.02 (n = 3)
C	3.26 ± 0.34 (n = 5)	3.93 ± 0.48 (n = 2)	2.97 ± 0.05 (n = 3)
D	5.91 ± 0.40 (n = 3)	6.06 ± 0.08 (n = 2)	5.98 ± 0.13 (n = 3)

[a] *The instrument does not allow the exact measurement of the air flow when the standard value of 33 mL/min is not chosen.*
From Isengard, H.-D., Kling, R., Reh, C.T., 2006b. Proposal of a new reference method to determine the water content of dried dairy products. Food Chem. 96, 418–422.

TABLE 23.3 Mass loss by "reference dying" at 87°C after 5 h (air flow 33 mL/min) for a dairy powder sample from Nestlé (see Table 23.2) using different sample sizes; two replicates for sample sizes 2, 3, 4, and 6 g, five replicates for sample size 5 g; water content by KFT 3.42 ± 0.02 g/100 g (n = 3).

Sample size (g)	2	3	4	5	6
Mass loss (g/100 g)	5.31 ± 0.27	4.21 ± 0.06	4.43 ± 0.35	4.10 ± 0.49	3.73 ± 0.08

From Isengard, H.-D., Kling, R., Reh, C.T., 2006b. Proposal of a new reference method to determine the water content of dried dairy products. Food Chem. 96, 418–422.

TABLE 23.4 Mass loss of a lactose sample by "reference drying" at 102°C in dependence of drying time; mass loss at standard conditions: 1.08 ± 0.14 g/100 g (n = 3); n = number of replicates; water content by KFT: 4.46 ± 0.10 g/100 g (n = 4).

Drying time (h)	2	3	4	5	6	7
Mass loss (g/100 g)	1.79 ± 0.69 (n = 2)	2.37 ± 0.56 (n = 4)	2.65 ± 0.51 (n = 6)	3.00 ± 0.32 (n = 6	3.17 ± 0.30 (n = 4)	3.35 ± 0.23 (n = 2)

From Isengard, H.-D., Kling, R., Reh, C.T., 2006b. Proposal of a new reference method to determine the water content of dried dairy products. Food Chem. 96, 418–422.

23.3.10 Concluding considerations

Results obtained for mass loss by drying and for water content by Karl Fischer titration can clearly differ. With increasing α-lactose content, the difference increases and is extreme for pure lactose. The drying techniques neither determine the total water nor the free water fraction alone. The defenders of the new reference method argue that the water of crystallization is of no practical importance, because it has no influence on the flowability of the powders and almost no importance for the microbiological stability and thus the shelf life of the product. This argumentation does not consider the fact that this "bound" water is set free when the product is dissolved. This has to be accounted for when recipes are designed. Therefore, the total water content (including water of crystallization) is important.

The results of the "reference method" depend very strongly on the drying time and also other parameters (de Knegt and van den Brink, 1998; Isengard et al., 2006a,b). Only for ordinary milk powders are they close to the water content values found by the Karl Fischer titration. For products with other compositions, other product-specific parameters would have to be chosen. This means that the method is limited and cannot be applied to dairy powders in general.

The Karl Fischer method detects the total water content selectively and is independent from the lactose content. The precision of the Karl Fischer results is very good, even though the sample sizes are much smaller than those of the drying techniques. The precision of the new reference method is no better than that of oven drying.

The drying techniques are more time consuming than the Karl Fischer method. Conventional drying takes several hours and real mass constancy is only rarely reached. Experience shows that terminating the measurement after a fixed time of 2 h is recommended. This makes results more comparable, as in many cases additional mass loss may be due to decomposition processes. The "reference drying" is very time consuming (practically 1 day for a set of eight samples). The Karl Fischer method is by far the most rapid method for one sample (a couple of minutes). However, a disadvantage of the Karl Fischer technique is the use of chemicals.

These and other investigations (Rückold et al., 2000; Isengard et al., 2006a,b) have shown that the "reference drying" method is correct only for ordinary milk powders but not necessarily for other dried dairy products. In contrast to this, the Karl Fischer titration can generally be applied to these products and would be a more reasonable reference method.

From the scientific point of view, these considerations are clear and straightforward and cannot be doubted. Nevertheless, attempts to introduce the Karl Fischer titration as reference method for water determination in dairy powders met the resolute resistance of the dairy industry. A reason might have been the economic interest in not "finding" all of the water in the product being sold. The detection of the true water content would have lowered the price of the product if it was still calculated on the basis of dry matter.

Interestingly enough, the same scientific group that had proposed the Karl Fischer method for determining the water content in dairy powders was asked by IDF to develop a method to analyze the water content in lactose. This was carried out (Isengard et al., 2011; Merkh et al., 2012) and was accepted by the IDF (ISO12779. IDF227, 2011, confirmed 2018). The parameters of this method were the same as for the proposed and refused method for dried milk on the basis of the Karl Fischer titration.

Scientific facts are, so far, not strong enough against economic arguments. Harmonizing scientific truth and accuracy with economic power and interest is necessary. Such a harmonization should, for ethical reasons, be aimed at in the interest of honest and correct trade.

23.4 Water content determination by near-infrared spectroscopy

23.4.1 Rapid water determination by near-infrared spectroscopy

Near-infrared (NIR) spectroscopy is one of the rapid techniques which allow even in-line measurements. Many properties of samples can be analyzed, once an appropriate calibration against a direct method has been established. As NIR spectroscopy is based on light absorption of chemical components in the sample, a connection between one of these components or a group of components and the property to be measured must exist. This property may be a physical characteristic of the sample or the concentration or mass concentration of a compound or group of compounds in the sample. Water is one of these compounds which can be analyzed by NIR spectroscopy.

23.4.2 Water determination in a whey powder by NIR spectroscopy (from Isengard et al., 2010)

Lactoserum Euvoserum (from Nestlé, Lausanne, Switzerland), which essentially contains whey powder, was analyzed for water content by NIR spectroscopy after calibration against mass loss by drying and against water content determination by Karl Fischer titration. Previous experiments had shown that the temperature of about 102°C usually applied for drying dairy powders in drying ovens is not sufficient to liberate the water from this product (Merkh, 2006). Higher temperatures were therefore applied. This example shows the effect at a drying temperature of 145°C. Karl Fischer titrations were carried out with a KF Titrando 841 from Metrohm, Herisau, Switzerland, using the two-component technique with Hydranal Titrant 2 as titrating solution and a mixture of Hydranal Solvent and formamide in a volume ration of 2:1 as working medium (Schmitt and Isengard, 1997), all chemicals being from Sigma–Aldrich (now Honeywell), Seelze, Germany. To shorten titration times, the analyses were carried out in a double-walled titration vessel at 40°C (Isengard and Schmitt, 1995). NIR spectra were registered using a Fourier transform NIR spectrometer NIRVIS from Bühler, Uzwil, Switzerland (now represented by Büchi, Essen, Germany).

23.4.3 Results and discussion of NIR measurements

Fig. 23.6 shows the calibration line based on mass loss.

The calibration is acceptable if the mass loss on drying is really the property to be analyzed. It is, however, useless if water content is to be analyzed. The reason is that the water content analyzed by Karl Fischer titration lies only in the range highlighted by a rectangle in Fig. 23.7.

If the calibration is based on the Karl Fischer results for real water content, the line in Fig. 23.8 is obtained.

The reason for this phenomenon is that the product continues to lose mass at the temperature applied. This is depicted in Fig. 23.9 (from Isengard et al., 2010). The dried product was analyzed for water content by Karl Fischer titration after various drying times. After some time, the residual water remains constant, although the mass loss still increased. The product obviously undergoes degradation, with formation of volatile substances. This can also be derived from its more and more darkening color. When the drying curve reaches the Karl Fischer value, there is still about half of the water in the sample. Further analyses showed that the relatively constant water content found in the dried fractions is to a great extent due to water which was taken up by the hygroscopic dried material during handling (cooling in a desiccator, weighing, and transferring into the titration vessel).

It is not possible to determine the water content of this product by NIR spectroscopy if the calibration is based on oven drying at 145°C. Such a calibration leads to mass loss results with good correlation. However, if water content is to be measured, the calibration has to be based on Karl Fischer titration. The temperatures for the liberation of water from α-lactose and the degradation of the product overlap. It is, therefore, not possible to choose a temperature at which water is evaporated completely without degradation of the product.

23.4.4 Concluding considerations

Establishing a connection between a sample property to be analyzed with one method and another property which is measured by another method includes the risk of yielding good calibration lines without giving correct results of the sample property. This is particularly true for techniques with chemometric evaluation of the measurements like NIR spectrometry.

The whole approach depends on the "correctness" of the reference method, because it is always possible to draw a regression line through data points. For every analytical measurement, a corresponding concentration of the analyte will be found and even a high precision is no proof of correctness. A possible error will not be detected because the results are based on the same wrong conditions and assumptions. It is therefore essential that the reference method is "correct." The importance of this prerequisite was shown at the example of water determination in a whey powder product.

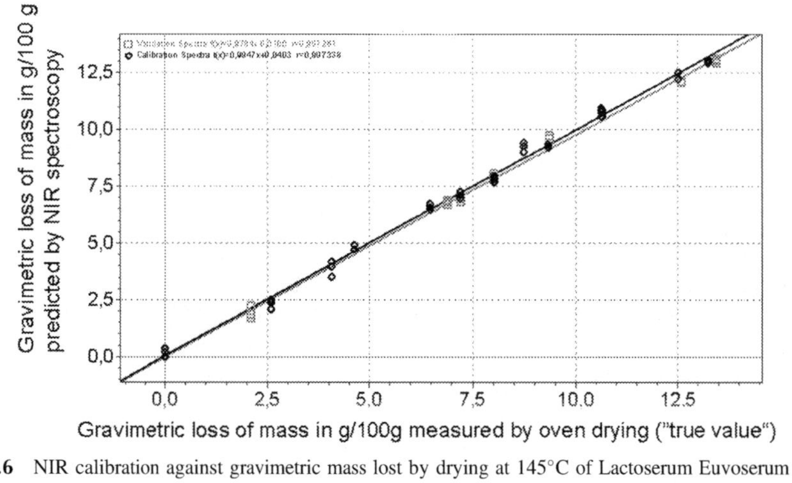

FIGURE 23.6 NIR calibration against gravimetric mass lost by drying at 145°C of Lactoserum Euvoserum from Nestlé.

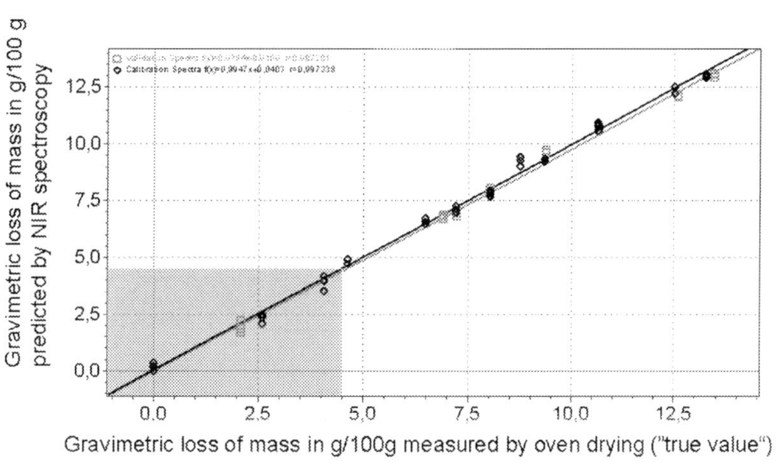

FIGURE 23.7 NIR calibration against gravimetric mass lost by drying at 145°C of Lactoserum Euvoserum from Nestlé. The water content occurring in the powders analyzed lies within the highlighted rectangle.

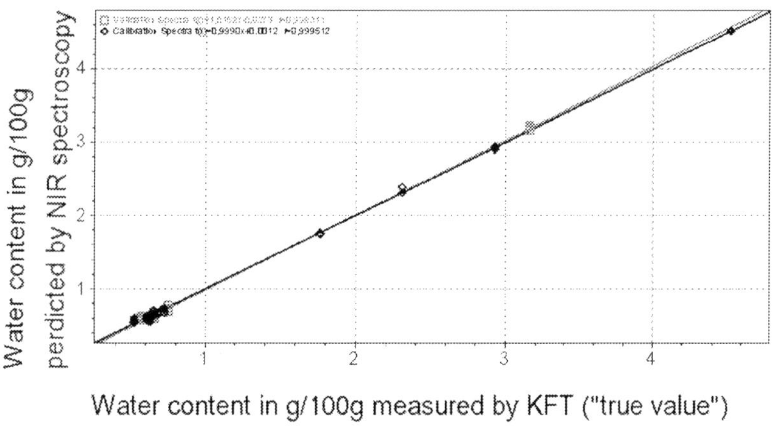

FIGURE 23.8 NIR calibration against water content determined by Karl Fischer titration of Lactoserum Euvoserum from Nestlé.

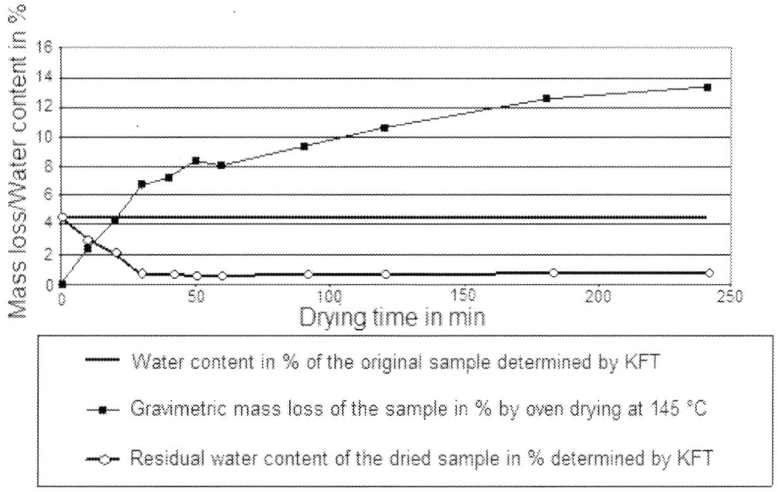

FIGURE 23.9 Drying curve of Lactoserum Euvoserum from Nestlé at 145°C and residual water content of the dried fractions. *From Isengard, H.-D., Merkh, G., Schreib, K, Labitzke, I., Dubois, C., 2010. The influence of the reference method on the results of the secondary method via calibration. Food Chem. 122, 429–435.*

23.5 Summary

To avoid disputes and problems in trade, analytical methods should be harmonized. The methods agreed upon must, however, be scientifically correct and correspond to the state of art. In order to be in a position to compare scientific findings, methods with which analyses have been carried out must be described in all essential details, because results depend on the methods and parameters used.

Political or economic interests may have an influence on the choice of the reference or standard methods and lead to a situation that a reference method is scientifically incorrect. On first sight this may seem harmless, since it applies to all concerned in the same way. This, however, is not the case. If the agreed analytical method does not yield correct results, this is ethically not acceptable. In addition, one of the trade partners is at an economic disadvantage. Also, the incorrect result may have negative consequences. Thus, in the example of dairy powders, the incomplete detection of the water contained in the product by using the official reference method for "determination of the moisture content" may lead the buyer of the product think that the indicated moisture content corresponds to the true water content. But when the product is dissolved in water, the undetected water is set free and the solution contains less dry mass of dairy powder than was foreseen in this recipe.

When incorrect reference methods are used for calibration of secondary methods, this may lead to good calibration lines with satisfactory correlation factors, when in fact the desired property was not calibrated. This does not become apparent and false results seem to be confirmed, even if they are completely senseless.

References

de Knegt, R.J., van den Brink, H., 1998. Improvement of the drying oven method for the determination of the moisture content of milk powder. Int. Dairy J. 8, 733–738.

Grobecker, K.H., Rückold, S., Anklam, E., 1999. Determination of the Water Content in Milk Powder: Report of a Collaborative Study Performed in the Period of June - July 1999. European Commission, Joint Research Centre - Institute for Reference Materials and Measurements, Geel, Belgium, and Joint Research Centre - Institute for Health and Consumer Protection. EUR 19005 EN, European Communities, Ispra, Italy.

Heinze, P., Isengard, H.-D., 2001. Determination of the water content in different sugar syrups by halogen drying. Food Control 12, 483–486.

IDF 26A, 1993. Dried Milk and Dried Cream, Determination of Water Content.

Isengard, H.-D., 1995. Rapid water determination in foodstuffs. Trends Food Sci. Technol. 6, 155–162.

Isengard, H.-D., 2001. Water content, one of the most important properties of food. Food Control 12, 395–400.

Isengard, H.-D., 2006. Harmonisation of analytical methods – best solution is same regulations everywhere. Food Eng. Ingredients 31 (2), 22–26.

Isengard, H.-D., 2008. Water determination – scientific and economic dimensions. Food Chem. 106, 1393–1398.

Isengard, H.-D., Färber, J.-M., 1999. Hidden parameters of infrared drying for determining low water contents in instant powders. Talanta 50, 239–246.

Isengard, H.-D., Felgner, A., Kling, R., Reh, C.T., 2006a. Water determination in dried milk: is the international standard reasonable? In: del Pilar Buero, M., Welti-Chanes, J., Lillford, P.J., Corti, H.R. (Eds.), Water Properties of Food, Pharmaceutical, and Biological Materials. CRC Press, Taylor & Francis Group, Boca Raton, FL, ISBN 0-8493-2993-0, pp. 631–637.

Isengard, H.-D., Kling, R., Reh, C.T., 2006b. Proposal of a new reference method to determine the water content of dried dairy products. Food Chem. 96, 418–422.

Isengard, H.-D., Merkh, G., Schreib, K., Labitzke, I., Dubois, C., 2010. The influence of the reference method on the results of the secondary method via calibration. Food Chem. 122, 429–435.

Isengard, H.-D., Haschka, E., Merkh, G., 2011. Determination of water content in lactose by Karl Fischer titration – interlaboratory study. Bull. Int. Dairy Fed. 450, 1–15.

Isengard, H.-D., Präger, H., 2003. Water determination in products with high sugar content by infrared drying. Food Chem. 82, 161–167.

Isengard, H.-D., Schmitt, K., 1995. Karl Fischer titration at elevated temperatures. Mikrochim. Acta 120, 329–337.

Isengard, H.-D., Walter, M., 1998. Can the true water content in dairy products be determined accurately by microwave drying? Zeitschrift für Lebensmittel-Untersuchung und -Forschung A 207, 377–380.

ISO 5537. IDF 26, 2004. Dried Milk – Determination of Moisture Content (Reference Method).

ISO 12779. IDF 227, 2011, confirmed 2018. Lactose – Determination of Water Content – Karl Fischer Method (International Standard).

Merkh, G., 2006. Vergleichende Untersuchungen von Methoden zur Wassergehaltsbestimmung in einigen pulverförmigen Milchprodukten (Master thesis (Diplomarbeit)). University of Hohenheim, Institute of Food Science and Biotechnology, Stuttgart, Germany.

Merkh, G., Haschka, E., Isengard, H.-D., 2012. Development of a method for water determination in lactose. Food Chem. 132, 1660–1663.

Rückold, S., Grobecker, K.H., Isengard, H.-D., 2000. Determination of the contents of water and moisture in milk powder. Fresenius J. Anal. Chem. 368, 522–527.

Rückold, S., Grobecker, K.H., Isengard, H.-D., 2001. Water as a source of errors in reference materials. Fresenius J. Anal. Chem. 370, 189–193.

Rückold, S., Isengard, H.-D., Hanss, J., Grobecker, K.H., 2003. The energy of interaction between water and surfaces of biological reference materials. Food Chem. 82, 51–59.

Rüegg, M., Moor, U., 1987. Die Bestimmung des Wassergehaltes in Milch und Milchprodukten mit der Karl-Fischer-Methode, V. Die Wasserbestimmung von getrockneten Milchprodukten. Mitteilungen Gebiete Lebensmitteluntersuchung und Hygiene 78, 309–319.

Schmitt, K., Isengard, H.-D., 1997. Method for avoiding the interference of formamide with the Karl Fischer titration. Fresenius J. Anal. Chem. 357, 806–811.

Scholz, E., 1984. Karl-Fischer-Titration. Springer Berlin, Heidelberg, New York, Tokyo, ISBN 3-540-12846-8.

Chapter 24

Global harmonization of analytical methods

Pamela L. Coleman, Anthony J. Fontana and John Szpylka
Mérieux NutriSciences, Chicago, IL, United States

24.1 Introduction

"In God we trust, all others bring data." This statement, attributed to W. Edward Deming, summarizes the importance of data and can be applied, equally well, to a myriad of different industries. The global food industry is certainly no exception. Consumer trust and satisfaction, the main concern of the food industry, is influenced and measured by application of analytical methods. Analytical data are needed for numerous and varied purposes. Data are used to quantify components, to determine value in the marketplace, to monitor key production parameters in a factory, to establish the nutrient content for labeling purposes, and to ensure the absence of harmful contaminants. The food industry relies heavily on data generated through the application of analytical chemistry methods in simple as well as complex food matrices.

Food analytical chemistry is complex and the lack of global harmonization of analytical methods and legislation makes it more complicated. The dizzying array of foods processed and manufactured around the world has led to the development of a wide variety of analytical methods for even the most basic of measurements. While the goal of each method is to generate accurate or "correct" data, the lack of consensus on the most appropriate method often leads to trade barriers, inefficient use of resources, and economic loss. So while the utility of applying analytical chemistry techniques such as gravimetric analysis, coulometry, acid–base titrimetry, chromatography, atomic absorption or emission, and mass spectroscopy to the task of generating data concerning food is clear, the wide range of potential techniques makes the food analytical laboratory a very complex environment.

There are many areas of concern where correct analytical measurements are important, such as establishing, making, and supporting labeling claims or testing for the presence of harmful substances in food. When different analytical methods are used, they may not align with each other, in particular for new rapid methods or test kits where limited validation and the lack of reference materials for some analytes exist. Layered over the large range of methods in use are the regulatory requirements of major producing and consuming countries, sometimes harmonized, but many times divergent. Consumers ultimately suffer from this inefficiency of different countries using different methods to test foods. Whether the goal is to distribute food commodities or to develop new processed foods with nutrition benefits, the current situation of varying analytical methods and government norms is untenable.

To address this situation and reduce method variation, major commodity groups have international associations and method committees, thereby simplifying trade between states, provinces, and countries. Methods for basic quality and product parameters are generally the original focus, but eventually, key contaminants and other analytes will be addressed. For example, the International Dairy Federation (IDF) has Standing Committees on *Main Components in Milk*, on *Minor Components and Characterization of Physical Properties in Milk*, on *Quality Assurance, Statistics of Analytical Data and Sampling*, and on *Analytical Methods for Additives and Contaminants in Milk* (IDF, 2007). These committees work with the International Organization for Standardization (ISO) to develop and harmonize standard methods.

While this chapter will not address the harmonization of regulations in any detail, it is important to note that there are numerous groups focused on this important effort. Regulatory bodies should refrain from inserting specific analytical methods into the regulations; rather they should focus on the scientific basis for the regulation. These bodies should adhere

Ensuring Global Food Safety. https://doi.org/10.1016/B978-0-12-816011-4.00009-4

to defining the performance parameters of an acceptable analytical method instead of discussing how to show compliance. Even if international regulations were harmonized through scientific consensus, duplicate testing may be required for trade purposes if countries included specific analytical methods in the regulations themselves.

At least two prominent groups work to harmonize and validate analytical methods for foods. The ISO and the Association of Analytical Communities (AOAC International) provide guidelines with minimum recommendations on procedures that should be employed to ensure adequate validation of analytical methods. The goal of these organizations is to establish a system which allows confidence in analytical results through expert reviews of collaborative studies and consensus of methods. These protocols/standards require a minimum number of laboratories and test materials to be included in the collaborative trial to validate fully the analytical method.

AOACI serves the communities of analytical sciences by providing the tools and processes necessary for community stakeholders to collaborate and, through consensus building, develop fit-for-purpose methods and services for assuring quality measurements. An example of this process is the pesticide contamination of soft drinks reported by a prominent laboratory in India which resulted in these drinks being banned across several Indian states.[1] Two major soft drink producers worked with AOAC, a coalition of several industry experts, and the Indian government to develop and validate methodology for detecting pesticides in soft drinks that could be utilized by both producers and regulators. Through this collaboration, no regulations were changed, but the compliance to the regulations was harmonized in that the producers and the regulators agreed on the analytical method.

Within the food industry, the concept of "fitness of purpose" is of key importance. Many have attempted to define this concept. Table 24.1 outlines the base considerations to be addressed when trying to choose a method that applies to a specific situation (Nielsen, 2010). The purpose for which the data are needed should help determine the method applied. In 2011, AOACI adopted an approach where industry stakeholders and regulators first define the *Standard Method Performance Requirements* (SMPR) needed for an analytical method to be fit-for-purpose. Analytical methods are then located and a vetted Expert Review Panel compares the method's performance to the SMPR. To date, over 85 methods have been recognized as Official Methods of Analysis using this protocol.

In this chapter, three main areas of food analysis will be discussed: (1) methods meant to define the basic composition, quality, or economic value of foods; (2) methods designed to determine the nutrient value of foods; and (3) methods meant to detect or confirm the absence of contaminants in foods. For each of these three areas, at least two examples will be reviewed which highlight the importance of method selection as well as current global disagreements or compromises.

24.2 Methods for establishing the basic composition, quality, or economic value of foods

Analytical methods meant to establish and monitor the basic composition and quality of foods are often used as economic value indicators. In this area of analytical concern, international agreement on what and how each component is to be measured is critical in determining the correct economic value of food products as well as in preventing harmful adulterations. The melamine incidents that occurred in 2007 and 2008 are examples of economic adulteration with dire consequences, first to animals and then to humans. While there is international standardization and regulatory agreement on the method for protein analysis, the agreed upon method is a nonspecific analysis. Rather than analyzing the protein content of foods by direct methods, which can be time consuming and costly, faster and less expensive in-direct methods are generally applied. The classical Kjeldahl or Dumas methods determine the total organic nitrogen followed by conversion of measured total nitrogen into crude protein content using a suitable conversion factor (Mermelstein, 2009). Conversion of nitrogen into protein can only occur accurately if the nitrogen content of the protein fraction is known and if the food product includes no other nitrogen-containing matter besides protein.

Fortunately, the development of specific, rapid, and less costly methods is underway (Mermelstein, 2009). Several different approaches for protein analysis are now available, with more to follow. Improving the speed and sensitivity of the analysis for amino acids is one promising option. Since the current method requires 24 h for the sample preparatory hydrolysis step and additional time for the detection and quantification, instrument companies are working to reduce the time required by incorporating microwave digestion. Other approaches move away from amino acid analysis by utilizing an azo-dye called Crocein Orange G, which tags by binding to the protein. The dye is added to a sample in excess. Then

1. Toxic Pesticides found in India's Soft Drinks: http://newfarm.rodaleinstitute.org/international/news/080103/081103/in_pest_drinks.shtml.

TABLE 24.1 Criteria for choice of food analysis methods.

Characteristic	Critical questions
Inherent properties	
Specificity	Is the property being measured the same as that claimed to be measured? What steps are being taken to ensure a high degree of specificity?
Precision	What is the precision of the method? Is there within-batch, batch-to-batch, or day-to-day variation? What steps in the procedure contribute the greatest variability?
Accuracy	How does the new method compare in accuracy to the old or a standard method? What is the percent recovery?
Applicability of method to laboratory	
Sample size	How much sample is needed? Is it too large or too small to fit your needs? Does it fit your equipment and/or glassware?
Reagents	Can you properly prepare them? Are they stable? For how long and under what conditions? Is the method sensitive to slight or moderate changes in the reagents?
Equipment	Do you have the appropriate equipment? Are personnel competent to operate equipment?
Cost	What is the cost in terms of equipment, reagents, and personnel?
Usefulness	
Time required	How fast is it? How fast does it need to be?
Reliability	How reliable is it from the standpoints of precision and stability?
Need	Does it meet a need or better meet a need? Is any change in method worth the trouble of the change?
Personnel	
Safety	Are special precautions necessary?
Procedures	Who will prepare the written description of the procedures and reagents? Who will do any required calculations?

From Nielsen, S.S., 2010. Introduction to food analysis. In: Food Analysis. Springer, Boston, MA.

colorimetry is used to detect the amount of dye not bound to protein in order to calculate the amount of protein present. By analyzing specifically for protein, the economic value of nitrogenous adulterants would no longer be present and thus prevent harmful adulterations.

When payment is based on the protein content, as it is in several commodities, the incentive for fraud is high and relatively easy to achieve with nitrogenous compounds that, in fact, contain no protein at all. Thus, melamine and similar compounds can be illegally added into products for economic gain. Since we have not seen the last of these types of economic adulterations, the food industry must invest in methods to make a more accurate, fast, inexpensive, and specific protein assessment a reality. In these authors' opinion, if the focus is only on the analysis of a single adulterant, the industry will be fooled again when the culprits adulterate with another nitrogenous compound in the future.

Another example of analytical challenge is the determination of fruit juice authenticity. A single test to determine juice authenticity does not yet exist. Several different analytical procedures, organic acids, mono- and disaccharides, and minerals are compared with known levels of authentic juice to determine authenticity. While the analytical methods employed are well established, the interpretation is not. Even experienced laboratories often do not agree on the meaning of the analytical results. Juices processed from fruit grown in different geographical regions can further confuse the issue.

For this reason, numerous attempts have been made in finding suitable methods for authenticity control and determination of the fruit content in fruit-based products. The major analytical problem is due to the complexity of the products and to the substantial variance of the fruit specific components (Fügel et al., 2005). Ongoing research into DNA and mass spectrometry-based detection systems shows promise. Knight (2000) developed an analytical approach for the detection of the adulteration of orange juice by PCR technique, providing a quantification of 2.5% mandarin juice in orange juice. Fraudulent addition of sugars to authentic fruit juices and concentrates was detected using isotope ratio mass spectrometry to characterize the specific natural isotope profile of sugars and organic acids (González et al., 1999).

In response to the rise of unforeseen economic adulterants, Non-Targeted Testing (NTT) is being more widely developed to assess the risk of a food or ingredient containing an unknown material. NTT employs either a single or group of analytical method(s) such as spectrometry, NMR, and LC/MS to create a "fingerprint" of the food or ingredient. Incoming lots of the food are then compared to the fingerprint and the "degree of difference" assesses the risk of the food being adulterated for economic gain. NTT is possible due to analytical instruments being more sensitive and accurate with the raw data being processed by improved chemometric software to determine the risk level. This approach does lower the risk of economic adulteration by looking at foods in a more holistic manner to locate unpredicted modifications.

Moisture content is another important food composition and quality measurement that lacks agreement on an appropriate method. Moisture is very important to food microbiological, physical, and chemical stability. Moisture content is measured by direct methods, where moisture is removed by drying, distillation, extraction, or other physicochemical techniques and is measured gravimetrically or by any other direct means. Direct methods are widely used for laboratory analysis, having high accuracy and even absolute values, but are generally time-consuming and require manual operations. Some of these methods are faster and lend themselves to continuous and automated measurement in industrial processes. Loss-on-drying methods are the most common and easiest procedures for routine analysis of moisture content. Practical application of a particular method sometimes outweighs what is correct from a scientific point of view. For the most part, the dairy industry everywhere employs the loss-on-drying method for measuring moisture content. Case studies show results can differ from the actual water content of the dairy matrices; Chapter 6 (Water Determination in Food) provides a more in-depth account of this analytical topic. However, the industry consistently applies this method; thus, within the industry there is equity on a per product basis. If one part of the industry adopted the more specific and accurate Karl Fischer titration method and the rest of the industry stayed with the loss-on-drying method, there would be inequity. Perhaps this is an example of "fitness of purpose" as the dairy industry wants to stay with a simple, chemical-free method rather than move to one that is more scientifically correct. The fact that economic decisions are made on the basis of the results adds a complication; however, within the industry all players abide by the same rules so at least there is equity, if not scientific accuracy.

It should also be noted that dairy products are not the only food matrix with this issue. To obtain the true moisture content, it is important to be aware of factors involved in the loss-on-drying technique. Factors pertaining to sample weighing, oven conditions, and postdrying operations will affect the accuracy and reproducibility of the results. The loss-on-drying methods really determine the moisture content at the specified temperature and time conditions, and not the true moisture content of many foods. However, for purposes of calculating caloric content, these methods are accepted. AOAC goes into great detail on the many permutations of the loss-on-drying method and how they apply to food matrices.

The outcome of method harmonization for the analysis of basic composition and quality tests should result in less adulteration and fewer food quality issues. Suppliers who monitor products using standard methods will help ensure buyers the value of those products. However, as shown by some examples, adulteration can have more far reaching and negative effects than economic damage, so harmonization to more specific methods would improve the situation. The industry would only have to intersperse the use of a more sophisticated measurement alongside the standard indirect methods. If only 1 per every 10 samples were tested using the direct methods and the results tracked per supplier, the threat of these specific methods would act as a deterrent. The international food community should work to collaborate and harmonize direct methods to remove the economic incentive for future adulteration issues. It is hoped that through the availability of novel analytical approaches the expense necessary for adulterating food will increase to a level which makes fraud extremely risky and increasingly uneconomical (Fügel et al., 2005).

24.3 Methods for establishing the nutrient content of foods

Most foods contain a mix of some or all of the nutrient classes. For a healthy diet, some nutrients are required regularly, while others are needed only occasionally. Poor health can be caused by an imbalance of nutrients, whether excessive or deficient. These nutrient classes can be categorized as either macronutrients (carbohydrates, fats, fiber, proteins, and water) or micronutrients (minerals and vitamins). Many of the macronutrient analyses have been discussed in the previous section in this chapter as they are often used for economic or production monitoring purposes. When used for nutrition value purposes, there are often differences in the analytical methods applied. The easiest method to estimate the carbohydrate content is to utilize the "by difference" method and then to calculate caloric content using Atwater factors.

Like proteins and lipids, carbohydrates are one of the major classes of food components. Although a well-defined group of compounds, some confusion often occurs as to what constitutes a carbohydrate and what does not. The measurement of carbohydrates in food ranges from the very simple "by difference" method to the intensive analysis of individual mono- and disaccharides, oligosaccharides, and other complex carbohydrates. When used for nutrition labeling purposes, the "by difference" method is allowed in several countries. However, it results in an overstatement of the actual content, especially when a significant level of fiber or sugar alcohols is present.

The Food and Agriculture Organization (FAO) and the World Health Organization (WHO) of the United Nations held a joint expert panel on carbohydrates (FAO/WHO, 1997), during which some important recommendations were issued regarding the analysis of food carbohydrates:

1. The terminology used to describe carbohydrates should be standardized by primarily classifying them by molecular size (degree of polymerization or DP) into sugar (DP 1−2), oligosaccharides (DP 3−9), and polysaccharides (DP10+). Further subdivision can be made on the basis of monosaccharide composition.
2. Food analysis laboratories should measure total carbohydrates in the diet as the sum of the individual carbohydrates and not "by difference."
3. The analysis and labeling of dietary carbohydrate, for whatever purpose, should be based on the chemical divisions recommended. Additional groupings such as polyols, resistant starch, nondigestible oligosaccharides, and dietary fiber can be used, provided that the included components are clearly defined.

The carbohydrate industry is rapidly changing with the advent of nonnutritive sweeteners and multiple fiber components. The reasons for determining the carbohydrate content are also evolving to now include quality control of drinks and foodstuffs, monitoring of food-labeling claims, analysis of sweeteners, and authentication. The analysis of carbohydrates sometimes involves not only the determination of the total amount of sugar present in the sample, but also the identity, configuration, and conformation of the carbohydrate components.

The American Dietetic Association (ADA) and the British Nutrition Foundation (BNF), respectively, recommend a minimum fiber intake of 20−35 g/day and 12−24 g/day for healthy adults. The term "dietary fiber" was originally synonymous with nondigestible plant matter. Even when defined in its most chemically precise form, as nonstarch polysaccharides, dietary fiber is very heterogeneous and includes pectin substances, hemicellulose, cellulose, lignin, etc. With the expansion of processed and functional foods, there has been an enormous increase in intakes of noncell wall sources of dietary fiber added to foods as ingredients, and an increasing need for regulation of nutrient claims for fiber. Therefore, the concept of dietary fiber has been extended to include oligosaccharides such as oligofructans and nondigestible polysaccharides, such as pectins, gums, mucilages, and resistant starch.

The definition of dietary fiber has been revised several times and continues to evolve. A number of approaches have been taken to dietary fiber analysis, reflecting the lack of agreement on the definition. The methods developed involved chemical measurement referred to as enzymatic−gravimetric, enzymatic−chemical, enzymatic−chemical−gravimetric, and nonenzymatic−gravimetric among other methods.

The most current definition that best captures both dietary fiber molecular structures and physiological behaviors was adopted by CODEX Alimentarius Commission (CAC) which defines dietary fiber as nondigestible carbohydrates having chain lengths DP > 3 and not digested in the small intestine. CAC recognizes AOAC 2009.01/AACCI 32-45.01 and AOAC 2011.25/AACCI 32-50.01 as Type I analytical methods CODEX Standard 234. In addition, FDA states in their regulations they will use analytical methods that best capture nondigestible carbs, and list AOAC 2009.01 and AOAC 2011.25. Earlier versions of analytical methods can still be used if the present nondigestible carbs can be accurately measured. As the definition of dietary fiber continues to be refined, it will require further changes in methods to ensure that they provide a true measure of polysaccharides that are not digested under physiological conditions.

The final class of macronutrients is lipids, which consist of a broad group of compounds that are generally soluble in organic solvents but only sparingly soluble in water. Fatty acids are key components of lipids. Similar to moisture content, much of the food industry applies a shorthand method involving solvent extraction to quickly determine the crude fat content of foods. What is extractable includes many components like pigments, phospholipids, glycolipids, fat-soluble vitamins, and phenols that may not meet the definition of fat. Thus, a food analyzed by solvent extraction would be biased higher than one by only fatty acid analysis. The European Union uses the definition of all lipids from foods (mono-, di- and triglycerides, phospholipids, etc.), while the US FDA has adopted the definition of lipids to only include the total lipid fatty acids expressed as triglycerides. These two definitions are different and thus many times result in different amounts of lipids. While this is close enough for many purposes, the differences between the crude fat methods and the more time and capital intensive fatty acid procedure may not warrant a move away from the crude fat methods.

However, for some food matrices, the differences will be sufficient to warrant this change. In fact, the food industry continues to utilize crude fat methods to satisfy specification requirements for economic purposes and even basic quality purposes, while then turning to the sum of the fatty acids for nutrition labeling and caloric content purposes.

Nutrition labels describe the nutrient content of a food and are intended to guide the consumer in making healthy food choices. This is especially important for single nutritional source foods such as infant formula, adult nutritionals, and pet food/animal feeds. As nutrition labeling efforts continue to evolve toward harmonization of nutritional information listing, the analytical methods employed to substantiate the claims will also harmonize. The Codex Guidelines on Nutrition Labeling play an important role in providing guidance to member countries when developing or updating their national regulations and encourage harmonization of national standards with international standards.

Products marketed in different countries are currently subject to multiple regulations and formats for nutrition labeling which require specific packaging materials for each country. Chapter 17 (Nutrition and Bioavailability: Sense and Nonsense of Nutrition Labeling) provides a more in-depth account of this topic. In some cases, the same analytical methods can be employed, but the results are utilized differently as with the calculation of calories and carbohydrates. Through the harmonization efforts in the European Union, some of these differences will disappear. These disparities complicate and prolong the product development process, which ultimately increases the price of the products.

24.4 Methods for detecting or confirming the absence of contaminants in foods

Preventing acute as well as chronic harmful effects from contaminants in foods requires that foods have little to no residue of chemical contaminants that are known to be toxic. From this very simple concept, a mammoth industry has evolved. Regulators determine what exactly can be allowed and at what levels. Instrument manufacturers design ever better separation and detection devices, which command higher and higher prices. Analytical testing laboratories compete to offer the lowest sensitivity and the broadest range of analytes to their clients. Is it enough to look for 200 compounds at the ppm level or 400 analytes at the ppb level or even more compounds at even lower limits of detection? Is all of this effort resulting in food that is safer? Or does this situation drive up the cost without true benefit to the world's population?

While these questions cannot be answered here, a risk—benefit analysis approach should be employed. In no other area is regulatory harmonization needed more. The "chasing zero" issue should be addressed and subjected to scientific consensus on risk; simply because a compound can be detected does not mean that this compound is harmful at that low detection level. Regulations should not specify methods but rather outline the method performance parameters that should be satisfied, leaving the choice of technology open.

When employing methods for the purpose of detecting contaminants, it is vitally important to consider the entire analytical process from a holistic viewpoint. The sampling (size and protocol), sample preparation, and analytical methods need to be considered as a whole to determine the total variability associated with determining the level of contaminants. An example is mycotoxin testing where sampling represents the largest source of variation in the entire analytical process. Mycotoxins may be present in small regions randomly distributed throughout the material in question and appropriate sampling is critical. Thus, sole focus on lowering the detection limit that does not incorporate appropriate sample size and preparation parameters will not result in more accurate data (EUROCHEM, 2007). Whitaker et al. (2006) and Ozay et al. (2006) investigated the uncertainty associated with sampling, sample preparation, and analysis for aflatoxin in almonds and hazelnuts, respectively. For hazelnuts, the sampling, sample preparation, and analytical steps of the aflatoxin test procedure accounted for 99.4%, 0.4%, and 0.2% of the total variability, respectively (Ozay et al., 2006). For almonds, the percentages of the total variance associated with sampling, sample preparation, and analytical steps were 96.2%, 3.6%, and 0.2%, respectively (Whitaker et al., 2006). As with the testing of other commodities for mycotoxins, the sampling step contributes the most variability, followed by the sample preparation step, and then the analytical step. The best use of resources to reduce the total variability of the aflatoxin test procedure would be to increase the size of the test sample (Whitaker et al., 2006). Due to the fact that sampling plays a critical role in the precision of the determination of the levels of contaminants in foods, the EU clearly defines the sampling methods (Commission Directive 98/53/EC). In Annex I of the Directive, the methods of sampling for official checking control of the levels of aflatoxins in certain foods are clearly defined. The following formula can be used as a guide for the sampling of lots:

$$\text{Sampling frequency (SF)} = \frac{\text{Weight (kg) of the lot} \times \text{weight of the incremental sample}}{\text{Weight of the aggregate sample} \times \text{weight of individual packing}}$$

where the weight of the incremental sample should be about 300 g, and the number of incremental samples to be taken depends on the weight of the lot, with a minimum of 10 and a maximum of 100. Annex II of the Directive discusses the sample preparation and criteria for methods of analysis used in official checking of the levels of aflatoxins in certain foods. This is an excellent example where a regulatory body has not defined a specific analytical method; rather it has defined the sampling of lots and the criteria which the method of analysis has to comply with to ensure that laboratories obtain comparable results.

In 2004, the California Attorney General's Office filed a Proposition 65 lawsuit against 34 manufacturers whose candies tested positive for dangerous levels of lead. California's Proposition 65 law requires warnings on products that can expose the public to carcinogens or reproductive toxins. Chili powder and tamarind are popular ingredients in Mexican-style candies that are sold in California. Data have shown that some of these Mexican-style candies are contaminated with the toxic metal lead. Research has determined that some of the lead in the candies comes from the chili powder and tamarind ingredients. The US FDA is recommending that lead levels in candy products likely to be consumed frequently by small children should not exceed 0.1 ppm because such levels are achievable under good manufacturing practices and would not pose a significant risk to small children for adverse effects.

The lawsuit was settled by seven large manufacturers in June 2006. As part of the settlement, a certification process and an educational out-reach fund were established. Among the companies that signed the settlement are some of the world's largest candy makers and the three leading sellers of popular spicy candies from Mexico. The certification process includes an audit of the manufacturing process to pinpoint possible sources of lead contamination, auditor's recommendations to eliminate possible sources of lead contamination, and regular testing to make certain the candy is safe. The California law does not mandate a specific technology that may change in the next 3−5 years; rather various analytical procedures may be used to ensure that the candy meets the required specification.

Analytical methods to detect residues and contaminants in food have significantly improved and will continue to improve with new technologies. Improving the sensitivity of the detection of the analytical method does little to ensure the safety of the food if an inappropriate sample is used or if the levels detected do not pose a safety or toxicity risk. If testing is done with the "wrong" sample, increasing the sensitivity will not lead to improving the efficiency of the surveillance system in place. However, a risk-based strategy will improve current surveillance systems that target at reducing residues and contaminants in foods.

24.5 Conclusion

Global harmonization of regulations as well as standard methods of analysis would benefit the food industry, but more important, would benefit the worlds consumers significantly. Analytical methods used to analyze the basic composition and quality of food products need to be more specific in order to reduce food safety concerns as well as the threat of economic adulterants. The analytical methods for establishing the nutrient content of foods should also be harmonized to avoid lengthening the product development process and increasing the cost to introduce new products, such as fortified foods into the global market. Finally, harmonizing the analytical as well as the regulatory approach to food contaminants would focus testing efforts on ensuring food safety and would reduce the destruction of foods that may be safe to consume.

References

EUROCHEM, 2007. Measurement Uncertainty Arising from Sampling: A Guide to Method and Approaches, p. iii. www.eurachem.org/guides/ufs.2007.pdf.

FAO/WHO, April 14−18, 1997. Expert Consultation on Carbohydrates in Human Nutrition. Rome.

Fügel, R., Carle, R., Schieber, A., 2005. Quality and authenticity control of fruit purees, fruit preparations and jams − a review. Trends Food Sci. Technol. 16, 433−441.

González, J., Remaus, G., Jamin, E., Naulet, N., Martin, G.G., 1999. Specific natural isotope profile studied by isotope ratio mass spectrometry (SNIP-IRMS): 13C/12C ratios of fructose, glucose, and sucrose for improved detection of sugar addition to pineapple juices and concentrates. J. Agric. Food Chem. 47, 2316−2321.

IDF, 2007. ISO 14501/IDF 171. Milk and Milk Powder— Determination of Aflatoxin M1 Content—Clean-Up by Immunoaffinity Chromatography and Determination by High-Performance Liquid Chromatography.

Knight, A., 2000. Development and validation of a PCR- based heteroduplex assay for the quantitative detection of Mandarin juice in processed orange juices. Agro Food Ind. Hi Tech. 11, 7−8.

Mermelstein, N.H., 2009. Analyzing for melamine. Food Technol. 63 (2), 70−75.

Nielsen, S.S., 2010. Introduction to food analysis. In: Food Analysis. Springer, Boston, MA.

Ozay, G., Seyhan, F., Yilmaz, A., et al., 2006. Sampling hazelnuts for aflatoxin: uncertainty associated with sampling, sample preparation, and analysis. J. Assoc. Off. Anal. Chem. 89 (4), 1004−1011.

Whitaker, T.B., Slate, A.B., Jacobs, M., et al., 2006. Sampling almonds for aflatoxin, Part I: estimation of uncertainty associated with sampling, sample preparation, and analysis. J. Assoc. Off. Anal. Chem. 89 (4), 1027−1034.

Chapter 25

Global harmonization of the control of microbiological risks

Cynthia M. Stewart[1], Frank F. Busta[2] and John Y.H. Tang[3]

[1]*Silliker Food Science Center, South Holland, IL, United States;* [2]*University of Minnesota, Minneapolis, St. Paul, MN, United States;* [3]*Universiti Sultan Zainal Abidin, Terengganu, Malaysia*

25.1 Introduction

Control of microbial risks, as related to food, involves procedures designed to eliminate or minimize the presence of specific microorganisms, their byproducts, and/or toxins. To harmonize these control measures, it is essential to define processes as well as desired outcomes. Consequently, metrics are required to monitor processes (e.g. time, temperature, pressure, concentration, pH, etc.) and determine outcomes (e.g. absence of pathogenic bacteria, inactivation of food enzymes, etc.). In many cases, these metrics serve as a motivation to implement one or more interventions with wider control programmes. Determination of these metrics requires standardization and validation of protocols used to verify outcomes and, wherever multiple approaches exist, methods and results need to be harmonized.

During a GHI Workshop in Lisbon, Portugal, during November 2007, the Microbiology Working Group discussed topics of interest. Priorities were standardization of microbiological methods, definition of "pasteurization" beyond use for dairy products, and harmonization of global regulations for *Listeria monocytogenes* in ready-to-eat (RTE) foods. The latter was discussed at the GHI symposium during EFFoST's First European Food Congress in Ljubljana, Slovenia (November 2008). The resulting working group, in collaboration with Dr. Ewen Todd (Michigan State University), hosted a workshop in Amsterdam, 5–7 May 2009, supported by Elsevier and EFFoST. The expected outcome of this workshop was a series of papers culminating in a consensus article, published in a special edition of *Food Control*. This chapter focuses on the need identified for harmonization of microbiological methods, criteria, and standards.

25.2 Microbiological food safety management

Control of foodborne pathogenic microorganisms is a worldwide public health issue. Prevention and/or reduction of foodborne disease continues to be a major goal of societies, dating back to when food was first preserved through drying and salting (ICMSF, 2005). Food safety issues are complex and, while appropriate policy responses are not always clear, international harmonization of regulations is highly desirable and would facilitate trade. Changes in consumer preferences and social values have a key role in policy issues and there is increasing emphasis on food quality, as incomes and educational attainment rise (OECD, 1998). Currently, there is a much greater public concern about and less tolerance of health risks associated with foods than from other manufactured products, because of the nature of foods and because such risks impact the whole population (OECD, 1998). Consumers also insist on higher safety standards when they perceive others are in control. Pressure for new and stricter enforcement of food safety regulations typically gains momentum following highly publicized incidents of food contamination, such as those that have been associated with bovine spongiform encephalopathy, and outbreaks of *Salmonella*, *Escherichia coli*, or *Listeria monocytogenes.*

The importance of foodborne diseases varies among countries, depending not only on the types of foods consumed, but also food processing, preparation, handling, storage, and vulnerability to a particular pathogenic microorganism. While elimination of foodborne diseases remains unattainable, both government public health managers as well as the food industry are committed to reducing outbreaks and mortality associated with contaminated food (ICMSF, 2005).

Ensuring Global Food Safety. https://doi.org/10.1016/B978-0-12-816011-4.00017-3

Responsibility for selling foods that are not harmful to consumers is that of food businesses along the food production and distribution chain. However, food safety, while vigorously discussed, is an elusive concept that is a source consternation for regulatory authorities and food business operators (FBOs). Assessing the public health (safety) status of a food is risk-based activity and what is an acceptable level of risk and for whom challenging. Food safety might be defined as the '*biological, chemical or physical status of a food that will permit its consumption without incurring excessive risk of injury, morbidity or mortality*' (Keener, 2005). Logically, it follows that food safety lies judging acceptability of risks, which can be a normative, qualitative or, frequently, political decision (Stewart et al., 2010).

Regarding food supply, the primary role of governments is to protect consumers, in parallel with facilitating trade. Over recent decades, new evidence-based outcome-based approaches have being implemented by governments and the food industry alike and, increasingly, food safety risks are communicated alongside other considerations (costs, culture, eating habits, etc.). The term "appropriate level of protection" (ALOP) has also been introduced, meaning "*the level of protection deemed appropriate by the Member (country) establishing a sanitary or phytosanitary measure to protect human, animal, or plant life or health within its territory*" (ICMSF, 2002). While ALOP is defined as the level of risk that a society is willing to accept, many countries have set goals to lower the incidence of foodborne disease and, therefore, lower limits for future ALOPs. For example, current outbreaks of listeriosis might be 6 per million people per year, but a country may want to reduce this to 3 per million people per year (ICMSF, 2005).

The Sanitary and Phytosanitary (SPS) and Technical Barriers to Trade (TBT) agreements, under the World Trade Organization (WTO), have provided an important momentum toward use of international food safety standards, while maintaining sovereign rights of governments to decide on levels of appropriate health protection (OECD, 1998). The SPS agreement, which was signed by more than 100 countries, states that "*while a country has the sovereign right to decide on the degree of protection it wishes for its citizens, it must provide, if required, the scientific evidence on which this level of protection rests*." The TBT agreement also requires that a country must not ask for a higher degree of food safety for imported foods than it does for food produced locally (ICMSF, 2005). Criteria used by a country to determine whether a food should be considered safe have to be conveyed clearly to exporting countries and should be scientifically justifiable (ICMSF, 2001).

While a government might express public health goals relative to the incidence of disease, this does not provide FBOs or trade partners with information about what they need to do to help achieve these goals (ICMSF, 2005). Instead, these goals need to be translated into parameters that can be used by FBOs to manufacture and distribute foods as well as parameters that can be assessed by government agencies. Food Safety Objectives (FSOs) are intended to be the link between public health-based goals and control measures. Good Agricultural Practices (GAPs), Good Manufacturing Practices (GMPs), Good Hygienic Practices (GHPs), and Hazard Analysis Critical Control Points (HACCPs) remain essential for food safety management systems to achieve FSOs or performance objectives (POs) (ICMSF, 2005).

Keener (1999) advocated and developed an integrated food-safety system that focused on collective risk-reducing contributions of subordinating elements to manufacturing supply chains in delivering product safety. This integrated approach to food safety management is reflected in the requirements of ISO standard 22000–2005, which was published in September 2005. A more in-depth discussion of how FSOs link governmental public health goals to control measures can be found in Chapter 6 (A Simplified Guide to Understanding and Using Food Safety Objectives and Performance Objectives). However, the cornerstones for modern food safety management were developed in the early 1970s based on HACCP.

HACCP was developed to address shortcomings and the lack of food safety assurance offered by traditional inspection and sampling (ICMSF, 2005). HACCP focuses on hazards associated with a particular food commodity that are likely to impact public health and monitors conditions to control such hazards. To be successful, HACCP must to be built on practices such as GAPs, GMPs, and GHPs, which can be viewed as basic sanitary conditions, and on practices that must be maintained to produce safe foods, including raw material selection, labeling, and coding (Stewart et al., 2002). GMPs form the basis for HACCP programs to develop and implement food safety management systems controlling significant hazards. In-depth discussions about GMPs, GHPs, GAPs, and HACCP can be found elsewhere.

25.3 Emerging foodborne pathogens

Foodborne pathogens can change significantly over time and new food processes and technology require different control measures. Studies have found novel food processing treatments can cause pathogens to mutate or adapt to new environments (Horn and Bhunia, 2018) and, in doing so, acquire virulent genes via gene transfer, develop antibiotic resistance, or enter into a new niche (Horn and Bhunia, 2018; Laird, 2015). Processing and preservation techniques aim to reduce or eliminate microbial contamination as well as prolong shelf life. These processes can, however, lead to changes in gene expression related to persistence and virulence (Esbelin et al., 2018). While advances in food processing, such as pasteurization, sanitation programs, and thermal and nonthermal treatments, have improved food safety greatly, certain technologies (e.g., chilling and freezing) are associated with pathogenic pathogens, namely *Listeria monocytogenes* (Laird, 2015).

Food pathogens continue to pose significant health risks to humans, resulting in food poisoning incidences. Apparent increased numbers of food poisoning cases can be attributed to several factors, which include more sensitive detection methods, increased surveillance, improved reporting, changes in food consumption habits, increased vulnerable and high risks populations, pathogens adaptations, and resistance to antimicrobial agents (Alvarez-Ordóñez et al., 2017; Begley and Hill, 2015; Bhunia, 2018).

Emerging pathogens are those harmful microorganisms that appear newly in a population, have a new vehicle of transmission, or increased numbers of incidences. These pathogens might be widespread for many years but remain undetected by conventional methods. Subsequently, they are discovered when new knowledge and methods become available (van de Venter, 2000). Global emergence of food pathogens is related to new food types or products, food source exposure, population growth, food standards violations, and climate change (Hoffmann et al., 2017; Laird, 2015). However, globalization of the food supply chain has also increased the spread the pathogens (Laird, 2015).

Continuous monitoring and research have increased our understanding about foodborne pathogens and their control. Some pathogens are fully characterized for growth, resistance, and behavioral patterns. With that knowledge, strategies and guidelines have been implemented with the aim of controlling and eliminating them. Many of these pathogens are controlled successfully, but ever evolving bacteria emerge as a new strain some of which are resistant to conventional treatments and processes, and capable of causing illnesses even at low doses. Current emerging pathogens worldwide include *Salmonella* spp., *Staphylococcus aureus, Campylobacter* spp., *Listeria monocytogenes,* and *Escherichia coli. L. monocytogenes* foodborne infection will be discussed at the end of this chapter together with the harmonization of regulations to control its presence and contamination in food.

25.3.1 *Salmonella* spp.

Salmonella are rod-shaped, Gram-negative, and facultative anaerobes. They are widely distributed in the environment and have been recognized as major pathogens in both humans and animals (Addis et al., 2011). They have been implicated frequently in foodborne disease outbreaks and are responsible for ca. 155,000 deaths per year worldwide (Heredia and Garcia, 2018). Although *Salmonella* have been identified as pathogens since the 1900s, recent increases in antibiotic resistant strains continue to pose a risk to human health. Emerging multidrug resistant (MDR) strains lead to ineffective treatment of severe salmonellosis with common antibiotics, such as azithromycin or ciprofloxacin. Emerging MDR strains are also more virulent, increasing mortality rates especially among vulnerable individuals (CDC, 2019; Eng et al., 2015).

Salmonella may colonize humans, birds, and reptiles. However, the major sources of human salmonellosis include poultry, pigs, and cattle. *Salmonella* colonize the intestinal tract of these farm animals that, although recognized as the reservoir for *Salmonella*, are also asymptomatic after infection. Products from livestock, such as meat, eggs, and milk, are commonly responsible for *Salmonella* outbreaks (Abebe et al., 2020).

The incubation period for *Salmonella* infection ranges from 12 to 72 h. Those infected with *Salmonella* may experience symptoms ranging from self-limiting gastroenteritis to septicemia, and severity depends on host immunity and strain virulence. Vulnerable groups, such as young children, elderly, and immunocompromised individuals, might experience severe infection (Abebe et al., 2020; Eng et al., 2015), the symptoms of which include enteric fever, headache, gastroenteritis, bloody diarrhea with mucous, bacteraemia, extraintestinal complications, and chronic carrier state (Darby and Sheorey, 2008).

25.3.2 *Staphylococcus aureus*

S. aureus are Gram-positive cocci bacteria that live mainly as commensal and opportunistic pathogens. *Staphylococcal* infection can range from minor skin infections to severe and potentially fatal disease (Kadariya et al., 2014). *S. aureus* pathogenicity is due to enterotoxin, invasive infection, and multidrug resistance. Although nonspore forming bacteria, *S. aureus* can contaminate and persist on inanimate surfaces, transferring to people and food products during preparation and processing (Chaibenjawong and Foster, 2011; Kadariya et al., 2014).

The onset of staphylococcal infection or intoxication is abrupt with symptoms of nausea, vomiting, abdominal cramping, and, in some cases, diarrhea. Generally, staphylococcal foodborne disease is self-limiting and resolves within 24–48 h (Argudín et al., 2010). In the events of *S. aureus* poisoning, the toxin cannot be detected in foods by culture methods (Murray, 2005) but requires instead bioassays, molecular, or immunological techniques (Hennekinne et al., 2012).

Studies have shown that animals are important reservoirs for human MRSA infections (Monaco et al., 2013). US and European farms and commercial meats have been found to have high prevalence of MRSA, which poses a risk to meat handlers and consumers (Pomba et al., 2009; Smith et al., 2009). Pigs (Pomba et al., 2009), poultry (Mulders et al., 2010), and cattle (Feßler et al., 2010) are frequently implicated with MRSA outbreaks in humans.

Jones et al. (2002) reported an outbreak in the United States and proved that MRSA-contaminated foods can also infect low-risk individuals. In the study, food handlers were identified as the source of outbreak rather than foods directly. Thus, poor and improper food handling is recognized as a cause of staphylococcal foodborne outbreaks. Temperature abuse during food storage also contributes to outbreaks, as the toxin is heat stable and cannot be rendered safe by cooking. Heat treatments remove *S. aureus*, but not the toxin that is responsible for the majority of staphylococcal food poisoning events (Kadariya et al., 2014).

In 2010, a new strain of *S. aureus* has emerged, identified as livestock-associated methicillin-resistant *Staphylococcus aureus* (LA-MRSA), which has caused significant numbers of human infections (Larson et al., 2011). LA-MRSA with sequence type 398 (ST398) lineages was highly prevalent in livestock-dense areas in Europe and its emergence was detected in France and The Netherlands in 2005 (Armand-Lefevre et al., 2005). LA-MRSA has also been detected in Asia (Guardabassi et al., 2009; Wagenaar et al., 2009), other parts of Europe (Pomba et al., 2009), and the United States (Smith et al., 2009).

25.3.3 *Campylobacter* spp.

Campylobacter spp. measure 0.2–0.8 μm × 0.5–5 μm and are Gram negative, spiral shaped, and motile. Most of the species move in a corkscrew-like motion by means of single polar unsheathed flagella (Silva et al., 2011). They are nonspore forming bacteria and can form viable but nonculturable cells under unfavorable conditions (Portner et al., 2007). Thermophilic *Campylobacter* species (*C. jejuni* and *C. coli*) grow between 37 and 42°C, but not below 30°C (Silva et al., 2011). They are most frequently implicated with human campylobacteriosis. One study found that as few as 800 campylobacter cells are enough to cause infection. However, the number of bacterial cells does not determine severity of symptoms (Black et al., 1988). Studies have reported lower numbers of *Campylobacter* (360 cells) that could possibly cause infection, but 9×10^4 cells were reported to be optimum (Kaakoush et al., 2015).

Symptoms after campylobacter infection range from watery diarrhea to bloody stool, abdominal pain, and vomiting. Usually, these symptoms will resolve in 5–7 days. Although campylobacteriosis rarely results in death, it is associated with complications in 1% of cases. These complications may include Guillain–Barré Syndrome (GBS), reactive arthritis, and irritable bowel syndrome (IBS) (Kaakoush et al., 2015; Nyati and Nyati, 2013).

Campylobacter spp. are commensal organisms in birds, cattle, sheep, and swine. Generally, birds have higher body temperatures and are considered common hosts for thermophilic *Campylobacter* spp. (Skirrow, 1977). Thus, due to high consumption rates of poultry, they present a significant risk of human campylobacteriosis (Humphrey et al., 2007).

Campylobacter infection has been recognized as a primary cause of human foodborne infection and, in certain areas, cases exceed those caused by *Salmonella* or *Yersinia* (Silva et al., 2011; Chlebicz and Ślizewska, 2018). In the past decades, studies have shown increased campylobacteriosis incidence in Europe, United States, and Australia. There are only limited epidemiological data available for Asia, Middle East, and Africa, but research data have shown *Campylobacter* infection is endemic in these regions (Kaakoush et al., 2015).

The European Food Safety Authority described several factors contributing to the frequency and severity of campylobacter infections. Young children, especially those younger than 5 years, are at high risk. Summer months record higher incidences and host immunity, demographics, and campylobacter strain are also important factors for campylobacter infection (EFSA, 2007).

Poultry, especially chicken, are a primary source of high-quality protein in many countries. Because birds are the hosts for *Campylobacter,* consumption of chicken has caused large outbreaks of acute campylobacteriosis in humans. Investigations have identified a range of factors leading to outbreaks, include poor handling raw poultry, consumption of undercooked meat, and cross-contamination of raw to cooked foods (Silva et al., 2011).

The cecum and colon of chicken harbor many *Campylobacter* spp. and chicken carcasses are easily contaminated with *Campylobacter* during processing (Berrang et al., 2001). *Campylobacter* spp. are thought to contaminate chicken skin, which serves as a potential sources for further cross-contamination (Chantarapanont et al., 2003) and previous studies have reported the potential for *Campylobacter* spp. to grow on chicken skin when stored at room temperature (Scherer et al., 2006).

25.3.4 *Escherichia coli*

Escherichia coli colonize intestines of human and animals. The majority of *E. coli* are harmless and support a balanced microbiome in the human gut. However, some *E. coli* are pathogenic and cause illnesses and food poisoning. Pathogenic

E. coli include Shiga-toxin producing *E. coli* (STEC), enterotoxigenic *E. coli* (ETEC), enteropathogenic *E. coli* (EPEC), enteroaggregative (EAEC), enteroinvasive *E. coli* (EIEC), and diffusely adherent *E. coli* (DAEC). STEC, also known as verocytotoxin producing *E. coli* (VTEC) or enterohemorrhagic *E. coli* (EHEC), is associated most frequently with foodborne poisoning and outbreaks (CDC, 2014).

An STEC (*E. coli* O157:H7 strain) outbreak was first reported in 1982, due to consumption of hamburgers from a fast food restaurant (Riley et al., 1983). While most food pathogens relate to food from animal origins, STEC is also implicated in fresh produce and water sources (Koutsoumanis et al., 2020). Two toxins are produced by STEC, type 1 (Stx1), and type 2 (Stx2), which can be further classified as Stx1/Stx2 subtypes and variants depending on the receptor preference and toxin potency (Melton-Celsa, 2014). Stx2 is reported to cause more severe infection than Stx1, which results in severe weight loss and renal injury (Pradhan et al., 2016). Stx covers a group of bacterial exotoxins produced by both STEC and *Shigella dysenteriae* serotype 1 (Sandvig, 2001). These toxins are highly potent, causing serious infections that include bloody diarrhea, hemolytic uremic syndrome, and central nervous system damages (Lee and Tesh, 2019).

Ruminants, especially postweaning calves and heifers, are recognized as principal asymptomatic reservoirs of STEC (Gyles, 2007). STEC has long been associated with eating undercooked contaminated ground (minced) meat, pork, egg, and dairy products (Koutsoumanis et al., 2020). In recent years, fresh produce has been associated with multiple STEC outbreaks. For example, Germany reported an STEC serotype O104:H4 outbreak linked to contaminated sprouted seeds (Luna-Guevara et al., 2019). Contamination of fresh produce can occur from farm to table, such as during cultivation, postharvest processing, and domestic kitchen handling (Ailes et al., 2008). Studies have identified several sources of contamination that include fertilizer (chemical or animal manure), irrigation water, soil, and pesticides. After harvest, fresh produce transportation, handling, washing, and degree of processing can lead to cross-contamination (Luna-Guevara et al., 2019).

25.4 Microbiological criteria

Microbiological limits that include methods and sampling plans are defined as "microbiological criteria" (ICMSF, 2002). Microbiological criteria (MC) should specify the number of sample units to be collected, analytical method, and numbers of analytical units that should conform to the limits. Additionally, MC should be accompanied by information such as specific food products, sampling plans, methods of examination, and microbiological limits (ICMSF, 2005). Under certain circumstances, MC may be established to determine the acceptability of specific production lots, particularly when conditions of production are not known. MC are measures against which decisions are based and, traditionally, they have been designed for testing lots or shipments for acceptance or rejection. In contrast, FSOs and POs are maximum levels and do not specify details needed for testing (ICMSF, 2005). However, MC can be based on POs, where testing of foods for a specific microorganism can be an effective means for verification. While there are several approaches that can be used (e.g., lot testing, process control testing), all compare results against predetermined limits. ICMSF (2002) has provided guidance on the establishment of MC. MC may be established for quality as well as safety (CAC, 1997) and are used in setting standards, guidelines, and purchase specifications, which are defined as follows (ICMSF, 2009):

- Microbiological standards are contained in international, federal, and regional laws. Relatively few standards exist in the United States, with most relating to a specific pathogen, such as *Listeria monocytogenes* in RTE foods, *Escherichia coli* O157: H7 in ground beef, and *Salmonella* and generic *E. coli* in meats relative to USDA's Pathogen Reduction Program. Exceeding a standard relating to a pathogen, such as *Salmonella* or *Listeria*, may lead to a product recall and/or punitive action.
- Microbiological guidelines are internal, advisory criteria established by a processor or a trade association. Failure to meet them serves as an alert to the processor, indicating that remedial action should be taken. A wide variety of criteria fit into this category, such as results on preop swabs from equipment, in-process samples of product or equipment, and environmental samples tested for pathogens.
- Microbiological specifications or purchase specifications are agreements between the vendor and buyer of a product as a basis for sale. These criteria can be looked upon as mandatory since failure of the vendor to meet specifications can be used as a basis for product rejection.

While safe food can be produced by adhering to GHP and HACCP programs, levels of safety these systems are expected to deliver have seldom been defined in quantitative terms. Establishment of FSOs and POs (see Chapter 6) provide the industry with quantitative targets (ICMSF, 2009). Although FSOs and POs are expressed in quantitative terms, they are not MC. MC are designed to determine adherence to GHPs and HACCP (i.e., verification) when more effective and efficient means are not available. FSOs and POs are targets to be met but, in this context, MC based on within-lot testing are meant to provide statistically designed means for determining whether these targets are being achieved (van Schothorst et al., 2009). A detailed description on setting MC is available elsewhere (ICMSF, 2009). Uses of MC include the following:

- Verifying compliance of POs and FSOs (within the limits of sampling and testing);
- Validating that an HACCP/GHP system(s) provide the desired level of control;
- Verifying control within HACCP/GHP systems;
- Demonstrating the utility (suitability) of a food or ingredient for a particular purpose;
- Establishing the keeping quality (shelf-life) of certain perishable foods;
- Driving industry improvement when used as a regulatory tool;
- Achieving market access;
- Identifying unacceptable from acceptable product defined by standards, guidelines, and specifications not directly related to the above applications.

MC for foods in international trade are addressed by the joint Food and Agriculture Organization/World Health Organization (FAO/WHO) food standards program, as implemented by the Codex Alimentarius Commission (CAC, 1997). This program was the direct result of conflict between national food legislation and general requirements of global food markets. It was established in 1962, the same year as the International Commission on Microbiological Specifications for Foods was established. Serious nontariff obstacles to trade were caused by differing national food legislation (ICMSF, 2002). At that time, the CAC's objectives were to develop international food standards, codes of practice, and guidelines, anticipating that their adoption would help remove and prevent nontariff barriers to the food trade.

The Codex Committee on Food Hygiene (CCFH) has responsibility for all provisions on food hygiene practice (effectively, GHP). CCFH requires expert advice in dealing with highly specialized microbiological matters, especially when developing microbiological criteria. Such advice has been provided frequently by ICMSF through its publications on sampling plans and principles for the establishment and application of MC for foods and several other discussion papers (CAC, 1997; ICMSF, 1998).

Problems with some current MC include application of sampling statistics based on random distribution to situations where contamination is not random; use of too few samples to draw valid conclusions; noncompliance; lack of meaningful negative results; resampling of products that failed initial tests; and regulatory standards that ignore principles of establishment of criteria, e.g., zero tolerance; and use of indicator tests in the absence of an established relationship of the indicator to the pathogen of interest (Carter, 2008).

25.5 Microbiological testing

Monitoring microbial contamination is critical to ensure the safety of products including foods, pharmaceuticals, cosmetics, and healthcare items. Many methods are used in different regions around the world, depending on local regulations. However, the lack of alignment or "harmonization" of methods for microbiological testing creates an extra burden for FBOs that intend to market products globally. Lack of harmonization also results in conflicts between regulatory agencies of trading partners.

Microbiological testing can be a useful tool in the management of food safety. However, microbiological tests should be selected and applied with knowledge about limitations and benefits, and their purpose (ICMSF, 2009). In many instances, other means of assessment are quicker and more effective than microbiological testing. The need for microbiological testing varies along the food chain and, therefore, points in the food chain where sampling and testing occur should be selected where information about microbiological risk will prove most useful for control. Several types of microbiological testing may be utilized by FBOs and governmental agencies. One of the most commonly used is within-lot testing, which compares microbiological hazards detected in a food against a prespecified limit, i.e., an MC (ICMSF, 2002).

ICMSF has written extensively on the principles of controlling microbial hazards in foods, but recognizes that those same principles are applicable to control of microorganisms that may be associated with spoilage, or general indicators of good hygienic and good manufacturing practices (ICMSF, 2009). For example, the combination of pasteurization and refrigeration of milk is effective means for reducing and controlling both pathogenic and spoilage microflora and, therefore, a safe product with extended shelf life. Good hygienic practices designed to prevent microbial contamination of the product during manufacture and packaging prevent recontamination of the product with pathogens and spoilage microorganisms. Microbiological testing of pasteurized milk for pathogens is not justified, because the process is effective in reducing microorganisms to well below levels that can be detected (ICMSF, 2009). However, pasteurized milk is tested routinely to determine the Standard Plate Count (SPC)/mL and Coliform Count/mL as indicators of the efficacy of pasteurization and prevention of postpasteurization contamination. Pasteurized milk is just one example of microbial hazards MC that also assure microbial quality. Further, microbiological testing to monitor or verify control is indicative of both quality and safety.

It is important to consider the purpose of microbiological tests because, typically, they are performed to reach a decision. Their purpose determines type of test (indicator or pathogen), method (rapidity, accuracy, repeatability, reproducibility, etc.), sampling (line residue, end-product), interpretation of the results, and actions to be taken (rejection of a lot, investigational sampling, readjustment of the process, etc.) (ICMSF, 2009). If the purpose cannot be defined, then analysis should not be done. The rationale for testing falls into four categories: to determine safety, adherence to GHPs and/or GMPs, and utility of a food or ingredient for a particular purpose, and to predict product stability (ICMSF, 2009).

Some of the basic steps in microbiological testing of food include collecting samples, preparing homogenates (as required), and conducting either quantitative or qualitative analyses. Each of these steps has numerous procedures, depending on the purpose of the test and the type of food being analyzed.

Quantitative analysis is usually conducted via colony counts or the most probable number (MPN) methodology, although other methods are available if concentrations of microorganisms are sufficiently high. In qualitative analysis, there is one or more enrichment steps for population amplification followed by detection and isolation of the target microorganism (confirmation, if required), with evidence of growth and identity measured by means, such as visual, biochemical, immunological, and genetic techniques (ICMSF, 2009). Modern detection technologies include enzyme immunoassay, immunocapture, immunoprecipitation, nucleic acid hybridization, use of DNA/RNA, PCR and other nucleic amplification methodologies, electrochemical techniques, enzymatic amplification, chromatographic separation, etc.

For quantitative microbial tests, approximate lower limits are as follows (ICMSF, 2009):

- $<10-100$ cfu/g MPN;
- $>10-100$ cfu/g viable counts;
- $>10^3-10^4$ cfu/g DEFT;
- $>10^4-10^5$ cfu/g ELISA, flow cytometry, quantitative PCR;
- $>10^5-10^6$ cfu/g direct microscopy, spectrophotometry.

Applications of microbiological testing, both nationally and internationally, are broad. Microbiological testing is used when gathering epidemiological data (e.g., during outbreaks, recalls, etc.); conducting baseline studies, international trade, trade association studies, such as surveys of products in the marketplace, and retail surveys (by government), comparative assays by FBOs across production facilities and production lines; customer/supplier requirements (purchase specifications); and HACCP or other prerequisite programs (GMPs, GHPs, facility/product specific) (ICMSF, 2009).

Additionally, microbiological testing has an essential role in food safety programs, such as hazard analysis; validating processes; monitoring critical ingredients and high-risk finished products; verifying critical control points; determining potential for postprocess contamination; establishing the adequacy and frequency of cleaning and sanitation; detecting difficult areas to clean and sanitize; compliance testing; complying with mandatory regulatory programs; confirming purchase specifications; documenting situations in case of litigation; problem solving, etc (ICMSF, 2009).

Often microbiological data exist but are only used for acceptance on a given unit of production (e.g., batch, lot, 1 day's production). Trend analysis of these data can assist in identifying the most likely sources of a problem, or pinpointing areas for further investigation. A detailed discussion about the uncertainty of microbiological data, sampling plans, method validation, and use of lab proficiency testing, as part of a quality program, is given in ICMSF (2009).

Many limitations of microbiological testing revolve around sampling, i.e., not practical to test a sufficient numbers of samples, sampling might cause incorrect conclusions to be drawn, and no feasible sampling plan can ensure absence of a pathogen (Carter, 2008). In addition, results identify outcomes, not causes or controls. Irrespective of considerations or scenarios, specific methods are required for assessing and validating the fact that accepted and risks are being controlled. Despite the limitations, microbiological testing is almost always an important component of any integrated program to assure the safety of foods at all levels of government and industry.

25.6 Validation of microbiological methods

AOAC and ISO provide similar, yet different, procedures for validation and certification of microbiological methods. While these organizations have assisted in acceptance of test methods by regulatory bodies, methods are not harmonized across the organizations (i.e., methods may be accredited by ISO, but not AOAC and vice versa), thus impeding trade.

While these institutions have not harmonized methodology, they have taken steps in the right direction by creating procedures for determining if different microbial methods provide equivalent results, especially in regard of limits of detection, false positive and negative results, and impact of food matrices on results.

25.6.1 Association of analytical communities

AOAC has had official observer status in *Codex Alimentarius* since its inception, giving AOAC the ability to input on development of international standards for foods and agriculture: if an AOAC method is available, it will generally be accepted into Codex standards. The majority of analytical methods cited in Codex standards are those of AOAC, and many AOAC methods are required specifically in the enforcement of some state, provincial, municipal, and local laws as well as many federal food standards worldwide.

AOAC's "Official Methods of Analysis" have been designated as official by regulations promulgated for enforcement of the Food, Drug, and Cosmetic Act (21 CFR 2.19), recognized in Title 9 of the USDA-FSIS Code of Federal Regulations and, in some cases, by the US Environmental Protection Agency (Stewart and Busta, 2010). In 2002, the AOAC published guidelines on validation of qualitative and quantitative food microbiological methods (Feldsine et al., 2002). The guidelines define steps involved in method validation, including selection of study director, ruggedness, methods comparison, or precollaborative study, which includes inclusivity/exclusivity, interlaboratory collaborative study, and the AOAC approval process (Feldsine et al., 2002). These guidelines were provided by the Methods Committee on Microbiology and Extraneous Materials as part of an initiative to specify validation criteria for methods comparison/precollaborative studies and interlaboratory studies and to harmonize validation methodology with ISO standard 16140 "Protocol for the Validation of Alternative Methods." They are also applicable to the validation of any alternative methods, whether proprietary or nonproprietary, which are submitted to AOAC for OMA status recognition. It is the intent of the guidelines to harmonize validation procedures with ISO standard 16140. Data produced for an alternative method that satisfies ISO 16140 protocol requirements and acceptance criteria may be recognized reciprocally and apply for validation, although further data may be required (Feldsine et al., 2002).

AOAC recommends single laboratory validation, based on a study protocol that is reviewed by the Expert Review Panel (ERP) and the appropriate AOAC methods committee. As suggested by the name, single laboratory validation shows how a method performs within one laboratory. A full collaborative study shows how methods perform in multiple laboratories. The value of single laboratory validation is that it gives a good indication of performance and provides some measure of likely performance in a collaborative study. To ensure the success of a full collaborative study, involving 8–10 sites, the protocol is designed using AOAC Official Methods^{SM} Program guidelines and must be approved by the appropriate AOAC methods committee and general referee(s); ERP also provides review comments. AOAC committees are composed of seven or more experts in the topic areas who review recommendations of the general referee(s), study director and ERP, and provide written review of the study. After the study is complete, the study director submits the results for review by the relevant committee. If completed successfully, the study is submitted to the AOAC Official Methods Board for further review and approval.

25.6.2 International organization for standardization

ISO 16140:2016 "*Microbiology of the food chain— Method validation— Part 2: Protocol for the validation of alternative (proprietary) methods against a reference method*" defines the general principles and technical protocols for validation of alternative methods for microbiological analysis of food, animal feed, and environmental and veterinary samples used for the validation. These can be used in the framework of official control, and international acceptance of results obtained using these method (ISO, 2016). It also establishes the principles for certification of these methods, based on validation defined in ISO 16140:2016. Where an alternative method is used on a routine basis for internal laboratory use without the requirement to meet (higher) external criteria, less stringent validation of the method may be appropriate. Alternative methods that have not been validated against a reference method are acceptable under this standard provided:

- Validation protocols have been approved by a recognized panel of technical reviewers and the results have been accepted;
- Technical reviewers operated under the sponsorship of internationally recognized organizations performing method validations (e.g., AFNOR, NordVal, AOAC International, AOAC Research Institute);
- Validations include studies that conform to at least the total sample number and food matrix requirements of this standard (ISO, 2016).

According to ISO 16140:2016, when alternative methods have been compared with an internationally recognized reference method (e.g., AOAC International) that differs in minor aspects from the reference method, and if the protocol is similar, then the results can be accepted. If the method is substantially different from the reference method, an assessment must be made to determine if these differences would have a minor or major impact on method performance. An assessment is made regarding supplementary data for the resolution of procedural and/or reference method differences (e.g., primary enrichment broth). Decisions that reference methods contain major differences must be substantiated by the organizing laboratory with documented data. Data required to resolve perceived differences have to be stipulated.

25.7 Harmonization of global regulations for *Listeria monocytogenes* in ready-to-eat foods

An example of the challenges associated with harmonization of global regulations is control of public health threats arising from *Listeria monocytogenes*. Measurements (presence and concentration) of *Listeria monocytogenes* have a significant role in the control of this microorganism in RTE foods.

Listeria monocytogenes is a bacterium that can contaminate foods and cause a mild noninvasive illness (listerial gastroenteritis) or a severe, sometimes life-threatening, illness (invasive listeriosis). Invasive listeriosis is characterized by a high case fatality rate, ranging from 20% to 30% (FDA, 2017). The main target populations for listeriosis are the following (FDA, 2000):

- Pregnant women/fetus—perinatal and neonatal infections;
- Individuals who are immunocompromised by corticosteroids, anticancer drugs, graft suppression therapy, AIDS;
- Cancer patients, particularly those with leukemia
- Diabetic, cirrhotic, asthmatic, and ulcerative colitis patients;
- Elderly;
- Individuals taking antacids or cimetidine

Manifestations of listeriosis include sepsis, meningitis (or meningoencephalitis), encephalitis, and intrauterine or cervical infections in pregnant women, which may result in spontaneous abortion (second/third trimester) or stillbirth. Onset is usually preceded by influenza-like symptoms including fever (FDA, 2000). Gastrointestinal symptoms, such as nausea, vomiting, and diarrhea, may precede more serious forms of listeriosis or may be the only symptoms. Onset times for serious forms of listeriosis are unknown but may range from a few days to 3 weeks. Onset times for gastrointestinal symptoms are also unknown but is probably more than 12 h. An infective dose of *L. monocytogenes* is unknown but thought to vary with strain and susceptibility. From cases contracted via raw or supposedly pasteurized milk, in susceptible individuals, fewer than 1000 microorganisms can lead to illness (FDA, 2000).

L. monocytogenes is widespread in the environment, being found in soil, water including sewage, and decaying vegetation. It can be isolated from humans, domestic animals, raw agricultural commodities, and food processing environments (particularly cool damp areas). Control of *L. monocytogenes* in food processing environments has been the subject of a number of scientific publications (FDA, 2017). *L. monocytogenes* can grow slowly at refrigeration temperatures and, therefore, refrigerated RTE foods can support growth of *L. monocytogenes*, meaning these products must be managed appropriately.

Most cases of human listeriosis occur sporadically. However, much of what is known about its epidemiology is derived from outbreak-associated cases. With rare exceptions, foods that have been reported to be associated with outbreaks or sporadic cases of listeriosis have those that can support growth including RTE products such as coleslaw, fresh soft cheese made with unpasteurized milk, frankfurters, deli meats, and butter, and are often associated with failures during production or processing (FDA, 2017). Data from a variety of sources of information show that *L. monocytogenes* has been detected to varying degrees in unpasteurized and pasteurized milks, high-fat dairy products, soft unripened cheeses (cottage cheese, cream cheese, ricotta), cooked RTE crustaceans, smoked seafood, fresh soft cheese (queso fresco), semisoft cheese (blue, brick, Monterrey), soft-ripened cheese (brie, camembert, feta), deli-type salads, sandwiches, fresh-cut fruits and vegetables, and raw molluscan shellfish (FDA, 2017). However, these data also show that most RTE foods do not contain detectable numbers of *L. monocytogenes*. For many RTE foods, contamination can be avoided—e.g., through application of current GMPs, listericidal and listeristatic processes, segregation of foods that have been cooked from those that have not, and sanitation controls with effective environmental monitoring programs (FDA, 2017).

Rates of listeriosis are about the same in developed countries, independent of policies in place. For example, in the United States, there are an estimated 3.4−4.4 cases/million people, with a 20%−30% mortality rate; in Australia, the rate is 3 cases/million, with a 23% mortality rate; in New Zealand a rate of 5 cases/million, 17% mortality rate, and in the EU a rate of 0.3−7.5 cases/million is estimated (Todd, 2006).

There have been three risk assessments completed for *L. monocytogenes* in RTE foods, all of which are excellent sources of information for determining scientifically based MCs for this microorganism. In 2003, the US FDA published "*Quantitative Assessment of the Relative Risk to Public Health from Foodborne Listeria monocytogenes Among Selected Categories of Ready-to-Eat Foods*" (HHS/USDA, 2003) and the USDA Food Safety and Inspection Service published "*Risk Assessment of Listeria monocytogenes in Deli Meat*" (FSIS, 2003) in the same year. The following year, FAO and WHO published "*Risk Assessment of Listeria monocytogenes in Ready-to-Eat Foods*" (FAO, 2004). Conclusions from the USDA FSIS risk assessment for deli meats stated that increased frequency of food contact surface testing and sanitation was estimated to lower the risk of listeriosis; and combinations of interventions (i.e., microbiological testing and sanitation of food contact surfaces, pre- and postpackaging treatments, and the use of growth inhibitors/product reformulation) were more effective than any single intervention step (Todd, 2006). For example, estimated numbers of deaths annually due to listeriosis could drop from 250 to <100 if FBOs used a combination of growth inhibitors and postpackaging pasteurization of products. Based on some of the information from this risk assessment, under an interim final rule released on June 6, 2003 (FSIS, 2004), USDA FSIS afforded RTE products different regulatory treatment; products had to be produced using one of three control programs to reduce or eliminate *L. monocytogenes.*

These programs established means that may be used, specifically:

- Postlethality treatment that reduces or eliminates *L. monocytogenes* AND an antimicrobial agent or process that suppresses or limits growth throughout shelf life (Alternative 1);
- Postlethality treatment that reduces or eliminates *L. monocytogenes* OR an antimicrobial agent or process that suppresses or limits its growth throughout shelf life (Alternative 2);
- Sanitation procedures to prevent *L. monocytogenes* contamination; those using Alternative 3 are tested most frequently by regulators.

Under Alternative 1 or 2, FSIS would apply relatively less sampling to a product undergoing a postlethality treatment giving >2-log_{10} reduction of *L. monocytogenes*, relatively more sampling to a product receiving a postlethality treatment giving between 1- and 2-log_{10} reduction and would consider <1-log_{10} reduction not eligible for these Alternatives, unless there was supporting documentation demonstrating an adequate safety margin (FSIS, 2004).

HHS/USDA risk assessment (2003) estimated that foods posing the highest risk support growth of *L. monocytogenes*. In contrast, foods with the lowest risk of listeriosis are those that have intrinsic or extrinsic factors preventing growth of *L. monocytogenes*, or are processed in such a way as to alter the normal characteristics of the food (FDA, 2008). For example, it is well established that *L. monocytogenes* does not grow when (i) pH is less than or equal to 4.4; (ii) water activity is less than or equal to 0.92; or (iii) foods are frozen.

Currently, there is no international agreement on what numbers of *L. monocytogenes* in foods are acceptable to protect the consumer. In several countries, different criteria or recommendations for tolerable levels of *L. monocytogenes* in RTE foods have been established, but the rationale is not always clear (Todd, 2006). In the United States, a major outbreak of listeriosis occurred in 1985, when a Mexican-style soft cheese caused over 142 cases of illness with 48 fatalities. In the same year, a survey conducted by the FDA found *L. monocytogenes* in both imported and domestic fresh cheeses. It was recognized that *L. monocytogenes* had caused foodborne diseases in food products regulated by the FDA and USDA, or had the potential to do so (Todd, 2006). Thus, the FDA established a policy of "zero tolerance" for *L. monocytogenes* in RTE foods, the MC being absent in 25 g (<0.04 cfu/g).

The European Union (EU) has different MC for *L. monocytogenes* in RTE foods, based on the categories of food—whether products can support growth of the microorganism or not and if the food is intended for the general population or special medical purposes. The EU also describes stages where criteria apply, either for products on the market and during their shelf-life or before the food has left the immediate control of FBOs (Table 25.1; Food Safety Authority of Ireland, 2011).

In Canada, RTE foods are categorized based upon health risk. Category 1 foods are those with a causally link to documented outbreaks and/or have been placed in the "high risk" category, based on HHS/USDA risk assessment (Health Canada, 2011). Products in Category 1 should receive the highest priority for inspection and compliance activities. Category 2 considers all RTE foods that are capable of supporting growth of *L. monocytogenes* and have a shelf-life exceeding 10 days. The presence of *L. monocytogenes* in these products triggers a Health 2 concern, with possible need for a public alert. These products are given the second highest priority in inspection and compliance activities.

TABLE 25.1 European Union microbiological criteria for *Listeria monocytogenes* in ready-to-eat foods.

Food category	Sampling plan	Limit	Analytical reference method	Stage where the criterion applies
RTE foods intended for infants and RTE foods for special medical purposes	$n = 10$	Absence in 25 g	EN/ISO	Products placed on the market during their shelf life
	$c = 0$		11290–1	
RTE foods able to support the growth of *L. monocytogenes* other than those intended for infants and for special medical purposes	$n = 5$	100 cfu/g	EN/ISO	Products placed on the market during their shelf life
	$c = 0$		11290–2	
	$n = 5$	Absence in 25 g	EN/ISO	Before the food has left the immediate control of the FBO[a] who has produced it
	$c = 0$		11290–1	
RTE foods unable to support the growth of *L. monocytogenes* other than those intended for infants and for special medical purposes	$n = 5$	100 cfu/g	EN/ISO	Products placed on the market during their shelf life
	$c = 0$		11290–2	

[a]*Food business operator, n = number of units comprising the sample, c = number of sample units giving values over the limit.*

Category 3 contains two types of RTE food products; those supporting growth with a <10-day shelf-life and those not supporting growth. These products receive the lowest priority in terms of inspection and compliance. For Category 3 RTE foods, factors, such as adherence to GMP, levels of *L. monocytogenes* in foods (action level 100 cfu/g), and/or a health risk, are considered in determining compliance actions.

In the United States, in 2003, 15 trade associations submitted a petition to the FDA to amend the regulatory limit to 100 cfu/g *L. monocytogenes* in foods that do not support growth, thereby establishing a scientific basis for actions related to *L. monocytogenes* in such foods.

In response, the FDA (2008) issued two draft documents: "*Draft Guidance for Industry: Control of L. monocytogenes in Refrigerated or Frozen RTE Foods*" and "*Draft Compliance Policy Guideline.*" The guidance document divides foods into two categories: foods that support the growth of *L. monocytogenes* high-risk and MC would remain absence in 25 g (<0.04 cfu/g) and foods that do not support growth low risk, where the MC would be < 100 cfu/g. If the food contains *L. monocytogenes* at higher levels, the food would be considered contaminated. MCs in these draft guidelines were alignment with those of the EU and Canada. The enforcement policy in the draft guidelines clarify which foods support the growth of *L. monocytogenes,* ensuring FDA resources are focused on high-risk foods. FDA anticipated that it might be able to increase numbers of samples that it collects and tests for *L. monocytogenes* to verify compliance with limits in low-risk RTE foods, while it continued to focus on inspection and outreach efforts on high-risk RTE foods. There was a 60-day comment period and, at the time of writing, no further updates are available.

Harmonization of food regulations can only be done through agreement among governments and CAC one international bodies of member countries that may facilitate this. Discussions continue with subgroups in parallel with studies, including FAO/WHO assessments, and it is hoped that science-based criteria may be agreed. For example, that 100 cfu/g *L. monocytogenes* at the point of consumption presents little risk to the healthy population. How governments ask the food industry to achieve this can differ. For example, some foods have interventions, such as low pH or in-package pasteurization (i.e., for sliced deli meats), but these approaches will not work for all products (i.e., cheese and some fruits and vegetables) (Todd, 2006).

Acceptable levels of risk also have a cultural component. For example, certain foods, such as raw milk cheeses, are culturally important in Europe. A "zero tolerance" regulation might not be effective in reducing numbers of cases. Consumer education and labeling have had limited success (i.e., education of risk to pregnant women), but do have influence, and risk communication efforts should continue (Todd, 2006). There are certain subpopulations who are at greater risk of contracting listeriosis, e.g., those undergoing cancer therapy, with AIDS and transplant recipients. Therefore, a different strategy/standard might be required. Different strategies are being considered for these populations. For example, the EU is considering more stringent policies, but most policies are likely to be internal developed and implemented by institutions supporting these populations, rather than governments imposing national standards for *L. monocytogenes* in food delivered to these subpopulations (Todd, 2006).

25.8 Conclusion

Processes and procedures to eliminate or minimize the presence of specific microorganisms or their byproducts and toxins must be harmonized. Metrics required to measure these procedures and their outcomes must also be synchronized. Implementation of one or more effective interventions as control programs requires standardization and validation of methods and protocols that will be used to verify outcomes. Our goal, therefore, must be that the existing plethora of methods, procedures, processes, protocols, and sampling approaches are reconciled through harmonization to deliver meaningful results.

References

Abebe, E., Gugsa, G., Ahmed, M., 2020. Review on major foodborne zoonotic bacterial pathogens. J. Trop. Med. 3, 4674235. https://doi.org/10.1155/2020/4674235.

Addis, Z., Kebede, N., Worku, Z., Gezahegn, H., Yirsaw, A., Kassa, T., 2011. Prevalence and antimicrobial resistance of *Salmonella* isolated from lactating cows and in contact humans in dairy farms of addis ababa: a cross sectional study. BMC Infect. Dis. 11 (1), 1−7.

Ailes, E.C., Leon, J.S., Jaykus, L.-A., Johnston, L.M., Clayton, H.A., Blanding, S., Kleinbaum, D.G., Backer, L.C., Moe, C.L., 2008. Microbial concentrations on fresh produce are affected by postharvest processing, importation, and season. J. Food Protect. 71 (12), 2389−2397.

Alvarez-Ordóñez, A., López, M., Prieto, M., 2017. Role of stress response on microbial ecology of foods and its impact on the fate of food'borne microorganisms. In: de Souza Sant'Ana, A. (Ed.), Quantitative Microbiology in Food Processing: Modeling the Microbial Ecology. Wiley, Hoboken, NJ, pp. 631−648.

Argudín, M.A., Mendoza, M.C., Rodicio, M.R., 2010. Food poisoning and *Staphylococcus aureus* enterotoxins. Toxins 2 (7), 1751−1773.

Armand-Lefevre, L., Ruimy, R., Andremont, A., 2005. Clonal comparison of *Staphylococcus* from healthy pig farmers, human controls, and pigs. Emerg. Infect. Dis. 11 (5), 711−714.

Begley, M., Hill, C., 2015. Stress adaptation in foodborne pathogens. Annu. Rev. Food Sci. Technol. 6, 191−210. https://doi.org/10.1146/annurev-food-030713-092350.

Berrang, M.E., Buhr, R.J., Cason, J.A., Dickens, J.A., 2001. Broiler carcass contamination with *Campylobacter* from feces during defeathering. J. Food Prot. 64, 2063−2066.

Bhunia, A., 2018. Foodborne Microbial Pathogens: Mechanisms and Pathogenesis. Springer, Berlin. https://doi.org/10.1007/978-1-4939-7349-1.

Black, R.E., Levine, M.M., Clements, M.L., Hughes, T.P., Blaser, M.J., 1988. Experimental *Campylobacter jejuni* infection in humans. J. Infect. Dis. 157, 472−479.

CAC, 1997. Recommended International Code of Practice for the General Principles of Food Hygiene. Codex Alimentarius Commission. CAC/RCP 1-1969, Rev. 3.

Carter, M., June 24, 2008. The steps before PCR: sampling and enrichment, concentration procedures, compositing, true real time PCR without enrichment. In: Presented during the Colorado State University Workshop on Molecular Methods in Food Microbiology. CPU Department of Animal Science.

Center for Disease Control and Prevention (CDC), 2019. Emerging Strain of *Salmonella* Raises Concern. www.cdc.gov/foodsafety/newsletter/emerging-strain-salmonella-8-22-19.html#:~:text=An%20investigation%20into%20an%20emerging,Morbidity%20and%20Mortality%20Weekly%20Report (Accessed 7 September 2020).

Center for Disease Control and Prevention (CDC), 2014. *E. coli* (*Escherichia coli*). https://www.cdc.gov/ecoli/general/index.html (Accessed 10 September 2020).

Chaibenjawong, P., Foster, S.J., 2011. Desiccation tolerance in *Staphylococcus aureus*. Arch. Microbiol. 193 (2), 125−135.

Chantarapanont, W., Berrang, M., Frank, J.F., 2003. Direct microscopic observation and viability determination of *Campylobacter* jejuni on chicken skin. J. Food Protect. 66, 2222−2230.

Chlebicz, A., Ślizewska, K., 2018. Campylobacteriosis, salmonellosis, yersiniosis, and listeriosis as zoonotic foodborne diseases: a review. Int. J. Environ. Res. Publ. Health 15, 863. https://doi.org/10.3390/ijerph15050863.

Darby, J., Sheorey, H., 2008. Searching for *Salmonella*. Aust. Fam. Physician 37, 806−810.

Eng, S.K., Pusparajah, P., Ab Mutalib, N.S., Leng, S.H., Chan, K.G., Han, L.L., 2015. *Salmonella*: A review on pathogenesis, epidemiology and antibiotic resistance. Front. Life Sci. 8 (3), 284−293.

Esbelin, J., Santos, T., Hébraud, M., 2018. Desiccation: an environmental and food industry stress that bacteria commonly face. Food Microbiol. 69, 82−88.

European Food Safety Authority (EFSA), 2007. The community summary report on trends and sources of zoonoses, zoonotic agents, antimicrobial resistance and foodborne outbreaks in the European Union in 2006. EFSA J. 130, 130−155.

FAO, 2004. Risk Assessment of *Listeria monocytogenes* in Ready-To- Eat Foods. World Health Organization Food and Agriculture Organization and of the United Nations. http://www.fao.org/3/a-y5394e.pdf (Accessed 16 September 2020).

FDA, 2017. Guidance for Industry: Control of *Listeria monocytogenes* in Refrigerated or Frozen Ready-To-Eat Foods Draft Guidance. Food and Drug Administration, Center for Food Safety and Applied Nutrition. https://www.fda.gov/regulatory-information/search-fda-guidance-documents/draft-guidance-industry-control-listeria-monocytogenes-ready-eat-foods (Accessed 16 September 2020).

FDA, 2000. Foodborne Pathogenic Microorganisms and Natural Toxins Handbook: *Listeria monocytogenes*. Food and Drug Administration. https://pdf.usaid.gov/pdf_docs/pnado152.pdf (Accessed 16 September 2020).

Feßler, A., Scott, C., Kadlec, K., Ehricht, R., Monecke, S., Schwarz, S., 2010. Characterization of methicillin-resistant *Staphylococcus aureus* ST398 from cases of bovine mastitis. J. Antimicrob. Chemother. 65 (4), 619−625.

Feldsine, P., Abeyta, C., Andrews, W.H., 2002. AOAC international methods committee guidelines for validation of qualitative and quantitative food microbiological official methods of analysis. J. AOAC Int. 85, 1187−1200.

Food Safety Authority of Ireland, 2011. *Listeria monocytogenes*. FSAI. https://www.fsai.ie/search-results.html?searchString=listeria%20monocytogenes (Accessed 16 September 2020).

FSIS, 2003. Risk Assessment for *Listeria monocytogenes* in Deli Meat. Food Safety and Inspection Service. https://www.fsis.usda.gov/shared/PDF/Lm_Deli_Risk_Assess_Final_2003.pdf (Accessed 16 September 2020).

FSIS, 2004. Compliance guidelines to control *Listeria monocytogenes* in post- lethality exposed ready-to-eat meat and poultry products. Food Safety and Inspection Service. http://www.fsis.usda.gov/OPPDE/rdad/FRPubs/97-013F/Lm_Rule_Compliance_Guidelines_ 2004.pdf (Accessed 20 April 2005). − Note updated May 2006 https://www.fsis.usda.gov/wps/wcm/connect/8cf5e6a1-1f52-406c-bd8b-e3608a5a3c7e/Lm_Rule_Compliance_Guidelines_May_2006.pdf?MOD=AJPERES (Accessed 16 September 2020).

Guardabassi, L., O'Donoghue, M., Moodley, A., Ho, J., Boost, M., 2009. Novel lineage of methicillin-resistant *Staphylococcus aureus*, Hong Kong. Emerg. Infect. Dis. 15 (12), 1998−2000.

Gyles, C.L., 2007. Shiga toxin-producing *Escherichia coli*: an overview. J. Anim. Sci. 85, E45−E62. https://doi.org/10.2527/jas.2006-508.

Health Canada, 2011. Policy on *Listeria monocytogenes* in Ready-To-Eat Foods. Health Canada. https://www.canada.ca/content/dam/hc-sc/migration/hc-sc/fn-an/alt_formats/pdf/legislation/pol/policy_listeria_monocytogenes_2011-eng.pdf (Accessed 16 September 2020).

Hennekinne, J.-A., de Buyser, M.-L., Dragacci, S., 2012. *Staphylococcus aureus* and its food poisoning toxins: characterization and outbreak investigation. FEMS Microbiol. Rev. 36, 815−836.

Heredia, N., García, S., 2018. Animals as sources of food-borne pathogens: a review. Animal Nutr. 4, 250−255.

HHS/USDA, 2003. Quantitative Assessment of the Relative Risk to Public Health from Food-Borne Listeria Monocytogenes Among Selected Categories of Ready-To-Eat Foods, vols. 23−28. HHS Food and Drug Administration and USDA Food Safety and Inspection Service. Available in Docket No. 1999N-1168. https://www.fda.gov/media/77947/download (Accessed 16 September 2020).

Hoffmann, S., Devleesschauwer, B., Aspinall, W., et al., 2017. Attribution of global foodborne disease to specific foods: findings from a World Health Organization structured expert elicitation. PLoS One 12 (9), e0183641.

Horn, N., Bhunia, A.K., 2018. Food associated stress primes foodborne pathogens for gastrointestinal phase of infection. Front. Microbiol. 9, 1962. https://doi.org/10.3389/fmicb.2018.01962.

Humphrey, T., O'Brien, S., Madsen, M., 2007. Campylobacters as zoonotic pathogens: a food production perspective. Int. J. Food Microbiol. 117, 237−257.

ICMSF, 1998. International Commission on Microbiological Specifications for Foods. Microbial ecology of food commodities. In: Microorganisms in Foods. London: Blackie Academic and Professional. ICMSF.

ICMSF, March 2001. The Role of Food Safety Objective in the Management of Microbiological Safety of Food According to Codex Documents. Document Prepared for the Codex Committee on Food Hygiene. International Commission on Micro- Biological Specifications for Foods.

ICMSF, 2002. Microorganisms in foods, 7. In: Microbiological Testing in Food Safety Management. Kluwer Academic/Plenum Publishers, International Commission on Microbiological Specifications for Foods, New York.

ICMSF, 2005. A Simplified Guide to Understanding and Using Food Safety Objectives and Performance Objectives. International Commission on Microbiological Specifications for Foods. http://www.icmsf.iit.edu/main/articles_papers.html (Accessed 23 March 2009).

ICMSF, 2009. Microorganisms in Foods, 8. Use of Data for Assessing Process Control and Product Acceptance. Springer in press, International Commission on Microbiological Specifications for Foods, New York.

ISO, 2016. ISO 16140:2016 Microbiology of the Food Chain — Method Validation — Part 2: Protocol for the Validation of Alternative (Proprietary) Methods against a Reference Method. International Organization for Standardization. https://www.iso.org/standard/54870.html (Accessed 16 September 2020).

Jones, T.F., Kellum, M.E., Porter, S.S., Bell, M., Schaffner, W., 2002. An outbreak of community-acquired foodborne illness caused by methicillin-resistant *Staphylococcus aureus*. Emerg. Infect. Dis. 8 (1), 82−84.

Kaakoush, N.O., Castaño-Rodríguez, N., Hazel, M., Mitchell, H.M., Man, S.M., 2015. Global epidemiology of *Campylobacter* infection. Clin. Microbiol. Rev. 8 (3), 687−720.

Kadariya, J., Smith, T.C., Thapaliya, D., 2014. *Staphylococcus aureus* and staphylococcal food-borne disease: an ongoing challenge in public health. BioMed Res. Int. 2014, 827965. https://doi.org/10.1155/2014/827965.

Keener, L., 1999. Is HACCP enough for ensuring food safety. Food Test. Anal. 5 (9), 17−19.

Keener, L., 2005. Maximizing Food Safety Return on Investment. FI Food Safety and Innovation Seminar, Paris, France.

Koutsoumanis, K., Allende, A., Alvarez-Ordóñez, A., Bover-Cid, S., Chemaly, M., Davies, R., De Cesare, A., Herman, L., Hilbert, F., Lindqvist, R., Nauta, M., Peixe, L., Ru, G., Simmons, M., Skandamis, P., Suffredini, E., Jenkins, C., Pires, S.M., Morabito, S., Niskanen, T., Scheutz, F., da Silva Felício, M.T., Messens, W., Bolton, D., 2020. Pathogenicity assessment of Shiga toxin-producing *Escherichia coli* (STEC) and the public health risk posed by contamination of food with STEC. EFSA J. 18 (1), e05967. https://doi.org/10.2903/j.efsa.2020.5967.

Laird, K., 2015. New and emerging bacterial food pathogens. In: Food Safety: Emerging Issues, Technologies and Systems. Elsevier, US, pp. 309−316.

Larson, K.R.L., Harper, A.L., Hanson, B.M., Male, M.J., Wardyn, S.E., Dressler, A.E., Wagstrom, E.A., Tendolkar, S., Diekema, D.J., Donham, K.J., Smith, T.C., 2011. Methicillin resistant *Staphylococcus aureus* in pork production shower facilities. Appl. Environ. Microbiol. 77 (2), 696–698.

Lee, M.S., Tesh, V.L., 2019. Roles of Shiga toxins in immunopathology. Toxins 11, 212. https://doi.org/10.3390/toxins11040212.

Luna-Guevara, J.J., Arenas-Hernandez, M.M.P., de la Peña, C.M., Silva, J.L., Luna-Guevara, M.L., 2019. The role of pathogenic *E. coli* in fresh vegetables: behavior, contamination factors, and preventive measures. Int. J. Microbiol. 2019, 2894328. https://doi.org/10.1155/2019/2894328.

Melton-Celsa, A.R., 2014. Shiga toxin (Stx) classification, structure and function. Microbiol. Spectr. 2. https://doi.org/10.1128/microbiolspec.EHEC-0024-2013. EHEC-0024-2013.

Monaco, M., Pedroni, P., Sanchini, A., Bonomini, A., Indelicato, A., Annalisa Pantosti, A., 2013. Livestock-associated methicillin-resistant *Staphylococcus aureus* responsible for human colonization and infection in an area of Italy with high density of pig farming. BMC Infect. Dis. 13 article 258.

Mulders, M.N., Haenen, A.P.J., Geenen, P.L., Vesseur, P.C., Poldervaart, E.S., Bosch, T., Huijsdens, X.W., Hengeveld, P.D., Dam-Deisz, W.D.C., Graat, E.A.M., Mevius, D., Voss, A., Van De Giessen, A.W., 2010. Prevalence of livestock-associated MRSA in broiler flocks and risk factors for slaughterhouse personnel in The Netherlands. Epidemiol. Infect. 138 (5), 743–755.

Murray, R.J., 2005. Recognition and management of *Staphylococcus aureus* toxin-mediated disease. Intern. Med. J. 35 (S2), S106–S119.

Nyati, K.K., Nyati, R., 2013. Role of *Campylobacter jejuni* infection in the pathogenesis of Guillain-Barré Syndrome: an update. BioMed Res. Int. 2013, 852195.

OECD, 1998. Organisation for Economic Co-operation and Development. Regulatory Reform in the Global Economy: Asian and Latin American Perspectives. OECD Publishing.

Pradhan, S., Pellino, C., Macmaster, K., Coyle, D., Weiss, A.A., 2016. Shiga toxin mediated neurologic changes in murine model of disease. Fron. Cell. Infect. Microbiol. 6, 114. https://doi.org/10.3389/fcimb.2016.00114.

Pomba, C., Hasman, H., Cavaco, L.M., da Fonseca, J.D., Aarestrup, F.M., 2009. First description of meticillin-resistant *Staphylococcus aureus* (MRSA) CC30 and CC398 from swine in Portugal. Int. J. Antimicrob. Agents 34 (2), 193–194.

Portner, D.C., Leuschner, R.G.K., Murray, B.S., 2007. Optimising the viability during storage of freeze-dried cell preparations of *Campylobacter jejuni*. Cryobiology 54, 265–270.

Riley, L.W., Remis, R.S., Helgerson, S.D., Okott, E.S., Johnson, L.M., Hargratt, N.T., Blake, P.A., Cohen, M.L., 1983. Haemorrhagic colitis associated with a rare *E.coli* serotype. N. Engl. J. Med. 308, 681–685.

Sandvig, K., 2001. Shiga toxins. Toxicon 39, 1629–1635. https://doi.org/10.1016/S0041-0101(01)00150-7.

Scherer, K., Bartelta, E., Sommerfelda, C., Hildebrandt, G., 2006. Comparison of different sampling techniques and enumeration methods for the isolation and quantification of *Campylobacter* spp. in raw retail chicken legs. Int. J. Food Microbiol. 108, 115–119.

Silva, J., Leite, D., Fernandes, M., Mena, C., Gibbs, P.A., Teixeira, P., 2011. *Campylobacter* spp. as a foodborne pathogen: a review. Front. Microbiol. 2. https://doi.org/10.3389/fmicb.2011.00200 article 200.

Skirrow, M.B., 1977. *Campylobacter enteritis*: a "new" disease. Br. Med. J. 2, 9–11.

Smith, T.C., Male, M.J., Harper, A.L., Kroeger, J.S., Tinkler, G.P., Moritz, E.D., Capuano, A.W., Herwaldt, L.A., Diekema, D.J., 2009. Methicillin resistant *Staphylococcus aureus* (MRSA) strain ST398 is present in midwestern U.S. swine and swine workers. PLoS One 4 (1), e4258.

Stewart, M.S., Busta, F.F., 2010. Global harmonization of the control of microbiological risks. In: Ensuring Global Food Safety. Academic Press, US, pp. 177–191.

Stewart, C.M., Tompkin, R.B., Cole, M.B., 2002. Food safety: new concepts for the new millennium. Innovat. Food Sci. Emerg. Technol. 3, 105–112.

Stewart, C.M., Cole, M.B., Hoover, D.G., Keener, L., 2010. New tools for microbiological risk assessment, risk management and process validation methodology. In: Non-thermal Processing Technologies for Food. Blackwell Publishing, UK, pp. 550–561.

Todd, E., 2006. Harmonizing International Regulations for *Listeria monocytogenes* in Ready-To-Eat Foods: Use of Risk Assessments for Helping Make Science-Based Decisions. Food Safety and Inspection Service. http://www.fsis.usda.gov/PDF/Slides_092806_ ETodd3.pdf (Accessed 16 September 2020).

van de Venter, T., 2000. Emerging Foodborne Diseases: A Global Responsibility. http://www.fao.org/tempref/docrep/fao/003/X7133m/x7133m01.pdf (Accessed 5 September 2020).

van Schothorst, R., Zwietering, M.H., Ross, T., Buchanen, R.L., Cole, M.B., 2009. Relating microbiological criteria to food safety objectives and performance objectives. Food Contr. 20 (11), 967–979.

Wagenaar, J.A., Yue, H., Pritchard, J., Broekhuizen-Stins, M., Huijsdens, X., Mevius, D.J., Bosch, T., Van Duijkeren, E., 2009. Unexpected sequence types in livestock associated methicillin-resistant *Staphylococcus aureus* (MRSA): MRSA ST9 and a single locus variant of ST9 in pig farming in China. Vet. Microbiol. 139 (3–4), 405–409. https://doi.org/10.1016/j.vetmic.2009.06.014.

Chapter 26

Testing for food safety using human competent liver cells (HepG2): a review

Firouz Darroudi

Global Harminization Initiaitve (GHI), Section of Genetic Toxicology and Genomics, Oegstgeest, The Netherlands

26.1 Introduction

It is well-documented that diet plays a crucial role in the etiology of cancer in humans (Sugimura, 1982, 2000; Knasmüller and Verhagen, 2002; Steck et al., 2007; Gonzales et al., 2014).

A number of epidemiological studies indicated that 40%–70% of the cancer incidence in humans is due to nutritional factors (Steck et al., 2007; Stepien et al., 2016).

The nature of the food we consume and our total diet greatly influence our health and well-being. Food should be tasty, nutritious, and safe. The safety of our food supply is a shared responsibility, from farm to fork, the food producing industry, regulatory authorities, and consumers. As part of this safety assurance, it is essential to assess the potential risks posed by food and food ingredients.

The risk assessment process, defined as the determination of the probability of harm resulting from exposure to a food component, must be based on sound scientific data and carried out to internationally agreed standards in a transparent manner (Verhagen et al., 2003). This is essential to ensure consumer confidence in the safety of the food supply, particularly when foods are traded on an international basis. Furthermore, as the world beat of globalization maintains its fast tempo, the challenges of ensuring food safety, security, and nutrition on a global scale continue to grow in complexity.

In this review, the use of a human hepatoma cell line, and in particular HepG2 cell system that is established and validated in genetic toxicology more than 30 years ago (Natarajan and Darroudi, 1991; Darroudi and Natarajan, 1993) in detecting and discriminating mutagens/carcinogens and nonmutagens/noncarcinogens in human dietary components is being discussed.

26.2 Assessment of human food safety and the current problems using existing *in vitro* and *in vivo* assays

The traditional risk assessment process applied to food additives relies on toxicology testing *in vitro* and *in vivo* assays. These models, however, have important imperfections, such as:

a. In "*in vitro*" assays with metabolically incompetent cells require the addition of exogenous enzyme homogenates (S9-fraction from rat liver is commonly being used) to catalyze the activation of genotoxic food derived carcinogens for indirectly acting chemicals that may have mutagenic/carcinogenic potential. This process is generally dependent on Cytochrome P450 enzymatic activities present in rat liver. However, they may reflect only few of the mechanisms that modulate the genotoxic effects of food carcinogens. Consequently, certain modes of action (such as to detect protective and synergistic effects) are not represented adequately in the *in vitro* models (Kassie et al., 2003a,b; Knasmüller et al., 1998, 2003, 2004a; Mersch-Sundermann et al., 2004). Furthermore, high rates of false-positive results were obtained when undertaken *in vitro* genotoxicity testing (Kirkland et al., 2005, 2007).

b. In "*in vivo*" assays in animals use intake levels many times higher than is likely in humans. Employing safety factors then carries out by extrapolating of the data to determine the safe level for humans. Such an approach does not quantitatively assess the relationship between exposure and adverse health effects (Knasmüller et al., 1998, 2003, 2004a).

Ensuring Global Food Safety. https://doi.org/10.1016/B978-0-12-816011-4.00020-3

c. At present, there is also abundant evidence that various plant constituents decrease the genotoxic and carcinogenic effects of food specific toxins (Kassie et al., 2003a,b; Knasmüller et al., 1998; Mersch-Sundermann et al., 2004; Izquierdo-Vega et al., 2017).

Synergistic and antagonistic effects may have a strong impact on the cancer risk in human, but they cannot be monitored adequately in conventional *in vitro* test systems. So far, the only reliable approaches that enable the prediction of co- and antimutagenic effects of compounds that interfere with metabolic activation/detoxification pathways of carcinogens are *in vivo* models with laboratory animals. However, they are, in general, time consuming, and costly, and the requirement of large numbers of animals argues against their use in screening trials. Furthermore, it became evident that there is a significant difference between human and laboratory animals on enzymatic profiles for both phase I and II cytochrome P450 enzymes.

Therefore, it becomes of a great importance to develop and to validate more suitable *in vitro* assays as a model to assess the genotoxic and/or antigenotoxic potential of human dietary component (Kirkland et al., 2007; Corvi and Madia, 2017; Corvi et al., 2017).

26.3 Human HepG2 cell system

Darroudi and Natarajan (Natarajan and Darroudi, 1991; Darroudi and Natarajan, 1993) have started more than 30 years ago to establish and validate human HepG2 cell system as an *in vitro* model and an alternative to use of animals in genotoxicity studies by generating a wide range of cytogenetic assays (such as micronucleus, chromosome aberrations, fluorescence *in situ* hybridization for detecting human chromosomes and centromeres and for whole genome analysis of structural and numerical abnormalities) and DNA molecular biology assays (gamma-H2AX and Rad 51) for testing different classes of chemicals in food chains and namely those human are being exposed to on a regular and daily basis.

In addition, the application of HepG2 cells was explored further by isolating S9-fraction from HepG2 cells (Darroudi and Natarajan, 1993) and was applied in combination with indirectly acting chemicals in other eu- and prokaryotic cells, such as Chinese hamster ovary (CHO), by using micronuclei (MN), sister-chromatid exchanges, cell survival and gene mutation analysis at HPRT locus assays, as well as Ames *Salmonella* assay by analyzing both point- and frame shift-mutations.

In addition, these updated, and state-of-the-art cytogenetic assays were complimented with Comet, reactive oxygen species (ROS) assays, and microarray to assess gene expression profiles.

Currently, HepG2 cells are the only human hepatoma cell line that has been extensively incorporated into basic and applied science research programs worldwide and in particular toward identification of genotoxicants and anti-genotoxicants in human dietary components.

26.4 Specific features of human HepG2 cells

1. This human hepatoma cell line was originally established from a human liver tumor biopsy and was supplied by G. Dalner (Stockholm, Sweden) (Natarajan and Darroudi, 1991).

It is worthy to note that this cell line cannot be cultured for an unlimited time. In order to preserve its enzymatic profiles that are important to assess genotoxicity and antigenotoxicity potential of selected chemicals and to maintain reproducibility of data, it should be always used after cell culturing and -propagation at low passage number (≤ 15).

1. Mutagenic effects can be studied directly in HepG2 cells that can also activate the test compound if it is needed. This simplifies the test. It also allows to study the mutagenic activity of very short-lived metabolites.
2. Subcellular fraction of HepG2 cells can be obtained and used as an exogenous source for metabolic activation in CHO cells, human lymphocytes, and Ames *Salmonella* assay (Darroudi and Natarajan, 1993; Knasmüller et al., 2004a; Darroudi et al., 1998).
3. Phase I and II profiles in HepG2:

Gaining knowledge on the metabolism of a drug, the enzymes involved, and its inhibition or induction potential is a necessary step in pharmaceutical development of new compounds, screening of human dietary components, and environmental mutagens. Primary human hepatocytes are considered to be a cellular model of reference, as they express the majority of drug metabolizing enzymes, respond to enzyme inducers, and are capable of generating *in vitro* a metabolic profile similar to that found *in vivo*. However, hepatocytes show phenotypic instability and have a restricted accessibility.

In HepG2 cells, though transcripts of series of phase I enzymes were found, mRNA levels for most of these Cytochrome P450 phase I enzymes (1A1, 1A2, 2C9, and 3A4) are significantly lower than in primary hepatocytes (Wilkening et al., 2003; Westerink and Schoonen, 2007a). This imbalance might explain why some compounds with known toxicity in primary hepatocytes remain nontoxic in HepG2 cells.

For phase II enzymes (responsible for detoxification), HepG2 cells retain the activities of many relevant enzymes (SULT1A1, 1A2, 1E1, 1A2, 2A1, microsomal GST1, NAT1, and EPHX1) almost similar to levels found in primary human hepatocytes (Westerink and Schoonen, 2007b).

1. Various biological assays (cytogenetic and DNA molecular biology) for detecting DNA damage induction, repair kinetics, and biological consequences following exposure to different classes of human dietary components and environmental mixture (in air and water) were developed and successfully validated using HepG2 cells (Natarajan and Darroudi, 1991; Darroudi et al., 1996, 1998; Uhl et al., 1999, 2000, 2001, 2003a,b; Bezrookove et al., 2003; Lu et al., 2004; Filipic and Hei, 2004; Hreljac et al., 2008; Yuan et al., 2005; Jondeau et al., 2006; Buchmann et al., 2007).
2. In addition, state-of-the-art techniques such as fluorescence *in situ* hybridization with a pan-centromeric probe to detect aneugen and clastogens (Darroudi et al., 1996), as well as microarray and proteomic were adapted to these cells and used to define the modes of action of various classes of human dietary carcinogens, noncarcinogens, and anticarcinogens (Harries et al., 2001; Gerner et al., 2002; Breuza et al., 2004; van Delft et al., 2004; Hockley et al., 2006, 2009).
3. This *in vitro* model is found to have the potential to be applied in mutagenicity testing to Refine, Reduce, and Replace animal use (3Rs) (Russell, 1995) while ensuring protection of human and animal health and the environment (Kirkland et al., 2007; Rueff et al., 1996).

26.5 Validation and application of human HepG2 cells and their S9-fractions in genetic toxicology studies for assessing food safety

26.5.1 Assessment of the genotoxic potential of known carcinogen and noncarcinogens

HepG2 cells were used as metabolic activation systems as well as targets for evaluating DNA damage (Natarajan and Darroudi, 1991; Knasmüller et al., 1998, 1999, 2004a). In addition, in order to expand application of this *in vitro* model and to compare with the exisiting model (rat liver), S9-fractions from HepG2 cells were isolated to investigate their ability to activate promutagenic carcinogens using CHO cells *in vitro* and Ames *Salmonella* assay (Darroudi and Natarajan, 1993). Various biological endpoints, such as sister-chromatid exchanges, MN in binucleated cells, aneuploidy using MN assay in combination with a pan-centromeric probe, cytotoxicity, gene mutation in CHO cells (at HPRT locus), Comet assay as well as gamma-H2AX, and Rad 51 foci formation, were developed and validated (Natarajan and Darroudi, 1991; Darroudi and Natarajan, 1993; Knasmüller et al., 1998, 2003, 2004a; Mersch-Sundermann et al., 2004; Darroudi et al., 1996, 1998; Lamy et al., 2004).

The results (Table 26.1) indicate that the human HepG2 cell system reflects the activation/detoxification of genotoxic carcinogens better than other indicator *in vitro* assays (such as CHO cells and Ames *Salmonella* test) that are currently being used, and the outcomes are positively correlated to the known carcinogenicity data *in vivo* using animal models (Kassie et al., 2003a,b; Knasmüller et al., 2004a; Mersch-Sundermann et al., 2001, 2004; Schmeiser et al., 2001; Wu et al., 2005).

In addition, HepG2 cell system was shown to be capable of detecting and discriminating between structurally related chemicals carcinogens and noncarcinogens (Table 26.1).

Potent *in vivo* carcinogens such as 2-acetylaminofluorene, benzo(a)pyrene (B(a)P), and ochratoxin A have shown genotoxic potential in HepG2 cells (*in vitro*). In contrast, structurally related chemicals such as 4-acetylaminofluorene, pyrene, and ochratoxin B, respectively, reported to be noncarcinogen *in vivo* have shown no genotoxic effect in HepG2 cells using MN and Comet assays (Natarajan and Darroudi, 1991; Knasmüller et al., 2004a,b).

Interestingly, when HepG2 cells were applied to assess the genotoxicity potential of two known carcinogens *in vivo* tests, HMPA and safrole (Table 26.1) that so far, no positive genotoxicity effect is reported in other *in vitro* assays, in HepG2 cells as well as in the presence of S9-microsome derived from HepG2 cells in CHO cells and Ames assay (TA98), a positive effect was evident.

The observed positive correlation between HepG2 cells *in vitro* model with *in vivo* data opens the possibility of applying this model as a suitable alternative to animal studies. Furthermore, the potential of detecting and discriminating accurately genotoxicants and structurally related nongenotoxicants has an advantage of exploring further the mode of action of these chemicals and consequently to reduce and/or eliminate the hazardous effects to human.

TABLE 26.1 A comparative study between genotoxicity data *in vitro* using human HepG2 cell system, S9-microsomal fraction from HepG2 and rat liver, and carcinogenicity/noncarcinogenicity data *in vivo*.

Chemical	Carcinogen (*in vivo* test)	HepG2 cells (*in vitro*)	CHO cells (*in vitro*)		Ames *Salmonella* test	
			With S9-fraction derived from			
			HepG2	Rat liver	HepG2	Rat liver
2-AAF	+	+	+	−	N.T.	+
4-AAF	−	−	−	−		
B(a)P	+	+	+	+	N.T.	+
Pyrene	−	−	−	−		
Ochratoxin A	+	+	+	−	N.T.	−
Ochratoxin B	−	−	−	−		
CP	+	+	+	+	N.T.	+
DMN	+	+	+	+	N.T.	−
Ethanol	+	+	+	−	N.T.	−
Methyl carbamate	+	+	+			
Ethyl carbamate	+	+	+		N.T.	−
HMPA	+	+	+	−	+	−
Safrole	+	+	+	−	+	−

Abbreviations: −, Results were negative; +, Results were positive; *2-AA F*, 2-acetylaminofluorene; *4-AAF*, 4-acetylaminofluorene; *B(a)P*, benzo(a)pyrene; *CHO*, Chinese hamster ovary; *CP*, cyclophosphamide; *DMN*, dimethylnitrosamine; *HMPA*, hexamethylphosphoramide; *N.T.*, Not tested.

26.5.2 Assessment of the genotoxic potential of mycotoxins

Several studies were designed for validating the application of human HepG2 cells in assessing the genotoxic potential of different classes of mycotoxins, mainly those present in food as human dietary components (Knasmüller et al., 2004b; Ehrlich et al., 2002a,b). These mycotoxins (Table 26.2) are known to cause cancer of different origins in animals, but so far, no positive effects (except for aflatoxin B1) had been reported, neither in any mammalian tests *in vitro* nor in Ames *Salmonella* assay.

Mycotoxins of interest in naturally contaminated human foods and feeds are nivalenol, 2-deoxynivalenol, T2-toxin, ochratoxin A (these can be found in wheat, barley, and oats); fusarenone X, fumonisin B1, ochratoxin A (can be found in maize); and aflatoxin B1 (can be found in maize and peanuts).

All mycotoxins are very stable during storage and milling processing, and it is even difficult to be destroyed at high temperatures (IARC, 1993). Primarily, most of the data were generated in animal studies. Ochratoxin A was found to be a potent nephrotoxin. Deoxynivalenol caused general toxicity. Fumonisins ingestion via feed prepared from corn contaminated with *Fusarium* could cause leukoencephalomalacia (horse), pulmonary edema (pig), and liver cancer (rat) (Knasmüller et al., 2004b; Ehrlich et al., 2002a,b). Among the tested mycotoxins, aflatoxin B1 was found to be the most potent carcinogen *in vivo* and was the only one for which its genotoxic potential has been reported in other *in vitro* assays. In humans, there are no genotoxicity data reported; however, epidemiological studies showed that presence of mycotoxins in food can enhance the risk of esophageal cancer.

TABLE 26.2 Genotoxic potential of mycotoxins in HepG2 cells.

Substance	Nivalenol	Deoxynivalenol	Fumonisin B1	Ochratoxin A	Aflatoxin B1
M.E.D (μg/mL)	50	50	25	25	0.2
Ranking order	3	3	2	2	1

M.E.D., Minimum effective dose. It is defined and estimated as a dose that could increase micronucleus frequency by twofold in comparison to the control (untreated) sample.

Chemicals such as nivalenol, deoxynivalenol, fumonisin B1, citrinin, and ochratoxin A were tested in human HepG2 cells and were found to be potent genotoxicants (Table 26.2) (Knasmüller et al., 2004b; Ehrlich et al., 2002a,b).

In a comparative study, based on a dose that could enhance the biological assay (micronuclei) by a factor of 2 in comparison to the untreated samples, a ranking order was made for tested mycotoxins (Table 26.2). The known carcinogen aflatoxin B1 was by far the most potent chemical; the effective concentration was found to be 0.2 μg/mL compared to 25 μg/mL for fumonisin B1. These sets of data are the first to elucidate the genotoxic potential of the selected mycotoxins in an *in vitro* assay and, therefore, can open the possibility of further assessments of the origin and mechanisms of genotoxicity of these chemicals in order to enhance human health. Interestingly, the noncarcinogen ochratoxin B that is a structurally related chemical to the carcinogen ochratoxin A, under identical conditions, was found to be nongenotoxic in HepG2 cells (Knasmüller et al., 2004b).

26.5.3 Assessment of the genotoxic potential of heterocyclic aromatic amines

Heterocyclic aromatic amines (HAAs) can be found in protein-rich food, after baking, cooking and barbecuing (IARC, 1993) and are known to be mutagens in prokaryotes (Sugimura, 1982, 2000). The most common ones, IQ, MeIQ, MeIQx, PhIP, Trp-p-1, and Trp-p-2, were studied in human HepG2 cells. Existing data revealed that HAAs are mutagenic in pro- and eukaryotic organisms. Experiments with laboratory rodents also showed that these food-derived compounds can cause cancer (Turesky, 2007; Zheng and Lee, 2009). Furthermore, epidemiological studies indicated that they might be involved in the etiology of colon cancer in humans (Rohrmann et al., 2009).

HAAs have been tested extensively in microbial *in vitro* assays with rat liver microsomal fractions in Ames *Salmonella* assay and were found to be potent mutagens (Sugimura, 1982, 2000). However, the results obtained in genotoxicity tests with mammalian cell lines are highly divergent (Knasmüller et al., 1995, 1999, 2004a,c; Majer et al., 2004a,b).

Several HAAs such as IQ, MeIQ, MeIQx, Trp-p-1, Trp-p-2, and PhIP were tested in human HepG2 cells in MN as well as Comet or single cell gel electrophoresis (SCGE) assays (Knasmüller et al., 2004a,c; Russell, 1995). With all of them, positive results were obtained (Table 26.3). It is notable that, in contrast to experiments with CHO cells and *Salmonella*/microsome assay, the ranking order of the genotoxic potencies of these HAAs observed in HepG2 cells correlated with that of their carcinogenic activities in rodents. Moreover, the sensitivity of SCGE assay toward most HAAs was found to be similar to that of MN tests.

26.5.4 Antigenotoxic potential of glycine betaine on a heterocyclic aromatic amine Trp-p-2 in HepG2 cells

In an earlier study, glycine betaine is found to inhibit the mutagenicity of sanma fish mutagen, 2-chloro-4-methylthiobutanoic acid in *Salmonella* typhimurium TA100, and TA1535 (Kimura et al., 1999).

Attempts were made to assess the antigenotoxic potential of glycine betaine in HepG2 cells following treatment with Trp-p-2, that is found to be among all HAA tested the most genotoxicant (Table 26.3). In addition, the levels of phase II enzyme Uridine 5′-diphospho-glucuronosyltransferase (UGT2) in all experimental conditions were determined. UGTs are important enzymes responsible for glucuronidation of a serum bilirubin, which is a natural antioxidant. UGTs can also transform toxic compounds to less or nontoxic hydrosoluble forms.

The results are presented in Table 26.4. HepG2 cells were exposed to Trp-p-2 for 24 h with two doses of 5 and 10 μM. The MN frequency increased significantly in comparison to the level in control samples and it was as well in proportion to the dose. This implicates and confirms the genotoxic potential of Trp-p-2 in HepG2 cells. Glycine betaine significantly reduced MN frequency and it reached to control level. Its antigenotoxicity effect mainly was related to enhancing UGTs

TABLE 26.3 Genotoxic potential of heterocyclic aromatic amines (cooked food mutagens) in HepG2 cells.

Substance	IQ	MeIQ	MeIQx	PhIP	Trp-p-1	Trp-p-2
M.E.D (μM)	80	50	30	30	15	5
Ranking order	5	4	3	3	2	1

Abbreviations: *IQ*, 2-Amino-3methylimidazol[4,5-f]quinoline; *M.E.D.*, Minimum effective dose. It is defined and estimated as a dose that could increase micronucleus frequency by twofold in comparison to the control (untreated) sample; *MeIQ*, 2-Amino-3,4-dimethylimidazol[4,5-f]quinoline; *MeIQx*, 2-Amino-3,8-dimethylimidazol[4,5-f]quinoxaline; *PhIP*, 2-Amino-1-methyl-6-phenylimiazo[4,5-b]pyridine; *Trp-p-1*, 3.Amino-1,4-dimethyl-5H-pyrido[4,3-b]indole; *Trp-p-2*, 3.Amino-1-methyl-5H-pyrido[4,3-b]indole.

TABLE 26.4 Genotoxic potential of Trp-p-2[a] and modulating effect of glycine betaine (GB) in HepG2 cells.

Chemical	Number of binucleated	Number of micronuclei (MN)	Induced MN	UTG_2 level[b]	
(Dose)	cells (BNC) scored	in 1000 BNC analyzed ± S_d	in 1000 BNC	% ↓	% ↑
Controls[c]:					
PBS	1000	23 ± 1.2	0		
GB (1 mM)	1000	24 ± 1.3	1		
Trp-p-2[a]					
(5 μM)	1000	52 ± 3.3	29		
Trp-p-2					
(5 μM) +					
GB (1 mM)	1000	23 ± 2.1	−1		
Trp-p-2[a]					
(10 μM)	1000	81 ± 6.1	58	40	
Trp-p-2					
(10 μM) +					
GB (1 mM)	1000	24 ± 1.6	0		14

[a]*Trp-p-2 = 3Amino-1-methyl-5H-pyrido[4,3-b]indole.*
[b]*UTG_2 = UDP-glucuronosyl-transferase. The percentage (%) of ↓ (decrease) or ↑ (increase) in comparison to the level found in untreated (control) samples was estimated.*
[c]*PBS = Phosphate buffered saline.*

level in HepG2. It is useful to follow up this work by analyzing different types of phase I- and phase II- enzymes in order to reveal mode of action(s) of both genotoxic Trp-p-2 as well as antigenotoxic potential of glycine betaine.

26.5.5 Toxicity studies of compounds and mechanistic assays on NAD(P)H, ATP, DNA contents (cell proliferation), glutathione depletion, calcein uptake, and radical oxygen assay using human HepG2 cells

An approach to reduce the cost aspects of analyzing large numbers of synthetic compounds within the pharmaceutical industry is the introduction of medium and high throughput *in vitro* screening for toxicity measurements using Alamar Blue (AB) and Hoechst 33342 coloration, and luminometric assays, using Cyto-Lite and AT P-Lite, dichlorofluorescein diacetate, monochlorobimane, and calcein-AM (Schoonen et al., 2005a,b).

With AB, AT P-Lite, and Cyto-Lite, the energy status of the cell is measured and with Hoechst 33342 the amount of DNA. Dichlorofluorescein diacetate, monochlorobimane, and calcein-AM are fluorophores for the measurement of the formation of ROS, the quantification of glutathione, and the membrane stability, respectively (Schoonen et al., 2005a,b; Miret et al., 2006).

Further developments in this area may lead to an earlier prediction of toxic effects of compounds in cellular assay, *in vitro*, and animal studies can be largely reduced in this assay. Due to the high attrition rate of toxicity of drug candidates in development (50% of the compounds), there is increased interest within the pharmaceutical industry to examine compounds at earlier stages and with relatively small quantities of compound.

For a first glance on toxicity, simple assays are needed to identify general aspects of cellular toxicity. Interference with normal cell physiology, such as energy metabolism and cellular proliferation, can be simple cell toxicity markers.

The human HepG2 cell system has been examined using the above-mentioned assays for 100−110 reference compounds with different modes of action. Also, 60 up to 100 were tested on HeLa cells, CHO cells, and human endometrium (ECC-1). The highest dose tested varied between 3.16×10^{-3} M and 3.16×10^{-5} M, and the minimal toxicity dose obtained for different assays was found to be between 3.16×10^{-3} M and 3.16×10^{-8} M (Schoonen et al., 2005a).

The outcome of these studies revealed that all four cell lines were responsive to the same set of drugs (classified as directly acting); however, for some drugs, HepG2 cells appeared to be more sensitive, as compared to the other three cell lines. In general, in HepG2 cells, it was possible to predict toxicity up to 75%. This implies that a high throughput toxicological screening can be set up with HepG2 cells and can also be applied when tested drugs are indirectly acting and need an exogenous activation system. Moreover, these cells originated from human liver cells, which can increase the predictability for assessing human risk compared to the predictability based on rat, mice, hamster, or monkey cell lines.

Furthermore, it was reported that glutathione depletion and calcein uptake assays are almost equally potent and both assays are much more sensitive than the ROS (reactive oxygen species) measurement. Consequently, it was concluded that *in vitro* screening appears to be a realistic and reliable alternative to the use of *in vivo* studies with vertebrate animals.

Miret and coworkers used other sets of *in vitro* models of cytotoxicity in human HepG2 cells. Those included AB for the measurement of cellular adenosine triphosphate (ATP); ToxiLight as an indicator of cellular necrosis by measurement of released AK; and Caspase-3 Fluorometric Assay, Apo-ON E Caspase-3/7 Homogenous Assay, and Caspase-Glo for the determination of caspase-3/7 activity.

In order to evaluate the assays, several known cytotoxic compounds such as dimethyl sulfoxide, butyric acid, carbonyl cyanide 4-(trifluoromethoxy)phenylhydrazone (FCCP), and camptothecin were examined (Miret et al., 2006). Data revealed that the best way to evaluate the potential cytotoxicity of a compound is to employ a battery of assays. The use of ATP levels, cell necrosis, and caspase-3/7 resulted in the most useful combination in HepG2 cells (Miret et al., 2006).

26.5.6 The genotoxic potential of heavy metals in HepG2 cells

Arsenic, cadmium, and selenium were examined in HepG2 cells and induction of MN is used as biological endpoint (Darroudi, 2021a). When genotoxic effect was evident, then a pan-centromeric probe for whole genome is used by combining with MN assay (Darroudi et al., 1996), to elucidate the clastogenic or aneugenic potential of these selected chemicals.

Arsenic and cadmium are frequent environmental pollutants, both are present in human diet, and were classified as human carcinogens by IARC (IARC, 1993). Selenium is an oligoelement with essential biological functions. Diet is the most important source of selenium intake, and it is dependent on its concentration in food and amount of food consumed. Among the essential human micronutrients, selenium is peculiar due to its both beneficial physiological activity and toxicity effect. It may have anticarcinogenic effects at low concentrations, whereas at concentrations higher than those necessary for nutrition, it can be genotoxic and carcinogenic. In the last decade, there has been increasing interest in several nutritional selenium compounds because of their environmental, biological, and toxicological properties, particularly for their cancer- and disease-preventing activities (Valdiglesias et al., 2010).

Arsenic at a dose level of 5 μM and following 24 h treatment could increase MN frequency in HepG2 cells by twofold in comparison to untreated (control) samples. HepG2 cells were treated for 24 h with cadmium in a dose range of 0.5−5 μM. Using this experimental protocol, the genotoxicity of cadmium was evident and MN frequency enhanced proportional to the dose. In exposed HepG2 cells, the clastogenic activity of cadmium was evident and more than 60% of MN were found to be centromere negative (derived from a part of chromosomes) (Darroudi et al., 1996; Darroudi, 2021a). HepG2 cells were exposed to selenium for 24 h in a dose range of 1−10 μM. Frequency of MN slightly (that was found to be significant $P < 0.05$) increased at doses of 5 and 10 μM. Furthermore, the antigenotoxic potential of selenium was investigated against B(a)P at a dose of 2.5 μM. The frequency of induced MN decreased significantly ($P \leq 0.01$) in HepG2 cells treated with both B(a)P and selenium in comparison with MN frequency obtained after B(a)P treatment (Darroudi, 2021a). These data implicate that though selenium can have antigenotoxicity potential, a caution should be made on consumption (dose) of this compound as one of the human dietary components.

26.5.7 To assess the genotoxic potential of human dietary components in fermented food and in alcoholic beverages using HepG2 cells

Taking into account that dietary exposure to both food components and alcoholic beverages is of concern to assess human risk, the genotoxic potential of both vinyl carbamate and ethyl carbamate (Urethane) was studied in human HepG2 cells as well as in CHO cells in the presence of S9-fractions derived from HepG2 cells (Table 26.5).

Methyl- and Ethyl-carbamate (Urethane) are naturally formed in fermented food, such as bread, soy sauce, and alcoholic beverages such as grape wines, sherries, and whiskies (in the range of 69−247 mg/kg) (Shin and Yang, 2012; Ryu et al., 2015).

In IARC Monograph on the evaluation of carcinogenic risk to human ethyl carbamate is classified in group 2A, as probably carcinogenic. The toxic effect of ethyl carbamate is found primarily in lung and other organs as well, such as the heart, brain, mammary glands, and liver (Gewd et al., 2018).

TABLE 26.5 Genotoxic potential of methyl carbamate and ethyl carbamate in HepG2 and CHO[a] cells.

Chemicals	Dose (mg/mL)	Number of binucleated cells (BNC) scored	Number of micronuclei (MN) in 1000 BNC ± S_d	Induced MN in 1000 BNC
In HepG2 cells:				
Methyl carbamate				
	0	1000	28 ± 2.8	0
	0.025	1000	85 ± 9.0	57
	0.05	1000	125 ± 21	97
	0.1	1000	225 ± 27	197
Ethyl carbamate				
(Urethane)	0	1000	27 ± 2.5	0
	0.025	1000	96 ± 9.5	69
	0.05	880	159 ± 24	132
	0.1	720	201 ± 27	174
In CHO[a] cells + S9-fractions isolated from HepG2 cells:				
Methyl carbamate				
	0	1000	12 ± 2.1	0
	0.05	1000	65 ± 4.4	53
	0.1	1000	118 ± 10.5	106
	0.2	855	180 ± 22.6	168

These cells are treated with Methyl carbamate in the presence of S9-fractions derived from HepG2 cells (Darroudi and Natarajan, 1993). In the absence of HepG2 S9-fraction, when CHO cells are treated with a dose of 0.2 mg/mL, MN frequency remained at the control level.
[a]*CHO = Chinese hamster ovary cells.*

Methyl carbamate is not listed; however, unlike in mice, liver carcinogenesis with Methyl carbamate treatment is reported in rats.

HepG2 cells are treated with these chemicals in a dose range of 0.025−0.1 mg/mL for 1 h (Table 26.5).

MN frequency in binucleated cells was used to assess genotoxic potential of these chemicals. Both chemicals increased MN frequency proportional to the dose. At highest dose of 0.1 mg/mL with ethyl carbamate due to toxicity, the frequency of binucleated cells decreased significantly. This led to a deviation from a linear dose−response curve (Table 26.5).

In addition, CHO cells were exposed to methyl carbamate in the presence of S9-fraction derived from HepG2 cells (Darroudi and Natarajan, 1993). MN frequency increased in a dose-dependent manner and the genotoxic potential of methyl carbamate was evident.

These data reveal a new set of information for these chemicals and can be useful to extrapolate the risk to humans. Furthermore, it is feasible to proceed with analyzing the mode of action of these chemicals under these experimental protocols in both HepG2 and CHO cells.

26.5.8 To assess DNA damage induction, repair kinetics, and biological consequences of chemical mutagens/carcinogens in HepG2 cells

It is known that for ionizing radiation of different qualities as well as chemicals different repair processes are involved at different stages of cell cycle. Two prominent repair processes involved are nonhomologous end-joining and homologous recombinations, which are mainly involved at G1 and G2 stage of cell cycle in mammalian cells, respectively (Shrivastav et al., 2008). In order to assess induced DNA damages and their repair kinetics in HepG2 cells with different classes of physical and chemical agents, Darroudi and coworkers established and applied gamma-H2AX and Rad51 assays (Darroudi, 2021b). On the basis of mode of action of the agent of interest, it was evident that chemicals which are having radiomimic potential, such as bleomycin and Etoposide (VP16), can be easily detected by using gamma-H2AX assay.

The induction of DNA damage is happening fast within 5 min after treatment; approximately 50% are repaired within first 2 h, afterward frequency decreased gradually. In contrast, chemicals which need S-phase to exert their genotoxic potential (such as Mitomycin C and cyclophosphamide) showed no effect using gamma-H2AX, but when Rad51 foci used as biological endpoint, initially number of foci increased, and further increase in number of Rad51 foci was evident with increasing postexposure time (Darroudi, 2021b). These assays are very useful for classifying the mode of action of unknown chemicals rapidly and to assess the genotoxic potential as well as may pave the way toward developing drugs against cancer of different origins.

26.5.9 Application of human HepG2 cell system to detect dietary antigenotoxicants

There is increasing evidence that chemicals cannot only have adverse effects. There are many substances (such as human dietary components, vitamins) that can have a beneficial effect on health, and may even inhibit the effect of carcinogens in human (Knasmueller et al., 2002).

However, the conventional *in vitro* assay outcomes are clearly not conclusive because many studies do not fully reflect the complex activation and detoxification processes that take place in mammals (*in vivo*). Consequently, it is not possible to extrapolate the results of such experiments to humans. Several dietary constituents that are DNA protective under *in vitro* conditions in experiments with indicator cells that require addition of exogenous activation system (liver S9-mixture) were inactive in rodent bioassays, or even led to an enhancement of the DNA damaging properties of the tested HAAs (Schwab et al., 1999, 2000). Furthermore, certain conventional *in vivo* experiments with laboratory rodents are not adequate tools for identifying protective constituents in human food since only marginal or negative effects are obtained in these models with HAAs (IARC, 1993).

Several HAAs (Table 26.3) were examined using MN assay in HepG2 cells, and a ranking order was made. It is notable that, in contrast to experiments with CHO cells and Ames *Salmonella*/microsome assay, the ranking order made with HepG2 cells for the genotoxic potencies of the HAAs positively correlated with that of their carcinogenic activities in rodents. In the follow-up studies, attempts were made to study the antigenotoxic potential of series of human dietary components (including vitamins) against HAAs (IQ, MeIQ, MeIQx, Trp-p-1, Trp-p-2, and PhIP) in human HepG2 cells by using MN and Comet assays (Kassie et al., 2003a,b; Knasmüller et al., 2004a,c; Mersch-Sundermann et al., 2004; Uhl et al., 2003a,b; Sanyal et al., 1997; Dauer et al., 2003; Steinkellner et al., 2001; Laky et al., 2002; Majer et al., 2005; Lhoste et al., 2004).

Table 26.6 summarizes the data of antimutagenicity studies with HAA s and B(a)P using HepG2 cells. The HepG2 cell system also proved to be a useful *in vitro* model for detecting human dietary antigenotoxicants (Kassie et al., 2003a,b; Knasmüller et al., 2004a,c; Mersch-Sundermann et al., 2004; Majer et al., 2005). Interestingly, the results revealed that the antioxidants, such as ascorbic acid and beta-carotene, do give genotoxic responses in HepG2 cells at high concentration, 20 and 10 mg/mL, respectively. The genotoxic potential of selenium in HepG2 cells was found at a significantly lower dose of 5 μM.

26.5.10 The use of genomic and proteomic technologies in HepG2 cells

The ability to monitor variations at the transcriptional and translational levels using DNA microarrays and proteomics is essential to improve our understanding of physiologically relevant processes at the molecular level (Gerner et al., 2002; Yokoo et al., 2004; Breuza et al., 2004; Staal et al., 2006). These are certainly integrated approaches to identify the alteration of biological/cellular pathways in HepG2 cells upon treating with different classes of chemicals (Gandhi et al., 2015).

The DNA damage caused by chemical carcinogens is important in the initiation of carcinogenesis. However, for promotion and progression of an initiated cell to occur, other events within the cell need to take place and such events are likely to involve gene expression changes induced by the carcinogen. A broader understanding of the impact of carcinogen treatment in specific cells can be mechanistically informative and may enlarge the number of candidate genes contributing to variations in individual susceptibility to carcinogens. Microarray technology offers an attractive method to analyze globally profiles of genes (suppressed and/or expressed by carcinogen exposure). Consequently, this may give key insights into the carcinogenic and anticarcinogenic potential of various classes of chemicals that humans are being exposed to via food or environment (Tien et al., 2003).

Gene expression profiling is also used in human HepG2 cell systems and could discriminate genotoxic from non-genotoxic carcinogens (van Delft et al., 2004). Furthermore, gene expression changes in human HepG2 and MCF-7 cell lines were studied following treatment with carcinogen B(a)P or its noncarcinogenic isomer benzo(e)pyrene (B(e)P)

TABLE 26.6 Antigenotoxicity studies of dietary constituents which protect against heterocyclic aromatic amines (HAA) and benzo(a)pyrene [B(a)P] in HepG2 cells.

Putative antimutagen	Dose range	End-point[a]	HAA	Results	Remarks
Ascorbic acid (Vit. C)	20 μg-20 mg/mL	SCGE	IQ, PhIP	+, +	Genotoxic at ≥20 mg/mL
Brussels sprout	1.0 μL/mL	SCGE	IQ	+	
Caffeine	1.0–500 μg/mL	MN	IQ, MeIQ	+, +	
			MeIQx, PhIP	+, +	
Beta-carotene	10 μg–10 mg/mL	SCGE	IQ, PhIP	+, +	Genotoxic at ≥20 mg/mL
Chrysin	1.3–33 μg/mL	MN	PhIP	+	
Coumarin	1.0–500 μg/mL	MN	IQ	+	
Glycine betaine	1.0 mM	MN	Trp-P-2	+	Mode of action:
		Ames assay[b]	Trp-P-2	+	Enhancing
					Phase II enzymes
Alpha-naphtoflavone	20 μg–20 mg/mL	MN	IQ	+	
Beta-naphtoflavone	20 μg–20 mg/mL	MN	PhIP	+	
Phenethylisocyanate	0.25–1.0 μg/mL	MN	PhIP	+	
		Ames assay[b]	PhIP	+	
Tannic acid	5.0–500 μg/mL	MN	IQ	+	
Vanillin	1.0–500 μg/mL	MN	IQ, MeIQ,	+, +	
			MeIQx, PhIP	+, +	
			Trp-P-2	+	
Selenium	0.25–10 μM	MN	B(a)P	+	Genotoxic at ≥5 μM
Antigenotoxic potential of flavonols					**Effective at**
Fistein	10–100 μM	SCGE	B(a)P	+	[50 μM]
Kaempferol	10–100 μM	SCGE	B(a)P	+	[50 μM]
Myricetin	10–100 μM	SCGE	B(a)P	+	[50 μM]
Quercetin	10–100 μM	SCGE	B(a)P	+	[50 μM]

+, Indicates that dietary component revealed antimutagenic potential against selected HAA or B(a)P.
[a]*Two biological end-points were used; SCGE (Single Cell Gel Electrophoresis) and MN (micronuclei in binucleated cells).*
[b]*For Ames Salmonella assay using both point- and frame-shift mutations, S9-fractions isolated from HepG2 cells were used.*

(Hockley et al., 2006, 2009). The overall response to B(a)P consisted of upregulation of tumor suppressor genes and downregulation of oncogenes promoting cell cycle arrest and apoptosis. Antiapoptotic signaling that may increase cell survival and promote tumorigenesis was also evident (Hockley et al., 2006). In contrast, B(e)P did not induce consistent gene expression changes at the same concentrations.

Interestingly, in another study, a number of the genes identified have been induced in normal human mammary epithelial cells by B(a)P (CYP1B1 and NQO1) (Keshava et al., 2005). This promises that the expression changes observed in these two cell systems are not likely to be artifacts of their cancer phenotype.

Carcinogenesis is also an important chronic toxicity of metals and metalloids, although their mechanisms of action are still unclear. Comparison of gene expression patterns induced by carcinogenic metals, metalloids, and model carcinogens would give an insight into understanding of their carcinogenic mechanisms. HepG2 was exposed to two metals [cadmium (Cd) and nickel (Ni)], a metalloid (arsenic, As), and three model carcinogenic chemicals *N*-dimethylnitrosamine (DMN), 12-*O*-tetradecanoylphorbol-13-acetate (TPA), and tetrachloroethylene (TCE). Afterward, DNA microarrays with 8795 human genes were applied to elucidate the mode(s) of action of these classes of chemicals (Kawataki et al., 2009). Of the genes altered by As, Cd, and Ni exposures, 31%–55% were overlapped with those altered by three model carcinogenic chemical exposures in these studies. In particular, the metals and metalloid shared certain characteristics with TPA and TCE in remarkable upregulations of the genes associated with progression of cell cycle, which might play a central role in As, Cd, and Ni carcinogenesis. This characteristic of gene expression alteration was partially counteracted by intracellular accumulation of vitamin C in As-exposed cells, whereas the number of cell cycle—associated genes was increased in Cd- and Ni-exposed cells. ROS might have an accelerative effect on the cell proliferation mechanisms of As, but have an inhibitory effect on those of other two heavy metals (Kawataki et al., 2009).

26.6 Conclusion

The HepG2 cell system has been demonstrated to be a useful *in vitro* model for detecting environmental and human dietary genotoxicants and antigenotoxicants. Furthermore, this *in vitro* cell system has the potential to discriminate between structurally related carcinogens and noncarcinogens, as well as between genotoxic and nongenotoxic carcinogens, and could discriminate between clastogens and aneugens.

Using gene expression profiling and proteomics will open the way to elucidate the kinetics and mode of action of various classes of human dietary carcinogens, such as mycotoxins, HAAs, and water contaminants, in which earlier studies using conventional tests mostly failed to reveal any DNA damaging effect.

Furthermore, certain characteristics of the commonly used rodent cell lines (CHO, CHL, V79, L5178Y, etc.), such as lack of p53, karyotype instability, DNA repair deficiencies, are recognized as possibly contributing to the high rate of false positives. In addition, in those cell types, the need for exogenous metabolism is also expected to contribute to the false-positive/negative rate. In contrast, it appears that the possibility of getting false-positive results is low in tests using HepG2 cells, and a wide variety of biological assays that are developed could enhance the potential of this cell system in genotoxicity studies as a model for human risk assessment, with additionally the potential to be used as an alternative to vertebrate animals in mutagenicity/carcinogenicity studies.

Recent studies are oriented toward successfully generating 3D spheroid models for HepG2 cells. These cells cultivated in 3D resemble organ structures better than 2D monolayer cultures (Elje et al., 2020). They were applied to detect genotoxic potential of different classes of nanoparticles (Elje et al., 2020).

The cells derived from human hepatoma cell system (such as HepG2) have shown to possess the potential of being incorporated in research programs that are dealing with global harmonization of food safety regulations and are focused to assess the genotoxic and antigenotoxic potential of human dietary components at biologically relevant doses.

Acknowledgments

These studies were initiated and performed by the author at the former department of Toxicogenetic of Leiden University Medical Centre, Leiden, The Netherlands.

The author is grateful to Prof. Knasmueller, Prof. Mersch-Sundermann, Prof. Bader, and Dr. Filipic and their coworkers for their contributions in HepG2 research programs (its application in genetic toxicology and chemical mutagenesis for examining human dietary components).

References

Bezrookove, V., Smits, R., Moeslein, G., et al., 2003. Premature chromosome condensation revisited: a novel chemical approach permits efficient cytogenetic analysis of cancers. Gene Chromosome Cancer 38, 177—186.

Breuza, L., Halneisen, R., Jan, P., et al., 2004. Proteomics and endoplasmic reticulum-golgi intermediate compartment (ERGIC) membrane from Brefeldin A-treated HepG2 cells identifies ERGIC-32, a new cycling protein that interacts with human Erv-46. J. Biol. Chem. 279, 47242—47253.

Buchmann, C.A., Nersesyan, A., Kopp, B., et al., 2007. DIMBOA and DIBOA, two naturally occurring benzoxazinones contained in sprouts are potent aneugens in HepG2 cells. Cancer Lett. 246, 290—299.

Corvi, R., Madia, F., 2017. In vitro genotoxicity testing- Can the performance be enhanced? Food Chem. Toxicol. 106, 600—608.

Corvi, R., Madia, F., Guyton, K.Z., Kasper, P., Rudel, R., Colacci, A., Kleinjans, J., Jennings, P., 2017. Moving forward in carcinogenicity assessment: report of an EURL ECVAM/ESTIV workshop. Toxicol in Vitro 45 (Pt 3), 278—286.

Darroudi, F., Natarajan, A.T., 1993. Metabolic activation of chemicals to mutagenic carcinogens by human hepatoma microsomal extracts in Chinese hamster ovary cells (*in vitro*). Mutagenesis 8, 11—15.

Darroudi, F., Meijers, C.M., Hadjidekova, V., Natarajan, A.T., 1996. Detection of aneugenic and clastogenic potential of X-rays, directly and indirectly acting chemicals in human hepatoma (Hep G2) and peripheral blood lymphocytes, using the micronucleus assay and fluorescent *in situ* hybridization with a DNA centromeric probe. Mutagenesis 11, 425—433.

Darroudi, F., Knasmüller, S., Natarajan, A.T., 1998. Use of metabolically competent human hepatoma cells for the detection and characterization of mutagens and antimutagens: an alternative system to the use of vertebrate animals in mutagenicity testing, Netherlands Centre Alternative to Animal Use. News Lett. 6, 6—7.

Darroudi, F., 2021a. Genotoxic and Anti-genotoxic Effects of Heavy Metals in HepG2 Cells (Manuscript in Preparation).

Darroudi, F., 2021b. Assessment of DNA Damage Induction, Repair Kinetics and Biological Conseuqences Following Exposure of HepG2 Cells to Genotoxicants (Manuscript in Preparation).

Dauer, A., Hensel, A., Lhoste, E., et al., 2003. Genotoxic and antigenotoxic effects of tannins from the bark of *Hamamelis virginiana* L. in metabolically competent, human hepatoma cells (HepG2) using single cell gel electrophoresis. Phytochemistry 63, 199–207.

Ehrlich, V., Darroudi, F., Uhl, M., et al., 2002a. Fumonisin B1 is genotoxic in human derived hepatoma (HepG2) cells. Mutagenesis 17, 257–260.

Ehrlich, V., Darroudi, F., Uhl, M., et al., 2002b. Genotoxic effects of ochratoxin A in human-derived hepatoma (HepG2) cells. Food Chem. Toxicol. 40, 1085–1090.

Elje, E., Mariussen, E., Moriones, O.H., Bastús, N.G., Puntes, V., Kohl, Y., Dusinska, M., Rundén-Pran, E., 2020. Hepato(Geno)Toxicity assessment of nanoparticles in a HepG2 liver spheroid model. Nanomaterials 10, 545.

Filipic, M., Hei, T.K., 2004. Mutagenicity of cadmium in mammalian cells: implication of oxidative DNA damage. Mutat. Res. 546 (1–2), 81–91.

Gandhi, D., Tarale, P., Naoghare, P.K., Bafane, A., Krishnamurthi, K., Arrigo, P., Saravanadevi, S., 2015. An integrated genomic and proteomic approach to identify signatures of endosulfan exposure in hepatocellular carcinoma cells. Pestic. Biochem. Physiol. 125, 8–16.

Gerner, C., Vejda, S., Gelbmann, D., et al., 2002. Concomitant determination of absolute values of cellular protein amounts, synthesis rates and turnover rates by quantitative proteome profiling. Mol. Cell. Proteomics 1 (7), 528–537.

Gonzales, J.F., Barnard, N.D., Jenkins, D.J., Lanou, A.J., Davis, B., Saxe, G., Levin, S., 2014. Applying the precautionary principle to nutrition and cancer. J. Am. Coll. Nutr. 33, 239–246.

Harries, H.M., Fletcher, S.T., Duggen, C.M., Baker, V.A., 2001. The use of genomics technology to investigate gene expression changes in cultured human liver cells. Toxicol In Vitro 15, 399–405.

Hockley, S.L., Arlt, V.M., Brewer, D., et al., 2006. Time and concentration-dependent changes in gene expression induced by benzo(a)pyrene in two human cell lines, MCF-7 and HepG2. BMC Genom. 7, 260–283.

Hockley, S.L., Mathijs, K., Staal, Y.C.M., et al., 2009. Interlaboratory and interplatform comparison of microarray gene expression analysis of HepG2 cells exposed to benzo(a)pyrene. OMICS 2, 115–118.

Hreljac, L., Zaic, I., Lah, T., Filipic, M., 2008. Effects of model organophosphorous pesticides on DNA damage and proliferation of HepG2 cells. Environ. Mol. Mutagen. 49, 360–367.

IARC, 1993. Some Naturally Occurring Substances: Food Items and Constituents, Heterocyclic Aromatic Amines and Mycotoxins, vol. 56. IARC Press, Lyon.

Izquierdo-Vega, J.A., Morales-González, J.A., SánchezGutiérrez, M., Betanzos-Cabrera, G., Sosa-Delgado, S.M., Sumaya-Martínez, M.T., Morales-González, Á., Paniagua-Pérez, R., Madrigal-Bujaidar, E., Madrigal-Santillán, E., 2017. Evidence of some natural products with antigenotoxic effects. Part 1: fruits and polysaccharides. Nutrients 9, 102–129.

Jondeau, A., Dahbi, L., Bani-Estivals, M.H., Chagnon, M.C., 2006. Evaluation of the sensitivity of three sublethal cytotoxicity assays in human HepG2 cell line using water contaminants. Toxicology 226, 218–228.

Kassie, F., Laky, B., Gminski, R., et al., 2003a. Effects of garden and water cress juices and their constituents, benzyl and phenethyl isothiocyanates, towards benzo(a)pyrene-induced DNA damage: a model study with the single cell gel electrophoresis/Hep G2 assay. Chem. Biol. Interact. 142, 285–296.

Kassie, F., Mersch-Sundermann, V., Edenharder, R., et al., 2003b. Development and application of test methods for the detection of dietary constituents which protect against heterocyclic aromatic amines Review. Mutat. Res. 523–524, 183–192.

Kawataki, K., Shimazaki, R., Satoshi, O., 2009. Comparison of gene expression profiles in HepG2 cells exposed to arsenic, cadmium, nickel, and three model carcinogens for investigating the mechanisms of metal carcinogenesis. Environ. Mol. Mutagen. 50, 45–59.

Keshava, C., Whipkey, D., Weston, A., 2005. Transcriptional signatures of environmentally relevant exposures in normal human mammary epithelial cells: benzo(a)pyrene. Cancer Lett. 221, 201–211.

Kimura, S., Hayatsu, H., Arimoto-Kobayashi, S., 1999. Glycine betaine in beer as an antimutagenic substance against 2-chloro-4-methylthiobutanoic acid, the sanma-fish mutagen. Mutat. Res. 439, 267–276.

Kirkland, D., Aardema, M., Henderson, L., Mueller, L., 2005. Evaluation of the ability of a battery of three *in vitro* genotoxicity tests to discriminate rodent carcinogens and non-carcinogens I. sensitivity, specificity and relative predictivity. Mutat. Res. 584, 1–256.

Kirkland, D., Pfuhler, S., Tweats, D., et al., 2007. How to reduce false positive results when undertaking *in vitro* genotoxicity testing and thus avoid unnecessary followup animal tests: report of an ECVAM Workshop. Mutat. Res. 628, 31–55.

Knasmueller, S., Steinkellner, S., Majer, B., et al., 2002. Search for dietary antimutagens and anticarcinogens: methodological aspects and extrapolation problems. Food Chem. Toxicol. 40, 1051–1062.

Knasmüller, S., Verhagen, H., 2002. Impact of dietary factors on cancer causes and DNA integrity: new trends and aspects. Food Chem. Toxicol. 40, 1047–1050.

Knasmüller, S., Sanyal, R., Kassie, F., Darroudi, F., 1995. Induction of cytogenetic effects by cooked food mutagens and their inhibition by dietary constituents in human hepatoma cells. Mutat. Res. 335, 62–63.

Knasmüller, S., Parzefall, W., Sanyal, R., et al., 1998. Use of metabolically competent human hepatoma cells for the detection of mutagens and antimutagens Review. Mutat. Res. 402, 185–202.

Knasmüller, S., Schwab, C.E., Land, S.J., et al., 1999. Genotoxic effects of heterocyclic aromatic amines in human derived hepatoma (HepG2) cells. Mutagenesis 14, 533–539.

Knasmüller, S., Uhl, M., Pfau, W., et al., 2003. Use of human cell lines in toxicology. Toxicology 191, 15–16.

Knasmüller, S., Mersch-Sundermann, V., Kevekordes, S., et al., 2004a. Use of human-derived liver cell lines for the detection of environmental and dietary genotoxins; current state of knowledge. Toxicology 198, 315–328.

Knasmüller, S., Cavin, C., Chakraborty, A., et al., 2004b. Structurally related mycotoxins, ochratoxin A, ochratoxin B, citrinin differ in their genotoxic activities, and in their mode of action in human derived liver (HepG2) cells: implication for risk assessment. Nutr. Cancer 50, 190–197.

Knasmüller, S., Murkovic, M., Pfau, W., Sontag, G., 2004c. Heterocyclic aromatic amines — still a challenge for scientist. J. Chromatogr. B 802, 1–2.

Laky, B., Knasmüller, S., Gminski, R., et al., 2002. Protective effects of Brussels sprout towards BaP induced DNA damage: a model study with the single cell gel electrophoresis (SCGE)/Hep G2 assay. Food Chem. Toxicol. 40, 1077–1083.

Lamy, E., Kassie, F., Gminski, R., et al., 2004. 3-Nitrobenzanthrone (3-NBA) induced micronucleus formation and DNA damage in human hepatoma (HepG2) cells. Toxicol. Lett. 146, 103–109.

Lhoste, E.F., Gloux, K., De Waziers, I., et al., 2004. The activities of several detoxication enzymes are differentially induced by juices of garden cress, water cress and mustard in human HepG2 cells. Toxicol. Lett. 146, 103–109.

Lu, W.-Q., Chen, D., Wu, X.J., et al., 2004. DNA damage caused by extracts of chlorinated drinking water in human derived liver cells (Hep G2). Toxicology 198, 351–357.

Majer, B.J., Kassie, F., Sasaki, Y., et al., 2004a. Investigation of the genotoxic effects of 2-amino–9H-pyrido[2,3-b]indole (AalphaC) in different organs of rodents and in human derived cells. J. Chromatogr. B 802, 167–173.

Majer, B.J., Mersch-Sundermann, V., Darroudi, F., et al., 2004b. Genotoxic effects of dietary and lifestyle related carcinogens in human derived hepatoma (Hep G2, Hep 3B) cells. Mutat. Res. 551, 153–166.

Majer, B.J., Hofer, E., Cavin, C., et al., 2005. Coffee diterpenes prevent the genotoxic effects of 2-amino–1-methyl–6- phenylimidazo[4,5-b]pyridine (PhIP) and N-nitrosodimethylamine in a human derived liver cell line (HepG2). Food Chem. Toxicol. 43, 433–441.

Mersch-Sundermann, V., Schneider, H., Freywald, C., et al., 2001. Musk ketone enhances benzo[a]pyrene induced mutagenicity in human derived Hep G2 cells. Mutat. Res. 495, 89–96.

Mersch-Sundermann, V., Knasmüller, S., Wu, X., et al., 2004. Use of a human derived liver cell line for the detection of cytoprotective, antigenotoxic and cogenotoxic agents Review. Toxicology 198, 329–340.

Miret, S., De Groene, E.M., Klaffke, W., 2006. Comparison of *in vitro* assays of cellular toxicity in the human hepatic cell line HepG2. J. Biomol. Screen 11, 184–193.

Natarajan, A.T., Darroudi, F., 1991. Use of human hepatoma cells for *in vitro* metabolic activation of chemical mutagens/carcinogens. Mutagenesis 6, 399–403.

Rohrmann, S., Hermann, S., Linseisen, J., 2009. Heterocyclic aromatic amine intake increases colorectal adenoma risk: findings from a prospective European cohort study. Am. J. Clin. Nutr. 89, 1418–1424.

Rueff, J., Chiapella, C., Chipman, J.K., et al., 1996. Development and validation of alternative metabolic systems for mutagenicity testing in short-term assays. Mutat. Res. 353, 151–176.

Russell, W.M., 1995. The development of the three Rs concept. Altern Lab Anim 23, 298–304.

Ryu, D., Choi, B., Kim, E., et al., 2015. Determination of ethyl carbamate in alcoholic beverages and fermented foods sold in Korea. Toxicol. Res. 31, 289–297.

Gewd, V.S., Su, H., Karlovsky, P., Chen, W., 2018. Ethyl carbamate: an emerging food and environmental toxicant. Food Chem. 248, 312–321.

Sanyal, R., Darroudi, F., Parzefall, W., et al., 1997. Inhibition of the genotoxic effects of heterocyclic amines in human derived hepatoma cells by dietary bioantimutagens. Mutagenesis 12, 297–303.

Schmeiser, H., Gminski, R., Mersch-Sundermann, V., 2001. Evaluation of health risks caused by musk ketone. Int. J. Hyg. Environ. Health 203, 293–299.

Schoonen, W.G.E.J., Westerink, W.M.A., De Roos, J.A.D.M., Debiton, E., 2005a. Cytotoxic effects of 100 reference compounds on HepG2 and HeLa cells and for 60 compounds on ECC–1 and CHO cells. I. Mechanistic assays on ROS, glutathione depletion and calcein uptake. Toxicol. In Vitro 19, 505–516.

Schoonen, W.G.E.J., De Roos, J.A.D.M., Westerink, W.M.A., Debiton, E., 2005b. Cytotoxic effects of 110 reference compounds on HepG2 and HeLa cells and for 60 compounds on ECC–1 and CHO cells. II. Mechanistic assays on NAD(P)H, AT P and DNA contents. Toxicol. In Vitro 19, 491–503.

Schwab, C., Kassie, F., Qin, H.M., et al., 1999. Development of test systems for the detection of compounds that prevent the genotoxic effects of heterocyclic aromatic amines: preliminary results with constituents of cruciferous vegetables and other dietary constituents. J. Environ. Pathol. Toxicol. Oncol. 18, 109–118.

Schwab, C.E., Huber, W.W., Parzefall, W., et al., 2000. Search for compounds which inhibit the genotoxic and carcinogenic effects of heterocyclic aromatic amines. Crit. Rev. Toxicol. 30, 1–69.

Shin, H.S., Yang, E.U., 2012. Simultaneous determination of methylcarbamate and ethylcarbamate in fermented foods and beverages by derivatization and GC-MS analysis. Chem. Cent. J. 6, 157–164.

Shrivastav, M., De Haro, L.P., Nickoloff, J.A., 2008. Regulation of DNA double-strand break repair pathway choice. Cell Res. 18, 34–147.

Staal, Y.C.M., van Herwijnen, M.H.M., van Schooten, F.J., Delft van, J.H.M., 2006. Modulation of gene expression and DNA adduct formation in HepG2 cells by polycyclic aromatic hydrocarbons with different carcinogenic potencies. Carcinogenesis 3, 646–655.

Steck, S.E., Gaudet, M.M., Eng, S.M., et al., 2007. Cooked meat and risk of breast cancer — Lifetime versus recent dietary intake. Epidemiology 18, 373–382.

Steinkellner, H., Rabot, S., Frewald, C., et al., 2001. Effect of cruciferous vegetables and their constituents on drug metabolizing enzymes involved in the bioactivation of DNA-reactive dietary carcinogens. Mutat. Res. 480–481, 285–297.

Stepien, M., Chajes, V., Romieu, I., 2016. The role of diet in cancer: the epidemiologic link. Aalud Publica Mex 58, 261–273.

Sugimura, T., 1982. Mutagens, carcinogens, and tumor promoters in our daily food. Cancer 49, 1970–1984.

Sugimura, T., 2000. Nutrition and dietary carcinogens. Carcinogenesis 21, 387–395.

Tien, E.S., Gray, J.P., Peters, J.M., van den Heuvel, J.P., 2003. Comprehensive gene expression analysis of peroxisome proliferator-treated immortalized hepatocytes: identification of peroxisome proliferator-activated receptor alpha-dependent growth regulatory genes. Cancer Res. 63, 5767–5780.

Turesky, R.J., 2007. Formation and biochemistry of carcinogenic heterocyclic aromatic amines in cooked meats. Toxicol. Lett. 168, 219–227.

Uhl, M., Helma, C., Knasmüller, S., 1999. Single-cell gel electrophoresis assays with human-derived hepatoma (Hep G2) cells. Mutat. Res. 441, 215–224.

Uhl, M., Helma, C., Knasmüller, S., 2000. Evaluation of the single-cell gel electrophoresis assays with human hepatoma (HepG2) cells. Mutat. Res. 468, 213–225.

Uhl, M., Darroudi, F., Seybel, A., et al., 2001. Development of new experimental models for the identification of DNA protective and anticarcinogenic plant constituents. In: Kreft, I., Skrabanja, V. (Eds.), Molecular and Genetic Interactions Involving Phytochemicals, pp. 21–33.

Uhl, M., Ecker, S., Kassie, F., et al., 2003a. Effect of chrysin, a flavonoid compound, on the mutagenic activity of 2-amino–1-methyl–6-phenylimidazo [4,5-b]pyridine (PhIP) and benzo(a)pyrene (P(a)P in bacterial and human hepatoma (HepG2) cells. Arch. Toxicol. 77, 477–484.

Uhl, M., Laky, B., Lhoste, E., et al., 2003b. Effects of mustard sprouts and allylisothiocyanate on benzo(a)pyrene-induced DNA damage in human-derived cells: a model study with the single cell gel electrophoresis/Hep G2 assay. Teratog. Carcinog. Mutagen. 273–282.

Valdiglesias, V., Pasaro, E., Mendez, J., Laffon, B., 2010. In vitro evaluation of selenium genotoxic, cytotoxic, and protective effects: a review. Arch. Toxicol. 84, 337–351.

van Delft, J.H.M., Agen van, E., Breda van, S.G.J., et al., 2004. Discrimination of genotoxic from non-genotoxic carcinogens by gene expression profiling. Carcinogenisis 25 (7), 1265–1276.

Verhagen, H., Aruoma, O.I., van Delft, J.H.M., et al., 2003. The 10 basic requirements for a scientific paper reporting antioxidant, antimutagenic or anticarcinogenic potential of test substances in *in vitro* experiments and animal studies *in vivo*. Food Chem. Toxicol. 41, 603–610.

Westerink, W.M.A., Schoonen, W.G.E.J., 2007a. Cytochrome P450 enzyme levels in HepG2 cells and cryopreserved primary human hepatocytes and their induction in HepG2 cells. Toxicol. In Vitro 21, 1581–1591.

Westerink, W.M.A., Schoonen, W.G.E.J., 2007b. Phase II enzyme levels in HepG2 cells and cryopreserved primary human hepatocytes and their induction in HepG2 cells. Toxicol. In Vitro 21, 1592–1602.

Wilkening, S., Stahl, F., Bader, A., 2003. Comparison of primary human hepatocytes and hepatoma cell line HepG2 with regard to their biotransformation properties. Drug Metabol. Dispos. 31, 1035–1042.

Wu, X., Lu, W.Q., Roos, P.H., Mersch-Sundermann, V., 2005. Vinclozolin, a widely used fungicide, enhanced BaP-induced micronucleus formation in human derived hepatoma cells by increasing CYP1A1 expression. Toxicol. Lett. 159, 83–88.

Yokoo, H., Kondo, T., Fujii, et al., 2004. Proteomic signature corresponding to alpha fetoprotein expression in liver cancer cells. Hepatology 3, 609–617.

Yuan, J., Lu, W.Q., Dai, W.T., et al., 2005. Chlorinated drinking water caused oxidative damage, DNA migration and cytotoxicity in human cells. Int. J. Hyg. Environ. Health 208, 481–488.

Zheng, W., Lee, S.A., 2009. Well-done meat intake, heterocyclic amine exposure, and cancer risk. Nutr. Cancer 61, 437–446.

Chapter 27

Capacity building

Harmonization and achieving food safety in an era of unilateral legislation

Larry Keener[1] **and Tatiana Koutchma**[2]
[1]*International Product Safety Consultants, Seattle, WA, United States;* [2]*Agriculture and Agri Foods, Canada*

27.1 Introduction

Global trade involving food and foodstuffs expanded rapidly during the latter decades of 20th century. Developing nations, newly enrolled supply chain partners, significantly benefited from the opening of new market opportunities as well as from the opportunity of participating in the international economy. Between 1980 and 1994, for example, the contribution to the food sector from developing countries to the overall world value of international trade increased by 3.5%, while that of the European Union increased by 4.3% and that of North America by only 2.4% (UNIDO, 1997). Much of the observed growth resulted from the aggressive sourcing practices of the industrialized nations. The CEOs of multinational corporations based in Europe, the United States, and Japan and other "Group of Eight (G8) nations" extended their supply chains into new geographic areas with a view to improved margins and in cultivating new markets. Miranda-da-Cruz et al. (2009) of the United Nations Industrial Development Organization (UNIDO) estimate the present value of the global food industry in excess of 5 trillion U.S. dollars.

The much heralded stories of India, China, and Brazil's successful emergence as low cost producers and source countries are reported widely by contemporary news outlets. The economies of these nations benefited enormously from this explosion in food trade and poverty rates in each declined precipitously, according to the World Bank (Thompson, 2005). Moreover, there was also a serendipitous improvement in the public health status of the foods and foodstuffs offered by each nation for both domestic consumption and international trade. These collective gains appear to have been hard learned and also expensive.

Concurrent with the reported successes were a number of sensational transnational food safety scares. The majority of the incidents have been assigned classical modes of failure, e.g., bacterial, viral, or protozoan agents, and industrial, agricultural, or environmental chemical contaminants. Consider, for example, the 2007 and 2008 episodes of melamine contamination in dairy products, infant formula, and assorted other foodstuffs sourced from China. Likewise, in 1994, enterotoxin tainted, mushrooms exported to the United States from China were implicated in a major foodborne illness outbreak (Ballentine, 1989). Similarly, in 2004 one of the largest transnational food scares in history, attributed to the industrial dye (prohibited in foods intended for human use) Sudan Red 1, is reported to have had its origins in food ingredients sourced from India (Mishra et al., 2007). By no means do China, India, and Brazil comprise a complete listing of the countries that have placed unsafe food into the global food supply chain. In fact, the list of culpable nations, underdeveloped or otherwise, is a long one and growing. Curtailing the occurrence of such failures is paramount for preserving the public health and for sustaining the global trade in food.

"As our world transforms and becomes increasingly globalized, we must come together in new, unprecedented, even unexpected, ways to build a public health safety net for consumers around the world," (FDA Commissioner Margaret A. Hamburg, M.D. (FDA, 2012).[5] The safety net of which Dr. Hamburg envisions can only be erected, and sustained, when each member of the supply chain network have the capacity, both scientific and regulatory, to confirm that the foods they offer in commerce are safe. Dr. Hamburg's statement implies further that participating countries require equivalent standards as well as the capacity to enforce them.

Ensuring Global Food Safety. https://doi.org/10.1016/B978-0-12-816011-4.00025-2

Unilateral food safety legislation promulgated, primarily, by the developed nations are increasingly written with the view to transnational protections for public health, by way of forced capacity building among complex and disparate networks of supply chain partners. The Food Safety Modernization Act (FSMA) of the USA Food and Drug Administration (2011) is an excellent example of unilateral legislation that aims to improve food safety within the United States by demanding that its supply chain partners are capable of adhering to, and enforcing the food safety standards codified in this broadly written legislation. Other noteworthy examples of unilateral food safety legislation include the European Union's General Food Law (2002) and The Safe Foods Canadians Act (2012).

The impact of these unilateral initiatives has yet to be fully realized, but it is clear that individual nations are taking the initiative to write domestic legislation with purposeful and deliberate transnational intentions. Because of the anticipated transnational implications, questions have been raised by WTO members as to whether or not the unilateral standards codified by the FSMA, for example, are discriminatory and thereby create an economic hardship for lesser developed nations; A violation of the WTO agreement.

China has raised several trade concerns about the FSMA. Other WTO members, such as Brazil, Belize, and Korea, supported China's concerns. One of the major points of contention was the FSMA's import certification requirements and whether its outsourced third-party auditors will conduct food safety inspections in a manner consistent with the SPS Agreements. Member states must engage in risk assessments and regulate food imports in a manner that is "no more restrictive than necessary" to protect against the health risks identified by scientific evidence. Under the WTO regime, food safety laws that could restrict the free movement of food commodities must be sufficiently justified by scientific evidence (McNeil, 2012). Title III, of the FSMA, which regulates imported food, may create extra burdens for importers and therefore act as a barrier to trade.

The FSMA undoubtedly expands the FDA's regulatory authority over imported food products, but the law outsources the implementation authority to third party auditors. When such third party audits in effect constitute a more stringent requirement on imported food products than that on U.S. products, there might be WTO inconsistencies (such as the violation of the principles of nondiscrimination). The FSMA improves oversight of America's food safety system. What will be on trial before the World Trade Organization (WTO), however, is not the law's content, but the science supporting it (McNeil, 2012).

Updating and upgrading the scientific capabilities and regulatory framework of all nations participating in global food supply chains is fundamental. Collectively, these processes are referred to as capacity building.

27.2 Capacity building

In the process of assigning cause to food safety failures, the epidemiologists and regulatory officials are beginning to acknowledge that insufficiency in both scientific and regulatory capability is a major contributing factor. Lacking capacity frequently translates into an inability to provide the surveillance mechanisms necessary for ensuring the safety of foodstuffs bound for international commerce. Lacking capacity is increasingly critical and problematic as global food supply chains are projected deeper into economically underdeveloped regions of the planet. Countries in Africa, Asia, and the Americas, in which the per capita income is less than $1000.00 annually, are increasingly targeted by industrialized states as potential sources for food and food ingredients (see Table 27.1).

There is an abundance of data in the contemporary literature suggesting a correlation between per capita income and the likelihood of a nation state having both scientific capability and regulatory infrastructure required to sustain the safe exportation of food products. This finding is supported, for example, by a 2003 Pan American Health Organization/World Health Organization's (WHO) food safety assessment of its regional members. The data presented in Table 27.2 show a large disparity in the food safety systems between the wealthy and poor nations in the region. Brazil, Canada, Chile, and the United States of America (clusters 1 and 7), for example, received excellent scores in the PAHO/WHO assessments (see Table 27.2). By contrast, the food safety systems of the remaining 29 countries that participated in the study, less wealthy nations, did not achieve the minimum international standard (PAHO/WHO, 2003). These facts are daunting when one considers the contributions of the poor nations in this region to the global trade in food.

The PAHO/WHO study concluded the following with regard to the main weakness in food safety system development as highlighted by the study:

1. Absence or scarcity of provisions establishing ***punishing mechanisms*** and control mechanisms
2. Absence or scarcity of ***transparency*** in the drafting of food regulations (no integration of entities related to the food area, provisions drafted by the government without consulting all sectors involved)

TABLE 27.1 Potential supply chain partners with per capita daily income and population data.

Country	Population (000)[b]	% < $1/day[a]	% < $2/day
China	1299	16.6	46.7
India	1065	34.7	79.9
Indonesia	239	7.5	52.4
Brazil	184	8.2	22.4
Pakistan	159	13.4	65.6
Russia	144	6.1	23.8
Bangladesh	141	36.0	82.8
Nigeria	126	70.2	90.8
Mexico	105	9.9	26.3

[a]*Source: World Bank. World Development Indicators database (Bob Thompson, UI).*
[b]*Source: U.S. CIA -World 6,536,473,538 Aug. 2006.*

TABLE 27.2 Pan American Health Association/World Health Organization's Food Safety Assessment of American Regional Members (2003).

Cluster (representative member)	Number of countries/Cluster	Performance rating
1 (Brazil, Canada, USA)	3	96%–100%
2	14	25%–60%
3 (Mexico)	5	58%–81%
4	4	25%–60%
5	3	58%–81%
6	3	58%–81%
7 (Chile)	1	58%–81%

33 of 35 Nations in the Americas participated in the study (Cuba and Haiti were excluded due to insufficiency of Food Safety system data). Food safety systems in 4 nations (clusters 1 and 7) were defined as excellent. By contrast, the remaining 29 countries received ratings reflective of low levels of food safety system development.

3. **Insufficient harmonization with international standards** and lack of provisions referred to follow-up and confiscation of food in case of problems
4. Insufficient definition and delimitation of functions of governmental authorities (***overlapping of functions or loopholes*** in the appointment of competent authorities for the control of certain activities)

Similar findings have been reported by the UNIDO in assessment of food safety system development in 25 African nations (Ouaouich, 2005). As can be seen in Table 27.3 even the best performing nations in the UNIDO assessments suffered deficiencies in their food safety programs that would cause concern for the public health status of foods products emanating from those sources. The main deficiencies (Table 27.4) in food safety system development as reported by the UNIDO workers were also remarkably similar to those reported by the PAHO/WHO studies. The African nations, like those in the Americas, suffered from insufficient harmonization with international standards, poor cooperation among regulatory bodies, and a lack of national food safety policy.

Food safety experts have historically assigned food safety failures to one of three principle modes, i.e., microbiological, chemical, and foreign body contamination. Increasingly, however, lacking food safety capacity is cited as a primary or contributing factor in outbreak investigations.

Building capacity across the transnational expanses of complex food supply chains is regarded by regulatory authorities and health officials as a major contributor in protecting the public health and preserving the integrity of the global food supply.

TABLE 27.3 United Nations Industrial Development Organization (UNIDO) Capacity Building Assessment in 25 African Nations (2005).

Country	Food safety Policy	Food safety information	Regulations	Food safety management	Auditing	Scientific support (monitoring)	Labs
Senegal	3	3	3	3	3	3	3
Mauritania	2	3	3	2	3	2	3
Guinea	2	3	2	2	3	1	1
Uganda	3	3	3	3	3	3	3
Tanzania	3	3	3	3	3	3	3
Kenya	3	3	3	3	3	3	
Angola	2	3	2	2	3	2	2

The data in the table for the 7 best performing nations of 25 involved in UNIDO food safety survey. (Key 3 = Satisfactory).
Source: Ahmidou Ouaouich UNIDO-Vienna.

TABLE 27.4 UNIDO summary of major findings from African Survey (2005).

Frequency of food safety Element's inclusion in UNIDO programs (%)	Key finding on implementation/Efficacy from participating nations
Food safety policy 68%	Few countries with established national food safety policy
Food safety information/Awareness **100%**	High level of awareness and many training opportunities
Legislation, regulations, and standards **76%**	• 50% of nations have harmonized with international standards • 25% nations in process of developing standards • 25% of nations with no development activity
National food safety management framework **75%**	• Poor corporation between agencies • 7/25 nations in study have national framework
Inspection and auditing programs **96%**	• 18/25 nations have high competent inspectors • 750 inspectors trained in last 10 years
88% scientific support for surveillance and monitoring	**Low level of development 5/25 with risk assessment capability**
Laboratory programs and standards **84%**	• 13/25 nations with international accredited labs • 7/25 nations making progress with GLPs • 5/25 nations lack infrastructure and staff

Source: Ahmidou Ouaouich Industrial Development Officer UNIDO-Vienna.

The term "globalization of public health" has emerged in policy discourses to express the transnational or globalized nature of public health threats, including those represented by food and waterborne disease, in an increasingly interdependent world. Because foodborne diseases and chemical contaminants have no regard for the geopolitical boundaries of nation states, and because notions of state sovereignty are unknown concepts in the microbial realm, all of humanity is now (owing in part to the global trade in food) vulnerable to the emerging and reemerging threats associated with foodstuffs improperly handled or processed.

27.3 The role of multilateral agreements in achieving food safety

In the past 10 or so years, food safety has increasingly become an important topic in international law. Its implications are far reaching and impact a number of multilateral regimes such as WHO's International Health Regulations (IHR), WTO's Trade-Related Aspects of Intellectual Property Rights (TRIPS) and Sanitary and Phytosanitary (SPS) Agreements, and the Joint FAO/WHO Codex Alimentarius Commission standards on food safety. Achieving public health objectives related to

TABLE 27.5 Types of food safety governance systems and schemes.

Governance system	Example
Unilateral or National Food Safety Governance	General Food Law – European Union 2002 The Food Safety Modernization Act – USA 2011 The Safe Food for Canadian's Act – Canada 2012
Multilateral Food Safety or International Schemes and Agreements	World Trade Organization (WTO); Food Agriculture Organization/World Health Organization (FAO/WHO); Codex Alimentarious (Codex); Regional Trade Agreements (RTAs)
Private Food Safety Governance Schemes	Global Food Safety Initiative (GFSI); International Standards Organization (ISO); British Retailers Consortium (BRC Global Standards); NSF and Bureau Veritas

preserving the integrity of the food supply present enormous global challenges that frequently transcend the capabilities of individual governments. Achieving food safety requires harmonized and global governance strategies. Historically, international law has played an important role in this dynamic, because states have used treaties and conventions to solve public health problems that are transnational in nature. See Table 27.5 for examples of food safety governance schemes and strategies.

27.3.1 Historical developments in food safety management and multilateral agreements

The 1924 Pan-American Sanitary Code is an excellent example of a multilateral agreement that was constructed with the intent of curtailing the spread of communicable disease and other "dangerous contagion liable to spread through international commerce" (Pan American Health Organization). It is noteworthy that surveillance and sharing epidemiological data were important elements of Code's mandate. The 1924 sanitary code supplanted a narrowly written 1905 version that had been agreed for the expressed purpose of monitoring a specific but limited set of communicable diseases plaguing the region at that time. The success of these regional multilateral agreements sets the stage for the adoption, in 1951, of the International Sanitary Regulations (ISR) of the WHO. The ISRs were renamed the IHR in 1969 and modified in 1973 and 1981. According to WHO, the IHRs are among the earliest multilateral regulatory mechanisms focusing on global surveillance for communicable diseases. As of 1997, the IHRs were legally binding on all WHO's Member States except Australia. The effectiveness and impact of the IHRs are contentious and the subject of much debate among WHO members. It is ironic but one of the main points cited in making the case against the IHRs has to do with reporting of surveillance data. For example, in 1994, India after reporting a plague outbreak suffered nearly two billion dollar losses in trade resulting from the excessive embargoes that were imposed by other nations. Similarly in 1997, the economies of several East African nations were harmed by a ban against the importation of fresh fish into the European Community following a reported outbreak of cholera (Aginam, 2002). Another frequently cited reason for the ineffectiveness of the IHRs is the regulations' inflexibility and inability to adapt to changing circumstance in international trade and public health. Indubitably much has changed in international trade related to food and foodstuffs since the 1981 modification of the IHRs that would call into questions the regulations usefulness in terms of controlling the spread of foodborne diseases. Promoting food safety, it appears, was never the explicit intent of the IHR regulations. The regulations were promulgated and subsequently modified with the intent of providing a mechanism for the control and sharing of epidemiological information on the transboundary spread of cholera, plague, and yellow fever, diseases that, with the exception of cholera, are not transmitted by either food or water.

27.3.1.1 Sanitary and Phytosanitary agreement of the World Trade Organization

The 1995 "Sanitary and Phytosanitary" (SPS) agreement of the WTO covers risk measures such as quarantine restrictions on imported agricultural products, intended to prevent the introduction of pests or diseases which could harm domestic industries or the natural environment within the territory of a WTO Member, as well as bans on imported food products given that they contain contaminants or additives which pose a risk to public health. The SPS measures potentially impact a broad range of national regulations designed to protect against risk to human, animal, or plant life or health. The SPS agreement was driven by a desire to minimize the impacts of national SPS measures on trade by harmonizing the WTO Members' SPS measures. Article 3.1 of the SPS Agreement requires that Members base their SPS measures on the international standards, guidelines, and recommendations developed by the Codex Alimentarius Commission (food safety), the International Office of Epizootics (animal health), and the International Plant Protection Convention (plant health). While harmonization was the objective, it was recognized that this goal could not be achieved in all cases. Consider, for example, where Members' SPS measure could not be harmonized because of a lack of an agreed international standard, or because certain WTO Members opted for more stringent regulations. Article 3.3 of the SPS agreement specifically allows Members to adopt more stringent national regulations provided that they have a scientific basis. There are two provisions of the Agreement that are important in defining the criteria for measures to have a basis in science. The first is Article 2.2 which stipulates that WTO Members must base any SPS measure they wish to introduce on "scientific principles" and to ensure that their SPS measures are "not maintained without sufficient scientific evidence." Article 2.2, however, is subject to an exception set out at Article 5.7. The exemption allows Members to adopt SPS measures "on the basis of available pertinent information" in circumstance where "relevant scientific evidence is insufficient."

The second provision is Article 5.0 which establishes obligations for Members to ensure that their SPS measures are "based on" risk assessment. Moreover, in conducting an assessment of risks, Members must take into account "available scientific evidence" and risk assessment techniques developed by the international organizations whose standards are referenced in the Agreement (SPS Article 5.1).

Considering that the SPS measures are risk based, it is worthwhile noting that nowhere in the WTO documents is there mention of risk management. It is generally accepted that the components of the risk analysis process involve the following operations: risk assessment, risk communication, and risk management.

The SPS Agreement has given rise to conflicts among WTO Members. Developing nations have complained that they are held to the more onerous SPS measures, as allowed by WTO, of the developed nations and that this disparity has the effect of inhibiting the participation of the poor in remunerative economic activities. The proliferation and increased level of standards pose challenges for developing countries. For example, how to cost-effectively meet external regulatory or supply chain requirements? Many in the developing world are seeking external assistance as called for under the Agreement. The provision of technical assistance and other capacity-building measures in this area are under the World Bank—supported projects and there appears to be increasing demand in this area from the Bank's clients. To date, there has not been a comprehensive assessment to compare or contrast the assistance that has been provided in terms of its efficacy, efficiency, or sustainability in addressing the needs of the developing nations.

As mentioned previously "capacity," in terms of providing adequate surveillance of the public health issues attendant to the global food supply, consists of two separate but interrelated elements. Ideally, foods safety laws and regulations are developed and underpinned by sound objective science. Capacity in terms of food safety is the construct that results from the amalgamation of science and laws. Both are fundamental for building a nation's capability to protect and preserve public health. WTO's SPS Agreement sits squarely at that intersection where scientific method meets jurisprudence.

27.3.1.2 FAO/WHO and Codex Alimentarius

In 1961, the Eleventh Session of the Conference of the Food and Agriculture Organization (FAO) of the United Nations passed a resolution to set up the Codex Alimentarius Commission. In May 1963, the Sixteenth World Health Assembly approved the establishment of the Joint FAO/WHO Food Standards Program and adopted the statutes of the Codex Alimentarius Commission. According to Codex documents, a driving force for the establishment of the Commission was the need to harmonize standards that could be adopted in advancing public health and providing for the safe trade in food products. As previously noted, the SPS Agreement of the WTO specifically references the Codex Alimentarius and stipulates further that WTO Members adhere to the guidelines and standards of Codex in developing national SPS measures. Moreover, WTO has increasingly used Codex as an international reference standard for the resolution of disputes concerning food safety and consumer protection.

Codex Alimentarius is framed and directed by the following five guiding principles:

1. protecting the health of consumers and ensuring fair practices in the food trade;
2. promoting coordination of all food standards work undertaken by international governmental and nongovernmental organizations;
3. determining priorities and initiating and guiding the preparation of draft standards through and with the aid of appropriate organizations;

4. finalizing standards elaborated under (3) above, and after acceptance by governments, publishing them in a Codex Alimentarius either as regional or worldwide standards, together with international standards already finalized by other bodies under (2) above, wherever this is practicable;
5. amending published standards, after appropriate survey in the light of developments.

To adopt Codex standards, countries require an adequate food law as well as a technical and administrative infrastructure with the capacity to implement it and ensure compliance. It is clear then that Codex recognizes in a fundamental way the need for those adopting its voluntary guidelines and standards to have the "capacity" for both validations and enforcement. For many years, FAO and WHO have provided assistance to developing countries to enable them to take full advantage of the Commission's work.

According to the Commission, Codex's assistance to developing countries has included the following:

- *convening expert meetings*, including JECFA and the Joint FAO/WHO Meeting on Pesticide Residues (JMPR), to advise the Codex Alimentarius Commission;
- *establishing and strengthening national food control systems*, including the formulation and revision of food legislation (acts and regulations) and food standards in accordance with Codex standards;
- *conducting workshops and training courses*, not only for transferring information, knowledge, and skills associated with food control, but also to increase awareness of the Codex Alimentarius and activities carried out by the Commission.
- *strengthening laboratory analysis and food inspection capabilities*;
- *providing training in all aspects of food control* associated with protecting the health of consumers and ensuring honest practices in the sale of food;
- *presenting papers at conferences, meetings, and symposia* on the relevance of Codex activities to the provision of safe food of acceptable quality;
- *extending guidance on matters directly related to Codex activities*, such as safety assessment of food produced using biotechnology;
- *developing and publishing manuals and texts* that are associated with food quality control and provide recommendations for the development and operation of food quality and safety systems;
- *helping with the establishment and strengthening of food control agencies* as well as with training in the necessary technical and administrative skills to ensure their effective operation;
- *developing and publishing training manuals* on food inspection and quality and safety assurance, particularly with respect to the application of the Hazard Analysis Critical Control Point (HACCP) system in the food-processing industry.

Since its inception in 1963, the Commission's demographics have changed rather dramatically. As can be seen in Fig. 27.1, there has been a radical shift in the numbers of participating countries from developing nations versus those from developed nations. By 1997, membership of the developing nations outnumbered developed nations by approximately 3−1 (Codex Commission). While this trend has continued paradoxically, it is reported that key problems facing smaller and less developed countries in their efforts to improve food safety capacity have to do with a lack of representation in international standard setting bodies including Codex Alimentarius. The facts are that, for many poor nations, the cost of sending a delegation to Commission meetings is simply prohibitive.

27.4 Unilateral food safety legislation for promoting capacity building

Unilateral food safety legislation are those measures written and enacted by individual nation states without consultation or agreements with others. The regulations are intended to provide for the safety of the domestic food supply. Unilateral food safety standards and regulations are typically written on the back of existing public health measures. For example the Safe Food for Canadians Act (SFCA) consolidated and updated 13 of those countries long-standing public health (food safety) regulations. The same can be said of the U.S.'s FSMA, another classic example of Unilateral Food Safety legislation. While it is the aim of these unilateral rules to protect national populations, it is also true that in the case of SFCA and FSMA that they both include explicit language that is designed to have a transnational impact. SFCA and FSMA legislation are written with the view to improving food safety assurance among their supply chain partners.

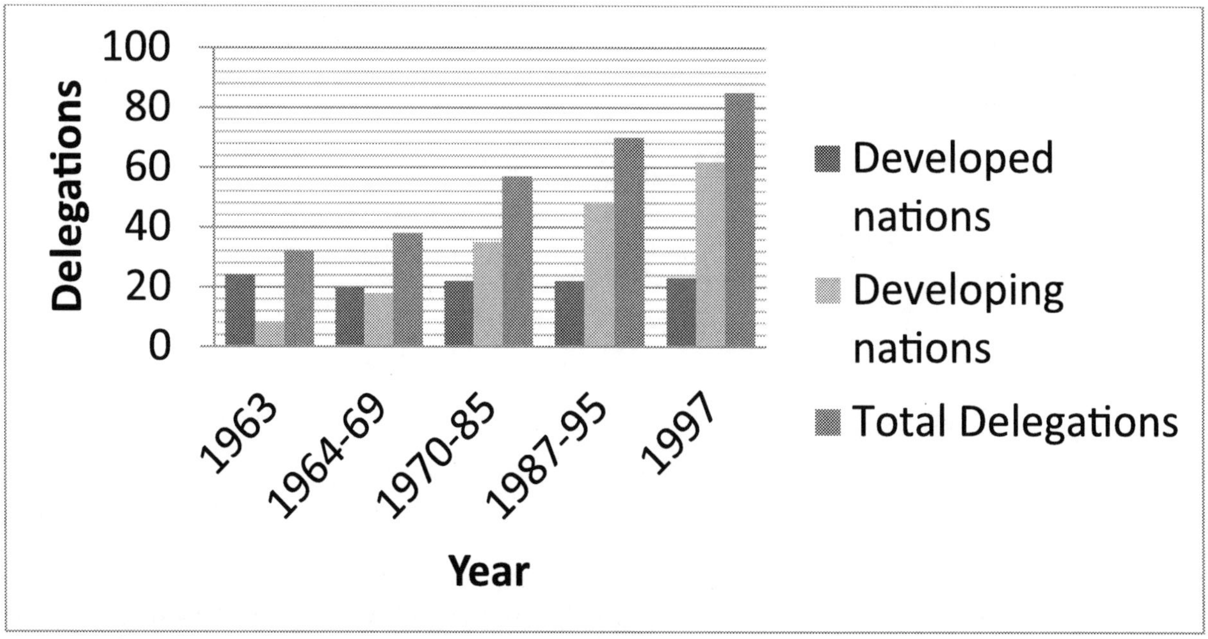

FIGURE 27.1 Codex's changing delegation demographics 1963–1997. Shift in the composition and number of delegations from developing nations v. developed nations attending Codex between 1963 and 1997 (Codex Commission).

27.4.1 U.S. FDA Food Safety Modernization Act

The burden of foodborne illness in the United States (US) is substantial. According to the U.S. Centers for Disease Control and Prevention (CDC), one in six Americans experiences a foodborne illness every year. Of those, nearly 130,000 are hospitalized and 3000 die (CDC, 2012). It is generally agreed and accepted among U.S. academics, medical epidemiologist, and food safety scientists that these official estimates of morbidity and mortality are likely underreported. Most assuredly, these CDC data do not include viral mediated foodborne illnesses.

The total societal cost of foodborne illnesses in the U.S. may be as much as $152 billion per year, according to a recent report (2010) from the Produce Safety Project at Georgetown University. CDC and U.S. Food and Drug Administration (FDA) officials report that preventing foodborne illnesses will improve public health, reduce medical costs, and avoid costly disruptions of the food system caused by outbreaks and large-scale product recalls. Accordingly, preventing foodborne illnesses will help promote trade, result in improved public health, and contribute to the nations' prosperity.

Foodborne disease outbreaks caused by imported food appeared to rise in 2009 and 2010, and nearly half of the outbreaks associated with imported food and implicated foods imported from areas which previously had not been associated with outbreaks (CDC, 2011). CDC experts reviewed outbreaks reported to CDC's Foodborne Disease Outbreak Surveillance System from 2005 to 2010 for implicated foods that were imported into the United States. During that 5-year period, 39 outbreaks and 2348 illnesses were linked to imported food from 15 countries. Of those outbreaks, nearly half (17) occurred in 2009 and 2010. Overall, fish (17 outbreaks) were the most common source of implicated imported foodborne disease outbreaks, followed by spices (six outbreaks, including five from fresh or dried peppers). Nearly 45% of the imported foods causing outbreaks came from Asia (CDC, 2011).

The FDA estimates that U.S. consumers purchased $2 trillion worth of imported products in 2007. These products came from 825,000 importers through more than 300 U.S. ports of entry. And the volume of imports could double every 5 years, according to FDA staff. FDA Commissioner Margaret Hamburg has reported an estimated 24 million food import entries into the United States during 2011 compared to about 12 million entries in 2007.

The FDA's office of regulatory affairs has reported that, in 2010, its inspectors physically examined 2.06% of all food-related imports. As reported previously, the FDA anticipated 24 million agency-regulated products entering the United States in 2011. While actual inspection rates are not available for these shipments, FDA had previously projected that it would inspect only 1.59% of them. Further, this downward trend in inspection rates was expected to continue in 2012 as FDA indicated that only 1.47% of all food imports will be examined (FDA, 2012).

In 2011, the U.S. FDA, at the direction of then president Barak H. Obama, enacted the FSMA. This legislation was heralded at that time as the most sweeping, comprehensive rewrite of U.S. Food Safety law in nearly 70 years. The agency has said that in writing this legislation that its intent was to impact, affirmatively, the food safety behavior and practices of those countries exporting food and food ingredients to the United States. Succinctly, the FSMA stipulates that the food safety requirements for imported food should be equivalent to those of foods produced domestically. This notion of equivalence is contentious and has been long debated by U.S. trading partners. Succintly then, FSMA demands that countries exporting foods to the United States have the capacity, regulatory, and technical, to assure the safety of its exported food products. U.S. regulatory officials have come to the realization that achieving food safety was not compatible with its archaic at the port entry inspection-based schemes. The agency has reported that less than 2% of all food imported in to the United States are subjected to agency inspection programs. FSMA's intent is to push the onus for food safety assurance with the exporting country or with those causing food to be imported into the United States. The United States is demanding its trading partners to embrace accepted scientific methods for confirming and documenting the safety of the food they were placing into international commerce.

Building capacity among its supply chain partners was seen as the best strategy for protecting the public health. The argument is easily made that with the enactment of the FSMA that the United States of America was attempting to force capacity building among its many and disparate supply chain partners.

It is also interesting to note but also anticipate that the transnational reach of the FSMA regulations should result in complaints to the WTO. The WTO is the international organization that governs trade between its member nations, and covers the majority of world trade. Imported food must be treated the same as domestically produced food to remain in compliance with the WTO agreements. If implementation of FSMA results in a foreign supplier being at a disadvantage relative to domestic producers, there could be a cause for a WTO complaint. For example, implementation of the Foreign Supplier Verification (FSV) provisions of the FSMA demand that the hazard analysis for foreign sourced produce includes microbial, chemical, and physical risks, while only microbial hazard analysis is required for domestically produced products under the Produce Safety Rule. This is an example where hazard analysis under the FSV program is more extensive and imposes additional costs on imported foods that are not incurred by domestic firms, thereby allowing for a potential WTO complaint. Furthermore, the requirements for FSV activities may be particularly challenging in developing countries, while imports from developed countries with food safety regulations, deemed equivalent to those of the United States, may be exempt from FSVP (Humphrey, 2012). Again, this may potentially put developing country producers at a great disadvantage relative to foreign suppliers from developed countries. China, Korea, and Belize have raised concerns about the FSVP with the WTO (Ching-Fu, 2013).

The goal of the FSMA is to ensure the safety of the U.S. food supply, including food produced both domestically and abroad by moving to food safety regulations that are preventative measures to avert food contamination before outbreaks occur. Accordingly, the FSVP has been created to govern the standards that imported food must adhere to. Those countries lacking scientific capacity and regulatory enforcement capacity will suffer disproportionately the effect of the FSMA regulation. Furthermore, the new food safety standards will cause an increase in the cost of imported food, particularly for imports from developing countries without domestic food safety policies (capacity) that are similar to those in the United States. It should be anticipated that in many developing countries, where the gross domestic product (GDP) is dependent on exportation of food and food ingredients, that there will be a significant disruption of their economies resulting from loss of trade with the U.S. markets. This then is the collateral damage, as it were, of building the scientific and regulatory capacity globally to protect the integrity and safety of the food supply.

Anticipating the possible blow-back and collateral economic damage caused by the FSMA regulations, the FDA has begun a policy of deploying agency personnel for duty in support of its mission at U.S. embassies and consulates around the world. At present, the agency reports that it has established post in 13 countries as an integral part of its global food safety strategy. The primary objectives of FDA staff at the overseas posting are as follows (FDA, 2012):

1. **Relationship Building:** Engage with foreign counterpart regulatory authorities to establish and maintain collaborative and cooperative working relationships that ensure timely exchange of information regarding the manufacturing and distribution of food/feed and medical products that are exported to the United States.
2. **Obtaining Information on the Regulation, Production, Manufacture, and Transport of Products:** Interact routinely with counterpart regulatory agencies and industry in country to: (a) learn more about a country's regulatory capacity and (b) leverage their information and resources regarding the conduct of clinical trials and the manufacture of products in the country that will be exported to the United States.
3. **Providing Information on FDA Policies, Laws, Regulations, Standards, and Expectations:** Work with the regulated industries that wish to export their products to the United States to help ensure and increase their understanding

of FDA's requirements and expectations regarding FDA-regulated products. This work includes serving as a resource: (a) for industry, academia, professional organizations, trade groups, local governments, and in-country international organizations for information about FDA; and (b) to identify needed documents and technical experts within FDA, when appropriate, including triaging and prioritizing requests.

4. **Engaging in Technical Cooperation and Capacity-Building:** Partner with foreign regulatory counterparts to address regulatory issues of mutual concern and priority, and leverage each other's information and activities, as appropriate. This partnering includes capacity-building to strengthen regulatory data and systems to help ensure the safety and quality of FDA-regulated products. Capacity-building, in turn, supports a data-driven approach to FDA decision-making.
5. **Conducting Inspections and Investigations:** As directed by the Office of Regulatory Affairs (ORA), FDA investigators posted in China and India conduct inspections of a routine, priority, or emergency nature and collect information about the manufacture of FDA-regulated products destined for the United States. This information-gathering increases FDA capacity for better targeting of inspections of high-risk facilities and high-risk products, and better analyses of risks around imported products.
6. **Environmental Scanning:** Assess conditions and events in-country that might affect the safety, quality, efficacy, security, and availability of FDA-regulated products exported to the United States. FDA's Office of International Programs (OIP) country/regional Offices are responsible for "scanning" the environment within their geographic locales to obtain information that may be helpful to the FDA Centers, ORA, and other FDA Offices, and senior executive leadership in their decision-making, while increasing the Agency's knowledge of a country's/region's regulatory landscape.
7. **Cooperation with Other U.S. Government Agencies:** Coordinate and collaborate routinely on product quality and safety issues with other U.S. Government agencies at post that have complementary missions. These include, for example, the U.S. Departments of State, Agriculture, Homeland Security, Justice, and Commerce and their relevant agencies, the United States Trade Representative, and Health and Human Services Agencies, such as the CDC.

The impact and reach of the new U.S. food safety regulations are intended to be transformative while simultaneously transcending national boundary considerations. Ultimately, the FSMA regulations coming out of Washington, D.C., must withstand WTO scrutiny. The United States will be compelled to demonstrate that its FSMA regulations do not adversely impact trade and commerce. The demand for equivalence in standards and public health measures for assuring food safety, as these are currently written in the FSMA regulations, may be viewed as a hardship for many of the U.S.'s trading partners, especially the developing nations. Section 1003(b)(3) of the Federal Food, Drug, and Cosmetic Act states that, as part of its statutory mission, FDA will "participate through appropriate processes with representatives of other countries to reduce the burden of regulation, harmonize regulatory requirements, and achieve appropriate reciprocal arrangements." It appears that the FDA is seeking to comply with this statutory mandate by use of an integrated strategy that includes new rules and standards as well as direct oversight of its supply chain adherents.

27.4.2 European Union General Food Law

A series of food safety failures, in Europe, during the late 1990s drew attention to the need to establish general principles and requirements concerning food and feed law **at the Union level**. Accordingly, the European Commission developed an **integrated approach to food safety "from farm to table**," as it was called at the time. The EU food safety strategy encompassed all sectors of the food supply chain and was applicable to all EU member states. Compelling reasons for the new law were expressed by EU leadership at that time; "European citizens need to have access to safe and wholesome food of the highest standard."

The 2002 General Food Law of the European Union covers all sectors of the **food** chain, including feed production, primary production, **food** processing, storage, transport, and retail sale. In 2002, the **European Parliament** and the Council adopted Regulation (EC) No 178/2002 laying down the **general** principles and requirements of food law (General Food Law Regulation, 2002).

The, unilateral, General Food Law Regulations are the foundation of food and feed law. It sets outs an overarching and coherent framework for the development of food and feed legislation both at Union and national levels. To this end, it lays down general principles, requirements, and procedures that underpin decision-making in matters of food and feed safety, covering all stages of food and feed production and distribution. It also sets up an independent agency responsible for scientific advice and support, the European Food Safety Authority (EFSA). The formation of EFSA appears to be a hedge against future inquiries from WTO about the risk assessments and the scientific basis of the General Food Law's standards and regulations.

The General Food Law Regulations reported ensure a high level of protection of human life and consumers' interests in relation to food safety while simultaneously ensuring the effective functioning of the European market.

Unlike the FSMA regulations of the United States of America and in particular the FSV requirements of that legislation, the General Food Law of the EU does not impose its requirements on other non-EU supply chain partners. The law as written is a great incentive for EU members to build regulatory and scientific capacity necessary for ensuring safe food but dose little by way of promoting capacity building for its many supply chain partners outside of the Union.

27.4.3 Safe Food for Canadians Act

To address and manage modern risks to food safety and to protect Canadian families from potentially unsafe food, the Government of Canada tabled the SFCA on June 7, 2012. The Safe Food for Canadians Act, S-11, was adopted by the Senate on October 17, 2012 and passed by the House of Commons on November 20, 2012. On November 22, 2012, it received Royal Assent.

The new *SFCA* consolidates the authorities of the *Fish Inspection Act*, the *Canada Agricultural Products Act*, the *Meat Inspection Act*, and the food provisions of the *Consumer Packaging and Labeling Act*.

The *Safe Food for Canadians Act*:

- makes food as safe as possible for Canadian families;
- protects consumers by targeting unsafe practices;
- implements tougher penalties for activities that put health and safety at risk;
- provides better control over imports;
- institutes a more consistent inspection regime across all food commodities; and
- Strengthens food traceability.

Similar to the FSMA, it virtually overhauls the existing system, in this case, by consolidating the food provisions of four statutes into one. The *SFSA*, S-11, was adopted by the Senate on October 17, 2012 and passed by the House of Commons on November 20, 2012. On November 22, 2012, it received Royal Assent. The SFCA also provides for the promulgation of regulations to implement the intent of the act, but to date there has been no publication of any proposed regulations.

The SFSA focuses on three important areas:

1. **Improved food safety oversight to better protect consumers:** The SFSA provides new authorities to address immediate food safety risks and builds additional safety into the system, from producer or importer to consumer. For example, the new prohibitions provide a stronger deterrent against deceptive practices, tampering and hoaxes. Related penalties and fines also increase to deter willful or reckless threats to health and safety. The legislation also provides the Canadian Food Inspection Agency (CFIA) with strengthened authorities to develop regulations related to tracing and recalling food, and the appropriate tools to take action on potentially unsafe food commodities. This includes a prohibition against selling food commodities that have been recalled.
2. **Streamlined and strengthened legislative authorities:** Existing CFIA food commodity statutes contain inconsistencies in inspection and enforcement authorities. Consolidating diverse provisions and authorities into one Act aligns inspection and enforcement powers, making them consistent across all food commodities, enabling inspectors to be more efficient, and fostering even higher rates of compliance for industry. As a result, consumers will enjoy a safer food supply.
3. **Enhanced international market opportunities for Canadian industry:** In recent years, more countries have required that the foods they import be certified, reflecting an international effort to ensure food safety. The legislation provides the authority to certify all food commodities for export, allowing the CFIA to treat exported food commodities consistently.

With the adoption of the Act, the Government of Canada has new authorities that will result in clearer rules for Canadian food commodity exporters and for trading partners importing food commodities to Canada. The legislation provides the CFIA the authority to certify all food commodities for export, allowing for a consistent approach to Canadian export certification. The new import control measures builds on Canada's existing suite of import safety measures by clearly prohibiting the entry of potentially unsafe food commodities into Canada. In addition, the Act includes provisions to register or license importers, holding them accountable for the safety of the food commodities they bring into the country.

On January 21, 2017, the Government of Canada published the second version of its proposed *Safe Food for Canadians Regulations* ("SFCR") in Canada Gazette Part I (Canada Gazette, 2017). The proposed regulations were created under authority of the *SFCA*. The SFCR will establish a set of regulations to govern all food sectors in Canada by consolidating 13 existing regulations and the food labeling provisions of the *Consumer Packaging and Labeling Act*. The end result is that, once implemented, all foods subject to CFIA oversight (i.e., federally registered sectors, and food that is destined for import, export, or interprovincial trade) will be regulated under the *SFCA* and the *Food and Drugs Act*.

These regulations will have a direct impact throughout the food supply chain, and will affect the way many manufacturers and importers conduct business in Canada. Those importers without a place of business in Canada would still be eligible for a license to import, provided they have a fixed place of business in a state with a food safety regime that provides equivalent protections to Canada's. Hence, Canada, like the U.S. FSMA, has unilaterally established domestic regulations that force capacity building among the countries supply chain partners.

Elements covered by the SFCR include the following:

- Licensing requirements for the import, export, and interprovincial trade of food
- Traceability requirements
- Reporting requirements and timelines
- Preventative controls and a preventative control plan
- Export Certificate request process
- Vegetable Dispute Resolution Corporation ("DRC") membership requirements for buyers and sellers or fresh fruit and vegetables
- Changes to the imported meat inspection requirements
- Establishment of standards for foreign systems of inspection for meat and shellfish
- Prescribed container sizes and weights for certain food products
- Labeling regulations and standards of identity
- Grade requirements for certain foods
- Expansion of organic certification to service providers and additional products

27.5 Conclusion

Achieving and sustaining capacity is difficult, even for wealthy, technology rich nations. Consider that the United States of America, one of the wealthiest nations on the planet and certainly among the world's leaders in its technological prowess, has recently embarked upon a campaign to modernize and improve its food safety systems. The U.S. President in announcing the campaign indicated that the overall number of foodborne illness outbreaks in the United States had increased from 100/year during the decade of the 90s to upwards of 350 by 2008. The cause for the increase was attributed to inadequate surveillance of the food manufacturing and distribution supply chains (Obama, 2009). President Barack Obama indicated that a failure to adequately fund the agencies responsible for food safety oversight had created a "hazard to the public health." Accordingly, the United States, it is reported, will invest 1 billion dollars in modernizing its food safety surveillance systems. China experienced similar failures in the last decades of the 20th century during its rapid ascent to prominence as a low cost, global supplier of food and foodstuffs. Cognizant of the threat to its international reputation and its future economic development, China deployed an aggressive national program to raise the level of food safety awareness and to bolster both its scientific capability and regulatory framework to limit future foodborne illness outbreaks. There are no financial data available relating to the cost of these programs, but it is reasonable to conclude that China's investment was also in the billions of dollars.

Imagine the hardship that such expenditure would cause for developing nations desirous of participating in the global trade in food. The proposition is simply untenable and cost prohibitive. The Republic of Chad, for example, according to the CIA's World Factbook, in 2004 reported a GDP of just over 15 billion dollars. It is difficult to conceive that this central African country of approximately 11 million people and a contributor to the global food supply could make a proportional investment in attaining the capacity to properly monitor its food production and distribution systems.

FAO and WHO reports concluded that the food systems of developing countries are extremely diverse and tend to be less organized, comprehensive, and effective than those in developed countries. The reports go on to say that the food safety systems in these countries are challenged by problems of rapid growing population, urbanization, and natural environments that expose consumers to a wide range of potential food safety risks. It is interesting to note that FAO also concluded that food safety standards in developing countries might actually attain those of international standards, but the lack of technical and institutional capacity to control and ensure compliance essentially makes the standards less effective. According to FAO, this apparent paradox results from a lack of technical and scientific infrastructure, e.g., testing laboratories, human and financial resources, national legislative and regulatory frameworks, enforcement capacity, management, and coordination of the food safety system. These findings are confirmed or supported by PAHO/WHO workers in their assessment of food safety systems in the Americas and again by the outcomes of UNIDO's report from its study of food safety system development in 25 African nations.

Widespread changes in the global food economy and the rapidly evolving environment in which food safety must be considered have demonstrated the profound interdependence of all elements of the supply chain and highlighted their contribution to achieving food safety. The historical record is replete with bilateral and multilateral conventions that have been agreed between and among nations for the purpose of preventing the transnational spread of contagion and disease causing agents. The 1924 Sanitary Agreement among the nations in the Americas, for example, offers a good case study in both the successes and failing of such agreements. The 1924 agreement was too narrowly crafted and overly ridged to keep pace with the rapid advances in science and technology. Likewise WHO's IHR have been reported to suffer similar shortcomings. These observations withstanding it remain a fact that achieving food safety involves all elements of the food supply chain including food production, food processing, and food distribution networks. Science and science-based regulatory constructs are also required to support and sustain the supply chain. Moreover, risk-based ex-ante (risk avoidance) measures are preferred over the more conventional ex-post strategies for achieving food safety.

Food safety has been defined as "the biological, chemical, or physical status of a food that will permit its consumption without incurring excessive risk of injury, morbidity, or mortality" (Keener, 2004). The inequality in food safety capacity, scientific, and regulatory, among various supply chain contributors, places the global trade in food at risk. Building the scientific capability and regulatory framework corresponding with food safety capacity is an expensive proposition. For many developing and underdeveloped nations, the costs of acquiring food safety capacity are simply prohibitive. Yet international trading data suggest that these nations will increasingly participate as future supply chain partners.

Developing nations, to be successful in building scientific and regulatory capability and therefore capacity to provide for the adequate surveillance of their food and agricultural resources, will require the continued financial support of their trading partners in the "West." Building analytical capacity is critical for supporting and elaborating a comprehensive food safety surveillance system. Analytical capability demands people with expert skills as well as facilities and equipment. Acquiring these components represents an enormous challenge for many developing nations. Recall that both the UNIDO and PAHO/WHO studies, discussed earlier in this chapter, concluded that there was a lack of analytical skill in many of the countries involved with their surveys. Babu and Rhoe (2001) provide an excellent summary of the typical challenges facing developing countries in the organization and modernization of food safety systems. According to these workers, the problems are often related to lack of human capacity and personnel, the lack of financial capacity to establish basic infrastructure, to conduct training and make the system sustainable over time. These stark realities of increased food safety failures and the implication for increased morbidity and mortality, associated with such failures, are the impetus for erecting safety nets around imported foods from the lesser developed countries. Unilaterally, both the United States of America and Canada have subsequently passed aggressive domestic legislation that has embedded explicit intent to force capacity building across their extensive and disparate supply chains. These unilateral regulations intend transnational actions by which all importers will be required to demonstrate technical and regulatory food safety capacity equivalent to that of the United States of America or Canada, as a requirement of doing business.

Similar trends run through FSMA and SFCA acts: both focus on prevention of food safety problems and both add requirements for importers. Both acts plan to achieve this is by requiring members of the food chain to design and implement prevention plans, and by policing these plans by requiring facilities to register with or be licensed by the government.

In regards to importers, both Acts specifically focus on food moving across borders and put the onus firmly on importers to verify that there are systems in place to prevent unsafe food from entering the country. The SFCA prohibits the sending or conveying of a food commodity from one province to another, or to import or export, unless the person is authorized to do so by a registration specifically made under the Act or by a license issued under the Act. These registrations are nontransferable. Conditions of registration and licensing will likely define in regulations. Both the Canadian and U.S. governments have acknowledged the need to work with foreign governments to harmonize the registration and inspection of food facilities.

There is a definite emphasis on harmonization and modernization that runs through both the FSMA and SFCA. Both increase the roles and responsibilities of all supply chain members and the food industry, and give enhanced powers to government to police the system.

References

Aginam, O., 2002. International Law and Communicable Diseases, vol. 80. Bulletin of the World Health Organization.

Babu, S., Rhoe, V., 2001. Food Security, Regional Trade, and Food Safety in Central Asia-Case Studies from Kyrgyz Republic and Kazakhstan. International Food Policy Research Institute, Washington, D.C.

Ballentine, C., 1989. FDA Consumer Magazine 23.

Canada Gazette Part I, 2017. Canada Gazette, Part I: Volume 151.

Ching-Fu, L., November 6, 2013. FDA Food Safety Modernization Act Might Raise Trade Concerns? Harvard Law.

CDC, 2011. CDC's Foodborne Disease Outbreak Surveillance System Report.

CDC, 2012. https://www.cdc.gov/media/releases/2012/p0314_foodborne.html.

Food and Agriculture Organization. Sanitary/Phytosanitary Agreement, Articles 2.2, 5.0,5.1 and 5.2,5.5 and 5.7.

Food and Drug Administration, USA, 2012. FDA Commissioner Margaret A. Hamburg, M.D.

General Food Law Regulation. https://ec.europa.eu/food/horizontal-topics/general-food-law_en.

Humphrey, J., 2012. Food Safety, Private Standards, Schemes and Trade: The Implications of the FDA Food Safety Modernization Act. Institute of Development Studies. http://www.ids.ac.uk/publication/food-safety-private-standards-schemes-and-trade-the-implications-of-the-fda-food-.

Keener, L., 2004. Nonthermal Processing Technologies and the Global Harmonization Initiative, IFT/EFFoST NPD Conference Cork, Ireland.

Mishra, K.K., Dixit, S., Purshottam, S.K., Pandey, R.C., Das*, M., Khanna, S.K., 2007. Exposure assessment to Sudan dyes through consumption of artificially coloured chilli powders in India. Int. J. Food Sci. Technol. 42 (11).

McNeill, N., 2012. The Food Safety Modernization Act: a barrier to trade? Only if the science says so. Food Drug Law J. 67 (2), 177−190.

Miranda-da-Cruz, S., Schebesta, K., 2009. Global Development of the Food Industry-Perspective of UNIDO. http://www.futurefood6.com/files/1.Global%20development%20of%20the%20food%20industry.pdf.

Obama, B., 2009. U.S. President's Statement on Food Safety Modernization, March 14. Associated Press International.

Ouaouich, A., 2005. A Review of the Capacity Building Efforts in Developing Countries: Case Study Africa; Presented 6th World Congress on Seafood Safety, Sydney Australia.

PAHO/WHO, 2003. National Food Safety Systems in the AMERICAS and the Caribbean - A SITUATION ANALYSIS. http://www.fao.org/docrep/meeting/010/a0394e/A0394E20.htm.

Thompson, R.L., 2005. World Ag Trade Negotiations: Doha Development Agenda. IFT annual convention, New Orleans, LA USA. http://ift.confex.com/ift/2005/techprogram/paper_27890.htm.

UNIDO, 1997. Trade Developments in Food and Contributions of Developing Countries, Tapia, M. IFT Presentation 2004 Las Vegas.

Chapter 28

Capacity building: building analytical capacity for microbial food safety

Debdeep Dasgupta[1], Mandyam C. Varadaraj[2] and Paula Bourke[3]

[1]*Department of Microbiology, Surendranath College-Kolkata, Kolkata, West Bengal, India;* [2]*Department of Human Resource Development, Central Food Technological Research Institute, Mysore, Karnataka, India;* [3]*School of Biosystems and Food Engineering, University College Dublin, Dublin, Ireland*

28.1 Introduction

There is no capsule definition for food safety although, in general, it would mean "the biological, chemical, or physical status of a food that will permit its consumption without incurring excessive risk of injury, morbidity, or mortality."

Food safety issues are closely associated with changes evolving in society, economy, lifestyle, and eating habits (Doores, 1999) and can arise in response to environmental, climate, and geographical location factors. Notwithstanding temporary halts associated with societal lockdowns, there has been increasing globalization, with significant travel and movement by people. These influences have driven changes to lifestyle, food consumption trends with little limitation on what can be available in different global regions, as well as a consistent trend of processed foods for convenience as well as health and nutrition in addition to increased animal protein trade and consumption. International trade in food and ingredients continues to increase and often raw materials are obtained from one or more regions or countries and the food is processed through the use of varied unit operations in either centralized or compartmentalized scenarios. The increased globalization of the food supply chain combined with knowledgeable consumers led to the demand for a safer food supply (Stringer, 2005), which remains and is consistently challenged with traditional and new risks that evolve.

The joint FAO/WHO World Declaration on Nutrition of 1992 (FAO and WHO, 1992) declared that access to nutritionally adequate and safe food is a right of each individual. The WHO has urged the government health agencies, the entire food industry, and the consumers to assume greater responsibility in the area of food safety. The World Health Assembly adopted a resolution on food safety calling upon countries to integrate food safety as one of their essential public health functions through capacity building programs (Miliotis and Bier, 2003). This resolution was considered as a milestone in the history of public health since, for the first time in the WHO's 50 plus years of existence, it identified food safety as an important objective and as an essential function of the public health community (Miliotis and Bier, 2003). Globally, many regulatory agencies and organizations continue to strengthen the implementation of such programs.

28.2 Significance of microbial food safety

In the food safety scenario, the most important objective and function is microbiological food safety. Microorganisms replaced chemical adulterants as the major recognized agent of food poisoning in the latter part of the 20th century which continues through the 21st century. The situation is driven through changes on a global scale, including population explosion, urbanization and changes in lifestyle, consumption of minimally processed ready-to-eat foods, resistance to antimicrobials and plant protection products, short supply of potable drinking water in some regions, international trade in food and animal feed, as well as the international movement of people. Foods are prone to microbial contaminants that are known for the undesirable changes they bring about in foods, causing either spoilage and/or health hazards. The emergence of new food production paradigms to address the increasing demand will likely bring unknown issues for the retention of microbiological food safety. A serious aspect of microbial contamination due to the presence of potential food poisoning

Ensuring Global Food Safety. https://doi.org/10.1016/B978-0-12-816011-4.00023-9

microorganisms and their toxic metabolites leading to food poisoning outbreaks persists. The recognized foodborne pathogens include multicellular animal parasites, protozoa, fungi, bacteria, and viruses, and the strains causing the most prevalent or serious risks to human health can ebb and flow on a global and regional level. Deciding on capacity building programs requires knowledge of the major bacterial species that play a significant role in maintenance of food safety and retention of human health.

Bacterial food poisoning comprises two broad categories: infection and intoxication. Infections are diseases caused by the presence of viable, usually multiplying microorganisms at the site of inflammation. In this case, the viable bacterial cells have been ingested with food. The dose required to produce an infection varies with the type of microorganism, although the microorganism may multiply in the gastrointestinal tract or some other organ of the body to produce the infectious disease. Common bacteria involved in foodborne infections are *Salmonella* sp., *Shigella dysenteriae*, *Campylobacter jejuni*, *Clostridium perfringens*, *Escherichia coli*, *Yersinia enterocolitica*, *Listeria monocytogenes*, and *Vibrio parahaemolyticus*. Intoxications are strictly poisoning events, implying the ingestion of a toxin produced or preformed in the ingested food. Common bacteria implicated are *Staphylococcus aureus*, *Bacillus cereus*, and *Clostridium botulinum*. In most of these poisoning events, the food serves only as a vehicle of transmission. The role of food here is significant, since the product may not only permit the survival of the pathogen, but may also serve as the suitable medium for the rapid proliferation of microorganisms and production of toxin, as in the case of exotoxin producing microorganisms. It is at this important juncture that the challenge for global food safety and integrity programs lies on capacity building approaches. These create and retain the intellectual workforce and networks to realize the benefits of the accurate and reliable approved methods now available for detecting potent pathogenic/virulent bacterial species.

Food microbiologists used the well-established culture-based approach of isolating organism(s) of interest in pure culture and performing predetermined biochemical tests as a means to identify the cultures, including pathogenicity/virulence (Mossel, 1986). All of the current methods have limitations in either sensitivity or specificity. Considering the changing global scenario of food safety, it is desirable to have analytical techniques that are rapid, accurate, simple to use, cost-effective, reliable, and reproducible, and which could provide results in "Real-Time." These techniques complement rather than eliminate the need for culture-based assessment of microbiological risks, but qualitative and quantitative detection methods and their performance and interpretation require harmonization for effective global food integrity. The detection of microorganisms, particularly those known to cause health hazards, occurs in very low numbers and the focus of all approaches has been to achieve a complete scenario of safety aspects of food samples through detection methods. Within the international and local food chains, there is an increasing public awareness of microbiological food safety. Rapid methods and automation in microbiology are dynamic fields of study that address the utilization of microbiological, chemical, biochemical, biophysical, immunological, and serological methods to improve isolation, the early detection, characterization and enumeration of microorganisms, and their products in clinical, food, industrial, and environmental samples.

The following sections present methods of detection with respect to three significant bacterial pathogenic species that are consistently and commonly implicated in food related health hazards over recent decades. The bacterial species focused on here include *Staphylococcus aureus*, *Listeria monocytogenes*, and *Bacillus cereus*.

28.3 *Staphylococcus* and its species

28.3.1 Characteristics

Staphylococcus is a very significant organism because of its ability to cause a number of diseases and infections in human and animals. It causes localized suppurative lesions and abscesses on skin, or deep-seated infections like osteomyelitis, endocarditis, or more serious skin infections like furunculosis. Because of their ability to develop resistance to many antibiotics, *Staphylococcus aureus* and *Staphylococcus epidermidis* form a major group of nosocomial pathogens infecting surgical wounds and causing infections associated with indwelling medical devices. *Staphylococcus aureus* is a major cause of intoxication, resulting in food poisoning due to the release of enterotoxins in foods. *Staphylococcus saprophyticus* is known to cause urinary tract infections and other staphylococci are infrequent pathogens.

Staphylococci are nonmotile, nonspore-forming, catalase-positive, facultatively anaerobic cocci, except for *Staphylococcus saccharolyticus*, which is a true anaerobe. Growth is more rapid and abundant under aerobic conditions and acetoin is formed as a product of glucose metabolism. Fermentation of mannitol is found to be characteristic of *S. aureus*. Colonies appear as smooth, circular and convex on agar plates (Baird-Parker, 1965). Staphylococci produce several extracellular proteins, among which are the characteristic enzymes and toxins produced by *S. aureus;* hemolysins, nuclease, lipase, coagulases, staphylokinase (fibrinolysin), as well as a large variety of cytotoxins and cytolysins and enterotoxins.

Production of thermostable (heat resistant nuclease) is a characteristic of *S. aureus*, *S. intermedius*, and *S. hyicus*, while S. *carnosus* shows a delayed reaction and *S. epidermidis*, *S. simulans*, *and S. hyicus* subsp. *achromogenes* are negative to weakly positive (Genigeorgis, 1989). Detection of Staphylococcus and its enterotoxins in foods is of prime concern, which can be hampered or interfered with by the complexity of the food matrices. However, there has been considerable improvement in detection methods that increase the sensitivity, specificity, and response times of the tests.

28.3.2 Methods of detection

The earliest method based on serological/immunological principles was the microslide method (Micro-Ouchterlony slidetest), an AOAC-approved method, used by FDA for the detection of enterotoxin. This has been widely used, due to its efficiency in detecting low amounts (<0.1−0.2 μg) of enterotoxins in the food extract (Casman and Bennett, 1965). Reiser et al. (1974) modified the assay procedure so that within 1−3 days as little as 0.05 μg/mL of enterotoxin could be detected in food extracts. The disadvantages involved with microslide assay, like time consumption and lack of sensitivity, were overcome by other methods like Laurell electro-immunodiffusion (Gasper et al., 1973), but the sensitivity was not adequate.

Improved serological methods such as Reverse passive hemagglutination assay (RPHA) and solid phase radioimmunoassay evolved in due course. These were able to detect 1.5 ng of enterotoxin per ml (Bergdoll et al., 1976). The development of enzyme immunoassays for enterotoxin detection and typing, in particular the enzyme-linked immunosorbent assay (ELISA) methods and Reverse passive latex agglutination test (RPLA), evolved as most sensitive detection methods (Baird-Parker, 2000). An enzyme-linked immune-filtration assay for the detection of staphylococcal enterotoxin was shown to detect 1 ng/mL of enterotoxin B in spiked milk samples <1 h (Valdivieso-Garcia et al., 1996). The RPLA method uses latex particles sensitized with purified antistaphylococcal enterotoxin immunoglobulins that agglutinate in the presence of homologous enterotoxins. Although occasional nonspecific reactions have been reported, the method is simple and easy to perform (Marin et al., 1992).

Freed et al. (1982) developed a detection method for staphylococcal enterotoxins based on the principle of ELISA, where a very high sensitivity and specificity has been achieved. An indirect double sandwich ELISA was applied in detection of enterotoxin, B, C1, and D in food samples using monoclonal antibodies (Lapeyre et al., 1988). However, one limitation of ELISA was its inability to identify and characterize the antigen that reacts with the antibody to give a positive signal. Given the variety and the complexity of foods, it is difficult to rule out a cross reactivity of the antibodies with the food matrix even with proper controls.

This led to the development of nucleic acid-based detection systems, which can overcome some of the problems commonly encountered with other methods of detection. These techniques facilitate rapid, sensitive, and specific analysis and are therefore an appropriate choice to record foodborne disease outbreaks (Hill and Keasler, 1991). Nucleic acid probe−based methods have been developed for the detection and enumeration of foodborne pathogens. Several kits applying DNA probes are commercially available. Methods involving a polymerase chain reaction (PCR) have proved promising of the rapid microbiological methods. The following section will detail the background and advances in the application of these methods to the detection of Staphylococcus and its enterotoxins in a food system.

28.3.2.1 Nucleic acid probes

An ideal probe is a single-stranded molecule with a short sequence of nucleotide bases that can hybridize only if a target is present. A probe can be either DNA or RNA. Most of the applications so far have used DNA probes (Wolcott, 1991). In order to detect the hybridization, DNA fragments have to be labeled. The label can be a radioisotope or a chemical moiety. Because of problems encountered during handling of radioisotopes, probes proved to be less popular. The availability of nonisotopic labels and efficient chemicals like digoxigenin, psoralen, and others has improved the utility of a DNA probe as a diagnostic tool. Probes can be produced either by screening of restriction genomic DNA fragments or targeting the unique sequences in genes coding for the known products or virulence genes.

Two-phase DNA hybridization formats are used in DNA probe assays. The dot-blot hybridization format requires the immobilization of crude or purified DNA containing the target on nitrocellulose or nylon membrane. The colony hybridization method involves the impression transfer of bacterial cells from colonies on a primary isolation plate to nitrocellulose or nylon membrane (Swaminathan and Feng, 1994). Anon-isotopic DNA hybridization assay was developed by Wilson et al. (1994). A DIG-labeled total genomic DNA probe was used to identify *S. aureus* in their study. The GENE-TRAK and AccuProbe (GEN-PROBE) DNA probe assays have been developed for *S. aureus*. GENE-TRAK assay is a semiquantitative assay, which detects *S. aureus* colorimetrically and includes a dipstick solid phase system.

28.3.2.2 Polymerase chain reaction

In the commercial set up, DNA probes could not gain the projected dominance. However, the basic principle has been utilized when developing many technologies (Feng, 2001). PCR technology has proven to be one of the most promising of the rapid microbiological methods for the detection and identification of bacteria in many food samples. This method is based on an enzymatic amplification of target nucleic acid sequences using a specific primer pair and a heat stable DNA polymerase (Lantz et al., 1994).

Various approaches have been developed using PCR for the detection of single or multiple target genes in a single reaction. The availability of nucleotide sequences for the enterotoxin genes (A-K) has enabled the detection of enterotoxigenic staphylococci using PCR. Jones and Khan (1986) gave the first complete nucleotide sequence of enterotoxin B. Later, a number of researchers worked toward the same end, and as a result, the nucleotide sequences of staphylococcal enterotoxins A, C, D, E, G, H, I, J, and K were published and made available in public databases.

As a means of enhancing the sensitivity of detection, Wilson et al. (1991) employed nested PCR, which uses internal primers for the specific gene amplicon. The nested primers for nuc, entB, and entC genes of *S. aureus* could detect one fg of purified DNA, while 100 pg was required for the detection of these genes, when a single primer pair was used. A decrease in the sensitivity was observed when applied in skimmed milk, which could be due to the presence of PCR inhibitors in milk. The problem of sensitivity was resolved by Tsen and Chen (1992), who detected enterotoxigenic staphylococci at the level of 100−101 cells per gram of various naturally contaminated food samples like beef, pork, chicken, and fish, using the primers targeted to enterotoxin A, D, or E genes. In another study, the same group (Tsen et al., 1994) developed a PCR-based method for the direct detection of enterotoxigenic *S. aureus* in various foods, without prior enrichment and could achieve a sensitivity of <100 cfu per gram of food sample.

Khan et al. (1998) developed a PCR-based method targeting the nuc gene encoding thermonuclease of *S. aureus*. A modified rapid boil method for the isolation of DNA directly from artificially contaminated milk samples provided a sensitivity of 100%. The nuc gene has been widely used as a taxonomic tool to identify *S. aureus* in food and clinical samples (Kim et al., 2001). The wide application of PCR in clinical microbiology has led to a large number of reports particularly targeting the mec gene, coding for the methicillinase gene of methicillin resistant *S. aureus* (MRSA) in clinical isolates (Mehrotra et al., 2000).

The major disadvantage of uniplex PCR is that a series of separate reactions is needed to identify a single gene or subset of genes. However, multiplex PCR has the advantage of simultaneous detection and differentiation of multiple genes in a single reaction. In multiplex PCR, multiple pairs of primers specific for different DNA segments are included in the same reaction to enable amplification of multiple target sequences in one reaction. The primers used in multiplex PCR should have similar annealing temperatures or Tm. The major advantage of multiplex PCR over conventional PCR is reduced preparation and analysis time (Phuektes et al., 2001). The sensitivity of PCR is dependent on the quality of the template DNA. The extraction of DNA from *S. aureus* generally requires an enzymatic treatment. The isolation of DNA from food samples has been simplified in order to overcome the inhibitors and enhance the sensitivity of detection of staphylococci (Tamarapu et al., 2001; Ramesh et al., 2002).

PCR-ELISA is an efficient quantification tool, which combines PCR with ELISA for post-PCR analysis. The assays utilize the internal biotinylated probes to capture the amplified PCR products of toxin encoding sequences and produce quantifiable signals through enzymatic amplification of a colorimetric detection system. PCR-ELISA, sometimes called enzyme-linked oligosorbent assay, has been used for the detection of *S. aureus* and its enterotoxins (Gilligan et al., 2000). Many reports are available wherein PCR-ELISA was used for the quantitative detection of bacterial pathogens like *Escherichia coli*, *Campylobacter* spp., *Listeria monocytogenes*, *Streptococcus* spp., in food and water systems (Daly et al., 2002; Ge et al., 2002; Sails et al., 2002). The sensitivity of PCR was found to be enhanced when applying PCR-ELISA.

Real-Time PCR and biprobe assays have been used to identify and distinguish bacterial species by many researchers (Logan et al., 2001; Vishnubhatla et al., 2001). Another alternative rapid detection method is Real-Time quantitative PCR (RTQ-PCR), which quantifies DNA and thus has the potential for accurate enumeration of microorganisms. This system combines air thermocycler and fluorometer enabling rapid cycle PCR. Hein et al. (2001) used two different approaches to the RTQ-PCR-based quantification of the nuc gene of *S. aureus* in cheese. The first approach was by using the dsDNA-binding fluorescent probe called the TaqMan probe. Quantification studies proved SYBR Green I to be less sensitive (60 nuc gene copies/μL) than using TaqMan probe (6 nuc gene copies/μL). Subsequently, *S. aureus* was also detected from milk and meat samples (10^3 CFU/mL) without preenrichment using primers for heat shock protein (Chiang et al., 2007), whereas in another study, a two-step enrichment method was used for identification of *S. aureus* by multiplex real-time PCR from raw meat, animal nasal swabs, and deli meat (Velasco et al., 2014). Similarly, real-time immune-quantitative

PCR method, which has approximately 1000 times more sensitivity (<10 pg mL^{-1}), was developed for detection of *Staphylococcus aureus* enterotoxin B (SEB) from both pure culture and food samples (Rajkovic et al., 2006). Detection of *Staphylococcus aureus* at a very low concentration in milk samples was made possible by real-time PCR where standard curves were prepared with 50 ng to 50 fg for purified DNA (Goto et al., 2007). Similarly, multiplex RT PCR was employed for detection of *Staphylococcus aureus* (12cfu/25 mL) in post preenrichment steps from milk (Ding et al., 2017).

28.3.2.3 Molecular typing

Intraspecies differentiation of *S. aureus* and other staphylococci is essential during investigation into epidemiological outbreaks. Various methods of typing staphylococcal isolates are available. These include phage typing, ribotyping, comparison of plasmid profiles, restriction fragment length polymorphism (RFLP), randomly amplified polymorphic DNA (RAPD), and pulsed field gel electrophoresis (PFGE). These methods are based on the premise that clonally related organisms share traits that can differentiate them from other unrelated organisms (Fueyo et al., 2001).

PFGE as shown by the number of workers is a convenient discriminative tool not only for strain identification, but also for the resolution of the clonal relationships of *S. aureus* strains in the clinical environment and in foods (Shimizu et al., 2000). Staphylococci responsible for food poisoning outbreaks in China and Japan were studied individually for coagulase type VII and enterotoxin A profiles using PFGE. In both cases, chromosomal DNA was digested with SmaI to characterize the strains of *S. aureus* and considerable genetic diversity was observed, which could be linked to the geographic relatedness and source of the strain (Shimizu et al., 2000). RFLP, in combination with PCR, has been applied for the identification of staphylococcal species (Yugueros et al., 2001). For example, RFLP study on camel meat showed significant positive signal in 37 MRSA strains (Ziad et al., 2013). A similar method of identification conducted on bovine raw milk and recently on camel milk samples, in two separate studies, also exhibited the presence of novel *S. aureus* species (Hakimi et al., 2018; Aljahani et al., 2020). The RFLP studies along with PCR reported nearly 12 *S. aureus* isolates from different food samples (Zare et al., 2019). The use of PCR based RFLP was demonstrated while finding the clonal relationship among the isolates of *S. aureus* from raw poultry produced by different broiler chicken producers in south Brazil (Pinto et al., 2015).

28.3.2.4 Microarrays or biochips

DNA microarrays provide a powerful tool for the parallel analysis of many genes and their expression levels (Cho and Tiedje, 2002). In one study, microarrays were designed for the identification of *Staphylococcus*, its species and methicillin resistance. The microarray contained five selective DNA capture probes for simultaneous and differential identification of five clinically important staphylococcal species (*S. aureus*, *S. epidermidis*, *S. haemolyticus*, *S. hominis*, *and S. saprophyticus*), with a consensus capture probe to detect all femA sequences, allowing the identification of genus *Staphylococcus*. The hybridization and identification processes were completed in less than 2 h. Therefore, the application of low-density microarrays as a powerful multigenotypic post-PCR analysis could compete with conventional bacterial identification (Hamels et al., 2001).

Considering microarray applications to food studies, a survey of fresh chicken and turkey meat found 86 positive samples of MRSA employing DNA microarray analysis (Fessler et al., 2011). Microarray-based genetic profiling revealed a low prevalence of *Staphylococcus aureus* (8%) within abattoir byproducts of pork origin (Morach et al., 2019), while for fresh produce, 24 different probes were developed for the detection of *S. Aureus* pathogens directly from fresh-Cut Cantaloupe and Lettuce (Sarengaowa et al., 2020).

28.3.2.5 Staphylococcal enterotoxin identification using chromatography methods

Liquid chromatography coupled with mass spectrometry (LC-MS/MS) has long been used for the purpose of yielding information on molecular weight of amino acids in peptide chains digested from basic proteins that can be enzymatically degraded. Kientz et al. (1997) first reported the successful detection of SEB in juices, milk, green beans, crackers, chicken meat, and shrimp at concentrations as low as 3 pmol/mL by using LC-MS/MS (Callahan et al., 2006; Sospedra et al., 2012a,b; Bao et al., 2012; Muratovic et al., 2015).

28.3.2.6 Chemiluminescence immunoassays

With a large surface area and electrical conductivity, carbon nanoparticles such as carbon nanotubes (CNTs) are successfully employed for antibody immobilization. Moreover, due to its low absorption in visible range, the CNTs are recently applied for SEB detection in soy milk, baby food, and apple juices (Yang et al., 2008, 2010).

28.3.2.7 Surface plasmon resonance immunoassays

In the case of surface plasmon resonance (SPR), the antibodies are coated on a sensor chip where the reflection intensity changes as the target binds the antibody. The changes in the refractive index at the thin metal layer or film can be detected and therefore the method is suitable for real-time measurement in a label-free manner. This real-time approach was employed for the detection of Staphylococcal enterotoxins to monitor the quantity of toxin (as low as 10–100 ng/g) in hotdog, mushrooms, potato salad, and milk (Rasooly and Rasooly, 1999). Subsequently an SPR sandwich immunosensor suitable for detection of SEB spiked in potted meat at a minimum of 10 ppb concentration (Rasooly, 2001), and SEB was detected down to 2.5 ppb when spiked in ham using sandwich immunoassay method (Medina, 2003). The detection of SEA and SEB in whole egg samples spiked at concentrations between 1 and 40 ng/mL and SEA in fresh milk as low as 1 ng/mL has been subsequently achieved (Medina, 2005, 2006).

28.3.2.8 Aptamer-based bioassays

To achieve higher affinity and selectivity, short ssDNA or RNA molecules have been used for in vitro selection and use as biosensors. These sensors are chemically synthesized and have many advantages over the antibody-based assays due to the small molecular weight and superior stability (Nimjee et al., 2005). The detection of SEB in butter and skim milk was made possible using aptamers as recognition element in piezoresistive cantilever-based approach (Zhao et al., 2014).

28.4 Listeria monocytogenes

The genus *Listeria* is a gram-positive, microaerophilic, nonspore-forming rod. The collective knowledge generated by investigation and research done over the last number of decades in the field of *Listeria* has made this organism the most important investigated gram-positive bacterium. There are six recognized species of *Listeria*: *Listeria monocytogenes*, *Listeria innocua*, *Listeria seeligeri*, *Listeria welshimeri*, *Listeria ivanovii*, and *Listeria grayi*. *Listeria monocytogenes* was not a major concern from the point of food safety until the 1980s; the speed with which *L. monocytogenes* emerged as the etiological agent of "listeriosis" is beyond comparison with any other foodborne pathogen.

28.4.1 Conventional isolation methods

Isolation of *Listeria* species including that of *L. monocytogenes*, from foods and other environments, offers a challenging task for microbiologists. Its detection, isolation, and identification from inoculated and naturally contaminated foods and the recovery of sublethally injured *Listeria* from foods is quite challenging and interesting. Direct plating, cold enrichment, selective enrichment, and several rapid methods have been used singly or in different combinations to detect *L. monocytogenes in food*, as well as clinical and environmental samples. Early attempts to isolate small numbers of Listeria from samples containing large populations of indigenous microflora relied on direct plating, and often ended in failure.

Conventionally, culture methods used for the detection of *Listeria* spp. are based on preenrichment, selective enrichment, and plating followed by colony morphology, sugar fermentation, and hemolytic properties to identify different species of *Listeria* (Janzten et al., 2006). These methods are sensitive and remain considered as "gold standards" compared to other methods even today. However, the time required for confirmation of the positive result is usually 5–7 days from the time of sample analysis (Paoli et al., 2005). Various factors such as the presence of high populations of other competitive bacteria, low numbers of the test pathogen, and inhibitory components of food matrices hamper the detection of *L. monocytogenes* in foods (Janzten et al., 2006).

Against this background, several isolation protocols have been developed over the years for the improved detection of *L. monocytogenes* in foods. Other efficient methods based on antibodies or molecular techniques (DNA hybridization or PCR) are developed which are equally sensitive and rapid so as to give a positive confirmation within 48 h. Real-time PCR is increasingly used in food diagnostics for the detection of *L. monocytogenes*. Microarrays and biosensors are other newer techniques being used to detect *Listeria* spp. in foods.

The cold enrichment developed by Gray involves homogenization of samples in tryptose broth, incubation at 4°C and weekly and biweekly plating on tryptose agar during 3 months of storage. This was adopted as a standard method to detect *L. monocytogenes*, as the organism could be recovered after prolonged incubation at 4°C, even in the presence of other contaminants. Usually, cold enrichment is followed by selective enrichment, wherein the use of selective or inhibitory agents during enrichment at elevated temperatures (30–37°C) selectively inhibits other microflora, while at the same time

TABLE 28.1 Selective agents used in isolation of *Listeria*.

Selective/Antimicrobial agent	Conc. mg/L	Uses	References
Potassium tellurite	5–15	Selective/differential for *Listeria* that reduce tellurite to tellurium producing black colonies	Gray et al. (1950)
Lithium chloride/Phenyl ethanol	0.5 mg/L–15 g	Amplification of *Listeria* in presence of gram-negative bacteria	Hao et al. (1989)
Nalidixic acid	20–40	Inhibitory to gram-negative bacteria by interfering DNA gyrase except Pseudomonas and Proteus	Farber et al. (1988), Ortel (1972)
Acriflavine/trypaflavine (acridine dyes)	5–25	Inhibitory to gram positives including *Lactobacillus bulgaricus* and *Streptococcus thermophilus*	Ralovich et al. (1971), Ortel (1972)
Polymyxin B	$1-10^6$ U	Inhibits growth of gram-negative rods and streptococci	Doyle and Schoeni (1986), Rodriguez et al. (1984)
Moxalactam	20	Broad-spectrum antibiotic, inhibits many gram-positive and gram-negative bacteria including *Staphylococcus, Proteus,* and *Pseudomonas*	Lee and McClain (1986)
Ceftazidime	4–50	Broad-spectrum cephalosporin antibiotic	Lovet et al. (1987), Lovet et al. (1988)
Cycloheximide	50	Inhibits fungi	Curtis et al. (1989)

A few of the selected enrichment broths and isolation media are presented briefly.

allowing the growth of *Listeria*. The selective agents included chemicals, antimicrobials, and dyes which could inhibit the growth of indigenous microorganisms and allow *L. monocytogenes* to grow. The various selective agents used and their functions are listed in Table 28.1.

A few of the selected enrichment broths and isolation media used in the recovery of *L. monocytogenes* are presented in brief. The University of Vermont (UVM) selective broth was originally recommended by the FDA and USDA-FSIS for the enrichment of food samples, and subsequently, certain modifications were included into the medium. The modified medium designated as *Listeria* enrichment broth (LEB) by Donnelly and Baigent (1986) was used to selectively enrich *L. monocytogenes* in raw milk from other contaminants. Fraser broth (Fraser and Sperber, 1988) was a modification of UVM broth, wherein the culture of Listeria turns Fraser broth black due to esculin hydrolysis within 48 h of incubation. Hence, this media has now replaced LEB in the USDA protocol as a secondary enrichment medium for meat, poultry, and environmental samples (USDA/FSIS, 2002). The PALCAM enrichment broth developed by Van Netten et al. (1989) gave better results when compared to USDA-LEB as well as tryptose broth-based antibiotic medium of Beckers et al. (1987) in detecting *L. monocytogenes* from naturally contaminated cheese, meat, fermented sausage, raw chicken, and sausage.

McBride Listeria agar (MLA) was the first plating medium widely used for selective isolation of *L. monocytogenes*. It was first introduced by McBride and Girard (1960) and was prepared from phenyl ethanol agar to which lithium chloride, glycine, and sheep blood were added. Subsequently, the medium has undergone several modifications. The plates with grown colonies were observed under oblique illumination for bluish to bluish-green *Listeria* colonies (Lovett et al., 1987). The lithium chloride-phenyl ethanol-moxalactam (LPM) agar was a modification of MLA medium formulated by Lee and McClain (1986). This LPM medium-plus esculin and ferric iron have been in use as one of the selective agars in the FDA procedure. The Oxford agar developed by Curtis et al. (1989) eliminated oblique illumination. Modified Oxford agar (MOXA) is the modification of Oxford agar by the inclusion of moxalactam and has been a recommended plating medium in the USDA-FSIS procedure. Oxford agar has been one of the selective media in the FDA method (Carnevale and Johnston, 1989).

PALCAM (Polymyxin acriflavine lithium chloride ceftazidime aesculin mannitol) agar is a selective plating medium used after primary or secondary enrichments. PALCAM agar plates are incubated at 30°C for 48 h under microaerophilic conditions (5% oxygen, 7.5% CO_2, 7.5% hydrogen, 80% nitrogen). The colonies of *Listeria* appear as grey-green with black sunken centers. The PALCAM medium, along with L-PALCAMY enrichment broth, forms the basis of a method used for the isolation and detection of *Listeria* by the Netherlands Government Food Inspection Service. Gunasinghe et al. (1994) made a comparative study of two plating media (PALCAM and Oxford) to detect *Listeria* spp. in meat products.

They found that the PALCAM medium was more effective in suppressing other background microflora, wherein isolation and identification of *Listeria* spp. were easier in this medium.

In the background of several enrichment broths and selective isolation agar media, an attempt has been made to shortlist a few of these broths and agar media based on efficacy studies. Those recommended are being routinely used in the food industry worldwide; even though sensitive and reliable, they are time consuming, almost taking 5–6 days for the results to become available. Difficulties have also been observed in these procedures, such as the inability to isolate *Listeria* from all positive samples and recover sublethally injured cells encountered during food processing.

The ISO 11290 method developed by the International Organization for Standardization (1996, 1998) is a two-stage enrichment process using Fraser broth. The presence of *Listeria* is indicated by the blackening of the medium. Samples of primary and secondary enrichment broths are then plated on Oxford and PALCAM agar plates for detection of *L. monocytogenes*. The FDA method with only one enrichment step was originally developed by Lovett (1988) and has been frequently used for the isolation and detection of *L. monocytogenes* in milk and milk products (particularly ice cream and cheese), seafood, and vegetables. Food samples in 25 g quantities are preenriched at 30°C for 4 h in buffered *Listeria* enrichment broth. After 4 h of nonselective enrichment, selective agents (acriflavine, nalidixic acid with or without cycloheximide) are added and further incubated for 48 h for selective enrichment. The enriched sample is then plated on one of the esculin containing selective agars such as OXA or MOXA or PALCAM or LPM. The plates are incubated at 30–35°C for 24–48 h for the development of *Listeria* colonies.

The International Dairy Federation method was developed as a reference method to recover *L. monocytogenes* from dairy products (Terplan, 1988). The present AOAC-approved IDF method resembles the FDA procedure (Association of Official Analytical Chemists, 1996). The enriched samples are streaked on Oxford agar. The presumptive colonies are then confirmed by other confirmatory tests. This method requires a minimum of 4 days to obtain presumptive results and is a popular method in European countries for detecting *Listeria* in dairy products. The USDA-FSIS method developed by the United States Department of Agriculture, Food Safety, and Inspection Service (USDA-FSIS) is used to isolate the organism from meat and poultry products as well as environmental samples (USDA/FSIS, 2002). It involves a two-stage enrichment process. The enriched samples are then streaked onto MOXA plates for presumptive identification of black *Listeria* colonies. The presumptive Listeria cultures are further confirmed by the use of specific biochemical characteristics such as hemolysis, phospholipase C production, sugar fermentation, oblique illumination, and growth in the presence of chromogenic substrates. Also, sensitivity and rapidity have been focal points in the detection and isolation of *L. monocytogenes* from foods. Attempts to achieve these objectives have led to the development of immunological-based and nucleic acid–based methods.

28.4.2 Immunological detection methods

These methods are based on antibodies specific for *Listeria* and that play an important role in the host defense mechanism. Certain unique properties of antibodies include binding to epitopes, present in living cells. Antibodies directed against surface antigens do not require that they penetrate inside the cell in order to reach the target, making them suitable for detecting live cells. ELISA is the most common immunoassay used for pathogen detection in foods. Ky et al. (2004) have developed monoclonal antibodies against the protein p60 encoded by the iap gene for the detection of *L. monocytogenes*. Several commercial kits like Transia plate *Listeria monocytogenes* by Diffchamb AB and VIDAS LMO by bio-Mérieux have been developed based on ELISA to detect *L. monocytogenes* in foods (Hitchins, 2003). Immuno-capture on immunomagnetic separation is another technique, wherein magnetic beads coated with specific antibodies are used to separate target organisms from other competing microflora and inhibitory food components. This approach has been used to capture and concentrate *Listeria* spp. directly from foods and environmental samples (Mitchell et al., 1994; Hudson et al., 2001). The rapid detection of *L. monocytogenes* was further improved by a comparative study of ELISA, immunochromatography (ICG) strip test, and immunomagnetic bead separation (IMBS) system (Shim et al., 2008). This study demonstrated that ICG-IMBS–based methods were rapid, cost effective, and suitable for onsite screening of food sample such as pork, beef, mutton, and chicken within 14–15 h.

28.4.3 Nucleic acid–based methods

The availability of complete genome sequences of *L. monocytogenes* (serotypes 1/2a, 4b, and 6a) and *L. innocua* has led to a greater understanding of molecular features involved in the pathogenesis of *L. monocytogenes* (Buchrieser et al., 2003). In addition, they have also made available new diagnostic targets to detect *L. monocytogenes*. Several nucleic acid–based methods such as DNA hybridization, PCR and nucleic acid–based amplification are available to detect *L. monocytogenes* and other *Listeria* spp. from innumerable samples.

28.4.3.1 Polymerase chain reaction

A PCR provides an exponential amplification of a specific DNA sequence present in the target organism. The components of a buffered PCR reaction include oligonucleotide primers, deoxyribonucleotide triphosphates, the DNA to be amplified, and a thermostable DNA polymerase. Different types of DNA extractions, cell lysis, and DNA purification techniques are applied directly to foods or to enrichment broths with varying sensitivity. PCR methods are highly sensitive, rapid, and reproducible.

Various genes have been targeted for the detection of *L. monocytogenes.* They include gene hlyA coding for listeriolysin O (Mengaud et al., 1988); iap gene coding for invasive associated protein p60 (Kohler et al., 1990); actA gene, a surface protein required for intracellular bacterial propulsion and cell to cell invasion (Kocks et al., 1992); internalin inlA and inlB, surface proteins conferring invasiveness to human erythrocytes (Gaillard et al., 1991); lmaA gene (also known as Dth-18) responsible for delayed-type hypersensitivity response in *Listeria*-immune mice (Gohmaan et al., 1990); flaA gene for flagellin which is specific to all species of *Listeria* (Dons et al., 1992); plcB gene encoding phospholipase C (Geoffroy et al., 1991); prfA gene responsible for the regulation of expression of a cascade of virulence factors including listeriolysin O (Leimeister-Wachter et al., 1990); and fibronectin-binding protein (Gilot and Content, 2002). In addition, genus and species-specific 16S rRNA and 23S rRNA sequences have also been targeted (Wesley et al., 2002; Rodriguez- Lazaro et al., 2004). A brief overview of selected genes targeted, the specific PCR primers used, generated amplicon size, sensitivity, and types of foods samples tested by several researchers have been presented in Table 28.2.

Conventional PCR is the amplification of the specific gene sequence using a single primer pair for the detection of one pathogen at a time. This method is used to detect the pathogen either directly or indirectly from foods (Herman et al., 1995). One of the advantages of enrichment, prior to PCR, would be the elimination of false-positive results, which may arise due to DNA of nonviable bacteria in the food sample. A variety of DNA extraction methods have been employed to prepare the template for PCR amplification, which makes the method more suitable and sensitive to detecting the target organism, even if present in low numbers (Ramesh et al., 2002; Liu, 2008). Nested PCR involves two sets of primers and two rounds of thermal cycling. The major advantage of nested PCR is increased sensitivity of detection by several orders of magnitude than that achieved by primary amplification and enhanced specificity, wherein it is unlikely that any nonspecific amplification at the primary round will give a positive result in the second round (Levin, 2003). In multiplex PCR, two or more gene loci are simultaneously amplified in one reaction. This technique is widely used to characterize pathogens based on their antigenic traits and virulence factors. Although designing a robust multiplex assay for food is quite challenging, once optimized for specific pathogens and food products, this method has the advantage of being cost-effective and highly efficient (Ramesh et al., 2002). Multiplex PCR was employed for identification and differentiation of *Listeria monocytogenes* and *Listeria* species in deli meats (Liu et al., 2015).

TABLE 28.2 Overview of selected genes targeted for PCR detection of *L. monocytogenes.*

Gene targeted	Primer sequence 5′→3′	Size of amplicon (bp)	Sensitivity (detection limit)	Samples tested	References
hlyA	F – TTG CCA GGA ATG ACT AAT CAA G R – ATT CAC TGT AAG CCA TTT CGTC	172	10 cfu/mL	Milk	Amagliani et al. (2004)
iap	F – GGGCTTTATCCATAAAATA R – TTGGAAGAACCTTGATTA	453	10^{-1}–10^{-2} cells/mL or g	Meat, sausages, cheese	Manzano et al. (1997)
actA	F – GCTGATTTAAGAGATAGAGGAACA R – TTTATGTGGTTATTTGCTGTC	827	10^1 cfu/25 g	Pork, milk	Zhou and Jiao (2006)
16S rRNA	F – GCTAATACCGAATGATAAGA R – GGCTAATACCGAATGATGAA		4×10^{-2} to 4×10^{-1} cfu/g	Fresh and ready to eat meat and fish, potato salads, vegetable salads, pasta, ice cream	Somer and Kashi (2003)
prfA	F – GATACAGAAACATCGGTTGGC R – TGACCGCAAATAGAGCCAAG	215	1 cfu/25 g	cooked ham	Jofre et al. (2005)

Real-time PCR technology is based on the ability to detect and quantify PCR products as the reaction cycle progresses. The DNA is quantified by measuring the fluorescence with respect to the binding of an intercalating dye or the binding of a fluorescent hybridization probe. SYBR green I have been the most frequently used DNA-binding dye in Real-Time PCR (Fairchild et al., 2006). Results are obtained in an hour or less, which is faster than conventional PCR. The convenience and rapidity have made Real-Time PCR very pertinent as an alternative to a conventional culture-based or immuno-based assay for the detection of *L. monocytogenes* (Norton, 2002). Various Real-Time PCR assays for the detection of *L. monocytogenes* in food have been documented in the literature (Berrada et al., 2006). Various commercial Real-Time PCR kits like Bax-PCR (Dupont-Qualicon, Delaware, USA), Probelia (Bio-Rad, Hercules, California, USA), and others are available. The reader is directed to two insightful reviews on the isolation and rapid identification methods of *L. monocytogenes* which provide a comprehensive update (Law et al., 2015a,b; Chen et al., 2017).

28.4.4 Other methods

Enterobacterial repetitive intergenic consensus sequence—based PCR assays were used to compare *Listeria* spp. isolated from different sources. This method was used to generate DNA fingerprints, by which *Listeria* spp. were divided into three major clusters and allowed differentiation among the serotypes 1/2a, 4b, 6a, and 6b within each cluster (Laciar et al., 2006). Restriction enzyme analysis and PFGE were employed to detect *Listeria* species in raw whole milk and farm bulk tanks. This method allowed the isolated organisms to be differentiated into 16 clonal types, with the majority of them belonging to serovar 1/2a (Waak et al., 2002). DNA microarrays are composed of many discretely located probes which are composed of a sequence that is complementary to a pathogen-specific gene sequence. This allows the analysis of thousands of gene sequences in a relatively short time. This technology has been used for the detection of *L. monocytogenes* in environmental samples (Call et al., 2003). Microarrays along with PFGE and serotyping have been used to study the genetic diversity of *L. monocytogenes* strains in dairy farms (Borucki et al., 2005). However, the disadvantage of serotyping alone with antisera includes limited discriminatory power, cross-reactivity, and lower accuracy (Jadhav et al., 2012).

DNA microarrays are composed of many discretely located probes on a solid substrate such as glass, wherein each probe is composed of a sequence that is complementary to a pathogen-specific gene sequence. Spectroscopic methods, such as Fourier Transform Infrared (FT-IR) and Raman spectroscopies, provide unique spectral fingerprints of specific cell types, so they can discriminate between bacteria at genus, species, or strain levels. These are whole-cell nondestructive methods; destructive methods, such as matrix-assisted laser desorption/ionization mass spectrometry, can also be used. As early as 1995, FT-IR spectrometry was used to differentiate all of the six species of *Listeria* (Holt et al., 1995).

The major players in the food supply chain benefit enormously from the easy availability of rapid, reliable, and sensitive detection methods, which enable the implementation and monitoring of food safety on a global platform. Even though newer methods emerge and are valuable tools for routinely screening food and environmental samples, they have still not entirely replaced some standard techniques, although cross-validation of methods is ongoing. Rapid methods capable of differentiating living and dead cells and recovering sublethally injured cells are still needed. Nevertheless, nucleic acid—based methods do allow for efficient and reliable results within a short timeframe and demonstrate the detection of *L. monocytogenes* in foods.

28.5 Bacillus cereus

Among the important bacterial pathogens, *Bacillus cereus* is of significance as it is an opportunistic organism and can dominate in any given situation, because of its ubiquitous nature and ability to occur in a wide range of foods (Reyes et al., 2007; Roy et al., 2007). The study of *B. cereus* in relation to foods retains significance in the light of its ability to form heat-resistant endospores and its capacity to grow and produce toxins in a wide variety of foods (Ehling-Schulz et al., 2004; Ouoba et al., 2008). *Bacillus cereus* food poisoning is underreported, as both types of illnesses are relatively mild and usually last for less than 24 h. The combined properties of heat-resistant endospore-forming ability, toxin production, and psychrotrophic nature give ample scope for this organism to be considered a prime cause of public health hazard (Griffiths and Schraft, 2002). Strains of *B. cereus* cause two different types of foodborne illnesses in humans, namely diarrheal and emetic (Schoeni and Wong, 2005). Three types of heat-labile enterotoxins involved in food poisoning have been characterized at the molecular level (Tsen et al., 2000). Molecular diagnostic methods based on the PCR have been described for the detection of potent toxigenic cultures of *B. cereus* (Radhika et al., 2002; Abriouel et al., 2007; Ngamwongsatit et al., 2008).

28.5.1 Detection methods

Several selective and differential enumeration media have been formulated to achieve the maximum recovery of *B. cereus* from foods (Varadaraj, 1993). One of the earliest methods developed in 1955 involved surface plating on blood agar, followed by incubation for 18 h at 37°C and the plates were observed for colonies surrounded by a halo (lecithinase activity). Later, a peptone beef extracts egg yolk agar containing lithium chloride and polymyxin B as selective agents was formulated. Typical B. *cereus* colonies were surrounded by an opaque zone after 18 h at 30°C. A further improvement in this medium was through the addition of mannitol, which enabled non-*B. cereus* colonies to be differentiated from mannitol negative *B. cereus* colonies (Mossel, 1986). Kim and Goepfert (1971) formulated egg yolk containing medium (KG medium), which could result in sporulation within 24 h of incubation. This medium gave a good recovery of *B. cereus* from foods with inherent background microflora. Polymyxin was used for selectivity, egg yolk to record lecithinase activity, and low levels of peptone enabled sporulation.

As a means of achieving a better recovery of the organisms from food samples, Holbrook and Anderson (1980) developed a medium known as polymyxin pyruvate egg yolk mannitol bromothymol blue agar (PEMBA) based on the characteristics of mannitol utilization (peacock blue colored colonies) and lecithinase activity (precipitation of egg yolk) of *B. cereus* cultures. The presence of pyruvate in the medium restricts the characteristic spreading nature of spore formers. This medium supports the growth of even small numbers of *B. cereus* cells and spores. For this reason, the medium found wider application in research investigations relating to the enumeration of *B. cereus* from foods. The use of PEMBA enabled easy differentiation of *B. cereus* from other *Bacillus* species present in Indian snack and lunch foods (Varadaraj et al., 1992). A later modification of PEMBA was the replacement of bromothymol blue with bromocresol blue in the medium (Szabo et al., 1984).

Although identified *B. cereus* isolates exhibited a diversified pattern with respect to virulence/toxigenic traits being harbored by them, it is important to observe that phospholipase activity was more predominant among the isolates. Phospholipase is known to cause degradation of cell and mucous membranes, which are rich in phospholipids, leading to necrosis and considered virulence factors. In view of the importance of this character in *B. cereus*, our earlier studies resulted in obtaining a patent for the phospholipase (Pl) oligonucleotide primer designed to detect *B. cereus* among the cluster of other related species of *B. cereus* (Padmapriya et al., 2004). Another virulence factor associated with *B. cereus* isolates was that of sphingomyelinase activity, which was recorded in 42% isolates by PCR. An earlier study concluded that phospholipase C and sphingomyelinase encoded by two tandemly arranged genes with close genetic linkage constitute a biologically functional cytolytic determinant and act in natural concert, causing lysis of target cells and are thus regarded as an effective cytolysin (Beecher and Wong, 2000).

Although diarrheal food poisoning arising from *B. cereus* is due to three enterotoxins—hemolysin BL (HBL), nonhemolytic enterotoxin (NHE), and cytotoxin K (CytK)—the major health hazard attributed to *B. cereus* is mostly related to the HBL enterotoxin complex (Hsieh et al., 1999). The enterotoxigenic among isolates of *B. cereus* has been attributed to the presence of all three components: a binding component B and two lytic components L1 and L2 in the HBL complex (Schoeni and Wong, 2005). Studies have shown that the potential virulence factors (hemolysis, sphingomyelinase, and phospholipase C) are under the control of a pleiotropic regulator PlcR and are expressed simultaneously. A complex interaction among these virulence factors exists, which may be cooperative, synergistic, and antagonistic (Slamti et al., 2004). Another study has designed eight sets of novel PCR primers to detect the wide distribution pattern of enterotoxin genes among *B. cereus* clusters (Ngamwongsatit et al., 2008).

28.5.1.1 Real-time PCR

Real-time PCR was successfully employed for the detection of *Bacillus* species from food samples such as lettuce, vegetable salad, apple juice, liquid whole egg, cow milk, hamburger, boiled rice, and boiled spaghetti with concentration as high as 10^4–10^5 cfu/g (Ueda et al., 2013). Similarly, strains of *B. cereus* were also identified from UHT-treated milk samples with qRT-PCR using propidium monoazide treatment (Cattani et al., 2016). The detection limit for *B. cereus* was tested using artificially contaminated liquid egg and reconstituted infant formula and the results confirmed the presence of *B. cereus* at a concentration of about 3cfu per reaction or 60 cfu/mL of food (Martínez-Blanch et al., 2009). Artificial inoculation in a rice and pasta sample was applied in another study, where the emetic *B. cereus* was identified as the primary cause for an outbreak in Germany (Fricker et al., 2007). Recently, multiplex real-time PCR showed a positive signal for detection of *B. cereus* from milk samples compared to culture-based methods (Wei et al., 2019). The detection of spores was achieved by measuring the expression of pc-plc gene encoding for phosphatidylcholine-specific phospholipase in *B. cereus* (Martinez-Blanch et al., 2010), and this method was used for detection of spores in artificially contaminated or reconstituted infant formula at an efficiency of detection at 4 spores per reaction or 60 spores per mL.

28.5.1.2 Aptamer-based spore trapping

Due to several drawbacks in detection of spores from milk by PCR, alternative methods such as aptamer-based trapping method have been optimized for quality control purposes. The PCR method can be time consuming and tedious due to the presence of ions and fat molecules in milk samples for qRT-PCR analysis (Vidic et al., 2020). Therefore, magnetic capture of spores had gained interest as an excellent strategy for concentrating bacteria from complex food samples or even highly diluted solutions (Fisher et al., 2015; Kotsiri et al., 2019). The magnetic bead-based aptamers were used for spore trapping from milk in higher affinity (sixfold increase in enrichment) suitable for further PCR-based identification at detection limit of 10^3 CFU/mL for *B. cereus* in milk.

28.5.1.3 Loop-mediated isothermal amplification

An innovative approach for detection of *B. cereus* spores was developed with rotate and react SlipChip (RnR-SlipChip), where the visual identification time is nearly 60 min with a success rate of 100% (Xia et al., 2016). This system was based on the principle of loop-mediated isothermal nucleic acid amplification or loop-mediated isothermal amplification (Vidic et al., 2019).

28.5.1.4 DNA microarray

Milk samples spiked with *Bacillus anthracis* were used for identification of strain specifically by microarray focusing on any bioterrorism event (Goji et al., 2012). Similarly in another study, the absolute sensitivity of *Bacillus anthracis* was determined, and the value was close to as low as 10 fg per gram DNA (Wilson et al., 2002). Similarly, Wang et al. (2007) developed a DNA microarray assay for 22 different strains including *Bacillus* species from food samples, and from infant formula (Wang et al., 2009).

28.5.1.5 Biosensors

In the last 2 decades, the detection of foodborne pathogens advanced with the use of biosensors. Pal et al. (2008) identified *Bacillus* strains at a limit of 35−88 CFU/mL from vegetables including alfalfa sprouts, strawberries, lettuce, tomatoes, fried rice, and cooked corn using electrochemical biosensors. A combined PCR-ELISA (Bacillus Diarrheal Enterotoxin Visual Immunoassay) was developed for detection of toxin directly from foods such as milk, infant formula, chocolate dietary supplements, canned beef, vegetable soup, parboiled rice, and dehydrated food products like beef gravy mix, whey powder, pancake mix, and mashed potatoes (Tallent et al., 2017).

ELISA-based identification gains interest for detection of pathogens contaminating food samples such as *Bacillus* strains in egg, meat, and dairy products. Laboratory grown strains of *Bacillus* species spiked with various other strains procured were artificially inoculated in food samples such as comminuted meat and a sensitive double antibody sandwich ELISA was performed. When a concentration of $1 \times 10^4 - 2.8 \times 10^6$ cells/mL was inoculated in foods, this was detected at a limit (LoD) of 0.9×10^3 cells/mL (Zhu et al., 2016). ELISA was often used coupled with PCR earlier for the detection of pathogens in food samples such as milk and dairy products (Gutiérrez et al., 1997). However, with the advent of monoclonal antibody-based assays, the rapid detection of pathogens can be solely done by ELISA in laboratory scale and food industries worldwide.

28.5.1.6 Raman spectroscopy

Optical methods such as spectrophotometry, Raman/surface-enhanced Raman scattering (SERS) and infrared spectroscopies, fluorescence, and photoluminescence had been an important tool for identification of spore-forming species from milk and dairy products. The spore-specific dipicolinic acid of *B. cereus* was targeted as a biomarker for its identification and quantification. This molecule that is present as a Ca^{2+} chelate complex can be targeted as it is responsible for 5%−15% of dry weight (Janssen and Anderson, 1958). Various new techniques were designed to glutathione-capped gold nanoparticles complexed with Ca2+−ions (Ca2+-AuNP@GSH) through GSH-Ca2+ chelation as a sensing agent for the complex matrices (Baig et al., 2018). In another method, terbium chloride was used to detect calcium dipicolinate where the terbium−(III)−dipicolinate anion produces photoluminescence for identification (Rosen et al., 2006).

In the presence of dipicolinic acid, Ca2+-AuNP@GSH is dissociated to the Ca2+-dipicolinic complex and AuNP@GSH, which desegregates nanoparticles and induces a color change of the solution. The feasibility of using this sensing method for quantitative detection of dipicolinic acid was demonstrated in lysate samples prepared from the *B. cereus* spore suspension. Due to an intrinsic Raman signature, spores of *B. cereus* can be identified as single particle

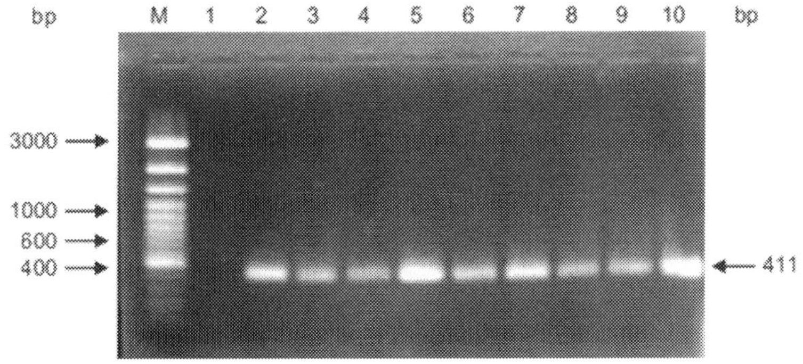

FIGURE 28.1 Agarose gel electrophoretic pattern of PCR products showing amplicons with entB primers in native food isolates of S. aureus and S. epidermidis. Lane M, 100 bp marker; Lanes 1 – 10, native food isolates of S. aureus and S. Epidermidis.

(Chan et al., 2004). This was further developed using SERS-active gold nanoparticles tailored for discrimination of spores of *Bacillus* species (He et al., 2008). Interestingly, a portable device was designed in 2013, where a SERS driven system had the potential to differentiate spores of *B. cereus* and *B. subtilis* (Cowcher et al., 2013). The method had an accuracy near to qRT-PCR and approximate LoD of 1100 spores in 200 μL sample volume. The dipicolinic acid detection was possible in milk samples by following studies such as one by Han et al. (2011), where the solution absorbance of spores was directly correlated with endospore containing pure culture diluted into solution and the obtained LoD was nearly 1.46 × 103 CFU/mL.

28.6 Capacity building in India

At the Central Food Technological Research Institute, Mysore, India, an initiative toward developing PCR-based detection methods for foodborne pathogens in the late 1990s has enabled the development of sensitive methods for detecting *S. aureus*, both in culture broth (Fig. 28.1) and food systems. Further, multiplex PCR has been optimized for the simultaneous detection of *S. aureus* and *Y. enterocolitica* (Fig. 28.2) in milk through the optimization of DNA extraction protocols from milk and PCR conditions (Ramesh et al., 2002; Padmapriya et al., 2003). Through the use of PCR and colony hybridization protocols, targeted isolates of *B. cereus* from food samples (Figs. 29.3 and 29.4) were detected (Radhika et al., 2002). Considering the expertise available at CFTRI, Mysore, several collaborative programs, both at national and international levels, have been organized over the past 2 decades. CFTRI has been associated with international organizations of the United Nations and other related bodies in conducting training programs, workshops, symposia, one-to-one interactions, technical support, and advice in setting up of infrastructural facilities for analysis and

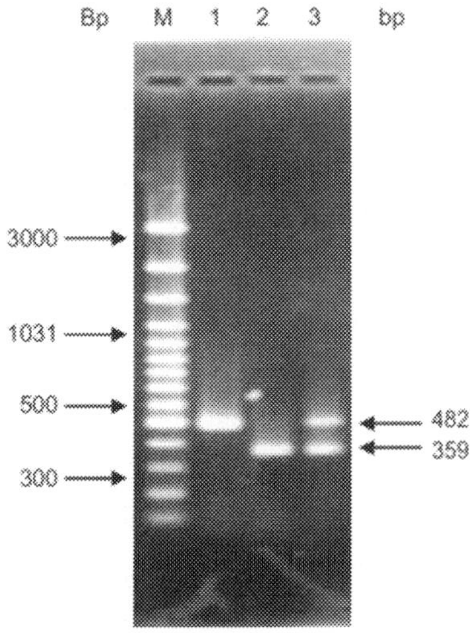

FIGURE 28.2 Agarose gel electrophoretic pattern of multiplex PCR products showing amplicons with nuc primers of S. aureus and ail primers of Y. enterocolitica. Lane M, 100 bp marker; Lanes 1, S. aureus ; 2, Y. enterocolitica ; 3, S. aureus and Y. Enterocolitica.

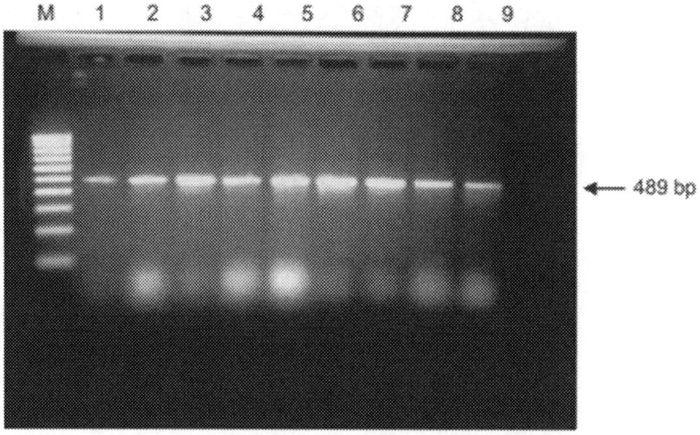

FIGURE 28.3 Agarose gel electrophoretic pattern of PCR products showing amplicons with Ha-1 primers in cultures of B. cereus. Lane M, 100 bp marker; Lanes 1 − 9, potent toxigenic native food isolates of B. Cereus.

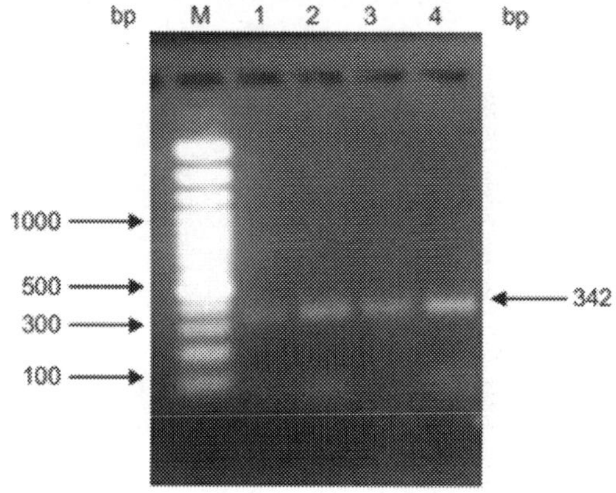

FIGURE 28.4 Agarose gel electrophoretic pattern of PCR products showing amplicons with PI primers of B. cereus in food samples. Lane M, 100 bp marker; Lanes 1, mustard powder; 2, turmeric powder; 3, chilli powder; 4, positive control of B. Cereus.

testing, and many other focal issues related to global food safety. The effort for capacity building for food safety and pathogen detection in India has been initiated in many other laboratories. Specifically, three organisms including *S. aureus*, *B. cereus*, and *L. monocytogenes* were targeted for their critical role in the spread of foodborne infection and poisoning.

Recently, aptamer-linked sandwich ELISA was developed for in house identification of enterotoxigenic *Staphylococcus aureus* from Indian food samples such as bakery products (Sundararaj et al., 2019). Another study demonstrated the isolation of *Staphylococcus aureus* from a total of 120 retail pork samples where the overall prevalence of the strain was 76.67% (Savariraj et al., 2019). The isolation of *S. aureus* from 450 bovine milk samples with and without mastitis was carried out in a study where 9.6% of strains were methicillin resistant in nature (Mahantia et al., 2020). The virulent strains of *S. aureus* were also isolated from 430 street food samples such as animal products such as egg, meat, and milk samples sold by vendors in India (Sivakumar et al., 2019). The same strain was also isolated from different raw meat samples such as pork, beef, and chicken where femA gene was screened (Sharma et al., 2018). The pathogenic strain *S. aureus* was subsequently isolated from buffalo meat with the simultaneous detection of femA and mecA gene by duplex PCR (Ruban et al., 2018).

In India, food pathogens such as *Listeria monocytogenes* were also detected in many studies as it is of primary interest. Das et al. (2013) reported multiplex PCR-based identification of *Listeria monocytogenes* from seafood in Kerala, India. Similarly, *L. monocytogenes* was isolated from a various food source such as chicken, pork, milk, egg, beef, and vegetables (Dhanashreea et al., 2003). The techniques employed are mostly culture-based methods rather than direct rapid detection. *L. monocytogenes* were isolated from slaughterhouses and food samples such as meat and milk, and then identified by PCR-based methods (Khan et al., 2014).

Numerous studies also report isolation and identification of *Bacillus cereus* from restaurants and food outlets while testing food samples such as milk, meat, water, and beverages in Himachal Pradesh, India (Rana et al., 2020). Previous research reported the isolation of *B. cereus* from food samples such as mutton tikka and chutney (32.5%) and raw milk (Altah et al., 2012). The strain was also reported from poultry-based street foods sold at Hyderabad (Sudershan et al., 2012). Up to 30.9% of the overall incidence of *B. cereus* was observed in raw and cooked meat products (Tewari et al., 2015). Interesting work from Machaiaha and Krishnan (2014) showed hemolytic enterotoxin detection by an immunological method of ELISA where the limit was as low as 1−250 pg.

The diversity of methods developed to detect and identify microorganisms and undesirable microbial metabolites in food products provides technical advantages in the maintenance of food safety and environmental health. However, the harmonization of such methods would make it easier to build the analytical capacity required for food safety surveillance. It would increase the recognition of data and reduce costs. At the same time, it would reduce the restriction of imports due to the lack of availability of data produced by internationally recognized methods.

References

Abriouel, H., Ben Omar, N., Lucas López, R., et al., 2007. Differentiation and characterization by molecular techniques of *Bacillus cereus* group isolates from poto poto and dégué, two traditional cereal-based fermented foods of Burkina Faso and Republic of Congo. J. Food Protect. 70, 1165−1173.

Aljahani, A.H., Alarjani, K.M., Hassan, Z.K., Elkhadragy, M.F., Ismail, E.A., Al-Masoud, A.H., Yehia, H.M., 2020. Molecular detection of methicillin heat-resistant *Staphylococcus aureus* strains in pasteurized camel milk in Saudi Arabia. Biosci. Rep. 40 (4). BSR20193470.

Altaf, M.S., Iqbal, A., Ahmad, M., Hussain, S.A., Ahmad, R., Willayat, M.M., 2012. Study of enterotoxigenicity of B. cereus emetic strain by skin vasopermeability reaction in rabbits and poultry. Int. J. Pharma Bio Sci. 3 (2), 166−172.

Amagliani, G., Brandi, G., Omiccioli, E., et al., 2004. Direct detection of *Listeria monocytogenes* from milk by magnetic based DNA isolation and PCR. Food Microbiol. 21, 597−603.

Association of Official Analytical Chemists, 1996. 17.10.01 AOAC official method 993.12. *Listeria monocytogenes* in milk and dairy products. In: Official Methods of Analysis of the Association of Official Analytical Chemists. AOACInternational, Gaithersburg, MD.

Baig, M.M.F., Chen, Y.-C., 2018. Gold nanoparticle-based colorimetric sensing of dipicolinic acid from complex samples. Anal. Bioanal. Chem. 410, 1805−1815.

Baird-Parker, A.C., 1965. The classification of staphylococci and micrococci from world-wide sources. J. Gen. Microbiol. 38, 363−387.

Baird-Parker, T.C., 2000. Staphylococcus aureus. In: Lund, M.B., Baird-Parker, T.C., Gould, G.W. (Eds.), The Microbiological Safety and Quality of Food:, vol. 2. Aspen Publishers Inc, Maryland, pp. 1317−1335.

Bao, K.D., Letellier, A., Beaudry, F., 2012. Analysis of Staphylococcus enterotoxin B using differential isotopic tags and liquid chromatography quadrupole ion trap mass spectrometry. Biomed. Chromatogr. 26, 1049−1057.

Beckers, H.J., Soentoro, P.S.S., Asc, E.H. M.D. -v., 1987. The occurrence of *Listeria monocytogenes* in soft cheeses and raw milk and its resistance to heat. Int. J. Food Microbiol. 4, 249−256.

Beecher, D.J., Wong, A.C.L., 2000. Cooperative, synergistic and antagonistic haemolytic interactions between haemolysin BL, phosphatidylcholine phospholipase C and sphingomyelinase from *Bacillus cereus*. Microbiology 146, 3033−3039.

Bergdoll, M.S., Reiser, R., Spintz, J., 1976. Staphylococcal enterotoxins—detection in food. Food Technol. 5, 80−84.

Berrada, H., Soriano, J.M., Pico, Y., Manes, J., 2006. Quantification of *Listeria monocytogenes* in salads by Real-Time quantitative PCR. Int. J. Food Microbiol. 107, 202−206.

Borucki, M.K., Gay, C.C., Reynolds, J., et al., 2005. Genetic diversity of *Listeria monocytogenes* strains from a high- prevalence dairy farm. Appl. Environ. Microbiol. 71, 5893−5899.

Buchrieser, C., Rusniok, C., Kunst, F., et al., 2003. The *Listeria* Consortium, Comparison of the genome sequences of *Listeria monocytogenes* and *Listeria innocua*: clues for evolution and pathogenicity. FEMS Immunol. Med. Microbiol. 35, 207−213.

Call, D.R., Borucki, M.K., Loge, F.J., 2003. Detection of bacterial pathogens in environmental samples using DNA microarrays. J. Microbiol. Methods 53, 235−243.

Callahan, J.H., Shefcheck, K.J., Williams, T.L., Musser, S.M., 2006. Detection, confirmation, and quantification of Staphylococcal enterotoxin B in food matrixes using liquid chromatography-mass spectrometry. Anal. Chem. 78, 1789−1800.

Carnevale, R.A., Johnston, R.W., 1989. Method for the Isolation and Identification of *Listeria Monocytogenes* from Meat and Poultry products.US Department of Agriculture Food Safety and Inspection Service. Laboratory Communication, Washington, DC. No.57.

Casman, E.P., Bennett, R.W., 1965. Detection of staphylococcal enterotoxin in Food. Appl. Microbiol. 13, 181−189.

Cattani, F., Barth Jr., V.C., Nasário, J.S.R., Ferreira, C.A.S., Oliveira, S.D., 2016. Detection and quantification of viable Bacillus cereus group species in milk by propidium monoazide quantitative real-time PCR. J. Dairy Sci. 99 (4), 2617−2624.

Chan, J.W., Esposito, A., Talley, C., Hollars, C., Lane, S., Huser, T., 2004. Reagentless identification of single bacterial spores in aqueous solution by confocal laser tweezers Raman spectroscopy. Anal. Chem. 76, 599−603.

Chen, J.Q., Regan, P., Laksanalamai, P., Healey, S., Hu, Z., 2017. Prevalence and methodologies for detection, characterization and subtyping of *Listeria monocytogenes* and *L. ivanovii* in foods and environmental sources. Food Sci. Hum. Wellness 6 (3), 97−120.

Chiang, Y.C., Fan, C.M., Liao, W.W., Lin, C.K., Tsen, H.Y., 2007. Real-time PCR detection of *Staphylococcus aureus* in milk and meat using new primers designed from the heat shock protein gene htrA sequence. J. Food Protect. 70 (12), 2855–2859.

Cho, J.-C., Tiedje, J.M., 2002. Quantitative detection of microbial genes by using DNA microarrays. Appl. Environ. Microbiol. 68, 1425–1430.

Cowcher, D.P., Xu, Y., Goodacre, R.P., 2013. Quantitative detection of Bacillus bacterial spores using surface-enhanced Raman scattering. Anal. Chem. 85, 3297–3302.

Curtis, G.D.W., Mitchell, R.G., King, A.F., Griffen, E.J., 1989. A selective differential medium for the isolation of *Listeria monocytogenes*. Lett. Appl. Microbiol. 8, 95–98.

Daly, P., Collier, T., Doyle, S., 2002. PCR-ELISA detection of *Escherichia coli* in milk. Lett. Appl. Microbiol. 34, 222–226.

Das, S., Lalitha, K.V., Thampuran, N., Surendran, P.K., 2013. Isolation and characterization of Listeria monocytogenes from tropical seafood of Kerala, India. Ann Microbiol 63, 1093–1098.

Dhanashreea, B., Ottab, Karunasagarb, I., Goebelc, W., Karunasagarb, I., 2003. Incidence of Listeria spp. in clinical and food samples in Mangalore, India. Food Microbiol. 20, 447–453.

Ding, T., Suo, Y., Zhang, Z., Liu, D., Ye, X., Chen, S., Zhao, Y., 2017. A multiplex RT-PCR assay for *S. aureus*, *L. Monocytogenes*, and Salmonella spp. detection in raw milk with pre-enrichment. Front. Microbiol. 31 (8), 989.

Donnelly, C.W., Baigent, G.J., 1986. Method for flow cytometric detection of *Listeria* monocytogenes in milk. Appl. Environ. Microbiol. 52, 689–695.

Dons, I., Rasmussen, O.F., Olsen, J.E., 1992. Cloning and characterization of a gene encoding flagellin of *Listeria monocytogenes*. Mol. Microbiol. 6, 2919–2929.

Doores, S., 1999. Food Safety: Current Status and Future Needs. Critical Issues Colloquia Reports, American Academy of Microbiology. http://academy.asm.org/index.php?option=com_content&task=view&id=44&I temid=55.

Doyle, M.P., Schoeni, J.L., 1986. Selective-enrichment procedure for isolation of *Listeria monocytogenes* in soft, surface-ripened cheese. J. Food Protect. 50, 4–6.

Ehling-Schulz, M., Fricker, M., Scherer, S., 2004. *Bacillus cereus*, the causative agent of an emetic type of food- borne illness. Mol. Nutr. Food Res. 48, 479–487.

Fairchild, A., Lee, M.D., Maurer, J.J., 2006. PCR basis. In: Maurer, J. (Ed.), PCR Methods in Foods. Springer-Science Business Media Inc, USA, pp. 1–25.

Farber, J.M., Sanders, G.W., Speirs, J.I., 1988. Methodology for isolation of *Listeria* from foods—a Canadian perspective. J. Assoc. Off. Anal. Chem. 71, 675–678.

Feng, P., 2001. Development and impact of rapid methods for detection of foodborne pathogens. In: Doyle, M.P., Beuchat, L.R., Montaville, T.J. (Eds.), Food Microbiology: Fundamentals and Frontiers, second ed. ASM Press, Washington DC, pp. 775–793.

Fessler, A.T., Kadlec, K., Hassel, M., Hauschild, T., Eidam, C., Ehricht, R., Monecke, S., Schwarz, S., 2011. Characterization of methicillin-resistant *Staphylococcus aureus* isolates from food and food products of poultry origin in Germany. Appl. Environ. Microbiol. 77 (20), 7151–7157.

Fraser, J.A., Sperber, W.H., 1988. Rapid detection of *Listeria* spp. in food and environmental samples by esculin hydrolysis. J. Food Protect. 51, 762–765.

Food and Agricultural Organization of the United Nations and World Health Organization, December 1992. Joint World Declaration and Plan of Action for Nutrition. Rome, Italy.

Freed, R.C., Evenson, M.L., Reiser, R.F., Bergdoll, M.S., 1982. Enzyme-linked immunosorbent assay for detection of *staphylococcal enterotoxins* in foods. Appl. Environ. Microbiol. 44, 1349–1355.

Fricker, M., Messelhäusser, U., Busch, U., Scherer, S., Ehling-Schulz, M., 2007. Diagnostic real-time PCR assays for the detection of emetic *Bacillus cereus* strains in foods and recent food-borne outbreaks. Appl. Environ. Microbiol. 73 (6), 1892–1898.

Fueyo, J.M., Martin, M.C., Gonzalez-Hevia, M.A., Mendoza, M.C., 2001. Enterotoxin production and DNA fingerprinting in *Staphylococcus aureus* isolated from human and food samples. Relations between genetic types and enterotoxins. Int. J. Food Microbiol. 67, 139–145.

Gaillard, J.L., Berche, P., Frehel, C., et al., 1991. Entry of *Listeria monocytogenes* into cells is mediated by internalin, a repeat protein reminiscent of surface antigens from Gram-positive cocci. Cell 65, 1127–1141.

Gasper, E., Heimsch, R.C., Anderson, A.W., 1973. Quantitative detection of type A *staphylococcal entero- toxin* by Laurell electro immuno diffusion. Appl. Microbiol. 25, 421–426.

Ge, B., Zhao, S., Hall, R., Meng, J., 2002. APCR-ELISA for detecting Shiga toxin-producing *Escherichia Coli*. Microb Infect 4, 285–290.

Genigeorgis, C.A., 1989. Present state of knowledge of staphylococcal intoxication. Int. J. Food Microbiol. 9, 327–360.

Geoffroy, C., Raveneau, J., Beretti, J.L., et al., 1991. Purification and characterization of an extracellular 29 kDa phospholipase C from *Listeria monocytogenes*. Infect. Immun. 59, 2382–2388.

Gilligan, K., Shipley, M., Stiles, B., et al., 2000. Identification of *Staphylococcus aureus* enterotoxins A and B genes by PCR-ELISA. Mol. Cell. Probes 14, 71–78.

Gilot, P., Content, J., 2002. Specific identification of *Listeria welshimeri* and *Listeria monocytogenes* by PCR assays targeting a gene encoding a fibronectin-binding protein. J. Clin. Microbiol. 40, 698–703.

Gohmann, S., Leimester-Wachter, M., Schlitz, E., et al., 1990. Characterization of a *Listeria monocytogenes*- specific protein capable of inducing delayed hypersensitivity in *Listeria*-immune mice. Mol. Microbiol. 4, 1091–1099.

Goji, N., MacMillan, T., Amoako, K.K., 2012. A new generation microarray for the simultaneous detection and identification of *Yersinia pestis* and Bacillus anthracis in food. J. Pathog. (8), 627036.

Goto, M., Takahashi, H., Segawa, Y., Hayashidani, H., Takatori, K., Hara-Kudo, Y., 2007. Real-time PCR method for quantification of *Staphylococcus aureus* in milk. J. Food Protect. 70 (1), 90−96.

Gray, M.L., Stafseth, H.J., Thorp Jr., F., 1950. The use of potassium tellurite, sodium azide and acetic acid in a selective medium for the isolation of *Listeria monocytogenes*. J. Bacteriol. 59, 443−444.

Griffiths, M.W., Schraft, H., 2002. *Bacillus cereus* food poisoning. In: Cliver, D.O., Riemann, H.P. (Eds.), Food- Borne Diseases. Academic Press, London, pp. 261−270.

Gunasinghe, C.P.G.L., Henderson, C., Rutter, M.A., 1994. Comparative study of two plating media (PALCAM and Oxford) for detection of *Listeria* species in a range of meat products following a variety of enrichment procedures. Lett. Appl. Microbiol. 18, 156−158.

Gutiérrez, R., García, T., González, I., Sanz, B., Hernández, P.E., Martín, R., 1997. A quantitative PCR-ELISA for the rapid enumeration of bacteria in refrigerated raw milk. J. Appl. Microbiol. 83 (4), 518−523.

Hakimi, A.R., Mohammadzadeh, A., Mahmoodi, P., 2018. Molecular typing of Staphylococcus aureus of different origins based on the polymorphism of the spa gene: characterization of a novel spa type. 3. Biotech 8 (1), 58.

Hamels, S., Galo, J.-L., Dufour, S., et al., 2001. Consensus PCR and Microarray for diagnosis of the genus *Staphylococcus*, species and methicillin resistance. Biotechniques 31, 1364−1372.

Han, X., Zhang, L.W., Zhen, F., Yi, H.X., Du, M., Zhang, L.L., Li, Y.H., Wang, W.J., 2011. Dipicolinic acid contents used for estimating the number of spores in raw milk. Adv. Mater. Res. 183−185, 1467−1471.

Hao, D.Y.Y., Beuchat, L.R., Brackett, R.E., 1989. Comparison of media and methods for detecting and enumerating *Listeria monocytogenes* in refrigerated cabbage. Appl. Environ. Microbiol. 53, 955−957.

He, L., Liu, Y., Lin, M., Mustapha, A., Wang, Y., 2008. Detecting single Bacillus spores by surface enhanced Raman spectroscopy. Sens. Instrum. Food Qual. Saf. 2, 247.

Hein, I., Lehner, A., Rieck, P., et al., 2001. Comparison of different approaches to quantify *Staphylococcus aureus* cells by Real-Time quantitative PCR and application of this technique for examination of cheese. Appl. Environ. Microbiol. 67, 3122−3126.

Herman, L.M.F., DeBlock, J.H.G.E., Moermans, R.J.B., 1995. Direct detection of *Listeria monocytogenes* in 25 milliliters of raw milk by a two-step PCR with nested primers. Appl. Environ. Microbiol. 61, 817−819.

Hill, W.E., Keasler, S.P., 1991. Identification of food- borne pathogens by nucleic acid hybridization. Int. J. Food Microbiol. 12, 67−76.

Hitchins, A.D., 2003. Listeria monocytogenes. In: Bacterio- Logical Analytical Manual, eighth ed. AOAC International, pp. 10.01−10.13.

Holbrook, R., Anderson, J.M., 1980. An improved selective and diagnostic medium for the isolation and enumeration of *Bacillus cereus* in foods. Can. J. Microbiol. 26, 753−759.

Holt, C., Hirst, D., Sutherland, A., MacDonald, F., 1995. Discrimination of species in the genus *Listeria* by Fourier transform infrared spectroscopy and canonical variate analysis. Appl. Environ. Microbiol. 61, 377−378.

Hsieh, Y.M., Sheu, S.J., Chen, Y.L., Tsen, H.Y., 1999. Enterotoxigenic profiles and polymerase chain reaction detection of *Bacillus cereus* group cells and *B. cereus* strains from foods and foodborne outbreaks. J. Appl. Microbiol. 87, 481−490.

Hudson, J.A., Lake, R.J., Savill, M.G., 2001. Rapid detection of *Listeria monocytogenes* in ham samples using immunomagnetic separation followed by polymerase chain reaction. J. Appl. Microbiol. 90, 614−621.

International Organization for Standardization, D., 1996, 1998. Website: https://www.iso.org/obp/ui/#iso:std:60313:en.

Jadhav, S., Bhave, M., Palombo, E.A., 2012. Methods used for the detection and subtyping of Listeria monocytogenes. J. Microbiol. Methods 88 (3), 327−341.

Janssen, F.L.A.J., Anderson, L.E., 1958. Colorimetric assay for dipicolinic acid in bacterial spores. Science 127, 26−27.

Janzten, M.M., Navas, J., Corujo, A., et al., 2006. Review: specific detection of *Listeria* monocytogenes in foods using commercial methods: from chromogenic media to real-time PCR. Spanish J. Agric. Res. 4, 235−247.

Jofre, A., Martin, B., Garriga, M., et al., 2005. Simultaneous detection of *Listeria monocytogenes* and *Salmonella* by multiplex PCR in cooked ham. Food Microbiol. 22, 109−115.

Jones, C.L., Khan, S.A., 1986. Nucleotide sequence of the enterotoxin B gene from *Staphylococcus aureus*. J. Bacteriol. 166, 29−33.

Khan, J.A., Rathore, R.S., Khan, S., Ahmad, I., 2014. In vitro detection of pathogenic Listeria monocytogenes from food sources by conventional, molecular and cell culture method. Braz. J. Microbiol. 44 (3), 751−758.

Khan, M.A., Kim, C.H., Kaoma, I., et al., 1998. Detection of *Staphylococcus aureus* in milk by use of polymerase chain reaction analysis. Am. J. Vet. Res. 59, 807−813.

Kientz, C.E., Hulst, A.G., Wils, E.R., 1997. Determination of Staphylococcal enterotoxin B by on-line (micro) liquid chromatography-electrospray mass spectrometry. J. Chromatogr. A 757, 51−64.

Kim, H.U., Goepfert, J.M., 1971. Enumeration and identification of *Bacillus cereus* in foods. I, 24-hour presumptive test medium. Appl. Microbiol. 22, 581−587.

Kim, C.−H., Khan, M., Morin, D.E., et al., 2001. Optimization of the PCR for detection of *Staphylococcus aureus nuc* gene in Bovine milk. J. Dairy Sci. 84, 74−83.

Kocks, C., Gouin, E., Tabouret, M., et al., 1992. *Listeria monocytogenes* induced actin assembly requires the *actA* gene product, a surface protein. Cell 68, 521−531.

Kohler, S., Leimeister-Wachter, M., Chakraborty, T., et al., 1990. The gene coding for protein p60 of *Listeria monocytogenes* and its use as a specific probe for *Listeria monocytogenes*. Infect. Immun. 58, 1943−1950.

Kotsiri, Z., Vantarakis, A., Rizzotto, F., Kavanaugh, D., Ramarao, N., Vidic, J., 2019. Sensitive detection of *E. coli* in artificial seawater by aptamer-coated magnetic beads and direct PCR. Appl. Sci. 9, 5392.

Ky, Y.U., Noh, Y., Park, H.J., et al., 2004. Use of mono- clonal antibodies that recognize p60 for identification of *Listeria monocytogenes*. Clin. Diagn. Lab. Immunol. 1, 446–451.

Laciar, A., Vaca, L., Lopresti, R., et al., 2006. DNA finger-printing by ERIC-PCR for comparing *Listeria* spp. strains isolated from different sources in San Luis, Argentina. Rev. Argent. Microbiol. 38, 55–60.

Lantz, P.-G., Hahn-Hägerdal, B., Radstrom, P., 1994. Sample preparation methods in PCR-based detection of food pathogens. Trends Food Sci. Technol. 5, 384–389.

Lapeyre, C., Janin, F., Kaveri, S.V., 1998. Indirect double sandwich ELISA using monoclonal antibodies for the direct detection of *Staphylococcal enterotoxins* A, B, C and D in food samples. Food Microbiol. 5, 25–32.

Law, J.W., Ab Mutalib, N.S., Chan, K.G., Lee, L.H., 2015a. An insight into the isolation, enumeration, and molecular detection of Listeria monocytogenes in food. Front. Microbiol. 3 (6), 1227.

Law, J.W., Ab Mutalib, N.S., Chan, K.G., Lee, L.H., 2015b. Rapid methods for the detection of foodborne bacterial pathogens: principles, applications, advantages and limitations. Front. Microbiol. 12 (5), 770.

Lee, W.H., McClain, D., 1986. Improved *Listeria monocytogenes* selective agar. Appl. Environ. Microbiol. 52, 1215–1217.

Leimeister-Wachter, M., Haffner, C., Domann, E., et al., 1990. Identification of a gene that positively regulates expression of *Listeriolysin*, the major virulence factor of *Listeria monocytogenes*. Proc. Natl. Acad. Sci. U. S. A 87, 8336–8340.

Levin, R.E., 2003. Application of the polymerase chain reaction for detection of Listeria monocytogenes in foods: a review of methodology. Food Biotechnol 17, 99–116.

Liu, D., 2008. Preparation of *Listeria monocytogenes* specimens for molecular detection and identification. Int. J. Food Microbiol. 122, 229–242.

Liu, H., Lu, L., Pan, Y., Sun, X., Hwang, C.A., Zhao, Y., et al., 2015. Rapid detection and differentiation of *Listeria monocytogenes* and *Listeria* species in deli meats by a new multiplex PCR method. Food Control 52, 78–84.

Logan, J.M.J., Edwards, K.J., Saunders, N.A., Stanley, J., 2001. Rapid identification of Campylobacter spp. by melting peak analysis of biprobes in real-time PCR. J. Clin. Microbiol. 39, 2227–2232.

Lovett, J., 1988. Isolation and identification of *Listeria monocytogenes* in dairy products. J. Assoc. Off. Anal. Chem. 71, 658–660.

Lovett, J., Francis, D.W., Hunt, J.M., 1987. *Listeria monocytogenes* in raw milk: detection, incidence and pathogenicity. J. Food Protect. 50, 188–192.

Machaiaha, M.I., Krishnan, M.V., 2014. Immunodetection of Bacillus cereus haemolytic enterotoxin (HBL) in food samples. Anal. Methods 6, 1841–1847.

Mahantia, A., Narayan, S., Samiran, J., Bandyopadhyay, J., Banerjee, J., Ghosh, S., Batabyal, K., Sar, T.K., Samantaa, I., 2020. Characterization of methicillin-resistant and enterotoxins producing *Staphylococcus aureus* in bovine milk in India. J. Agric. Food Res. 2, 100017.

Manzano, M., Cocolin, L., Cantoni, C., Comi, G., 1997. Detection and identification of Listeria monocytogenes from milk and cheese by a single-step PCR. Mol Biotechnol 7 (1), 85–88. https://doi.org/10.1007/BF02821546. PMID: 9163724.

Marin, M.E., Rosa, M.C., Cornejo, I., 1992. Entertoxigenicity of *Staphylococcus* strains isolated from Spanish dry-cured hams. Appl. Environ. Microbiol. 58, 1067–1069.

Martínez-Blanch, J.F., Sánchez, G., Garay, E., Aznar, R., 2009. Development of a real-time PCR assay for detection and quantification of enterotoxigenic members of *Bacillus cereus* group in food samples. Int. J. Food Microbiol. 135 (1), 15–21.

Martinez-Blanch, J.F., Sanchez, G., Garay, E., Aznar, R., 2010. Evaluation of a real-time PCR assay for the detection and quantification of *Bacillus cereus* group spores in food. J. Food Prot 73, 1480–1485.

McBride, M.E., Girard, K.F., 1960. A selective method for the isolation of *Listeria monocytogenes* from mixed bacterial populations. J. Lab. Clin. Med. 55, 153–157.

Medina, M.B., 2003. Detection of Staphylococcal enterotoxin B (SEB) with surface plasmon resonance biosensor. J. Rapid Methods Autom. Microbiol 11, 225–243.

Medina, M.B., 2005. A biosensor method for a competitive immunoassay detection of Staphylococcal enterotoxin B (SEB) in milk. J. Rapid Methods Autom. Microbiol. 13, 37–55.

Medina, M.B., 2006. A biosensor method for detection of Staphylococcal enterotoxin A in raw whole egg. J. Rapid Methods Autom. Microbiol. 14, 119–132.

Mehrotra, M., Wang, G., Johnson, W.M., 2000. Multiplex PCR for detection of genes for *Staphylococcus aureus* enterotoxins, exfoliative toxins, toxic shock syndrome toxin 1 and methicillin resistance. J. Clin. Microbiol. 38, 1032–1035.

Mengaud, J., Vicente, M.F., Chenevert, J., et al., 1988. Expression in *Escherichia coli* and sequence analysis of the *listeriolysin O* determinant of *Listeria monocytogenes*. Infect. Immun. 56, 766–772.

Miliotis, M.D., Bier, J.W., 2003. International Handbook of Foodborne Pathogens. Marcel Dekker, Inc.

Mitchell, B.A., Milbury, J.A., Brookins, A.M., et al., 1994. Use of immunomagnetic capture on beads to recover *Listeria* from environmental samples. J. Food Protect. 57, 743–745.

Morach, M., Käppeli, N., Hochreutener, M., Johler, S., Julmi, J., Stephan, R., Etter, D., 2019. Microarray based genetic profiling of *Staphylococcus aureus* isolated from abattoir byproducts of pork origin. PLoS One 14 (9), e0222036.

Mossel, D.A.A., 1986. Developing methodology for food- borne microorganisms: fundamentals of analytical techniques. In: Pierson, M.D., Stern, N.J. (Eds.), Foodborne Microorganisms and Their Toxins: Developing Methodology. Marcel Dekker Inc, New York, pp. 1–22.

Muratovic, A.Z., Hagström, T., Rosén, J., Granelli, K., Hellenäs, K.E., 2015. Quantitative analysis of Staphylococcal enterotoxins A and B in food matrices using ultra high-performance liquid chromatography tandem mass spectrometry (UPLC-MS/MS). Toxins 7, 3637–3656.

Ngamwongsatit, P., Buasri, W., Pianariyanon, P., et al., 2008. Broad distribution of enterotoxin genes (hblCDA, nhe- ABC, cytK and entFM) among *Bacillus thuringiensis* and *Bacillus cereus* as shown by novel primers. Int. J. Food Microbiol. 121, 352–356.

Nimjee, S.M., Rusconi, C.P., Sullenger, B.A., 2005. Aptamers: an emerging class of therapeutics. Annu. Rev. Med. 56, 555–583.

Norton, D.M., 2002. Polymerase chain reaction-based methods for detection of *Listeria monocytogenes*: toward real-time screening for food and environmental samples. J. Assoc. Off. Anal. Chem. 85, 505–515.

Ortel, S., 1972. Experience with nalidixic acid-trypaflavine agar. Acta Microbiol. Acad. Sci. Hungar. 19, 363–365.

Ouoba, L.I., Thorsen, L., Varnam, A.H., 2008. Enterotoxins and emetic toxins production by *Bacillus cereus* and other species of *Bacillus* isolated from Soumbala and Bikalga, African alkaline fermented food condiments. Int. J. Food Microbiol. 124, 224–230.

Padmapriya, B.P., Ramesh, A., Chandrashekar, A., et al., 2004. Oligo Nucleotide Primers for Phosphotidylinositol in *Bacillus Cereus*. US Patent, US 6713620 dt.30/03/2004.

Padmapriya, B.P., Ramesh, A., Chandrashekar, A., Varadaraj, M.C., 2003. Staphylococcal accessory gene regulator (sar) as a signature gene to detect enterotoxigenic staphylococci. J. Appl. Microbiol. 95, 974–981.

Pal, S., Ying, W., Alocilja, E.C., Downes, F.P., 2008. Sensitivity and specificity performance of direct-charge transfer biosensor for detecting Bacillus cereus in selected food matrices. Biosyst. Eng. 99, 461–468. https://doi.org/10.1016/j.biosystemseng.2007.11.015.

Paoli, G.C., Bhunia, A.K., Bayles, D.O., 2005. *Listeria* monocytogenes. In: Fratamico, P.M., Bhunia, A.K., Smith, J.L. (Eds.), Foodborne Pathogens: Microbiology and Molecular Biology. Caister Academic Press, Wymondham, Norfolk, UK, pp. 295–325.

Phuektes, P., Mansell, P.D., Browning, G.F., 2001. Multiplex polymerase chain reaction assay for simultaneous detection of *Staphylococcus aureus* and Streptococcal causes of bovine mastitis. J. Dairy Sci. 84, 1140–1148.

Pinto, J.B., Rossatto, F.C.P., Martins, P.D., et al., 2015. Genetic relationships and virulence factors in *Staphylococcus aureus* isolated from raw poultry in. South Brazil. Ann Microbiol 65, 1933–1940.

Radhika, B., Padmapriya, B.P., Chandrashekar, A., et al., 2002. Detection of *Bacillus cereus* in foods by colony hybridization using PCR-generated probe and characterization of isolates for toxins by PCR. Int. J. Food Microbiol. 74, 131–138.

Rajkovic, A., El-Moualij, B., Uyttendaele, M., Brolet, P., Zorzi, W., Heinen, E., Foubert, E., Debevere, J., 2006. Immunoquantitative real-time PCR for detection and quantification of *Staphylococcus aureus* enterotoxin B in foods. Appl. Environ. Microbiol. 72 (10), 6593–6599.

Ralovich, B., Forray, A., Mérö, E., Málovics, H., Százados, I., 1971. New selective medium for isolation of L. monocytogenes. Zentralbl Bakteriol Orig 216 (1), 88–91. PMID: 4994622.

Ramesh, A., Padmapriya, B.P., Chandrashekar, A., Varadaraj, M.C., 2002. Application of a convenient DNA extraction method and multiplex PCR for direct detection of Staphylococcus aureus and Yersinia enterocolitica in milk samples. Mol. Cell. Probes. 16, 307–314.

Rana, N., Panda, A.K., Pathak, N., et al., 2020. Bacillus cereus: public health burden associated with ready-to-eat foods in Himachal Pradesh, India. J. Food Sci. Technol. 57, 2293–2302.

Rasooly, A., 2001. Surface plasmon resonance analysis of Staphylococcal enterotoxin B in food. J. Food Protect. 64, 37–43.

Rasooly, L., Rasooly, A., 1999. Real time biosensor analysis of Staphylococcal enterotoxin A in food. Int. J. Food Microbiol 49, 119–127.

Reiser, R., Conaway, D., Bergdoll, M.S., 1974. Detection of staphylococcal enterotoxins in foods. Appl. Microbiol. 27, 83–85.

Reyes, J.E., Bastias, J.M., Gutierrez, M.R., Rodriguez, M.O., 2007. Prevalence of *Bacillus cereus* in dried milk products used by Chilean school feeding program. Food Microbiol. 24, 1–6.

Rodriguez, D.L., Fernandez, G.S., Garayzabal, J.F.F., et al., 1984. New methodology for the isolation of *Listeria* microorganisms from heavily contaminated environments. Appl. Environ. Microbiol. 47, 1188–1190.

Rodriguez-Lazaro, D., Hernandez, M., Pla, M., 2004. Simultaneous quantitative detection of *Listeria* spp. and *Listeria monocytogenes* using a duplex real-time PCR- based assay. FEMS Microbiol. Lett. 233, 257–267.

Rosen, D.L., 2006. Airborne bacterial endospores detected by use of an impinger containing aqueous terbium chloride. Appl. Opt. 45, 3152–3157.

Roy, A., Moktan, B., Sarkar, P.K., 2007. Characteristics of *Bacillus cereus* isolates from legume-based Indian fermented foods. Food Control 18, 1555–1564.

Ruban, S.W., et al., 2018. Prevalence of methicillin resistant *Staphylococcus aureus* in retail buffalo meat in Chennai, India. Buffalo Bull. 37 (1), 51–58. ISSN 2539-5696.

Sails, A.D., Bolton, F.J., Fox, A.J., et al., 2002. Detection of *Campylobacter jejuni* and *Campylobacter coli* in environmental waters by PCR enzyme-linked immunosorbent assay. Appl. Environ. Microbiol. 68, 1319–1324.

Sarengaowa, H.W., Feng, K., Jiang, A., Xiu, Z., Lao, Y., Li, Y., Long, Y., 2020. An in situ-synthesized gene chip for the detection of food-borne pathogens on fresh-cut Cantaloupe and lettuce. Front. Microbiol. 5 (10), 3089.

Savariraj, W.R., Ravindran, N.B., Kannan, P., Paramasivam, R., Senthilkumar, T.M.A., Kumarasamy, P., Rao, V.A., 2019. Prevalence, antimicrobial susceptibility and virulence genes of *Staphylococcus aureus* isolated from pork meat in retail outlets in India. J. Food Saf. 39 (1), e12589.

Schoeni, J.L., Wong, A.C.L., 2005. *Bacillus cereus* food poisoning and its toxins. J. Food Protect. 68, 636–648.

Sharma, B.J.S., Rathnapraba, S., Meignanalakshmi, S., Manoharan, S., Saranya, S., Vijayarani, K., 2018. Incidence and molecular characterization of *Staphylococcus aureus* isolated from meat products. Int. J. Curr. Microbiol. App. Sci, 7 (9), 3163–3169.

Shim, W.B., Choi, J.G., Kim, J.Y., Yang, Z.Y., Lee, K.H., Kim, M.G., Ha, S.D., Kim, K.S., Kim, K.Y., Kim, C.H., Eremin, S.A., Chung, D.H., 2008. Enhanced rapidity for qualitative detection of Listeria monocytogenes using an enzyme-linked immunosorbent assay and immunochromatography strip test combined with immunomagnetic bead separation. J. Food Protect. 71 (4), 781–789. https://doi.org/10.4315/0362-028x-71.4.781. PMID: 18468033.

Shimizu, A., Fujita, M., Igarashi, H., et al., 2000. Characterization of *Staphylococcus aureus* coagulase type VII isolates from staphylococcal food poisoning outbreaks (1980–1995) in Tokyo, Japan, by pulsed-field gel electrophoresis. J. Clin. Microbiol. 38, 3746–3749.

Sivakumar, M., Dubal, Z.B., Kumar, A., Bhilegaonkar, K., Vinodh Kumar, O.R., Kumar, S., Kadwalia, A., Shagufta, B., Grace, M.R., Ramees, T.P., Dwivedi, A., 2019. Virulent methicillin resistant *Staphylococcus aureus* (MRSA) in street vended foods. J. Food Sci. Technol. 56 (3), 1116–1126.

Slamti, L., Perchat, S., Gominet, M., et al., 2004. Distinct mutations in Plc R explain why some strains of the *Bacillus cereus* group are non hemolytic. J. Bacteriol. 186, 3531–3538.

Somer, L., Kashi, Y., 2003. A PCR method based on 16S rRNA sequence for simultaneous detection of the genus *Listeria* and the species *Listeria monocytogenes* in food products. J. Food Protect. 66, 1658–1665.

Sospedra, I., Marín, R., Mañes, J., Soriano, J.M., 2012a. Rapid whole protein quantification of Staphylococcal enterotoxin B by liquid chromatography. Food Chem. 133, 163–166.

Sospedra, I., Soler, C., Manes, J., Soriano, J.M., 2012b. Rapid whole protein quantitation of Staphylococcal enterotoxins A and B by liquid chromatography/mass spectrometry. J. Chromatogr. A 1238, 54–59.

Stringer, M., 2005. Food safety objectives—role in microbiological food safety management. Food Control 16, 775–794.

Sudershan, R.V., Naveen Kumar, R., Kashinath, L., Bhaskar, V., Polasa, K., 2012. Microbiological hazard identification and exposure assessment of poultry products sold in various localities of Hyderabad, India. Sci. World J. 2012, 736040.

Sundararaj, N., Kalagatur, N.K., Mudili, V., Krishna, K., Antonysamy, M., 2019. Isolation and identification of enterotoxigenic *Staphylococcus aureus* isolates from Indian food samples: evaluation of in-house developed aptamer linked sandwich ELISA (ALISA) method. J. Food Sci. Technol. 56 (2), 1016–1026.

Swaminathan, B., Feng, P., 1994. Rapid detection of food- borne pathogenic bacteria. Annu. Rev. Microbiol. 48, 401–426.

Szabo, R.A., Todd, E.C.D., Rayman, M.K., 1984. Twenty-four hour isolation and confirmation of *Bacillus cereus* in foods. J. Food Protect. 47, 856–860.

Tallent, S.M., Hait, J.M., Knolhoff, A.M., Bennett, R.W., Hammack, T.S., Croley, T.R., 2017. Rapid testing of food matrices for Bacillus cereus enterotoxins. J. Food Saf. 37 (1), e12292.

Tamarapu, S., McKillip, J.L., Drake, M., 2001. Development of a multiplex Polymerase chain reaction assay for detection and differentiation of *Staphylococcus aureus* in dairy products. J. Food Protect. 64, 664–668.

Terplan, G., 1988. Provisional IDF-Recommended Method: Milk and Milk Products—Detection of Listeria Monocytogenes. Brussels: International Dairy Federation.

Tewari, A., Singh, S.P., Singh, R., 2015. Incidence and enterotoxigenic profile of Bacillus cereus in meat and meat products of Uttarakhand, India. J. Food Sci. Technol. 52 (3), 1796–1801.

Tsen, H.-Y., Chen, T.R., 1992. Use of the polymerase chain reaction for specific detection of type A, D and E enterotoxigenic *Staphylococcus aureus* in foods. Appl. Microbiol. Biotechnol. 37, 685–690.

Tsen, H.-Y., Chen, M.L., Hsieh, Y.M., et al., 2000. *Bacillus cereus* group strains, their haemolysin BL activity and their detection in foods using a 16S RNA and haemolysin BL gene-targeted multiplex polymerase chain reaction system. J. Food Protect. 63, 1496–1502.

Tsen, H.-Y., Chen, T.R., Yu, G.-K., 1994. Detection of B and C types enterotoxigenic *Staphylococcus aureus* using polymerase chain reaction. J. Chin. Agric. Chem. Soc. 32, 322–331.

Ueda, S., Yamaguchi, M., Iwase, M., Kuwabara, Y., 2013. Detection of emetic Bacillus cereus by real-time PCR in foods. Biocontrol Sci. 18 (4), 227–232.

USDA/FSIS, 2002. Isolation and identification of *Listeria monocytogenes* from red meat, poultry, egg and environ- mental samples. In: Microbiology Laboratory Guidebook, third ed. Revision 3, (Chapter 8).

Valdivieso-Garcia, A., Surujballi, K.D., Habib, D., et al., 1996. Development and evaluation of a rapid enzyme linked immunofiltration assay (ELIFA) and enzyme linked immunosorbent assay (ELISA) for the detection of staphylococcal enterotoxin B. J. Rapid Methods Autom. Microbiol. 4, 285–295.

Van Netten, P., Perales, I., Van de Moosdijk, A., et al., 1989. Liquid and solid selective differential media for the detection and enumeration of *Listeria monocytogenes* and other *Listeria* spp. Int. J. Food Microbiol. 8, 299–316.

Varadaraj, M.C., 1993. Methods for detection and enumeration of foodborne bacterial pathogens: a critical evaluation. J. Food Sci. Technol. 30, 1–13.

Varadaraj, M.C., Keshava, N., Devi, N., et al., 1992. Occurrence of *Bacillus cereus* and other Bacillus species in Indian snack and lunch foods and their ability to grow in a rice preparation. J. Food Sci. Technol. 29, 344–347.

Velasco, V., Sherwood, J.S., Rojas-García, P.P., Logue, C.M., 2014. Multiplex real-time PCR for detection of *Staphylococcus aureus*, mecA and Panton-Valentine Leukocidin (PVL) genes from selective enrichments from animals and retail meat. PLoS One 9 (5), e97617.

Vidic, J., Chaix, C., Manzano, M., Heyndrickx, M., 2020. Food sensing: detection of Bacillus cereus spores in dairy products. Biosensors 10 (3), 15.

Vidic, J., Vizzini, P., Manzano, M., Kavanaugh, D., Ramarao, N., Zivkovic, M., Radonic, V., Knezevic, N., Giouroudi, I., Gadjanski, I., 2019. Point-of-need DNA testing for detection of foodborne pathogenic bacteria. Sensors 19, 1100.

Vishnubhatla, B., Oberst, R., Fung, D., Wonglumsom, W., Hays, M., Nagaraja, T., 2001. Evaluation of a 5-nuclease (TaqMan) assay for the detection of virulent strains of Yersinia enterocolitica in raw meat and tofu samples. J. Food Prot. 64, 355–360.

Waak, E., Tham, W., Danielsson-Tham, M.L., 2002. Prevalence and fingerprinting of *Listeria monocytogenes* strains isolated from raw whole milk in farm and in dairy plant receiving tanks. Appl. Environ. Microbiol. 68, 3366–3370.

Wang, M., Cao, B., Gao, Q., Sun, Y., Liu, P., Feng, L., et al., 2009. Detection of Enterobacter sakazakii and other pathogens associated with infant formula powder by use of a DNA microarray. J. Clin. Microbiol. 47, 3178–3184.

Wang, X.W., Zhang, L., Jin, L.Q., Jin, M., Shen, Z.Q., An, S., et al., 2007. Development and application of an oligonucleotide microarray for the detection of food-borne bacterial pathogens. Appl. Microbiol. Biotechnol. 76, 225–233.

Wei, S., Daliri, E.B.-M., Chelliah, R., Park, B-j, Lim, J.S., Baek, M.A., Nam, Y.S., Seo, K.H., Jin, Y.G., Oh, D.H., 2019. Development of a multiplex real-time PCR for simultaneous detection of Bacillus cereus, Listeria monocytogenes, and *Staphylococcus aureus* in food samples. J. Food Saf. 39 (1), e12558.

Wesley, I.V., Harmon, K.M., Dickson, J.S., Schwartz, A.R., 2002. Application of multiplex polymerase chain reaction assay for the simultaneous confirmation of *Listeria monocytogenes* and other *Listeria* species in Turkey sample surveillance. J. Food Protect. 65, 780–785.

Wilson, I.G., Gilmour, A., Cooper, J.E., et al., 1994. A non- isotopic DNA hybridization assay for the identification of *Staphylococcus aureus* isolated from foods. Int. J. Food Microbiol. 22, 43–54.

Wilson, W.J., Strout, C.L., DeSantis, T.Z., Stilwell, J.L., Carrano, A.V., Andersen, G.L., 2002. Sequence-specific identification of 18 pathogenic microorganisms using microarray technology. Mol. Cell. Probes 16, 119–127.

Wilson, G.I., Cooper, J.E., Gilmour, A., 1991. Detection of enterotoxigenic *Staphylococcus aureus* in dried skimmed milk: use of the polymerase chain reaction for amplification and detection of staphylococcal enterotoxin genes *entB* and *entC*1 and the thermonuclease gene *nuc*. Appl. Environ. Microbiol. 57, 1793–1798.

Wolcott, M., 1991. DNA-based rapid methods for the detection of foodborne pathogens. J. Food Protect. 54, 387–401.

Xia, Y., Liu, Z., Yan, S., Yin, F., Feng, X., Liu, B.-F., 2016. Identifying multiple bacterial pathogens by loop-mediated isothermal amplification on a rotate & react slipchip. Sensor. Actuator. B Chem. 228, 491–499.

Yang, M.H., Kostov, Y., Bruck, H.A., Rasooly, A., 2008. Carbon nanotubes with enhanced chemiluminescence immunoassay for CCD-based detection of Staphylococcal enterotoxin B in food. Anal. Chem. 80, 8532–8537.

Yang, M.H., Sun, S., Kostov, Y., Rasooly, A., 2010. Lab-on-a-chip for carbon nanotubes based immunoassay detection of Staphylococcal enterotoxin B (SEB). Lab Chip 10, 1011–1017.

Yugueros, J., Temprano, A., Sanchez, M., et al., 2001. Identification of Staphylococcus spp by PCR-restriction fragment length polymorphism of gap gene. J. Clin. Microbiol. 39, 3693–3695.

Zare, S., Derakhshandeh, A., Haghkhah, M., Naziri, Z., Broujeni, A.M., 2019. Molecular typing of *Staphylococcus aureus* from different sources by RAPD-PCR analysis. Heliyon 5 (8), e02231.

Zhao, R., Wen, Y.Z., Yang, J.C., Zhang, J.L., Yu, X.M., 2014. Aptasensor for Staphylococcus enterotoxin B detection using high SNR piezoresistive microcantilevers. J. Microelectromech. Syst 23, 1054–1062.

Zhou, X., Jiao, X., 2006. Prevalence and lineages of Listeria monocytogenes in Chinese food products. Lett. Appl. Microbiol. 43 (5), 554–559.

Zhu, L., He, J., Cao, X., et al., 2016. Development of a double-antibody sandwich ELISA for rapid detection of Bacillus Cereus in food. Sci. Rep. 6, 16092.

Ziad, J., Akram al, A., Mahmoud, S., Qotaibah, A., 2013. *Staphylococcus aureus* isolates from camels differ in coagulase production, genotype and methicillin resistance gene profiles. J. Microbiol. Biotechnol. Food Sci. 2 (6), 2455–2461.

Further reading

Blais, B.W., Phillippe, L.M., 1993. A simple RNA probe system for analysis of *Listeria monocytogenes* polymerase chain reaction products. Appl. Environ. Microbiol. 59, 2795–2800.

Chapter 29

Role of education and training of food handlers in improving food safety and nutrition: the Indian experience

Jamuna Prakash
Global Harmonization Initiative, Austria

The past century has been an era of great changes in the lives of human beings. The human race has witnessed an unprecedented rapid rate of transition in all walks of life. The population demographics have changed the impact on all other aspects of life in a major way, whether it is transport, communication, technology, mechanizations, education, or ecology. There are consequent changes in lifestyle too, increasing the level of comfort and luxury and lifespan for mankind owing to medical advances. However, the same cannot be said about the quality of life, as the changing food environment has not improved it as expected. The dietary and nutrition transitions have been detrimental for most of the population and the reduction in the incidences of malnutrition has been very slow with newer emerging challenges of overweight and obesity.

29.1 Food environment: dietary and nutrition transition as prime determinants of food behavior

A considerable change in food environment is seen. As what one eats depends on what is grown and what is available, agricultural resources have defined the food environment, whether from plant, animal, or marine sources. Food processing operations have moved out of individual households. Advances in food processing and mechanization of operations have reduced the hours spent in the kitchen and brought in a revolutionary entry of packaged foods in everyday meals. This has obviously influenced the dietary intakes. Two major changes associated with dietary and nutrition transition are replacement of home cooked food with catered or packaged foods and inclusion of varieties of processed foods in everyone's meals, starting from an infant up to an elderly person. Food choices are dictated by taste, availability, affordability, and convenience rather than health, nutrition, or safety concerns. Consumer's knowledge and awareness levels vary; hence, even informed choices may not be appropriate in all cases. From the public health point of view, safety of food should be the first priority, though nutrition is important too!!

Dietary and nutrition transition in Indian context has resulted in a reduction in coarse millets, whole grains, fresh seasonal fruits and vegetables, and natural foods, and an increase in highly processed foods, fats, sweets, sugar, salt, animal foods, and with chemicals laden foods. The change in food behavior is consequential to the easy access to fast food vendors, processed foods, sweets, bakeries and soft drinks, increased dietary diversity with an improper food selection, altered purchasing patterns, exposure to mass media, and availability of unsafe food. Eating away from home is a major change in food behavior observed both in urban and rural population on account of social and economic changes happening in communities. There is an increase in disposable income, in number of food business operators, more working women, easy access to processed and catered foods, online order and home deliveries of food, and exposure to advertisements.

A study in urban area of Bangalore city on determinants of food behavior in school children ($n = 1200$ families) revealed that processed and convenient foods topped the list with traditional meals getting a low score (Table 29.1). The practice of storing foods in refrigerator was also very common and practiced by 96% of families. Television was the most

Ensuring Global Food Safety. https://doi.org/10.1016/B978-0-12-816011-4.00029-X

TABLE 29.1 Food behavior in urban children: selection and influencing environment (percent of subjects).

Types of food selected		Sources of nutrition information/influence of media	
Ready to eat foods	83.0	Television	89.5
Branded foods	87.0	Professional advice	45.0
Convenience foods	98.0	Print media	65.0
Foods stored in refrigerator	96.0	Interest in TV ads	99.3
Traditional meals	63.0	Ask for foods seen in ads	95.4

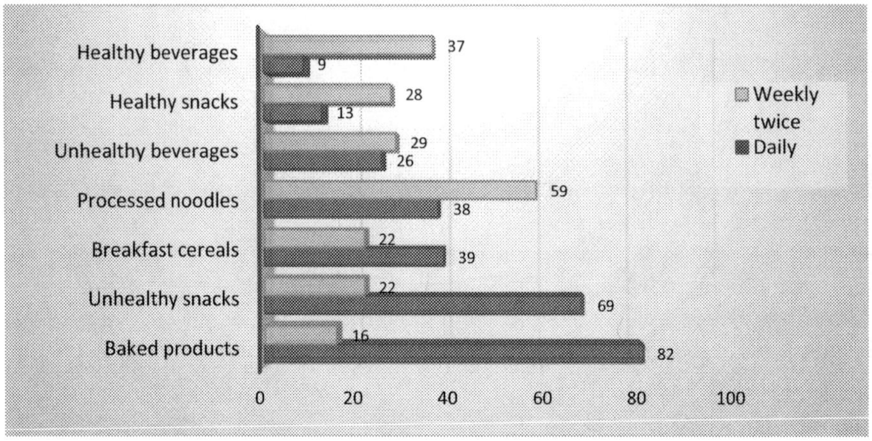

FIGURE 29.1 Consumption pattern of processed foods in children (%).

prominent source of nutrition information and most of the children asked for foods seen on the television. Baked products, unhealthy snacks, and processed noodles were consumed more frequently than healthy snacks or beverages (Fig. 29.1) revealing the impact of dietary and nutrition transition in the young generation (Prakash and Prakash, 2012).

29.1.1 Food and waterborne infections are one of the leading causes of illness among young and old alike, especially in developing countries

Globally, the loss caused by unsafe food is unknown. This is because there are very little data available on the documented evidence of the problems and the damage it has caused. The food safety loss has been mostly underestimated. Every year, many adults are affected by food borne infections causing loss of working days and productivity, and many children who contract a disease die. The key facts given by WHO (2020) reproduced in Box 29.1 speak for themselves and present the alarming situation.

In India, foodborne diseases cost around $28 billion every year (Rs. 178,100 crore) (Food for all partnership of World Bank Group & Netherlands Govt.). However, this may only be the tip of the iceberg as many incidences are not reported. Mild cases are not brought to the notice of healthcare provider, and treated at home. Diagnostic tools are not applied to investigate the origin of infections, and due to poor awareness, illnesses are not associated with intake of a particular food.

It may also be noted that eating out episodes are not confined only to urban areas but have moved out to villages. As a case study, in a rural population [sample size - 500 families, target population - 2500–3000, from 5 villages in southern India], a survey of processed foods consumption was conducted and the results are compiled in Table 29.2. As can be seen, a large number of subjects were consuming street vended foods or baked products on a daily basis. Most of these were rich in fat, salt, or sugar and prepared with refined flours. Moreover, since these come from food handlers who have no basic training on food safety, foodborne infections were common.

BOX 29.1 Key facts: food safety

- Access to sufficient amounts of safe and nutritious food is key to sustaining life and promoting good health.
- Unsafe food containing harmful bacteria, viruses, parasites, or chemical substances causes more than 200 diseases—ranging from diarrhea to cancers.
- An estimated 600 million—almost 1 in 10 people in the world—fall ill after eating contaminated food and 420,000 die every year, resulting in the loss of 33 million healthy life years (DALYs).
- Children under 5 years of age carry 40% of the foodborne disease burden, with 125,000 deaths every year.
- Diarrheal diseases are the most common illnesses resulting from the consumption of contaminated food, causing 550 million people to fall ill and 230,000 deaths every year.
- Food safety, nutrition, and food security are inextricably linked. Unsafe food creates a vicious cycle of disease and malnutrition, particularly affecting infants, young children, elderly, and the sick.
- Foodborne diseases impede socioeconomic development by straining health care systems, and harming national economies, tourism, and trade.
- Food supply chains now cross multiple national borders. Good collaboration between governments, producers, and consumers helps ensure food safety.

Reproduced from WHO, 2020. World Health Organisation. Available from: www.who.int/en/news-room/fact-sheets/detail/food-safety.

TABLE 29.2 Consumption of street vended foods, baked, and fried products (daily intake, % of subjects).

Type of food	Percent of subjects	Type of food	Percent of subjects	Type of food	Percent of subjects
Spiced snacks		Fried foods		Baked products	
Masala puri	13	*Seviya*	75	Bun	62
Churmuri	45	*Kodubale*	32	Bread	61
Pani poori	43	*Chakli*	26	Cake	13
Chinese food		*Pakora*	54	Rusk	7
Noodles	51	*Bonda*	47	Puffs	6
Gobi manchoorian	12	*Vada*	58	–	–
Fried rice	19	Chips	58	–	–

Description of ethnic snacks

Masala puri	Fried small wheat *poories* [flat breads] eaten with finely cut tomatoes, onions and spiced chutneys
Churmuri	Puffed rice eaten with finely cut tomatoes, onions, fresh coriander, roasted peanuts, and spiced chutneys
Pani poori	Fried small wheat *poories* stuffed with boiled potatoes and eaten with finely cut tomatoes, onions, and spiced water
Gobi manchoorian	Fried cauliflower seasoned with highly spiced onions, ginger, garlic, and soy sauce
Seviya	Deep fried extruded snack prepared with chick pea flour
Kodubale	Deep fried snack prepared with rice flour, coconut, and spices
Chakli	Deep fried extruded snack prepared with rice and black gram flour, and spices
Pakora	Deep fried vegetables dipped in chick pea flour
Bonda	Boiled spiced potato balls dipped in chick pea flour batter and deep fried.
Vada	Deep fried savoury snack prepared with soaked and ground decorticated legumes.

Computed from original data Puttarathnamma, D., Prakash, J., Prabhavathi, S.N., 2015. Consumption trends of processed foods among rural population selected from South India. Int. J. Food Nutr. Sci. 6 (2), 1–6, https://doi.org/10.15436/2377-0619.15.039.

Consequences of eating unsafe food: Let us examine the adverse impact of consuming unsafe food.

- There is increased morbidity and the resultant impact is increased health care cost, loss of productivity in adults, and learning time in children. Sometimes, it can be fatal too.
- Frequent infections compromise nutrition security of an individual increasing nutrient requirements and augmenting existing deficiencies. Children are particularly vulnerable to this situation.
- This gives rise to vicious cycle of malnutrition in children, exacerbated on account of unhygienic food, and contributes toward increased morbidity and mortality.
- This adds on to the burden of food waste. Unsafe food is lost on account of poor handling thereby wasting the related resources used for production and processing of that food.
- It also poses an extra burden on environmental resources.

Food safety issues in Indian situation: Though there is a definite food authority in place and there are stringent laws against any kind of fraud, there are some food safety—related challenges observed now and then. These are not uniform to all situations and regional variations are seen. There are frequent episodes of food poisoning reported in mass feeding situations on account of mishandling of food and unhygienic surroundings. Food and waterborne infections are frequently reported. There is indiscriminate overuse of pesticides on agricultural crops, the residue of which is harmful for human consumption. Meat and poultry are contaminated with antibiotic residues used for prolonging shelf life of meat. Nonapproved fruit ripening agents, such as calcium carbide, are used by traders for quick ripening of fruits. Arsenic and fluoride contamination of water is encountered in some regions. Cases of adulterated milk as well as adulterated fats and oils [animal fat in clarified butter] have been reported. Use of nonpermitted colors and additives is not uncommon. There are other fraudulent practices such as selling of expired date products or mishandling of processed products.

Both safety and nutritional composition of foods are hidden qualities. If a processed food comes from an organized sector, the composition is declared on the label and consumer can make an informed choice. However, food coming from the unorganized sector has no label of composition; hence, the nutritional value is not known. Same is true for any kind of catered food, where the nutrient density is not known. And, in general, the nutrient density would vary depending upon the ingredients used and the portion size which varies from one facility to other.

In comparison to processed and packaged foods, the catering sector of the food establishment is mostly unorganized. Also, the sector is partially automated which stresses the need for trained food handlers to ensure safety. The challenge therefore lies in convincing the unorganized food business operators to invest in food safety. This would mean they invest in better infrastructure, technology, resources, training, and documentation. It is therefore important to understand why they are unorganized and how they can be motivated to adopt systems. This requires a study of their systems, followed by strategizing the appropriate solutions, implementing them, and finally confirming what has worked best with the given conditions (Mishra and Prakash, 2016, 2017; Prakash et al., 2017).

Therefore, adequate education and training of food handlers is a requirement for the food industry and the catering facilities. Whether small or large, all food-related operations from farm to fork should be based on sound principles of safety. Our experiences with education programs in the area of food safety and nutrition have revealed that the impact is always positive and considerable improvement can be seen on post-assessments of trainings. The study description and major observation of some of these programs are summarized in Table 29.3.

A sustainable program in food safety requires the support of management and appropriate training of personnel to create and manage the prerequisites, operational prerequisites, and critical control point design in accordance with the business. The FBOs need to understand the importance of food safety and the potential risk of serving unsafe food to customers. The training process involves identification of gap areas in infrastructure facilities, food handling, and personal hygiene of workers, based on which the program objectives are framed. Appropriate training modules are designed based on the education and skill level of workers and then implemented. The last element of training is assessment and evaluation of the program with follow-up and retrainings for reinforcement.

To summarize, there has been a dramatic rise in consumption of processed and catered foods across Indian population with two major concerns, safety of foods and nutritional composition. Food safety can be ensured with adequate education and training of food handlers across all levels. This is a continuous process, and needs frequent reinforcement to ensure quality and compliance to food regulations.

TABLE 29.3 A bird's eye view of community educational programs undertaken for nutrition and safety of food.

Study description	Observation/results
Community: Training program for women on utilizing green leafy vegetables (Hemalatha and Prakash, 2002)	
Participants: 60 women **Tools and Techniques**: Lectures, demonstration, and workshops **Duration**: 6 weeks for training/educating. **Assessment:** Pre- and post-assessment of knowledge, attitude, and practice with follow up for 9 months. Maintenance of food record for 6 months.	• Increase in knowledge levels of all participants • Interest and awareness regarding healthy eating • Increase in greens consumption by family from 1 to 5 times a week in substantial quantity • Positive feedback.
Community: Role of water in health - Hygienic habits related to water (unpublished data, project taken up by authors team)	
Participants: School children from 5 villages. **Tools and techniques**: Education integrated through schools regarding role of water in health, specifically, hygiene related to water, washing hands, drinking clean water. **Assessment:** Part of ongoing project for mapping morbidity in children for a year followed by various intervention programs. **Initial assessment** – Health status, nutritional status, and morbidity in children for a year. Follow up for 6 months after intervention.	• Significant improvement in nutritional status of children • Increase in knowledge and awareness regarding hygiene and using clean water. • Significant reduction in overall morbidity, specifically in helminthic infection by over 70% as evidenced by fecal examination.
Community: Awareness creation program: Quality of packed school lunch and parental nutritional awareness of preschool children (Prakash et al., 2013)	
Participants: Preschool children from 3 schools and their parents. **Tools and techniques**: Assessment of quality of packed school lunch and parent's knowledge levels through a structured questionnaire. Education of parents through lecture and demonstration of healthy school meals. **Assessment:** Pre- and postassessment of quality of packed lunch and parental knowledge through follow-up.	• Improvement in nutritional quality of packed lunch by inclusion of healthy foods, fruits and vegetables, home cooked meals and less of baked, deep-fried, high fat, and high sugar snacks. • Improvement of knowledge and awareness of parents.
Foodservice Providers: Mid-day meal program - operational constraints and impact of training on food handlers (Dachana et al., 2010)	
Area: 20 schools, each 300–600 children [10 rural and 10 urban] in each school. Initial assessment with structured questionnaire • Location of kitchen and dining area. • An assessment of kitchen and storage facilities - area, space, ventilation, fuel used, cleaning facilities. • Water supply, hygiene, and sanitation, toilet facilities. • Handling of food waste. • Training program for cooks and teachers in multiple sessions in food safety and nutrition Follow-up assessment after 3 months Education program: Specially designed nutrition education program for children on 5–7th grade to sensitize them regarding nutritional needs.	**Infrastructural facilities -** Construction of kitchens, provision of store room, proper storage area, storage containers. **Hygienic practices -** significant improvement in general cleanliness. **Food storage -** food grains and spices cleaned and stored in appropriate containers. **Serving of food -** children served lunch in cleaner surroundings avoiding open fields thus reducing dust contamination. Proper serving vessels used for serving liquid curry thus reducing spillage. **Nutrition, kitchen garden -** improvement in nutritional quality of foods—increased diversity, use of local ingredients—greens, legumes. **Water supply and washing facilities -** Improvement in hygiene and cleanliness - children washed their hands properly before eating and kept the premises clean. The cooks used head scarf while cooking. Extra taps for hand washing were fixed in schools. **Handling food waste** - waste water used for kitchen garden. Solid food waste used as manure for growing plants.

Continued

TABLE 29.3 A bird's eye view of community educational programs undertaken for nutrition and safety of food.—cont'd

Study description	Observation/results
Assessment of risk determinants of sustainable food safety status in food court (Mishra and Prakash, 2017)	
The status of food safety at a food court in Bangalore city studied through risk determinants which defined the relationship between prerequisites (PRP), operational prerequisites (OPRP), and critical control point (CCP). **Participants:** Six food business operators (FBO) **Tools and techniques:** Data collected through visual observation and verification of the flow of operations with structured questionnaires based on FSSAI [Food Safety and Standards Authority of India] guidelines. **Assessment** using an ordinal scale of four for rating. • Kitchen location and design • Status of equipment and facilities • Critical control points • Sanitation and personal hygiene	Results analyzed as the compliance rating for the PRP and OPRP of the FBOs with FSSAI guidelines. • Location: 50% compliance. • Lack of facilities in all six FBOs like drain, hand wash, and space constraint: 43% compliance, affecting the CCPs of kitchen and personnel hygiene. • Potable water, drainage, sewage disposal, ventilation and lighting: 51% compliance. • Personnel hygiene, cross-contamination of food, the operational area and food handling methods: 39% compliance. • Storage area: 47.5% compliance [space constraint leading to overload of food in operations, freezers, and refrigerator without proper segregation.] • Traceability of incoming raw material: Only 40% were traceable. • Pest control and waste management: Very poor. • Health status of the personnel: Not assessed. • Training in food safety - No training provided.

References

Dachana, K.B., Prakash, D.J., Prakash, J., 2010. Mid-day meal program – a comprehensive analysis of operational constraints and impact of training on the personnel involved in selected schools. Indian J. Nutr. Diet. 47 (10), 435–443.

Hemalatha, M.S., Prakash, J., 2002. Impact of an awareness creation program for women on nutrition through green leafy vegetables. Indian J. Nutr. Diet. 39, 17–25.

Mishra, M., Prakash, J., 2016. Impact of awareness creation program on food safety management system in institutional catering. MOJ Food Process. Technol. 3 (2), 00068. https://doi.org/10.15406/mojfpt.2016.03.00068.

Mishra, M., Prakash, J., 2017. Assessment of risk determinants of sustainable food safety status in food court. Indian J. Nutr. Diet. 54 (2), 161–171.

Prakash, D., Prakash, J., 2012. Impact of nutrition education on nutrition knowledge of children and their parents and food behaviour of children. Indian J. Nutr. Diet. 49 (8), 341–351.

Prakash, D., Shilpa, M.S., Prakash, J., 2013. Impact of nutrition education of parents of preschool children on quality of packed school lunch. Int. J. Food Nutr. Diet. 1 (2), 61–68.

Prakash, D.J., Siddiqui, A.A., Sandeep, P.G., Prakash, J., 2017. Qualitative analysis of street foods and identification of constraints for ensuring their safety. In: Komala, M., Prakash, J. (Eds.), The Book, Emerging Health Issues Across Life Stages. Aayu Publications, New Delhi, pp. 291–300.

Puttarathnamma, D., Prakash, J., Prabhavathi, S.N., 2015. Consumption trends of processed foods among rural population selected from South India. Int. J. Food Nutr. Sci. 6 (2), 1–6. https://doi.org/10.15436/2377-0619.15.039.

WHO, 2020. World Health Organisation. Available on: www.who.int/en/news-room/fact-sheets/detail/food-safety, 4th June, 2019. (Accessed 11 February 2020).

Index

Note: 'Page numbers followed by "f" indicate figures and "t" indicate tables.'

G

H

Printed in the United States
by Baker & Taylor Publisher Services